# STUDY GUIDE AND STUDENT SOLUTIONS MANUAL

## DOUGLAS BRANDT

# PHYSICS

for

## SCIENTISTS & ENGINEERS

### Third Edition

# DOUGLAS C. GIANCOLI

Prentice
Hall

Upper Saddle River, NJ 07458

Executive Editor: Alison Reeves
Project Manager: Elizabeth Kell
Special Projects Manager: Barbara A. Murray
Supplement Cover Manager: Paul Gourhan
Supplement Cover Designer: PM Workshop Inc.
Manufacturing Manager: Trudy Pisciotti
*Cover Photo Credit: Oone van der Wal/ Young America*

Printed in the United States of America

10  9

ISBN 0-13-021475-2

Prentice-Hall International (UK) Limited, London
Prentice-Hall of Australia Pty. Limited, Sydney
Prentice-Hall Canada, Inc., Toronto
Prentice-Hall Hispanoamericana, S.A., Mexico
Prentice-Hall of India Private Limited, New Delhi
Pearson Education Asia Pte. Ltd., Singapore
Prentice-Hall of Japan, Inc., Tokyo
Editora Prentice-Hall do Brazil, Ltda., Rio de Janeiro

# Contents

# Preface

This study guide has been prepared to accompany the textbook *Physics for Scientists & Engineers* by Douglas C. Giancoli.

Each chapter begins with a brief overview of what the chapter covers and a list of objectives for the chapter. The list of objectives is comprehensive and your course may omit some of these objectives. Pay attention to what your instructor is emphasizing and wants you to be able to gain knowledge of from each chapter.

The list of equations for each chapter includes both the fundamental laws of physics in equation form and equations that result from application of the laws or principles to specific situations. Some instructors want their students to memorize equations, others don't. More than memorizing the equation, it is important to know what the equation is stating and to what situations it applies. If you are not required to memorize the equations, you should be aware of the existence of each of the relationships in the list of equations and whether they are definitions, statements of laws of physics, or derived equations that apply to particular situations.

The chapter summary is brief review of the principles covered in the chapter. In some places it includes supplementary discussion of topics covered in the text.

A sample quiz has been included in each chapter. The quiz consists of ten conceptual questions and five quantitative problems. The level of difficulty of the both the conceptual questions and quantitative problems vary from relatively easy to difficult. The answers to the quiz questions are in the back of the study guide.

The solutions to alternate odd numbered problems from the text have been included in each chapter. These well written solutions are from the *Instructor's Solution Manual* prepared by Irving A. Brown. To make the best use of these solutions, make an honest attempt at solving the problems directly from the text before looking at the solutions. Use the solutions to verify that you have correctly solved the problems or to find the step in the reasoning used to solve the problem that is giving you some difficulty.

I have tried to take great care assuring that everything in this study guide is correct. I am solely responsible for any errors or omissions that remain in this study guide. I am happy to receive any feedback that readers send, both positive and negative, that may help me improve any future edition or other work that I prepare in the future.

Finally, I would like to thank everyone that assisted me directly or indirectly while in preparing this study guide. Thank you to Liz Kell at Prentice Hall for asking me to attempt this work, for all her assistance while the study guide was a work in progress, and for her patience during the times my teaching schedule was preventing rapid progress. Thank you to Irving A. Brown for preparing the solutions for the *Instructor's Solution Manual* for the text, some of which have been included in this study guide. My deepest thanks to my mother-in-law and father-in law, Mitsue and Takao Sato, who have been so good and generous to my family and me while I completed most of this study guide on their dining room table in Hawaii. And, finally, thanks to my wife Charlene and my children Stephanie and Donovan for their patience and support and for giving me the time and space to complete this work.

# Chapter 1: Introduction, Measurement, Estimating

## Chapter Overview and Objectives

This chapter describes the nature of science. It defines the terms theory, model and law. It describes the measurement process and uncertainties in measurement, units, and estimation. It also introduces the concept of dimensional analysis in analyzing relationships between physical quantities.

After completing study of this chapter, you should:

- Understand the nature of the process of science.
- Know the difference between models, theories, and laws.
- Know that measurements always have some uncertainty and how that uncertainty can be reported.
- Know the SI units of length, mass, and time and their abbreviations.
- Know most of the common metric prefixes.
- Know the procedure for converting physical quantities from one set of units to another.
- Be able to make order of magnitude estimates of things encountered in everyday life.
- Understand how dimensional analysis can be used to analyze relationships between physical quantities.

## Summary of Equations

There are no equations in this chapter.

## Chapter Summary

### Section 1-1  The Nature of Science

One of the basic components of science is **observation** of events. This is done in an organized way by designing and performing experiments in the laboratory or other appropriate environment. A second component of science is the creation of **theories** or rules that can be applied to explain the results of observations. Finally, a component of science that sets it apart from other human endeavors is further observation to **test** the theories that have been created. This further testing can often prove a theory false or of limited applicability in which case new theories may be developed to replace the older theories, but no theory can be proven absolutely true.

### Section 1-2  Models, Theories, and Laws

Scientists create **models** of particular phenomena to help understand that phenomenon. Models are often simple explanations of phenomena using ideas that can explain those phenomena in terms of more well understood and familiar phenomena that are similar in behavior in some way. One purpose of a model is to help visualize phenomena in some way that is useful. Models are often suggestive as to what further investigation scientists should make about particular phenomena to gain a deeper understanding.

The words theory and model are often used as synonyms. The word **theory**, though, is usually used to mean a more general rule that applies to a variety of phenomena and is able to explain the phenomena with quantitative accuracy and precision or in great detail.

The word **law** is used for a very general concise rule that appears to describe the behavior of nature in general. To keep science in perspective, it is important to remember that the statements made by scientists concerning the behavior of the universe are descriptive statements. The statements are not absolutes that dictate the behavior of the universe. The theories and laws are general statements about the way the universe seems to behave.

### Section 1-3  Measurement and Uncertainty; Significant Figures

Observations in physics involve the measurement of physical quantities. A measurement is a comparison to a standard amount of that physical quantity. Because measurement is a comparison process, it is of limited precision and accuracy. For a measurement to be useful, its **estimated uncertainty** due to limited precision and accuracy must also be known. That precision can be stated explicitly, as in $d = 34.6 \pm 0.1$ cm. This notation means the quantity $d$ is somewhere between 34.5 cm and 34.7 cm. Sometimes the percentage the uncertainty is of the quantity is reported. For the above example, the percentage uncertainty is

$$\text{percentage uncertainty} = \frac{\text{uncertainty}}{\text{value}} \times 100\% = \frac{\pm 0.1 \text{ cm}}{34.6 \text{ cm}} \times 100\% = 0.3\%$$

Often, particularly in textbooks, the uncertainty in a number is given implicitly by using **significant figures**. A physical quantity reported by a number such as 78.5 means 78.5 ± 0.05. The uncertainty is implied to be one half of the least significant figure reported. When significant figures are used to represent uncertainty in quantities, it is very important to use the correct number of significant figures in reporting quantities.

To keep consistency when significant figures are used to represent the uncertainty in quantities, it is important to follow some rules regarding the number of significant figures in the results of calculations. The result of a multiplication or division has as many significant figures as the number with the fewest number of significant figures used in the calculation. The result of an addition or subtraction has a least significant figure the same as the largest decimal place of the least significant figures of the operands. Intermediate results during calculations should keep an additional digit beyond the significant figures to minimize cumulative round-off error in the final result.

Numbers such as 3900 create an ambiguity in determining the number of significant figures. Are the two zero digits significant figures or are they only place holders to get the 3 and the 9 into the proper powers of ten places? To avoid this ambiguity, numbers can be written in **scientific notation**. Scientific notation expresses numbers as a decimal number greater than or equal to one and less than 10 multiplied by a power of ten. If this were applied to the number 3900 above, it would be written as $3.9 \times 10^2$ if only the 3 and the 9 were significant figures. It would be written as $3.900 \times 10^2$ if all the digits were significant figures.

### Section 1-4  Units, Standards, and the SI System

As measurements are comparisons to a standard amount of a quantity, a number representing a quantity is meaningless unless the standard amount of the quantity used for comparison is known. Writing all quantities with units appended to the number representing the quantity provides this information.

• The scientific standard for length is the **meter** (abbreviated as **m**). The meter is defined as the distance light travels in a time of 1/299,792,458 of a second.

• The standard for time is the **second** (abbreviated **s**). The second is defined as the time for 9,192,631,770 periods of the radiation from a cesium atom when they change between two particular states.

• The standard unit of mass is the **kilogram** (abbreviated **kg**). The kilogram is defined as the mass of a platinum-iridium cylinder kept in Paris, France.

In the metric system of units, larger and smaller units are built from the units of the standards by adding prefixes that represent powers of ten. The following table gives the prefix name, its abbreviation, and the power of ten it represents:

| Prefix | Abbreviation | Power | Prefix | Abbreviation | Power |
|---|---|---|---|---|---|
| exa | E | $10^{18}$ | centi | c | $10^{-2}$ |
| peta | P | $10^{15}$ | milli | m | $10^{-3}$ |
| tera | T | $10^{12}$ | micro | $\mu$ | $10^{-6}$ |
| giga | G | $10^{9}$ | nano | n | $10^{-9}$ |
| mega | M | $10^{6}$ | pico | p | $10^{-12}$ |
| kilo | K | $10^{3}$ | femto | f | $10^{-15}$ |

The most common system of units used in scientific work is the **Systeme International** or **SI** system. The standard of length is the meter, the standard of time is the second, and the standard of mass is the kilogram in the SI unit system. A second metric based system is the **cgs** system of units. The centimeter is the standard of length, the second is the standard of time, and the gram is the standard of mass in the cgs unit system. A third system of units in use today is the **British engineering system**. The standard of length is a foot, the standard of time is the second, and the standard of force is the pound in the British engineering unit system.

### Section 1-5  Converting Units

Quantities are often given in one type of units, but need to be converted to a different type of units. This is done by the use of conversion factors.

**Example 1-5-A**

What is the area in square meters of a one foot by one foot floor tile?

**Solution:**

The area of a one foot by one foot floor tile is 1.00 square feet. It can be converted to square meters by using the conversion factor 1 ft = 0.3048 m:

$$1\,\text{ft}^2 = 1\,\text{ft}^2 \times \frac{0.3048\,\text{m}}{1\,\text{ft}} \times \frac{0.3048\,\text{m}}{1\,\text{ft}} = 0.0929\,\text{m}^2$$

### Section 1-6  Order of Magnitude: Rapid Estimating

There are times that we want to make rough estimates of a quantity based on our intuition and experience. We may want to do this to avoid the time it takes for a detailed calculation or possibly we may not really know how to make a detailed calculation. Other times we might want to check a detailed calculation by making a rough estimate. Involved in learning how to do this rough estimation is knowledge of the size of the units that we use.

**Example 1-6-A**

Estimate the number of grains of sand needed to make a beach volleyball court.

**Solution:**

A volleyball court is approximately 10 m × 20 m. The sand must be about 20 cm or 0.2 m deep. This means the volume, $V$, of sand needed is about 10 m × 20 m × 0.2 m = 40 m$^3$. The volume of a single grain of sand depends on how fine the sand is. Assuming each sand grain is a millimeter ($10^{-3}$ m) on a side gives a grain volume, $V_{\text{grain}}$, of $(10^{-3}\text{ m})^3 = 10^{-9}$ m$^3$. Dividing the total volume of sand by the volume of one grain gives the number of grains:

$$N = V/V_{\text{grain}} = 40\text{ m}^3/10^{-9}\text{ m}^3 = 4 \times 10^{10}$$

Try counting the grains of sand on a court sometime and see how close this estimate is!

### Section 1-7  Dimensions and Dimensional Analysis

The dimensions of a physical quantity specify the type of physical quantity it is. In an equation involving physical quantities, the dimensions on each side of the equation must be identical. A quantity with dimensions of time on one side of the equation cannot be equal to a quantity with dimensions of length divided by time. This, in principle, is the same as "you can't add apples and oranges" that you may have heard many times in your life.

Applying dimensional analysis to an equation can sometimes find incorrect relationships between quantities, but it cannot be used to show that a relationship is correct.

**Example 1-7-A**

Determine which of the following relationships cannot possibly be correct expressions for the volume of a cone with a regular hexagonal base with side $s$ and height $h$:

a) $\pi s h^2$      b) $\frac{3}{2}\sqrt{3}s^2 h$      c) $\sqrt{3}sh/2$      d) $\pi s^2 h \cos(s/h)$      e) $s^3/h^2$

**Solution:**

Volume has dimensions of $[\text{L}^3]$. The length of a side of the base, $s$, and the height of the cone, $h$, each have dimensions of $[\text{L}]$. For expression of the form $s^n h^m$, we must have

$$[\text{L}^3] = [\text{L}]^n[\text{L}]^m \quad \text{or} \quad 3 = n + m$$

In the list of expressions above, only a, b, and d meet this criteria so c and e cannot be possible expressions for the volume of the cone.  Note that the expression in answer d also has a factor that depends on the dimensionless combination $s/h$.  Dimensional analysis is unable to determine any dependence on dimensionless combinations of the physical quantities involved in the expression.

## Practice Quiz

1.    Estimate the number of revolutions a car tire must make for a car to travel one mile.

   a)   10
   b)   100
   c)   1000
   d)   10,000

2.    Given a measurement of the length of a stick to be 0.3 m, what would be an appropriate expression for the length in feet?

   a)   1 ft
   b)   0.98 ft
   c)   0.984 ft
   d)   0.1 ft

3.    What is the conversion factor between cubic light-years ($ly^3$) and cubic meters ($m^3$)?

   a)   $8.5 \times 10^{47}$ $m^3/ly^3$
   b)   $9.5 \times 10^{15}$ $m^3/ly^3$
   c)   $1.2 \times 10^{-48}$ $m^3/ly^3$
   d)   $9.0 \times 10^{31}$ $m^3/ly^3$

4.    An expression for the position of a certain car as a function of time is given by $At^3/d$ where $t$ is time and $d$ has dimensions of length.  What are the dimensions of $A$?

   a)   $[L/T^2]$
   b)   $[L^2/T^2]$
   c)   $[L^2/T^3]$
   d)   $[L^3/T^4]$

5.    Estimate the number of crayons sold in the United States each year.

   a)   $10^5$
   b)   $10^7$
   c)   $10^9$
   d)   $10^{11}$

6.    If the fractional uncertainty in $x$ is 1%, what is the approximate fractional uncertainty in $x^2$?

   a)   10%
   b)   2%
   c)   1%
   d)   0.01%

7.    If the fractional uncertainty in $x$ is 1%, what is the approximate fractional uncertainty in $x^n$?

   a)   $(1 + n\%$
   b)   $n \times 1\%$
   c)   1%
   d)   $(1/n)\%$

8.      Estimate the number of word entries in a typical college dictionary of length 500 pages.

     a)  1000
     b)  10,000
     c)  100,000
     d)  1,000,000

9.      One year is often approximated as $\pi \times 10^7$ s.  What fractional uncertainty is made by using this approximation?

     a)  0.005%
     b)  0.5%
     c)  1.4%
     d)  $1.4 \times 10^5$%

10.     What would your hourly wages in dollars be if you earned a megadollar/gigasecond?

     a)  $3.60
     b)  $0.001
     c)  $1.00
     d)  $3600

11.     Divide 1.546 by 0.05673 and report the result with the appropriate number of significant figures.

12.     What is the percent uncertainty in the area of a rug that is a rectangle with dimensions 3.44 m by 4.67 m?

13.     What percentage error is being made if a liter container is mistaken for a quart container?

14.     Estimate the number of baseballs used in major league baseball during a season.

15.     Quantity $A$ has dimensions [ML/T$^3$], quantity $B$ has dimensions [M/T], quantity $C$ has dimensions [L/T] and quantity $D$ has dimensions [M$^2$].  Determine a possible expression for quantity $A$ in terms of quantities $B$, $C$, and $D$.

## Problem Solutions

Solutions to problems with a number equal to $4n - 1$ where $n$ is an integer:

3.      (a)  $1,156 = 1.156 \times 10^3$.
     (b)  $21.8 = 2.18 \times 10^1$.
     (c)  $0.0068 = 6.8 \times 10^{-3}$.
     (d)  $27.635 = 2.7635 \times 10^1$.
     (e)  $0.219 = 2.19 \times 10^{-1}$.
     (f)  $22 = 2.2 \times 10^1$.

7.      We assume an uncertainty of 0.2 s.
     (a)  % uncertainty = [(0.2 s)/(5 s)] 100 = 4%.
        Because the uncertainty has 1 significant figure, the % uncertainty has 1 significant figure.
     (b)  % uncertainty = [(0.2 s)/(50 s)] 100 = 0.4%.
        Because the uncertainty has 1 significant figure, the % uncertainty has 1 significant figure.
     (c)  % uncertainty = [(0.2 s)/(5 min)(60 s/min)] 100 = 0.07%.
        Because the uncertainty has 1 significant figure, the % uncertainty has 1 significant figure.

11.     We compare the volume with the specified radius to the volume for the extreme radius.
$$V_1 = \tfrac{4}{3}\pi R_1^3 = \tfrac{4}{3}\pi (2.86\text{ m})^3 = 98.0\text{ m}^3;$$
$$V_2 = \tfrac{4}{3}\pi R_2^3 = \tfrac{4}{3}\pi (2.86\text{ m} + 0.08\text{ m})^3 = 106.45\text{ m}^3,$$
so the uncertainty in the volume is $\Delta V = V_2 - V_1 = 8.5$ m$^3$; and the % uncertainty is
     % uncertainty = [(8.5 m$^3$)/(98.0 m$^3$)](100 %) = 9%.
We could also treat the change as a differential:
$$dV = 4\pi R^2\, dR = 4\pi (2.86\text{ m})^2(0.08\text{ m}) = 8.22\text{ m}^3.$$

15.    If we assume a height of 5 ft 10 in, we have
$$\text{height} = 5 \text{ ft } 10 \text{ in} = 70 \text{ in} = (70 \text{ in})[(1 \text{ m})/(39.37 \text{ in})] = 1.8 \text{ m}.$$

19.    (*a*)  $1.0 \times 10^{-10} \text{ m} = (1.0 \times 10^{-10} \text{ m})[(1 \text{ in})/(2.54 \text{ cm})][(100 \text{ cm})/(1 \text{ m})] = 3.9 \times 10^{-9} \text{ in}.$
       (*b*)  We let the units lead us to the answer:
$$(1.0 \text{ cm})[(1 \text{ m})/(100 \text{ cm})][(1 \text{ atom})/(1.0 \times 10^{-10} \text{ m})] = 1.0 \times 10^{8} \text{ atoms}.$$

23.    (*a*)  $1.00 \text{ ly} = (2.998 \times 10^{8} \text{ m/s})(1.00 \text{ yr})[(365.25 \text{ days})/(1 \text{ yr})][(24 \text{ h})/(1 \text{ day})][(3600 \text{ s})/(1 \text{ h})]$
$$= 9.46 \times 10^{15} \text{ m}.$$
       (*b*)  $1.00 \text{ ly} = (9.46 \times 10^{15} \text{ m})[(1 \text{ AU})/(1.50 \times 10^{8} \text{ km})][(1 \text{ km})/(1000 \text{ m})] = 6.31 \times 10^{4} \text{ AU}.$
       (*c*)  speed of light $= (2.998 \times 10^{8} \text{ m/s})[(1 \text{ AU})/(1.50 \times 10^{8} \text{ km})][(1 \text{ km})/(1000 \text{ m})][(3600 \text{ s})/(1 \text{ h})]$
$$= 7.20 \text{ AU/h}.$$

27.    We assume a rectangular house 40 ft × 30 ft, 8 ft high; so the total wall area is
$$A_{\text{total}} = [2(40 \text{ ft}) + 2(30 \text{ ft})](8 \text{ ft}) \approx 1000 \text{ ft}^{2}.$$
       If we assume there are 12 windows with dimensions 3 ft × 5 ft, the window area is
$$A_{\text{window}} = 12(3 \text{ ft})(5 \text{ ft}) \approx 200 \text{ ft}^{2}.$$
       Thus we have
$$\% \text{ window area} = [A_{\text{window}}/A_{\text{total}}](100) = [(200 \text{ ft}^{2})/(1000 \text{ ft}^{2})](100) \approx 20 \%.$$

31.    We assume that each dentist sees 10 patients/day, 5 days/wk for 48 wk/yr, for a total number of
       visits of
$$N_{\text{visits}} = (10 \text{ visits/day})(5 \text{ days/wk})(48 \text{ wk/yr}) \approx 2400 \text{ visits/yr}.$$
       We assume that each person sees a dentist 2 times/yr.
       (*a*)  We assume that the population of San Francisco is 700,000.  We let the units lead us to the answer:
$$N = (700,000)(2 \text{ visits/yr})/(2400 \text{ visits/yr}) \approx 600 \text{ dentists}.$$
       (*b*)  Left to the reader to estimate the population.

35.    For the right triangle shown on the diagram we have
$$r^{2} + d^{2} = (r + h)^{2};$$
$$d^{2} = 2rh + h^{2} = 2(6.4 \times 10^{6} \text{ m})(200 \text{ m}) + (200 \text{ m})^{2}, \text{ which gives}$$
$$d = 5.1 \times 10^{4} \text{ m} = 51 \text{ km}.$$

39.    (*a*)  $1.0 \text{ Å} = (1.0 \times 10^{-10} \text{ m})/(10^{-9} \text{ m/nm}) = 0.10 \text{ nm}.$
       (*b*)  $1.0 \text{ Å} = (1.0 \ 10^{-10} \text{ m})/(10^{-15} \text{ m/fm}) = 1.0 \times 10^{5} \text{ fm}.$
       (*c*)  $1.0 \text{ m} = (1.0 \text{ m})/(10^{-10} \text{ m/Å}) = 1.0 \times 10^{10} \text{ Å}.$
       (*d*)  From the result for Problem 23, we have
$$1.0 \text{ ly} = (9.5 \times 10^{15} \text{ m})/(10^{-10} \text{ m/Å}) = 9.5 \times 10^{25} \text{ Å}.$$

43.    (*a*)  The maximum number of buses is needed during rush hour.  If we assume that at any time there are 40,000
       persons commuting by bus and each bus has 30 passengers, we have
$$N = (40,000 \text{ commuters})/(30 \text{ passengers/bus}) \approx 1000 \text{ buses} \approx 1000 \text{ drivers}.$$
       (*b*)  Left to the reader.

47.    We will convert all units to meters.  The volume used in one year is
$$V = [(40,000 \text{ persons})/(4 \text{ persons/family})](1200 \text{ L/family a day})(365 \text{ days/yr})(10^{-3} \text{ m}^{3}/\text{L})$$
$$= 4.4 \times 10^{6} \text{ m}^{3}.$$
       If we let *d* represent the loss in depth, we have
$$d = V/\text{area} = (4.4 \times 10^{6} \text{ m}^{3})/[(50 \text{ km}^{2})(10^{3} \text{ m/km})^{2}] \approx 0.09 \text{ m} \approx 9 \text{ cm}.$$

51.    We assume that we can walk an average of 15 miles a day.  If we ignore the impossibility of walking on water
       and travel around the equator, the time required is
$$\text{time} = 2\pi \, r_{\text{Earth}}/\text{speed} = 2\pi \, (6 \times 10^{3} \text{ km})(0.621 \text{ mi/km})/(15 \text{ mi/day})(365 \text{ days/yr})$$
$$\approx 4 \text{ yr}.$$

55.    The trigonometric function is not a linear function of the angle, so we can find the uncertainty in the sine by calculating two values.

(a)  Percent uncertainty in $\theta$ = $(0.5°/15.0°)$ 100% = 3%.

We find the percent uncertainty in the sine from

$\sin 15.0° = 0.2588$;

$\sin 15.5° = 0.2672$;

Percent uncertainty in $\sin \theta$ = $[(0.2672 - 0.2588)/0.2588]$ 100% = 3%.

(b)  Percent uncertainty in $\theta$ = $(0.5°/75.0°)$ 100 = 0.7%.

We find the percent uncertainty in the sine from

$\sin 75.0° = 0.9659$;

$\sin 75.5° = 0.9681$;

Percent uncertainty in $\sin \theta$ = $[(0.9681 - 0.9659)/0.9659]$ 100 % = 0.2%.

Note that it is possible to approximate uncertainties with differential quantities, if the angle is not very small (i. e., $\sin \theta$ is not small ).  We have

$d(\sin \theta) = \cos \theta \, d\theta$,

$d(\sin \theta)/(\sin \theta) = d\theta/(\tan \theta)$.

The angle must be in radians, so we get

$d(\sin \theta)/(\sin \theta) = [(0.5°)( \pi /180°)]/\tan 15.0° = 0.03 = 3\%$;

$d(\sin \theta)/(\sin \theta) = [(0.5°)( \pi /180°)]/\tan 75.0° = 0.002 = 0.2\%$.

# Chapter 2: Describing Motion: Kinematics in One Dimension

## Chapter Overview and Objectives

This chapter defines the quantities that are used to describe motion of objects in one-dimensional motion. These quantities include position, displacement, velocity, and acceleration. The relationships between these quantities are also developed in this chapter.

After completing study of this chapter, you should:

- Know and understand the definitions of the kinematical variables used to describe motion.
- Know the equations of constant accelerated motion.
- Know the integral relationships between the different kinematical variables.
- Know the SI units of kinematical quantities.
- Be able to apply the definitions of kinematical variables and the equations of constant accelerated motion to solve problems about the one-dimensional translational motion of objects.

## Summary of Equations

Definition of average velocity: $\bar{v} = \dfrac{\Delta x}{\Delta t}$

Definition of instantaneous velocity: $v = \dfrac{dx}{dt}$

Definition of average acceleration: $\bar{a} = \dfrac{\Delta v}{\Delta t}$

Definition of instantaneous acceleration: $a = \dfrac{dv}{dt}$

Relationships between position, velocity, acceleration, and time for constant accelerated motion:

$$x = x_0 + v_0 t + \tfrac{1}{2} a t^2 \qquad\qquad v^2 - v_0^2 = 2a(x - x_0)$$

$$v = v_0 + at \qquad\qquad x = x_0 + \tfrac{1}{2}(v + v_0)t$$

General integral relationships between position, velocity, acceleration, and time:

$$\int_{v_1}^{v_2} dv = \int_{t_1}^{t_2} a \, dt \qquad\qquad \int_{x_1}^{x_2} dx = \int_{t_1}^{t_2} v \, dt = \int_{t_1}^{t_2} \left( v_1 + \int_{t_1}^{t'} a \, dt' \right) dt$$

## Chapter Summary

### Section 2-1 Reference Frames and Displacement

Measurements of position, velocity, speed, and acceleration depend on the frame of reference of the observer. A coordinate system is used to describe positions within a frame of reference. Change in position from one point to another is defined as **displacement**. Displacement is a vector quantity, a quantity that has magnitude ( or size ) and direction.

### Section 2-2 Average Velocity

There are two closely related physical quantities used to describe motion, **speed** and **velocity**. In every day language, the names of those two quantities are often used synonymously, but in physics they have distinct meanings. **Average speed** is the distance traveled by an object along its path divided by the amount of time it takes to travel that distance.

$$average\ speed = \frac{distance\ traveled}{time\ elapsed}$$

**Average velocity** is a vector quantity that is defined as the displacement of an object divided by the time to make that displacement. The direction of the average velocity vector is in the same direction as the displacement vector.

$$average\ velocity = \bar{v} = \frac{\Delta x}{\Delta t} = \frac{x_2 - x_1}{t_2 - t_1}$$

## Example 2-2-A

A car travels along a road that runs east-west. It first travels east at 24.2 m/s for 562 s. It then turns around and travels back to the place it started in a time of 634 s. What is the average velocity for the entire trip? What is the average speed for the entire trip? What is the average velocity for the return trip?

**Solution:**

The car returns to the position that it starts from. That means that the total displacement for the entire trip is zero. Average velocity is the displacement divided by the time that displacement takes, so the average velocity for the entire trip is zero.

To determine the average speed for the trip, we need to know the total path length traveled by the car during the trip. We need to use the information about the eastward part of the trip to determine how far the car traveled east.

$$\Delta x = v\Delta t = (24.2\ \text{m/s})(562\ \text{s}) = 1.36 \times 10^4\ \text{m}$$

The total path length is twice this distance because the car travels this distance again when returning to the starting position. The time the trip takes is 562 s + 634 s = 1196 s. The average speed for the entire trip is

$$average\ speed = \frac{distance\ traveled}{time\ elapsed} = \frac{2.72 \times 10^4\ \text{m}}{1196\ \text{s}} = 22.7\ \text{m/s}$$

The average velocity during the second leg of the trip is the displacement during the second leg of the trip, $1.36 \times 10^4$ m west, divided by the time that leg takes, 634 s:

$$\bar{v} = \frac{\Delta x}{\Delta t} = \frac{1.36 \times 10^4\ \text{m west}}{634\ \text{s}} = 21.5\ \text{m/s west}$$

## Section 2-3 Instantaneous Velocity

The **instantaneous velocity** is defined as the time derivative of the position:

$$instantaneous\ velocity = v = \lim_{t \to 0} \frac{\Delta x}{\Delta t} = \frac{dx}{dt}$$

The **instantaneous speed** is the magnitude of the instantaneous velocity.

## Example 2-3-A

The position of an object is given as x = 2.3 m + 1.4 m/s $t$ + 0.65 m/s $t^3$. What is the velocity of the object as a function of time?

**Solution:**

$$v = \frac{dx}{dt} = \frac{d}{dt}\left(2.3\ m + 1.4\ \text{m/s}\,t + 0.65\ \text{m/s}^3\,t^3\right) = 1.4\ \text{m/s} + 1.95\ \text{m/s}^3\,t^2$$

**Section 2-4 Acceleration**

The quantity **acceleration** measures the rate of change of velocity. **Average velocity** is the change in velocity divided by the amount of time it takes to make the change:

$$average\ acceleration = \frac{change\ in\ velocity}{time\ elapsed} = \frac{v_2 - v_1}{t_2 - t_1} = \frac{\Delta v}{\Delta t}$$

The **instantaneous acceleration** is the limit of the average velocity as the time elapsed goes to zero or the time derivative of the acceleration:

$$(instantaneous)\ acceleration = \lim_{t \to 0} \frac{\Delta v}{\Delta t} = \frac{dv}{dt} = \frac{d^2 x}{dt^2}$$

**Example 2-4-A**

An object has a position given by $x = 1.24$ m/s $t^4$. What is the acceleration of the particle?

**Solution:**

$$a = \frac{d^2 x}{dt^2} = \frac{d}{dt}\frac{dx}{dt} = \frac{d}{dt}\frac{d}{dt}\left\{1.24\,\text{m/s}^4\,t^4\right\} = \frac{d}{dt}\left\{4.96\,\text{m/s}^4\,t^3\right\} = 14.9\,\text{m/s}^4\,t^2$$

**Section 2-5 Motion at Constant Acceleration**

A common situation of accelerated motion is the case where the acceleration is constant (or approximated by a constant). In this case of constant acceleration, the relationships between position, velocity, acceleration, and time take the form:

$$x = x_0 + v_0 t + \tfrac{1}{2} a t^2 \qquad\qquad v^2 - v_0^2 = 2a(x - x_0)$$

$$v = v_0 + at \qquad\qquad x = x_0 + \tfrac{1}{2}(v + v_0)t$$

Be careful that these equations are only used under the conditions that the acceleration is constant.

**Example 2-5-A**

An automobile is traveling with a speed of 28.5 m/s when the driver sees a stop sign in front of him a distance 93.7 m away. a) What constant acceleration must the automobile have so that the car comes to rest at the stop sign? b) How long does it take the automobile to reach the stop sign with this acceleration?

**Solution:**

First, read the problem carefully to determine that it is a problem with constant acceleration and that the following information is given:

$$v_0 = 28.5\ \text{m/s} \qquad\qquad v = 0 \qquad\qquad x - x_0 = 93.7\ \text{m}$$

In part a), you are trying to find the acceleration. Looking at the relationships between kinematical variables that apply under conditions of constant acceleration, we find that the relationship

$$v^2 - v_0^2 = 2a(x - x_0)$$

relates the known kinematical variables, ($v_0$, $v$, $x - x_0$), to the unknown, $a$. Algebraically solving for $a$:

$$a = \frac{v^2 - v_0^2}{2(x - x_0)} = \frac{(0)^2 - (28.5\,\text{m/s})^2}{2(93.7\,\text{m})} = -4.33\ \text{m/s}^2$$

In part b), you are trying to find the time it takes to reach the stop sign, $t$. Any of the constant acceleration kinematical relationships that include $t$ can now be used to solve for $t$. For example, using

$$x = x_0 + \tfrac{1}{2}(v + v_0)t$$

and solving for $t$ results in

$$t = \frac{2(x - x_0)}{v + v_0} = \frac{2(93.7\,\text{m})}{0 + 28.5\,\text{m/s}} = 6.58\,\text{s}$$

### Example 2-5-B

A car is traveling at a speed of 18.6 m/s and is entering a 12.3 m wide intersection when the light has 0.582 s more time remaining yellow. What does the acceleration of the car need to be so that its front is out of the intersection when the light turns red? What will be the velocity of the car as it leaves the intersection if it has this acceleration

**Solution:**

We are given that the initial velocity, $v_0$, of the car is 18.6 m/s, the amount of time, $t$, the car travels must be 0.582 s, and the displacement the car during that time must be 12.3 m. We are asked to find the acceleration, $a$, of the car. We use the constant acceleration kinematic relationship that relates initial velocity, displacement, time, and acceleration:

$$x = x_0 + v_0 t + \tfrac{1}{2}at^2$$

and solve for the acceleration:

$$a = \frac{2(x - x_0 - v_0 t)}{t^2} = \frac{2(12.3\,\text{m} - (18.6\,\text{m/s})(0.582\,\text{s}))}{(0.582\,\text{s})^2} = 8.71\,\text{m/s}$$

To then find the velocity of the car as it exits the intersection, the relationship between initial velocity, final velocity, acceleration and time can be used:

$$v = v_0 + at = 18.6\,\text{m/s} + \left(8.71\,\text{m/s}^2\right)(0.582\,\text{s}) = 23.7\,\text{m/s}$$

### Example 2-5-C

A car is traveling with velocity of 10.2 m/s and is accelerating at 1.68 m/s$^2$ in the direction it is traveling. How long does it take to travel 389 m?

**Solution:**

The initial velocity, $v_0$, is given as 10.2 m/s, the acceleration, $a$, is given as 1.68 m/s$^2$, and the displacement, $x - x_0$, is given as 389 m. The question asks for the amount of time to make this displacement. The constant acceleration kinematical equation that relates these quantities is

$$x = x_0 + v_0 t + \tfrac{1}{2}at^2$$

which written in the standard form for a quadratic equation in $t$ is

$$\tfrac{1}{2}at^2 + v_0 t - (x - x_0) = 0$$

Using the quadratic formula to solve for $t$

$$t = \frac{-v_0 \pm \sqrt{v_0^2 + 2a(x - x_0)}}{a} = \frac{-10.2 \text{ m/s} \pm \sqrt{(10.2 \text{ m/s})^2 + 2(1.68 \text{ m/s}^2)(389 \text{ m})}}{1.68 \text{ m/s}^2}$$

$$t = 16.3 \text{ s} \quad \text{or} \quad -28.4 \text{ s}$$

Which of these solutions is the desired one and why are there two solutions? The solution we are looking for is 16.3 s because the time we are looking for occurs after the initial conditions. The second solution exists because if the situation were traced backward in time and the condition of constant acceleration in the current direction of travel existed at those times, the car would be found at the same position as our final position traveling in the opposite direction at a time −28.4 s prior to the initial conditions of the problem.

## Section 2-6 Solving Problems

It is very tempting to try to solve physics problems by searching for an equation that seems to apply and blindly plug in numbers to get an answer. This method of solving problems might produce some correct answers for relatively simple problems, but is sure to fail for even moderately complicated problems. Just getting the right answers is not a measurement of your understanding physics, you must understand the logical process of getting to the answer. To assist you in learning this logical thought process, an outline of a general approach to problems is given:

1.  Read the problem carefully, forming an image in your mind of the problem. Reread the problem to see if you missed any details and gain further understanding of what the situation is and what is being asked for.
2.  Draw a diagram of the situation. Include a set of coordinate axes whenever needed.
3.  Write down the quantities that are given in the problem and the quantities that are to be determined.
4.  Think about what physical principles apply to the problem.
5.  Determine which relationships that you know of relate the quantities that are known to the unknowns in the problem. A single equation may not directly relate the knowns to the unknowns. You may need to use several different equations or multiple applications of the same relationship to solve the problem.
6.  Carry out calculations if a numerical result is needed. Results of intermediate calculations should be kept to a precision of one or two significant figures more than the final solution to prevent cumulative round-off error. Round off the final result to the appropriate number of significant figures.
7.  Try to decide if the result is reasonable. Does it make sense to your intuition and experience? Make a rough estimate of your calculations using powers of ten (order of magnitude calculation). This may not catch all errors, but often catches silly mistakes that have been made.
8.  Check to see that units have been kept track of correctly. If the same units do not occur on both sides of the equation, you have made an error.

## Section 2-7 Falling Objects

Near the surface of the earth, objects in free fall have constant accelerated motion in the vertical direction. The acceleration is 9.8 m/s$^2$ downward. The constant $g$ is the magnitude of this acceleration, 9.8 m/s$^2$.

## Example 2-7-A

An object is thrown upward. It reaches a height of 10.7 m above its release point and then reaches a height of 35.3 m a time 3.24 s after it was at a height of 10.7 m. What was the velocity of the object as it passed the height of 10.7 m? What was the initial upward velocity of the object at the time it was released?

## Solution:

We are given the position of the object at two different times we will call $t_1$ and $t_2$, and we are given the difference, $t_2 - t_1$. We will label the corresponding displacements from the release point $x_1 - x_0$ and $x_2 - x_0$, which are given. The acceleration is $g$ in the downward direction. The question asks for the initial velocity, $v_0$. We can use the constant acceleration kinematic relationship that relates position, initial velocity, final velocity, acceleration, and time to each other for each of the given positions:

$$v_2 - v_1 = -g(t_2 - t_1) \qquad \text{and} \qquad x_2 - x_1 = \tfrac{1}{2}(v_2 + v_1)(t_2 - t_1)$$

where the upward direction has been taken as positive and $v_1$ is the velocity at time $t_1$ and $v_2$ is the velocity at time $t_2$. Eliminating $v_2$ from these equations algebraically gives

$$v_1 = \tfrac{1}{2} g(t_2 - t_1) + \frac{2(x_2 - x_1)}{(t_2 - t_1)} = \tfrac{1}{2}(9.8 \text{ m/s}^2)(3.24 \text{ s}) + \frac{2(35.3 \text{ m} - 10.7 \text{ m})}{3.24 \text{ s}} = 31.1 \text{ m/s}$$

Knowing the velocity at $t_1$ allows us to use a relationship between initial velocity, final velocity, acceleration and displacement to the velocity at the time of the release of the object, $v_0$:

$$v_1^2 - v_0^2 = -2g(x_1 - x_0)$$

Solving this for $v_0$ results in

$$v_0 = \sqrt{v_1^2 + 2g(x_1 - x_0)} = \sqrt{(31.1 \text{ m/s})^2 + 2(9.8 \text{ m/s}^2)(10.7 \text{ m})} = 34.3 \text{ m/s}$$

## Section 2-8 Use of Calculus; Variable Acceleration

From the definition of velocity, we can write $dv = a\, dt$. Both sides of this equation can be integrated:

$$\int_{v'=v_0}^{v} dv' = \int_{t'=0}^{t} a\, dt'$$

Completing the integration on the left side of the equation:

$$v - v_0 = \int_{t'=0}^{t} a\, dt' \qquad \text{or} \qquad v = v_0 + \int_{t'=0}^{t} a\, dt'$$

Similarly, from the definition of velocity, we can write $dx = v\, dt$. Again, both sides of this equation can be integrated:

$$\int_{x'=x_0}^{x} dx' = \int_{t'=0}^{t} v\, dt'$$

Completing the integration on the left side of the equation and substituting the above expression for $v$:

$$x - x_0 = \int_{t'=0}^{t}\left\{ v_0 + \int_{t''=0}^{t'} a\, dt'' \right\} dt' \qquad \text{or} \qquad x = x_0 + v_0 t + \int_{t'=0}^{t}\int_{t''=0}^{t'} a\, dt''\, dt'$$

## Example 2-8-A

An object starts from rest and has an acceleration given by $a = 3.07 \text{ m/s}^4\, t^2 - 4.21 \text{ m/s}^3\, t$.  a) Determine the velocity of the object at the end of 4.55 s.  b) Determine how far the object is from its starting point at the end of 4.55 s.

## Solution:

First recognize that this is *not* motion with constant acceleration!  That means the more general integral relationships between position, velocity, and acceleration must be used to determine velocity and position from acceleration.

a)  $v = v_0 + \int_{t'=0}^{t} a\, dt' = 0 + \int_{t'=0}^{4.55 \text{ s}} \left\{ 3.07 \text{ m/s}^4\, t'^2 - 4.21 \text{ m/s}^3\, t' \right\} dt'$

$v = \left\{ 3.07 \text{ m/s}^4\, \dfrac{t'^3}{3} - 4.21 \text{m/s}^3\, \dfrac{t'^2}{2} \right\}\Big|_{t'=0}^{4.55 \text{ s}} = 96.39 \text{ m/s} - 43.58 \text{ m/s} = 52.8 \text{ m/s}$

b) $x = x_0 + v_0 t + \int_{t'=0}^{t} \int_{t''=0}^{t'} a\, dt''\, dt' = 0 + 0 + \int_{t'=0}^{4.55\,s} \int_{t''=0}^{t'} \left\{ 3.07 \text{ m/s}^4\, t''^2 - 4.21 \text{ m/s}^3\, t'' \right\} dt''\, dt'$

$x = \int_{t'=0}^{4.55\,s} \left\{ 3.07 \text{ m/s}^4\, \frac{t'^3}{3} - 4.21 \text{ m/s}^3\, \frac{t'^2}{2} \right\} dt' = \left\{ 3.07 \text{ m/s}^4\, \frac{t'^4}{12} - 4.21 \text{ m/s}^3\, \frac{t'^3}{6} \right\} \Big|_{t'=0}^{4.55\,s}$

$x = 109.6 \text{ m} - 66.1 \text{ m} = 43.5 \text{ m}$

## Practice Quiz

1.  Over a particular time interval $t$, an object has an average speed of 10.67 m/s. What can you conclude about the magnitude of the average velocity of the object during that same time interval?

    a)  The magnitude of the average velocity is equal to 10.67 m/s.
    b)  The magnitude of the average velocity could be any value.
    c)  The magnitude of the average velocity is less than or equal to 10.67 m/s.
    d)  The magnitude of the average velocity is zero.

2.  At a given instant, the instantaneous speed of a car is 23.45 m/s. What can you conclude about the magnitude of the instantaneous velocity?

    a)  The magnitude of the instantaneous velocity is equal to 23.45 m/s.
    b)  The magnitude of the instantaneous velocity could be any value.
    c)  The magnitude of the instantaneous velocity is less than or equal to 23.45 m/s
    d)  The magnitude of the instantaneous velocity is zero.

3.  The velocity and the acceleration of an object are in opposite directions. What is the motion of the object at the time those conditions are true?

    a)  The object is traveling with constant speed and direction.
    b)  The object is traveling with constant speed but is changing direction.
    c)  The object is speeding up.
    d)  The object is slowing down.

Questions 4 through 7 refer to the following graph of position vs. time.

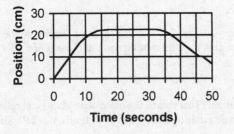

4.  During which time interval on the graph is the average speed the greatest?

    a)  0 to 7 s
    b)  7 s to 15 s
    c)  15 s to 33 s
    d)  33 s to 37 s
    e)  37 s to 50 s

5.      During which time interval on the graph is the average speed the least?

        a)  0 to 7 s
        b)  7 s to 15 s
        c)  15 s to 33 s
        d)  33 s to 37 s
        e)  37 s to 50 s

6.      During which time interval is the acceleration opposite in direction to the velocity?

        a)  0 to 7 s
        b)  7 s to 15 s
        c)  15 s to 33 s
        d)  33 s to 37 s
        e)  37 s to 50 s

7.      During which time interval is the object slowing down?

        a)  0 to 7 s
        b)  7 s to 15 s
        c)  15 s to 33 s
        d)  33 s to 37 s
        e)  37 s to 50 s

8.      A car changes velocity from 12.6 m/s to a velocity of 26.7 m/s in a time such that the average acceleration is 4.8 m/s$^2$.  If the time for the velocity change to occur was twice as much, what would the acceleration be?

        a)  4.8 m/s$^2$
        b)  2.4 m/s$^2$
        c)  1.2 m/s$^2$
        d)  9.6 m/s$^2$
        e)  19.2 m/s$^2$

9.      Car A moves from point $x_1$ to point $x_2$ with constant *velocity* in a time $t$. Car B moves from point $x_1$ to point $x_2$ with constant *acceleration* after starting from rest in the same time $t$.  Which car had the greater average velocity in going from $x_1$ to $x_2$?

        a)  A
        b)  B
        c)  Neither; the average velocity is the same for both.
        d)  Not enough information is given to determine which had the greater average velocity.

10.     Which car in problem 9 reached the greater instantaneous speed in going from point $x_1$ to $x_2$?

        a)  A
        b)  B
        c)  Neither; the average velocity is the same for both.
        d)  Not enough information is given to determine which had the greater average velocity.

11.     A ball is thrown upward vertically in the air with a speed of 24.2 m/s and is released from a height of 2.18 m above the ground.  What is the maximum height that the ball reaches above the ground?

12.     How long does it take the ball in question number 11 to hit the ground from the time it is released?

13.     A race car accelerates from rest at 5.31 m/s$^2$ for 8.05 s and then travels with constant velocity.  A second race car accelerates from rest at the same time as the first car at 4.78 m/s$^2$ for 8.95 s and then travels with constant velocity.  Which car wins the race if the race is 400 m long?

14.    A car travels with a constant speed of 30 mph for a distance of 45 miles. The car then travels at a speed of 60 mph for another 45 miles. What is the average speed of the car for the 90 mile trip?

15.    A ball is thrown downward with an initial speed of 30 m/s from a height of 135 m from the ground. How long does it take to hit the ground?

## Problem Solutions

Solutions to problems with a number equal to $4n - 1$ where $n$ is an integer:

3.    Problem asks for distance traveled, $d$, in a time, $t$, of 2.0 s while traveling with a speed, $v$, of 110 km/h.

First, change the speed, $v$, into SI units:        110 km/h (1000 m/km ) (1 h / 3600 s )= 30.56 m/s

Then:    $d = vt = (30.56 \text{ m/s}) \times 2.0 \text{ s} = 61 \text{ m}$

7.     Because there is no elapsed time when the light arrives, the sound travels one mile in 5 seconds.
We find the speed of sound from
        speed = $d/t$ = (1 mi)(1610 m/1 mi)/(5 s) = 300 m/s.

11.    (a) We find the instantaneous velocity from the slope of the straight line from $t = 0$ to $t = 10.0$ s:
        $v_{10} = \Delta x/\Delta t$ = (2.8 m – 0)/(10.0 s – 0) = 0.28 m/s.
(b) We find the instantaneous velocity from the slope of a tangent to the line at $t = 30.0$ s:
        $v_{30} = \Delta x/\Delta t$ = (22 m – 10 m)/(35 s – 25 s) = 1.2 m/s.
(c) The velocity is constant for the first 17 s (position vs. time is a straight line), so the velocity is the same as the velocity at $t = 10$ s:
        $\bar{v}_{0 \to 5}$ = 0.28 m/s.
(d) For the average velocity we have
        $\bar{v}_{25 \to 30} = \Delta x/\Delta t$ = (16 m – 8 m)/(30.0 s – 25.0 s) = 1.6 m/s.
(e) For the average velocity we have
        $\bar{v}_{40 \to 50} = \Delta x/\Delta t$ = (10 m – 20 m)/(50.0 s – 40.0 s) = –1.0 m/s.

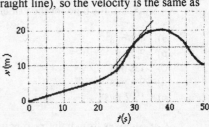

15.    Because the velocities are constant, we can use the relative speed of the car to find the time:
        $t = d/v_{\text{rel}}$ = [(0.100 km)/(90 km/h – 75 km/h)](60 min/h) = 0.40 min = 24 s.

19.    We find the time for the sound to travel the length of the lane from
        $t_{\text{sound}} = d/v_{\text{sound}}$ = (16.5 m)/(340 m/s) = 0.0485 s.
The time for the ball to travel the length of the lane is $T - t_{\text{sound}}$. We find the speed of the ball from
        $v = d/(T - t_{\text{sound}})$
            = (16.5 m)/(2.50 s – 0.0485 s) = 6.73 m/s.

23.    We find the acceleration (assumed to be constant) from
        $v^2 = v_0^2 + 2a(x_2 - x_1)$;
        0 = [(100 km/h)/(3.6 ks/h)]$^2$ + 2a(55 m), which gives $a$ = –7.0 m/s$^2$.
The number of $g$'s is
        $N = |a|/g$ = (7.0 m/s$^2$)/(9.80 m/s$^2$) = 0.72.

27.    We find the velocity and acceleration by differentiating $x$ = (6.0 m/s)$t$ + (8.5 m/s$^2$)$t^2$:
        $v = dx/dt$ = (6.0 m/s) + (17 m/s$^2$)$t$;
        $a = dv/dt$ = 17 m/s$^2$.

31.     We find the length of the runway from
        $v^2 = v_0^2 + 2aL$;
        (32 m/s)$^2$ = 0 + 2(3.0 m/s$^2$) $L$, which gives $L$ = 1.7 $\times$ 10$^2$ m.

35.    For constant acceleration the average speed is ½($v + v_0$), thus
        $x$ = ½($v + v_0$)$t$
            = ½(0 + 22.0 m/s)(5.00 s) = 55.0 m.

39.    We convert the units for the speed:  $(90 \text{ km/h})/(3.6 \text{ ks/h}) = 25 \text{ m/s}$.
       With the origin at the beginning of the reaction, the location when the brakes are applied is
$$x_0 = v_0 t = (25 \text{ m/s})(1.0 \text{ s}) = 25 \text{ m}.$$
       (a)  We find the location of the car after the brakes are applied from
$$v^2 = v_0^2 + 2a_1(x_1 - x_0)$$
$$0 = (25 \text{ m/s})^2 + 2(-4.0 \text{ m/s}^2)(x_1 - 25 \text{ m}), \text{ which gives } x_1 = 103 \text{ m}.$$
       (b)  We repeat the calculation for the new acceleration:
$$v^2 = v_0^2 + 2a_2(x_2 - x_0);$$
$$0 = (25 \text{ m/s})^2 + 2(-8.0 \text{ m/s}^2)(x_2 - 25 \text{ m}), \text{ which gives } x_2 = 64 \text{ m}.$$

43.    (a)  We assume constant velocity of $v_0$ through the intersection.   The time to travel at this speed is
$$t = (d_S + d_I)/v_0 = t_R - (v_0/2a) + (d_I/v_0).$$
       (b)  For the two speeds we have
$$t_1 = t_R - (v_{01}/2a) + (d_I/v_{01})$$
$$= 0.500 \text{ s} - (30.0 \text{ km/h})/[(3.60 \text{ ks/h})2(-4.00 \text{ m/s}^2)] + [(14.4 \text{ m})(3.60 \text{ ks/h})/(30.0 \text{ km/h})]$$
$$= 3.27 \text{ s}.$$
$$t_2 = t_R - (v_{02}/2a) + (d_I/v_{02})$$
$$= 0.500 \text{ s} - (60.0 \text{ km/h})/[(3.60 \text{ ks/h})2(-4.00 \text{ m/s}^2)] + [(14.4 \text{ m})(3.60 \text{ ks/h})/(60.0 \text{ km/h})]$$
$$= 3.45 \text{ s}.$$
       Thus the chosen time is 3.45 s.

47.    We use a coordinate system with the origin at the top of the cliff and down positive.
       To find the time for the object to acquire the velocity, we have
$$v = v_0 + at;$$
$$(100 \text{ km/h})/(3.6 \text{ ks/h}) = 0 + (9.80 \text{ m/s}^2)t, \text{ which gives } t = 2.83 \text{ s}.$$

51.    We use a coordinate system with the origin at the ground and up positive.
       We can find the initial velocity from the maximum height (where the velocity is zero):
$$v^2 = v_0^2 + 2ah;$$
$$0 = v_0^2 + 2(-9.80 \text{ m/s}^2)(2.55 \text{ m}), \text{ which gives } v_0 = 7.07 \text{ m/s}.$$
       When the kangaroo returns to the ground, its displacement is zero.  For the entire jump we have
$$y = y_0 + v_0 t + \tfrac{1}{2}at^2;$$
$$0 = 0 + (7.07 \text{ m/s})t + \tfrac{1}{2} (-9.80 \text{ m/s}^2)t^2,$$
       which gives $t = 0$ (when the kangaroo jumps), and $t = 1.44$ s.

55.    We use a coordinate system with the origin at the ground and up positive.  When the package returns to the
       ground, its displacement is zero, so we have
$$y = y_0 + v_0 t + \tfrac{1}{2}at^2;$$
$$0 = 115 \text{ m} + (5.60 \text{ m/s})t + \tfrac{1}{2}(-9.80 \text{ m/s}^2)t^2.$$
       The solutions of this quadratic equation are $t = -4.31$ s and $t = 5.44$ s.
       Because the package is released at $t = 0$,
       the positive answer is the physical answer: 5.44 s.

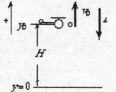

59.    We use a coordinate system with the origin at the release point and down positive.  On paper the apple measures
       6 mm, which we will call 6 mmp.  If its true diameter is 10 cm, the conversion is 0.10 m/6 mmp.  The images of
       the apple immediately after release overlap.  We will use the first clear image which is 9 mmp below the release
       point.  The final image is 62 mmp below the release point, and there are 7 intervals between these two images.
       The position of the apple is given by
$$y = y_0 + v_0 t + \tfrac{1}{2}at^2 = 0 + 0 + \tfrac{1}{2}gt^2.$$
       For the two selected images we have
$$y_1 = \tfrac{1}{2}gt_1^2; \quad (9 \text{ mmp})(0.10 \text{ m/6 mmp}) = \tfrac{1}{2} (9.8 \text{ m/s}^2)t_1^2, \text{ which gives } t_1 = 0.175 \text{ s};$$
$$y_2 = \tfrac{1}{2}gt_2^2; \quad (62 \text{ mmp})(0.10 \text{ m/6 mmp}) = \tfrac{1}{2} (9.8 \text{ m/s}^2)t_2^2, \text{ which gives } t_2 = 0.459 \text{ s}.$$
       Thus the time interval between successive images is
$$\Delta t = (t_2 - t_1)/7 = (0.459 \text{ s} - 0.175 \text{ s})/7 = 0.041 \text{ s}.$$

63.    If the height of the cliff is $H$, the time for the sound to travel from the ocean to the top is
$$t_{sound} = H/v_{sound}.$$
The time of fall for the rock is $T - t_{sound}$. We use a coordinate system with the origin at the top of the cliff and down positive. For the falling motion we have
$$y = y_0 + v_0 t + \tfrac{1}{2}at^2;$$
$$H = 0 + 0 + \tfrac{1}{2}a(T - t_{sound})^2 = \tfrac{1}{2}(9.80 \text{ m/s}^2)[3.4 \text{ s} - H/(340 \text{ m/s})]^2.$$
This is a quadratic equation for $H$:
$$4.24 \times 10^{-5} H^2 - 1.098H + 56.64 = 0, \text{ with } H \text{ in m; which has the solutions}$$
$$H = 5 \times 10^1 \text{ m}, 2.58 \times 10^4 \text{ m}.$$
The larger result corresponds to $t_{sound}$ greater than 3.4 s, so the height of the cliff is $5 \times 10^1$ m.

67.    (a) If we make the suggested change of variable, we have
$$u = g - kv, \text{ and } du = -k\,dv, \text{ and } a = g - kv = u.$$
Thus from the definition of acceleration, we have
$$a = dv/dt;$$
$$u = (-du/k)/dt, \text{ or } du/u = -k\,dt.$$
When we integrate, we get
$$\int_{g}^{g-kv} \frac{du}{u} = \int_{0}^{t} -k\,dt$$
which gives
$$\ln\left(\frac{g - kv}{g}\right) = -kt$$
or    $g - kv = -ge^{-kt}$
Thus the velocity as a function of time is
$$v = (g/k)(1 - e^{-kt}).$$
(b) When the falling body reaches its terminal velocity, the acceleration will be zero, so we have
$$a = g - kv_{term} = 0, \text{ or } v_{term} = g/k.$$
Note that this is the terminal velocity from the velocity expression because as $t \to \infty$, $e^{-kt} \to 0$.

71.    We assume that the seat belt keeps the occupant fixed with respect to the car. The distance the occupant moves with respect to the front end is the distance the front end collapses, so we have
$$v^2 = v_0^2 + 2a(x - x_0);$$
$$0 = [(100 \text{ km/h})/(3.6 \text{ ks/h})]^2 + 2(-30)(9.80 \text{ m/s}^2)(x - 0), \text{ which gives } x = 1.3 \text{ m}.$$

75.    (a) At the top of the motion the velocity is zero, so we find the maximum height of the second child from
$$v^2 = v_{02}^2 + 2ah_2;$$
$$0 = (5.0 \text{ m/s})^2 + 2(-9.80 \text{ m/s}^2)h_2, \text{ which gives } h_2 = 1.28 \text{ m} = 1.3 \text{ m}.$$
(b) If the first child reaches a height $h_1 = 1.5h_2$, we find the initial speed from
$$v^2 = v_{01}^2 + 2ah_1 = v_{01}^2 + 2a(1.5h_2) = v_{01}^2 + 1.5v_{02}^2;$$
$$v_{01}^2 = (1.5)(5.0 \text{ m/s})^2, \text{ which gives } v_{01} = 6.1 \text{ m/s}.$$
(c) We find the time for the first child from
$$y = y_0 + v_0 t + \tfrac{1}{2}at^2$$
$$0 = 0 + (6.1 \text{ m/s})t + \tfrac{1}{2}(-9.80 \text{ m/s}^2)t^2,$$
which gives $t = 0$ (when the child starts up), and t = 1.2 s.

79.    We convert the maximum speed units: $v_{max} = (90 \text{ km/h})/(3.6 \text{ ks/h}) = 25$ m/s.
(a) There are $(36 \text{ km})/0.80 \text{ km}) = 45$ trip segments, which means 46 stations (with 44 intermediate stations). In each segment there are three motions.
Motion 1 is the acceleration to $v_{max}$. We find the time for this motion from
$$v_{max} = v_{01} + a_1 t_1;$$
$$25 \text{ m/s} = 0 + (1.1 \text{ m/s}^2)t_1, \text{ which gives } t_1 = 22.7 \text{ s}.$$
We find the distance for this motion from
$$x_1 = x_{01} + v_{01}t + \tfrac{1}{2}a_1 t_1^2;$$
$$L_1 = 0 + 0 + \tfrac{1}{2}(1.1 \text{ m/s}^2)(22.73 \text{ s})^2 = 284 \text{ m}.$$
Motion 2 is the constant speed of $v_{max}$, for which we have
$$x_2 = x_{02} + v_{max}t_2;$$
$$L_2 = 0 + v_{max}t_2.$$
Motion 3 is the acceleration from $v_{max}$ to 0. We find the time for this motion from
$$0 = v_{max} + a_3 t_3;$$
$$0 = 25 \text{ m/s} + (-2.0 \text{ m/s}^2)t_3, \text{ which gives } t_3 = 12.5 \text{ s}.$$

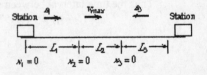

We find the distance for this motion from
$$x_3 = x_{03} + v_{max}t + \tfrac{1}{2}a_3t_3^2\ ;$$
$$L_3 = 0 + (25\text{ m/s})(12.5\text{ s}) + \tfrac{1}{2}(-2.0\text{ m/s}^2)(12.5\text{ s})^2 = 156\text{ m}.$$
The distance for Motion 2 is
$$L_2 = 800\text{ m} - L_1 - L_3 = 800\text{ m} - 284\text{ m} - 156\text{ m} = 360\text{ m}, \text{ so the time for Motion 2 is}$$
$$t_2 = L_2/v_{max} = (359.6\text{ m})/(25\text{ m/s}) = 14.4\text{ s}.$$
Thus the total time for the 45 segments and 44 stops is
$$T = 45(t_1 + t_2 + t_3) + 44(20\text{ s}) = 45(22.7\text{ s} + 14.4\text{ s} + 12.5\text{ s}) + 44(20\text{ s}) = 3112\text{ s} = 52\text{ min}.$$
(b)  There are (36 km / 3.0 km) = 12 trip segments, which means 13 stations (with 11 intermediate stations ).
The results for Motion 1 and Motion 3 are the same:
$$t_1 = 22.7\text{ s},\ L_1 = 284\text{ m},\ t_3 = 12.5\text{ s},\ L_3 = 156\text{ m}.$$
The distance for Motion 2 is
$$L_2 = 3000\text{ m} - L_1 - L_3 = 3000\text{ m} - 284\text{ m} - 156\text{ m} = 2560\text{ m}, \text{ so the time for Motion 2 is}$$
$$t_2 = L_2/v_{max} = (2560\text{ m})/(25\text{ m/s}) = 102\text{ s}.$$
Thus the total time for the 12 segments and 11 stops is
$$T = 12(t_1 + t_2 + t_3) + 11(20\text{ s}) = 12(22.7\text{ s} + 102\text{ s} + 12.5\text{ s}) + 11(20\text{ s}) = 1870\text{ s} = 31\text{ min}.$$
This means there is a higher average speed for stations farther apart.

83.  We use a coordinate system with the origin at the top of the cliff and up positive.
(a)  For the motion of the stone from the top of the cliff to the ground, we have
$$y = y_0 + v_0t + \tfrac{1}{2}at^2;$$
$$-65.0\text{ m} = 0 + (10.0\text{ m/s})t + \tfrac{1}{2}(-9.80\text{ m/s}^2)t^2.$$
This is a quadratic equation for $t$, which has the solutions $t = -2.76$ s, 4.80 s.
Because the stone starts at $t = 0$, the time is 4.80 s.
(b)  We find the speed from
$$v = v_0 + at$$
$$= 10.0\text{ m/s} + (-9.80\text{ m/s}^2)(4.80\text{ s}) = -37.0\text{ m/s}.$$
The negative sign indicates the downward direction, so the speed is 37.0 m/s.
(c)  The total distance includes the distance up to the maximum height, down to the top of the cliff, and down to the bottom.  We find the maximum height from
$$v^2 = v_0^2 + 2ah;$$
$$0 = (10.0\text{ m/s})^2 + 2(-9.80\text{ m/s}^2)h, \text{ which gives } h = 5.10\text{ m}.$$
The total distance traveled is
$$d = 5.10\text{ m} + 5.10\text{ m} + 65.0\text{ m} = 75.2\text{ m}.$$

87.  In each case we use a coordinate system with the origin at the beginning of the putt and the positive direction in the direction of the putt.  The limits on the putting distance are 6.0 m < $x$ < 8.0 m.
For the downhill putt we have:
$$v^2 = v_{0down}^2 + 2a_{down}(x - x_0);$$
$$0 = v_{0down}^2 + 2(-2.0\text{ m/s}^2)x.$$
When we use the limits for $x$, we get 4.9 m/s < $v_{0down}$ < 5.7 m/s, or $\Delta v_{0down} = 0.8$ m/s.
For the uphill putt we have:
$$v^2 = v_{0up}^2 + 2a_{up}(x - x_0);$$
$$0 = v_{0up}^2 + 2(-3.0\text{ m/s}^2)x.$$
When we use the limits for $x$, we get 6.0 m/s < $v_{0up}$ < 6.9 m/s,  or $\Delta v_{0up} = 0.9$ m/s.
The smaller spread in allowable initial velocities makes the downhill putt more difficult.

# Chapter 3: Kinematics In Two Dimensions; Vectors

## Chapter Overview and Objectives

This chapter concerns kinematics when objects move along more than just one direction. It introduces the concepts of vectors and applies all the kinematical concepts learned in Chapter 2 to positions described by vectors. The chapter applies these concepts to the special cases of projectile motion and uniform circular motion.

After completing study of this chapter, you should:

- Know the properties of a vector quantity as opposed to a scalar quantity.
- Know how to resolve two dimensional vectors into Cartesian components.
- Know how to add and subtract vectors, both graphically and by components.
- Know the vector definitions of velocity and acceleration.
- Know how to solve two-dimensional constant acceleration kinematical problems.
- Know that an object in uniform circular motion has centripetal acceleration.
- Know how to relate the velocity measurements made by two observers in relative motion to each other.

## Summary of Equations

Resolution of vector into Cartesian components:
$$\mathbf{V} = \mathbf{V}_x + \mathbf{V}_y + \mathbf{V}_z$$

Relationships between component magnitudes, vector magnitude, and direction:

$$V_x = V \cos\theta$$
$$V_y = V \sin\theta$$
$$V = \sqrt{V_x^2 + V_y^2}$$
$$\theta = \arctan\frac{V_y}{V_x}$$

Vector written as linear combination of Cartesian unit vectors:

$$\mathbf{V} = V_x\mathbf{i} + V_y\mathbf{j} + V_z\mathbf{k}$$

Vector definition of velocity:
$$\mathbf{v} = \frac{d\mathbf{r}}{dt}$$

Vector definition of acceleration:
$$\mathbf{a} = \frac{d\mathbf{v}}{dt}$$

Vector equations of motion for constant acceleration:

$$\mathbf{v} = \mathbf{v}_0 + \mathbf{a}t$$
$$\mathbf{r} = \mathbf{r}_0 + \mathbf{v}_0 t + \tfrac{1}{2}\mathbf{a}t^2$$
$$\bar{\mathbf{v}} = \frac{\mathbf{v} + \mathbf{v}_0}{2}$$

Centripetal acceleration:
$$a_R = \frac{v^2}{r}$$

Relationship between period and frequency:
$$T = \frac{1}{f}$$

Relationship between speed and radius of uniform circular motion and the period of the motion:

$$v = \frac{2\pi r}{T}$$

Relationship between velocities measured by different observers and the relative velocity of the observers:

$$\mathbf{v}_{AC} = \mathbf{v}_{AB} + \mathbf{v}_{BC}$$

## Chapter Summary

### Section 3-1  Vectors and Scalars

Quantities that have magnitude and direction are **vector** quantities.  Quantities that are completely specified by only a number without any dependence on the coordinate system used are called **scalars**.

### Section 3-2 Addition of Vectors -Graphical Methods

A graphical representation can be made of vector quantities.  An arrow is drawn with the direction of the arrow pointing in the corresponding direction of the vector quantity.  The length of the arrow is drawn such that the magnitude of the vector is proportional to the length of the arrow.

The operation of vector addition is a well-defined operation.  This operation can be introduced by defining a graphical method of adding vectors.  To add two vectors graphically:

1.   Decide on the scale for the length of the vectors.
2.   Draw a set of Cartesian axes to set up a direction reference.
3.   Draw the first of the two vectors to be added.  Start the vector at the origin of the coordinate system and draw it to the correct length and in the correct direction relative to the Cartesian axes.
4.   Draw the second vector to be added beginning from the end of the first vector.  Draw the vector of the correct length and in the appropriate direction.
5.   Draw an arrow from the origin of the coordinate axes to the end of the second vector.  This arrow's length is representative of the vector sum magnitude and its direction is the direction of the vector sum.

### Example 3-2-A

An airplane flies 300 km in a direction northeast.  It then flies 200 km north.  What is the plane's final displacement vector from its originating point?

### Solution:

We will solve this problem graphically.  Following the step by step procedures above:

1.   Determine a scale for the drawing:  We will let 1 cm on our drawing represent 100 km

```
├────┼────┼────┼────┼────┤
0   100  200  300  400  500  km
```

2.   Draw a coordinate axes for a direction reference:

North

East

3.   Draw the first vector in the appropriate direction and of length to scale:

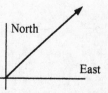

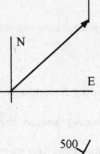

4.  Draw the second vector with its beginning at the tip of the first vector.   Draw it in the
    appropriate direction and of length to scale.

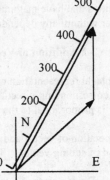

5.  Draw a vector from the beginning of the first vector drawn to the tip of the second
    arrow drawn.  This represents the sum of the two vectors being added.  Use the scale
    to determine the magnitude of the sum vector.  Here it appears the magnitude of the
    sum is about 460 km.  A protractor can be used to determine the direction of the
    resultant vector.

## Section 3-3  Subtraction of Vectors, and Multiplication of a Vector by a Scalar

A vector can be multiplied by a scalar.  Multiplying a scalar $c$ times vector $\mathbf{V}$ results in a vector that has a magnitude that
is the absolute value of $c$ times as big as the magnitude of $\mathbf{V}$. The resulting vector is in the same direction as $\mathbf{V}$ if $c$ is
positive and it is in the opposite direction of $\mathbf{V}$ if $c$ is negative.

Subtraction of two vectors, $\mathbf{A} - \mathbf{B}$, is defined as the addition of $\mathbf{A}$ and the vector $(-1)\mathbf{B}$

$$\mathbf{A} - \mathbf{B} = \mathbf{A} + (-1) \cdot \mathbf{B}$$

### Example 3-3-A

Graphically determine the difference of the first displacement and the second displacement of the airplane in
Example 3-2-A.

### Solution:

We will subtract the 200 km northward displacement from the 300 km northeast displacement.  To do so, we add the
negative of the 200 km displacement to the 300 km northeast displacement.  The negative of 200 km northward is 200
km southward.  The graphical addition appears as

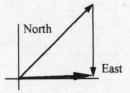

**Section 3-4  Adding Vectors by Components**

Graphical addition or subtraction of vectors has limited precision because of finite thickness of lines and limited precision of scales and protractors.  Sometimes greater precision than graphical methods can provide is necessary.  This additional precision can be obtained by using an algebraic method of vector addition.

To use the algebraic method of vector addition, we first need to choose a set of Cartesian coordinates to work with.  This means choosing a set of three orthogonal (perpendicular) directions.  We will restrict ourselves to only two dimensions here to help simplify the discussion.  We will usually choose the positive x-axis horizontal and to the right and the positive y-axis vertical and upward, but we are free to choose the directions as long as they are orthogonal.  Any vector, **V**, that lies in the plane of these two axes can be written uniquely as the sum of one vector in the x-axis direction, **V**$_x$, and one vector in the y-axis direction, **V**$_y$, as shown below

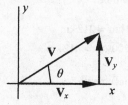

As a vector equation we write

$$\mathbf{V} = \mathbf{V}_x + \mathbf{V}_y$$

or if we needed to use all three spatial dimensions

$$\mathbf{V} = \mathbf{V}_x + \mathbf{V}_y + \mathbf{V}_z$$

There are some simple relations between the magnitudes of the component vectors, $V_x$ and $V_y$, the magnitude of the vector, $V$, and some trigonometric functions of the angle $\theta$:

$$V_x = V \cos\theta$$
$$V_y = V \sin\theta$$
$$V = \sqrt{V_x^2 + V_y^2}$$
$$\theta = \arctan\frac{V_y}{V_x}$$

To apply this to the addition of two vectors, it is easy to see that the magnitudes of the components of the sum of two vectors are simply the sums of the magnitudes of the corresponding components of the vectors being added.  See Figure 3-12 in the text.  If $\mathbf{V} = \mathbf{V}_1 + \mathbf{V}_2$, then

$$V_x = V_{1x} + V_{2x}$$
$$V_y = V_{1y} + V_{2y}$$

If the magnitude and the direction of the sum are desired, then the above relations between component magnitudes and sum magnitude and direction can be used.

**Example 3-4-A**

Add the vectors in Example 3-2-A by using vector components.

**Solution:**

We use a coordinate system with the positive x-axis toward the east and the positive y-axis toward the north.  We then express the two vectors as a sum of a vector in the x direction and a vector in the y direction.

For the first displacement, $\mathbf{V}_1$, 300 km northeast:

$$V_{1x} = 300\,\text{km}\cos 45° = 212\,\text{km}$$
$$V_{1y} = 300\,\text{km}\sin 45° = 212\,\text{km}$$

For the second displacement, $\mathbf{V}_1$, 300 km northeast:

$$V_{1x} = 0$$
$$V_{1y} = 200\,\text{km}$$

The component magnitudes for the sum of the vectors will be

$$V_x = V_{1x} + V_{2x} = 212\,\text{km} + 0\,\text{km} = 212\,\text{km}$$
$$V_y = V_{1y} + V_{2y} = 212\,\text{km} + 200\,\text{km} = 412\,\text{km}$$

We can determine the magnitude of the sum of the displacements

$$V = \sqrt{V_x^2 + V_y^2} = \sqrt{(212\,\text{km})^2 + (412\,\text{km})^2} = 463\,\text{km}$$

and the direction of the sum of the displacements:

$$\theta = \arctan\frac{V_y}{V_x} = \arctan\frac{412\,\text{km}}{212\,\text{km}} = 62.8°$$

where this angle is the direction toward north measured away from east.  Note the agreement between these results and the graphical results in Example 3-2-A.

### Section 3-5  Unit Vectors

A unit vector is a vector of magnitude one.  Any vector can be written as a scalar multiple of a unit vector.  Any vector can also be written uniquely as the sum of scalar multiples of a set of three orthogonal unit vectors (in three dimensions).  We commonly use the three unit vectors that lie along the direction of the positive x, y, and z axes of our three dimensional Cartesian coordinate system.  These three unit vectors are called $\mathbf{i}$, $\mathbf{j}$, and $\mathbf{k}$ respectively.

We can write any vector in terms of its components and the three Cartesian unit vectors

$$\mathbf{V} = V_x\mathbf{i} + V_y\mathbf{j} + V_z\mathbf{k}$$

### Example 3-5-A

A vector, $\mathbf{A}$, has a magnitude of 10.5 and a direction pointing 37.6° counter-clockwise from the positive x-axis.  Express this vector in terms of the Cartesian unit vectors.

### Solution:

We know how to calculate the components of a vector using simple trigonometry when the direction and magnitude of the vector are given.

$$A_x = A\cos\theta = (10.5)\cos 37.6° = 8.32$$
$$A_y = A\sin\theta = (10.5)\sin 37.6° = 6.41$$

We then express the vector as a linear combination of the Cartesian unit vectors:

$$\mathbf{A} = A_x\mathbf{i} + A_y\mathbf{j} = 8.32\mathbf{i} + 6.41\mathbf{j}$$

## Section 3-6  Vector Kinematics

The position of an object is a vector quantity. We will denote the position of an object by the vector $\mathbf{r}$. The definitions of our kinematical quantities velocity and acceleration remain as they were for one-dimensional motion. Velocity, $\mathbf{v}$, is the time derivative of the position and acceleration, $\mathbf{a}$, is the time derivative of the velocity:

$$\mathbf{v} = \frac{d\mathbf{r}}{dt} \quad and \quad \mathbf{a} = \frac{d\mathbf{v}}{dt}$$

The velocity and acceleration can be written in terms of the Cartesian unit vectors

$$\mathbf{v} = \frac{dx}{dt}\mathbf{i} + \frac{dy}{dt}\mathbf{j} + \frac{dz}{dt}\mathbf{k} = v_x\mathbf{i} + v_y\mathbf{j} + v_z\mathbf{k}$$

$$\mathbf{a} = \frac{dv_x}{dt}\mathbf{i} + \frac{dv_y}{dt}\mathbf{j} + \frac{dv_z}{dt}\mathbf{k} = a_x\mathbf{i} + a_y\mathbf{j} + a_z\mathbf{k}$$

The equations of constant accelerated motion also hold for the vector kinematical quantities:

$$\mathbf{v} = \mathbf{v}_0 + \mathbf{a}t$$

$$\mathbf{r} = \mathbf{r}_0 + \mathbf{v}_0 t + \tfrac{1}{2}\mathbf{a}t^2$$

$$\bar{\mathbf{v}} = \frac{\mathbf{v} + \mathbf{v}_0}{2}$$

where $\mathbf{r}$ is the position at time $t$, $\mathbf{r}_0$ is the position at time zero, $\mathbf{v}$ is the velocity at time $t$, $\mathbf{v}_0$ is the velocity at time zero, $\mathbf{v}$ is the average velocity from time zero to time $t$, and $\mathbf{a}$ is the constant acceleration. The remaining constant accelerated motion equation involving $v^2$ cannot be written as a vector equation because we don't yet have any meaning to assign to the square of a vector such as $\mathbf{v}^2$. However, we can write it in terms of the components of the vectors:

$$v_x^2 = v_{x0}^2 + 2a_x(x - x_0)$$

$$v_y^2 = v_{y0}^2 + 2a_y(y - y_0)$$

$$v_z^2 = v_{z0}^2 + 2a_z(z - z_0)$$

## Example 3-6-A

An object has an initial velocity $\mathbf{v}_0 = 2.0$ m/s $\mathbf{i} + 3.0$ m/s $\mathbf{j} - 1.5$ m/s $\mathbf{k}$ and a constant acceleration $\mathbf{a} = -1.0$ m/s$^2$ $\mathbf{i} + 1.0$ m/s$^2$ $\mathbf{j} + 2.5$ m/s$^2$ $\mathbf{k}$. What is the object's displacement after 3.0 s?

## Solution:

The displacement, $\Delta\mathbf{r}$, is the difference in the final position, $\mathbf{r}$, and initial position, $\mathbf{r}_0$:

$$\Delta\mathbf{r} = \mathbf{r} - \mathbf{r}_0 = \mathbf{v}_0 t + \tfrac{1}{2}\mathbf{a}t^2$$

$$= (2.0\,\text{m/s}\,\mathbf{i} + 3.0\,\text{m/s}\,\mathbf{j} + 1.5\,\text{m/s}\,\mathbf{k})(3.0\,\text{s}) + \tfrac{1}{2}(-1.0\,\text{m/s}^2\,\mathbf{i} + 1.0\,\text{m/s}^2\,\mathbf{j} + 2.5\,\text{m/s}^2\,\mathbf{k})(3.0\,\text{s})^2$$

$$= 1.5\,\text{m}\,\mathbf{i} + 13.5\,\text{m}\,\mathbf{j} + 15.8\,\text{m}\,\mathbf{k}$$

## Section 3-7  Projectile Motion

If we consider motion near the surface of the Earth for objects free to fall in the Earth's gravity, the motion undergoes constant acceleration in the downward vertical direction with a magnitude 9.80 m/s$^2$. This magnitude is called the local acceleration due to gravity and is given the symbol $g$. Because the direction of the acceleration of gravity is vertical, horizontal motion occurs with constant velocity under these conditions. Motion under these conditions is called **projectile motion**.

If we use a coordinate system with the x-axis parallel to the initial horizontal velocity, if any, and the y-axis positive in the upward vertical direction, we can write the kinematical equations for this constant accelerated motion as

<table>
<tr><td>Horizontal motion</td><td>Vertical motion</td></tr>
</table>

$$v_y = v_{y0} - gt$$

$$v_x = v_{x0}$$

$$x = x_0 + v_{x0}t$$

$$y = y_0 + v_{y0}t - \tfrac{1}{2}gt^2$$

$$v_y^2 = v_{y0}^2 - 2g(y - y_0)$$

In some cases, we are given the initial speed, $v$, and angle above horizontal, $\theta$, of the initial motion. The components of the initial horizontal velocity, $v_{x0}$, and the initial vertical velocity, $v_{y0}$, are related to the initial speed and angle by

$$v_{x0} = v_0 \cos\theta$$

$$v_{y0} = v_0 \sin\theta$$

Note that $\theta$ is negative if the initial velocity is below horizontal.

## Section 3-8  Solving Problems Involving Projectile Motion

Applying the general rules of problem solving to projectile motion problems:

1.  Read the problem carefully, trying to form a picture of the problem in your head. Reread the problem to be sure you have read it correctly.
2.  Choose a set of coordinate axes to use and draw a diagram.
3.  Make a list of known and unknown quantities, remembering that for projectile motion problems the horizontal acceleration is zero, the horizontal velocity is constant, and the vertical acceleration is downward with the magnitude $g = 9.80$ m/s$^2$.
4.  Think before selecting which of the equations of projectile motion to use before beginning to solve equations. Use the equations that allow you to solve for the unknown quantities in terms of the known quantities.

## Example 3-8-A

An airplane is climbing at a $10°$ angle above horizontal at an altitude of 366 m and a speed of 256 km/hr when a package is released from the airplane. How far does the package travel horizontally from the release point before hitting the ground? Assume air resistance is negligible.

## Solution:

Reading the problem carefully, we recognize that we are given the initial velocity of the package, which will be the same as the velocity of the airplane, 256 km/hr at $10°$ above horizontal. We are given the initial vertical position of the package (and airplane) 366 m above the ground. We know the final vertical position will be on the ground. The package is in free-fall so the acceleration is $g$ in the downward direction. We are asked to find the horizontal distance traveled. We choose a coordinate system with the x-axis horizontal and the positive y-axis vertical upward. Writing this information symbolically:

Horizontal
$v_{x0} = 256$ km/hr cos $10° = 252$ km/hr = 70.0 m/s
$x - x_0 = ?$
$a_x = 0$

Vertical
$v_{y0} = 256$ km/hr sin $10° = 44.4$ km/hr = 12.3 m/s
$y_0 = 366$ m
$y = 0$
$a_y = -9.8$ m/s$^2$

We know from looking at the information about the horizontal motion; we could answer the question if we knew the time to reach the ground. The information about the vertical motion is enough for us to find the time it takes the package to reach the ground:

$$y = y_0 + v_{y0}t + \tfrac{1}{2}a_y t^2$$

We can use the quadratic formula to solve for $t$:

$$t = \frac{-v_{y0} \pm \sqrt{v_{y0}^2 - 2a_y(y_0 - y)}}{a_y} = \frac{-12.3\,\text{m/s} \pm \sqrt{(12.3\,\text{m/s})^2 - 2(-9.8\,\text{m/s}^2)(366\,\text{m} - 0)}}{-9.80\,\text{m/s}^2}$$

$$= -7.48\,\text{s} \quad or \quad 9.99\,\text{s}$$

The negative solution is the time before the object was released that a projectile would have left the ground to reach the release point with the release velocity of the package. The positive solution for the time is the time at which the package will hit the ground after being released, which is what we are looking for. We can now easily solve for the horizontal position of the package:

$$x = x_0 + v_{x0}t \quad \Rightarrow \quad x - x_0 = v_{x0}t = (70.0\,\text{m/s})(9.99\,\text{s}) = 699\,\text{m}$$

## Example 3-8-B

A projectile is launched from a height of 28 m above the ground with a speed of 68 m/s. What angle must it be launched at to reach a target on the ground a distance of 184 m away horizontally?

**Solution:**

We are given the initial speed of the projectile, the initial and final vertical positions, and the acceleration of the object. We are looking for the direction of the initial velocity. Choosing a coordinate system with the x-axis horizontal and the positive y-axis vertical and upward:

Horizontal
$v_{x0} = v_0 \cos\theta = 68\,\text{m/s}\,\cos\,\theta$
$x - x_0 = 184\,\text{m}$
$a_x = 0$

Vertical
$v_{y0} = v_0 \sin\theta = 68\,\text{m/s}\,\sin\,\theta$
$y_0 = 28\,\text{m}$
$y = 0$
$a_y = -9.8\,\text{m/s}^2$

Although the time does not appear in the known quantities or unknown quantities, it ties the two sets of kinematical quantities together. The final $x$ occurs at the same time as the final $y$. We can solve for the time as a function of the other kinematical variables for both the horizontal and the vertical motion. We write down the equation of constant velocity motion in the x direction:

$$x = x_0 + v_{x0}t \quad \Rightarrow \quad t = \frac{x - x_0}{v_{x0}} = \frac{x - x_0}{v_0 \cos\theta}$$

and using the equation of constant accelerated motion for the y direction:

$$y = y_0 + v_{y0}t + \tfrac{1}{2}a_y t^2 = y_0 + v_0(\sin\theta)t + \tfrac{1}{2}a_y t^2$$

We substitute the expression for time above into the equation of vertical motion:

$$y = y_0 + (v_0 \sin\theta)\frac{x - x_0}{v_0 \cos\theta} + \tfrac{1}{2}a_y\left(\frac{x - x_0}{v_0 \cos\theta}\right)^2$$

Using the trigonometric identity $1 + \tan^2\theta = 1/\cos^2\,\theta$:

$$0 = y_0 - y + \tfrac{1}{2}a_y\left(\frac{x - x_0}{v_0}\right)^2 + (x - x_0)\tan\theta + \tfrac{1}{2}a_y\left(\frac{x - x_0}{v_0}\right)^2 \tan^2\theta$$

$$0 = -7.9 + 184\tan\theta - 35.9\tan^2\theta$$

This has solutions for tan $\theta$:

$$\tan \theta = 0.043 \quad \text{or} \quad \tan \theta = 5.08$$

Solving for $\theta$:

$$\theta = 2.5° \quad \text{or} \quad \theta = 78.9°$$

## Section 3-9  Uniform Circular Motion

**Uniform circular motion** is motion along a circular path at constant speed.  Note that even though the speed is constant, the velocity is not constant because it is continuously changing direction.  The direction of the acceleration of an object in uniform circular motion is toward the center of the circle, and  therefore is called **centripetal acceleration**.  The magnitude of the centripetal acceleration, $a_R$, is

$$a_R = \frac{v^2}{r}$$

where $v$ is the speed of the object and $r$ is the radius of the circular path.

The time to complete one complete circle is called the period, $T$, of the circular motion.  The number of complete circles per unit of time is called the frequency, $f$, of the circular motion.  The period and the frequency are related by

$$T = \frac{1}{f}$$

The period is related to the speed of the circular motion by

$$v = \frac{2\pi r}{T}$$

where $r$ is the radius of the circular motion.

## Example 3-9-A

A child sits on a merry-go-round that makes 10 revolutions in one minute.  The child is sitting 1.2 m from the rotation axis of the merry-go-round.  What is the acceleration of the child?

**Solution:**

The child makes 10 revolutions in one minute, so the period of the circular motion is 6.0 s.  We can then calculate the speed of the circular motion:

$$v = \frac{2\pi r}{T} = \frac{2\pi (1.2 \, \text{m})}{(6.0 \, \text{s})} = 1.26 \; \text{m/s}$$

Knowing the speed and the radius of the circle, we can calculate the centripetal acceleration:

$$a_R = \frac{v^2}{r} = \frac{1.26 \, \text{m/s}}{1.2 \, \text{m}} = 1.0 \, \text{m/s}$$

This acceleration is always directed toward the center of the merry-go-round from the location of the child.

## Section 3-10  Relative Velocity

Velocity measurements made in different inertial reference frames are related by

$$\mathbf{v}_{AC} = \mathbf{v}_{AB} + \mathbf{v}_{BC}$$

where $\mathbf{v}_{AB}$ is the velocity of the object A measured in reference frame B, $\mathbf{v}_{AC}$ is the velocity of the object A measured in reference frame C, and $\mathbf{v}_{BC}$ is the velocity of reference frame B measured in reference frame C.

We will often use the fact that the velocity of reference frame A as measured by an observer in reference frame B is the negative of the velocity of reference frame B as measured by an observer in reference frame A:

$$\mathbf{v}_{BA} = -\mathbf{v}_{AB}$$

## Example 3-10-A

An airplane flies at a speed of 300 km/hr directly northward relative to the air. The wind is blowing at 20 m/s toward the southwest. What is the velocity of the plane relative to the ground?

## Solution:

We will use the symbols $\mathbf{v}_{PA}$ for the velocity of the plane relative to the air, $\mathbf{v}_{AG}$ for the velocity of the air relative to the ground, and $\mathbf{v}_{PG}$ for the velocity of the plane relative to the ground. These velocities are related by

$$\mathbf{v}_{PG} = \mathbf{v}_{PA} + \mathbf{v}_{AG}$$

To determine the velocity of the plane relative to the ground, we need to add the velocity of the plane relative to the air to the velocity of the air relative to the ground (the wind velocity). To do so we set up a coordinate system with the positive x-axis pointing east and the positive y-axis pointing north. We can then express the two known velocities in terms of the Cartesian unit vectors:

$$\mathbf{v}_{PA} = -(20\,\text{m/s})\cos 45^{o}\,\mathbf{i} + -(20\,\text{m/s})\sin 45^{o}\,\mathbf{j} = -14.1\,\text{m/s}\,\mathbf{i} - 14.1\,\text{m/s}\,\mathbf{j}$$
$$\mathbf{v}_{AG} = (300\,\text{km/hr})/(3.6\,[\text{km/hr}]/[\text{m/s}])\mathbf{j} = 83.3\,\text{m/s}\,\mathbf{j}$$

This implies that the velocity of the plane relative to the ground is

$$\mathbf{v}_{PG} = -14.1\,\text{m/s}\,\mathbf{i} - 14.1\,\text{m/s}\,\mathbf{j} + 83.3\,\text{m/s}\,\mathbf{j}$$
$$= -14\,\text{m/s}\,\mathbf{i} + 69\,\text{m/s}\,\mathbf{j}$$

## Practice Quiz

1.      Which of the following is not a vector quantity?

   a)   Velocity
   b)   Acceleration
   c)   Displacement
   d)   Time

2.      Which inequality does the vector sum of vectors **A** and **B** necessarily satisfy?

   a)   The magnitude of **A** + **B** is greater than or equal to the magnitude of **A** and the magnitude of **B**.
   b)   The magnitude of **A** + **B** is greater than or equal to the magnitude of **A** plus the magnitude of **B**.
   c)   The magnitude of **A** + **B** is less than or equal to the magnitude of **A** and the magnitude of **B**.
   d)   The magnitude of **A** + **B** is less than or equal to the magnitude of **A** plus the magnitude of **B**.

3.      Given vector **A**, what vector added to it results in the zero vector?

   a)   **A**
   b)   −**A**
   c)   There is no such vector.
   d)   Zero

4.    A projectile is launched with an initial speed $v$. At a point in its trajectory above the initial launch position of the projectile, which of the following statements is necessarily true?

   a)   The speed of the projectile is greater than the launch speed.
   b)   The speed of the projectile is equal to the launch speed.
   c)   The speed of the projectile is less than the launch speed.
   d)   The angle of the velocity is steeper than the angle of the launch velocity.

5.    What are the conditions on the motion of a projectile at its highest point on its trajectory?

   a)   The speed and the acceleration are zero.
   b)   The horizontal velocity and the acceleration are zero.
   c)   The vertical velocity and the acceleration are zero.
   d)   The vertical velocity is zero.

6.    You are rounding a right hand turn in a car at constant speed. Which direction is your acceleration?

   a)   No direction. Your acceleration is zero because your speed is constant.
   b)   To your right.
   c)   To your left.
   d)   Straight ahead.

7.    A car travels around a curve of radius $r$ at constant speed $v$. Its acceleration has a magnitude $a$. If the car rounds the same curve so that its acceleration is $2a$, what will its speed be?

   a)   $v/2$
   b)   $2v$
   c)   $4v$
   d)   $\sqrt{2}\,v$

8.    A car travels around a curve of radius $r$ at constant speed $v$. Its acceleration has a magnitude $a$. If the car rounds a different curve at the same speed such that its acceleration is $2a$, what will the radius of the curve be?

   a)   $2r$
   b)   $r/2$
   c)   $4r$
   d)   $\sqrt{2}\,r$

9.    To observer A, an object is moving with a velocity $\mathbf{v}$. What would the velocity of observer B be relative to observer A, if observer B sees the object at rest?

   a)   $\mathbf{v}$
   b)   $-\mathbf{v}$
   c)   zero
   d)   That is impossible.

10.    Observer B is moving with a constant velocity relative to observer A. Observer A sees an object accelerate with an acceleration $\mathbf{a}$. What does observer B see as the acceleration of the object?

   a)   $\mathbf{a}$
   a)   $-\mathbf{a}$
   b)   Zero
   c)   Depends on the direction of the relative velocity of B to A.

11.    Graphically find the difference between a vector of magnitude 3.65 pointing 32° north of east and a vector of magnitude 5.88 pointing 12° north of west.

12.    Find the vector sum of a vector of magnitude 3.45 pointing 43° south of east and a vector of magnitude 7.63 pointing 21° west of south using vector components.

13. A diver leaps off the end of a diving board 3.00 m above the water with an initial velocity of 5.8 m/s at an angle 80° above horizontal. How long is the diver in the air before striking the water? How far horizontally from the end of the diving board does the diver hit the water? Ignore any changes in the orientation of the diver's body.

14. The earth moves in a nearly circular orbit around the sun once a year. Assuming the earth's orbit is circular, what is the centripetal acceleration of the earth in its orbit around the sun?

15. An airplane flies at a speed of 224 km/hr in a direction 12° east of north. An observer on the ground observes the plane moving at a speed of 241 km/hr in a direction 2° west of north. What is the wind velocity?

## Problem Solutions

Solutions to problems with a number equal to $4n - 1$ where $n$ is an integer:

3. From Fig. 3-6c, if we write the equivalent vector addition, we have
$$\mathbf{V}_1 + \mathbf{V}_{wrong} = \mathbf{V}_2, \quad \text{or} \quad \mathbf{V}_{wrong} = \mathbf{V}_2 - \mathbf{V}_1.$$

7. (a)

    (b) For the components of the vector we have
$$V_x = -V \cos \theta = -14.3 \cos 34.8° = -11.7;$$
$$V_y = V \sin \theta = 14.3 \sin 34.8° = 8.16.$$
    (c) We find the vector from
$$V = (V_x^2 + V_y^2)^{1/2} = [(-11.7)^2 + (8.16)^2]^{1/2}$$
$$= 14.3;$$
$$\tan \theta = V_y/V_x = (8.16)/(11.7) = 1.42, \text{ which gives}$$
$$\theta = 34.9° \text{ above} - x\text{-axis}.$$
This is within significant figures. Note that we have used the magnitude of $V_x$ for the angle indicated on the diagram.

11. The vectors are $\mathbf{V}_1 = 4\mathbf{i} - 8\mathbf{j}$, $\mathbf{V}_2 = \mathbf{i} + \mathbf{j}$, $\mathbf{V}_3 = -2\mathbf{i} + 4\mathbf{j}$.
    (a) For the sum $\mathbf{V}_1 + \mathbf{V}_2 + \mathbf{V}_3$ we have
$$\mathbf{V}_1 + \mathbf{V}_2 + \mathbf{V}_3 = 3\mathbf{i} - 3\mathbf{j}.$$
For the magnitude of $\mathbf{V}_1 + \mathbf{V}_2 + \mathbf{V}_3$ we have
$$|\mathbf{V}_1 + \mathbf{V}_2 + \mathbf{V}_3| = [(3)^2 + (-3)^2]^{1/2} = 4.2.$$
We find the direction from
$$\tan \theta_a = (-3)/(3) = -1.0.$$
From the signs of the components, we have $\theta_a = 45°$ below $+ x$-axis.
    (b) For $\mathbf{V}_1 - \mathbf{V}_2 + \mathbf{V}_3$ we have
$$\mathbf{V}_1 - \mathbf{V}_2 + \mathbf{V}_3 = \mathbf{i} - 5\mathbf{j}.$$
For the magnitude of $\mathbf{V}_1 - \mathbf{V}_2 + \mathbf{V}_3$ we have
$$|\mathbf{V}_1 - \mathbf{V}_2 + \mathbf{V}_3| = [(1)^2 + (-5)^2]^{1/2} = 5.1.$$
We find the direction from
$$\tan \theta_b = (-5)/(1) = -5.0.$$
From the signs of the components, we have $\theta_b = 79°$ below $+ x$-axis.

15.     (a)  For the components we have

$R_x = B_x - 2A_x$

$= -26.5 \cos 56.0° - 2(44.0 \cos 28.0°)$

$= -92.5;$

$R_y = B_y - 2A_y$

$= 26.5 \sin 56.0° - 2(44.0 \sin 28.0°)$

$= -19.3.$

We find the resultant from

$R = (R_x^2 + R_y^2)^{1/2} = [(-92.5)^2 + (-19.3)^2]^{1/2}$

$= 94.5;$

$\tan \theta = R_y/R_x = (19.3)/(92.5) = 0.209,$ which gives

$\theta = 11.8°$ below $-x$-axis.

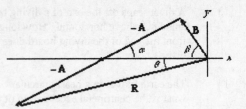

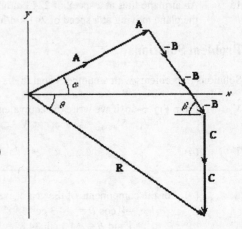

(b)  For the components we have

$R_x = 2A_x - 3B_x + 2C_x$

$= 2(44.0 \cos 28.0°) - 3(-26.5 \cos 56.0°) + 2(0)$

$= 122.2;$

$R_y = 2A_y - 3B_y + 2C_y$

$= 2(44.0 \sin 28.0°) - 3(26.5 \sin 56.0°) + 2(-31.0)$

$= -86.6.$

We find the resultant from

$R = (R_x^2 + R_y^2)^{1/2} = [(122.2)^2 + (-86.6)^2]^{1/2}$

$= 150;$

$\tan \theta = R_y/R_x = (86.6)/(122.2) = 0.709,$ which gives

$\theta = 35.3°$ below $+x$-axis.

19.     The positions of the particle at the two times are

$\mathbf{r}_1 = (7.60 \text{ m/s})(1.00 \text{ s})\mathbf{i} + (8.85 \text{ m})\mathbf{j} - (1.00 \text{ m/s}^2)(1.00 \text{ s})^2\mathbf{k} = (7.60 \text{ m})\mathbf{i} + (8.85 \text{ m})\mathbf{j} - (1.00 \text{ m})\mathbf{k};$

$\mathbf{r}_3 = (7.60 \text{ m/s})(3.00 \text{ s})\mathbf{i} + (8.85 \text{ m})\mathbf{j} - (1.00 \text{ m/s}^2)(3.00 \text{ s})^2\mathbf{k} = (22.8 \text{ m})\mathbf{i} + (8.85 \text{ m})\mathbf{j} - (9.00 \text{ m})\mathbf{k}.$

The average velocity is

$\mathbf{v}_{av} = \pi\mathbf{r}/\pi t = [(15.2 \text{ m})\mathbf{i} - (8.00 \text{ m})\mathbf{k}]/(3.00 \text{ s} - 1.00 \text{ s}) = (7.60 \text{ m/s})\mathbf{i} - (4.00 \text{ m/s})\mathbf{k}.$

The instantaneous velocity is given by $d\mathbf{r}/dt$:

$v(t) = (7.60 \text{ m/s})\mathbf{i} - (2.00 \text{ m/s}^2)t\mathbf{k}$

At the midpoint of the interval this is

$\mathbf{v}(2.00 \text{ s}) = (7.60 \text{ m/s})\mathbf{i} - (2.00 \text{ m/s}^2)(2.00 \text{ s})\mathbf{k} = (7.60 \text{ m/s})\mathbf{i} - (4.00 \text{ m/s})\mathbf{k}.$

Note that this is the same as the average velocity because the acceleration is constant. The magnitude of the instantaneous velocity at $t = 2.00$ s is

$v_2 = [(7.60 \text{ m/s})^2 + (4.00 \text{ m/s})^2]^{1/2} = 8.59 \text{ m/s}.$

23.     The acceleration is $\mathbf{a} = (4.0 \text{ m/s}^2)\mathbf{i} + (3.0 \text{ m/s}^2)\mathbf{j}.$

(a)  We find the velocity by integrating:

$$\int_0^v d\mathbf{v} = \int_0^t \mathbf{a}\, dt$$

$$\mathbf{v} = \int_0^t [(4.0 \text{ m/s}^2)\mathbf{i} + (3.0 \text{ m/s}^2)\mathbf{j}]dt = [(4.0 \text{ m/s}^2)t\,\mathbf{i} + (3.0 \text{ m/s}^2)t\,\mathbf{j}]$$

(b)  The speed of the particle is

$|\mathbf{v}| = \{[(4.0 \text{ m/s}^2)t]^2 + [(3.0 \text{ m/s}^2)t]^2\}^{1/2} = (5.0 \text{ m/s}^2)t.$

(c)  We find the position by integrating:

$$\int_0^r d\mathbf{r} = \int_0^t \mathbf{v}\, dt$$

$$\mathbf{r} = \int_0^t [(4.0 \text{ m/s}^2)t\,\mathbf{i} + (3.0 \text{ m/s}^2)t\,\mathbf{j}]dt = [(2.0 \text{ m/s}^2)t^2\,\mathbf{i} + (1.5 \text{ m/s}^2)t^2\,\mathbf{j}]$$

(*d*)  For the given time we have
$$\mathbf{v} = (4.0 \text{ m/s}^2)t\mathbf{i} + (3.0 \text{ m/s}^2)t\mathbf{j} = (4.0 \text{ m/s}^2)(2.0 \text{ s})\mathbf{i} + (3.0 \text{ m/s}^2)(2.0 \text{ s})\mathbf{j}$$
$$= (8.0 \text{ m/s})\mathbf{i} + (6.0 \text{ m/s})\mathbf{j}.$$
$$|\mathbf{v}| = (5.0 \text{ m/s}^2)t = (5.0 \text{ m/s}^2)(2.0 \text{ s}) = 10.0 \text{ m/s}.$$
$$\mathbf{r} = (2.0 \text{ m/s}^2)t^2\mathbf{i} + (1.5 \text{ m/s}^2)t^2\mathbf{j} = (2.0 \text{ m/s}^2)(2.0 \text{ s})^2\mathbf{i} + (1.5 \text{ m/s}^2)(2.0 \text{ s})^2\mathbf{j}$$
$$= (8.0 \text{ m})\mathbf{i} + (6.0 \text{ m})\mathbf{j}.$$

27.   We choose a coordinate system with the origin at the takeoff point, with *x* horizontal and *y* vertical, with the positive direction down.  We find the height of the cliff from the vertical displacement:
$$y = y_0 + v_{0y}t + \tfrac{1}{2}a_yt^2;$$
$$y = 0 + 0 + \tfrac{1}{2}(9.80 \text{ m/s}^2)(3.0 \text{ s})^2 = \ 44 \text{ m}.$$
The horizontal motion will have constant velocity.
We find the distance from the base of the cliff from
$$x = x_0 + v_{0x}t;$$
$$x = 0 + (2.1 \text{ m/s})(3.0 \text{ s}) = 6.3 \text{ m}.$$

31.   We find the time of flight from the vertical displacement:
$$y = y_0 + v_{0y}t + \tfrac{1}{2}a_yt^2;$$
$$0 = 0 + (18.0 \text{ m/s})(\sin 32.0°)t + \tfrac{1}{2}(-9.80 \text{ m/s}^2)t^2, \text{ which gives } t = 0, 1.95 \text{ s}.$$
The ball is kicked at *t* = 0, so the football hits the ground 1.95 s later.

35.   To plot the trajectory, we need a relationship between *x* and *y*, which can be obtained by eliminating *t* from the equations for the two components of the motion:
$$x = v_{0x}t = v_0 (\cos \theta)t;$$
$$y = y_0 + v_{0y}t + \tfrac{1}{2}a_yt^2 = 0 + v_0 (\sin \theta)t + \tfrac{1}{2}(-g)t^2.$$
The relationship is
$$y = (\tan \theta)x - \tfrac{1}{2}g[x/(v_0 \cos \theta)^2].$$

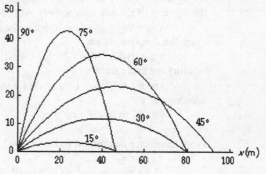

39.   (*a*)  At the highest point, the vertical velocity $v_y = 0$. We find the maximum height *h* from
$$v_y^2 = v_{0y}^2 + 2a_y(y - y_0);$$
$$0 = [(51.2 \text{ m/s}) \sin 44.5°]^2 + 2(-9.80 \text{ m/s}^2)(h - 0),$$
which gives *h* = 65.7 m.
      (*b*)  Because the projectile returns to the same elevation,
we have      $$y = y_0 + v_{0y}t + \tfrac{1}{2}a_yt^2;$$
$$0 = 0 + (51.2 \text{ m/s})(\sin 44.5°)t + \tfrac{1}{2}(-9.80 \text{ m/s}^2)t^2,$$
which gives *t* = 0, and 7.32 s.
Because *t* = 0 was the launch time, the total time in the air was 7.32 s.
      (*c*)  We find the horizontal distance from
$$x = v_{0x}t = (51.2 \text{ m/s})(\cos 44.5°)(7.32 \text{ s}) = 267 \text{ m}.$$
      (*d*)  The horizontal velocity will be constant: $v_x = v_{0x} = (51.2 \text{ m/s}) \cos 44.5° = 36.5 \text{ m/s}.$
      We find the vertical velocity from
$$v_y = v_{0y} + a_yt = (51.2 \text{ m/s}) \sin 44.5° + (-9.80 \text{ m/s}^2)(1.50 \text{ s}) = 21.2 \text{ m/s}.$$
      The magnitude of the velocity is
$$v = (v_x^2 + v_y^2)^{1/2} = [(36.5 \text{ m/s})^2 + (21.2 \text{ m/s})^2]^{1/2} = 42.2 \text{ m/s}.$$
      We find the angle from
$$\tan \theta = v_y/v_x = (21.2 \text{ m/s})/(36.5 \text{ m/s}) = 0.581, \text{ which gives } \theta = 30.1° \text{ above the horizontal.}$$

43.   The ball passes the goal posts when it has traveled the horizontal distance of 36.0 m.  From this we can find the time when it passes the goal posts:
$$x = v_{0x}t; \quad 36.0 \text{ m} = (20.0 \text{ m/s}) \cos 37.0° \ t, \text{ which gives } \ t = 2.25 \text{ s}.$$
To see if the kick is successful, we must find the height of the ball at this time:
$$y = y_0 + v_{0y}t + \tfrac{1}{2}a_yt^2 = 0 + (20.0 \text{ m/s}) \sin 37.0° (2.25 \text{ s}) + \tfrac{1}{2}(-9.80 \text{ m/s}^2)(2.25 \text{ s})^2$$
$$= 2.24 \text{ m}.$$
Thus the kick is *unsuccessful* because it passes 0.76 m below the bar.
To have a successful kick, the ball must pass the goal posts with an elevation of at least 3.00 m.  We find the time when the ball has this height from
$$y = y_0 + v_{0y}t + \tfrac{1}{2}a_yt^2;$$
$$3.00 \text{ m} = 0 + (20.0 \text{ m/s}) \sin 37.0° \ t \ + \tfrac{1}{2}(-9.80 \text{ m/s}^2)t^2.$$

The two solutions of this quadratic equation are $t = 0.28$ s, $2.17$ s. The horizontal distance traveled by the ball is found from

$x = v_{0x}t = (20.0 \text{ m/s}) \cos 37.0° t.$

For the two times, we get $x = 4.5$ m, $34.7$ m.

Thus the kick must be made no farther than $34.7$ m from the goal posts (and no nearer than $4.5$ m).

If the vertical velocity is found at these two times from

$v_y = v_{0y} + a_y t = (20.0 \text{ m/s}) \sin 37.0° + (-9.80 \text{ m/s}^2)t = +9.3 \text{ m/s}, -9.3 \text{ m/s},$

we see that the ball is falling at the goal posts for a kick from $34.7$ m and rising at the goal posts for a kick from $4.5$ m.

47.    We will take down as the positive direction. The direction of motion is the direction of the velocity. For the velocity components, we have

$v_x = v_{0x} = v_0$ .

$v_y = v_{0y} + a_y t = 0 + gt = gt.$

We find the angle that the velocity vector makes with the horizontal from

$\tan \theta = v_y/v_x = gt/v_0$ , or $\theta = \tan^{-1}(gt/v_0)$ below the horizontal.

51.    The centripetal acceleration is

$a_R = v^2/r = (500 \text{ m/s})^2/[(3.50 \times 10^3 \text{ m})(9.80 \text{ m/s}^2/g)] = 7.29g$ up.

55.    To complete an orbit in time $T$, the speed of the shuttle must be

$v = 2\pi r/T.$

Thus the centripetal acceleration in terms of $g$ is

$a_R/g = v^2/rg = (2\pi r/T)^2/rg = 4\pi^2 r/gT^2$

$= 4\pi^2(6.38 \times 10^6 \text{ m} + 0.40 \times 10^6 \text{ m})/[(9.80 \text{ m/s}^2)(90 \text{ min})(60 \text{ s/min})]^2 = 0.94g.$

59.    If $\mathbf{v}_{HR}$ is the velocity of Huck with respect to the raft, $\mathbf{v}_{HB}$ the velocity of Huck with respect to the bank, and $\mathbf{v}_{RB}$ the velocity of the raft with respect to the bank, then

$\mathbf{v}_{HB} = \mathbf{v}_{HR} + \mathbf{v}_{RB}$ , as shown in the diagram.

From the diagram we get

$v_{HB} = (v_{HR}^2 + v_{RB}^2)^{1/2} = [(1.0 \text{ m/s})^2 + (2.5 \text{ m/s})^2]^{1/2} = 2.7 \text{ m/s}.$

We find the angle from

$\tan \theta = v_{HR}/v_{RB} = (1.0 \text{ m/s})/(2.5 \text{ m/s}) = 0.40,$

which gives

$\theta = 22°$ from the river bank.

63.    From the vector diagram of Example 3–13, we have

$v_{BW}^2 = v_{BS}^2 + v_{WS}^2$ ;

$(1.85 \text{ m/s})^2 = v_{BS}^2 + (1.20 \text{ m/s})^2$ ,

which gives

$v_{BS} = 1.41 \text{ m/s}.$

67.    If $\mathbf{v}_{SB}$ is the velocity of the swimmer with respect to the bank, $\mathbf{v}_{SW}$ the velocity of the swimmer with respect to the water, and $\mathbf{v}_{WB}$ the velocity of the water with respect to the bank, then

$\mathbf{v}_{SB} = \mathbf{v}_{SW} + \mathbf{v}_{WB}$ , as shown in the diagram.

(*a*)  We find the angle from

$\tan \theta = v_{WB}/v_{SW} = (0.80 \text{ m/s})/(1.00 \text{ m/s}) = 0.80,$ which gives $\theta = 38.7°.$

Because the swimmer travels in a straight line, we have

$\tan \theta = d_{shore}/d_{river}$ ;  $0.80 = d_{shore}/(75 \text{ m}),$ which gives $d_{shore} = 60$ m.

(*b*)  We can find how long it takes by using the components across the river:

$t = d_{river}/v_{SW} = (75 \text{ m})/(1.00 \text{ m/s}) = 75$ s.

71.    If $\mathbf{v}_{CG}$ is the velocity of the car with respect to the ground, $\mathbf{v}_{MG}$ the velocity of the motorcycle with respect to the ground, and $\mathbf{v}_{MC}$ the velocity of the motorcycle with respect to the car, then

$\mathbf{v}_{MC} = \mathbf{v}_{MG} - \mathbf{v}_{CG}$ .

Because the motion is in one dimension, for the initial relative velocity we have

$v_{MC} = v_{MG} - v_{CG} = (95.0 \text{ km/h} - 75.0 \text{ km/h})/(3.6 \text{ ks/h}) = 5.56 \text{ m/s}.$

For the linear motion, in the reference frame of the car we have

$x = x_0 + v_0 t + \tfrac{1}{2}at^2;$

$60.0 \text{ m} = 0 + (5.56 \text{ m/s})(10.0 \text{ s}) + \tfrac{1}{2}a(10.0 \text{ s})^2,$ which gives $a = 0.0889 \text{ m/s}^2.$

Note that this is also the acceleration in the reference frame of the ground.

75.    The horizontal velocity is constant, and the vertical velocity will be zero when the pebbles hit the window. Using the coordinate system shown, we find the vertical component of the initial velocity from

$$v_y^2 = v_{0y}^2 + 2a_y(h - y_0) \; ;$$
$$0 = v_{0y}^2 + 2(-9.80 \text{ m/s}^2)(8.0 \text{ m} - 0), \text{ which gives } v_{0y} = 12.5 \text{ m/s}.$$

(We choose the positive square root because we know that the pebbles are thrown upward.)
We find the time for the pebbles to hit the window from the vertical motion:

$$v_y = v_{0y} + a_y t;$$
$$0 = 12.5 \text{ m/s} + (9.80 \text{ m/s}^2)t, \text{ which gives } t = 1.28 \text{ s}.$$

For the horizontal motion we have

$$x = x_0 + v_{0x}t;$$
$$9.0 \text{ m} = 0 + v_{0x}(1.28 \text{ s}), \text{ which gives } v_{0x} = 7.0 \text{ m/s}.$$

Because the pebbles are traveling horizontally when they hit the window, this is their speed.

79.    If $\mathbf{v}_{PW}$ is the velocity of the airplane with respect to the wind, $\mathbf{v}_{PG}$ the velocity of the airplane with respect to the ground, and $\mathbf{v}_{WG}$ the velocity of the wind with respect to the ground, then

$$\mathbf{v}_{PG} = \mathbf{v}_{PW} + \mathbf{v}_{WG}, \text{ as shown in the diagram}.$$

Because the plane has covered 180 km in 1.00 hour, $v_{PG} = 180$ km/h.
We use the diagram to write the component equations:

$$v_{WGE} = v_{PGE} = v_{PG} \sin 45° = (180 \text{ km/h}) \sin 45° = 127 \text{ km/h};$$
$$v_{WGN} = v_{PGN} - v_{PWN} = -v_{PG} \cos 45° - v_{PW}$$
$$= -(180 \text{ km/h}) \cos 45° - (-240 \text{ km/h}) = 113 \text{ km/h}.$$

For the magnitude we have

$$v_{WG} = (v_{WGE}^2 + v_{WGN}^2)^{1/2} = [(127 \text{ km/h})^2 + (113 \text{ km/h})^2]^{1/2} = 170 \text{ km/h}.$$

We find the angle from

$$\tan \theta = v_{WGN}/v_{WGE} = (113 \text{ km/h})/(127 \text{ km/h}) = 0.886,$$

which gives  $\theta = 41.5°$ N of E.

83.    We choose a coordinate system with the origin at the takeoff point, with $x$ horizontal and $y$ vertical, with the positive direction down.  We find the time for the diver to reach the water from the vertical motion:

$$y = y_0 + v_{0y}t + \tfrac{1}{2}a_y t^2;$$
$$35 \text{ m} = 0 + 0 + \tfrac{1}{2}(9.80 \text{ m/s}^2)t^2, \text{ which gives } t = 2.7 \text{ s}.$$

The horizontal motion will have constant velocity.
We find the minimum horizontal initial velocity needed to land beyond the rocky outcrop from

$$x = x_0 + v_{0x}t;$$
$$5.0 \text{ m} = 0 + v_0(2.7 \text{ s}), \text{ which gives } v_0 = 1.9 \text{ m/s}.$$

87.    If $\mathbf{v}_{AG}$ is the velocity of the automobile with respect to the ground, $\mathbf{v}_{HG}$ the velocity of the helicopter with respect to the ground, and $\mathbf{v}_{HA}$ the velocity of the helicopter with respect to the automobile, then

$$\mathbf{v}_{HA} = \mathbf{v}_{HG} - \mathbf{v}_{AG}.$$

For the horizontal relative velocity we have

$$v_{HA} = v_{HG} - v_{AG} = (200 \text{ km/h} - 150 \text{ km/h})/(3.6 \text{ ks/h}) = 13.9 \text{ m/s}.$$

This is the initial (horizontal) velocity of the document, so we can find the time of fall from

$$y = y_0 + v_{0y}t + \tfrac{1}{2}a_y t^2;$$
$$78.0 \text{ m} = 0 + 0 + \tfrac{1}{2}(+ 9.80 \text{ m/s}^2)t^2, \text{ which gives } t = 3.99 \text{ s}.$$

During this time, we find the horizontal distance the document travels with respect to the car from

$$x = v_{HA}t = (13.9 \text{ m/s})(3.99 \text{ s}) = 55.4 \text{ m}.$$

Because the helicopter is always directly above the document, this is how far behind the automobile the helicopter must be when the explosive is dropped.  Thus we find the angle from

$$\tan \theta = y/x = (78.0 \text{ m})/(55.4 \text{ m}) = 1.41, \text{ which gives } \theta = 54.6° \text{ below the horizontal}.$$

91.    We use the coordinate system shown in the diagram.  If Agent Logan heads downstream at speed $v_B$ at an angle $\theta$, the time required to cross the river is

$$t_1 = D/v_B \cos \theta.$$

The distance he will travel in the $y$-direction in this time is

$$y = (v_B \sin \theta + v_W)t_1 = (D/v_B)[(v_B \sin \theta + v_W)/\cos \theta].$$

Because he will be below the point directly across the river, he must run this distance, which, at speed $v_L$, will take a time

$$t_2 = y/v_L = (D/v_B v_L)[(v_B \sin \theta + v_W)/\cos \theta].$$

Thus the total time is

$$t = t_1 + t_2 = (D/v_B v_L)[(v_L + v_B \sin \theta + v_W)/\cos \theta].$$

To find the angle that produces the minimum time, we set $dt/d\theta = 0$:

$$\begin{aligned} dt/d\theta &= (D/v_Bv_L)\{v_B - [(v_L + v_B\sin\theta + v_W)(-\sin\theta)/\cos^2\theta]\} \\ &= (D/v_Bv_L)\{[v_B\cos^2\theta + (v_L + v_W)\sin\theta + v_B\sin^2\theta]/\cos^2\theta\} \\ &= (D/v_Bv_L\cos^2\theta)[v_B + (v_L + v_W)\sin\theta] = 0, \text{ which gives} \end{aligned}$$

$$\sin\theta = -v_B/(v_L + v_W) = -(1.50\text{ m/s})/(3.00\text{ m/s} + 0.80\text{ m/s}) = -0.395, \theta = -23.2°.$$

The time to reach the shore is

$$t_1 = D/v_B\cos\theta = (1600\text{ m})/(1.50\text{ m/s})\cos(-23.2°) = 1160\text{ s}.$$

The distance he must run is

$$y = (v_B\sin\theta + v_W)t_1 = [(1.50\text{ m/s})\sin(-23.2°) + 0.80\text{ m/s}](1160\text{ s}) = 241\text{ m}.$$

The running time is

$$t_2 = y/v_L = (241\text{ m})/(3.00\text{ m/s}) = 80\text{ s},$$

so the total time is

$$t = t_1 + t_2 = 1160\text{ s} + 80\text{ s} = 1240\text{ s} = 20.7\text{ min}.$$

Thus Agent Logan must row at an angle of 23° upstream and run 241 m in a total time of 20.7 min.

# Chapter 4: Dynamics: Newton's Laws of Motion

## Chapter Overview and Objectives

This chapter defines the quantities that are used to describe motion of objects in one-dimensional motion. These quantities include position, displacement, velocity, and acceleration. The relationships between these quantities are also developed in this chapter.

After completing study of this chapter, you should:

- Know and understand Newton's three laws of motion.
- Have an understanding of the concepts of mass and force.
- Understand the difference between mass and weight.
- Know the SI units of mass and force.
- Be able to apply Newton's laws of motion to solve for unknown forces, accelerations, or masses.

## Summary of Equations

Newton's Second Law of Motion: $\qquad\qquad \sum \vec{F} = m\vec{a}$

Newton's Third Law of Motion: $\qquad\qquad \vec{F}_{12} = -\vec{F}_{21}$

Weight: $\qquad\qquad\qquad\qquad\qquad \vec{F}_G = m\vec{g}$

## Chapter Summary

### Section 4-1 Force

**Force** is an interaction which in everyday life we would describe as a push or pull. For an object to accelerate, it must have a force applied to it. Force is a vector quantity because force has a magnitude and direction. Spring scales can be used to measure forces. The net force acting on a body is the vector sum of each of the forces acting on the body.

### Section 4-2 Newton's First Law of Motion

Every body continues in its state of rest or of uniform speed in a straight line unless acted on by a nonzero net force. This law of motion is only valid in inertial reference frames and can be viewed as a definition of inertial reference frames.

### Section 4-3 Mass

**Mass** is a measurement of a body's inertia against a change in motion. The standard of mass in the SI unit system is the **kilogram** (kg). Mass is a property of an object. **Weight** is the force of gravity acting on the object; it depends on both the object's mass and the location of the object near a gravitating body.

### Section 4-4 Newton's Second Law of Motion

The acceleration of a body is proportional to the net force acting on it and inversely proportional to the body's mass. The direction of the acceleration is the direction of the net force acting on the body. Written as a mathematical equation, Newton's second law is

$$\sum \vec{F} = m\vec{a}.$$

### Example 4-4-A

A runner with a mass of 80.6 kg runs a 50 m dash starting from rest with constant acceleration. If the runner finishes in a time of 6.2 s, what was the force of the ground acting on the runner?

**Solution:**

If you read this problem carefully, you should recognize that there is both kinematical information (distance, time, constant acceleration) and dynamical information (force) mentioned in the problem. As acceleration (which is not given) is the only quantity that ties kinematics to dynamics, you should realize that one kinematical relationship and Newton's second law are necessary to solve this problem.

The given quantities are:

$$v_0 = 0 \qquad\qquad t = 6.2\,\text{s} \qquad\qquad x - x_0 = 50\,\text{m}$$

We recognize that this is enough information to solve for the acceleration using the constant accelerated motion relationship:

$$x - x_0 = v_0 t - \tfrac{1}{2} a t^2$$

which implies:

$$a = \frac{2(x - x_0 - v_0 t)}{t^2} = \frac{2(50\,\text{m} - 0)}{(6.2\,\text{s})^2} = 2.60\,\text{m/s}^2$$

Knowing the acceleration and the mass, Newton's second law can be used to determine the force:

$$F = ma = (80.6\,\text{kg})(2.60\,\text{m/s}^2) = 210\,\text{N}$$

**Example 4-4-B**

A truck pulls a car on level ground. Starting from rest, the truck and car accelerate at 2.34 m/s². The horizontal force that the ground pushes on the truck with is $1.41 \times 10^4$ N and the mass of the truck is 3400 kg. a) What is the mass of the car? b) How much tension is in the cable that tows the car?

**Solution:**

You should recognize that this is a Newton's second law problem because force and acceleration are mentioned and possibly a Newton's third law problem because the car and the truck interact through the cable. When you draw free-body diagrams you might draw them as:

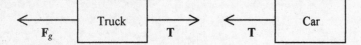

The symbol $\mathbf{F}_g$ is for the force of the ground on the car and $\mathbf{T}$ is the tension in the cable. Or, because the car and truck move with the same acceleration, they can be treated as the same body and the free-body diagram could be drawn as:

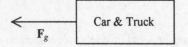

where the tension doesn't appear because it is an internal force. Both diagrams are correct, but you can see that the tension doesn't appear in the latter diagram, so it would not help you solve for the tension in the cable.

Using the upper diagrams, we write Newton's second law equations for each object:

$$\sum_{\text{on truck}} \mathbf{F} = \mathbf{F}_g + \mathbf{T} = m_t \mathbf{a}_t \qquad\qquad \sum_{\text{on car}} \mathbf{F} = \mathbf{T} = m_c \mathbf{a}_c$$

These are vector equations. We need to rewrite them as scalar equations in terms of the magnitude of the vectors as

$$-F_g + T = m_t(-a_t) \qquad\qquad -T = m_c(-a_c)$$

where we have chosen a coordinate system such that the positive direction is to the right in our free-body diagrams. Notice that we include a negative sign on the magnitude of all vectors that point to the left or negative direction. We notice that we can solve the first equation for the tension in terms of the remaining variables in the equation, for which we know the values:

$$T = m_t(-a_c) + F_g = (3400\,\text{kg})(-2.34\,\text{m/s}^2) + 1.4 \times 10^4\,\text{N} = 6.14\,\text{N}$$

Now we can use the second of the pair of equations above to solve for the mass of the car:

$$m_c = \frac{-T}{-a_c} = \frac{6.14\,\text{N}}{2.34\,\text{m/s}^2} = 2.62\,\text{kg}$$

## Section 4-5 Newton's Third Law of Motion

Force is the result of an interaction between two bodies. Each interacting body is acted on by a force during the interaction. The magnitude of the force on each body is identical and the forces on the two bodies are in opposite directions to each other. This is called Newton's third law of motion. If we use the convention that $\mathbf{F}_{AB}$ means the force on object A from object B, then Newton's third law can be written:

$$\mathbf{F}_{BA} = -\mathbf{F}_{AB}$$

### Example 4-5-A

A jet airplane's engines accelerated 50 kg of air at rest to a speed of 200 m/s as it passed through the engine a distance of 3.67 m. What was the force of the air on the airplane?

### Solution:

First, recognize that kinematic information about the air is given and you are asked about the force on the airplane. If the kinematic information is sufficient to solve for the acceleration of the air, then the force on the air from the airplane can be calculated using Newton's second law and then the force on the airplane from the air can be calculated using Newton's third law.

The following kinematical information about the air is given:

$$v_0 = 0 \qquad\qquad v = 200\,\text{m/s} \qquad\qquad x - x_0 = 3.67\,\text{m}$$

We can determine the acceleration using the constant accelerated motion relationship:

$$v^2 - v_0^2 = 2a(x - x_0)$$

which implies:

$$a = \frac{v^2 - v_0^2}{2(x - x_0)} = \frac{(200\,\text{m/s})^2 - 0^2}{2(3.67\,\text{m})} = 5.50 \times 10^3\,\text{m/s}^2$$

Now, Newton's second law can be used to determine the force on the air:

$$F_{on\ air} = ma = (50\,\text{kg})(5.50 \times 10^3\,\text{m/s}^2) = 2.8 \times 10^5\,\text{N}$$

Finally, Newton's third law can be used to determine the force on the airplane:

$$F_{on\ airplane} = -F_{on\ air} = -2.8 \times 10^5\ \text{N}$$

where the minus sign means the force on the airplane is opposite in direction to the force on the air.

### Section 4-6 Weight—the Force of Gravity; and the Normal Force

### Example 4-6-A

A box with a mass of 8.7 kg sits on a table. A person pushes down on the box with a force of 80 N at an angle $30^0$ below horizontal. Determine the normal force of the table on the box.

### Solution:

The table surface provides a constraint on the acceleration of the object in the direction perpendicular to the table surface. The acceleration can never be into the table's surface. The table pushes upward on the box with force that balances the other downward forces on the box. The weight of the box is a downward force on the box with a magnitude *mg*. The force the person applies to the box also has a downward component of force on the box:

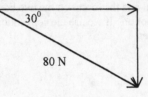

### Section 4-7 Solving Problems with Newton's Laws: Free-Body Diagrams

The **net force** on a body is the *vector* sum of all the forces acting on the body. An aid in correctly including all the forces acting on a given body is the construction of a free body diagram. A free body diagram is a simple representation of an object with an arrow drawn in the direction of each force acting on the body. Each force arrow is labeled in a way that makes it clear which of the forces involved in the problem that arrow represents. For example, suppose an object has three forces acting on it. The first force has a magnitude of 56 N and is in a direction $30°$ above horizontal to the right. The second force has a magnitude of 45 N and is in a direction of vertically upward. The third force has a magnitude of 62 N and is in a direction to the left. The free body diagram for this object is:

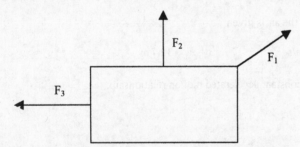

where we have labeled the forces with an index which represents whether it is the first, second, or third force.

The following are the steps to solving problems using Newton's Laws of Motion:

1. Draw a sketch of the situation.

2. For each object in the problem, draw a free-body diagram. Make sure that you include only the forces that act on that particular object the diagram represents.

3. Choose some directions for a set of coordinate axes and resolve each force on a particular object into components along the coordinate axes directions.

4. For each body, write down Newton's second law for each component direction for each of the objects in the problem.

5.  Solve the set of equations resulting from this procedure for the unknowns.

**Example 4-7-A**

A satellite of mass 156 kg is traveling horizontally directly toward the east.  The controllers need to divert the satellite so that it has an acceleration at an angle of 14.7° north of east.  The orientation of the satellite is such that one rocket on the satellite exhausts gas to the west with a force of 12.8 N.   A second rocket exhausts gas toward the south with an adjustable force.  To achieve the desired acceleration direction, what should be the force on the exhaust gas for the south-pointing rocket?  What will the magnitude of the satellite's acceleration be?

**Solution:**

First draw a simple sketch of the situation:

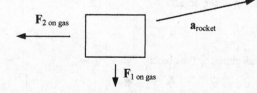

Using Newton's third law, we know the forces acting on the satellite will be equal in magnitude but opposite in direction to the forces on the exhaust gases.  We can now draw the free body diagram for the satellite:

Choosing coordinate directions of +$x$ to the east and +$y$ to the north, the two forces acting on the satellite can be resolved into components in the $x$ and $y$ directions:

$$F_{1x} = F_1 = 12.8 \, \text{N}$$
$$F_{1y} = 0$$
$$F_{2x} = 0$$
$$F_{2y} = F_2 = ?$$

where the subscripts should have obvious meanings.   Determine the sum of the forces in the $x$ and $y$ directions:

$$\sum F_x = F_{1x} + F_{2x} = 12.8 \, \text{N} + 0 = 12.8 \, \text{N}$$

$$\sum F_y = F_{1y} + F_{2y} = 0 + F_2 = F_2$$

Now writing Newton's second law for each direction:

$$x: \qquad 12.8 \, \text{N} = ma_x = ma \cos 14.7°$$

$$y: \qquad F_2 = ma_x = ma \sin 14.7°$$

where $a$ is the magnitude of the acceleration.  We recognize that we have two equations with two unknowns, $F_2$ and $a$. We can solve these equations by dividing the $y$ equation by the $x$ equation:

$$\frac{F_2}{12.8 \, \text{N}} = \frac{\sin 14.7°}{\cos 14.7°} = \tan 14.7°$$

$$F_2 = (12.8 \, \text{N}) \tan 14.7° = 3.36 \, \text{N}$$

Now either of the $x$ or $y$ direction Newton's second law equations can be used to solve for the magnitude of the acceleration. Using the $x$ direction equation:

$$a = \frac{12.8\,\text{N}}{m \cos 14.7^\circ} = \frac{12.8\,\text{N}}{(156\,\text{kg})(0.967)} = 8.48 \times 10^{-2}\ \text{m/s}^2$$

**Example 4-7-B**

Three forces act on an object. The first force has a magnitude 24.6 N and acts in a direction 23.6° east of north. The second force has a magnitude 38.7 N and acts in a direction 87.2° north of west. The third force has a magnitude 18.6 N and acts in a direction 45.7° south of west. What is the net force acting on the object? If the mass of the object is 16.8 kg, what is the acceleration of the object?

**Solution:**

First draw a free body diagram for the object:

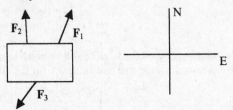

Choosing East as the $+x$ direction and north as the $+y$ direction, we resolve the forces into components in the two coordinate directions:

$$F_{1x} = F_1 \sin 23.6^\circ = (24.6\ \text{N})(0.4003) = 9.849\ \text{N}$$
$$F_{1y} = F_1 \cos 23.6^\circ = (24.6\ \text{N})(0.9164) = 22.54\ \text{N}$$
$$F_{2x} = -F_2 \cos 87.2^\circ = -(38.7\ \text{N})(0.04885) = -1.890\ \text{N}$$
$$F_{2y} = F_2 \sin 87.2^\circ = (38.7\ \text{N})(0.9988) = 38.65\ \text{N}$$
$$F_{3x} = -F_3 \cos 45.7^\circ = -(18.6\ \text{N})(0.6984) = -12.99\ \text{N}$$
$$F_{3y} = -F_3 \sin 45.7^\circ = -(18.6\ \text{N})(0.7157) = -13.31\ \text{N}$$

Notice that one extra significant figure has been kept because these are intermediate results. Now calculate the sum of the forces in the $x$ and $y$ directions:

$$\sum F_x = F_{1x} + F_{2x} + F_{3x} = 9.849\,\text{N} + (-1.890\,\text{N}) + (-13.29\,\text{N}) = -5.3\,\text{N}$$

$$\sum F_y = F_{1y} + F_{2y} + F_{3y} = 22.54\,\text{N} + (38.65\,\text{N}) + (-13.31\,\text{N}) = 47.9\,\text{N}$$

Notice the significant figures in the results. Why are these the appropriate numbers of significant figures? We are done calculating the net force on the object, but it is often desirable or required to express the force's magnitude and direction rather than the components in the coordinate axes directions. The magnitude is determined by the Pythagorean theorem:

$$F = \sqrt{\sum F_x + \sum F_y} = \sqrt{(-5.3\,\text{N})^2 + (47.9\,\text{N})^2} = 48.2\,\text{N}$$

and the direction can be found using the arctangent function:

$$\theta = \arctan \frac{\sum F_y}{\sum F_x} = \arctan \frac{47.9^\circ}{-5.3^\circ} = -84^\circ$$

What angle is this? This means that the force is in a direction 84° north of west. Draw a sketch of the vector if you are unable to see this.

The acceleration can now be determined using Newton's Second Law:

$$a = \frac{F}{m}$$

which for the magnitude of $a$ gives:

$$a = \frac{F}{m} = \frac{48.2\,\text{N}}{16.8\,\text{kg}} = 2.87\,\text{m/s}^2$$

The direction of the acceleration will be in the same direction as the net force, 84° north of west.

### Example 4-7-C

A sign of mass 6.34 kg is supported by two ropes as shown in the diagram. Determine the tension in each rope.

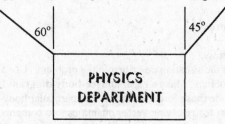

### Solution:

First, realize that the situation implies that the acceleration is zero. Although this is not definitely stated in the problem statement, the reasonable expectation that a sign hang stationary can be inferred. This of course implies that the net force on the sign is zero by Newton's second law. Another fact inferred by the situation described in the problem statement is that the force of gravity acts on the sign.

Start by drawing a free-body diagram. In this case, it looks a lot like the original sketch:

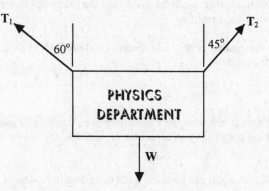

Next, choose a coordinate system with which to resolve the forces into components. As usual, we will choose horizontal to the right as the +x direction and upward as the +y direction. The components of the forces are

$$W_x = 0$$
$$W_y = -W = -mg = -(6.34\,\text{kg})(9.8\,\text{m/s2}) = -62\,\text{N}$$
$$T_{1x} = -T_1 \sin 60°$$
$$T_{1y} = T_1 \cos 60°$$
$$T_{2x} = T_2 \sin 45°$$
$$T_{2y} = T_2 \cos 45°$$

Finding the sum of the $x$ direction and $y$ direction components of the forces and setting them equal to the acceleration results in

$$\sum F_x = 0 + -T_1 \sin 60° + T_2 \sin 45° = 0$$

$$\sum F_y = -62\,\text{N} + T_1 \cos 60° + T_2 \cos 45° = 0$$

We recognize this as a set of two equations in two unknowns.  We are able to solve for the unknowns:

$$T_2 = \frac{T_1 \sin 60°}{\sin 45°}$$

$$T_1 = \frac{62\,\text{N}}{\cos 60° + \dfrac{\sin 60° \cos 45°}{\sin 45°}} = 45\,\text{N}$$

$$T_2 = \frac{45\,\text{N}\sin 60°}{\sin 45°} = 55\,\text{N}$$

### Section 4-8 Problem Solving—A General Approach

1.  Read and reread written problems carefully.
2.  Draw an accurate picture or diagram of the situation described in the problem.  Use arrows to represent all of the vector quantities involved in the problem.  Make a separate **free-body diagram** for each of the objects in the problem, being careful to only include those forces acting on that particular body.
3.  Choose a convenient **coordinate system** for resolving vector quantities into components so that vector additions can be carried out.  The choice of coordinates doesn't change the solution to the problem, but a good choice of coordinate systems will minimize the algebra that needs to be done to reach the final solution.  Pick the coordinate axes so that as many of the vector quantities as possible lie along the coordinate axes.
4.  Determine which quantities are known and which are unknowns that need to be solved for.  Let the conceptual principles that relate the knowns and unknowns guide you to writing down the **equations** that relate the knowns and unknowns.  Make sure the **relationships** you use are valid in the circumstances described in the problem.
5.  Make a **rough calculation** to see if the approach you intend to take is sufficient for solving the problem.  You may discover that an additional equation or more information is necessary.  Also, rough calculations can be used to check the values of a careful calculation.
6.  **Solve** the equation or system of equations that relate the knowns to the unknowns in the problem.  Solve them symbolically before entering numbers for quantities.
7.  Keep track of **units**.
8.  Consider whether your answer is **reasonable** or not.  Use **dimensional analysis** to determine if you have the correct dimension in the units of the solution.

## Practice Quiz

1.  A box with a weight of 300 N remains at rest while on a level floor even though you push on it with a horizontal force of 200 N.  Which statement is a physically correct explanation of why the box remains at rest even though you are applying a force to it?

    a)  The object is pushing back with an equal and opposite force according to Newton's third law and so the total force is zero.
    b)  The weight of the box exceeds the force you are pushing on it with.
    c)  The floor also pushes horizontally on the object with a force of 200 N in a direction opposite to your push.
    d)  The sum of your push, the normal force of the floor, and the weight of the box is zero.

2.  Which of the following objects has a mass of about 1.0 kg?

    a)  A marble
    b)  A book
    c)  A car
    d)  The Earth

3.    An airplane flies straight in level flight at constant speed. Newton's second law implies that the net force on the airplane is zero. How can this be if the gravitational force of the Earth is pulling downward on the plane?

a)    There is no gravitational force at the altitude planes fly at.
b)    The airplane is flying fast enough that it is actually falling around the curvature of the earth.
c)    The downward gravitational pull of the earth is balanced by the upward gravitational pull of the moon.
d)    There is an upward force on the plane from the air beneath the wings of the plane.

4.    Knowing that an object accelerates at $3.000 \text{ m/s}^2$ when the force on it is 396 N, how much force must be applied to the object so that the magnitude of its acceleration is $6.000 \text{ m/s}^2$?

a)  198 N          b)  99 N          c)  792 N          d)  1584 N

5.    What are the SI units of force divided by acceleration?

a) kg     b) m/s     c) kg m/s$^2$     d) m/s$^2$

6.    If only two forces act on an object but the object has no acceleration, what must be true about the two forces?

a)    Both forces are zero.
b)    The forces are equal in magnitude, but opposite in direction.
c)    The forces are equal in magnitude, but act in perpendicular directions.
d)    The forces can be any magnitude and direction.

7.    An object starts from rest and travels a distance $d$ in time $t$ when a constant force $F$ acts on the object. What is the constant force that acts on the object if it starts from rest and travels the same distance in a time $t/2$?

a)    $4F$
b)    $F/4$
c)    $2F$
d)    $F/2$

8.    When a person with a mass of 98.4 kg stands on a spring scale in an elevator, the scale reads 0 N. Which statement about the elevator is true?

a)    The elevator is stationary.
b)    The elevator is moving downward with a velocity of 9.8 m/s.
c)    The elevator is accelerating upward with an acceleration of $9.8 \text{ m/s}^2$.
d)    The elevator is accelerating downward with an acceleration of $9.8 \text{ m/s}^2$.

9.    When you jump upward from the ground, the Earth is applying an upward force on you. Newton's third law implies that you are applying an equal magnitude but opposite direction force on the Earth. Why doesn't everyone around you feel the Earth accelerate downward when you jump?

a)    It happens in too short a time for the people to notice.
b)    The Earth's mass is so great that its acceleration is unnoticeable.
c)    Everyone does notice and your jump is recorded on seismographs around the world.
d)    This is an incorrect application of Newton's third law.

10.    Two forces act on an object of mass 3.249 kg and cause it to accelerate with an acceleration of $1.000 \text{ m/s}^2$ toward the east. What is the magnitude and direction of a third force applied to the object so that its acceleration is zero?

a)    Impossible. Two forces are needed: one to cancel each of the other two forces.
b)    Need to know the other two forces to determine the third force.
c)    $1.000 \text{ m/s}^2$ toward the west.
d)    3.249 N toward the west.

11.    An object with a mass of 3.411 kg has three forces acting on it: 240 N in a direction east,  322 N in a direction 47° north of west, and 568 N in a direction 12° west of south.  What is the acceleration of the object?

12.    An object with a mass of 3.42 kg has a time-varying force acting on it given by

$$F(t) = 1.0 \text{ N/s } t - 2.5 \text{ N/s}^2 \, t^2$$

in a given direction.  If the object starts from rest at $t = 0$, how far from its initial position is it after 10 s?

13.    A 28.4 kg mass block rests on a frictionless inclined plane that is tilted $26.5^0$ from the horizontal.  A rope is attached to the block that pulls parallel to and up the incline.  The acceleration of the block is 0.893 m/s$^2$ down the plane.  What is the tension in the rope?

14.    What is the tension in the rope in question 13 if the acceleration is 0.893 m/s$^2$ up the plane instead of down the plane?

15.    An elevator is at rest and begins descending with constant acceleration.  After it descends 27.2 m, it has a downward velocity of 19.2 m/s.  If the weight of the elevator is 9340 N, what is the tension in the cable that is lowering the elevator?

## Problem Solutions

Solutions to problems with a number equal to $4n - 1$ where $n$ is an integer:

3.    We apply Newton's second law to the object:
      $\sum F = ma$;
      $F = (7.0 \times 10^{-3} \text{ kg})(10{,}000)(9.80 \text{ m/s}^2) = 6.9 \times 10^2 \text{ N}.$

7.    The required average acceleration can be found from the one-dimensional motion:
      $v^2 = v_0{}^2 + 2a(x - x_o)$;
      $(155 \text{ m/s})^2 = 0 + 2a(0.700 \text{ m} - 0)$, which gives $a = 1.72 \times 10^4 \text{ m/s}^2$.
      We apply Newton's second law to find the required average force
      $\sum F = ma$;
      $F = (6.25 \times 10^{-3} \text{ kg})(1.72 \times 10^4 \text{ m/s}^2) = 107 \text{ N}.$

11.    Because the line snapped, the tension $F_T > 25$ N.
      We write $\sum \mathbf{F} = m\mathbf{a}$ from the force diagram for the fish:
          $y$-component: $F_T - mg = ma$,  or  $F_T = m(a + g)$.
      We find the minimum mass from the minimum tension:
          $25 \text{ N} = m_{min}(3.5 \text{ m/s}^2 + 9.80 \text{ m/s}^2)$, which gives $m_{min} = 1.9$ kg.
      Thus we can say $m > 1.9$ kg.

15.    We write $\sum \mathbf{F} = m\mathbf{a}$ from the force diagram for the bucket:
          $y$-component: $F_T - mg = ma$;
          $63.0 \text{ N} - (7.50 \text{ kg})(9.80 \text{ m/s}^2) = (7.50 \text{ kg})a$,
      which gives  $a = -1.40 \text{ m/s}^2$ (down).

19.    The maximum tension will be exerted by the motor when the elevator has
      the maximum acceleration.  We write $\sum \mathbf{F} = m\mathbf{a}$ from the force diagram for the elevator:
          $y$-component: $F_{Tmax} - mg = ma_{max}$;
          $21{,}750 \text{ N} - (2100 \text{ kg})(9.80 \text{ m/s}^2) = (2100 \text{ kg})a_{max}$,
      which gives  $a_{max} = 0.557 \text{ m/s}^2.$

23.    We find the velocity necessary for the jump from the motion when the
      person leaves the ground to the highest point, where the velocity is zero:
          $v^2 = v_{jump}{}^2 + 2(-g)h$;
          $0 = v_{jump}{}^2 + 2(-9.80 \text{ m/s}^2)(0.80 \text{ m})$, which gives $v_{jump} = 3.96$ m/s.
      We can find the acceleration required to achieve this velocity during the
      crouch from
          $v_{jump}{}^2 = v_0{}^2 + 2a(y - y_0)$;
          $(3.96 \text{ m/s})^2 = 0 + 2a(0.20 \text{ m} - 0)$, which gives $a = 39.2 \text{ m/s}^2.$

Using the force diagram for the person during the crouch, we can write $\sum \mathbf{F} = m\mathbf{a}$:

$F_N - mg = ma$;

$F_N - (61\text{ kg})(9.80\text{ m/s}^2) = (61\text{ kg})(39.2\text{ m/s}^2)$, which gives $F_N = 3.0 \times 10^3$ N.

From Newton's third law, the person will exert an equal and opposite force on the ground:

$3.0 \times 10^3$ N downward.

27.  In order for the resultant to have no northerly component, the second force must have a northerly component equal in magnitude to that of the first force but pointing toward the south. Because the magnitudes of both the forces are equal and their northerly components are equal, the westerly components must also be equal. From the symmetry, the second force must be in the southwesterly direction.

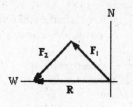

31. (*a*)  For the components of the net force we have

$F_{ax} = -F_1 = -20.2$ N;

$F_{ay} = -F_2 = -26.0$ N.

We find the magnitude from

$F_a^2 = F_{ax}^2 + F_{ay}^2 = (-20.2\text{ N})^2 + (-26.0\text{ N})^2$,

which gives $F_a = 32.9$ N.

We find the direction from

$\tan \alpha = |F_{ay}|/|F_{ax}| = (26.0\text{ N})/(20.2\text{ N}) = 1.29$,

which gives $\alpha = 52.2°$ below $-x$-axis.

The acceleration will be in the direction of the net force:

$a_a = F_a/m = (32.9\text{ N})/(29.0\text{ kg}) = 1.13$ m/s$^2$, 52.2° below $-x$-axis.

(*b*)  For the components of the net force we have

$F_{bx} = F_1 \cos \theta = (20.2\text{ N}) \cos 30° = 17.5$ N;

$F_{by} = F_2 - F_1 \sin \theta = 26.0\text{ N} - (20.2\text{ N}) \sin 30° = 15.9$ N.

We find the magnitude from

$F_b^2 = F_{bx}^2 + F_{by}^2 = (17.5\text{ N})^2 + (15.9\text{ N})^2$,

which gives $F_b = 23.6$ N.

We find the direction from

$\tan \beta = |F_{by}|/|F_{bx}| = (15.9\text{ N})/(17.5\text{ N}) = 0.909$,

which gives $\beta = 42.3°$ above $+x$-axis.

The acceleration will be in the direction of the net force:

$a_b = F_b/m = (23.6\text{ N})/(29.0\text{ kg}) = 0.814$ m/s$^2$, 42.3° above $+x$-axis.

(*a*)
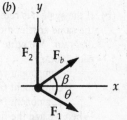

35. (*a*)  Because the buckets are at rest, the acceleration is zero.

We write $\sum \mathbf{F} = m\mathbf{a}$ from the force diagram for each bucket:

lower bucket:  $F_{T2} - m_2 g = m_2 a = 0$, which gives

$F_{T2} = m_2 g = (3.5\text{ kg})(9.80\text{ m/s}^2) = 34$ N.

upper bucket:  $F_{T1} - F_{T2} - m_1 g = m_1 a = 0$, which gives

$F_{T1} = F_{T2} + m_1 g = 34\text{ N} + (3.5\text{ kg})(9.80\text{ m/s}^2) = 68$ N.

(*b*)  The two buckets must have the same acceleration.

We write $\sum \mathbf{F} = m\mathbf{a}$ from the force diagram for each bucket:

lower bucket:  $F_{T2} - m_2 g = m_2 a$, which gives

$F_{T2} = m_2(g + a)$

$= (3.5\text{ kg})(9.80\text{ m/s}^2 + 1.60\text{ m/s}^2) = 40$ N.

upper bucket:  $F_{T1} - F_{T2} - m_1 g = m_1 a$, which gives

$F_{T1} = F_{T2} + m_1(g + a)$

$= 40\text{ N} + (3.5\text{ kg})(9.80\text{ m/s}^2 + 1.60\text{ m/s}^2) = 80$ N.

Note that we could have taken both buckets as the system to find the tension in the upper cord.

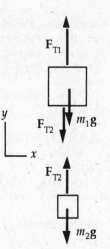

39. There is no acceleration perpendicular to the line L.
    We write $\sum \mathbf{F} = m\mathbf{a}$ from the force diagram:
    $\sum F_x = ma_x$;
    $F_B \sin \theta_B - F_A \sin \theta_A = 0$;
    $F_B \sin 30° - (4500 \text{ N}) \sin 50° = 0$, which gives $F_B = 6890$ N.
    The resultant force is in the $y$-direction:
    $\sum F_y = F_B \cos \theta_B + F_A \cos \theta_A = (6890 \text{ N}) \cos 30° + (4500 \text{ N}) \cos 50° = 8860$ N.

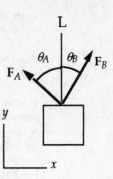

43. We find the two accelerations:
    $a_1 = F_0/m$;
    $a_2 = 2F_0/m$.
    We choose the origin of coordinates at the initial position. For the motion up to time $t_0$ we have
    $x_1 = x_{01} + v_{01}t + \frac{1}{2}a_1t^2 = 0 + 0 + \frac{1}{2}(F_0/m)t_0^2 = \frac{1}{2}(F_0/m)t_0^2$;   and
    $v_1 = v_{01} + a_1t_0 = 0 + (F_0/m)t_0 = (F_0/m)t_0$.
    These values become the initial ones for the motion after $t_0$:
    $x_2 = x_{02} + v_{02}t + \frac{1}{2}a_2t^2 = \frac{1}{2}(F_0/m)t_0^2 + (F_0/m)t_0(2t_0 - t_0) + \frac{1}{2}(2F_0/m)(2t_0 - t_0)^2 = \frac{5}{2}(F_0/m)t_0^2$.

47.   a)

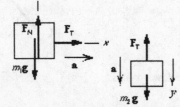

   (b) If the system is released from rest, the blocks will have
       the same acceleration in the directions indicated on the
       diagram. We write $\sum \mathbf{F} = m\mathbf{a}$ from the force diagram for
       each block:
           $x$-component ($m_1$):  $F_T = m_1 a$;
           $y$-component ($m_1$):  $F_N - m_1 g = 0$;
           $y$-component ($m_2$):  $m_2 g - F_T = m_2 a$.
       When we add the first and third equations, we get
           $m_2 g = (m_1 + m_2)a$, which gives  $a = m_2 g/(m_1 + m_2)$.
       When we use this result in the first equation, we get
           $F_T = m_1 m_2 g/(m_1 + m_2)$.

51.      The blocks and the cord will have the same
         acceleration. If we select the two blocks and
         cord as the system, we have
             $\sum F_x = ma_x$:  $F_P = (m_1 + m_2 + m_C)a$,
         which gives
             $a = F_P/(m_1 + m_2 + m_C) = (40.0 \text{ N})/(10.0 \text{ kg} + 12.0 \text{ kg} + 1.0 \text{ kg}) = 1.74 \text{ m/s}^2$.
         For block 1 we have $\sum F_x = ma_x$:
             $F_P - F_{T1} = m_1 a$;
             $40.0 \text{ N} - F_{T1} = (10.0 \text{ kg})(1.74 \text{ m/s}^2)$, which gives $F_{T1} = 22.6$ N.
         For block 2 we have $\sum F_x = ma_x$:
             $F_{T2} = m_2 a$;
             $F_{T2} = (12.0 \text{ kg})(1.74 \text{ m/s}^2) = 20.9$ N.
         Note that we can see if these agree with the analysis of $\sum F_x = ma_x$ for block 3:
             $F_{T1} - F_{T2} = m_3 a$:
             $22.6 \text{ N} - 20.9 \text{ N} = (1.0 \text{ kg})a$, which gives $a = 1.7 \text{ m/s}^2$.

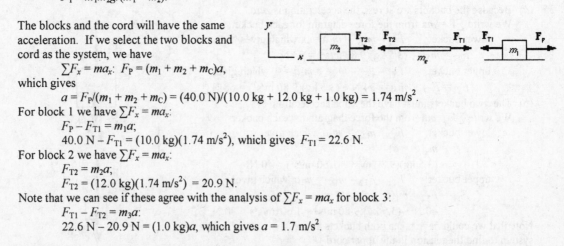

55.      The force diagrams for each of the masses and the movable pulley are shown. Note that we take down as
         positive and the indicated accelerations are relative to the fixed pulley. A downward acceleration of $m_3$ means
         an upward acceleration of the movable pulley. If we call $a_r$ the (downward) acceleration of $m_1$ with respect to
         the movable pulley, we have
             $a_1 = a_r - a_3$   and   $a_2 = -a_r - a_3$,

because the acceleration of $m_2$ with respect to the pulley must be the negative of $m_1$'s acceleration with respect to the pulley. If the mass of the pulley is negligible, for the movable pulley we write $\sum F_y = ma_y$:

$2F_{T1} - F_{T3} = (0)(-a_3)$, so  $2F_{T1} = F_{T3}$.

For each of the masses, for $\sum F_y = ma_y$ we get

mass $m_1$: $m_1g - F_{T1} = m_1a_1 = m_1(a_r - a_3)$,

mass $m_2$: $m_2g - F_{T1} = m_2a_2 = m_2(-a_r - a_3)$,

mass $m_3$: $m_3g - F_{T3} = m_3a_3$.

We have four equations for the four unknowns:

$F_{T1}$, $F_{T3}$, $a_r$, and $a_3$.

After some careful algebra, we get

$a_3 = [(m_1m_3 + m_2m_3 - 4m_1m_2)/(m_1m_3 + m_2m_3 + 4m_1m_2)]g$;

$a_r = [2(m_1m_3 - m_2m_3)/(m_1m_3 + m_2m_3 + 4m_1m_2)]g$;

$F_{T1} = [4m_1m_2m_3/(m_1m_3 + m_2m_3 + 4m_1m_2)]g$;  and

$F_{T3} = [8m_1m_2m_3/(m_1m_3 + m_2m_3 + 4m_1m_2)]g$.

We can now find the other accelerations:

$a_1 = [(m_1m_3 - 3m_2m_3 + 4m_1m_2)/(m_1m_3 + m_2m_3 + 4m_1m_2)]g$;

$a_2 = [(-3m_1m_3 + m_2m_3 + 4m_1m_2)/(m_1m_3 + m_2m_3 + 4m_1m_2)]g$.

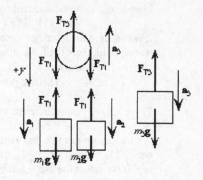

59.     The acceleration can be found from the blood's one-dimensional motion:

$v = v_0 + at$;

$0.35$ m/s $= (0.25$ m/s$) + a(0.10$ s$)$, which gives $a = 1.00$ m/s$^2$.

We apply Newton's second law to find the required force

$\sum F = ma$;

$F = (20 \times 10^{-3}$ kg$)(1.00$ m/s$^2) = 2.0 \times 10^{-2}$ N.

63.     For the motion until the elevator stops, we have

$v^2 = v_0^2 + 2a(x - x_0)$;

$0 = (3.5$ m/s$)^2 + 2a(3.0$ m$)$, which gives $a = -2.04$ m/s$^2$.

We write $\sum \mathbf{F} = m\mathbf{a}$ from the force diagram for the elevator:

$mg - F_T = ma$;  or

$(1300$ kg$)(9.80$ m/s$^2) - F_T = (1300$ kg$)(-2.04$ m/s$^2)$,

which gives

$F_T = 1.5 \times 10^4$ N.

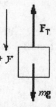

67.     (a)  The forces and coordinate systems are shown in the diagram. Note that we take up the incline as the positive direction. From the force diagram, with the block $m_2$ as the system, we can write $\sum \mathbf{F} = M\mathbf{a}$:

$y$-component: $m_2g - F_T = m_2a$.

From the force diagram, with the block $m_1$ as the system, we can write $\sum \mathbf{F} = m\mathbf{a}$:

$x$-component: $F_T - m_1g \sin \theta = m_1a$.

When we eliminate $F_T$ between these two equations, we get

$a = (m_2 - m_1 \sin \theta)g/(m_1 + m_2)$.

$= [1.00$ kg $- (1.00$ kg$)(\sin 30°)](9.80$ m/s$^2)/(1.00$ kg $+ 1.00$ kg$)$

$= 2.45$ m/s$^2$ (up the incline).

(b)  If the system remains at rest, the acceleration is zero, so we have

$a = (m_2 - m_1 \sin \theta)g/(m_1 + m_2) = 0$,  or

$m_2 = m_1 \sin \theta = (1.00$ kg$)(\sin 30°) = 0.50$ kg.

(c)  From the force equations, we have

$F_T = m_2(g - a)$.

For part (a), we get

$F_T = (1.00$ kg$)(9.80$ m/s$^2 - 2.45$ m/s$^2) = 7.35$ N.

For part (b), we get

$F_T = (0.50$ kg$)(9.80$ m/s$^2 - 0) = 4.9$ N.

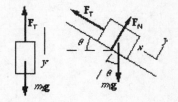

71.  We find the acceleration of the car on the level from
$$v = v_0 + at;$$
$$(21 \text{ m/s}) = 0 + a(14.0 \text{ s}), \text{ which gives } a = 1.5 \text{ m/s}^2.$$
This acceleration is produced by the net force:
$$F_{net} = ma = (1100 \text{ kg})(1.5 \text{ m/s}^2) = 1650 \text{ N}.$$
If we assume the same net force on the hill, with no acceleration on the
steepest hill, from the force diagram we have

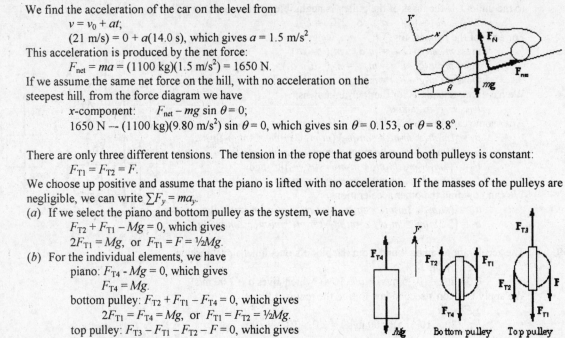

$x$-component:    $F_{net} - mg \sin \theta = 0$;
$$1650 \text{ N} - (1100 \text{ kg})(9.80 \text{ m/s}^2) \sin \theta = 0, \text{ which gives } \sin \theta = 0.153, \text{ or } \theta = 8.8°.$$

75.  There are only three different tensions.  The tension in the rope that goes around both pulleys is constant:
$$F_{T1} = F_{T2} = F.$$
We choose up positive and assume that the piano is lifted with no acceleration.  If the masses of the pulleys are
negligible, we can write $\sum F_y = ma_y$.
   (*a*)  If we select the piano and bottom pulley as the system, we have
$$F_{T2} + F_{T1} - Mg = 0, \text{ which gives}$$
$$2F_{T1} = Mg, \text{ or } F_{T1} = F = \tfrac{1}{2}Mg.$$
   (*b*)  For the individual elements, we have
         piano: $F_{T4} - Mg = 0$, which gives
$$F_{T4} = Mg.$$
         bottom pulley: $F_{T2} + F_{T1} - F_{T4} = 0$, which gives
$$2F_{T1} = F_{T4} = Mg, \text{ or } F_{T1} = F_{T2} = \tfrac{1}{2}Mg.$$
         top pulley: $F_{T3} - F_{T1} - F_{T2} - F = 0$, which gives
$$F_{T3} = 3F_{T1} = {}^3/2Mg.$$

79.  There will be two horizontal forces on the train: the force exerted against the track, $F_{track}$, and the drag force
from the air, $F_{drag}$, which depends on the speed.  We select the train as the system and write $\sum \mathbf{F} = m\mathbf{a}$:
$$F_{track} - F_{drag} = ma.$$
   (*a*)  The maximum acceleration will occur when the force on the track is maximum and the drag
force is minimum.  This occurs when the speed is zero, that is, when the train is starting to move.
Thus we have $400 \times 10^3 \text{ N} - 0 = (660,000 \text{ kg})a_{max}$, which gives $a_{max} = 0.606 \text{ m/s}^2.$
   (*b*)  At top speed the acceleration is zero, so we have
$$150 \times 10^3 \text{ N} - F_{drag} = 0, \text{ which gives } F_{drag} = 150 \times 10^3 \text{ N} = 150 \text{ kN}.$$

# Chapter 5: Further Applications of Newton's Laws

## Chapter Overview and Objectives

This chapter describes the application of Newton's second law to some particular situations and types of forces. It introduces friction, motion along circular paths, and velocity dependent forces.

After completing study of this chapter, you should:

- Be able to determine frictional forces on objects.
- Know the dynamics of motion in a circle.
- Be able to solve problems involving circular motion.
- Know that drag forces in fluids depend on the relative speed of objects to the fluid.

## Summary of Equations

Relationship between kinetic friction and normal force: $\qquad F_{fr} = \mu_k F_N$

Relationship between static friction and normal force: $\qquad F_{fr} \leq \mu_s F_N$

Centripetal acceleration of object in circular motion: $\qquad a_R = \dfrac{v^2}{r}$

Magnitude of tangential acceleration: $\qquad a_{\tan} = \dfrac{dv}{dt}$

Acceleration of object in circular motion: $\qquad \mathbf{a} = \mathbf{a}_R + \mathbf{a}_{\tan}$

Viscous drag force on an object moving in a fluid: $\qquad F_D = -bv$

Inertial drag force on an object moving in a fluid: $\qquad F_D = -Cv^2$

## Chapter Summary

### Section 5-1  Applications of Newton's Laws Involving Friction

Whenever there is relative motion between two surfaces in contact, there is force on each object parallel to the surface. This force acts on each object in a direction opposite to the relative motion of that object. This is the force of **kinetic friction**. The magnitude of the kinetic friction force is proportional to the normal force of contact of one surface on the other:

$$F_{fr} = \mu_k F_N$$

where $F_{fr}$ is the magnitude of the friction force and $F_N$ is the magnitude of the normal force. The **coefficient of kinetic friction**, $\mu_k$, depends on the materials the two surfaces in contact are made of, the roughness of those surfaces, and on conditions such as temperature and humidity. A table of typical values of the coefficient of friction between two different types of materials is in Table 5-1 of the textbook.

### Example 5-1-A

A box of mass 17.2 kg is slid across a level floor with constant velocity. The coefficient of friction between the floor and the box is 0.32. What force must the box be pushed with to maintain this constant velocity?

**Solution:**

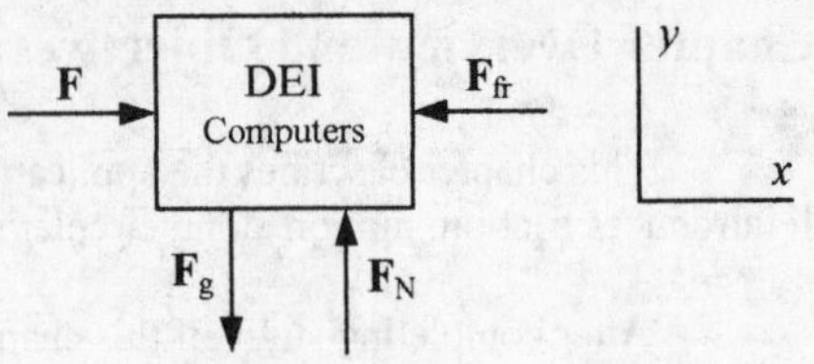

First we draw a free body diagram as shown to the right. We will use a
coordinate system with the *x*-axis horizontal and positive to the right.
The *y*-axis will be vertical and positive upward. Then we write
Newton's second law for the vertical direction and the horizontal
direction:

$$\sum F_x = F - F_{fr} = ma_x \qquad and \qquad \sum F_y = F_N - F_g = ma_y$$

We know the magnitude of the force of gravity is *mg*. We also know that both accelerations are zero. The object does
not move into or off of the floor, and the problem states the horizontal motion is at constant velocity. We recognize that
the two objects in contact, the box and the floor, are in relative motion. This means that the friction force is kinetic
friction. We have an expression for the magnitude of the kinetic friction force

$$F_{fr} = \mu_k F_N$$

Rewriting the Newton's second law equations with this additional information

$$\sum F_x = F - \mu_k F_N = 0 \qquad and \qquad \sum F_y = F_N - mg = 0$$

Notice that we cannot completely solve the horizontal direction equation, the direction parallel to the surfaces in contact,
until we use the vertical direction equation to solve for the normal force. This will usually be the case. Newton's second
law for the direction perpendicular to the surface of contact will need to be solved before the Newton's second law
parallel to the surface can be completely solved. The solution is now straightforward. Solving the *y*-direction equation
gives $F_N = mg$ and then using this in the *x*-direction equation results in $F = \mu_k mg$.

$$F = \mu_k mg = (0.32)(17.2\,\text{kg})(9.80\,\text{m/s}) = 54\,\text{N}$$

There can also be a friction force between two objects in contact not in relative motion if net forces on at least one of the
objects would cause relative acceleration of the objects if friction were not present We call this force **static friction**.
The rule that tells us how to calculate the magnitude of the static friction force is very different than that for the other
forces we have encountered thus far. The static friction force is equal to the opposite of the total of all other forces
acting on the object parallel to the surface of contact. However, there is an upper limit to this behavior. The magnitude
of the limit of the static friction force

$$F_{fr} \leq \mu_s F_N$$

where $\mu_s$ is the coefficient of static friction. This means that with static friction acting on the object, the net force acting
on the object will be zero unless the total of non-frictional forces exceeds in magnitude the upper limit of the static
friction force. The coefficients of static friction of some common surfaces in contact are also found in Table 5-1 of the
textbook.

**Example 5-1-B**

A box of mass 3.23 kg rests on a ramp with a slope of 34.2°. The coefficient of static friction between the box and the
ramp is 0.774. What addition force directed directly down the ramp will start the box sliding down the ramp?

**Solution:**

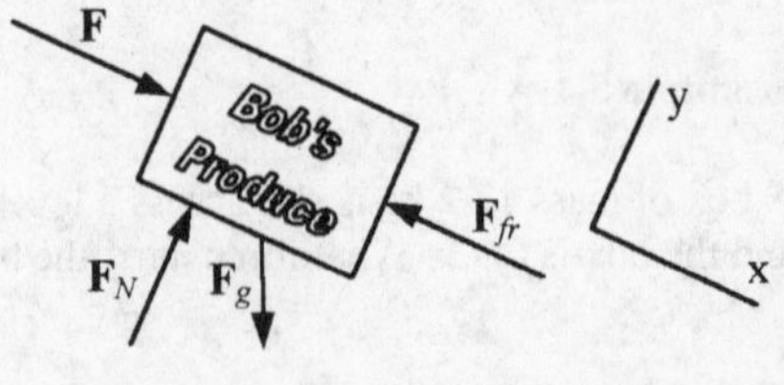

First we draw a free-body diagram of the situation where **F** is the
additional applied force needed to start the box moving, $\mathbf{F}_g$ is the
gravitational force on the box, $\mathbf{F}_N$ is the normal force of the ramp on the
box, and $\mathbf{F}_{fr}$ is the static friction force on the box. We will use a
coordinate system with the *x*-direction parallel and down the ramp and
the *y*-direction perpendicular to the ramp and angled upward. We write
down Newton's second law for each direction:

$$\sum F_x = ma_x \qquad \sum F_y = ma_y$$

Before we start solving these equation, we have to think about what the problem is asking. The problem asks for the minimum force that will start the box moving. The force necessary will be that force that just overcomes the maximum static friction force. We set the static friction force to its maximum value in magnitude, $\mu_s F_N$, and solve for the force $\mathbf{F}$ that would make the acceleration down the ramp zero.

$$\sum F_x = F_x + F_{gx} + F_{Nx} + F_{frx} = F + mg \sin 34.2° + 0 - F_{fr}$$

$$\sum F_y = F_y + F_{gy} + F_{Ny} + F_{fry} = 0 - mg \cos 34.2° + F_N + 0$$

Placing these net force expressions into Newton's second law

$$F + mg \sin 34.2° - \mu_s F_N = 0 \qquad -mg \cos 34.2° + F_N = 0$$

We recognize this as a set of two equations with two unknowns, $F_N$ and $F$. We first solve the second equation for $F_N$

$$F_N = mg \cos 34.2°$$

then put this expression into the first equation and solve for $F$

$$F = \mu_s mg \cos 34.2° - mg \sin 34.2°$$
$$= (0.774)(3.23 \text{ kg})(9.80 \text{ m/s}) \cos 34.2° - (3.23 \text{ kg})(9.80 \text{ m/s}) \sin 34.2° = 2.47 \text{ N}$$

## Example 5-1-C

Two boxes are in contact with each other on a horizontal surface. The larger box has a mass 4.55 kg and a coefficient of kinetic friction of 0.365 with the surface. The smaller block has a mass 3.21 kg and a coefficient of kinetic friction of 0.822 with the surface. A force of 90.6 N pushes on the larger box at an angle 10.2° below horizontal as shown in the diagram. What is the acceleration of the blocks?

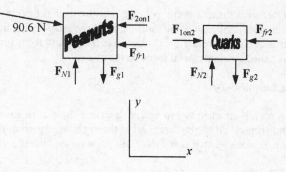

## Solution:

Draw a free-body diagram for each box:

We have used the symbols with subscript 1 for the larger box and subscript 2 for the smaller box. All of the other subscripts on the forces should be clear from pass usage except $\mathbf{F}_{1on2}$ and $\mathbf{F}_{2on1}$ which are respectively, the force of the first box on the second box and the force of the second box on the first. We now write down Newton's second law for each box

$$m_1 \mathbf{a}_1 = \sum \mathbf{F}_1 = \mathbf{F} + \mathbf{F}_{N1} + \mathbf{F}_{g1} + \mathbf{F}_{fr1} + \mathbf{F}_{2on1}$$
$$m_2 \mathbf{a}_2 = \sum \mathbf{F}_2 = \mathbf{F}_{N2} + \mathbf{F}_{g2} + \mathbf{F}_{fr2} + \mathbf{F}_{1on2}$$

We write each vector equation in terms of its $x$ and $y$ components (we choose the $x$-direction as horizontal and $y$-direction as vertical):

$$m_1 a_{1x} = (90.6 \text{ N}) \cos 10.2° + 0 + 0 - \mu_{k1} F_{N1} - F_{2on1}$$
$$m_1 a_{1y} = -(90.6 \text{ N}) \sin 10.2° + F_{N1} - m_1 g + 0 + 0$$
$$m_2 a_{2x} = 0 + 0 - \mu_{k2} F_{N2} + F_{1on2}$$
$$m_2 a_{2y} = F_{N2} - m_2 g + 0 + 0$$

We know both accelerations in the *y*-direction are zero; the boxes neither fly off nor sink into the surface.   We can now solve the *y*-direction equations for the normal forces:

$$F_{N1} = m_1 g + (90.6\,\text{N})\sin 10.2^\circ \qquad F_{N2} = m_2 g$$

These expressions for the normal forces can be substituted into the *x*-direction equations.  We also recognize that the *x*-direction accelerations of the two boxes must be the same, because the two boxes remain in contact with each other.  Calling the common acceleration of the two boxes *a*, we again write down the two *x*-direction equations:

$$m_1 a = (90.6\,\text{N})\cos 10.2^\circ + 0 + 0 - \mu_{k1}\big(m_1 g + (90.6\,\text{N})\sin 10.2^\circ\big) - F_{1on2}$$
$$m_2 a = 0 + 0 - \mu_{k2} m_2 g + F_{1on2}$$

where we have also used Newton's third law to set the magnitude of the force of each box on the other to be the same.  This is a set of two equations in the two unknowns *a* and $F_{1on2}$.  We solve for *a*:

$$a = \frac{(90.6\,\text{N})\cos 10.2^\circ - \mu_{k1}\big(m_1 g + (90.6\,\text{N})\sin 10.2^\circ\big) - \mu_{K2} m_2 g}{m_1 + m_2}$$

$$= \frac{(90.6\,\text{N})\cos 10.2^\circ - (0.365)\big((4.55\,\text{kg})(9.80\,\text{m/s}) + (90.6\,\text{N})\sin 10.2^\circ\big) - (0.822)(3.21\,\text{kg})(9.80\,\text{m/s})}{(4.55\,\text{kg}) + (3.21\,\text{kg})}$$

$$= 5.31\,\text{m/s}$$

### Section 5-2 Dynamics of Uniform Circular Motion

We know that an object following a circular path  of radius *r* at a constant speed *v* has an acceleration of magnitude

$$a_R = \frac{v^2}{r}$$

directed toward the center of the circular path.  This is called a centripetal acceleration, centripetal meaning toward the center.   Applying Newton's second law to this motion then requires the net force on the object to satisfy

$$\sum \mathbf{F} = -m\frac{v^2}{r}\hat{\mathbf{r}} \qquad \text{or} \qquad \sum F_R = m\frac{v^2}{r}$$

where $\hat{\mathbf{r}}$ is the unit vector pointing outward from the center of the circle to the position of the object and the $F_R$'s are the component of each of the forces in the direction pointing from the object to the center of the circle.  An object that moves in a circle at constant speed must have this inward force acting on it to continue in circular motion.  This necessary force is called the **centripetal force** on the object.

### Example 5-2-A

A rock is attached to a string and spun in a circle of radius *r*.  The plane of the circle is inclined at an angle $\theta$ to the horizontal.  At the highest point of the circle, the speed of the rock is *v*.  With what magnitude force does the string act on the rock when it is at the highest point on the circle?

### Solution:

We know that because the motion is in a circle tilted at angle $\theta$ to horizontal, the direction of the acceleration will be at an angle $\theta$ to horizontal also.  When the object is at the highest point, the centripetal direction will be angle $\theta$ below the horizontal.  We draw a free body diagram:

Note that the force of the string is shown at an angle $\alpha$ below horizontal. This will be different than the angle $\theta$ of the acceleration direction.  We write down Newton's second law in the *x* and *y* directions:

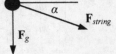

$$\sum F_x = ma_x \quad \Rightarrow \quad F_{string}\cos\alpha = m\frac{v^2}{r}\cos\theta$$

$$\sum F_y = ma_y \quad \Rightarrow \quad -F_{string}\sin\alpha - mg = -m\frac{v^2}{r}\sin\theta$$

This is a set of two equations in the two unknowns $F_{string}$ and $\alpha$. We solve by moving the $mg$ term in the second equation to the right hand side of that equation, squaring both sides of each equation and adding the two resulting equations together to get

$$F_{string}^2\left(\cos^2\alpha + \sin^2\alpha\right) = \left(m\frac{v^2}{r}\cos\theta\right)^2 + \left(-m\frac{v^2}{r}\sin\theta + mg\right)^2$$

Using the trigonometry identity $\cos^2\beta + \sin^2\beta = 1$, this can be simplified to

$$F_{string}^2 = \left(\frac{mv^2}{r}\right)^2 - \frac{2m^2v^2g}{r}\sin\theta + m^2g^2$$

Taking the square root of each side of this equation

$$F_{string} = \sqrt{\left(\frac{mv^2}{r}\right)^2 - \frac{2m^2v^2g}{r}\sin\theta + m^2g^2}$$

### Section 5-3  Highway Curves, Banked and Unbanked

A car traveling around a curve on a horizontal road must have a centripetal force provided by the friction force with the road. As the point of contact of the tires with the road is at rest with respect to the road, this is a static friction force. There is therefore a maximum centripetal force and a limit on the centripetal acceleration. If the speed exceeds a critical value, the maximum static friction force of the tires on the road is exceeded and the car can no longer travel without its tires skidding. Therefore, there is always a maximum speed at which a given curve can be rounded by a car without skidding. This maximum speed changes with a change in the coefficient of friction between the tires and the road. Under wet or icy road conditions the coefficient of friction is reduced and so is the maximum speed the car can round the curve without skidding.

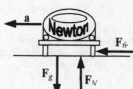

Similar arguments apply if the roadway of the turned is banked rather than horizontal. However, for a banked curve there will be only a component of the frictional force in the centripetal direction and an additional component of the centripetal force from the normal force of the roadway. Certainly, in these cases there is the possibility that no frictional force is required. That happens when the component of the normal force in the centripetal direction is equal to the required centripetal force.

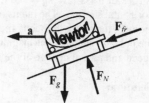

### Example 5-3-A

A car goes around a curve that is banked outward at 13.4° instead of inward as it should be. The radius of the curve is 65.4 m and the speed of the car is 17.3 m/s. What is the minimum coefficient of static friction that will enable the car to stay on the road?

### Solution:

Draw a free-body diagram. The curve is banked outward which implies that the direction of the acceleration is horizontal and to the right. Using a coordinate system with the x-axis at a 13.4° angle above horizontal and writing down Newton's second law for each coordinate direction:

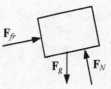

$$ma_x = \sum F_x \quad \Rightarrow \quad m\frac{v^2}{r}\cos 13.4° = \mu_s F_N - mg\sin 13.4°$$

$$ma_y = \sum F_y \quad \Rightarrow \quad -m\frac{v^2}{r}\sin 13.4° = F_N - mg\cos 13.4°$$

where we have inserted the maximum static frictional force into the expression. These are two equations for the two unknowns $F_N$ and $\mu_s$. Solving this pair of equations for $\mu_s$

$$\mu_s = \frac{g\sin 13.4° + \dfrac{v^2}{r}\cos 13.4°}{g\cos 13.4° - \dfrac{v^2}{r}\sin 13.4°} = \frac{\left(9.80\,\text{m/s}^2\right)\sin 13.4° + \dfrac{(17.3\,\text{m/s})^2}{65.4\,\text{m}}\cos 13.4°}{\left(9.80\,\text{m/s}^2\right)\cos 13.4° - \dfrac{(17.3\,\text{m/s})^2}{65.4\,\text{m}}\sin 13.4°} = 0.793$$

## Section 5-4  Nonuniform Circular Motion

Nonuniform circular motion means motion in a circle of fixed radius, but with varying speed. At any instant, the acceleration in the centripetal direction is still equal to $v^2/r$. However, there can now be a component of acceleration perpendicular to the centripetal direction or tangent to the circle. We call this acceleration the tangential acceleration. The magnitude of the tangential acceleration, $a_{\text{tan}}$, is equal to

$$a_{\text{tan}} = \frac{dv}{dt}$$

where $v$ is the speed of the object. The direction of the tangential acceleration is in the direction of the velocity of the object. The acceleration vector of the object, **a**, is the sum of the centripetal acceleration and the tangential acceleration:

$$\mathbf{a} = \mathbf{a}_R + \mathbf{a}_{\text{tan}}$$

This concept can be applied to an object traveling along an arbitrary curved path. At each point along any curved path there is a unique circle that is tangent to the path with the same radius of curvature as the path and in the plane of the path.

### Example 5-4-A

A rock of mass 1.27 kg is tied to a string and spun in a circle as it slides on a frictionless horizontal surface. The radius of the circle the rock follows is 1.04 m. At a given moment the string lies along the direction of the arrow in the diagram when the rock is in the position shown. The magnitude of the force of the string is 18.1 N. What are the speed and rate of change of the speed of the rock at that moment?

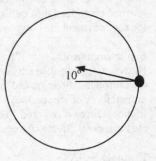

### Solution:

We can resolve the force vector into a component in the centripetal direction and a component in the tangential direction:

$$F_R = F\cos 10° = (18.1\,\text{N})\cos 10° = 17.8\,\text{N}$$

$$F_{\text{tan}} = F\sin 10° = (18.1\,\text{N})\sin 10° = 3.14\,\text{N}$$

Newton's second law in the centripetal direction gives

$$F_R = ma_R = \frac{mv^2}{r} \quad \Rightarrow \quad v = \sqrt{\frac{F_R r}{m}} = \sqrt{\frac{(17.8\,\text{N})(1.04\,\text{m})}{1.27\,\text{kg}}} = 3.82\,\text{m/s}$$

Newton's second law in the tangential direction gives

$$F_{\tan} = ma_{\tan} = m\frac{dv}{dt} \quad \Rightarrow \quad \frac{dv}{dt} = \frac{F_{\tan}}{m} = \frac{3.14\,\text{N}}{1.27\,\text{kg}} = 2.47\,\text{m/s}^2$$

## Section 5-5  Velocity-Dependent Forces; Terminal Velocity

When objects move through fluids, such as water or air, they feel a frictional force called a **drag force**. The drag force depends on the velocity of the object. The kinetic friction of two surfaces in contact depends on velocity in that the direction of the kinetic friction force is always opposite in direction to the relative velocity of the object it acts on. This is true for drag forces in fluids also, but the magnitude of the drag force also depends on the speed of the object relative to the fluid. This speed dependence is not a "simple" function of the speed.

For certain ranges of speed, the magnitude of the drag force can be successfully approximated by simple models. In the limit as the speed of the object goes to zero, the magnitude of the drag force, $F_D$, is proportional to the speed of the object

$$F_D = -bv$$

where the constant $b$ depends on the shape of the object, orientation of the object, and the viscosity of the fluid. This type of drag force is called viscous drag.

At higher speeds, the speed dependence of the drag force becomes proportional to the square of the speed of the object

$$F_D = -Cv^2$$

where the constant $C$ depends on the shape and orientation of the object and the density of the fluid. The drag force is called inertial drag when it has this dependence on speed.

### Example 5-5-A

An object enters a fluid at a speed $v$ where it is acted on only by the inertial drag force, $F_D = -Cv^2$. As long as the inertial drag dependence is valid, how does the speed of the object depend on time?

**Solution:**

We write Newton's second law in terms of drag force and write the acceleration as the time derivative of the velocity

$$\sum F = ma \quad \Rightarrow \quad -Cv^2 = m\frac{dv}{dt} \quad \Rightarrow \quad -\frac{C}{m}dt = \frac{1}{v^2}dv$$

Both sides of this expression can be integrated

$$\int_0^T \frac{-C}{m}dt = \int_{v(0)}^{v(T)} \frac{1}{v^2}dv \quad \Rightarrow \quad \frac{-C}{m}t\Big|_0^T = -\frac{1}{v}\Big|_{v(0)}^{v(T)} \quad \Rightarrow \quad \frac{-CT}{m} = -\frac{1}{v(T)} + \frac{1}{v(0)}$$

Solving this for $v(T)$ we get:

$$v(T) = \frac{1}{\dfrac{1}{v(0)} + \dfrac{CT}{m}}$$

## Practice Quiz

1.    Can kinetic friction ever cause an object to speed up?

   a)   Yes
   b)   No
   c)   Only if the coefficient of kinetic friction is greater than one
   d)   Only if the normal force is zero

2.    Three identical cars are connected to each other, one to the next by a towrope. The car in front starts to accelerate, pulling all the cars behind it. How does the minimum coefficient of static friction between the car's tires and the road compare to that for a single car with the same acceleration?

   a)   The same minimum coefficient of static friction is required.
   b)   A coefficient of static friction three times greater is required.
   c)   A minimum coefficient of static friction one third as great is required.
   d)   Not enough information is given to answer the question.

3.    Many of the problems in this chapter involved a car going around a curve at constant speed. What would happen to the required minimum coefficient of static friction, $\mu_s$, if the car were changing speed?

   a)   The minimum $\mu_s$ is bigger if speeding up and is less if slowing down.
   b)   The minimum $\mu_s$ is less if speeding up and is bigger if slowing down.
   c)   The minimum $\mu_s$ is the same whether the car is changing speed or not.
   d)   The minimum $\mu_s$ is bigger than if the car were not changing speed.

4.    Your uncle tries to impress you by pulling the tablecloth out from under the place settings on the table without everything crashing to the floor. Which of the following statements about the coefficients of friction between the tablecloth and the dinnerware is required to be true for the trick to work successfully?

   a)   The coefficient of static friction must be equal to the coefficient of kinetic friction.
   b)   The coefficient of static friction must be less than the coefficient of kinetic friction.
   c)   The coefficient of static friction must be greater than the coefficient of kinetic friction.
   d)   Both coefficients of friction must be zero.

5.    An object is speeding up as it goes around a circle. Which statement is necessarily true about the net force acting on the object?

   a)   The force on the object acts directly toward the center of the circle.
   b)   The force on the object acts directly away from the center of the circle.
   c)   The net force on the object acts in a direction tangent to the circle at the position of the object.
   d)   The net force cannot be in any of the three directions above.

6.    Anti-lock braking systems on cars are designed to keep the tires rolling on the road rather than sliding. Why is this advantageous to stopping and controlling a vehicle?

   a)   It isn't advantageous. The car will take longer to stop.
   b)   The car manufacturer can charge more money for the car.
   c)   The coefficient of static friction of the tires on the road is higher than the coefficient of kinetic friction.
   d)   The tires won't squeal when you stop.

7.    Two blocks made of different materials are placed on a horizontal board. Gradually the board is lifted at one end. Block A begins sliding at a gentler slope of the board than block B. What can you conclude?

   a)   Block A is heavier than block B.
   b)   Block B is heavier than block A.
   c)   The coefficient of kinetic friction of block A is less than that of block B.
   d)   The coefficient of static friction of block A is less than that of block B.

8.    If you place a toy boat in a stream, it begins to move. Why doesn't the drag force stop the toy boat?

   a)   The drag force is what makes the boat move.
   b)   The force of gravity on the boat overcomes the drag force.
   c)   The drag force is small and eventually the drag force will stop the boat.
   d)   The boat's hull is designed to make the drag force negligible.

9.    An object of mass 5.61 kg is dropped from an airplane. At the highest speeds attained by the object in free fall, the magnitude of the drag force on the object is given by $F_D = (0.634 \text{ Ns/m}) \, v$. What is the terminal speed of the object in free-fall?

      a)   86.7 m/s
      b)   9.8 m/s$^2$
      c)   34.9 m/s
      d)   8.85 m/s

10.   Two objects are moving through a fluid at the same speed and they experience the same drag force at a given moment. If object A is experiencing viscous drag ($F_D \propto v$) and object B is experiencing inertial drag ($F_D \propto v^2$), which object will come to a rest more quickly?

      a)   A
      b)   B
      c)   Neither; the velocity will be the same for both.
      d)   Not enough information is given to determine which will slow down more quickly.

11.   A car of mass 680 kg travels around a 3.05° banked curve with a speed of 12.6 m/s. The radius of the curve is 61.2 m. What is the force of friction acting on the car?

12.   A box of mass 3.44 kg rests on a sloped ramp angled at 29.1° above horizontal. The coefficient of friction between the ramp and the box is 0.697. At what horizontal acceleration of the ramp will the box begin sliding down the ramp?

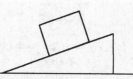

13.   Two boxes rest in contact each other while resting on the ramp as shown in the diagram. The slope of the ramp is 14.0° above horizontal. The lower box has a mass of 2.77 kg and a coefficient of static friction with the ramp of 0.718. The upper box has a coefficient of static friction 0.152 with the ramp. What is the upper limit on the mass of the upper box so the boxes will not begin sliding down the ramp?

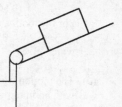

14.   A racetrack is circular and has a banking of 18.5° and a radius of 140 m. What is the minimum time that a car can complete a lap around the track if the coefficient of static friction between tires and road is 0.794?

15.   Two masses are connected by a string as shown in the diagram. The mass on the slope is 10.5 kg and its coefficient of kinetic friction with the surface of the slope is 0.559. The slope is at an angle of 22.6° above horizontal. The mass hanging on the vertical part of the rope is 15.8 kg. What is the acceleration of the blocks?

## Problem Solutions

Solutions to problems with a number equal to $4n - 1$ where $n$ is an integer:

3.   (a)                      (b)                      (c)

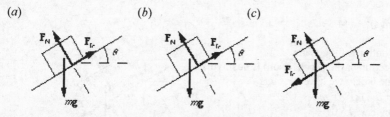

In (a) the friction is static and opposes the impending motion down the plane.
In (b) the friction is kinetic and opposes the motion down the plane.
In (c) the friction is kinetic and opposes the motion up the plane.

7.  While the box is sliding down, friction will be up the plane, opposing the motion. From the force diagram for the box, we have $\Sigma \mathbf{F} = m\mathbf{a}$:

   x-component:    $mg \sin \theta - F_{\text{fr}} = ma$.
   y-component:    $F_N - mg \cos \theta = 0$.

   From the x-equation, we have
   $$F_{\text{fr}} = mg \sin \theta - ma = m(g \sin \theta - a)$$
   $$= (15.0 \text{ kg})[(9.80 \text{ m/s}^2) \sin 30.0° - (0.30 \text{ m/s}^2)]$$
   $$= 69 \text{ N}.$$
   Because the friction is kinetic, we have
   $$F_{\text{fr}} = \mu_k F_N = \mu_k mg \cos \theta,$$
   $69 \text{ N} = \mu_k (15.0 \text{ kg})(9.80 \text{ m/s}^2) \cos 30.0°$, which gives $\mu_k = 0.54$.

11. The friction is kinetic, so $F_{\text{fr}} = \mu_k F_N$. Because the push is no longer there, the only horizontal force is the friction force. Using the force diagram for the box, we can write $\Sigma \mathbf{F} = m\mathbf{a}$:

   x-component: $-\mu_k F_N = Ma$;
   y-component: $F_N - Mg = 0$.

   Thus we have
   $$a = -\mu_k g.$$
   For the motion with constant acceleration, we have
   $$v^2 = v_0^2 + 2ax = v_0^2 + 2(-\mu_k g)x;$$
   $0 = (2.5 \text{ m/s})^2 + 2(-0.24)(9.80 \text{ m/s}^2)x$, which gives $x = 1.3 \text{ m}$.

15. (a)  We write $\Sigma \mathbf{F} = m\mathbf{a}$ from the force diagram for the snow while it is stationary on the roof:

   x-component: $mg \sin \theta - F_{\text{fr}} = 0$;
   y-component: $F_N - mg \cos \theta = 0$.

   When we combine the equations, we get
   $$F_{\text{fr}} = mg \sin \theta = \mu_s F_N = \mu_s mg \cos \theta.$$
   Thus we have
   $$\mu_s = \tan \theta = \tan 30° = 0.58.$$

   (b)  We write $\Sigma \mathbf{F} = m\mathbf{a}$ from the force diagram for the snow while it is sliding on the roof:

   x-component: $mg \sin \theta - \mu_k F_N = ma$;
   y-component: $F_N - mg \cos \theta = 0$.

   Thus $a = g(\sin \theta - \mu_k \cos \theta)$
   $$= (9.80 \text{ m/s}^2)[\sin 30° - (0.20) \cos 30°] = 3.20 \text{ m/s}^2.$$
   For the motion on the roof, we have
   $$v_1^2 = v_0^2 + 2a(x_1 - x_0) = 0 + 2(3.20 \text{ m/s}^2)(5.0 \text{ m}),$$
   which gives
   $$v_1 = 5.7 \text{ m/s}.$$

   (c)  The motion when the snow leaves the roof is projectile motion, with an initial velocity of $v_1 = 5.7$ m/s at 30° below the horizontal. If we use the new coordinate system shown, we have
   $$v_x = v_1 \cos \theta = (5.7 \text{ m/s}) \cos 30° = 4.9 \text{ m/s};$$
   $$v_y^2 = (-v_1 \sin \theta)^2 + 2gh = [-(5.7 \text{ m/s}) \sin 30°]^2 + 2(9.80 \text{ m/s}^2)(10.0 \text{ m}),$$
   which gives $v_y = 14.3$ m/s.
   The speed of the snow is
   $$v = (v_x^2 + v_y^2)^{1/2} = [(4.9 \text{ m/s})^2 + (14.3 \text{ m/s})^2]^{1/2} = 15 \text{ m/s}.$$

19. We choose the origin for x at the bottom of the plane.
   Note that down the plane (the direction of the acceleration) is positive.

   (a)  From the force diagram for the block, we have $\Sigma F = ma$:

   x-component:    $mg \sin \theta + \mu_k F_N = ma$.
   y-component:    $F_N - mg \cos \theta = 0$.

   When we combine these, we have
   $$a = g(\sin \theta + \mu_k \cos \theta)$$
   $$= (9.80 \text{ m/s}^2)[\sin 22.0° + (0.17) \cos 22.0°] = 5.22 \text{ m/s}^2.$$
   For the motion on the block until it stops, we have
   $$v^2 = v_0^2 + 2a(x - x_0);$$
   $0 = (-3.0 \text{ m/s})^2 + 2(5.22 \text{ m/s}^2)(x - 0)$, which gives $x = -0.86 \text{ m}$.

Thus the block travels 86 cm up the plane.

(*b*) We find the time to reach the highest point from

$v = v_0 + a_{up}t_{up}$ ;

$0 = (- 3.0 \text{ m/s}) + (5.22 \text{ m/s}^2)t_{up}$ , which gives $t_{up} = 0.575$ s.

When the block slides down the plane, the friction force will reverse, so the acceleration is

$a_{down} = g(\sin \theta - \mu_k \cos \theta)$

$= (9.80 \text{ m/s}^2)[\sin 22.0° - (0.17) \cos 22.0°] = 2.13 \text{ m/s}^2$ .

We find the time to slide down from

$x = x_0 + vt + \frac{1}{2}a_{down}t_{down}^2$ ;

$0 = - 0.862 \text{ m} + 0 + \frac{1}{2}(2.13 \text{ m/s}^2)t_{down}^2$ , which gives $t_{down} = 0.900$ s.

Thus the total time is

$t = t_{up} + t_{down} = 0.575 \text{ s} + 0.900 \text{ s} = 1.5$ s.

23. We choose the coordinate system shown in the force diagram. If the blocks are connected by a rod, which can support a tension or a compression, they must have the same acceleration. From the force diagram for each block, we have $\Sigma F = ma$: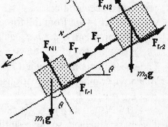

*y*-component (1)  $F_{N1} - m_1g \cos \theta = 0$,  or

$F_{N1} = m_1g \cos \theta$,

*x*-component (1): $m_1g \sin \theta - \mu_1F_{N1} - F_T = m_1a$;

$m_1 \sin \theta - \mu_1m_1 \cos \theta)g - F_T = m_1a$;

*y*-component (2)  $F_{N2} - m_2g \cos \theta = 0$,  or

$F_{N2} = m_2g \cos \theta$,

*x*-component (2): $m_2g \sin \theta - \mu_2F_{N2} + F_T = m_2a$;

$(m_2 \sin \theta - \mu_2m_2 \cos \theta)g + F_T = m_2a$.

When we add the two *x*-equations, we have

$(m_1 + m_2)a = [(m_1 + m_2) \sin \theta - (\mu_1m_1 + \mu_2m_2) \cos \theta]g$,  or

$a = \{\sin \theta - [(\mu_1m_1 + \mu_2m_2)/(m_1 + m_2)] \cos \theta\}g$.

We find the tension from

$F_T = m_1[(\sin \theta - \mu_1 \cos \theta)g - a] = m_1g\{\sin \theta - \mu_1 \cos \theta - \sin \theta + [(\mu_1m_1 + \mu_2m_2)/(m_1 + m_2)] \cos \theta\}$;

$F_T = [m_1m_2(\mu_2 - \mu_1)/(m_1 + m_2)]g \cos \theta$.

If $\mu_1 < \mu_2$, there will be tension in the rod; if $\mu_1 > \mu_2$, there will be compression in the rod.

27. The direction of the kinetic friction force is determined by the direction of the velocity, not the direction of the acceleration. We assume the block on the plane is moving up, so the friction force is down. The forces and coordinate systems are shown in the diagram. From the force diagram, with the block $m_2$ as the system, we can write $\Sigma \mathbf{F} = M\mathbf{a}$: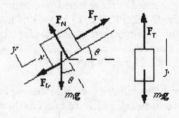

*y*-component:    $m_2g - F_T = m_2a$.

From the force diagram, with the block $m_1$ as the system, we can write $\Sigma \mathbf{F} = m\mathbf{a}$:

*x*-component:    $F_T - F_{fr} - m_1g \sin \theta = m_1a$;

*y*-component:    $F_N - m_1g \cos \theta = 0$; with $F_{fr} = \mu_kF_N$.

When we eliminate $F_T$ between these two equations, we get

$a = (m_2 - m_1 \sin \theta - \mu_km_1 \cos \theta)g/(m_1 + m_2)$.

(*a*) For a mass $m_1 = 5.0$ kg, we have

$a = [5.0 \text{ kg} - (5.0 \text{ kg}) \sin 30° - (0.10)(5.0 \text{ kg}) \cos 30°](9.80 \text{ m/s}^2)/(5.0 \text{ kg} + 5.0 \text{ kg})$

$= 2.0 \text{ m/s}^2$ up the plane.

The acceleration is up the plane because the answer is positive. This agrees with our assumption for the direction of motion.

(*b*) For a mass $m_1 = 2.0$ kg, we have

$a = [5.0 \text{ kg} - (2.0 \text{ kg}) \sin 30° - (0.10)(2.0 \text{ kg}) \cos 30°](9.80 \text{ m/s}^2)/(2.0 \text{ kg} + 5.0 \text{ kg})$

$= 5.4 \text{ m/s}^2$ up the plane.

The acceleration is up the plane because the answer is positive. This agrees with our assumption for the direction of motion.

31. The velocity is constant, so the acceleration is zero.

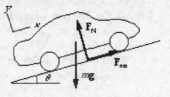

   (*a*)  From the force diagram for the bicycle, we can write $\Sigma \mathbf{F} = m\mathbf{a}$:

       *x*-component:  $mg \sin \theta - F_D = 0$,  or

       $mg \sin \theta = cv^2$;

       $(80.0 \text{ kg})(9.80 \text{ m/s}^2) \sin 7.0° = c[(9.5 \text{ km/h})/(3.6 \text{ ks/h})]^2$, which gives

       $c = 14 \text{ kg/m}$.

   (*b*)  We have an additional force in $\Sigma \mathbf{F} = m\mathbf{a}$:

       *x*-component:  $F + mg \sin \theta - F_D = 0$,  so

       $F = cv^2 - mg \sin \theta$

       $= (13.7 \text{ kg/m})[(25 \text{ km/h})/(3.6 \text{ ks/h})]^2 - (80 \text{ kg})(9.80 \text{ m/s}^2) \sin 7.0° = 5.7 \times 10^2 \text{ N}.$

35. If the car does not skid, the friction is static, with $F_{fr} \Sigma \mu_s F_N$.

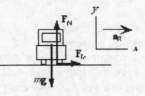

   This friction force provides the centripetal acceleration.  We take a
coordinate system with the *x*-axis in the direction of the
centripetal acceleration.

   We write $\Sigma \mathbf{F} = m\mathbf{a}$ from the force diagram for the auto:

       *x*-component:  $F_{fr} = ma = mv^2/R$;

       *y*-component:  $F_N - mg = 0$.

   The speed is maximum when $F_{fr} = F_{fr,max} = \mu_s F_N$.

   When we combine the equations, the mass cancels, and we get

       $\mu_s g = v_{max}^2/R$;

       $(0.55)(9.80 \text{ m/s}^2) = v_{max}^2/(80.0 \text{ m})$, which gives $v_{max} = 21 \text{ m/s}$.

   The mass canceled, so the result is independent of the mass.

39. Yes. If the bucket is traveling fast enough at the top of the
circle, in addition to the weight of the water a force from the
bucket, similar to a normal force, is required to provide the
necessary centripetal acceleration to make the water go in the
circle.  From the force diagram, we write

       $F_N + mg = ma = mv_{top}^2/R$.

   The minimum speed is that for which the normal force is zero:

       $0 + mg = mv_{top,min}^2/R$,  or  $v_{top,min} = (gR)^{1/2}$.

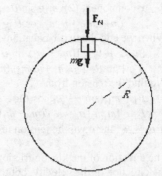

43. We find the speed of the skaters from the period of rotation:

       $v = 2\pi r/T = 2\Sigma(0.80 \text{ m})/(3.0 \text{ s}) = 1.68 \text{ m/s}$.

   The pull or tension in their arms provides the centripetal acceleration:

       $F_T = mv^2/R$;

       $= (60.0 \text{ kg})(1.68 \text{ m/s})^2/(0.80 \text{ m}) = 2.1 \times 10^2 \text{ N}.$

47. At the top of the hill, the normal force is upward and the
weight is downward, which we select as the positive direction.

   (*a*)  We write $\Sigma \mathbf{F} = m\mathbf{a}$ from the force diagram for the car:

       $m_{car}g - F_{Ncar} = mv^2/R$;

       $(1000 \text{ kg})(9.80 \text{ m/s}^2) - F_{Ncar} = (1000 \text{ kg})(20 \text{ m/s})^2/(100 \text{ m})$,

       which gives $F_{Ncar} = 5.8 \times 10^3 \text{ N}$.

   (*b*)  When we apply a similar analysis to the driver, we have

       $(70 \text{ kg})(9.80 \text{ m/s}^2) - F_{Npass} = (70 \text{ kg})(20 \text{ m/s})^2/(100 \text{ m})$,

       which gives $F_{Npass} = 4.1 \times 10^2 \text{ N}$.

   (*c*)  For the normal force to be equal to zero, we have

       $(1000 \text{ kg})(9.80 \text{ m/s}^2) - 0 = (1000 \text{ kg})v^2/(100 \text{ m})$,

       which gives  $v = 31 \text{ m/s}$  (110 km/h or 70 mi/h).

51. We convert the speed: $(90 \text{ km/h})/(3.6 \text{ ks/h}) = 25.0 \text{ m/s}$.

   At the speed for which the curve is banked perfectly,
there is no need for a friction force.  We take the *x*-axis
in the direction of the centripetal acceleration.

   We write $\Sigma \mathbf{F} = m\mathbf{a}$ from the force diagram for the car:

       *x*-component:  $F_{N1} \sin \theta = ma_1 = mv_1^2/R$;

       *y*-component:  $F_{N1} \cos \theta - mg = 0$.

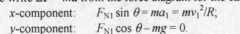

Combining these, we get

$\tan \theta = v_1^2/gR$

$= (25.0 \text{ m/s})^2/(9.80 \text{ m/s}^2)(60 \text{ m}) = 1.062$,    or    $\theta = 46.7°$.

At a higher speed, there is need for a friction force, which will be down the incline to help provide the greater centripetal acceleration. If the automobile does not skid, the friction is static, with $F_{fr} \Sigma \mu_s F_N$.

At the maximum speed, $F_{fr} = \mu_s F_N$. We write $\Sigma \mathbf{F} = m\mathbf{a}$ from the force diagram for the car:

x-component:    $F_{N2} \sin \theta + \mu_s F_{N2} \cos \theta = ma_2 = mv_{max}^2/R$;

y-component:    $F_{N2} \cos \theta - \mu_s F_{N2} \sin \theta - mg = 0$, or $F_{N2}(\cos \theta - \mu_s \sin \theta) = mg$.

When we eliminate $F_{N2}$ by dividing the equations, we get

$v_{max}^2 = gR(\sin \theta + \mu_s \cos \theta)/(\cos \theta - \mu_s \sin \theta) = gR(\tan \theta + \mu_s)/(1 - \mu_s \tan \theta) = gR(v_1^2 + \mu_s gR)/(gR - \mu_s v_1^2)$

$= (9.80 \text{ m/s}^2)(60 \text{ m})(1.062 + 0.30)/(1 - (0.30)(1.062))$,

which gives $v_{max} = 34.3 \text{ m/s} = 123 \text{ km/h}$.

At a lower speed, there is need for a friction force, which will be up the incline to prevent the car from sliding down the incline. If the automobile does not skid, the friction is static, with $F_{fr} \Sigma \mu_s F_N$.

At the minimum speed, $F_{fr} = \mu_s F_N$. The reversal of the direction of $F_{fr}$ can be incorporated in the above equations by changing the sign of $\mu_s$, so we have

$v_{min}^2 = gR[(\sin \theta - \mu_s \cos \theta)/(\cos \theta + \mu_s \sin \theta)]$

$= gR(\tan \theta - \mu_s)/(1 + \mu_s \tan \theta)$

$= (9.80 \text{ m/s}^2)(60 \text{ m})[(1.062 - 0.30)/(1 + (0.30)(1.062))]$,

which gives $v_{min} = 18.4 \text{ m/s} = 66 \text{ km/h}$.

Thus the range of permissible speeds is $66 \text{ km/h} < v < 123 \text{ km/h}$.

55. (a) We find the tangential acceleration from

$a_{tan} = dv/dt = d[3.6 \text{ m/s} + (1.5 \text{ m/s}^3)t^2]/dt = (3.0 \text{ m/s}^3)t = (3.0 \text{ m/s}^3)(3.0 \text{ s}) = 9.0 \text{ m/s}^2$.

(b) We find the radial acceleration from

$a_R = v^2/r = [3.6 \text{ m/s} + (1.5 \text{ m/s}^3)(3.0 \text{ s})^2]^2/(20 \text{ m}) = 15 \text{ m/s}^2$.

59. (a) For an initial velocity downward, we take the positive direction downward.

The drag force will be upward, so we have

$mg - bv = m \, dv/dt$, or $dv/[v - (mg/b)] = -(b/m) \, dt$.

When we integrate, we get

$$\int_{v_0}^{v} \frac{dv}{v - (mg/b)} = \int_0^t -\frac{b}{m} dt$$

$\ln\{[v - (mg/b)]/[v_0 - (mg/b)]\} = -bt/m$, or

$v = (mg/b) + [v_0 - (mg/b)] e^{-bt/m}$.

(b) For an initial velocity upward, we take the positive direction upward.

The drag force and gravity will be downward, so we have

$-mg - bv = m \, dv/dt$, or $dv/[v + (mg/b)] = -(b/m) \, dt$.

This is the same as in part (a), with g replaced by $-g$. Thus we have

$v = -(mg/b) + [v_0 + (mg/b)] e^{-bt/m}$.

For the motion after the object comes momentarily to rest, we use the result from part (a) with $v_0 = 0$.

63. When the engine is shut off, the only force on the boat is the drag force of the water, so we have

$F_D = -bv = m \, dv/dt$, or $dv/v = -b \, dt/m$.

When we integrate, we get

$$\int_{v_0}^{v} \frac{dv}{v} = \int_0^t -\frac{b}{m} dt$$

$\ln(v/v_0) = -bt/m$, or $v = v_0 e^{-bt/m}$.

We find the value of $b/m$ from the change in velocity in 3.0 s:

$\ln(½) = -(b/m)(3.0 \text{ s})$, which gives $b/m = 0.231 \text{ /s}$.

From the definition of $v = dx/dt$, we have

$dx/dt = v_0 e^{-bt/m}$, or $dx = v_0 e^{-bt/m} dt$.

When we integrate, we get

$$\int_0^x dx = \int_0^t v_0 e^{-bt/m} dt$$

$x = (-mv_0/b)(e^{-bt/m} - 1) = (mv_0/b)(1 - e^{-bt/m})$

$= [(2.4 \text{ m/s})/(0.231 \text{ /s})][1 - e^{-(0.231 \text{ /s})(3.0 \text{ s})}] = 10 \text{ m}$.

67. We can find the required acceleration, assumed constant, from
    $x = v_0t + \frac{1}{2}at^2$;
    $(0.250 \text{ mi})(1610 \text{ m/mi}) = 0 + \frac{1}{2}a (6.0 \text{ s})^2$, which gives $a = 22.4 \text{ m/s}^2$.
    If we assume that the tires are just on the verge of slipping, $F_{\text{fr,max}} = \mu_s F_N$, so we have
        $x$-component:  $\mu_s F_N = ma$;
        $y$-component:  $F_N - mg = 0$.
    Thus we have $\mu_s = ma/mg = a/g = (22.4 \text{ m/s}^2)/(9.80 \text{ m/s}^2) = 2.3$.

71. (*a*)  For the object to move with the ground, the static friction force must provide the same
        acceleration.  With the usual coordinate system, for $\Sigma\mathbf{F} = m\mathbf{a}$ we have
            $x$-component:  $F_{\text{fr}} = ma$;
            $y$-component:  $F_N - mg = 0$.
        For static friction, $F_{\text{fr}} \leq \mu_s F_N$,  or  $ma \leq \mu_s mg$ ; thus $\mu_s \geq a/g$.
        In order not to slide when the acceleration is maximum, the minimum required coefficient of static
        friction is $\mu_s = a_{\text{max}}/g$.
    (*b*)  For the greatest acceleration, the minimum required coefficient is
            $\mu_s = a_{\text{max}}/g = (4.0 \text{ m/s}^2)/(9.80 \text{ m/s}^2) = 0.41$.
        Because this is greater than 0.25, the chair will slide.

75. The horizontal force on the astronaut produces the centripetal acceleration:
        $F = ma_R = mv^2/r$;
        $(7.75)(2.0 \text{ kg})(9.80 \text{ m/s}^2) = (2.0 \text{ kg})v^2/(10.0 \text{ m})$, which gives $v = 27.6 \text{ m/s}$.
    The rotation rate is
        Rate $= v/2\pi r = (27.6 \text{ m/s})/2\pi(10.0 \text{ m}) = 0.439 \text{ rev/s}$.
    Note that the results are independent of mass, and thus are the same for all astronauts.

79. (*a*)  There will be two forces in the vertical direction.  We write $\Sigma\mathbf{F} = m\mathbf{a}$ from the force diagram.
        At point A the radial acceleration will be down, so we take that as the positive direction:
            $mg - F_{\text{NA}} = ma_A = mv^2/R$,  so  $F_{\text{NA}} = m(g - v^2/R)$.
        At point B the radial acceleration is zero, so we have
            $mg - F_{\text{NB}} = ma_B = 0$,  so  $F_{\text{NB}} = mg$.
        At point C the radial acceleration will be up, so we take that as the positive direction:
            $F_{\text{NC}} - mg = ma_C = mv^2/R$,  so  $F_{\text{NC}} = m(g + v^2/R)$.
        Thus we see that
            $F_{\text{NC}} > F_{\text{NB}} > F_{\text{NA}}$.

    (*b*)  The force diagram for the driver would look the same and there will be similar normal forces on
        the driver to provide the radial accelerations, so the drive will feel heaviest at C, and lightest at A.
    (*c*)  Because the normal force cannot be negative, the car will lose contact at A when
            $mg \leq mv^2/R$.
        Thus the maximum speed at A without losing contact is
            $v_{\text{Amax}} = (gR)^{1/2}$.

83. Before the mass slides, the friction is static, with $F_{\text{fr}} \leq \mu_s F_N$.
    The static friction force will be maximum just before the mass
    slides.  We write $\Sigma\mathbf{F} = m\mathbf{a}$ from the force diagram:
        $x$-component:  $mg \sin\phi - \mu_s F_N = 0$;
        $y$-component:  $F_N - mg \cos\phi = 0$.
    When we combine these, we get
        $\tan\phi = \mu_s = 0.60$,  or  $\phi = 31°$.

87. (*a*)  We can find the radius of the halfcircle from the radial acceleration.  We
        choose the $x$-direction in the radial direction.  We write $\Sigma\mathbf{F} = m\mathbf{a}$ from the force
        diagram:
            $F_{\text{lift}} \cos\theta - mg = 0$;
            $F_{\text{lift}} \sin\theta = mv^2/r$.
        When we combine these, we get
            $\tan\theta = v^2/gr$;
            $\tan 38° = [(520 \text{ km/h})/(3.6 \text{ ks/h})]^2/(9.8 \text{ m/s}^2)r$, which gives $r = 2.7 \times 10^3 \text{ m}$.
        We find the time to complete the halfcircle from
            $t = \pi r/v = \pi(2.7 \times 10^3 \text{ m})/[(520 \text{ km/h})/(3.6 \text{ ks/h})] = 59 \text{ s}$.
    (*b*)  The passengers will feel a greater normal force from the seat.

91. For the rising rocket we have
$$-mg - kv^2 = m\,dv/dt.$$
We can use the chain rule to replace the variable $t$ with the variable $y$:
$$-mg - kv^2 = m\,dv/dt = m\,(dv/dy)(dy/dt) = mv\,dv/dy, \quad \text{or} \quad v\,dv/[v^2 + (mg/k)] = -(k/m)\,dy.$$
When we integrate from the initial speed up to the highest point, we get

$$\int_{v_0}^{0} \frac{2v\,dv}{v^2 + (mg/k)} = \int_{0}^{h} -\frac{2k}{m}\,dy$$

$$\ln\left[v^2 + (mg/k)\right]\Big|_{v_0}^{0} = \ln\left[\frac{mg/k}{v_0^2 + (mg/k)}\right] = -\frac{2kh}{m}$$

$$h = -\frac{m}{2k}\ln\left[\frac{mg}{kv_0^2 + (mg)}\right] = -\frac{250\,kg}{2(0.65\,kg/m)}\ln\left[\frac{(250\,kg)(9.8\,m/s^2)}{(0.65\,kg/m)(120\,m/s)^2 + (250\,kg)(9.8\,m/s^2)}\right] = 302\,m$$

If there were no air resistance, $a = -g$, and we find the maximum height from
$$v^2 = v_0^2 + 2ay;$$
$$0 = (120\,m/s)^2 + 2(-9.80\,m/s^2)h', \quad \text{which gives } h' = 735\,m, \text{ almost } 2.5\times.$$

# Chapter 6: Gravitation and Newton's Synthesis

## Chapter Overview and Objectives

This chapter introduces Newton's law of universal gravitation. It also introduces Kepler's laws of planetary motion and Einstein's principle of equivalence.

After completing study of this chapter, you should:

- Know Newton's universal law of gravitation.
- Know how to obtain the surface gravitational acceleration of an object from Newton's universal law of gravitation.
- Know Kepler's laws of planetary motion.

## Summary of Equations

Magnitude of gravitational force:
$$F = \frac{Gm_1m_2}{r^2}$$

Vector form of Newton's law of gravitation:
$$\mathbf{F}_{12} = -\frac{Gm_1m_2}{r_{21}^2}\hat{\mathbf{r}}_{21}$$

Magnitude of acceleration due to gravity:
$$g = \frac{Gm}{r^2}$$

Vector form of acceleration due to gravity:
$$\mathbf{g} = \frac{Gm}{r^2}\hat{\mathbf{r}}$$

Kepler's Third Law:
$$\frac{T_1^2}{T_2^2} = \frac{s_1^3}{s_2^3}$$

## Chapter Summary

### Section 6-1 Newton's Law of Universal Gravitation

Every particle in the universe attracts every other particle with a force that is proportional to the product of their masses and inversely proportional to the square of the distance between them. This force acts along the line joining the two particles. The magnitude of this force, $F$, is

$$F = \frac{Gm_1m_2}{r^2}$$

where $G$ is the universal gravitation constant and is equal to $6.67 \times 10^{-11}$ Nm$^2$/kg$^2$. The masses of the two particles are $m_1$ and $m_2$ and the distance between the two particles is $r$.

### Example 6-1-A

Two small asteroids orbit each other in circular orbits. The distance between the centers of the two asteroids is 31.2 m. The mass of the larger asteroid, $m_{large}$, is $2.86 \times 10^6$ kg and the mass of the smaller asteroid, $m_{small}$ is $1.13 \times 10^5$ kg. What is the magnitude of the gravitational force of interaction between the two asteroids? What is the acceleration of each of the asteroids? What is the period of the circular orbits of each of the two asteroids? Ignore the force of the sun acting on the asteroids for the second and third parts of the question.

**Solution:**

First solve for the gravitational force of attraction for one asteroid on the other (equal in magnitude for each asteroid by Newton's third law):

$$F = \frac{Gm_{small}m_{large}}{d^2} = \frac{\left(6.67 \times 10^{-11} \ \text{Nm}^2/\text{kg}\right)\left(1.13 \times 10^5 \ \text{kg}\right)\left(2.86 \times 10^6 \ \text{kg}\right)}{(31.2 \ \text{m})^2} = 2.214 \times 10^{-2} \ \text{N}$$

Now the acceleration of each asteroid can be determined from Newton's second law:

$$a_{small} = \frac{F}{m_{small}} = \frac{2.214 \times 10^{-2} \ \text{N}}{1.13 \times 10^5 \ \text{kg}} = 1.960 \times 10^{-7} \ \text{m/s}^2$$

$$a_{large} = \frac{F}{m_{large l}} = \frac{2.214 \times 10^{-2} \ \text{N}}{2.86 \times 10^6 \ \text{kg}} = 7.741 \times 10^{-9} \ \text{m/s}^2$$

The two asteroids must have a common period to their orbits; otherwise they would not orbit around each other. Also, the sum of the orbital radii must add up to the distance between the centers of the asteroids. Using the centripetal acceleration is $v^2/r$, we end up with the following four relationships:

$$a_{small} = \frac{v_{small}^2}{r_{small}} \qquad\qquad a_{large} = \frac{v_{large}^2}{r_{large}}$$

$$r_{small} + r_{large} = d \qquad\qquad T = \frac{2\pi \, r_{small}}{v_{small}} = \frac{2\pi \, r_{large}}{v_{large}}$$

Eliminating the speeds and $r_{large}$ from this set of equations results in:

$$T^2 = \frac{4\pi^2 \, r_{small}}{a_{small}} = \frac{4\pi^2 \left(d - r_{small}\right)}{a_{large}}$$

Eliminating $r_{small}$ from these equations results in:

$$T = 2\pi\sqrt{\frac{d}{a_{large} + a_{small}}} = 2\pi\sqrt{\frac{31.2 \ \text{m}}{7.741 \times 10^{-9} \ \text{m/s}^2 + 1.960 \times 10^{-7} \ \text{m/s}^2}} = 7.78 \times 10^4 \ \text{s}$$

$$T = 7.78 \times 10^4 \ \text{s} = 21.6 \ \text{h}$$

## Section 6-2 Vector Form of Newton's Law of Universal Gravitation

Newton's law of universal gravitation can be written in vector form as:

$$\mathbf{F}_{12} = -\frac{Gm_1 m_2}{r_{21}^2}\hat{\mathbf{r}}_{21}$$

where $\mathbf{F}_{12}$ is the vector force on particle 1 of mass $m_1$ caused by particle 2 of mass $m_2$. The vector $\mathbf{r}_{12}$ is the unit vector that points from particle 2 along the line to particle 1. The minus sign appears because the force is attractive, which makes the force direction opposite to the direction of $\mathbf{r}_{12}$.

## Section 6-3  Gravity Near the Earth's Surface; Geophysical Applications

Comparing the expressions for the magnitude of the gravitational force in Newton's universal law of gravitation and the expression $mg$ which has been used for the weight of the object at the Earth's surface we get:

$$mg = \frac{Gm_E m}{r_E^2}$$

Where $m_E$ is the mass of the Earth and $r_E$ is the radius of the Earth. Solving for $g$, the acceleration due to gravity at the earth's surface:

$$g = \frac{Gm_E}{r_E^2}$$

This equation can also be used for calculating the acceleration due to gravity caused by other bodies by substituting the mass of those bodies for $m_E$ and for other distances from the center of the body by substituting those distances for $r_E$.

## Section 6-4 Satellites and "Weightlessness"

The only external force acting on satellites in orbit around a body or on any body in free fall is the force of gravity. Even though a gravitational force is acting on the body in these situations, this condition is called **weightlessness**. One reason this is called weightlessness is that objects in contact with one another under conditions of free fall have no contact force between them in the vertical direction. If you held a book in your hands while standing on the surface of the Earth you would need to apply an upward force on the book to prevent the weight of the book from accelerating the book downward. If you held the same book in your hands while in free fall, you would not need to apply any force to the book to keep it from accelerating relative to your body because both you and the book would be accelerating the same. Comparing it to the situation of standing stationary on the Earth; because there is no downward acceleration relative to you, the book appears to have no weight. This is often called **apparent weightlessness** because of the appearance of lack of downward acceleration observed by the freely falling observer. The word apparent is used because there actually is a weight force acting on the body.

Satellites that travel in circular orbits have a centripetal acceleration $v^2/r$, where $v$ is the speed of the satellite in its orbit and $r$ is the radius of its orbit. Because satellites are in free fall, Newton's second law applied to this situation results in:

$$\frac{Gm_E m_s}{r^2} = m_s \frac{v^2}{r}$$

where $m_s$ is the mass of the satellite and $r$ is the distance from the center of the Earth to the satellite.

## Section 6-5 Kepler's Laws and Newton's Synthesis

Kepler's laws of planetary motion are empirical laws describing the behavior of the motion of planets around the sun:

*Kepler's first law*: The path of each planet around the sun is an ellipse. The sun is located at one of the two foci of the ellipse.

*Kepler's second law*: Each planet moves so that the line joining the planet and the sun sweeps out area at a constant rate as the planet moves around the sun.

*Kepler's third law*: The ratio of the square of the period of a planet is proportional to the cube of the semi-major axis of the elliptical trajectory of the planet's orbit. This relationship can be written:

$$\left(\frac{T_1}{T_2}\right)^2 = \left(\frac{s_1}{s_2}\right)^3$$

## Example 6-5-A

A planet that is in a circular orbit a distance of $2.00 \times 10^8$ m from a star has an orbital period of 3.87 years. A second planet is observed to have an orbital period of 29.7 years. How far is the second planet from the star?

**Solution:**

Solving Kepler's third law relationship for the period of the second planet:

$$s_2 = \left(\frac{T_2}{T_1}\right)^{2/3} s_1 = \left(\frac{29.7\,\text{y}}{3.87\,\text{y}}\right)^{2/3} 2.00 \times 10^8\,\text{m} = 7.78 \times 10^8\,\text{m}$$

## Section 6-6  Gravitational Field

The law of universal gravitation implies "action at a distance," two objects not in contact with each other affecting the motion of each other.  The abstract notion of a gravitational field created by a massive body removes the action at a distance problem.  The gravitational field is a part of the massive body that extends through all space.  The gravitational field, **g**, at a point in space is defined to be the force that a test particle placed at that position experiences divided by the mass of the test particle:

$$\mathbf{g} = \frac{\mathbf{F}}{m}$$

The gravitational field of a point particle is given by

$$\mathbf{g} = -\frac{Gm}{r^2}\hat{\mathbf{r}}$$

where **r** is the vector displacement from the mass to the point at which the gravitational field is evaluated.  The gravitational field of an extended body is the vector sum of the gravitational field of point masses that make up the extended body.

## Section  6-7 Types of Forces in Nature

All of the interactions between objects that have been observed can be explained in terms of only four fundamental forces.  These fundamental forces are:

1.  The gravitational force
2.  The electromagnetic force
3.  The strong nuclear force
4.  The weak nuclear force

All other forces can be described in terms of these four forces.

## Section 6-8 Gravitational Versus Inertial Mass; the Principle of Equivalence

Einstein's principle of equivalence states that it is experimentally impossible to detect whether you are in an inertial reference frame in a uniform gravitational field or in an accelerating reference frame with no gravitational field.

## Section 6-9 Gravitation as Curvature of Space; Black Holes

When the principle of equivalence is applied to light, it implies that light should follow curved trajectories in space.  This is contrary to what we expect light to do in empty space; follow the shortest distance between two points, which is a straight line.  An alternative method for dealing with gravitational fields is to give up the notion that space is described by Euclidean geometry and assume that space can be curved by the presence of mass.  In curved space–time, the shortest distance between two points can be a curve.  Einstein's general theory of relativity describes the behavior of curved space–time.

If space has a strong enough curvature, the path of light may not allow it to escape from a region of space.  This occurs around very massive bodies called **black holes**.

## Practice Quiz

1.    Which gravitational pull is greater, the gravitational force of the Sun on the Earth or the gravitational force of the Earth on the Sun?

    a)    The force of the Sun on the Earth is greater.
    b)    The force of the Earth on the sun is greater.
    c)    The force of the Sun on the Earth is equal in magnitude to the force of the Earth on the Sun.
    d)    Can't tell from the information given.

2.    What happens to the magnitude of the gravitational force between two objects when the distance between them doubles?

    a)    The magnitude of the force doubles.
    b)    The magnitude of the force quadruples.
    c)    The magnitude of the force decreases by a factor of two.
    d)    The magnitude of the force decreases by a factor of four.

3.    A planet has the same mass as the earth, but its surface gravitational acceleration is $g/2$.  What is the radius of the planet?

    a)    $2 \times R_{Earth}$
    b)    $\sqrt{2} \times R_{Earth}$
    c)    $4 \times R_{Earth}$
    d)    $R_{Earth}/2$

4.    At which point is the gravitational force zero between two masses separated by a distance $d$ when one mass has a mass $m$ and the other has a mass $4m$?

    a)    $1/3$ $d$ away from the mass $m$
    b)    $1/3$ $d$ away from the mass $4m$
    c)    $1/4$ $d$ away from the mass $m$
    d)    $1/4$ $d$ away from the mass $4m$

5.    A mass $m$ is near a mass $3m$ as shown in the diagram.  At which point in the diagram is the gravitational field approximately zero?

    a)    A
    b)    B
    c)    C
    d)    D

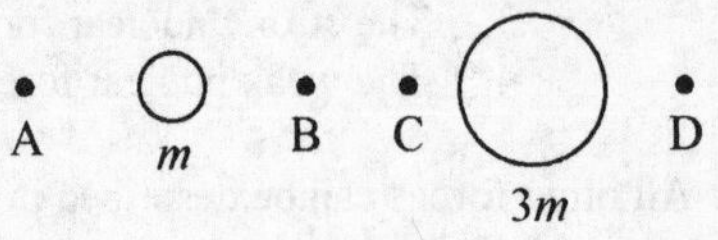

6.    For a given radius of circular orbit about a planet of mass $m$, the period is $T$.  What is the period of an orbit with the same radius about a planet of mass $2m$?

    a)    $2T$
    b)    $T/2$
    c)    $T/\sqrt{2}$
    d)    $T/4$

7.    Consider this statement: The gravitational field on the surface of a spherical planet with uniform density is constant.  What is wrong with that statement?

    a)    Nothing is wrong with that statement.
    b)    The magnitude of the gravitational field on the surface of the planet depends on the latitude.
    c)    The magnitude of the gravitational field on the surface of the planet depends on the longitude.
    d)    The magnitude of the gravitational field on the surface of the planet is constant, but the direction is not.

8.    Why do we get correct results using that the force of gravity is constant near the surface of the earth, but Newton's law of universal gravitation implies that the force of gravity should change with height?

a)    Newton's law of universal gravitation does not apply near the surface of the earth.
b)    The earth is not a point mass.
c)    The earth is flat.
d)    The change in height is small compared to the radius of the earth when we use that gravitational force is constant.

9.    Kepler's third law implies that the kinetic energy of an object in a circular orbit decreases as the radius of the circular orbit increases, but positive work must be done on the object to increase the radius of the orbit. How can this be true?

a)    The increase in gravitational potential energy is greater than the decrease in kinetic energy.
b)    The statement is not true, only the gravitational potential energy changes as the orbital radius increases.
c)    The statement is not true, negative work must be done on the object to increase the orbital radius.
d)    The statement is not true, the kinetic energy increases as the orbital radius increases.

10.    A spherical planet is made of uniform density material. How does the surface gravitational acceleration depend on the radius of the planet?

a)    $g \propto r$
b)    $g \propto r^2$
c)    $g \propto r^3$
d)    $g$ is constant

11.    Determine the period of the orbit for a satellite in a circular orbit that is just above the surface of the Sun. The mass of the Sun is $1.99 \times 10^{30}$ kg and the radius of the Sun is $6.96 \times 10^5$ km.

12.    Two planets are separated by a distance $r$. One planet has a mass that is three times the mass of the other planet. At what distance from the more massive planet is the sum of the gravitational forces of the two planets equal to zero along the line segment joining the two planets?

13.    Tidal forces can tear a satellite apart if those forces are large enough. For a simple model to understand this, consider a satellite that consists of two spheres of mass $m$ stacked on each other held together by only their gravitational attraction for each other. The orientation of the stacking is shown in the diagram.   Since the centers of the two spheres are different distances from the planet, $r$ and $r+d$, they have different gravitational forces acting on them. Because of the difference in gravitational forces from the planet, the spheres would separate if not held together by their gravitational attraction for each other. This difference can become great enough that the gravitational force of attraction between the two spheres is not enough to hold the two spheres together. Calculate the force necessary to hold the two spheres together in orbit around a planet of mass $M$. Set this equal to gravitational force of attraction between the two spheres and solve for the distance r at which the planets would just barely be held together. At a distance to the planet smaller than this, the spheres would separate from each other. Assume $M \gg m$ and $r \gg d$.

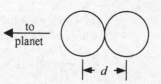

14.    Determine the radius of a satellite orbit around the earth that has a period of two days.

15.    Estimate the gravitational force between a 600 kg automobile and a 1200 kg truck parked 2 m apart.

## Problem Solutions

Solutions to problems with a number equal to $4n - 1$ where $n$ is an integer:

3.    The acceleration due to gravity on the surface of a planet is
        $g = F/M = GM/r^2$.
    If we form the ratio of the two accelerations, we have
        $g_{planet}/g_{Earth} = (M_{planet}/M_{Earth})/(r_{planet}/r_{Earth})^2$,   or
        $g_{planet} = g_{Earth}(M_{planet}/M_{Earth})/(r_{planet}/r_{Earth})^2 = (9.80 \text{ m/s}^2)(1)/(2.5)^2 = 1.6 \text{ m/s}^2$.

7. We choose the coordinate system shown in the figure and
find the force on the mass in the lower left corner.
Because the masses are equal, for the magnitudes of the
forces from the other corners we have

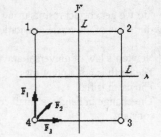

$$F_1 = F_3 = Gmm/r_1^2$$
$$= (6.67 \times 10^{-11}\ \text{N} \cdot \text{m}^2/\text{kg}^2)(8.5\ \text{kg})(8.5\ \text{kg})/(0.70\ \text{m})^2$$
$$= 9.83 \times 10^{-9}\ \text{N};$$
$$F_2 = Gmm/r_2^2$$
$$= (6.67 \times 10^{-11}\ \text{N} \cdot \text{m}^2/\text{kg}^2)(8.5\ \text{kg})(8.5\ \text{kg})/[(0.70\ \text{m})/\cos 45°]^2$$
$$= 4.92 \times 10^{-9}\ \text{N}.$$

From the symmetry of the forces we see that the resultant will be
along the diagonal. The resultant force is

$$F = 2F_1 \cos 45° + F_2$$
$$= 2(9.83 \times 10^{-9}\ \text{N}) \cos 45° + 4.92 \times 10^{-9}\ \text{N} = 1.9 \times 10^{-8}\ \text{N toward center of the square.}$$

11. The weight of objects is determined by $g$, which depends on the mass and radius of the Earth:
$$g = Gm_E/r_E^2.$$
If the density remains constant, the mass will be proportional to the radius cubed:
$$m_E'/m_E = (r_E'/r_E)^3.$$
If we form the ratio of $g$'s, we have
$$g'/g = (m_E'/m_E)/(r_E'/r_E)^2 = (m_E'/m_E)/(m_E'/m_E)^{2/3} = (m_E'/m_E)^{1/3} = 2^{1/3} = 1.26.$$

15. (a)  The acceleration due to gravity at a distance $r = r_E + \Delta r$ from the center is
$$g' = Gm/r^2 = Gm/(r_E + \Delta r)^2.$$
If we use the binomial expansion and keep only the first two terms, we get
$$g' = (Gm/r_E^2)[1 + (\Delta r/r_E)]^{-2} = g[1 - 2(\Delta r/r_E) + 3(\Delta r/r_E)^2 + \ldots] \sim g - 2g(\Delta r/r_E).$$
Thus we have
$$g' - g = \Delta g \sim -2g(\Delta r/r_E).$$
Note that this could also be obtained by treating the changes as differentials:
$$dg = -2(Gm/r^3)\ dr = -2g\ dr/r.$$
(b)  The negative sign means that $g$ decreases with an increase in height.
(c)  At a height of 100 km we get
$$\Delta g = -2g(\Delta r/r_E) = 2(9.80\ \text{m/s}^2)(100\ \text{km})/(6.38 \times 10^3\ \text{km}) = -0.307\ \text{m/s}^2.$$
Thus we have
$$g' = g + \Delta g = 9.800\ \text{m/s}^2 - 0.307\ \text{m/s}^2 = 9.493\ \text{m/s}^2.$$
If we use Eq. 6–1, we get
$$g' = Gm_E/(r_E + \Delta r)^2$$
$$= (6.67 \times 10^{-11}\ \text{N} \cdot \text{m}^2/\text{kg}^2)(5.98 \times 10^{24}\ \text{kg})/(6.38 \times 10^3\ \text{km} + 100\ \text{km})^2 = 9.499\ \text{m/s}^2.$$

19. The required centripetal acceleration of the circular orbit is provided by the gravitational attraction:
$$Gm_E M/r^2 = Mv^2/r,\ \text{so we have}$$
$$v^2 = Gm_E/(r_E + h)$$
$$= (6.67 \times 10^{-11}\ \text{N} \cdot \text{m}^2/\text{kg}^2)(5.98 \times 10^{24}\ \text{kg})/(6.38 \times 10^6\ \text{m} + 0.60 \times 10^6\ \text{m}),$$
which gives $v = 7.56 \times 10^3\ \text{m/s}.$

23. We take the positive direction upward. The spring scale reads the normal force expressed as an effective mass:
$F_N/g$.
We write $\Delta \mathbf{F} = m\mathbf{a}$ from the force diagram:
$$F_N - mg = ma,\ \text{or}\ m_{\text{effective}} = F_N/g = m(1 + a/g).$$
(a)  For a constant speed, there is no acceleration, so we have
$$m_{\text{effective}} = m(1 + a/g) = m = 56\ \text{kg}.$$
(b)  For a constant speed, there is no acceleration, so we have
$$m_{\text{effective}} = m(1 + a/g) = m = 56\ \text{kg}.$$
(c)  For the upward (positive) acceleration, we have
$$m_{\text{effective}} = m(1 + a/g) = m(1 + 0.33g/g) = 1.33(56\ \text{kg}) = 75\ \text{kg}.$$
(d)  For the downward (negative) acceleration, we have
$$m_{\text{effective}} = m(1 + a/g) = m(1 - 0.33g/g) = 0.67(56\ \text{kg}) = 38\ \text{kg}.$$
(e)  In free fall the acceleration is $-g$, so we have
$$m_{\text{effective}} = m(1 + a/g) = m(1 - g/g) = 0.$$

27. (a)  The attractive gravitational force between the stars is providing the required centripetal acceleration for the circular motion.

   (b)  We relate the orbital speed to the period of revolution from
   $v = 2\pi r/T$, where $r$ is the distance to the midpoint.
   The gravitational attraction provides the centripetal acceleration:
   $Gmm/(2r)^2 = mv^2/r = m(2\pi r/T)^2/r = m4\pi^2 r/T^2$, so we have
   $m = 16\pi^2 r^3/GT^2$
   $= 16\pi^2 (4.0 \times 10^{10}\ \text{m})^3/(6.67 \times 10^{-11}\ \text{N} \cdot \text{m}^2/\text{kg}^2)[(12.6\ \text{yr})(3.16 \times 10^7\ \text{s/yr})]^2 = 9.6 \times 10^{26}\ \text{kg}.$

31.  Each body will attract one of the other bodies with a force given by
   $F = Gmm/L^2$.
   From the symmetry we see that the net force on a body will be toward the center of the triangle and will provide the centripetal acceleration required for the circular orbit:
   $F_{\text{net}} = 2(Gmm/L^2) \cos 30° = mv^2/r = mv^2/(L/(2 \cos 30°))$, which gives
   $v = (Gm/L)^{1/2}$.

35.  From Kepler's third law, $T^2 = 4\pi^2 r^3/Gm_S$, we can relate the periods of the Earth and Neptune:
   $(T_{\text{Neptune}}/T_{\text{Earth}})^2 = (r_{\text{Neptune}}/r_{\text{Earth}})^3$;
   $(T_{\text{Neptune}}/1\ \text{yr})^2 = [(4.5 \times 10^{12}\ \text{m})/(1.50 \times 10^{11}\ \text{m})]^3$, which gives $T_{\text{Neptune}} = 1.6 \times 10^2\ \text{yr}.$

39.  From Kepler's third law, $T^2 = 4\pi^2 r^3/Gm_{\text{Jupiter}}$, we can relate the distances of the moons:
   $(r/r_{\text{Io}})^3 = (T/T_{\text{Io}})^2$.
   Thus we have
   $(r_{\text{Europa}}/422 \times 10^3\ \text{km})^3 = (3.55\ \text{d}/1.77\ \text{d})^2$, which gives $r_{\text{Europa}} = 6.71 \times 10^5\ \text{km}.$
   $(r_{\text{Ganymede}}/422 \times 10^3\ \text{km})^3 = (7.16\ \text{d}/1.77\ \text{d})^2$, which gives $r_{\text{Ganymede}} = 1.07 \times 10^6\ \text{km}.$
   $(r_{\text{Callisto}}/422 \times 10^3\ \text{km})^3 = (16.7\ \text{d}/1.77\ \text{d})^2$, which gives $r_{\text{Callisto}} = 1.88 \times 10^6\ \text{km}.$
   All values agree with the table.

43.  (a)  From Kepler's third law, $T^2 = 4\pi^2 r^3/Gm_S$, we can relate the periods of Hale–Bopp and the Earth:
   $(T_{\text{HB}}/T_E)^2 = (r_{\text{HB}}/r_E)^3$;
   $[(3000\ \text{y})/(1\ \text{yr})]^2 = (r_{\text{HB}}/(1\ \text{A.U.})]^3$, which gives $r_{\text{HB}} = 2.1 \times 10^2\ \text{A.U.}\ (3.1 \times 10^{13}\ \text{m}).$

   (b)  We find the farthest distance from
   $r_{\text{HB}} = \tfrac{1}{2}(d_N + d_F)$;
   $2.1 \times 10^2\ \text{A.U.} = \tfrac{1}{2}(1.0\ \text{A.U.} + d_F)$, which gives $d_F = 4.2 \times 10^2\ \text{A.U.}$

   (c)  From Problem 42 we have
   $v_N/v_F = d_F/d_N = (4.2 \times 10^2\ \text{A.U.})/(1.0\ \text{A.U.}) = 4.2 \times 10^2.$

47.  The acceleration due to gravity at a distance $r$ from the center of the Earth is
   $g = F/M = Gm_{\text{Earth}}/r^2$.
   If we form the ratio of the two accelerations for the different distances, we have
   $g/g_{\text{surface}} = [r_{\text{Earth}}/(r_{\text{Earth}} + h)]^2$;
   $1/2 = [(6400\ \text{km})/(6400\ \text{km} + h)]^2$, which gives $h = 2.7 \times 10^3\ \text{km}.$

51.  The acceleration due to gravity on the surface of the white dwarf star is
   $g = F/M = Gm/r^2 = (6.67 \times 10^{-11}\ \text{N} \cdot \text{m}^2/\text{kg}^2)(2.0 \times 10^{30}\ \text{kg})/(1.74 \times 10^6\ \text{m})^2 = 4.4 \times 10^7\ \text{m/s}^2.$

55.  We relate the orbital speed to the period of revolution from
   $v = 2\pi R/T$.
   The required centripetal acceleration is provided by the gravitational attraction:
   $Gm_S m/R^2 = mv^2/R = m(2\pi R/T)^2/R = m4\pi^2 R/T^2$, so we have
   $Gm_S = 4\pi^2 R^3/T^2$.
   For the two extreme orbits we have
   $(6.67 \times 10^{-11}\ \text{N} \cdot \text{m}^2/\text{kg}^2)(5.69 \times 10^{26}\ \text{kg}) = 4\pi^2 (7.3 \times 10^7\ \text{m})^3/T_{\text{inner}}^2$,
   which gives $T_{\text{inner}} = 2.0 \times 10^4\ \text{s} = 5\ \text{h}\ 35\ \text{min}$;
   $(6.67 \times 10^{-11}\ \text{N} \cdot \text{m}^2/\text{kg}^2)(5.69 \times 10^{26}\ \text{kg}) = 4\pi^2 (17 \times 10^7\ \text{m})^3/T_{\text{outer}}^2$,
   which gives $T_{\text{outer}} = 7.1 \times 10^4\ \text{s} = 19\ \text{h}\ 52\ \text{min}.$
   Because the mean rotation period of Saturn is between the two results, with respect to a point on the surface of Saturn, the edges of the rings are moving in opposite directions.

59. From Kepler's third law, $T^2 = 4\pi^2 R^3 / Gm_S$, we can relate the periods of Halley's comet and the Earth to find the mean distance of the comet from the Sun:

$(T_{Halley}/T_{Earth})^2 = (R_{Halley}/R_{Earth})^3$;

$(76 \text{ yr}/1 \text{ yr})^2 = [R_{Halley}/(1.50 \times 10^{11} \text{ m})]^3$, which gives $R_{Halley} = 2.68 \times 10^{12}$ m.

This mean distance is half the sum of the nearest and farthest distances. If we take the nearest distance to the Sun as zero, the farthest distance is

$d = 2R_{Halley} = 2(2.68 \times 10^{12} \text{ m}) = 5.4 \times 10^{12}$ m.

It is still orbiting the Sun and thus is in the Solar System. The planet nearest it is Pluto.

63. The acceleration due to gravity on the surface of a planet is

$g_P = F/M = Gm/r^2$, so we have

$m_P = g_P r^2 / G$.

67. The particle falls along a radial line. We take the positive direction upward. During the fall the gravitational attraction provides the acceleration:

$-Gm_E m/r^2 = ma = m \, dv/dt = m \, (dv/dr)(dr/dt) = mv \, (dv/dr)$, which we can write

$-(Gm_E/r^2) \, dr = v \, dv$.

We find the velocity by integration:

$$\int_{2r_E}^{r_E} -\left(\frac{Gm_E}{r^2}\right) dr = \int_0^{v_f} v \, dv \Rightarrow -\left(\frac{Gm_E}{r}\right)\Bigg|_{2r_E}^{r_E} = \tfrac{1}{2}v_f^2$$

$v_f^2 = 2Gm_E[(1/r_E) - (1/2r_E)] = Gm_E/r_E$

$= (6.67 \times 10^{-11} \text{ N} \cdot \text{m}^2/\text{kg}^2)(5.98 \times 10^{24} \text{ kg})/(6.38 \times 10^6 \text{ m})$,

which gives $v_f = 7.9 \times 10^3$ m/s.

# Chapter 7: Work and Energy

## Chapter Overview and Objectives

This chapter introduces the concepts of work and kinetic energy. It shows how these concepts are related to Newton's second law to derive the work-energy theorem. The expression for kinetic energy at relativistic speeds is introduced.

After completing study of this chapter, you should:

- Know the definitions of work done by both a constant force and a varying force.
- Know the definition of the scalar product of two vectors and the properties of the scalar product.
- Be able to calculate scalar products from knowledge of magnitudes and relative directions of the vectors.
- Be able to calculate scalar products from vectors expressed in Cartesian unit vectors.
- Know the dimensions and SI units of work and energy.
- Know the work-energy theorem and how to apply the work-energy theorem to problems.
- Know that the relativistic expression for kinetic energy differs from the classical expression.

## Summary of Equations

Defintion of work done by a constant force:
$$W = \mathbf{F} \cdot \mathbf{d} = F_\parallel d = Fd\cos\theta$$

Work done by varying force:
$$W = \int \mathbf{F} \cdot d\mathbf{l}$$

Definition of scalar product of two vectors:
$$\mathbf{A} \cdot \mathbf{B} = AB\cos\theta$$

Commutative property of scalar product:
$$\mathbf{A} \cdot \mathbf{B} = \mathbf{B} \cdot \mathbf{A}$$

Associative property of scalar product:
$$\mathbf{A} + (\mathbf{B} + \mathbf{C}) = (\mathbf{A} + \mathbf{B}) + \mathbf{C}$$

Distribution of scalar multiplication over vector addition:
$$\mathbf{A} \cdot (\mathbf{B} + \mathbf{C}) = \mathbf{A} \cdot \mathbf{B} + \mathbf{A} \cdot \mathbf{C}$$

Scalar product in terms of Cartesian components:
$$\mathbf{A} \cdot \mathbf{B} = A_x B_x + A_y B_y + A_z B_z$$

Definition of kinetic energy:
$$K = \tfrac{1}{2}mv^2$$

Work-energy theorem:
$$W = \Delta K = \tfrac{1}{2}mv'^2 - \tfrac{1}{2}mv^2$$

Relativistic kinetic energy:
$$K = mc^2 \left( \frac{1}{\sqrt{1 - \dfrac{v^2}{c^2}}} - 1 \right)$$

## Chapter Summary

### Section 7-1 Work Done by a Constant Force

The **work** done on an object by a force is defined to be the force on the object multiplied by the displacement of the object in the direction of the force

$$W = F_\parallel d = Fd\cos\theta$$

where $W$ is the work done on the object, $F_\parallel$ is the force component parallel to the displacement of the object, $d$ is the magnitude of the displacement of the object, $F$ is the magnitude of the force, and $\theta$ is the angle between the direction of the force and the direction of the displacement.

The dimensions of work are $[M][L]^2/[T]^2$ and the SI units of work are Newton·meters. A Newton·meter is defined to be a **joule** of work, $1\ J = 1\ N \cdot m$. The British unit system unit of work is the foot-pound. One foot-pound is equal to 1.36 J.

### Example 7-1-A

Your car of mass 652 kg runs out of gas. You need to push it up a hill to get to the gas station. The hill rises 2.00 m for every 100 m horizontally you travel. You push with a constant horizontal force on the car of 538 N on the car as you move the car 623 m to the gas station. How much work is done by your force on the car, the force of gravity acting on the car, and the normal force of the road acting on the car?

### Solution:

For each force we need to detemine the angle between the direction of the force and the direction of the displacement. To do this it is convenient to calculate the angle of the hill above horizontal. We know the hill rises 2 m for each 100 m horizontal run. Solving for angle $\theta$ using the tangent function:

$$\theta = \arctan \frac{2.00\ m}{100\ m} = 1.146°$$

The pushing force on the car is horizontal toward the uphill direction and the displacement is at an angle $1.146°$ above horizontal in the uphill direction. The work done by the pushing force on the car is

$$W = Fd\cos\theta = (538\ N)(623\ m)\cos(1.146°) = 3.35 \times 10^5\ J$$

The force of gravity points vertically downward. It should be easy to see that the angle between the direction of the force of gravity and the direction of the dispalcement is $1.146° + 90° = 91.146°$. The work done by the force of gravity on the car is

$$W = Fd\cos\theta = (538\ N)(623\ m)\cos(91.146°) = -6.70 \times 10^3\ J$$

The work done by the normal force is zero because the normal force points perpendicular to the surface of the hill by definition. The displacement is parallel to the surface of the hill. The displacement and the normal force are perpendicular to each other. The normal force does no work on the car.

### Section 7-2 Scalar Product of Two Vectors

To this point, the only multiplication operation involving vectors that has been introduced is multiplication of a vector by a scalar. It is useful to define two other vector multiplication operations involving vectors. The **scalar product** (also called **dot product**) of two vectors results in a scalar. The **vector product** of two vectors results in an axial vector. In this chapter, the scalar product is introduced.

The scalar product of vector **A** and vector **B** is denoted as **A·B** and is defined to be

$$\mathbf{A} \cdot \mathbf{B} = AB\cos\theta$$

where A and B are the magnitudes of vectors **A** and **B** respectively and $\theta$ is the angle between the direction of vector **A** and the direction of vector **B**.

The work, $W$, done on object acted on by a constant force **F** while making a displacement **d** can be written as

$$W = \mathbf{F} \cdot \mathbf{d} = Fd\cos\theta$$

There are several properties of scalar multiplication that may be useful to know. The scalar product is commutative, the order of the multiplication does not matter:

$$\mathbf{A} \cdot \mathbf{B} = \mathbf{B} \cdot \mathbf{A}$$

Scalar multiplication distributes over vector addition:

$$\mathbf{A} \cdot (\mathbf{B} + \mathbf{C}) = \mathbf{A} \cdot \mathbf{B} + \mathbf{A} \cdot \mathbf{C}$$

If vector $\mathbf{A}$ is in terms of the Cartesian unit vectors as $\mathbf{A} = A_x\mathbf{i} + A_y\mathbf{j} + A_z\mathbf{k}$ and vector $\mathbf{B}$ is expressed as $\mathbf{B} = B_x\mathbf{i} + B_y\mathbf{j} + B_z\mathbf{k}$, then the scalar product of $\mathbf{A}$ and $\mathbf{B}$ is given by

$$\mathbf{A} \cdot \mathbf{B} = A_xB_x + A_yB_y + A_zB_z.$$

It is useful to remember the scalar products of the Cartesian unit vectors

$$\mathbf{i} \cdot \mathbf{i} = \mathbf{j} \cdot \mathbf{j} = \mathbf{k} \cdot \mathbf{k} = 1$$
$$\mathbf{i} \cdot \mathbf{j} = \mathbf{j} \cdot \mathbf{k} = \mathbf{i} \cdot \mathbf{k} = 0$$

### Example 7-2-A

A vector $\mathbf{A}$ with magnitude 5.34 lies in the $xy$ plane. It makes an $43°$ angle from the $+x$-axis toward the $+y$-axis. A second vector $\mathbf{B}$ with magnitude 3.98 lies in the $yz$ plane and makes an angle $57°$ from the $+y$ axis towrd the $+z$ axis. Determine the scalar product of vectors $\mathbf{A}$ and $\mathbf{B}$. Determine the angle between the directions of vectors $\mathbf{A}$ and $\mathbf{B}$.

### Solution:

First, express each vector in terms of the cartesian unit vectors:

$$\mathbf{A} = A\cos 43°\,\mathbf{i} + A\sin 43°\,\mathbf{j}$$
$$= 5.34\cos 43°\,\mathbf{i} + 5.34\sin 43°\,\mathbf{j}$$
$$= 3.90\,\mathbf{i} + 3.64\,\mathbf{j}$$

$$\mathbf{B} = B\cos 57°\,\mathbf{j} + B\sin 57°\,\mathbf{k}$$
$$= 3.98\cos 57°\,\mathbf{j} + 3.98\sin 57°\,\mathbf{k}$$
$$= 2.17\,\mathbf{j} + 3.34\,\mathbf{k}$$

We can directly calculate the scalar product in terms of the cartesian components:

$$\mathbf{A} \cdot \mathbf{B} = A_xB_x + A_yB_y + A_zB_z = (3.90)(0) + (3.64)(2.17) + (0)(3.34)$$
$$= 7.90$$

To determine the angle between vecotrs A and B, we use the fact that the scalar product expressed in terms of the cosine of the angle between the two vectors:

$$\mathbf{A} \cdot \mathbf{B} = AB\cos\theta$$

or solving for $\theta$,

$$\theta = \arccos\left(\frac{A \cdot B}{AB}\right) = \arccos\left(\frac{7.90}{(5.34)(3.98)}\right) = 68°$$

### Section 7-3  Work Done by a Varying Force

When the force on an object is varying with position or the path the object follows is curved, calculus must be used to determine the work done one the object by the force:

$$W = \int \mathbf{F} \cdot d\mathbf{l}$$

where $d\mathbf{l}$ is an infinitesimal length vector pointing along the direction of travel parallel to the tangent to the path followed by the object. The integral is over the path from the initial position to the final position.

### Example 7-3-A

The electrostatic force, $\mathbf{F}$, between two charged particles has the magnitude $F = C/r^2$, where $C$ is a constant and $r$ is the distance between the charged particles. The force points along the line between the two particles and is repulsive. Determine the work done by this force on one particle as it directly approaches the other particle moving from an initial distance $r_1$ to a final distance $r_2$.

### Solution:

The force points along a line between the two particles. If the particle is approaching the other particle and the force is pointing away from the other particle, the force and displacement are in opposite directions. The work done is

$$W = \int_{r=r_1}^{r_2} \mathbf{F} \cdot d\mathbf{l} = \int_{r=r_1}^{r_2} F\, dr \cos 180^o = \int_{r=r_1}^{r_2} -\frac{C\, dr}{r^2} = \left.\frac{C}{r}\right|_{r=r_1}^{r_2} = C\left(\frac{1}{r_2} - \frac{1}{r_1}\right)$$

### Example 7-3-B

A spring with spring constant $k$ is attached to a mass that is constrained to slide along a straight rod as shown in the diagram. The natural length of the spring is the perpendicular distance from the rod to the fixed end of the spring, i.e., the force is zero when the spring is perpendicular to the rod. How much work is done by the spring on the mass if the mass slides from position $a$ to position $b$?

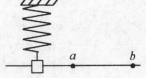

### Solution:

The magnitude of the force of the spring depends on the difference from its natural length. Using a coordinate system, $y$, along the rod for the position of the mass with the origin at the position for which the force is zero, the stretch of the spring, $x$, is given by

$$x = \sqrt{L^2 + y^2} - L$$

The magnitude of the force is then $F = kx$ and is directed along the spring away from the rod. We can write the force in vector form as

$$F = kx[(-\cos\theta)\mathbf{i} + (\sin\theta)\mathbf{j}]$$

with the angle $\theta$ as shown in the diagram. Moving from position $a$ to position $b$ along the rod, the displacement is always in the $\mathbf{j}$ direction. We can write $d\mathbf{l} = \mathbf{j}\, dy$. We can also write the angle as in terms of the position $y$ as

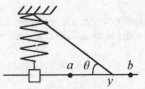

$$\sin\theta = \frac{y}{\sqrt{L^2 + y^2}} \qquad or \qquad \theta = \sin^{-1}\frac{y}{\sqrt{L^2 + y^2}}$$

Writing the expression for the work

$$W = \int \mathbf{F} \cdot d\mathbf{l} = \int_{y=a}^{b} k\left(\sqrt{L^2 + y^2} - L\right)\left[(-\cos\theta)\mathbf{i} + (\sin\theta)\mathbf{j}\right] \cdot \mathbf{j}\, dy$$

$$= \int_{y=a}^{b} k\left(\sqrt{L^2 + y^2} - L\right)\sin\theta\, dy = \int_{y=a}^{b} k\left(\sqrt{L^2 + y^2} - L\right)\frac{y}{\sqrt{L^2 + y^2}}\, dy$$

$$= \int_{y=a}^{b} ky\, dy - \int_{y=a}^{b} kL\frac{y}{\sqrt{L^2 + y^2}}\, dy = k\left[\tfrac{1}{2}y^2 - L\sqrt{L^2 + y^2}\right]\Big|_{y=a}^{b}$$

$$= \tfrac{1}{2}k\left(b^2 - a^2\right) - kL\left(\sqrt{L^2 + b^2} - \sqrt{L^2 + a^2}\right)$$

## Section 7-4  Kinetic Energy and the Work-Energy Principle

We define the translational kinetic energy, $K$, of an object to be

$$K = \tfrac{1}{2}mv^2$$

where $m$ is the mass of the object and $v$ is the speed of the object.  Newton's second law can be used to relate the work done one an object by the net force acting on the object and the change in kinetic energy of the object while that work is being done.  The resulting relationship is called the **work-energy theorem**:

$$W = \Delta K = \tfrac{1}{2}mv'^2 - \tfrac{1}{2}mv^2$$

where $m$ is the mass of the object, $v$ is the speed of the object before the work is done, and $v'$ is the speed of the object after the work has been done.

## Example 7-4-A

A 5.76 kg box is initially traveling with a speed of 3.43 m/s as it starts up a ramp on which the kinetic friction force has a magnitude of 37.6 N.  The ramp surface is tilted 30° to horizontal as shown.  What is the speed of the box after it has slid a distance 1.34 m along the ramp?

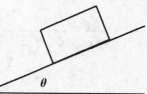

## Solution:

First, we draw a free-body diagram for the box:
The change in kinetic energy is equal to the net work done on the object by each of the forces acting on the object.  Clearly the normal force, $F_N$, does no work on the box as it is perpendicular to the surface.  Also, it is easy to see that the work done by the friction force on the box is

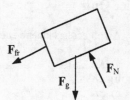

$$W_{fr} = \mathbf{F}_{fr} \cdot \mathbf{d} = F_{fr}d\cos\theta = F_{fr}d\cos\left(180°\right)$$
$$= -(37.6\,\text{N})(1.34\,\text{m}) = -50.4\,\text{J}$$

The work done by the force of gravity is equal to

$$W_g = F_g \cdot d = F_g d\cos\theta = mgd\cos\left(210°\right)$$
$$= (5.76\,\text{kg})(9.80\,\text{m/s}^2)(1.34\,\text{m})\cos\left(210°\right) = -65.5\,\text{J}$$

We can now use the work-energy theorem to solve for the final speed of the box:

$$W = \Delta K = \tfrac{1}{2}mv'^2 - \tfrac{1}{2}mv^2 \quad \Rightarrow$$

$$v' = \sqrt{\frac{2W}{m} + v^2} = \sqrt{\frac{2(0 + -50.4\,\text{J} - 65.5\,\text{J})}{5.76\,\text{kg}} + (8.43\,\text{m/s})^2} = 5.55\,\text{m/s}$$

### Section 7-5  Kinetic Energy at Very High Speed

The mechanics we have developed in this textbook has been consistent with Galilean invariance. Einstein's theory of special relativity is able to explain some phenomena better than classical physics is able to do. However, Einstein's special theory of relativity requires that measurements in different inertial reference be related by Lorentz transformations rather than by Galilean transformations. The quantity that behaves like kinetic energy and transforms correctly under Lorentz transformations is the relativistic kinetic energy

$$K = mc^2 \left( \frac{1}{\sqrt{1 - \frac{v^2}{c^2}}} - 1 \right)$$

where $m$ is the mass of the object, v is the speed of the object, and $c$ is the speed of light, $3.00 \times 10^8$ m/s. This expression is approximately equal to the classical kinetic energy expression, $\frac{1}{2}mv^2$, when the speed, $v$, is much less than the speed of light, $c$.

### Example 7-5-A

Determine the percentage difference between the classical kinetic energy and the relativistic kinetic energy when the speed of the object is one half the speed of light.

**Solution:**

The ratio of the classical kinetic energy to the relativistic kinetic energy is

$$\frac{K_{classical}}{K_{relativistic}} = \frac{\frac{1}{2}mv^2}{mc^2\left(\sqrt{\frac{1}{1-v^2/c^2}} - 1\right)} = \frac{\frac{1}{2}\left(\frac{1}{2}c\right)^2}{c^2\left(\sqrt{\frac{1}{1-\left(\frac{1}{2}c\right)^2/c^2}} - 1\right)} = \frac{\frac{1}{8}}{\left(\sqrt{\frac{1}{1-\frac{1}{4}}} - 1\right)} = 0.81$$

The classical kinetic energy is 81% of the relativistic kinetic energy or 19% lower than the relativistic energy.

## Practice Quiz

*1.*    An object has kinetic energy $K$. How much work do you have to do to double the speed of the object?

   a) $K$
   b) $2K$
   c) $3K$
   d) $4K$

2.    You're working on a construction job and the boss asks you to hold a piece of plywood up at a height of 1.27 m above the floor. The weight of the plywood is 34.1 N. You hold the piece of plywood up for a time of 18.2 minutes. How much work have you done on the sheet of plywood while holding it at this height?

   a) 43.3 J
   b) −43.3 J
   c) $6.00 \times 10^5$ J
   d) 0 J

3.  A force acting on an object accelerates it up to a speed *v* after starting from rest. Does accelerating to that speed in a shorter time increase or decrease the amount of work done by the force?

    a)  Shorter time means a bigger force so more work is done.
    b)  Shorter time means the car travels a shorter distance so less work is done.
    c)  Same change in kinetic energy regardless of time, so same amount of work is done.
    d)  More information is needed to answer the question.

4.  Why can't the work on an object due to friction be calculated using the constant force expressions for work in Section 7-1 if the object does not follow a straight line path?

    a)  The friction force would not be constant in magnitude.
    b)  The friction force would not be constant in direction.
    c)  The expressions in Section 7-1only apply to straight line paths.
    d)  Work is never done by friction.

5.  What is the direction of the scalar product of two vectors?

    a)  Half way between the direction of each of the vectors
    b)  Arcos[(A·B)/AB]
    c)  Arcos[AB/(A·B)]
    d)  The scalar product is a scalar; scalars don't have a direction.

6.  Work is done on an object to triple its kinetic energy. By what factor does the object's speed change?

    a)  9
    b)  1/9
    c)  $\sqrt{3}$
    d)  $1/\sqrt{3}$

7.  Determine which of the following is equivalent to $\mathbf{A \cdot B + C \cdot B - B \cdot B + (C + B) \cdot B}$.

    a)  $\mathbf{A \cdot B}$
    b)  $\mathbf{C \cdot B}$
    c)  $\mathbf{A \cdot B + 2C \cdot B}$
    d)  $\mathbf{A \cdot B + 2C \cdot B + 2B \cdot B}$

8.  An object moves around a closed path so its displacement for the entire motion is zero. Can you conclude that the work done is zero?

    a)  Yes, no displacement implies no work.
    b)  No, the work done in going around a closed path is always negative.
    c)  No, the work done in going around a closed path is always positive.
    d)  Not enough information is given to determine whether the work was zero, positive, or negative.

9.  Can an object have net negative work done on it of magnitude greater than the amount of kinetic energy it has?

    a)  Yes, there is no limit to the amount of work done on the object.
    b)  No, once the kinetic energy is brought to zero, no more negative work can be done on the object.
    c)  Yes, if the object has relativistic kinetic energy.
    d)  Only by frictional forces.

10. Going from a speed *v* to a speed 2*v*, the classical kinetic energy quadruples. How does the relativistic kinetic energy change in doubling the speed (assuming the doubled speed is less than the speed of light)?

    a)  Quadruples
    b)  Less than quadruples
    c)  More than quadruples
    d)  Doesn't change

11.    Calculate the work done on an object that moves from $x = 1$ m to $x = 2$ m when a force

$$F = 3.0 \, \text{N/m}^2 \, x^2$$

acts on the object.

12.    A box starts from rest and slides down a frictionless ramp that is tilted at $32°$ to horizontal. How fast is the box traveling after it slides a distance 3.39 m down the ramp?

13.    An object moves by a displacement $\mathbf{d} = 3.42$ m $\mathbf{i}$ + 2.31 m $\mathbf{j}$ + 1.78 m $\mathbf{k}$. While it is moving, it is acted upon by a constant force $\mathbf{F} = 56.3$ N $\mathbf{i}$ + 61.2 N $\mathbf{j}$ - 49.2 N $\mathbf{k}$. What is the work done by the force and what is the angle between the direction of the force and the direction of the displacement?

14.    A child with a mass 38.6 kg slides down a playground slide that is a length of 2.82 m. The slide slopes downward at an angle of $44.1°$. If the child starts from rest and reaches a speed of 4.34 m/s at the bottom of the slide, what is the frictional force acting on the child as the child goes down the slide?

15.    At what speed is the relativistic kinetic energy three times the classical kinetic energy? Express your answer as a fraction of the speed of light.

## Problem Solutions

Solutions to problems with a number equal to $4n - 1$ where $n$ is an integer:

3.    Because there is no acceleration, the contact force must have the same magnitude as the weight.
The displacement in the direction of this force is the vertical displacement. Thus,
$$W = F \, \Delta y = (mg) \, \Delta y = (65.0 \text{ kg})(9.80 \text{ m/s}^2)(20.0 \text{ m}) = 1.27 \times 10^4 \text{ J}.$$

7.    $1 \text{ J} = (1 \text{ kg} \cdot \text{m/s}^2)(1 \text{ m})(1000 \text{ g/kg})(100 \text{ cm/m})^2 = 1 \times 10^7 \text{ g} \cdot \text{m/s}^2 = 1 \times 10^7 \text{ erg}.$
$1 \text{ J} = (1 \text{ N} \cdot \text{m})(1 \text{ m})(0.225 \text{ lb/N})(3.28 \text{ ft/m})^2 = 0.738 \text{ ft} \cdot \text{lb}.$

11.    We assume that the input force is exerted perpendicular to the lever. When the lever rotates through a small angle $\theta$, the distance through which the input force acts is $l_1\theta$, and the distance the output force acts is $l_O\theta$. If the output work is equal to the input work, we have
$$W_{\text{I}} = F_{\text{I}}l_1\theta = W_{\text{O}} = F_{\text{O}}l_{\text{O}}\theta, \quad \text{or} \quad F_{\text{O}}/F_{\text{I}} = l_1/l_{\text{O}}.$$

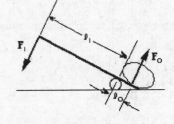

15.    Because the motion is in the $x$-direction, we see that the weight and normal forces do no work:
$$W_{\text{FN}} = W_{mg} = 0.$$
From the force diagram, we write $\Sigma\mathbf{F} = m\mathbf{a}$:
$x$-component: $F_{\text{P}} \cos \theta - F_{\text{fr}} = 0$, or $F_{\text{fr}} = F_{\text{P}} \cos \theta.$
For the work by these two forces, we have
$$W_{\text{FP}} = F_{\text{P}} \, \Delta x \, \cos \theta = (14 \text{ N})(15 \text{ m}) \cos 20° = 2.0 \times 10^2 \text{ J}.$$
$$W_{\text{fr}} = F_{\text{P}} \cos \theta \, \Delta x \, \cos 180° = (14 \text{ N}) \cos 20° (15 \text{ m})(- 1) = - 2.0 \times 10^2 \text{ J}.$$
As expected, the total work is zero: $W_{\text{FP}} = - W_{\text{fr}} = 2.0 \times 10^2 \text{ J}.$

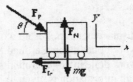

19.    $\mathbf{A} \cdot (-\mathbf{B}) = A_x(-B_x) + A_y(-B_y) + A_z(-B_z) = -(A_xB_x + A_yB_y + A_zB_z) = -\mathbf{A} \cdot \mathbf{B}.$

23.    Because $\mathbf{C}$ lies in the $xy$-plane and is perpendicular to $\mathbf{B}$, we have
$$\mathbf{B} \cdot \mathbf{C} = B_xC_x + B_yC_y = 9.6C_x + 6.7C_y = 0.$$
For the scalar product of A and C we have
$$\mathbf{A} \cdot \mathbf{C} = A_xC_x + A_yC_y = -4.8C_x + 7.8C_y = 20.0.$$
When we solve these two equations for the two unknowns, we get
$$C_x = -1.25, \quad \text{and} \quad C_y = 1.79.$$
Thus $\mathbf{C} = -1.3\mathbf{i} + 1.8\mathbf{j}.$

27. We find the angle between **A** and **B** by taking the scalar product:
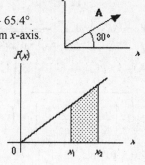

$$\mathbf{A} \cdot \mathbf{B} = A_x B_x + A_y B_y = AB \cos \theta,$$

$20.0 = (12.0)(4.0) \cos \theta$, which gives $\cos \theta = 0.417$, so $\theta = 65.4°, -65.4°$.

Thus the two angles that **B** may make with the $x$-axis are $95°$, $-35°$ from $x$-axis.

31. We obtain the forces at the beginning and end of the motion:

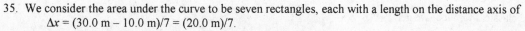

at $x_1 = 0.038$ m, $F_1 = kx_1 = (84 \text{ N/m})(0.030 \text{ m}) = 2.52$ N;

at $x_2 = 0.058$ m, $F_2 = kx_2 = (84 \text{ N/m})(0.055 \text{ m}) = 4.62$ N.

From the graph the work done in stretching the object is the area under the $F$ vs. $x$ graph:

$$\begin{aligned} W &= \tfrac{1}{2}(F_1 + F_2)(x_2 - x_1) \\ &= \tfrac{1}{2}(2.52 \text{ N} + 4.62 \text{ N})(0.055 \text{ m} - 0.030 \text{ m}) = 0.089 \text{ J}. \end{aligned}$$

35. We consider the area under the curve to be seven rectangles, each with a length on the distance axis of

$\Delta x = (30.0 \text{ m} - 10.0 \text{ m})/7 = (20.0 \text{ m})/7$.

We estimate the height of each segment from the graph to get

$$\begin{aligned} W &= \Sigma F_i\, \Delta x = (\Sigma F_i)\, \Delta x \\ &= (185 \text{ N} + 175 \text{ N} + 150 \text{ N} + 120 \text{ N} + 110 \text{ N} + 100 \text{ N} + 100 \text{ N})(20.0 \text{ m})/7 = 2.7 \times 10^3 \text{ J}. \end{aligned}$$

39. (a) If we use $r$ as the displacement, the force of gravity is negative and $\Delta r$ is negative. Thus we can plot the force as positive with a positive change in $r$, as shown. If we approximate the area as two rectangles, the average forces for the two are

at $r_E + \tfrac{1}{4}h = (6.38 \times 10^6 \text{ m}) + \tfrac{1}{4}(3.0 \times 10^6 \text{ m}) = 7.13 \times 10^6$ m,

$$\begin{aligned} F_1 &= GM_E m/r^2 \\ &= (6.67 \times 10^{-11} \text{ N} \cdot \text{m}^2/\text{kg}^2)(5.98 \times 10^{24} \text{ kg})(2500 \text{ kg})/(7.13 \times 10^6 \text{ m})^2 = 1.96 \times 10^4 \text{ N}. \end{aligned}$$

at $r_E + \tfrac{3}{4}h = 6.38 \times 10^6 \text{ m} + \tfrac{3}{4}(3.0 \times 10^6 \text{ m}) = 8.63 \times 10^6$ m,

$$\begin{aligned} F_2 &= GM_E m/r_E^2 \\ &= (6.67 \times 10^{-11} \text{ N} \cdot \text{m}^2/\text{kg}^2)(5.98 \times 10^{24} \text{ kg})(2500 \text{ kg})/(8.63 \times 10^6 \text{ m})^2 = 1.34 \times 10^4 \text{ N}. \end{aligned}$$

From the graph the work done is the area under the $F$ vs. $r$ graph:

$$\begin{aligned} W &= (F_1 + F_2)\tfrac{1}{2}h \\ &= \tfrac{1}{2}(1.96 \times 10^4 \text{ N} + 1.34 \times 10^4 \text{ N})(3.0 \times 10^6 \text{ m}) = 4.95 \times 10^{10} \text{ J} = 5.0 \times 10^{10} \text{ J}. \end{aligned}$$

(b) To find the work by integration, we have

$$W = \int F\, dr = \int_{r_E + h}^{r_E} -\frac{GM_E m}{r^2}\, dr = \left. \frac{GM_E m}{r} \right|_{r_E + h}^{r_E} = GM_E m\left(\frac{1}{r_E} - \frac{1}{r_E + h}\right) = mgr_E\left(1 - \frac{r_E}{r_E + h}\right)$$

$$W = (2500 \text{ kg})(9.80 \text{ m/s}^2)(6.38 \times 10^6 \text{ m})\left(1 - \frac{6.38 \times 10^6 \text{ m}}{9.38 \times 10^6 \text{ m}}\right) = 5.00 \times 10^{10} \text{ J}$$

Thus we see that our approximation in (a) is within 3%.

43. The work done on the car decreases its kinetic energy:

$$W = \Delta K = \tfrac{1}{2}mv^2 - \tfrac{1}{2}mv_0^2 = 0 - \tfrac{1}{2}(1300 \text{ kg})[(100 \text{ km/h})/(3.6 \text{ ks/h})]^2 = -5.02 \times 10^5 \text{ J}.$$

47. On a level road, the normal force is $mg$, so the kinetic friction force is $\mu_k mg$. Because it is the (negative) work of the friction force that stops the car, we have

$W = \Delta K$;

$-\mu_k mg\, d = \tfrac{1}{2}mv^2 - \tfrac{1}{2}mv_0^2$;

$-(0.38)m(9.80 \text{ m/s}^2)(78 \text{ m}) = -\tfrac{1}{2}mv_0^2$, which gives $v_0 = 24$ m/s (87 km/h or 54 mi/h).

Because every term contains the mass, the final answer does not depend on the mass.

51. On the level, the normal force is $F_N = mg$, so the friction force is $F_{fr} = \mu_k mg$.

The normal force and the weight do no work. The net work increases the kinetic energy of the mass:

$W_{net} = \Delta K = \tfrac{1}{2}mv_f^2 - \tfrac{1}{2}mv_i^2$;

$F(L_1 + L_2) - \mu_k mgL_2 = \tfrac{1}{2}mv_f^2 - 0$;

$(225 \text{ N})(11.0 \text{ m} + 10.0 \text{ m}) - (0.20)(66.0 \text{ kg})(9.80 \text{ m/s}^2)(10.0 \text{ m}) = \tfrac{1}{2}(66.0 \text{ kg})v_f^2$,

which gives $v_f = 10.2$ m/s.

59. For the proton we have

$$W_p = \Delta K_p = m_p c^2 \left[ \frac{1}{\sqrt{1 - (v_p/c)^2}} - 1 \right]$$

Solving for $v_p$

$$v_p = c\sqrt{1 - \left( \frac{1}{W_p/m_p c^2 + 1} \right)^2} = (3.0 \times 10^8 \text{ m/s})\sqrt{1 - \left[ \frac{1}{(3.2 \times 10^{-13} \text{ J})/(1.67 \times 10^{-27} \text{ kg})(3.0 \times 10^8 \text{ m/s})^2 + 1} \right]^2}$$

$$v_p = 2.0 \times 10^7 \text{ m/s}$$

Using the classical formula, we get
$W_p = \frac{1}{2}m_p v_{pc}^2$;
$3.2 \times 10^{-13}$ J $= \frac{1}{2}(1.67 \times 10^{-27}$ kg$)v_{pc}^2$, which gives $v_{pc} = 2.0 \times 10^7$ m/s,
the same as the relativistic formula.
For the electron we have

$$W_e = \Delta K_e = m_e c^2 \left[ \frac{1}{\sqrt{1 - (v_e/c)^2}} - 1 \right]$$

Solving for $v_e$

$$v_e = c\sqrt{1 - \left( \frac{1}{W_e/m_e c^2 + 1} \right)^2} = (3.0 \times 10^8 \text{ m/s})\sqrt{1 - \left[ \frac{1}{(3.2 \times 10^{-13} \text{ J})/(9.11 \times 10^{-31} \text{ kg})(3.0 \times 10^8 \text{ m/s})^2 + 1} \right]^2}$$

$$v_p = 2.9 \times 10^8 \text{ m/s}$$

Using the classical formula, we get
$W_e = \frac{1}{2}m_e v_{ec}^2$;
$3.2 \times 10^{-13}$ J $= \frac{1}{2}(9.1 \times 10^{-31}$ kg$)v_{ec}^2$, which gives $v_{ec} = 8.4 \times 10^8$ m/s,
not only much greater than the relativistic formula, but greater than $c$, which is impossible.

63. (*a*)  The initial kinetic energy of the block is
$K_i = \frac{1}{2}mv_i^2 = \frac{1}{2}(6.0$ kg$)(2.2$ m/s$)^2 = 14.5$ J $=$ 15 J.

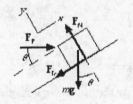

(*b*)  The work done by the applied force is
$W_P = F_P \cos \theta\, d = (75$ N$) \cos 37° (7.0$ m$) = 4.2 \times 10^2$ J.

(*c*)  The work done by friction is
$W_{fr} = -F_{fr}d$
$= -(25$ N$)(7.0$ m$) = -1.75 \times 10^2$ J $= -1.8 \times 10^2$ J.

(*d*)  The work done by gravity is
$W_{grav} = (-mg \sin \theta)d$
$= -(6.0$ kg$)(9.80$ m/s$^2) \sin 37° (7.0$ m$) = -2.5 \times 10^2$ J.

(*e*)  Because the normal force is perpendicular to the displacement,
it does no work: $W_N = 0$.

(*f*)   The net work done on the block changes its kinetic energy:
$W_{net} = W_P + W_{grav} + W_N + W_{fr} = K_f - K_i$;
$4.19 \times 10^2$ J $- 2.47 \times 10^2$ J $+ 0 - 1.75 \times 10^2$ J $= K_f - 14.5$ J, which gives $K_f = 1 \times 10^1$ J.
Note that there is only one significant figure in the result.

67. The work done by the force is
$W = \mathbf{F} \cdot \mathbf{d} = F_x d_x + F_y d_y + F_z d_z = (10.0$ kN$)(5.0$ m$) + (9.0$ kN$)(4.0$ m$) + (12.0$ kN$)(0) = 86$ kJ.
If we use the other expression for the scalar product, we have
$W = \mathbf{F} \cdot \mathbf{d} = Fd \cos \theta$,
86 kJ $= [(10.0$ kN$)^2 + (9.0$ kN$)^2 + (12.0$ kN$)^2]^{1/2}[(5.0$ m$)^2 + (4.0$ m$)^2]^{1/2} \cos \theta$, which gives
$\cos \theta = 0.746$ or $\theta = 42°$.

71. The work done by the air resistance decreases the kinetic energy of the ball:
$W = -F_{air}d = \Delta K = \frac{1}{2}mv^2 - \frac{1}{2}mv_0^2 = \frac{1}{2}m(0.90v_0)^2 - \frac{1}{2}mv_0^2 = \frac{1}{2}mv_0^2[(0.90)^2 - 1]$;
$-F_{air}(15$ m$) = \frac{1}{2}(0.25$ kg$)[(110$ km/h$)/(3.6$ ks/h$)]^2[(0.90)^2 - 1]$, which gives $F_{air} = 1.5$ N.

75. (a) Because the rider exerts a force on each side of the pedals, when the front sprocket has turned through $N_{front}$ revolutions, the work done will be

$W_{rider} = N_{front}\, 2FD_{front} = 2N_{front}\,(0.90mg)D_{front} = 1.80mgN_{front}D_{front}$ .

The number of points that the chain passes over must be the same for the front and back sprockets:

$N_{front}(42\ points/rev) = N_{back}(19\ points/rev)$,

so the number of revolutions of the rear wheel is

$N_{back} = (42/19)N_{front}$ .

After $N_{back}$ revolutions, the bike travels a distance

$d = N_{back}\,2\pi R_{wheel} = (42/19)N_{front}\,2\pi R_{wheel}$ .

From the work-energy principle we have

$W_{rider} + W_{grav} = \Delta K = 0$,  or

$1.80mgN_{front}D_{front} - (m + m_{bike})gd \sin\theta = 0$,  or

$1.80mgN_{front}D_{front} = (m + m_{bike})g(42/19)N_{front}\,2\pi R_{wheel} \sin\theta$,

$(1.80)(60\ kg)gN_{front}(0.36\ m) = (60\ kg + 12\ kg)g(42/19)N_{front}2\pi(0.34\ m)\sin\theta$,

which gives $\sin\theta = 0.114$, $\theta = 6.6°$.

(b) If the force is applied tangential to the pedal motion, the work done is

$W_{rider} = N_{front}\, F2\pi R_{front} = N_{front}\,(0.90mg)2\pi R_{front} = 1.80\pi mgN_{front}R_{front}$ .

From the work-energy principle we have

$W_{rider} + W_{grav} = \Delta K = 0$,  or

$1.80\pi mgN_{front}R_{front} - (m + m_{bike})gd \sin\theta = 0$,  or

$1.80\pi mgN_{front}R_{front} = (m + m_{bike})g(42/19)N_{front}2\pi R_{wheel} \sin\theta$,

$(1.80)\pi 60\ kg)gN_{front}(0.18\ m) = (60\ kg + 12\ kg)g(42/19)N_{front}2\pi(0.34\ m)\sin\theta$,

which gives $\sin\theta = 0.180$, $\theta = 10.3°$.

# Chapter 8: Conservation of Energy

## Chapter Overview and Objectives

This chapter defines the concept of potential energy for conservative forces and introduces the principle of conservation of energy. In addition, it introduces the concepts of power, equilibrium, and stability.

After completing study of this chapter, you should:

- Know the difference between conservative and nonconservative forces.
- Know how to calculate a potential energy function of a conservative force.
- Know how to determine the force from the potential energy function.
- Know how to apply the law of conservation of energy to problems.
- Know how to apply the definition of mechanical power to given situations.
- Know the conditions for mechanical equilibrium and the stability of the equibrium.

## Summary of Equations

Definition of change in potential energy:

$$\Delta U = U_2 - U_1 = -\int_1^2 \mathbf{F} \cdot d\mathbf{l}$$

Calculation of force from potential energy function:

$$\mathbf{F} = -\mathbf{i}\frac{dU}{dx} - \mathbf{j}\frac{dU}{dy} - \mathbf{k}\frac{dU}{dz}$$

Gravitational potential energy near Earth's surface:

$$U(y) = mgy$$

Elastic potential energy of spring:

$$U(x) = \tfrac{1}{2}kx^2$$

Principle of conservation of mechanical energy:

$$\Delta K + \Delta U = 0$$

$$K_1 + U_1 = K_2 + U_2$$

Definition of total mechanical energy:

$$E = K + U$$

Conservation of energy including nonconservative forces:

$$\Delta K + \Delta U = W_{NC}$$

General gravitational potential energy:

$$U(r) = -\frac{GMm}{r}$$

Definition of instantaneous power:

$$P = \frac{dW}{dt} = \frac{dE}{dt}$$

Average power:

$$\overline{P} = \frac{W}{t} = \frac{\Delta E}{t}$$

Power from force acting on moving object:

$$P = \mathbf{F} \cdot \mathbf{v}$$

Definition of efficiency of a machine:

$$e = \frac{P_{out}}{P_{in}}$$

Condition for equilibrium:

$$\frac{dU}{dx} = 0$$

Condition for stable equilibrium:

$$\frac{d^2U}{dx^2} > 0$$

Condition for unstable equilibrium:

$$\frac{d^2U}{dx^2} < 0$$

Condition for neutral equilibrium:

$$\frac{d^2U}{dx^2} = 0$$

## Chapter Summary

### Section 8-1 Conservative and Nonconservative Forces

Forces will be categorized into two types:

*Conservative force*: The work done by the force depends only on the intial and final position, not on the path taken. Or, alternatively, the work done on the object moving around any closed path is zero.

*Nonconservative force*: The work done on the object depends on the path the object follows.

### Section 8-2 Potential Energy

Potential energy functions can be defined for conservative forces. The change in potential energy of a system is defined to be the negative of the work done by a conservative force. Expressed, mathematically, this statement says

$$\Delta U = U_2 - U_1 = -\int_1^2 \mathbf{F} \cdot d\mathbf{l}$$

$\Delta U$ is the change in potential energy in moving from position 1 to position 2, $U_1$ and $U_2$ are the values of the potential energy function at position 1 and position 2 respectively, $\mathbf{F}$ is the conservative force acting on the object, $d\mathbf{l}$ is a infinitesimal displacement along the path followed by the objcct, and the limits on the integral mean the integral is along the path followed by the object from position1 to position 2. Because only the difference in potential energy is physically meaningful, the potential energy function is not unique for a given force. Any potential energy function can have a constant added to it and it remains a potential energy function for the given force.

From calculus we then know that given a function $U$ defined by the above:

$$\mathbf{F} = -\mathbf{i}\frac{dU}{dx} - \mathbf{j}\frac{dU}{dy} - \mathbf{k}\frac{dU}{dz}$$

### Example 8-2-A

Given a force as a function of position

$$\mathbf{F}(x) = \left(-\frac{C}{x^2} + B\sin sx\right)\mathbf{i}$$

Determine the change in potential energy in moving from $x_1$ to $x_2$.

**Solution:**

We should recognize that since there is only a component of the force in the **i** direction, only displacements along the **i** direction contribute to a change in the potential energy because of the dot product between **F** and *dl* in the integrand.

$$\Delta U = -\int_1^2 F \cdot dl = -\int_{x_1}^{x_2} \left( -\frac{C}{x^2} + B \sin sx \right) dx = -\left( \frac{C}{x} - \frac{B}{s} \cos sx \right)\Bigg|_{x_1}^{x_2}$$

**Example 8-2-B**

Given a potential energy function,

$$U(x,y) = Ax^3 + Bxy^2 + Cy^3$$

determine the force as a function of position.

**Solution:**

$$\mathbf{F} = -\mathbf{i}\frac{dU}{dx} - \mathbf{j}\frac{dU}{dy} - \mathbf{k}\frac{dU}{dz} = -\mathbf{i}\left( 3Ax^2 + By^2 \right) - \mathbf{j}\left( 2Bxy + 3Cy^2 \right)$$

The gravitational potential energy function for the gravitational force near the surface of the Earth is given by

$$U(y) = mgy$$

where $U$ is the gravitational potential energy function as a function of vertical position, $y$. $m$ is the mass of the object and $g$ is the local acceleration of gravity.

The elastic potential energy of a spring that follows Hooke's Law is given by

$$U(x) = \tfrac{1}{2}kx^2$$

where the potential energy is a function of $x$, the displacement of the end of the spring from its natural, unstretched length and $k$ is the spring constant of the spring.

**Example 8-2-C**

A spring with a spring constant of 120 N/m has a natural length of 24 cm. The spring is compressed until the potential energy stored in the spring is 1.8 J. How long is the spring when it is compressed?

**Solution:**

The potential energy when compressed, $U$, is 1.8 J and its spring constant, $k$, is 120 N/m. These are related to the compression distance by

$$U = \tfrac{1}{2}kx^2$$

This implies, on solving for $x$:

$$x = \sqrt{\frac{2U}{k}} = \sqrt{\frac{2(1.8\,\text{J})}{120\,\text{N/m}}} = 0.17\,\text{m} = 17\,\text{cm}$$

The question asks for the length of the spring when compressed, so it is 24 cm – 17 cm = 7 cm.

## Section 8-3 Mechanical Energy and Its Conservation

If we consider a system with only conservative forces acting, the work-energy theorem can be written in a new form:

$$\Delta K + \Delta U = 0$$

where $\Delta K$ is the change in total kinetic energy of the system and $\Delta U$ is the change in the total potential energy of the system. This relationship is an expression of the **Principle of Conservation of Energy**. The relationship is often written as

$$K_1 + U_1 = K_2 + U_2$$

where $K_1$ and $U_1$ are, respectively, the total kinetic and potential energy of the system in configuration 1 and $K_2$ and $U_2$ are the total kinetic and potential energy of the system in configuration 2.

We define the total mechanical energy, $E$, as the sum of the kinetic and potential energy:

$$E = K + U.$$

## Section 8-4 Problem Solving Using Conservation of Mechanical Energy

Problems that relate the initial and final positions to the initial and final speeds of an object are candidates for solving with the conservation of mechanical energy. When we use this principle to solve problems, we identify what types of potential energy may be changing during the process that occurs during the problem and include a term in the total energy for those particular types of potential energy. Remember that the form of conservation of energy above only applies if conservative forces act on the system.

### Example 8-4-A

A 2.98 kg box slides along a frictionless surface at a speed of 12.5 m/s. It moves toward a wall with a spring of spring constant 1.28 kN/m as shown. By what distance is the spring compressed when the box comes to a rest?

### Solution:

We recognize that the initial and final speeds are given to us in the problem and that the only potential energy that changes during the problem is the potential energy of the spring. We write down the law of conservation of energy for this system

$$K + U = K' + U' \quad \Rightarrow \quad \tfrac{1}{2}mv^2 + \tfrac{1}{2}kx^2 = \tfrac{1}{2}mv'^2 + \tfrac{1}{2}kx'^2$$

We recognize that the unknown is the final distortion of the spring, $x'$. We solve for $x'$:

$$x' = \sqrt{\frac{m\left(v^2 - v'^2\right)}{k} + x^2}$$

The initial speed is 12.5 m/s, the final speed is zero, and the initial compression of the spring is zero. Thus

$$x' = \sqrt{\frac{(2.98\ \text{kg})\left[(12.5\ \text{m/s})^2 - 0^2\right]}{1280\ \text{N/m}} + 0^2} = 0.603\ \text{m}$$

**Example 8-4-B**

A pendulum bob of mass 0.47 kg hangs vertically and rests against a spring compressed a distance of 1.7 cm. The length of the pendulum from the pivot to the center of mass of the bob is 83 cm. The spring constant of the spring is 340 N/m. If the pendulum bob is released, to what maximum angle will the pendulum bob swing upward?

**Solution:**

We see there will be a change in both the potential energy of the spring and a change in gravitational potential energy. Because the pendulum bob starts from rest and ends at rest, the change in kinetic energy will be zero. Writing down the conservation of energy for this situation

$$K + U_{grav} + U_{spring} = K' + U'_{grav} + U'_{spring} \quad \Rightarrow$$

$$\tfrac{1}{2}mv^2 + mgy + \tfrac{1}{2}kx^2 = \tfrac{1}{2}mv'^2 + mgy' + \tfrac{1}{2}kx'^2$$

where $y$ is the vertical position of the pendulum bob and $x$ is the displacement of the end of the spring from the uncompressed position. Reading the problem carefully, we know that $x = -1.7$ cm, $x' = 0$, $v = 0$, and $v' = 0$. If we choose the origin of the vertical coordinate to be zero at the pivot, then it is easy to see the initial vertical position is $y = -L$. We need to use simple trigonometry to determine that the final vertical position of the pendulum is $y' = -L \cos \theta$. Putting this information into the conservation of energy equation and solving for $\theta$:

$$0 + mg(-L) + \tfrac{1}{2}kx^2 = 0 + mg(-L\cos\theta) + 0 \quad \Rightarrow$$

$$\cos\theta = 1 - \frac{kx^2}{mgL} \quad \Rightarrow \quad \theta = \arcos\left(1 - \frac{kx^2}{mgL}\right)$$

$$= \arcos\left(1 - \frac{(340\ \text{N/m})(0.017\ \text{m})^2}{(0.47\ \text{kg})(9.80\ \text{m/s}^2)(0.83\ \text{m})}\right) = 13^o$$

### Section 8-5 The Law of Conservation of Energy

Nonconservative forces can be accounted for in the law of conservation of energy. The most common nonconservative force encountered in daily life is friction. The dissapative work done by frictional forces is always negative and acts to decrease the total macroscopic mechanical energy of a system. It has been recognized that the energy does not disappear, but acts to increase microscopic energy. This microscopic energy cannot be "seen" in the same way we can see that macroscopic objects have kinetic and potential energy. This **internal** or **thermal energy** is in the kinetic and potential energy of the individual atoms of the object. We can recognize some changes in the internal energy as changes in phase, such as the change from solid to liquid on melting, or changes in temperature.

The generalization of the principle of conservation of energy is:

> **The total energy in a closed system is neither increased nor decreased in any physical process. The energy can be changed from one form into another, but the total amount of energy remains constant.**

### Section 8-6 Energy Conservation with Dissipative Forces: Solving Problems

Work done by non-conservative forces can be included in problems which could be solved using the principle of conservation of energy if the non-conservative forces were not present. The principle of conservation of energy with nonconservative forces can be written:

$$\Delta K + \Delta U = W_{NC}$$

where $W_{NC}$ is the total work done by the nonconservative forces.

**Example 8-6-A**

A car of mass 733 kg is going down a 4.78° incline at 26.5 m/s. To avoid an accident the car must slow down to 18.3 m/s by the time it has gone another 123 m down the slope. What amount of work must be done on the car by nonconservative forces to slow the car down? What amount of work must be done on the car by nonconservative forces to slow the car down if it is headed up the hill instead of down?

**Solution:**

We recognize that as the car slows down there are changes in kinetic energy and gravitational potential energy and there must be some work done by nonconservative forces. Applying the law of conservation of energy

$$\Delta K + \Delta U = W_{NC}$$

We are able to determine the change in kinetic energy from the information given:

$$\Delta K = \tfrac{1}{2}mv'^2 - \tfrac{1}{2}mv^2 = \tfrac{1}{2}(733\,\text{kg})(18.3\,\text{m/s})^2 - \tfrac{1}{2}(733\,\text{kg})(26.5\,\text{m/s})^2$$
$$= -135 \times 10^5\,\text{J}$$

We can also determine the change in gravitational potential energy:

$$\Delta U = mgy' - mgy = mg\Delta y$$

To determine $\Delta y$, use simple trigonometry:

$$\Delta y = -(123\,\text{m})\sin 4.78° = -10.2\,\text{m}$$

Solving for $\Delta U$:

$$\Delta U = (733\,\text{kg})(9.80\,\text{m/s}^2)(-10.2\,\text{m}) = -7.33 \times 10^4\,\text{J}$$

Placing these values in the conservation of energy equation:

$$W_{NC} = \Delta K + \Delta U = -1.35 \times 10^5\,\text{J} + -7.33 \times 10^4\,\text{J}$$
$$= -2.08 \times 10^5\,\text{J}$$

If the car is headed up hill, the sign of the change in gravitational potential energy becomes positive:

$$W_{NC} = \Delta K + \Delta U = -1.35 \times 10^5\,\text{J} + 7.33 \times 10^4\,\text{J}$$
$$= -6.2 \times 10^4\,\text{J}$$

**Section 8-7 Gravitational Potential Energy and Escape Velocity**

The gravitational potential energy function associated with the universal gravitational force is given by:

$$U(r) = -\frac{GMm}{r}$$

where $U(r)$ is the gravitational potential energy as a function of $r$, the distance between the two bodies of mass $M$ and $m$, respectively. Note that the potential energy function, as written, is zero when the distance is infinite. This is the usual position chosen for the zero of the potential energy function when the force increases without bound as the distance decreases.

For a body with a very small mass $m$ interacting with a body with a very large mass $M$, such as man-made size objects of mass $m$ interacting with the earth of mass $M_E$, the change of the kinetic energy of the large mass object is negligible. In these circumstances we can write the principle of conservation of energy as:

$$\tfrac{1}{2}mv^2 - \frac{GMm}{r} = \text{Constant}$$

where $v$ is the speed of the object of mass $m$, $r$ is the distance between the object of mass $m$ and the object of mass $M$.

### Example 8-7-A

An asteroid undergoes a collision far from the earth that essentially brings it to rest relative to the Earth. Assuming gravitational forces from other bodies can be ignored, what would the speed of the asteroid be when it collides with the earth? If the mass of the asteroid is a mere $10^6$ kg, what will its kinetic energy be when it reaches the surface of the earth?

### Solution:

We are given that the initial speed of the asteroid is zero and it is far from the earth, which we will take to mean that the gravitational potential energy is approximately zero at that distance. Then

$$\tfrac{1}{2}mv^2 - \frac{GMm}{r} = 0 + 0 = 0$$

when the asteroid reaches the surface of the Earth, it will have the same total kinetic and potential energy, but now the speed is unknown and the distance will be the radius of the Earth:

$$\tfrac{1}{2}mv'^2 - \frac{GM_E m}{r_E} = 0$$

Solving this for the final speed

$$v' = \sqrt{\frac{2GM_E}{r_E}} = \sqrt{\frac{2\left(6.67\times10^{-11}\ \text{N}\cdot\text{m}^2/\text{kg}^2\right)\left(5.97\times10^{24}\ \text{kg}\right)}{6.38\times10^6\ \text{m}}} = 1.12\times10^5\ \text{m/s}$$

The kinetic energy of the asteroid would be

$$K = \tfrac{1}{2}mv'^2 = \tfrac{1}{2}\left(10^6\ \text{kg}\right)\left(1.12\times10^5\ \text{m/s}\right)^2 = 6.27\times10^{15}\ \text{J}$$

If the total energy of the system is zero, it is easy to see that the speed of the object will have to go to zero as the distance between the objects goes to infinity. The speed necessary to obtain this condition at a given separation between the objects is called the **escape speed**. We can solve for this speed as a function of distance between the bodies:

$$0 = \tfrac{1}{2}mv_{esc}^2 - \frac{GMm}{r} \Rightarrow v_{esc} = \sqrt{\frac{2GM}{r}}$$

where $v_{esc}$ is the speed necessary to satisfy the escape condition. Note that the **escape velocity** must have its component in the direction directly away from the large mass body equal to the escape speed!

### Example 8-7-B

The mass of the sun is $1.99 \times 10^{30}$ kg and its radius is $6.96 \times 10^5$ km. What is the escape speed at the surface of the sun?

**Solution:**

The escape velocity is given by

$$v_{esc} = \sqrt{\frac{2GM}{r}} = \sqrt{\frac{2\left(6.67 \times 10^{-11}\, \text{Nm}^2\,/\,\text{kg}^2\right)\left(1.99 \times 10^{30}\, \text{kg}\right)}{6.96 \times 10^{8}\, \text{m}}} = 6.18 \times 10^{5}\, \text{m/s}$$

## Section 8-8 Power

Instantaneous power is defined as the rate at which work is done:

$$P = \frac{dW}{dt}$$

or, more generally, the rate at which energy is transformed from one form into another:

$$P = \frac{dE}{dt}$$

The time average of the power can be calculated:

$$\overline{P} = \frac{W}{t} \quad or \quad \overline{P} = \frac{\Delta E}{t}$$

The SI unit of power is the **watt** (W).  One watt is equal to one Joule per second.  A common unit of power used from the British unit system is the **horsepower**.  One horsepower is equal to 746 watts.

## Example 8-8-A

A motor runs a conveyor system that raises 12 bags of rice that each weigh 50.0 pounds a distance of 10.0 ft every minute.  What average mechanical power is required from the motor to lift the bags of rice?

**Solution:**

The system changes the gravitational potential energy of the bags of rice.  The power needed is the change in potential energy of the bags divided by the time

$$\overline{P} = \frac{\Delta E}{t} = \frac{mg\Delta y}{t} = \frac{12\left(50.0\, \text{lbs} \times \dfrac{4.45\, \text{N}}{\text{lbs}}\right)\left(10.0\, ft \times \dfrac{0.305\, \text{m}}{\text{ft}}\right)}{60\, \text{s}} = 136\, \text{W}$$

A force, **F**, acting on an object with velocity **v** generates a power:

$$P = \mathbf{F} \cdot \mathbf{v}$$

## Example 8-8-B

A particular vehicle has a maximum speed of 94.2 mph.  It has an engine that can output 235 hp to the drive system. What is the magnitude of the total friction force acting against the vehicle's motion?

**Solution:**

The power generated by the vehicle's engine must be equal in magnitude but opposite in sign to the power of the resistive frictional forces.  The car travels opposite in direction to the frictional forces, so

$$P_{fr} = F_{fc}v\cos 180^\circ = -F_{fr}v$$

Solving this for $F_{fr}$.

$$F_{fr} = -\frac{P_{fr}}{v} = -\frac{-235\,hp \times \dfrac{746\,W}{hp}}{94.2\,mph \times \dfrac{0.447\,m/s}{mph}} = 4.16 \times 10^3 \, N$$

Efficiency of a machine is usually defined as the ratio of the rate of useful energy output to the rate of energy input or

$$e = \frac{P_{out}}{P_{in}}$$

**Example 8-8-C**

Determine the efficiency of the conveyor system in Example 8-8-A if the motor that runs the system uses one-half horsepower of power.

**Solution:**

The efficiency is easily calculated using the power output and the power input

$$e = \frac{P_{out}}{P_{in}} = \frac{136\,W}{\frac{1}{2}\,hp \times \dfrac{746\,W}{1\,hp}} = 0.365 = 36.5\%$$

**Section 8-9 Potential Energy Diagrams; Stable and Unstable Equilibrium**

A graph of potential energy as a function of position is useful in determining the qualitative nature of the motion of a system. Consider the diagram at the right of the potential energy of a system as a function of the position of one of the objects in the system. If the system has the energy $E_1$, then the motion of the system would carry the object back and forth between point $x_1$ and $x_2$. We say that the motion is bounded in this case. Any position of the system in which $dU/dx$ is zero is an **equilibrium** position. If the system has zero kinetic energy when located at the equilibrium position it will not move. If the system has exactly energy $E_2$, then no motion would occur because $F = dU/dx$ is zero at $x_3$. Also, if the system has exactly energy $E_3$ and the object is located at $x_4$, then no motion would occur because again $dU/dx$ is zero at the position of the object.

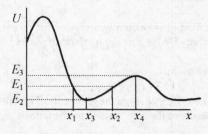

However, there is something very different about the two equilibrium positions at $x_3$ and $x_4$. The equilibrium position is a **stable equilibrium**, while that at $x_4$ is an **unstable equilibrium**. If the system is disturbed slightly from equilibriu... at position $x_3$, the forces act in a direction to return the system toward the equilibrium position. If the system at $x_4$ is disturbed slightly from equilbrium, the forces act in a direction to move the system away from equilibrium. A mathematical test for stability of equilibrium is to determine the sign of the second derivative of the potential energy function. If the second derivative $d^2U/dx^2$, is positive, the equilibrium position is stable. If the second derivative is negative, the equilibrium is unstable. If the second derivative is zero, the equilibrium is neutral.

**Example 8-9-A**

The potential energy as a function of position is given as

$$U(x) = Ax^3 - Bx$$

where A and B are both positive numbers. Determine the positions of the equilibrium positions and whether the equilibrium positions are stable, unstable, or neutral equilibrium positions.

**Solution:**

To determine equilibrium positions, we need to determine the positions at which $dU/dx$ is zero.

$$\frac{dU}{dx} = \frac{d}{dx}\left(Ax^3 - Bx\right) = 3Ax^2 - B$$

This expression is zero when

$$x = \pm\sqrt{\frac{B}{3A}}$$

There are two equilibrium positions, one at $+\sqrt{B/3A}$ and one at $-\sqrt{B/3A}$. To test which type of equilibrium positions these are, we use the second derivative test. The second derivative of the potential energy function is

$$\frac{d^2U}{dx^2} = 6Ax$$

The second derivative is positive when $x$ is positive and is negative when $x$ is negative (because $A$ is positive). This implies that the equilibrium position at $+\sqrt{B/3A}$ is a stable equilibrium position and the equilibrium position at $-\sqrt{B/3A}$ is an unstable equilibrium position. Draw a graph of the potential energy function as a function of position to see if it agrees with these statements.

## Practice Quiz

1.    What evidence is there that nonconservative forces act on an automobile while it is in motion? Identify as many of the nononservative forces as you can.

2.    Is the change in gravitational potential energy greater or lesser if one walks directly up the steep side of a hill compared to when one walks around the hill on a gently sloping path?

3.    If the potential energy and spring constant of a spring are known, does that imply that the displacement of the end of the spring from its equilibrium position is known?

4.    What is the form of the potential energy function for a force that is constant?

5.    If energy is always conserved, what do you think environmentally conscious people mean when they remind us to conserve energy when possible?

6.    Assume that frictional forces do not depend on speed. Car A travels from point X to point Y at a constant speed $v$. Car B travels from point X to point Y at a speed of $2v$. Does the fact that power is proportional to velocity imply that more energy is used by car B than car A?

7.    Explain Aristotle's views on motion and how they relate to conservation of energy. Aristotle's statements about motion are correct if what type of force acts on all systems?

8.    The conditions for equilibrium and stability discussed above and in the text only consider systems with one dimensional freedom. What would the conditions for equilibrium be in a system that has three dimensional freedom? What would the conditions for stability be in a system that has three dimensional freedom?

9.    The conditions for stable equilibrium were discussed in this chapter, but nothing was stated concerning why some equilibria are more stable than others. A house of cards must be stable if it is to stand, but why does it collapse much more readily than a textbook standing on edge? In other words, what determines how stable a configuration is?

10.   There have often been claims of perpetual motion, machines that effectively provide greater output power than input power. Explain how a machine of this type violates the principle of conservation of energy.

11.  Einstein's general theory of relativity includes a description of how light is effected by gravity. Suppose we simply assume that light cannot leave a place where the escape speed is greater than or equal to the speed of light, $3.00 \times 10^8$ m/s. If a body with the mass of the Earth had this escape speed, what would the radius of that body be?

12.  A non-linear spring has a force $F(x) = -kx - qx^3$, where x is the displacement of the end of the spring from its natural length. Determine the potential energy of the spring as a function of $x$.

13.  A ball of mass 0.0343 kg rests atop a spring that is compressed 3.67 cm. The spring has a spring constant 27.34 N/m. When the spring is released, how high will the ball be projected into the air?

14.  An object with a mass of 4.76 kg slides down a 30.0° slope. It starts with speed of 2.54 m/s and after sliding a distance of 3.29 m down the slope has a speed of 4.84 m/s. How large is the frictional force acting on the object?

15.  Determine the equilibrium positions and the stability of those equilibrium positions for a potential $U(x) = 2x + 3x^2 - x^3$.

## Problem Solutions

3.  For the potential energy change we have
$$\Delta U = mg\,\Delta y = (58 \text{ kg})(9.80 \text{ m/s}^2)(3.8 \text{ m}) = 2.2 \times 10^3 \text{ J}.$$

7.  (*a*)  Because the force $\mathbf{F} = (-kx + ax^3 + bx^4)\mathbf{i}$ is a function only of position, it is conservative.
    (*b*)  We find the form of the potential energy function from
$$U = -\int \mathbf{F} \cdot d\mathbf{r} = -\int (-kx + ax^3 + bx^4)\mathbf{i} \cdot (dx\mathbf{i} + dy\mathbf{j} + dz\mathbf{k})$$
$$= -\int (-kx + ax^3 + bx^4)dx = \tfrac{1}{2}kx^2 - \tfrac{1}{4}ax^4 - \tfrac{1}{5}bx^5 + \text{constant.}$$

11.  We choose the potential energy to be zero at the bottom ($y = 0$). Because there is no friction and the normal force does no work, energy is conserved, so we have
$$E = K_1 + U_1 = K_2 + U_2;$$
$$\tfrac{1}{2}mv_1^2 + mgy_1 = \tfrac{1}{2}mv_2^2 + mgy_2;$$
$$\tfrac{1}{2}m(0)^2 + m(9.80 \text{ m/s}^2)(105 \text{ m}) = \tfrac{1}{2}mv_2^2 + m(9.80 \text{ m/s}^2)(0), \text{ which gives } v_2 = 45.4$$
m/s.
This is 160 km/h. It is a good thing there is friction on the ski slopes.

15.  (*a*)  For the motion from the bridge to the lowest point, we use energy conservation:
$$K_i + U_{\text{gravi}} + U_{\text{cordi}} = K_f + U_{\text{gravf}} + U_{\text{cordf}};$$
$$0 + 0 + 0 = 0 + mg(-h) + \tfrac{1}{2}k(h - L_0)^2;$$
$$0 = -(60 \text{ kg})(9.80 \text{ m/s}^2)(31 \text{ m}) + \tfrac{1}{2}k(31 \text{ m} - 12 \text{ m})^2,$$
which gives $k = 1.0 \times 10^2$ N/m.
    (*b*)  The maximum acceleration will occur at the lowest point, where the upward restoring force in the cord is maximum:
$$kx_{\text{max}} - mg = ma_{\text{max}};$$
$$(1.0 \times 10^2 \text{ N/m})(31 \text{ m} - 12 \text{ m}) - (60 \text{ kg})(9.80 \text{ m/s}^2) = (60 \text{ kg})a_{\text{max}},$$
which gives $a_{\text{max}} = 22$ m/s$^2$.

19.  The potential energy is zero at $x = 0$.
    (*a*)  Because energy is conserved, the maximum speed occurs at the minimum potential energy:
$$K_i + U_i = K_f + U_f;$$
$$\tfrac{1}{2}mv_0^2 + \tfrac{1}{2}kx_0^2 = \tfrac{1}{2}mv_{\text{max}}^2 + 0, \text{ which gives } v_{\text{max}} = [v_0^2 + (kx_0^2/m)]^{1/2}.$$
    (*b*)  The maximum stretch occurs at the maximum potential energy or the minimum kinetic energy:
$$K_i + U_i = K_f + U_f;$$
$$\tfrac{1}{2}mv_0^2 + \tfrac{1}{2}kx_0^2 = 0 + \tfrac{1}{2}kx_{\text{max}}^2, \text{ which gives } x_{\text{max}} = [x_0^2 + (mv_0^2/k)]^{1/2}.$$

23. The maximum acceleration will occur at the lowest point,
where the upward restoring force in the spring is maximum:
$kx_{max} - Mg = Ma_{max} = M(5.0g)$, which gives $x_{max} = 6.0Mg/k$.
With $y = 0$ at the initial position of the top of the spring, for
the motion from the break point to the maximum compression
of the spring, we use energy conservation:
$K_i + U_{gravi} + U_{springi} = K_f + U_{gravf} + U_{springf}$ ;
$0 + Mgh + 0 = 0 + Mg(-x_{max}) + \tfrac{1}{2}kx_{max}^2$.
When we use the previous result, we get
$Mgh = -[6.0(Mg)^2/k] + \tfrac{1}{2}k(6.0Mg/k)^2$ , which gives $k = 12Mg/h$.

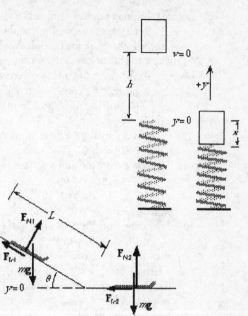

27. (a) We find the normal force from the force diagram for the ski:
$y$-component:    $F_{N1} = mg \cos \theta$,
which gives the friction force: $F_{fr1} = \mu_k mg \cos \theta$.
For the work-energy principle, we have
$W_{NC} = \Delta K + \Delta U = (\tfrac{1}{2}mv_f^2 - \tfrac{1}{2}mv_i^2) + mg(h_f - h_i)$;
$-\mu_k mg \cos \theta L = (\tfrac{1}{2}mv_f^2 - 0) + mg(0 - L \sin \theta)$;
$-(0.090)(9.80 \text{ m/s}^2) \cos 20° (100 \text{ m}) =$
$\tfrac{1}{2}v_f^2 - (9.80 \text{ m/s}^2)(100 \text{ m}) \sin 20°$,
which gives $v_f = 22$ m/s.

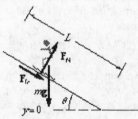

(b) On the level the normal force is $F_{N2} = mg$, so the
friction force is $F_{fr2} = \mu_k mg$.
For the work-energy principle, we have
$W_{NC} = \Delta K + \Delta U = (\tfrac{1}{2}mv_f^2 - \tfrac{1}{2}mv_i^2) + mg(h_f - h_i)$;
$-\mu_k mg D = (0 - \tfrac{1}{2}mv_i^2) + mg(0 - 0)$;
$-(0.090)(9.80 \text{ m/s}^2)D = -\tfrac{1}{2}(22.5 \text{ m/s})^2$ ,
which gives $D = 2.9 \times 10^2$ m.

31. We find the normal force from the force diagram for the skier:
$y$-component:    $F_N = mg \cos \theta$,
which gives the friction force: $F_{fr} = \mu_k mg \cos \theta$.
For the work-energy principle for the motion up the incline,
we have
$W_{NC} = \Delta K + \Delta U = (\tfrac{1}{2}mv_f^2 - \tfrac{1}{2}mv_i^2) + mg(h_f - h_i)$;
$-\mu_k mg \cos \theta L = (0 - \tfrac{1}{2}mv_i^2) + mg(L \sin \theta - 0)$;
$-\mu_k(9.80 \text{ m/s}^2) \cos 17° (12 \text{ m}) =$
$-\tfrac{1}{2}(11 \text{ m/s})^2 + (9.80 \text{ m/s}^2)(12 \text{ m}) \sin 17°$,
which gives $\mu_k = 0.23$.

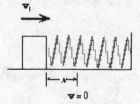

35. (a) On the level the normal force is $F_N = mg$, so the
friction force is $F_{fr} = \mu_k mg$. For the motion from the
impact point to where the block stops, for the
work-energy principle we have
$W_{NC} = \Delta K + \Delta U = (\tfrac{1}{2}mv_f^2 - \tfrac{1}{2}mv_i^2) + (\tfrac{1}{2}kx_f^2 - \tfrac{1}{2}kx_i^2)$;
$-\mu_k mgx = (0 - \tfrac{1}{2}mv_i^2) + \tfrac{1}{2}k(x^2 - 0)$;
$-(0.30)(2.0 \text{ kg})(9.80 \text{ m/s}^2)x = -\tfrac{1}{2}(2.0 \text{ kg})(1.3 \text{ m/s}) + \tfrac{1}{2}(120 \text{ N/m})x^2$.
This reduces to the quadratic equation
$60x^2 + 5.88x - 1.69 = 0$, which has the solutions $x = 0.126$ m, $-0.22$ m.
The negative solution corresponds to positive friction work, so the physical result is $x = 0.13$ m.

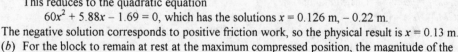

(b) For the block to remain at rest at the maximum compressed position, the magnitude of the
restoring force in the spring must equal the magnitude of the static friction force:
$kx = F_s = \mu_s mg$, or
$\mu_s = kx/mg = (120 \text{ N/m})(0.126 \text{ m})/(2.0 \text{ kg})(9.80 \text{ m/s}^2) = 0.77$.

(*c*)  Before the spring reaches its natural length, it is pushing on the block.  At the natural length, the force goes to zero.  Beyond the natural length the spring would want to pull on the block, but it is not attached; therefore the block leaves the spring.  If we consider the motion from the initial impact to the point where the block leaves the spring, for the work-energy principle, we have

$$W_{NC} = \Delta K + \Delta U = (\tfrac{1}{2}mv_f^2 - \tfrac{1}{2}mv_i^2) + (\tfrac{1}{2}kx_f^2 - \tfrac{1}{2}kx_i^2);$$
$$- \mu_k mg2x = (\tfrac{1}{2}mv_f^2 - \tfrac{1}{2}mv_i^2) + (0 - 0);$$
$$- (0.30)(9.80 \text{ m/s}^2)2(0.126 \text{ m}) = \tfrac{1}{2}v_f^2 - \tfrac{1}{2}(1.3 \text{ m/s}), \text{ which gives } v_f = 0.46 \text{ m/s}.$$

39.  The escape velocity from a mass $M$, as determined by energy conservation, is found from
$$v_{esc}^2 = 2GM/r.$$

(*a*)  To escape from the Sun's surface, we have
$$v_{esc}^2 = 2GM_S/r_S$$
$$= 2(6.67 \times 10^{-11} \text{ N} \cdot \text{m}^2/\text{kg}^2)(2.0 \times 10^{30} \text{ kg})/(7.0 \times 10^8 \text{ m}),$$
which gives $v_{esc} = 6.2 \times 10^5$ m/s.

(*b*)  To escape from the Sun when at the Earth's location, we have
$$v_{esc}^2 = 2GM_S/r$$
$$= 2(6.67 \times 10^{-11} \text{ N} \cdot \text{m}^2/\text{kg}^2)(2.0 \times 10^{30} \text{ kg})/(1.50 \times 10^{11} \text{ m}),$$
which gives $v_{esc} = 4.2 \times 10^4$ m/s.
Because the gravitational attraction provides the radial acceleration of the Earth, we have
$$GM_S M_E/r^2 = M_E v_{orbit}^2/r, \text{ or } v_{orbit}^2 = GM_S/r.$$
For the ratio we get
$$v_{esc}^2/v_{orbit}^2 = (2GM_S/r)/(GM_S/r), \text{ or } v_{esc}/v_{orbit} = \sqrt{2}.$$

43.  The escape velocity for a mass $m$ is the speed required so the mass can get infinitely far away with essentially zero velocity.  From energy conservation we have
$$K_1 + U_1 = K_2 + U_2;$$
$$\tfrac{1}{2}mv_{esc}^2 - GM_E m/r = 0 - GM_E m/\infty, \text{ or } v_{esc}^2 = 2GM_E/r.$$
Because the gravitational attraction provides the radial acceleration of the satellite, we have
$$GmM_E/r^2 = mv_{orbit}^2/r, \text{ or } v_{orbit}^2 = GM_S/r.$$
For the ratio we get
$$v_{esc}^2/v_{orbit}^2 = (2GM_E/r)/(GM_E/r), \text{ or } v_{esc}/v_{orbit} = \sqrt{2}.$$

47.  (*a*)  The escape velocity at a distance $r$ from the center of the Earth is
$$v_{esc} = (2GM_E/r)^{1/2}.$$
We find the rate at which $v_{esc}$ changes with a change in $r$ by differentiating:
$$dv_{esc}/dr = - \tfrac{1}{2}(2GM_E/r^3)^{1/2} = - v_{esc}/2r.$$

(*b*)  The escape speed from the surface of the Earth is found from
$$v_{esc0}^2 = 2GM_E/r_E = 2(6.67 \times 10^{-11} \text{ N} \cdot \text{m}^2/\text{kg}^2)(5.98 \times 10^{24} \text{ kg})/(6.38 \times 10^6 \text{ m}),$$
which gives $v_{esc0} = 1.12 \times 10^4$ m/s.
The approximate change at a height of 300 km is
$$\Delta v_{esc} \Delta (- v_{esc0}/2r_E) \Delta r = [- (1.12 \times 10^4 \text{ m/s})/2(6.38 \times 10^6 \text{ m})](300 \times 10^3 \text{ m}) = - 263 \text{ m/s}.$$
The new escape velocity is
$$v_{esc} = v_{esc0} + \Delta v_{esc} = 1.12 \times 10^4 \text{ m/s} - 263 \text{ m/s} = 1.09 \times 10^4 \text{ m/s}.$$

51.  If we neglect friction, we can apply conservation of energy:
$$K_1 + U_1 = K_2 + U_2;$$
$$\tfrac{1}{2}mv_1^2 - GM_E m/r_1 = \tfrac{1}{2}mv_2^2 - GM_E m/r_2, \text{ or }$$
$$v_2^2 = v_1^2 + 2GM_E[(1/r_2) - (1/r_1)];$$
$$v_2^2 = (600 \text{ m/s})^2 + 2(6.67 \times 10^{-11} \text{ N} \cdot \text{m}^2/\text{kg}^2)(5.98 \times 10^{24} \text{ kg})[(1/6.38 \times 10^6 \text{ m}) - (1/5.0 \times 10^9 \text{ m})]$$
which gives $v_2 = 1.1 \times 10^4$ m/s.

55.  We find the equivalent force exerted by the engine from
$$P = Fv;$$
$$(18 \text{ hp})(746 \text{ W/hp}) = F(90 \text{ km/h})/(3.6 \text{ ks/h}), \text{ which gives } F = 5.4 \times 10^2 \text{ N}.$$
At constant speed, this force is balanced by the average retarding force, which must be $5.4 \times 10^2$ N.

59. We find the average resistance force from the acceleration:

$F_R = ma = m \, \Delta v / \Delta t = (1000 \text{ kg})(70 \text{ km/h} - 90 \text{ km/h})/(3.6 \text{ ks/h})(6.0 \text{ s}) = -926 \text{ N}.$

If we assume that this is the resistance force at 80 km/h, the engine must provide an equal and opposite force to maintain a constant speed. We find the power required from

$P = Fv = (926 \text{ N})(80 \text{ km/h})/(3.6 \text{ ks/h}) = 2.1 \times 10^4 \text{ W} = (2.1 \times 10^4 \text{ W})/(746 \text{ W/hp}) = 28 \text{ hp}.$

63. The work done increases the potential energy of the player. We find the power from

$P = W/t = \Delta U/t = mg(h_f - h_i)/t$
$= (105 \text{ kg})(9.80 \text{ m/s}^2)[(140 \text{ m}) \sin 30° - 0]/(61 \text{ s}) = 1.2 \times 10^3 \text{ W}$ (about 1.6 hp).

67. Because the rate of work is $P = Fv$ and the applied force produces the acceleration, we find the velocity and acceleration as a function of time:

$x = (5.0 \text{ m/s}^3)t^3 - (8.0 \text{ m/s}^2)t^2 - (30 \text{ m/s})t;$
$v = dx/dt = (15.0 \text{ m/s}^3)t^2 - (16.0 \text{ m/s}^2)t - (30 \text{ m/s});$
$a = dv/dt = (30.0 \text{ m/s}^3)t - (16.0 \text{ m/s}^2).$

Thus the rate of work is

$P = Fv = mav = m[(30.0 \text{ m/s}^3)t - (16.0 \text{ m/s}^2)][(15.0 \text{ m/s}^3)t^2 - (16.0 \text{ m/s}^2)t - (30 \text{ m/s})].$

(a) At $t = 2.0$ s, we have

$P = (0.280 \text{ kg})[(30.0 \text{ m/s}^3)(2.0 \text{ s}) - (16.0 \text{ m/s}^2)][(15.0 \text{ m/s}^3)(2.0 \text{ s})^2 - (16.0 \text{ m/s}^2)(2.0 \text{ s}) - (30 \text{ m/s})]$
$= -25 \text{ W}.$

Note that the negative sign means there are times when the applied force is opposite to the motion.

(b) At $t = 4.0$ s, we have

$P = (0.280 \text{ kg})[(30.0 \text{ m/s}^3)(4.0 \text{ s}) - (16.0 \text{ m/s}^2)][(15.0 \text{ m/s}^3)(4.0 \text{ s})^2 - (16.0 \text{ m/s}^2)(4.0 \text{ s}) - (30 \text{ m/s})]$
$= +4.3 \times 10^3 \text{ W}.$

(c) Over a time interval, the average net power produces the change in kinetic energy:

$P = W/\Delta t = \Delta K/\Delta t = (\tfrac{1}{2}mv_f^2 - \tfrac{1}{2}mv_i^2)/\Delta t = \tfrac{1}{2}m(v_f^2 - v_i^2)/\Delta t.$

We find the velocities at the three times:

$v_0 = (15.0 \text{ m/s}^3)(0)^2 - (16.0 \text{ m/s}^2)(0) - (30 \text{ m/s}) = 30 \text{ m/s};$
$v_2 = (15.0 \text{ m/s}^3)(2.0 \text{ s})^2 - (16.0 \text{ m/s}^2)(2.0 \text{ s}) - (30 \text{ m/s}) = -2.0 \text{ m/s};$
$v_4 = (15.0 \text{ m/s}^3)(4.0 \text{ s})^2 - (16.0 \text{ m/s}^2)(4.0 \text{ s}) - (30 \text{ m/s}) = 146 \text{ m/s}.$

From $t = 0$ to $t = 2.0$ s, we have

$P = \tfrac{1}{2}(0.280 \text{ kg})[(-2.0 \text{ m/s})^2 - (30 \text{ m/s})^2]/(2.0 \text{ s} - 0) = -63 \text{ W}.$

From $t = 2.0$ s to $t = 4.0$ s, we have

$P = \tfrac{1}{2}(0.280 \text{ kg})[(146 \text{ m/s})^2 - (-2.0 \text{ m/s})^2]/(4.0 \text{ s} - 2.0 \text{ s}) = 1.5 \times 10^3 \text{ W}.$

71. For the binding energy, we have

$E_{binding} = U(\infty) - U(r_{min})$
$= 0 - [(-a/r_{min}^6) + (b/r_{min}^{12})] = [-a/(2b/a)] + [b/(2b/a)^2] = -a^2/4b.$

75. The work done increases the potential energy of the elevator. We find the power output from

$P = W/t = \Delta U/t = mg(h_f - h_i)/t = (850 \text{ kg})(9.80 \text{ m/s}^2)(32.0 \text{ m})/(11.0 \text{ s})(746 \text{ W/hp}) = 32.5 \text{ hp}.$

79. We choose the reference level for the gravitational potential energy at the lowest point. The tension in the cord is always perpendicular to the displacement and thus does no work.

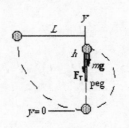

(a) With no air resistance during the fall, we have

$0 = \Delta K + \Delta U = (\tfrac{1}{2}mv_1^2 - \tfrac{1}{2}mv_0^2) + mg(h_1 - h_0),$ or
$\tfrac{1}{2}(v_1^2 - 0) = -g(0 - L),$ which gives $v_1 = (2gL)^{1/2}.$

(b) For the motion from release to the rise around the peg, we have

$0 = \Delta K + \Delta U = (\tfrac{1}{2}mv_2^2 - \tfrac{1}{2}mv_0^2) + mg(h_2 - h_0),$ or
$\tfrac{1}{2}(v_2^2 - 0) = -g[2(L - h) - L] = g(2h - L) = 0.60gL,$
which gives $v_2 = (1.2gL)^{1/2}.$

83. We choose the reference level for the gravitational potential energy at the lowest point.

(a) With no air resistance during the fall, we have

$0 = \Delta K + \Delta U = (\tfrac{1}{2}mv^2 - \tfrac{1}{2}mv_0^2) + mg(h - h_0),$ or
$\tfrac{1}{2}(v^2 - 0) = -g(0 - H),$ which gives
$v_1 = (2gH)^{1/2} = [2(9.80 \text{ m/s}^2)(80 \text{ m})] = 40 \text{ m/s}.$

(b)  If 60% of the kinetic energy of the water is transferred, we have

$P = (0.60)\tfrac{1}{2}mv^2/t = (0.60)\tfrac{1}{2}(m/t)v^2$
$= (0.60)\tfrac{1}{2}(550 \text{ kg/s})(40 \text{ m/s})^2 = 2.6 \times 10^5 \text{ W (about 350 hp)}.$

87.  We choose the potential energy to be zero at the lowest point ($y = 0$).

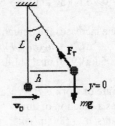

(a)  Because the tension in the rope does no work, energy is conserved, so we have

$K_i + U_i = K_f + U_f;$
$\tfrac{1}{2}mv_0^2 + 0 = 0 + mgh = mg(L - L \cos \theta) = mgL(1 - \cos \theta);$
$\tfrac{1}{2}m(5.0 \text{ m/s})^2 = m(9.80 \text{ m/s}^2)(10.0 \text{ m})(1 - \cos \theta)$

which gives $\cos \theta = 0.872$, or $\theta = 29°$.

(b)  The velocity is zero just before he releases, so there is no centripetal acceleration.  There is a tangential acceleration which has been decreasing his tangential velocity.  For the radial direction we have

$F_T - mg \cos \theta = 0;$  or
$F_T = mg \cos \theta = (75 \text{ kg})(9.80 \text{ m/s}^2)(0.872) = 6.4 \times 10^2 \text{ N}.$

(c)  The velocity and thus the centripetal acceleration is maximum at the bottom, so the tension will be maximum there.  For the radial direction we have

$F_T - mg = mv_0^2/L,$  or
$F_T = mg + mv_0^2/L = (75 \text{ kg})[(9.80 \text{ m/s}^2) + (5.0 \text{ m/s})^2/(10.0 \text{ m})] = 9.2 \times 10^2 \text{ N}.$

91.  If we consider a small length $L$ of the circular stream with radius $R$ that leaves the hose, the mass of this much water is $m = \rho \pi R^2 L$.  If the water leaves the hose with speed $v$, the time to leave the hose is $t = L/v$.  If we neglect air resistance and apply conservation of energy from leaving the hose to the highest point, we have

$K_1 + U_1 = K_2 + U_2;$
$\tfrac{1}{2}mv^2 + 0 = 0 + mgh,$  or  $v = (2gh)^{1/2} = [2(9.80 \text{ m/s}^2)(30 \text{ m})]^{1/2} = 24.2 \text{ m/s}.$

The minimal supplied power produces the kinetic energy of the water as it leaves the hose:

$P = W/t = \Delta K/t = \tfrac{1}{2}mv^2/t = \tfrac{1}{2}\rho \pi R^2 Lv^2/t = \tfrac{1}{2}\rho \pi R^2 v^3$
$= \tfrac{1}{2}(1.00 \times 10^3 \text{ kg/m}^3)\pi(1.5 \times 10^{-2} \text{ m})^2(24.2 \text{ m/s})^3/(746 \text{ W/hp}) = 6.8 \text{ hp}.$

95.  If 80% of the electrical power is used to increase the potential energy of the water, we have

$0.80P = mg(h_f - h_i)/t;$
$0.80P = (1.00 \times 10^3 \text{ kg/day})(9.80 \text{ m/s}^2)(400 \text{ m})/(24 \text{ h/day})(3600 \text{ s/h}),$

which gives $P = 5.7 \times 10^4 \text{ W} = 76 \text{ hp}.$

# Chapter 9: Linear Momentum and Collisions

## Chapter Overview and Objectives

This chapter introduces the concept of momentum, the law of conservation of momentum, and related topics.

After completing study of this chapter, you should:

- Know the definition of momentum.
- Know the definition of impulse.
- Know how Newton's laws can be expressed as the conservation of momentum.
- Know how to use the law of conservation of momentum to solve collision problems.
- Know how to determine the center of mass of an extended body and a system of particles.
- Know how to apply the law of conservation of momentum to a body that is changing mass.
- Know the SI units of momentum and impulse and their abbreviations.

## Summary of Equations

Definition of momentum:

$$\mathbf{p} = m\mathbf{v}$$

Newton's second law in terms of momentum:

$$\sum \mathbf{F} = \frac{d\mathbf{p}}{dt}$$

Integral form of Newton's second law:

$$\int_{t_1}^{t_2} \mathbf{F}\,dt = \int_{\mathbf{p}_1}^{\mathbf{p}_2} d\mathbf{p} = \mathbf{p}_2 - \mathbf{p}_1 = \Delta\mathbf{p}$$

Conservation of momentum:

$$\sum \frac{d\mathbf{p}}{dt} = \sum \mathbf{F}_{ext}$$

Definition of impulse:

$$\mathbf{J} = \int_{t_1}^{t_2} \mathbf{F}\,dt$$

Expression for average force in terms of impulse:

$$\overline{\mathbf{F}} = \frac{\mathbf{J}}{t}$$

Definition of center of mass of an extended body:

$$\mathbf{r}_{CM} = \frac{\int \mathbf{r}\,dm}{\int dm} = \frac{1}{M}\int \mathbf{r}\,dm$$

Definition of center of mass of a system of particles:

$$\mathbf{r}_{CM} = \frac{\sum m_i \mathbf{r}_i}{\sum m_i}$$

Newton's second law applied to center of mass of system:

$$\sum F_{ext} = Ma_{CM}$$

Principle of conservation of momentum applied to system with changing mass:

$$dP_{system} = (v - u)\,dm + m\,dv$$

# Chapter Summary

### Section 9-1 Momentum and Its Relation to Force

The linear momentum of an object is defined to be the product of its mass and its velocity:

$$\mathbf{p} = m\mathbf{v}$$

As the velocity of an object is a vector and its mass is a scalar, this product is the product of a scalar with a vector and results in a vector quantity. Thus, momentum is a vector quantity. The SI units of momentum are kg·m/s. Newton's second law of motion can be written in terms of momentum as:

$$\sum \mathbf{F} = \frac{d\mathbf{p}}{dt}$$

### Example 9-1-A

As an airplane flies through the air, it accelerates air downward at a rate of 1060 kg/s. The air starts at rest and is deflected downward with a velocity of 140 m/s. What is the force on the airplane?

### Solution:

The change of the momentum of the air is

$$dp = (v - v_0)dm \quad or \quad \frac{dp}{dt} = (v - v_0)\frac{dm}{dt} = (v - 0)\frac{dm}{dt} = (140 \text{ m/s})(1060 \text{ kg/s})$$
$$= 1.48 \times 10^5 \text{ N}$$

The force on the air is equal to the rate of change of the momentum. By Newton's third law, the airplane has an equal magnitude upward force so the force on the airplane is $1.48 \times 10^5$ N.

The above relationship can be multiplied on both sides by $dt$ and integrated

$$\int_{t_1}^{t_2} \mathbf{F}dt = \int_{\mathbf{p}_1}^{\mathbf{p}_2} d\mathbf{p} = \mathbf{p}_2 - \mathbf{p}_1 = \Delta\mathbf{p}$$

### Section 9-2 Conservation of Momentum

The relationship above can be applied to a system of interacting objects and summed over the momentum changes for all the objects in the system. Applying Newton's third law to the sum results in a total of zero momentum change for interactions between objects in the system. This results in the law of conservation of momentum. This law states that the sum of the changes in momentum of the objects in an **isolated system** is equal to zero.

$$\sum \Delta\mathbf{p} = 0 \quad or \quad \sum_{intial}\mathbf{p} = \sum_{final}\mathbf{p} \quad or \quad \sum \frac{d\mathbf{p}}{dt} = 0$$

An isolated system is a system with no forces acting on any component of the system arising from outside the system.

If external forces act on the system, this can be generalized to:

$$\sum \frac{d\mathbf{p}}{dt} = \sum \mathbf{F}_{ext}$$

where $\mathbf{F}_{ext}$ represents all of the forces acting on the system that arise from sources external to the system.

## Example 9-2-A

Two people run toward each other on ice and collide and head onto each other.  One person has a mass of 106 kg and runs with a speed of 2.67 m/s.  The other person has a mass of 78.4 kg and runs with a speed of 6.42 m/s.  What is the motion of the two people after they collide?

### Solution:

We will apply conservation of momentum to this collision.  We recognize that the two initial velocities are in opposite directions.  We will assign coordinates so that the initial velocity, $v_1$, of the 106 kg mass person is in the positive direction.  Applying conservation of momentum

$$\sum p_i = \sum p_f \quad \Rightarrow \quad m_1 v_1 + m_2 v_2 = m_1 v_1' + m_2 v_2'$$

We are given the masses, the initial velocities and that the final velocities are identical.  Solving for the final velocity

$$v' = \frac{m_1 v_1 + m_2 v_2}{m_1 + m_2} = \frac{(106\,\text{kg})(2.67\,\text{m/s}) + (78.4\,\text{kg})(-6.42\,\text{m/s})}{106\,\text{kg} + 78.4\,\text{kg}} = -1.19\,\text{m/s}$$

## Example 9-2-B

During a football game, a thrown ball with a mass of 0.214 kg hits a duck with a mass of 4.52 kg flying over the field.  Before the collision, the football is traveling with a velocity 4.5 m/s $\mathbf{i}$ +6.3 m/s $\mathbf{j}$+1.4 m/s $\mathbf{k}$, where $\mathbf{i}$ points to the east, $\mathbf{j}$ to the north, and $\mathbf{k}$ upward.  The duck's velocity before the collision is 8.2 m/s horizontal directly toward the west before the collision.  If the football is deflected so that its velocity is straight downward at 5.2 m/s after the collision, what is the velocity of the duck after the collision?

### Solution:

First write the duck's velocity in terms of the Cartesian unit vectors.  It should be easy to see that it is –8.2 m/s $\mathbf{i}$, as east is the $\mathbf{i}$ direction and the duck is flying west.  Writing down the law of conservation of momentum for this system:

$$\sum p_i = \sum p_f \quad \Rightarrow \quad m_{ball} v_{ball} + m_{duck} v_{duck} = m_{ball} v_{ball}' + m_{duck} v_{duck}'$$

Solving this for the final velocity of the duck:

$$\mathbf{v}_{duck}' = \frac{m_{ball} \mathbf{v}_{ball} + m_{duck} \mathbf{v}_{duck} - m_{ball} \mathbf{v}_{ball}'}{m_{duck}}$$

$$= \frac{(0.214\,\text{kg})[(4.5\,\text{m/s}\,\mathbf{i}+6.3\,\text{m/s}\,\mathbf{j}+1.4\,\text{m/s}\,\mathbf{k}) - (-5.2\,\text{m/s}\,\mathbf{k})] + (4.52\,kg)(-8.2\text{m/s}\,\mathbf{i})}{4.52\,\text{kg}}$$

$$= -8.0\,\text{m/s}\,\mathbf{i} + 0.30\,\text{m/s}\,\mathbf{j} + 0.31\,\text{m/s}\,\mathbf{k}$$

## Section 9-3 Collisions and Impulse

The integral of force over time that appears in the above relationships is called the **impulse, J**:

$$\mathbf{J} = \int_{t_1}^{t_2} \mathbf{F}\,dt$$

Note that impulse is a vector quantity with the same direction as the force.  The change in momentum of an object is equal to the total impulse that acts on the object:

$$\Delta \mathbf{p} = \mathbf{J}$$

There are times when the details of the force as a function of time are unknown. A useful concept when this is true is average force. The average force, $\overline{F}$, is the impulse divided by the time interval:

$$\overline{F} = \frac{J}{t}$$

## Example 9-3-A

A force has a time dependence given by $1.00 \text{ N} - 6.37 \text{ N/s}^2 \, t^2$. Determine the change in momentum of a particle that this force acts on during the time interval from $t = 0$ to $t = 3.42$ s.

**Solution:**

Using the relationship between impulse and change of momentum

$$\Delta p = J = \int_{t=0}^{3.42\,\text{s}} F \, dt = \int_{t=0}^{3.42\,\text{s}} 1.00 \text{ N} - 6.37 \text{ N/s}^2 \, t^2 \, dt$$

$$= \left( 1.00 \text{ N} \, t - \tfrac{1}{2} 6.37 \text{ N/s}^2 \, t^3 \right)_{t=0}^{3.42\,\text{s}} = -81.5 \text{ N} \cdot \text{s} = -81.5 \text{ kg} \cdot \text{m/s}$$

## Section 9-4 Conservation of Energy and Momentum in Collisions

In collisions between objects, the details of the forces that act on the objects are usually unknown. Newton's second law cannot be used to determine the motion resulting from the collision without this knowledge. However, using the laws of conservation of momentum and conservation of energy, the motions before and after the collision can be related to each other. In any collision in which external forces acting on the colliding bodies can be neglected, the law of conservation of momentum can be applied. In some collisions, the total kinetic energy of the colliding bodies remains constant. These collisions are called **elastic collisions**. For elastic collisions, a law of conservation of kinetic energy can be written down to related the motion before the collision to the motin after the collision. For a system of two object, 1 and 2, this can be written

$$\tfrac{1}{2} m_1 v_1^2 + \tfrac{1}{2} m_2 v_2^2 = \tfrac{1}{2} m_1 v_1'^2 + \tfrac{1}{2} m_2 v_2'^2$$

where the unprimed speeds are before the collision and the primed speeds are after the collision. The subscripts on the quantities refer to which particle, 1 or 2, they correspond to.

In the total kinetic energy does not remain constant during the collision, the collision is said to be an **inelastic collision**. During inelastic collisions, some of the intial kinetic energy is transformed into another form of energy:

$$\tfrac{1}{2} m_1 v_1^2 + \tfrac{1}{2} m_2 v_2^2 = \tfrac{1}{2} m_1 v_1'^2 + \tfrac{1}{2} m_2 v_2'^2 + \text{other energy}$$

## Example 9-4-A

A collision between two objects occurs. Initially, object A with a mass of 12 kg is traveling with a speed of 3.4 m/s and object B with a mass of 16 kg is traveling with a speed of 2.8 m/s. After the collision, object A is traveling with a speed of 2.2 m/s and object B is traveling with a speed of 3.0 m/s. Determine whether this is an elastic or inelastic collision. If it is an inelastic collision, how much kinetic energy was transformed into other forms of energy during the collision?

**Solution:**

The collision is elastic if the total kinetic energy of the objects is the same after the collision as before the collision. The kinetic energy before the collision, $K$, is

$$K = \tfrac{1}{2} m_1 v_1^2 + \tfrac{1}{2} m_2 v_2^2 = \tfrac{1}{2} (12 \text{ kg})(3.4 \text{ m/s})^2 + \tfrac{1}{2} (16 \text{ kg})(2.8 \text{ m/s})^2 = 1.3 \times 10^2 \text{ J}$$

The kinetic energy after the collision is

$$K = \tfrac{1}{2}m_1 v_1'^2 + \tfrac{1}{2}m_2 v_2'^2 = \tfrac{1}{2}(12\,\text{kg})(2.2\,\text{m/s})^2 + \tfrac{1}{2}(16\,\text{kg})(3.0\,\text{m/s})^2 = 1.0 \times 10^2\,\text{J}$$

The collision is not elastic and $0.3 \times 10^2$ J of kinetic energy were lost in the collision.

### Section 9-5 Elastic Collisons in One Dimension

Collisions in one dimension have all velocities, both intial and final, lying along the same line. If the collision is between two particles and is elastic, we can apply conservation of momentum and conservation of kinetic energy:

$$m_1 v_1 + m_2 v_2 = m_1 v_1' + m_2 v_2'$$

$$\tfrac{1}{2}m_1 v_1^2 + \tfrac{1}{2}m_2 v_2^2 = \tfrac{1}{2}m_1 v_1'^2 + \tfrac{1}{2}m_2 v_2'^2$$

Eliminating $m_1$ and $m_2$ from these two equation results in:

$$v_1 - v_2 = -(v_1' - v_2')$$

The result of the collision is to reverse the relative velocity of the two objects.

### Example 9-5-A

Collision problems are often solved in the center of mass reference frame. In this reference frame, the center of mass is at rest. What are the results of an elastic collision in one dimension if particle 1 has initial velocity $v_1$ and particle 2 has initial velocity $v_2$?

### Solution:

The position of the center of mass is given by

$$r_{CM} = \sum m_i x_i$$

If the center of mass is at rest, then the time derivative of the position of the center of mass is zero. This implies

$$0 = \frac{dr_{CM}}{dt} = \frac{d}{dt}\sum m_i x_i = \sum m_i \frac{dx_i}{dt} = \sum m_i v_i$$

This applies both before and after the collision. Writing down this condition along with the fact that an elastic collision reverses the relative velocity of the two objects

$$v_2 - v_1 = v_1' - v_2' \qquad m_1 v_1 + m_2 v_2 = 0 \qquad m_1 v_1' + m_2 v_2' = 0$$

Multiplying the first equation by $m_1$ and adding these three equations together

$$(m_1 + m_2)v_2 + m_1 v_1' + m_2 v_2' = m_1 v_1' - m_2 v_2' \;\Rightarrow\; v_2 = -v_2'$$

Similarly, mutiplying the first equation by $m_2$ and adding results in

$$v_1 = -v_1'$$

In the center of mass reference frame, an elastic collision simply reverses the directions of the colliding particles' velocities!

**Example 9-5-B**

One way a pool hustler can beat you is by substituting a cue ball whose mass is different from the other billiard balls. The directions of the balls after the collisions are not the same as for a cue ball with identical mass as the other balls, thus throwing off the opponent. The hustler has practiced with the heavy cue ball and knows what to expect; the other player doesn't. You have been playing pool and are having difficulty making shots. Your suspicoins are confirmed when you see an elastic head-on collision with a ball at rest in which the spin of the balls had no effect, but the cue ball continues to move after the collision with a velocity equal to one tenth its initial velocity. By what percentage is the cue ball bigger in mass that the ball it struck?

**Solution:**

We are given that the final velocity of the first ball is one tenth its initial velocity,

$$v_1' = \tfrac{1}{10} v_1$$

We also know the second ball is initially at rest, $v_2 = 0$.

Applying conservation of momentum to this system

$$\sum p = \sum p' \quad \Rightarrow \quad m_1 v_1 = \tfrac{1}{10} m_1 v_1 + m_2 v_2$$

Using the condition that the relative velocities reverse in a one dimensional elastic collision

$$v_2' - v_1' = v_1 - v_2 \quad \Rightarrow \quad v_2' = v_1 - v_2 + v_1' = v_1 + 0 + \tfrac{1}{10} v_1 = \tfrac{11}{10} v_1$$

Placing this expression for v′$_2$ in this conservation of momentum equation above

$$m_1 v_1 = \tfrac{1}{10} m_1 v_1 + \tfrac{11}{10} m_2 v_1 \quad \Rightarrow \quad \tfrac{9}{10} m_1 = \tfrac{11}{10} m_2 \quad \Rightarrow \quad m_1 = \tfrac{11}{9} m_2 = 1.22\, m_2$$

The cue ball is 22% more massive than the other ball.

**Section 9-6 Inelastic Collisions**

In inelastic collisions, the total kinetic energy of the objects is not conserved. To solve problems in which inelastic collisions are involved, usually some additional information will be necessary to solve the collision problem. One common type of problem encountered is the **completely inelastic** collision problem. In a completely inelastic collision, the colliding objects move with identical velocities after the collision.

$$v_1' = v_2'$$

**Example 9-6-A**

Two cars collide head on and lock together as they do so. The first car had a mass of 800 kg and was traveling north at 18 m/s. The second car had a mass of 600 kg and was traveling south. Evidence at the scene of the accident is consistent with the cars moving at velocity of 1.12 m/s north immediately after the collision before friction slowed them to a rest. The police believe the second car was exceeding the speed limit of 20 m/s just prior to the collision. Determine the speed of the second car prior to the collision to help the police determine whether the second car was speeding or not.

**Solution:**

Because the two cars lock together, we know the final velocity was common between the two cars. Writing down the conservation of momentum

$$\sum p_i = \sum p_f \quad \Rightarrow \quad m_1 v_1 + m_2 v_2 = (m_1 + m_2) v'$$

The first car's initial velocity, the final velocity, and the masses of the cars are given. We solve for the initial velocity of the second car

$$v_2 = \frac{(m_1 + m_2)v' - m_1 v_1}{m_2} = \frac{(800\,\text{kg} + 600\,\text{kg})(1.12\,\text{m/s}) - (800\,\text{kg})(18\,\text{m/s})}{600\,\text{kg}} = -21\,\text{m/s}$$

We have adopted a coordinate system such that north is positive. The minus sign on the result is consistent with the second car initially traveling toward the south. The car was exceeding the speed limit.

### Section 9-7 Collisions in Two or Three Dimensions

The law of conservation of momentum as written covers problems of any dimension because it is written in vector form. The mathematical details of the problem can be more complex in two or three dimensions as compared to one dimension. There will be one conservation of momentum equation for each component of the motion, two components in two dimensions and three components in three dimensions. If the final velocities of the particles are considered unknowns, then in a two particle collision there will be two unknown final velocities for each dimension. With one conservation of momentum equation for each dimension and two unknown velocities, there are not enough equations to solve for the unknown velocities in two or three dimensions even if the condition of an elastic collision is included. In two dimensions, one more parameter must be included and in three dimensions two more parameters are needed to make the elastic collision problem solvable. A completely inelastic problem will be solvable because the complete inelasticity condition reduces the number of unknown velocity components to the number of dimensions.

As an example, in two dimensions, the additional parameter given in an elastic collision problem is often one of the angles of one of the final velocities. Consider how this additional information provides enough information to make the problem solvable in the following example.

### Example 9-7-A

Two identical balls collide elastically. Initially, the two balls are traveling at right angles to one another, the first with speed of 8.73 m/s and the second with a speed of 2.97 m/s. After the collision, the first ball is deflected by an angle of 24.6° from its initial path. What are the speeds of the two balls after the collision and what is the direction of the final velocity of the second ball?

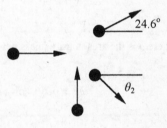

### Solution:

We have three relationships between the initial and final velocities and the masses: conservation of momentum in the $x$ direction, conservation of energy in the $y$ direction and conservation of kinetic energy.
Writing each of these down in terms of the initial and final velocities

$$x - direction\ momentum: \quad v_1 + 0 = v_1' \cos 24.6° + v_2' \cos\theta_2$$
$$y - direction\ momentum: \quad 0 + v_2 = v_1' \sin 24.6° + v_2' \sin\theta_2$$
$$kinetic\ energy: \quad v_1^2 + v_2^2 = v_1'^2 + v_2'^2$$

where we have divided through by the common mass in each equation and by the common factor of ½ in the kinetic energy equation. We recognize this as a set of three equations in three unknowns. This can be solved by moving the $v'_1$ terms on the left side to the right side of each of the first two equations, squaring each equation and adding the two together. This will eliminate angle $\theta_2$ and leave an expression for the final velocity of the second particle squared, which can be substituted into the final equation. The resulting solution is

$$v_1' = 9.17\,\text{m/s} \qquad v_2' = 0.93\,\text{m/s} \qquad \theta_2 = -65.2°$$

### Section 9-8 Center of Mass (CM)

General motion of rigid objects includes rotational motion as well as translational motion. Collections of particles can have very complex motions. However, the center of mass of a rigid body or collection of objects moves translationally according to Newton's second law of motion the same as a body with the total mass of the rigid object or collection of objects subject to the total external force on the rigid body or collection of objects.

The position of the center of mass, $\mathbf{r}_{CM}$, of an extended object is given by

$$\mathbf{r}_{CM} = \frac{\int \mathbf{r}\, dm}{\int dm} = \frac{1}{M}\int \mathbf{r}\, dm$$

where $\mathbf{r}$ is the position of the infinitesimal mass element $dm$ and $M$ is the total mass of the extended object. For a collection of objects, the center of mass is given by

$$\mathbf{r}_{CM} = \frac{\sum m_i \mathbf{r}_i}{\sum m_i}$$

where $\mathbf{r}_i$ is the postion of the $i$th particle of mass $m_i$.

In most cases, when the integral above is calculated it is written as an integral over the spatial extents of the body. The $dm$ is then written as an appropriate density times an appropriate infinitesimal volume. For a one dimensional object

$$dm = \lambda\, dl \quad \text{or} \quad \mathbf{r}_{CM} = \frac{\int \mathbf{r}\,\lambda\, dl}{\int \lambda\, dl} = \frac{1}{M}\int \mathbf{r}\,\lambda\, dl$$

where $\lambda$ is the linear density (mass per length) of the infinitesimal length element $dl$. For a two dimensional object

$$dm = \sigma\, dA \quad \text{or} \quad \mathbf{r}_{CM} = \frac{\int \mathbf{r}\,\sigma\, dA}{\int \sigma\, dA} = \frac{1}{M}\int \mathbf{r}\,\sigma\, dA$$

where $\sigma$ is the area density (mass per area) of the infinitesimal area element $dA$. For a three dimensional object

$$dm = \rho\, dV \quad \text{or} \quad \mathbf{r}_{CM} = \frac{\int \mathbf{r}\,\rho\, dV}{\int \rho\, dV} = \frac{1}{M}\int \mathbf{r}\,\rho\, dV$$

where $\rho$ is the volume density (mass per volume) of the infinitesimal area element $dV$.

### Example 9-8-A

Four masses are located at the corners of a square of side $a$. The masses, starting in the lower left hand corner and proceeding counter-clockwise around the square are 1.00 kg, 2.00 kg, 3.00 kg and 4.00 kg. What is the location of the center of mass?

### Solution:

Using a coordinate system with the origin located on the lower left corner of the square, the positions of the four masses are

$$r_1 = 0i + 0j \quad\quad r_2 = ai + 0j \quad\quad r_3 = ai + aj \quad\quad r_4 = 0i + aj$$

We determine the center of mass by

$$r_{CM} = \frac{\sum m_i r_i}{\sum m_i} = \frac{(1\,\text{kg})(0i + 0i) + (2\,\text{kg})(ai + 0i) + (3\,\text{kg})(ai + ai) + (4\,\text{kg})(0i + ai)}{1\,\text{kg} + 2\,\text{kg} + 3\,\text{kg} + 4\,\text{kg}}$$

$$r_{CM} = 0.50ai + 0.70aj$$

**Example 9-8-B**

A flat square piece of material with sides of length $s$ has an area density given by $\sigma(x,y) = Axy^2$ where x and y are both zero at one corner and increase along perpendicular sides of the square. Determine the position of the center of mass.

**Solution:**

This is a two dimensional object so the center of mass is given by

$$\mathbf{r}_{CM} = \frac{\int \mathbf{r}\, \sigma\, dA}{\int \sigma\, dA} = \frac{\int_{x=0}^{s}\int_{y=0}^{s}(x\mathbf{i}+y\mathbf{j})(Axy^2)dx\,dy}{\int_{x=0}^{s}\int_{y=0}^{s}(Axy^2)dx\,dy} = \frac{\int_{x=0}^{s}\int_{y=0}^{s}(x^2y^2)\mathbf{i}+(xy^3)\mathbf{j}\,dx\,dy}{\int_{x=0}^{s}\int_{y=0}^{s}(xy^2)dx\,dy}$$

$$\mathbf{r}_{CM} = \frac{\left(\frac{x^3}{3}\Big|_0^s \frac{y^3}{3}\Big|_0^s\right)\mathbf{i}+\left(\frac{x^2}{2}\Big|_0^s \frac{y^4}{4}\Big|_0^s\right)\mathbf{j}}{\left(\frac{x^2}{2}\Big|_0^s \frac{y^3}{3}\Big|_0^s\right)} = \frac{\frac{s^6}{9}\mathbf{i}+\frac{s^6}{8}\mathbf{j}}{\frac{s^5}{6}} = \frac{2}{3}s\mathbf{i}+\frac{3}{4}s\mathbf{j}$$

**Section 9-9 Center of Mass and Translational Motion**

As stated in the previous section, Newton's second law describes the motion of the center of mass of an extended object or system of particles if the mass in Newton's second law is the total mass and the force is the sum of all the external forces on the system:

$$\sum F_{ext} = Ma_{CM}$$

where $\mathbf{F}_{ext}$ are the external forces acting on the system, $M$ is the total mass of the system, and $\mathbf{a}_{CM}$ is the acceleration of the center of mass of the system.

**Example 9-9-A**

Two masses are connected by a spring as shown in the diagram. There are no external forces acting on the masses. The masses are at rest. When the masses are released, they move toward each other. Later, the 1.00 kg mass is found at $x = 0$. What is the position of the 2.00 kg mass at this time?

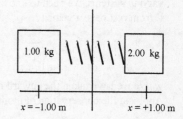

**Solution:**

There is no external force on the system of masses, so there is no acceleration of the center of mass of the system. The masses are initially at rest and so the center of mass must initially be at rest and must remain at rest as the masses move. The original center of mass is given by

$$x_{CM} = \frac{\sum m_i x_i}{\sum m_i} = \frac{(1.00\,\text{kg})(-1.00\,\text{m})+(2.00\,\text{kg})(+1.00\,\text{m})}{1.00\,\text{kg}+2.00\,\text{kg}} = 0.33\,\text{m}$$

The final center of mass must be in the same position

$$0.33\,m = x'_{CM} = \frac{(1.00\,\text{kg})(0.00\,\text{m})+(2.00\,\text{kg})x'_2}{1.00\,\text{kg}+2.00\,\text{kg}}$$

Solving for the final position of the second mass

$$x'_2 = \frac{(0.33\,\text{m})(1.00\,\text{kg}+2.00\,\text{kg})}{2.00\,\text{kg}} = 0.50\,\text{m}$$

## Section 9-10 Systems of Variable Mass; Rocket Propulsion

The law of conservation of momentum can be applied to systems that change mass. To see how changing mass of a system enters into the momentum, we use the product rule to differentiate the momentum expression of a given object in terms of $m$ and $\mathbf{v}$:

$$dP_{object} = d(mv) = v\,dm + m\,dv$$

where $dP$ is the change in the object's momentum, $\mathbf{v}$ is the velocity of the object, $d\mathbf{v}$ is the infinitesimal change in the objects's velocity, $m$ is the mass of the object, and $dm$ is the infinitesimal mass added to the object. This is only the change in momentum of the original object, but the object has now changed. To determine the change in momentum of the entire system, the original momentum of the infinitesimal mass element $dm$ before it was added to the object must be considered. If the infinitesimal mass element $dm$ had a velocity $u$ before being added to the object, its original momentum was $\mathbf{u}\,dm$ so the total change in momentum of the system is

$$dP_{system} = (v - u)dm + m\,dv$$

Applying Newton's second law to this system:

$$\sum F_{ext} = \frac{dP_{system}}{dt} = (v - u)\frac{dm}{dt} + m\frac{dv}{dt}$$

### Example 9-10-A

A beaker of mass M rests on a scale. A liquid of density $\rho$ leaves a tap with negligible velocity at a volume rate of Q from a distance L above the beaker. The liquid is caught in the beaker. What is the scale reading as a function of time?

### Solution:

We can break the force on the scale into three components: the weight of the beaker, the weight of the water that has been collected in the beaker, and the force necessary to stop the falling water. To apply the above equation for mass varying systems, we need to know the speed of the water as it reaches the beaker. We know that objects falling a distance L from rest have a velocity

$$u = -\sqrt{2gL}$$

where we have taken the upward direction as positive. The velocity of the liquid after being stopped in the beaker is zero. Applying the conservation of momentum rule for mass varying systems

$$\sum F_{ext} = -F_{grav} + F_{scale} = -mg + F_{scale} = \frac{dP}{dt} = (v - u)\frac{dm}{dt} + m\frac{dv}{dt}$$

$$\Rightarrow \quad F_{scale} = (v - u)\frac{dm}{dt} + m\frac{dv}{dt} + mg$$

where $m$ is the mass of the beaker and the water in it, $v$ is the velocity of the beaker, u is the velocity of the water just before reaching the beaker. The time derivative of $v$ is zero because the beaker remains at rest. It should be easy to see that $dm/dt$ is equalt to $\rho Q$ and $m$ as a function of time will be $M + \rho Qt$. The resulting expression for the scale reading is

$$F_{scale} = \left(0 - \left(-\sqrt{2gL}\right)\right)\rho Q + 0 + (M + \rho Qt)g = Mg + \rho Q\left(\sqrt{2gL} + gt\right)$$

## Practice Quiz

1.     A car is traveling down a road at a high speed which implies that it has momentum in the direction it is traveling. The car applies its brakes and comes to a rest. The momentum is now zero. If momentum is conserved, what happened to the original momentum of the car?

2.    A bomb is sitting at rest. When it explodes it breaks into two pieces, one being more massive than the other. Which piece will fly out at the highest speed?

3.    What can you say about the relative direction of the two pieces of bomb in the previous question?

4.    A rail transportation system makes many stops and starts. To get the train up to a given speed requires a certain impulse. Explain the trade-offs in creating the impulse over a relatively short time or over a relatively long time in terms of passenger comfort and total travel time between stops.

5.    In a one dimensional collision, applying conservation of energy and conservation of momentum completely determines the motion of two objects after the collision if the motion before the collision is known. Why isn't this true in two dimensions and three dimensions?

6.    Is it possible for the center of mass of an object to lie outside of the object? If so, give an example.

7.    In most of the examples of this chapter, we have been somewhat lax about enforcing the condition that momentum is conserved in a system *only if no net external force acts on the system*. What property of many collisions allows approximate momentum conservation even though a significant force may be acting during the collision?

8.    During a collision with an immovable object, the change in momentum of the colliding object is the same whether the surface is padded or not. Why does less damage occur if the surface is padded?

9.    There are occasionally accidents in Indy style car racing in which a driver strikes the wall at over 200 mph and the car is destroyed, but the driver receives only minor injuries. This is often explained in terms of energy considerations. Explain how the collapse of the car decreases the force on the driver in terms of impulse.

10.   Rubber bullets tend to bounce off of objects they are fired at rather than penetrating into them. Besides the obvious advantage that this may cause less damage than a bullet that penetrates the object that it hits, there is an additional advantage. Explain why a rubber bullet is more effective at knocking down what it hits than a bullet of equal mass and speed that penetrates the object it hits.

11.   A force depends on time as $F(t) = 2 N + (3 N/s) t$. What impulse does this force create during the time interval 0 s to 10 s?

12.   You jump off the surface of the Earth with enough velocity to reach a height of 0.732 m above the surface of the Earth. What was the velocity of the Earth at the moment you stopped being in contact with it?

13.   A bowling ball has a weight of 16 lbs. It is travelling at a speed of 8.42 m/s when it strikes a bowling pin with a weight of 30 ounces. Assuming the collision between the bowling ball and the pin is elastic and the direction of the pin after the collision is the same as the initial direction of the bowling ball, determine the velocity of the bowling pin after the collision.

14.   Determine the location of the center of mass of an object consisting of a uniform bar of length 1.08 m and mass 1.86 kg with a small sphere attached to one end with a mass of 2.95 kg.

15.   A 1000 kg car heading north at a speed of 18.5 m/s collides completely inelastically with an 800 kg car heading west at a speed of 13.8 m/s. What is the resulting velocity of the two cars immediately after the collision?

## Problem Solutions

3.    (a) $p = mv = (0.030 \text{ kg})(12 \text{ m/s}) = 0.36 \text{ kg} \cdot \text{m/s}$.
      (b) The force, opposite the direction of the velocity, changes the momentum:
      $$F = \Delta p/\Delta t;$$
      $$-2.0 \times 10^{-2} \text{ N} = (p_2 - 0.36 \text{ kg} \cdot \text{m/s})/(12 \text{ s}), \text{ which gives } p_2 = 0.12 \text{ kg} \cdot \text{m/s}.$$

7.    (a) We choose downward as positive. For the fall we have
      $$y = y_0 + v_0 t_1 + \tfrac{1}{2}at_1^2;$$
      $$h = 0 + 0 + \tfrac{1}{2}gt_1^2, \text{ which gives } t_1 = (2h/g)^{1/2}.$$
      To reach the same height on the rebound, the upward motion must be a reversal of the downward

motion. Thus the time to rise will be the same, so the total time is

$$t_{total} = 2t_1 = 2(2h/g)^{1/2} = (8h/g)^{1/2}.$$

(b) We find the speed from

$$v = v_0 + at_1 = 0 + g(2h/g)^{1/2} = (2gh)^{1/2}.$$

(c) To reach the same height on the rebound, the upward speed at the floor must be the same as the speed striking the floor. Thus the change in momentum is

$$\Delta p = m(-v) - mv = -2m(2gh)^{1/2} = -(8m^2gh)^{1/2} \text{ (up).}$$

(d) For the average force on the ball we have

$$F = \Delta p/\Delta t = -(8m^2gh)^{1/2}/(8h/g)^{1/2} = -mg \text{ (up).}$$

Thus the average force on the floor is *mg* (down), a surprising result.

11. The new nucleus and the alpha particle will recoil in opposite directions.
Momentum conservation gives us

$$0 = MV - m_\alpha v_\alpha,$$

$$0 = (57m_\alpha)V - m_\alpha(2.5 \times 10^5 \text{ m/s}), \text{ which gives } V = 4.4 \times 10^3 \text{ m/s}.$$

15. Momentum conservation gives

$$0 = m_1 v_1' + m_2 v_2', \text{ or } v_2'/v_1' = -m_1/m_2.$$

The ratio of kinetic energies is

$$K_2/K_1 = \tfrac{1}{2}m_2 v_2'^2 / \tfrac{1}{2}m_1 v_1'^2 = (m_2/m_1)(v_2'/v_1')^2 = 2.$$

When we use the result from momentum, we get

$$(m_2/m_1)(-m_1/m_2)^2 = 2, \text{ which gives } m_1/m_2 = 2.$$

The fragment with the lesser kinetic energy has the greater mass.

19. Because the initial momentum is zero, the momenta of the three products of the decay must add to zero. If we draw the vector diagram, we see that

$$\begin{aligned} p_{nucleus} &= (p_{electron}{}^2 + p_{neutrino}{}^2)^{1/2} \\ &= [(8.6 \times 10^{-23} \text{ kg} \cdot \text{m/s})^2 + (6.2 \times 10^{-23} \text{ kg} \cdot \text{m/s})^2]^{1/2} \\ &= 1.1 \times 10^{-22} \text{ kg} \cdot \text{m/s}. \end{aligned}$$

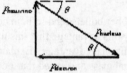

We find the angle from

$$\begin{aligned} \tan \theta &= p_{neutrino}/p_{electron} \\ &= (6.2 \times 10^{-23} \text{ kg} \cdot \text{m/s})/(8.6 \times 10^{-23} \text{ kg} \cdot \text{m/s}) \\ &= 0.721, \text{ so the angle is } 36° \text{ from the direction opposite to the electron's.} \end{aligned}$$

23. We find the average force on the ball from

$$F = \Delta p/\Delta t = m\Delta v/\Delta t = (0.0600 \text{ kg})[(65.0 \text{ m/s}) - 0]/(0.0300 \text{ s}) = 130 \text{ N}.$$

Because the weight of a 60-kg person is ~ 600 N, this force is not large enough.

27. (a) We find the average force on the molecule from

$$\begin{aligned} F &= \Delta p/\Delta t = m\Delta v/\Delta t \\ &= m[(-v) - (+v)]/\Delta t = -2mv/\Delta t. \end{aligned}$$

The average force on the wall is the reaction to this: $2mv/\Delta t$.

(b) If *t* is the average time between collisions, the number of collisions in time *T* is $N = T/t$. Thus in the time *T* the total momentum change is $N(2mv)$, so the average force on the wall is

$$F_{wall} = N(2mv)/T = 2mv/t.$$

31. We choose the upward direction as positive.

(a) If we consider a mass $\Delta m$ of water falling to the pan, we can find the speed just before hitting the pan from energy conservation:

$$0 = \Delta K + \Delta U;$$

$$0 = \tfrac{1}{2}mv^2 - 0 + (0 - mgh), \text{ or } v^2 = 2gh.$$

We find the average force required to stop the water from

$$F = \Delta p/\Delta t = (\Delta m/\Delta t) \Delta v = (\Delta m/\Delta t)[0 - (-v)] = (\Delta m/\Delta t)(2gh)^{1/2}.$$

The average force on the pan is the reaction to this: $-(\Delta m/\Delta t)(2gh)^{1/2}$ (down).

The scale reading is the increased normal force. After a time *t*, the water in the pan has a mass $m = (\Delta m/\Delta t)t$. If the acceleration of the water in the pan is negligible, we have

$$F_N - (\Delta m/\Delta t)(2gh)^{1/2} - mg = 0, \text{ or}$$

$$\begin{aligned} F_N &= (\Delta m/\Delta t)(2gh)^{1/2} + (\Delta m/\Delta t)gt = (\Delta m/\Delta t)[(2gh)^{1/2} + gt] \\ &= (0.12 \text{ kg/s})\{[2(9.80 \text{ m/s}^2)(2.5 \text{ m})]^{1/2} + (9.80 \text{ m/s}^2)t\} = (0.84 \text{ N}) + (1.2 \text{ N/s})t. \end{aligned}$$

(b) After 15 s we have
$$F_N = (0.84 \text{ N}) + (1.18 \text{ N/s})(15 \text{ s}) = 18.5 \text{ N}.$$

(c) After a time $t$, the water in the pan has a mass $m = (\Delta m/\Delta t)t$. The height of this water in the cylinder will be
$$h' = m/\rho A = (\Delta m/\Delta t)t/\rho A$$
$$= (0.12 \text{ kg/s})t/(1.0 \times 10^3 \text{ kg/m}^3)(20 \times 10^{-4} \text{ m}^2) = (6.0 \times 10^{-2} \text{ m/s})t.$$

The water falling into the cylinder at time $t$ will have fallen a distance $h - h'$. Thus the speed of the water will be given by $v^2 = 2g(h - h')$, and the impact force on the pan will be
$$F = -(\Delta m/\Delta t)[2g(h - h')]^{1/2} \quad \text{(down)}.$$
If the acceleration of the water in the pan is negligible, we have
$$F_N - (\Delta m/\Delta t)[2g(h - h')]^{1/2} - mg = 0, \quad \text{or}$$
$$F_N = (\Delta m/\Delta t)[2g(h - h')]^{1/2} + (\Delta m/\Delta t)gt = (\Delta m/\Delta t)\{[2g(h - h')]^{1/2} + gt\}$$
$$= (0.12 \text{ kg/s})(\{2(9.80 \text{ m/s}^2)[(2.5 \text{ m}) - (6.0 \times 10^{-2} \text{ m/s})t]\}^{1/2} + (9.80 \text{ m/s}^2)t)$$
$$= (0.12 \text{ kg/s})\{[(49 \text{ m}^2/\text{s}^2) - (1.12 \text{ m}^2/\text{s}^3)t]^{1/2} + (9.80 \text{ m/s}^2)t\}.$$
After 15 s we have
$$F_N = (0.12 \text{ kg/s})\{(49 \text{ m}^2/\text{s}^2) - (1.12 \text{ m}^2/\text{s}^3)t]\}^{1/2} + (9.80 \text{ m/s}^2)t\}$$
$$= (0.12 \text{ kg/s})\{[(49 \text{ m}^2/\text{s}^2) - (1.12 \text{ m}^2/\text{s}^3)(15 \text{ s})]^{1/2} + (9.80 \text{ m/s}^2)(15 \text{ s})\} = 18.3 \text{ N}.$$

35. (a) For the elastic collision of the two balls, we use momentum conservation for this one-dimensional motion:
$$m_1v_1 + m_2v_2 = m_1v_1' + m_2v_2';$$
$$(0.220 \text{ kg})(6.5 \text{ m/s}) + m_2(0) = (0.220 \text{ kg})(-3.8 \text{ m/s}) + m_2v_2'.$$
Because the collision is elastic, the relative speed does not change:
$$v_1 - v_2 = -(v_1' - v_2'), \quad \text{or} \quad 6.5 \text{ m/s} - 0 = v_2' - (-3.8 \text{ m/s}), \text{ which gives } v_2' = 2.7 \text{ m/s}.$$
(b) Using the result for $v_2'$ in the momentum equation, we get $m_2 = 0.84$ kg.

39. For the elastic collision, we use momentum conservation:
$$m_1v_1 + m_2v_2 = m_1v_1' + m_2v_2';$$
$$m_1v_1 + 0 = m_1v_1' + m_2v_2';$$
Because the collision is elastic, the relative speed does not change:
$$v_1 - v_2 = -(v_1' - v_2'), \quad \text{or} \quad v_1 - 0 = v_2' - v_1'.$$
When we combine the two equations, we get
$$v_1' = (m_1 - m_2)v_1/(m_1 + m_2).$$
The fraction of kinetic energy lost by the neutron is
$$\Delta K_1/K_1 = (\tfrac{1}{2}m_1v_1^2 - \tfrac{1}{2}m_1v_1'^2)/\tfrac{1}{2}m_1v_1^2 = 1 - [(m_1 - m_2)/(m_1 + m_2)]^2 = 4m_1m_2/(m_1 + m_2)^2.$$
(a) For $m_2 = 1.01$ u, we get
$$\Delta K_1/K_1 = 4m_1m_2/(m_1 + m_2)^2 = 4(1.01 \text{ u})(1.01 \text{ u})/(1.01 \text{ u} + 1.01 \text{ u})^2 = 1.00.$$
(b) For $m_2 = 2.01$ u, we get
$$\Delta K_1/K_1 = 4m_1m_2/(m_1 + m_2)^2 = 4(1.01 \text{ u})(2.01 \text{ u})/(1.01 \text{ u} + 2.01 \text{ u})^2 = 0.89.$$
(c) For $m_2 = 12.00$ u, we get
$$\Delta K_1/K_1 = 4m_1m_2/(m_1 + m_2)^2 = 4(1.01 \text{ u})(12.00 \text{ u})/(1.01 \text{ u} + 12.00 \text{ u})^2 = 0.29.$$
(d) For $m_2 = 208$ u, we get
$$\Delta K_1/K_1 = 4m_1m_2/(m_1 + m_2)^2 = 4(1.01 \text{ u})(208 \text{ u})/(1.01 \text{ u} + 208 \text{ u})^2 = 0.019.$$

43. (a) The velocity of the block and projectile after the collision is
$$v' = mv_1/(m + M).$$
The fraction of kinetic energy lost is
$$\text{fraction lost} = -\Delta K/K = -[\tfrac{1}{2}(m + M)v'^2 - \tfrac{1}{2}mv_1^2]/\tfrac{1}{2}mv_1^2$$
$$= -\{(m + M)[mv_1/(m + M)]^2 - mv_1^2\}/mv_1^2$$
$$= -[m/(m + M)] + 1 = +M/(m + M).$$
(b) For the data given we have
$$\text{fraction lost} = M/(m + M) = (380 \text{ g})/(14.0 \text{ g} + 380 \text{ g}) = 0.964.$$

47. (a) For a perfectly elastic collision, we use momentum conservation:
$$m_1v_1 + m_2v_2 = m_1v_1' + m_2v_2', \quad \text{or} \quad m_1(v_1 - v_1') = m_2(v_2' - v_2).$$
Kinetic energy is conserved, so we have
$$\tfrac{1}{2}m_1v_1^2 + \tfrac{1}{2}m_2v_2^2 = \tfrac{1}{2}m_1v_1'^2 + \tfrac{1}{2}m_2v_2'^2, \quad \text{or} \quad m_1(v_1^2 - v_1'^2) = m_2(v_2'^2 - v_2^2),$$
which can be written as
$$m_1(v_1 - v_1')(v_1 + v_1') = m_2(v_2' - v_2)(v_2' + v_2).$$

114    Giancoli, *Physics for Scientists & Engineers*: Study Guide

When we divide this by the momentum result, we get

$v_1 + v_1' = v_2' + v_2$, or $v_1' - v_2' = v_2 - v_1$.

If we use this in the definition of the coefficient of restitution, we get

$e = (v_1' - v_2')/(v_2 - v_1) = (v_2 - v_1)/(v_2 - v_1) = 1$.

For a completely inelastic collision, the two objects move together, so we have

$v_1' = v_2'$, which gives $e = 0$.

(b)  We find the speed after falling a height $h$ from energy conservation:

$\tfrac{1}{2}mv_1^2 = mgh$, or $v_1 = (2gh)^{1/2}$.

The same expression holds for the height reached by an object moving upward:

$v_1' = (2gh')^{1/2}$.

Because the steel plate does not move, when we take into account the directions we have

$e = (v_1' - v_2')/(v_2 - v_1) = [(2gh')^{1/2} - 0]/\{0 - [-(2gh)^{1/2}]\}$, so $e = (h'/h)^{1/2}$.

51.  For the collision we use momentum conservation:

x-direction:  $m_1v_1 + 0 = (m_1 + m_2)v' \cos\theta$,

(3.3 kg)(7.8 m/s) = (3.3 kg + 4.6 kg)$v'$ cos $\theta$, which gives

$v'$ cos $\theta = 3.26$ m/s.

y-direction:  $0 + m_2v_2 = (m_1 + m_2)v' \sin\theta$,

(4.6 kg)(10.2 m/s) = (3.3 kg + 4.6 kg)$v'$ sin $\theta$, which gives

$v'$ sin $\theta = 5.94$ m/s.

We find the direction by dividing the equations:

tan $\theta = (5.94$ m/s)/(3.26 m/s) = 1.82,

so $\theta = 61°$ from first eagle's direction.

We find the magnitude by squaring and adding the equations:

$v' = [(5.94$ m/s$)^2 + (3.26$ m/s$)^2]^{1/2} = 6.8$ m/s.

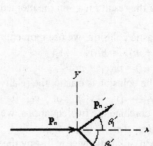

55.  Using the coordinate system shown, for momentum conservation we have

x:   $m_nv_n + 0 = m_nv_n' \cos\theta_1' + m_{He}v_{He}' \cos\theta_2'$;

$m_n(6.2 \times 10^5$ m/s$) = m_nv_n' \cos\theta_1' + 4m_nv_{He}' \cos 45°$, or

$v_n' \cos\theta_1' = (6.2 \times 10^5$ m/s$) - 4v_{He}' \cos 45°$.

y:   $0 + 0 = -m_nv_n' \sin\theta_1' + m_{He}v_{He}' \sin\theta_2'$;

$0 = -m_nv_n' \sin\theta_1' + 4m_nv_{He}' \sin 45°$, or

$v_n' \sin\theta_1' = 4v_{He}' \sin 45°$.

For the conservation of kinetic energy, we have

$\tfrac{1}{2}m_nv_n^2 + 0 = \tfrac{1}{2}m_nv_n'^2 + \tfrac{1}{2}m_{He}v_{He}'^2$;

$m_n(6.2 \times 10^5$ m/s$)^2 = m_nv_n'^2 + 4m_nv_{He}'^2$, or

$v_n'^2 + 4v_{He}'^2 = 3.84 \times 10^{11}$ m$^2$/s$^2$.

We have three equations in three unknowns: $\theta_1'$, $v_n'$, $v_{He}'$. We eliminate $\theta_1'$ by squaring and adding the two momentum results, and then combine this with the energy equation, with the results:

$\theta_1' = 76°$, $v_n' = 5.1 \times 10^5$ m/s, $v_{He}' = 1.8 \times 10^5$ m/s.

59.  We choose the origin at the carbon atom.  The center of mass will lie along the line joining the atoms:

$x_{CM} = (m_Cx_C + m_Ox_O)/(m_C + m_O)$

= $[0 + (16$ u$)(1.13 \times 10^{-10}$ m$)]/(12$ u $+ 16$ u$) = 6.5 \times 10^{-11}$ m  from the carbon atom.

63.  We choose the origin at the center of the raft, which is the center of mass of the raft:

$x_{CM} = (Mx_{raft} + m_1x_1 + m_2x_2 + m_3x_3)/(M + m_1 + m_2 + m_3)$

= $[0 + (1200$ kg$)(9.0$ m$) + (1200$ kg$)(9.0$ m$) +$

$(1200$ kg$)(-9.0$ m$)]/[6200$ kg $+ 3(1200$ kg$)]$

= 1.10 m (east).

$y_{CM} = (My_{raft} + m_1y_1 + m_2y_2 + m_3y_3)/(M + m_1 + m_2 + m_3)$

= $[0 + (1200$ kg$)(9.0$ m$) + (1200$ kg$)(-9.0$ m$) +$

$(1200$ kg$)(-9.0$ m$)]/[6200$ kg $+ 3(1200$ kg$)]$

= $-1.10$ m (south).

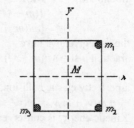

67. We know from the symmetry that the center of mass
lies on the $z$-axis: $x_{CM} = 0$, $y_{CM} = 0$.
For a differential element we use a circle at a
height $z$, thickness $dz$, and radius $r = (R/h)z$.
If $\rho$ is the mass density of the pyramid, the mass of
this element is $dm = \rho\Delta r^2\,dz$.
We integrate over the cone to find $z_{CM}$:

$$z_{CM} = \frac{\int z\,dm}{\int dm} = \frac{\int_0^h \rho z\pi r^2\,dz}{\int_0^h \rho\pi r^2\,dz} = \frac{\int_0^h (R/h)^2 z^3\,dz}{\int_0^h (R/h)^2 z^2\,dz} = \frac{h^4/4}{h^3/3} = \frac{3h}{4}$$

Thus the center of mass is at $x_{CM} = 0$, $y_{CM} = 0$, $z_{CM} = 3h/4$ above the point.

71. We choose the origin of our coordinate system at the man.
   (a) For their center of mass we have
   $$x_{CM} = (m_{woman}x_{woman} + m_{man}x_{man})/(m_{woman} + m_{man})$$
   $$= [(50\text{ kg})(11.0\text{ m}) + 0]/(50\text{ kg} + 70\text{ kg})$$
   $$= 4.6\text{ m}.$$
   (b) Because the center of mass will not move, we find the location of the woman from
   $$x_{CM} = (m_{woman}x_{woman}' + m_{man}x_{man}')/(m_{woman} + m_{man})$$
   $$4.6\text{ m} = [(50\text{ kg})x_{woman}' + (70\text{ kg})(2.8\text{ m})]/(50\text{ kg} + 70\text{ kg}),\text{ which gives}$$
   $$x_{woman}' = 7.1\text{ m}.$$
   The separation of the two will be 7.1 m – 2.8 m = 4.3 m.
   (c) The two will meet at the center of mass, so he will have moved 4.6 m.

75. We find the time for the man to reach the other end from
   $$t = L/v_{rel} = (25\text{ m})/(2.0\text{ m/s}) = 12.5\text{ s}.$$
   If we let $v_2$ be the speed of the flatcar while the man is walking, the speed of the man is $v_1 = v_2 + v_{rel}$.
   Because the velocity of the center of mass of the system of flatcar and man does not change, we have
   $$v_{CM} = (m_1v_1 + m_2v_2)/(m_1 + m_2)$$
   $$5.0\text{ m/s} = [(90\text{ kg})(v_2 + 2.0\text{ m/s}) + (200\text{ kg})v_2]/(90\text{ kg} + 200\text{ kg}),\text{ which gives }v_2 = 4.38\text{ m/s}.$$
   The flatcat will have moved
   $$x = v_2t = (4.38\text{ m/s})(12.5\text{ s}) = 55\text{ m}.$$

79. (a) Because chemical changes do not change the mass, the thrust from the ejected fuel is
   $$F_{fuel} = v_{rel}\,dM_{fuel}/dt = (-550\text{ m/s})(-4.2\text{ kg/s}) = 2.3\times 10^3\text{ N}.$$
   (b) The air is collected by the airplane at the speed of the airplane and ejected with the speed of the
   fuel, so the net thrust is
   $$F_{air} = v_{rel1}\,dM_{air}/dt + v_{rel2}\,dM_{air}/dt$$
   $$= (-270\text{ m/s})(100\text{ kg/s}) + (-550\text{ m/s})(-100\text{ kg/s}) = 2.8\times 10^4\text{ N}.$$
   (c) The power delivered by the two thrusts is
   $$P = (F_{fuel} + F_{air})v = (2.3\times 10^3\text{ N} + 2.8\times 10^4\text{ N})(270\text{ m/s})/(746\text{ W/hp}) = 1.1\times 10^4\text{ hp}.$$

83. We find the speed after being hit from the height $h$ using energy
conservation:
   $$\tfrac{1}{2}mv'^2 = mgh,\text{ or }v' = (2gh)^{1/2} = [2(9.80\text{ m/s}^2)(55.6\text{ m})]^{1/2} = 33.0\text{ m/s}.$$
   We see from the diagram that the magnitude of the change in momentum is
   $$\Delta p = m(v^2 + v'^2)^{1/2}$$
   $$= (0.145\text{ kg})[(35.0\text{ m/s})^2 + (33.0\text{ m/s})^2]^{1/2} = 6.98\text{ kg}\cdot\text{m/s}.$$

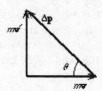

   We find the force from
   $$F\,\Delta t = \Delta p;$$
   $$F(0.50\times 10^{-3}\text{ s}) = 6.98\text{ kg}\cdot\text{m/s},\text{ which gives }F = 1.4\times 10^4\text{ N}.$$
   We find the direction of the force from
   $$\tan\theta = v'/v = (33.0\text{ m/s})/(35.0\text{ m/s}) = 0.943,\quad \theta = 43.3°.$$

87. For the elastic collision of the two balls, we use momentum conservation:

$mv_1 + m_2v_2 = mv_1' + m_2v_2'$;

$mv_1 + 0 = m(-0.600v_1) + m_2v_2'$, or $m_2v_2' = 1.600mv_1$.

Because the collision is elastic, the relative speed does not change:

$v_1 - 0 = -(v_1' - v_2')$; $v_1 = v_2' - (-0.600v_1)$, or $v_2' = 0.400v_1$.

Combining these two equations, we get

$m_2 = 4.00m$.

91. (a) No. The spring exerts equal but opposite forces on the blocks. For the system these forces are internal forces.

(b) Because there are no external forces on the system of the two blocks, momentum is conserved:

$0 = m_1v_1 + m_2v_2$, which gives $v_1/v_2 = -m_2/m_1$.

(c) For the ratio of kinetic energies, we have

$K_1/K_2 = \frac{1}{2}m_1v_1^2/\frac{1}{2}m_2v_2^2 = (m_1/m_2)(v_1/v_2)^2 = (m_1/m_2)(-m_2/m_1)^2 = m_2/m_1$.

(d) The center of mass is initially at rest. Because there are no external forces on the system of the two blocks, the center of mass does not move.

(e) The two friction forces will be in opposite directions, but they need not be equal. Thus there will be a net external force on the system. Momentum is not conserved, and the center of mass will move.

95. (a) We take the direction of the meteor for the positive direction.

For this perfectly inelastic collision, we use momentum conservation:

$M_{meteor}v_{meteor} + M_{Earth}v_{Earth} = (M_{meteor} + M_{Earth})V$;

$(10^8 \text{ kg})(15 \times 10^3 \text{ m/s}) + 0 = (10^8 \text{ kg} + 6.0 \times 10^{24} \text{ kg})V$, which gives $V = 2.5 \times 10^{-13}$ m/s.

(b) The fraction transformed was

fraction $= \Delta K_{Earth}/K_{meteor} = \frac{1}{2}m_{Earth}V^2/\frac{1}{2}m_{meteor}v_{meteor}^2$

$= (6.0 \times 10^{24} \text{ kg})(2.5 \times 10^{-13} \text{ m/s})^2/(10^8 \text{ kg})(15 \times 10^3 \text{ m/s})^2 = 1.7 \times 10^{-17}$.

(c) The change in the Earth's kinetic energy was

$\Delta K_{Earth} = \frac{1}{2}m_{Earth}V^2$

$= \frac{1}{2}(6.0 \times 10^{24} \text{ kg})(2.5 \times 10^{-13} \text{ m/s})^2 = 0.19$ J.

99. Obviously the spacecraft will have negligible effect on the motion of Saturn. In the reference frame of Saturn, we can treat this as the equivalent of a small mass "bouncing off" a massive object. The relative velocity of the spacecraft in this reference frame will be reversed.

The initial relative velocity of the spacecraft is

$v_{SpS} = v_{Sp} - v_S = 10.4$ km/s $- (-9.6$ km/s$) = 20.0$ km/s.

so the final relative velocity is $v_{SpS}' = -20.0$ km/s. Therefore, we find the final velocity of the spacecraft from

$v_{SpS}' = v_{Sp}' - v_S$;

$-20.0$ km/s $= v_{Sp}' - (-9.6$ km/s$)$, which gives $v_{Sp}' = -29.6$ km/s,

so the final speed of the spacecraft is 29.6 km/s.

103. (a) There is an obvious loss of kinetic energy, so this is an inelastic collision.

(b) If we assume a constant acceleration, we can find the time from

$x = \frac{1}{2}(v_0 + v)t$;

$0.70$ m $= \frac{1}{2}\{[(50 \text{ km/h})/(3.6 \text{ ks/h}) + 0\}t$, which gives $t = 0.10$ s.

(c) We find the average impulsive force from

$F = \Delta p/\Delta t = [m(v - v_0)]/\Delta t$

$= (1000 \text{ kg})[0 - (50 \text{ km/h})/(3.6 \text{ ks/h})]/(0.10 \text{ s}) = -1.4 \times 10^5$ N.

# Chapter 10: Rotational Motion About a Fixed Axis

## Chapter Overview and Objectives

This chapter describes the kinematics and dynamics of rotational motion about a fixed direction axis. It defines the terms angular velocity, angular acceleration, torque, moment of inertia, and angular momentum. It shows how Newton's laws of motion, the law of conservation of momentum, and the work-energy theorem can be applied to rotational motion about fixed direction axes.

After completing study of this chapter, you should:

- Know the definition of a radian of angle measure.
- Know the definitions of angular velocity and angular acceleration.
- Know how to apply the constant angular accelertion kinematical relationships to problems.
- Know how translational kinematic quantities of a point on a rotating object are related to the rotational kinematical quantities of the object.
- Know the condition for rolling without slipping.
- Know the definition of torque and angular momentum.
- Be able to apply Newton's second law to rotational dynamic problems.
- Know the definition of angular momentum.
- Be able to apply the law of conservation of angular momentum to problems.
- Know the expressions for work done by torque and rotational kinetic energy.
- Be able to apply the work-energy theorem to problems with rotational and translational motion .
- Know the dimensions and SI units of the quantities introduced in this chapter.

## Summary of Equations

Definition of angle in radians:
$$\theta = \frac{l}{R}$$

Definition of average angular velocity:
$$\overline{\omega} = \frac{\Delta\theta}{\Delta t}$$

Definition of instantaneous angular velocity:
$$\omega = \frac{d\theta}{dt}$$

Definition of average angular acceleration:
$$\overline{\alpha} = \frac{\Delta\omega}{\Delta t}$$

Definition of instantaneous angular acceleration:
$$\alpha = \frac{d\omega}{dt}$$

Relationship between angular velocity and velocity of point on object:
$$v = R\omega$$

Relationship between angular acceleration and tangential acceleration of point on object:
$$a = \frac{dv}{dt} = R\frac{d\omega}{dt} = R\alpha$$

Translational acceleration of point on rotating object: $a = -\dfrac{v^2}{r}r + R\alpha t$

Relationship between frequency and angular frequency:

$$f = \frac{\omega}{2\pi} \quad or \quad \omega = 2\pi f$$

Relationship between period and frequency:    $T = \frac{1}{f}$

Relationships between angular kinematical quantities for constant angular acceleration:

$$\omega = \omega_0 + \alpha t \qquad\qquad \theta = \omega_0 t + \alpha t^2$$

$$\omega^2 = \omega_0^2 + 2\alpha\theta \qquad\qquad \overline{\omega} = \frac{\omega + \omega_0}{2}$$

Condition for rolling without slipping:    $v_{axis} = \omega R_{contact}$

Definition of torque:    $\tau = R_\perp F = RF_\perp = RF\sin\theta$

Definition of moment of inertia:    $I = \sum m_i R_i^2$

Newton's second law for rotation:    $\sum \tau = I\alpha$

Moment of inertia of continuously-distributed mass:    $I = \int R^2 \, dm = \int R^2 \rho \, dV$

Parallel axis theorem:    $I = I_{CM} + Mh^2$

Perpendicular axis theorem:    $I_z = I_x + I_y$

Definition of angular momentum:    $L = I\omega$

Newton's second law written in terms of angular momentum:

$$\sum \tau = \frac{dL}{dt}$$

Rotational kinetic energy:    $K = \frac{1}{2} I\omega^2$

Work done by torque:    $W = \int \tau \, d\theta$

Work-energy theorem for rotational motion:    $\int \tau \, d\theta = \frac{1}{2} I\omega^2 - \frac{1}{2} I\omega_0^2$

Total kinetic energy of a rigid body:    $K = \frac{1}{2} mv_{CM}^2 + \frac{1}{2} I_{CM} \omega^2$

# Chapter Summary

### Section 10-1 Angular Quantities

The most simple mathematical expressions involving the relationship between angle measure and linear distance result from using **radians** as the unit for angle measure. One radian is the size of the angle that has a one unit length arc length at a radius of unit length. In other words, the angle measure in radians is the arc length divided by the radius:

$$\theta = \frac{l}{R}$$

where $\theta$ is the angle measure in radians, $l$ is the arclength, and $R$ is the radius of the arc. We will usually adopt the convention that angles are measured as positive counter-clockwise and negative clockwise.

Because there are $2\pi$ radians of arc and $360°$ of arc in a complete circle, the relationship between angle measure in radians and angle measure in degrees is

$$360° = 2\pi \text{ radians}$$

**Example 10-1-A**

How many radians correspond to one minute of arc? One second of arc?

**Solution:**

One minute of arc is 1/60 of a degree and one second of arc is 1/3600 degrees. Thus

$$1' = \left(\frac{1}{60}\deg\right)\frac{2\pi \text{ radians}}{360 \deg} = 2.91 \times 10^{-4} \text{ radians}$$

and

$$1'' = \left(\frac{1}{3600}\deg\right)\frac{2\pi \text{ radians}}{360 \deg} = 4.85 \times 10^{-6} \text{ radians}$$

The quantities that are used to describe rotational motion are analogous to the quantities used to describe translational motion. The rate of change of angular position is called angular velocity. Average angular velocity is

$$\overline{\omega} = \frac{\Delta\theta}{\Delta t}$$

where $\overline{\omega}$ is the average angular velocity, $\Delta\theta$ is the change in angular position which occurred in time $\Delta t$. The instantaneous angular velocity, $\omega$, is the limit of the average angular velocity as $\Delta t$ goes to zero. That is, the instantaneous angular velocity is the derivative of $\theta$ with respect to time,

$$\omega = \frac{d\theta}{dt}$$

Note that $\omega$ includes information about direction. An increasing angle with time results in a positive angular velocity and a decreasing angle with time results in a negative velocity, so the sign of the angular velocity has information about the direction of rotation of the object. If the standard convention for counter-clockwise angles being positive is adopted, then the rotation is counter-clockwise if the angular velocity is positive and clockwise if the angular velocity is negative.

The average angular acceleration, $\overline{\alpha}$, is defined to be

$$\overline{\alpha} = \frac{\Delta\omega}{\Delta t}$$

where $\Delta\omega$ is the change in angular velocity during the time $\Delta t$. The instantaneous angular acceleration is the limit of the average angular acceleration as the time $\Delta t$ goes to zero:

$$\alpha = \frac{d\omega}{dt}$$

Again, the sign of the angular acceleration is information about direction of the angular acceleration.

Any point on a rotating body has a velocity relative to the fixed axis that is tangential to a circle centered on the axis.  The speed, $v$, of a given point depends on the distance, $R$, that it is from the axis:

$$v = R\omega$$

Note that in this relationship the units of the result of the calculation appear to be m·radians/s if the radius is in meters and the angular velocity is in radians/s.  However, we know velocity should have units of m/s, not m·radians/s.  What happens to the radians?  The radians is just dropped because if we look at how the angle is calculated, we divided meters by meters to get the angle.  It is a dimensionless quantity.  This will be true for any dimensional quantity we calculate from an angular quantity; we always drop the radians from the units.

The acceleration of a point on a rotating body has a centripetal acceleration, as learned in Chapter 3, equal to $v^2/R$.  The point can also have an acceleration component in the tangential direction if the rotating object has an angular acceleration.  If the time derivative of the tangential velocity expression above is taken, we get

$$a = \frac{dv}{dt} = R\frac{d\omega}{dt} = R\alpha$$

The total acceleration of a point on the rotating object is the vector sum of the inward radial (centripetal) and tangential acceleration:

$$a = -\frac{v^2}{r}r + R\alpha\, t$$

where $r$ is a unit vector pointing from the axis of rotation to the point on the object and $t$ is a unit vector pointing tangent to the circle that the point moves on in the direction of positive angle measure.

The quantity **frequency**, $f$, of rotation is a measure of the number of revolutions per time that the object makes.  It is easy to see that frequency of rotation is related to angular velocity in radians per time by

$$f = \frac{\omega}{2\pi} \quad or \quad \omega = 2\pi f$$

The units for frequency are $s^{-1}$.  One $s^{-1}$ is also called a hertz (Hz).

The time required for one revolution of the object, called the **period**, $T$, of revolution.  The relationship between the frequency and the period is

$$T = \frac{1}{f}.$$

### Example 10-1-B

A bicyclist accelerates from rest to a speed of 5.38 m/s in a time of 8.93 s.  The diameter of his bicycle's tires is 27 inches.  What is the angular velocity of the bicycle wheels when he is moving at his final speed?  What was the average angular acceleration of the bicycle wheels?

### Solution:

The tires must rotate once for every circumference of the tires the bicycle moves forward (we assume the tires are not slipping).  The circumference of the tire is $C = \pi D$, where $D$ is the diameter of the tire.  One revolution is $2\pi$ radians, so the angular velocity of the tire will be

$$\omega = 2\pi \frac{v}{C} = 2\pi \frac{v}{\pi D} = \frac{2v}{D} = \frac{2(5.38\,\text{m/s})}{(27\,\text{in})(0.0254\,\text{m/in})} = 15.7\,\text{rad/s}$$

To determine the average angular acceleration, we divide the change in angular velocity by the time interval

$$\bar{\alpha} = \frac{\omega - \omega_0}{t} = \frac{15.7 \,\text{rad/s}}{8.93 \,\text{s}} = 1.76 \,\text{rad/s}^2$$

### Section 10-2 Kinematic Equations for Uniformly Accelerated Rotational Motion

Analogous to the kinematical equations for Chapter 2 are the kinematical equations for rotation about a fixed axis. Because the mathematical relationships between angular position, angular velocity, and angular acceleration is identical to the relationships between position, velocity, and acceleration, the kinematical equations have identical form for rotational quantities as for translational quantities.

For the case of constant angular or translational acceleration:

Rotation:

$$\omega = \omega_0 + \alpha t$$

$$\theta = \omega_0 t + \alpha t^2$$

$$\omega^2 = \omega_0^2 + 2\alpha\theta$$

$$\bar{\omega} = \frac{\omega + \omega_0}{2}$$

Translation:

$$v = v_0 + at$$

$$x = v_0 t + at^2$$

$$v^2 = v_0^2 + 2ax$$

$$\bar{v} = \frac{v + v_0}{2}$$

### Example 10-2-A

A winch is used to wind up cable. It is desired to wind up an additional 13.4 m of cable onto the drum of diameter 0.364 m. The current angular velocity of the drum is 10.4 rev/min. What must the constant angular acceleration of the drum be so that the drum comes to rest when the additional 13.4 m of cable is wound on the drum? Assume the additional cable on the drum does not change its diameter.

### Solution:

From the information given of the additional length of cable to wind up and the diameter of the drum, we can determine the angle that the drum must turn through

$$\theta = \frac{l}{r} = \frac{l}{D/2} = \frac{13.4 \,\text{m}}{0.364 \,\text{m} / 2} = 73.63 \,\text{rad}$$

The initial angular velocity and final angular velocity are given and we are trying to find the angular acceleration. We recognize that the equation

$$\omega^2 = \omega_0^2 + 2\alpha\theta$$

relates the known quantities to the quantity which we are to determine. Solving this for the angular acceleration

$$\alpha = \frac{\omega^2 - \omega_0^2}{2\theta} = \frac{\left[(10.4 \,\text{rev/min})(2\pi \,\text{rad/rev})(1 \,\text{min}/ \, 60 \,\text{s})\right]^2 - 0}{2(73.63 \,\text{rad})} = 8.05 \times 10^{-3} \,\text{rad/s}^2$$

### Section 10-3 Rolling Motion (without slipping)

Although a rolling object does not have a fixed axis in a frame of reference fixed to the surface it is rolling on, it does have a fixed axis in a frame of reference that is comoving with the axis of rotation. In the reference frame comoving with the axis of rotation, the surface that the object is rolling on is moving. If the object is rolling without slipping, the velocity of the surface must be the same as the velocity of the point in contact with the surface. This can be taken to define rolling without slipping in a mathematical way.

If we now change to the reference frame fixed to the surface the object is rolling on, we observe the axis of rotation of the object moving with velocity with an equal speed but opposite direction as the surface was moving relative to an observer fixed on the axis of rotation. This means that for an object rolling without slipping the speed of the axis of rotation of the object relative to the surface it is rolling on must be

$$v_{axis} = \omega R_{contact}$$

In determining the motion of any point on the rolling object, we can apply the relationships of section 1 and add the velocity of the axis to the velocity about the fixed axis of the given point on the object.

### Example 10-3-A

A spool of thread with outer radius $R_2$ rolls without slipping along a ramp that contacts the spool at radius $R_1$ as shown in the diagram. The person pulling the thread is pulling the thread at a constant speed $v_T$. What is the angular velocity of the spool?

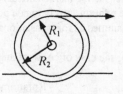

### Solution:

The thread that comes off the top of the spool must be moving at twice the speed of the center of the spool, so that $v_C = \frac{1}{2} v_T$, where $v_C$ is the speed of the center of the spool. To see this, consider what happens when the spool turns through one revolution. In one revolution the center of the spool advances a distance eqaul to the circumference of the spool. The length of the thread off the spool has also increased by one spool circumference. This means that the far end of the thread moves ahead by two spool circumferences for one revolution of the thread or twice as far as the center of the spool moves. This implies that the speed of the thread is twice the speed of the center of the spool.

Now we apply the rolling without slipping condition to this situation

$$v_C = \omega R_{contact} \quad \Rightarrow \quad \omega = \frac{v_C}{R} = \frac{v_T}{2R}$$

### Section 10-4 Vector Nature of Angular Quantities

It might seem to you that there is a directional nature to rotation; the axis has a direction to it and certainly rotation of an object about a vertical axis is very different motion that rotation about a horizontal axis. In fact, we can define a vector angular velocity and vector angular acceleration by defining the direction of these vectors along the axis of rotation with corresponding magnitudes defined in section 1. However, that definition is not complete as there are two opposite directions associated with each axis of rotation. An arbitrary choice between the two directions must be made and we must be consistent in choosing the direction. The direction has been chosen to be the direction defined by the **right-hand rule**. To determine the direction of the angular velocity vector, curl up the fingers of your right hand with the thumb sticking out away from your hand. Orient the curling fingers in the same direction the object rotates. Your thumb will point in the direction of the angular velocity vector.

Mathematically, the angular velocity vector defined in this manner is not a vector. Vectors are defined by how they transform under changes in the coordinate system. One of the properties vectors have is that if all the coordinate directions are reversed, the vector is unchanged. If all coordinate directions are reversed, then right and left handedness are reversed. This would result in the angular velocity vector being defined in the opposite direction so the definition is not invariant under coordinate reversal. It is not a vector. It is what is called an axial vector or pseudovector. However, for our purposes, the angular velocity and angular acceleration properties that we use can be treated the same as vector properties.

### Section 10-5 Torque

To cause a body to change its rotational motion requires a force. However, the point of application of the force and its direction are important in determining how the force changes the rotational motion. The perpendicular distance of the rotation axis to the line the force acts along is called the **moment arm**. The torque,$\tau$, is defined to be the product of the moment arm times the force:

$$\tau = R_\perp F$$

where $R_\perp$ is the moment arm of the force of magnitude $F$. An equivalent way of calcualting the torque is to write it as

$$\tau = RF_\perp$$

where $R$ is the distance from the axis of rotation to the point of application of the force on the object and $F_\perp$ is the component of the force perpendicular to the line from the axis of rotation to the point of contact of the force on the body. Another way to calculate the torque is

$$\tau = RF \sin \theta$$

where $R$ and $F$ are as given above and $\theta$ is the angle between the line joining the axis of rotation and the point of application of the force with the direction of the force.

The angular acceleration of the object is proportional to the torque:

$$\alpha \propto \tau$$

### Example 10-5-A

A wrench is used to loosen a bolt as shown in the diagram. The total frictional torque is equivalent to a 1.89 kN force acting with a moment arm of the radius of the bolt, 0.00432 m. The person loosening the bolt is pushing with a 132 N force in the direction shown in the diagram. The total torque is zero on the bolt. The bolt is moving at constant angular velocity. What is the distance from the axis of rotation to the point at which the person is applying the force to the wrench?

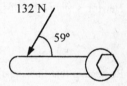

### Solution:

The sum of the torques acting on the bolt is zero. Clearly, from the diagram, the torque of the frictional force acts in a clockwise direction and the torque due to the person pushing on the wrench acts in the counterclockwise direction. Consistent with the right hand rule discussed in Section 4, the clockwise torque vector points into the page and the counterclockwise torque points out of the page. We choose to represent the direction out of the page as the positive direction and the direction into the page as the negative direction. The torque due to friction on the bolt is

$$\tau_{friction} = -R_\perp F = -(0.00432 \text{ m})(1.89 \times 10^3 \text{ N}) = -8.16 \text{ N} \cdot \text{m}$$

where the minus sign comes from the fact that the torque is clockwise.

The torque due to the applied force on the wrench is given by

$$\tau_{wrench} = +RF_\perp = +R(132 \text{ N}) \sin 59°$$

where the positive sign comes from the fact that the torque is counterclockwise.

The total torque sums to zero,

$$\tau = \tau_{friction} + \tau_{wrench} = -8.16 \text{ N} \cdot \text{m} + (132 \text{ N}) \sin 59° \, R = 0$$

Solving this equation for $R$:

$$R = \frac{8.16 \text{ N} \cdot \text{m}}{(132 \text{ N}) \sin 59°} = 0.072 \text{ cm}$$

## Section 10-6 Rotational Dynamics; Torque and Rotational Inertia

The previous equation needs a constant of proportionality to be complete.  Just as in the case of translational motion, the constant is a measure of the inertia for the corresponding motion. In the case of rotational motion, the corresponding inertial property depends on more than the mass of the object; it also depends on the distribution of mass.  This is easily seen by looking at the acceleration of the different points on the body for a given angular acceleration.  Points near the access of rotation will have a smaller linear acceleration than points far from the access of rotation.  This means a given torque on the body can accelerate a given amount of mass easily if it is near the axis of rotation, but will have a more difficult time acccelerating the same mass located far from the axis of rotation.  The appropriate measurement of the inertia of a body is called the **moment of inertia**, $I$, of the object about the axis of rotation.  It depends on how the mass of the object is distributed, but the *moment of inertia also depends on the position and direction of the axis of rotation*. It is incorrect to state that an object has a moment of inertia; it *does* have a moment of inertia about a given axis.   The moment of inertia about a fixed axis for a collection of discrete masses is equal to

$$I = \sum m_i R_i^2$$

where $m_i$ is the mass of the ith particle located a distance $R_i$ from the axis of rotation.  The sum is over each of the discrete masses that make up the object.

Newton's second law for rotation of an object is

$$\sum \tau = I\alpha$$

Remember that this equation is valid for a fixed axis of rotation.  The equation can be applied to a moving axis of rotation if the axis of rotation is through the center of mass of the body and it does not change direction in space during the motion.  Then

$$\left(\sum \tau\right)_{CM} = I_{CM}\alpha_{CM}$$

where the $CM$ subscript means about an axis through the center of mass.

## Section 10-7 Solving Problems in Rotational Dynamics

Remember the following general approach to rotational dynamics problems outlined in the text:

1.  As in any problem, draw a clear and complete diagram of the situation.  Identify the known and unknown quantities in the problem.

2.  Draw a free-body diagram for each object in the problem.  In rotational dynamics problems it is important to place the point of action of the force on the body correctly so that the torque is correctly calculated.

3.  Identify the axis of rotation of the problem.  Remember that we have adopted the convention that counter-clockwise rotation is positive and clockwise rotation is negative.  Similarly, any torque that tends to rotate the object counter-clockwise is positive and any torque that tends to rotate the object clockwise is negative.

4.  Apply Newton's second law of motion for rotation, $\Sigma\tau = I\alpha$.  Remember to express all quantities in a consistent set of units.

5.  Apply Newton's second law for translational motion also, if necessary.

6.  Solve the equations of motion for the unknown quantity.

7.  Check you answer by making order of magnitude estimates and asking yourself if the answer is sensible or not.

**Example 10-7-A**

A spool of rope is attached to a block of mass 6.45 kg that is free to slide down a 32° incline as shown in the diagram. The spool is on a fixed axis, its moment of inertia is 3.45 kg·m², and it has a radius of 0.187 m. What is the acceleration of the block down the plane.?

**Solution:**

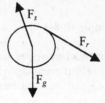

Draw a free-body diagram for the spool of rope where **F**r is the force of the rope on the spool, **F**g is the force of gravity on the spool, and **F**s is the force of the supports of the spool acting on the spool. The spool is fixed in position so the total force is zero. The only force that has a non-zero torque about the axis of the spool is the force of the rope. The other two forces have zero moment arms about the center of the spool. Newton's second law for rotation can be written as

$$\sum \tau = I\alpha \;\;\Rightarrow RF_r = I\alpha$$

where $R$ is the radius of the spool. The free-body diagram of the block sliding down the plane is also drawn where $\mathbf{F}_{rb}$ is the force of the rope on the block, $\mathbf{F}_N$ is the normal force of the plane on the block, and $\mathbf{F}_{gb}$ is the force of gravity on the block. The block only translates along the plane, so we write down Newton's second law for translation parallel to the plane

$$\sum F = ma \;\;\Rightarrow\;\; F_{gb}\sin 32° - F_{rb} = ma$$

Now we know the force of the rope on the block is equal in magnitude to the force of the rope on the spool. We also know when the spool turns clockwise by angle $\theta$, the block slides down the plane a distance $R\theta$ where $R$ is the radius of the spool. This implies that the angular acceleration of the spool, $\alpha$, and the acceleration of the block, $a$, are related by

$$\alpha = a / R$$

Using this to eliminate $\alpha$ from the Newton's second law equations results in

(spool)        $RF_r = Ia / R$

(block)        $mg\sin 32° - F_r = ma$

Eliminating $F_r$ from the equations

$$mg\sin 32° - \frac{Ia}{R^2} = ma$$

Solving this for the acceleration of the block

$$a = \frac{mg\sin 32°}{m + I/R^2} = \frac{(6.45\,\text{kg})(9.8\,\text{m/s}^2)\sin 32°}{(6.45\,\text{kg}) + (3.45\,\text{kg·m}^2)/(0.187\,\text{m})^2} = 0.239\,\text{m/s}^2$$

**Section 10-8 Determining Moments of Inertia**

The moment of inertia of a collection of point masses was given in section 6. To determine the moment of inertia about a given axis for an extended body, calculus must be used in a way similar to determining the center of mass of an extended body. We think of the extended body as a collection of infinitesimal masses of size $dm$. To determine the moment of inertia of the extended body, we add the moment of inertia of each piece, $R^2 dm$, by integrating over the body:

$$I = \int R^2\,dm$$

Here $R$ is the distance from the given moment of inertia to the location of the mass $dm$. As in the case of determining the center of mass of an extended body, we usually write the $dm$ as a density time an infinitesimal volume element, $\rho\, dV$. In this form the integral appears as

$$I = \int R^2 \rho\, dV$$

### Example 10-8-A

Three equal masses, $m$, lie on the corners of an equilateral triangle with sides of length $a$. Determine the moment of inertia about axes perpendicular to the plane of the triangle a) through the center of the triangle  b) through a mid-point of one edge of the triangle and c) through a vertex of the triangle.

### Solution:

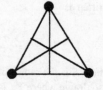

a)  We recall from geometry that the center of an equilateral triangle lies two thirds of the distance from the vertex to the base. Using that the altitude, $h$, of the equilateral triangle is

$$h = a \sin 60^\circ$$

we see that the distance each mass is from the axis of rotation will be

$$r = \tfrac{2}{3} a \sin 60^\circ = \sqrt{\tfrac{1}{3}}\, a$$

Adding up the moments of inertia of each mass gives

$$I = \sum mr^2 = 3m\left(\sqrt{\tfrac{1}{3}}a\right)^2 = ma^2$$

b)  For the axis located along one edge, it is easy to see that two of the masses are one half the side length from the axis, $r = a/2$ and one is an altitude of the triangle away, $r = a \sin 60^\circ$.

$$I = \sum mr^2 = m\left(\frac{a}{2}\right)^2 + m\left(\frac{a}{2}\right)^2 + m\left(a \sin 60^\circ\right)^2 = \tfrac{5}{4}ma^2$$

c)  For the axis located at one of the vertices of the triangle, one of the masses is on the axis, $r = 0$, and two are a side length away from the axis, $r = a$

$$I = \sum mr^2 = ma^2 + ma^2 + 0 = 2ma^2$$

### Example 10-8-B

A cylinder of radius 0.127 m and height 0.243 m has a density that varies with radius from the axis of the cylinder as $\rho(r) = 3200 \text{ kg/m}^3 - (1000 \text{ kg/m}^4)r$. What is the moment of inertia of this sphere about the axis of the cylinder?

### Solution:

The moment of inertia of a continuous distribution of mass is given by

$$I = \int \rho(r)\, r^2 dV = \int_0^{0.127\,\text{m}} \left[3200 \text{ kg/m}^3 - 1000 \text{ kg/m}^4\, r\right] r^2 \left(2\pi\, rh\, dr\right)$$

where we have used the density as given above and the infinitesimal volume unit is a thin cylindrical shell of radius $r$, height $h$, and thickness $dr$. Simplifying the integrand and completing the integral,

$$I = 2\pi h \int_0^{0.127\,m} \left[ 3200\,kg/m^3\ r^3 - 1000\,kg/m^4\ r^4 \right] dr$$

$$= 2\pi \left(0.243\,m\right) \left[ \tfrac{1}{4} \left(3200\,kg/m^3\ r^4\right) - \tfrac{1}{5}\left(1000\,kg/m^4\ r^5\right) \right]_0^{0.127\,m}$$

$$= 0.308\,kg \cdot m^2$$

There are two important theorems that can sometimes be used to assist in determining the moment of inertia of objects about particular axes. The parallel-axis theorem states that the moment of inertia about an axis that is displaced a distance $h$ from a parallel axis through the center of mass is the moment of inertia about the axis through the center of mass plus the mass of the body, $M$, multiplied by $h^2$:

$$I = I_{CM} + Mh^2$$

## Example 10-8-C

Determine the moment of inertia of a uniform solid sphere of radius $R$ and mass $M$ about an axis that is tangent to the surface of the sphere.

## Solution:

We already know from Figure 10-21 of the textbook that the moment of inertia of such a sphere about any axis through its center is $2/5 MR^2$. Now, any axis that is tangent to the surface of the sphere is parallel to some axis that passes through the center of the sphere, so the parallel axis theorem may be used:

$$I_{\text{tangent}} = I_{CM} + Mh^2 = \tfrac{2}{5}MR^2 + MR^2 = \tfrac{7}{5}MR^2$$

where we have used that the distance from the center of mass of the sphere to its surface is the radius of the sphere. Note that we have also used that the center of mass of the sphere is at the geometric center of the sphere. This is not necessarily the case if the density of the sphere is not uniform.

## Example 10-8-D

Obtain the answers for part b) and c) for Example 10-8-B using the answer to part a) and the parallel axis theorem.

## Solution:

The moment of inertia about the center of mass is determined in part a). The total mass of the system is $3m$. For part b), it is easy to determine that the mid-point of the side is a distance

$$h = \frac{1}{2\sqrt{3}}a$$

from the center of mass of the system. Using the parallel axis theorem results in a moment of inertia

$$I = I_{CM} + Mh^2 = ma^2 + 3m\left(\frac{1}{2\sqrt{3}}a\right)^2 = \tfrac{5}{4}ma^2$$

For part c) we know from the discussion of Example 10-8-B that the vertex is a distance

$$h = \frac{1}{\sqrt{3}}a$$

from the center of mass of the system. Applying the parallel axis theorem to this case results in a moment of inertia

$$I = I_{CM} + Mh^2 = ma^2 + 3m\left(\frac{1}{\sqrt{3}}a\right)^2 = 2ma^2$$

Giancoli, *Physics for Scientists & Engineers*: Study Guide

The second theorem is the perpendicular axis theorem. It applies only to effectively two dimensional objects, objects whose thickness is very small compared to their dimensions in the plane perpendicular to their thickness. The theorem states that the moment of inertia of the body about an axis perpendicular to the body is equal to the sum of the moments of inertia about any two perpendicular axes in the plane of the body that intersect with the first axis in the plane of the body. We write this as

$$I_z = I_x + I_y$$

where $I_z$ is the moment of inertia about the axis perpendicular to the plane of the body, and $I_x$ and $I_y$ are moments of inertia of the body about two perpendicular axes that lie in the plane of the body and intersect the $z$ axis in the plane of the body.

## Example 10-8-E

Use the perpendicular axis theorem to determine the moment of inertia of a flat metal washer of mass $M$ with an inner radius of $R_1$ and outer radius $R_2$ about an axis in the plane of the washer passing through a diameter of the washer.

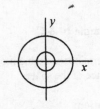

## Solution:

We recognize that we can take two perpendicular diameters of the washer in the plane of the washer and apply the perpendicular axis theorem to the moments of inertia about those diagonals. Because of the symmetry of the washer, the moments of inertia will be identical, $I_x = I_y$. These two axes intersect at the center of the washer. The moment of inertia through the center of washer perpendicular to the plane of the washer, $I_z$, is given in Figure 10-21 of the text as $\frac{1}{2}M(R_1^2+R_2^2)$. Applying these to the perpendicular axis theorem gives

$$I_z = I_x + I_y = 2I_x \quad \Rightarrow \quad I_x = \tfrac{1}{2}I_z = \tfrac{1}{2}\left[\tfrac{1}{2}M\left(R_1^2 + R_2^2\right)\right] = \tfrac{1}{4}M\left(R_1^2 + R_2^2\right)$$

## Section 10-9 Angular Momentum and Its Conservation

Angular momentum is defined as the product of the momentum of inertia of an object and its angular velocity:

$$L = I\omega$$

The dimensions of angular momentum are $ML^2/T$ and the SI units are kg·m²/s.

Just as Newton's second law for translational motion can be written in terms of translational momentum, Newton's second law for rotational motion can be written in terms of angular momentum:

$$\sum \tau = \frac{dL}{dt}$$

If total external torque acting on the system is zero, then the angular momentum remains constant.

$$I\omega = I_0\omega_0 = \text{constant}$$

## Example 10-9-A

A children's playground merry-go-round is rotating with an angular velocity of 1.43 rad/s. The merry-go-round is a flat uniform circular disk with a radius of 1.47 m and a mass of 165 kg. A child with a mass of 38.4 kg jumps onto the outer edge of the merry-go-round. The child has no velocity tangential to the edge of the merry-go-round as the child jumps onto it. What is the angular velocity of the merry-go-round after the child jumps on?

## Solution:

We will use conservation of angular momentum to solve this problem. We need to know the moment of inertia of the merry-go-round and the moment of inertia of the child on the merry-go-round. The merry-go-round is a flat circular disk, therefore the moment of inertia of the merry-go-round, $I_{mgr}$, about the rotation axis of the merry-go-round is

$$I_{mgr} = \tfrac{1}{2} m_{mgr} r_{mgr}^2 = \tfrac{1}{2}(165\,\text{kg})(1.47\,\text{m})^2 = 178.3\,\text{kg}\cdot\text{m}^2$$

The moment of inertia of the child, $I_{child}$, about this same axis while on the merry-go-round is

$$I_{child} = m_{child} r_{mgr}^2 = (38.4\,\text{kg})(1.47\,\text{m})^2 = 82.98\,\text{kg}\cdot\text{m}^2$$

Setting the total angular momentum before the child jumps on the merry-go-round equal to the total angular momentum after the child jumps on the merry-go-round

$$L_{mgr} + L_{child} = L'_{mgr} + L'_{child} \quad\Rightarrow\quad I_{mgr}\omega_0 + 0 = \left(I_{mgr} + I_{child}\right)\omega$$

Solving for $\omega$,

$$\omega = \frac{I_{mgr}\omega_0}{\left(I_{mgr} + I_{child}\right)} = \frac{\left(178.3\,\text{kg}\cdot\text{m}^2\right)\left(1.43\,\text{rad/s}\right)}{\left(178.3\,\text{kg}\cdot\text{m}^2 + 82.98\,\text{kg}\cdot\text{m}^2\right)} = 0.976\ \text{rad/s}$$

**Example 10-9-B**

The Earth is gradually slowing down in its rate of rotation which means the length of the day is getting longer (currently the length of a day is a small fraction of a second greater than 24 hours). Suppose, in the future, some fanatic group wants the length of the day to remain exactly 24 hours, but the length of the day has become 24 hours plus one second. They propose to put a large flywheel at the north pole and spin it so its angular momentum change on start up will result in an angular momentum change of the earth that will increase the angular velocity of the Earth. Suppose structural limitations limit the moment of inertia of the flywheel to about $10^{14}$ kg·m$^2$. Assume the Earth is a uniform sphere. What direction should the flywheel rotate? What angular velocity does the flywheel need to increase to in order to return the length of the day to 24 hours? Do you think this is practical?

**Solution:**

We will use conservation of angular momentum to solve this problem. We need to calculate the moment of inertia of the Earth:

$$I_{earth} = \tfrac{2}{5} m_{earth} r_{earth}^2 = \tfrac{2}{5}\left(5.97\times10^{24}\ \text{kg}\right)\left(6.38\times10^6\ \text{m}\right)^2 = 9.720\times10^{37}\ \text{kg}\cdot\text{m}^2$$

We need the initial angular velocity of the Earth and the flywheel

$$\omega_0 = \frac{2\pi}{T} = \frac{2\pi}{24\,\text{hr}+1\,\text{s}} = \frac{2\pi}{86401\,\text{s}} = 7.27212\times10^{-5}\ \text{s}^{-1}$$

We can also determine the final angular velocity of the Earth

$$\omega'_{earth} = \frac{2\pi}{T'} = \frac{2\pi}{24\,\text{hr}} = \frac{2\pi}{86400\,\text{s}} = 7.27221\times10^{-5}\ \text{s}^{-1}$$

We can now write down the conservation of angular momentum relation

$$L_{earth} + L_{flywheel} = L'_{earth} + L'_{flywheel}$$
$$\Rightarrow \left(I_{earth} + I_{flywheel}\right)\omega_0 = I_{earth}\omega'_{earth} + I_{flywheel}\omega'_{flyeheel}$$

Solving for the final angular velocity of the flywheel

$$\omega'_{flywheel} = \frac{(I_{earth} + I_{flywheel})\omega_0 - I_{earth}\omega'_{earth}}{I_{flywheel}}$$

$$= \frac{9.720 \times 10^{37} \text{ kg}(7.27212 \times 10^{-5} \text{ s}^{-1} - 7.27221 \times 10^{-5} \text{ s}^{-1})}{10^{14} \text{ kg} \cdot \text{m}^2} = -9 \times 10^{14} \text{ rad/s}$$

The rotation would be opposite in direction to the direction the earth rotates.  This is a ridiculously high angular velocity!

## Section 10-10 Rotational Kinetic Energy

The work-energy theorem can be written in terms of rotational quantities.  The rotational kinetic energy, $K$, of a body is equal to

$$K = \tfrac{1}{2} I \omega^2$$

The work done by torque on a body, $W$, is given by

$$W = \int \tau \, d\theta$$

The work-energy principle can be written in terms of these quantities as

$$\int \tau \, d\theta = \tfrac{1}{2} I \omega^2 - \tfrac{1}{2} I \omega_0^2$$

## Example 10-10-A

Stand a pencil on its point and let it go.  It will fall over and strike the table.  What is its angular velocity as it strikes the table?  Assume that it rotates about the point in contact with the table and its moment of inertia is that of a rod rotating about an axis through its end.

## Solution:

We will solve this problem using the work-energy theorem.  We know the center of mass of the pencil changes its height by dropping one half the length of the pencil as it falls.  The work done by gravity on the pencil is

$$W = Mg \frac{L}{2}$$

where $M$ is the mass of the pencil, $g$ is the acceleration of gravity, and $L$ is the length of the pencil.  The kinetic energy of the pencil when it is released is zero.  The kinetic energy of the pencil when it hits the table will be

$$K = \tfrac{1}{2} I \omega^2 = \tfrac{1}{2}\left(\tfrac{1}{3} M L^2\right)\omega^2$$

Applying the work-energy theorem

$$W = \Delta K \quad \Rightarrow \quad Mg \frac{L}{2} = \tfrac{1}{6} M L^2 \omega^2 - 0$$

Solving for $\omega$,

$$\omega = \sqrt{\frac{3g}{L}}$$

**Section 10-11 Rotational Plus Translational Motion; Rolling**

An object can have center of mass translational motion at the same time it rotates. If the axis of rotation of the object remains in a fixed direction, the total kinetic energy of the object can be written as the sum of translational and rotational kinetic energy

$$K = \tfrac{1}{2} m v_{CM}^2 + \tfrac{1}{2} I_{CM} \omega^2$$

where $v_{CM}$ is the velocity of the center of mass of the object, $I_{CM}$ is the moment inertia about an axis through the center of mass parallel to the axis of rotation of the object, $m$ is the mass of the object, and $\omega$ is the angular velocity of the object.

Be careful not to use $\Sigma\tau = I\alpha$ if either the axis of rotation is not fixed or the direction of the axis is not fixed and the axis does not pass through the center of mass of the object.

**Section 10-12 Why Does a Rolling Sphere Slow Down?**

If the only force acting on a rolling without slipping sphere were at the point of contact with the surface it is rolling on, no work would be done on the sphere because there is no displacement at the point of contact by definition of rolling without slipping. However, every real sphere and surface distort slightly on contact between the sphere and the surface. If the object is in motion, slippage has to occur at some points along the area of contact. The frictional force at the surface will do negative work on the object and slow it down.

## Practice Quiz

1.  What are the $x$ and $y$ components of the velocity of a point a distance $R$ from the axis of a rotating object that is rotating with a constant angular velocity $\omega$ if $\theta$ was zero at time $t = 0$.

2.  On a rotating object with a fixed axis, is it possible for the tangential and centripetal acceleration to add to each other so that a point a distance $R$ from the axis has zero total acceleration?

3.  On a rotating object with an axis with a fixed direction, is it possible for the acceleration of the center of mass to add to the centripetal acceleration so that a point a distance $R$ from the axis of rotation has zero total acceleration?

4.  What is the direction of the angular velocity of the second hand of a clock?

5.  The velocity of the point in contact with the ground of a wheel that is rolling without slipping is zero relative to the ground. What is the velocity of the point at the top of the wheel relative to the ground?

6.  What is incorrect about the statement "The moment of inertia of the object is 24 kg·m$^2$."

7.  Sketch the path of three different points on a wheel as it rolls without slipping: the center of the wheel, a point on the outside circumference of the wheel, and a point halfway from the center to the outside edge.

8.  Find the fallacy in the following. A perpetual motion apparatus can be constructed by hanging two masses on springs that slide on arms as shown in the diagram. If the apparatus is initially spun to some angular velocity, it will spin forever because if it starts to slow down, the masses will move inward due to reduced centripetal force. If the masses move inward, the angular momentum is reduced and the angular velocity must increase to conserve angular momentum. Thus the angular velocity can never decrease so the system will spin forever at constant speed.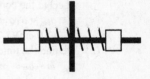

9.  For a given direction axis of rotation, through which point does the axis pass that minimizes the moment of inertia?

10. Some fancy doors have the door knob in the middle of the door rather than near the edge of the door opposite the hinges. Why does the knob in this position make it more difficult to pull the door open quickly?

11. A car accelerates from rest to 3.47 m/s in a time of 2.17 s while in first gear. If the car moves 43.7 cm per revolution of the car engine, what was the constant angular acceleration of the car engine during the car's acceleration?

12.  A cylindrical wheel with a diameter of 28.2 cm and a mass of 12.7 kg has a force acting on it as shown in the diagram. The point of application of the force is 19.5 cm from the axis of the wheel. What is the angular acceleration of the wheel?

13.  Three 2.4 kg masses are located in a plane as shown in the diagram. Determine the location of the center of mass and the moment of inertia about an axis perpendiculr to the page through the center of mass.

14.  The purpose of tail rotors on helicopters is to prevent conservation of angular momentum from rotating the body of the helicopter as the lift rotor changes angular speed. Inexpensive toy helicopters usually have no such tail rotor. A toy helicopter has its rotor with moment of inertia of 0.032 kg·m² about its rotation axis through the center of mass of the helicopter. The rest of the helicopter has a moment of inertia of 0.445 kg·m² about the same axis. The helicopter is initally at rest and a motor increases the angular velocity of the rotor to 800 revolutions per minute relative to the body of the helicopter. If no external torques act on the helicopter, what will the angular velocity of the body of the helicopter be?

15.  A yo-yo has a moment of inertia of $1.45 \times 10^{-6}$ kg·m² and a mass 0.0674 kg. Its string is wrapped around an axle of diameter 0.832 cm. What will the angular velocity of the yo-yo be when it has fallen 50.0 cm from its resting position?

## Problem Solutions

3.  We find the distance from
$$\theta = h/r;$$
$(7.5°)(\pi \text{ rad}/180°) = (300 \text{ m})/r;$ which gives $r = 2.3 \times 10^3$ m.

7.  All points will have the angular speed of the Earth:
$$\omega = \Delta\theta/\Delta t = (2\pi \text{ rad})/(1 \text{ day})(24 \text{ h/day})(3600 \text{ s/h}) = 7.27 \times 10^{-5} \text{ rad/s}.$$
Their linear speed will depend on the distance from the rotation axis.
(a) On the equator we have
$$v = R_{Earth}\omega = (6.38 \times 10^6 \text{ m})(7.27 \times 10^{-5} \text{ rad/s}) = 464 \text{ m/s}.$$
(b) At a latitude of 66.5° the distance is $R_{Earth} \cos 66.5°$, so we have
$$v = R_{Earth} \cos 66.5° \ \omega = (6.38 \times 10^6 \text{ m})(\cos 66.5°)(7.27 \times 10^{-5} \text{ rad/s}) = 185 \text{ m/s}.$$
(c) At a latitude of 40.0° the distance is $R_{Earth} \cos 40.0°$, so we have
$$v = R_{Earth} \cos 40.0° \ \omega = (6.38 \times 10^6 \text{ m})(\cos 40.0°)(7.27 \times 10^{-5} \text{ rad/s}) = 355 \text{ m/s}.$$

11.  In each revolution the ball rolls a distance equal to its circumference, so we have
$$L = N(\pi D);$$
$3.5 \text{ m} = (15.0)\pi D$, which gives $D = 0.074$ m = 7.4 cm.

15.  (a) If there is no slipping, the linear tangential acceleration of the pottery wheel and the rubber wheel at the contact point must be the same:
$$a_{tan} = R_1\alpha_1 = R_2\alpha_2;$$
$(2.0 \text{ cm})(7.2 \text{ rad/s}^2) = (25.0 \text{ cm})\alpha_2$, which gives $\alpha_2 = 0.58 \text{ rad/s}^2$.
(b) We find the time from
$$\omega = \omega_0 + \alpha t;$$
$(65 \text{ rev/min})(2\pi \text{ rad/rev})/(60 \text{ s/min}) = 0 + (0.58 \text{ rad/s}^2)t$, which gives $t = 12$ s.

19.  (a) The direction of $\omega_1$ is along the axle. At the instant shown,
$\omega_1$ is in the –x-direction.
The direction of $\omega_z$ is up. At any time, $\omega_2$ is in the +z-direction.
(b) At the instant shown, we have the vector diagram shown.
We find the magnitude from
$$\omega^2 = \omega_1^2 + \omega_2^2 = (50.0 \text{ rad/s})^2 + (35.0 \text{ rad/s})^2,$$
which gives $\omega = 61.0$ rad/s.
We find the angle from
$$\tan \theta = \omega_2/\omega_1 = (35.0 \text{ rad/s})/(50.0 \text{ rad/s}) = 0.700, \text{ so } \theta = 35.0°.$$

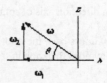

Thus the resultant angular velocity is
$$\omega = 61.0 \text{ rad/s, } 35.0° \text{ above } -x\text{-axis.}$$

(c) Because $\omega_2$ is constant, only $\omega_1$ is changing.  It will rotate at $\omega_2$ in the $xy$-plane.  If we let $t = 0$
at the instant shown,  we have
$$\omega_1 = \omega_1[-\cos(\omega_2 t)\mathbf{i} - \sin(\omega_2 t)\mathbf{j}].$$
From the definition of the angular acceleration we have
$$\alpha = d\omega/dt = d(\omega_1 + \omega_2)/dt = d\omega_1/dt = \omega_1\omega_2[+\sin(\omega_2 t)\mathbf{i} - \cos(\omega_2 t)\mathbf{j}].$$
At $t = 0$, we get
$$\alpha = -\omega_1\omega_2\mathbf{j} = -(50.0 \text{ rad/s})(35.0 \text{ rad/s})\mathbf{j} = -(1.75 \times 10^3 \text{ rad/s}^2)\mathbf{j}.$$

23. We assume clockwise motion, so the frictional torque is counterclockwise.
If we take the clockwise direction as positive, we have
$$\tau_{net} = rF_1 - RF_2 + RF_3 - \tau_{fr}$$
$$= (0.10 \text{ m})(35 \text{ N}) - (0.20 \text{ m})(30 \text{ N}) + (0.20 \text{ m})(20 \text{ N}) - 0.30 \text{ m} \cdot \text{N}$$
$$= 1.2 \text{ m} \cdot \text{N (clockwise).}$$

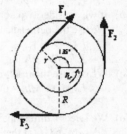

27. The torque produces the angular acceleration of the grindstone:
$$\tau_{motor} = I\alpha = \tfrac{1}{2}MR^2 \Delta\omega/\Delta t$$
$$= \tfrac{1}{2}(1.4 \text{ kg})(0.20 \text{ m})^2[(1800 \text{ rev/s})(2\pi \text{ rad/rev}) - 0]/(6.0 \text{ s}) = 53 \text{ m} \cdot \text{N}.$$

31. For the moment of inertia of the rotor blades we have
$$I = 3(\tfrac{1}{3} m_{blade}L^2) = m_{blade}L^2 = (160 \text{ kg})(3.75 \text{ m})^2 = 2.25 \times 10^3 \text{ kg} \cdot \text{m}^2.$$

We find the required torque from
$$\tau = I\alpha = I(\omega - \omega_0)/t$$
$$= (2.25 \times 10^{-3} \text{ kg} \cdot \text{m}^2)[(5.0 \text{ rev/s})(2\pi \text{ rad/rev}) - 0]/(8.0 \text{ s}) = 8.8 \times 10^3 \text{ m} \cdot \text{N}.$$

35. Because the force varies with time, the angular acceleration will vary with time:
$$\tau = F_T R_0 - \tau_{fr} = I\alpha,$$
$$[(3.00 \text{ N/s})t - (0.20 \text{ N/s}^2)t^2](0.330 \text{ m}) - 1.10 \text{ m} \cdot \text{N} = (0.385 \text{ kg} \cdot \text{m}^2)\alpha, \text{ which gives}$$
$$\alpha = -(2.86 \text{ rad/s}^2) + (2.57 \text{ rad/s}^3)t - (0.171 \text{ rad/s}^4)t^2.$$
We integrate this variable acceleration to find the angular velocity:
$$\int_0^\omega d\varpi = \int_0^t \alpha \, dt = \int_0^t \left[-(2.86 \text{ rad/s}^2) + (2.57 \text{ rad/s})t - (0.171 \text{ rad/s}^4)t^2\right]dt$$
$$\omega = -(2.86 \text{ rad/s}^2)t + (1.29 \text{ rad/s}^3)t^2 - (0.057 \text{ rad/s}^4)t^3$$
$$= -(2.86 \text{ rad/s}^2)(8.0 \text{ s}) + (1.29 \text{ rad/s}^3)(8.0 \text{ s})^2 - (0.057 \text{ rad/s}^4)(8.0 \text{ s})^3 = 30.2 \text{ rad.}$$
The linear speed of a point on the rim is the tangential speed:
$$v = R_0\omega = (0.330 \text{ m})(30.2 \text{ rad}) = 10 \text{ m/s.}$$

39. Thin hoop (through center): $Mk^2 = MR_0^2$, which gives $k = R_0$;
Thin hoop (through diameter): $Mk^2 = (MR_0^2/2) + (MW^2)/12$,
  which gives $k = [(MR_0^2/2) + (MW^2)/12]^{1/2}$;
Solid cylinder (through center): $Mk^2 = MR^2/2$, which gives $k = R^2/2$;
Hollow cylinder (through center): $Mk^2 = M(R_1^2 + R_2^2)/2$, which gives $k = [(R_1^2 + R_2^2)/2]^{1/2}$;
Uniform sphere (through center): $Mk^2 = 2Mr_0^2/5$, which gives $k = (2r_0^2/5)^{1/2}$;
Rod (through center): $Mk^2 = ML^2/12$, which gives $k = L^2/12$;
Rod (through end): $Mk^2 = ML^2/3$, which gives $k = L^2/3$;
Plate (through center): $Mk^2 = M(L^2 + W^2)/12$, which gives $k = [(L^2 + W^2)/12]^{1/2}$.

43. (a) Because the two diagonals are equivalent and perpendicular,
we can relate the moment of inertia about an axis through the
center perpendicular to the plate to the moments of inertia
about the diagonals:
$$I_c = I_a + I_a;$$
$$M(s^2 + s^2)/12 = 2I_a, \text{ or } I_a = Ms^2/12.$$

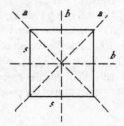

(*b*) Because the two axes parallel to the sides are equivalent
and perpendicular, we can relate the moment of inertia about
an axis through the center perpendicular to the plate to the
moments of inertia about the *b* axes:
$$I_c = I_b + I_b ;$$
$$M(s^2 + s^2)/12 = 2I_b , \text{ or } I_b = Ms^2/12.$$

47. (*a*) We use the parallel-axis theorem:
$$I_a = I_{CM} + Mh^2 = (MR_0^2/2) + M(R_0/4)^2 = 9MR_0^2/16.$$
(*b*) If we consider two horizontal axes, they must have the same moment of inertia.
We use the perpendicular-axis theorem:
$$I_{CM} = I_b + I_b;$$
$$MR_0^2/2 = 2I_b, \text{ or } I_b = MR_0^2/4.$$
(*c*) We can use the result from part (*b*) and the parallel-axis theorem:
$$I_c = I_b + Mh^2 = (MR_0^2/4) + MR_0^2 = 5MR_0^2/4.$$

51. (*a*) We select a differential element of the plate with
sides *dx* and *dy* at the location *x*, *y*. The element is
equivalent to a point mass with a mass of
*dm* = (*M*/¼*w*) *dx dy*.
We integrate from *x* = – ¼/2 to *x* = ¼/2 and *y* = – *w*/2 to
*y* = *w*/2 to find the moment of inertia of the plate:

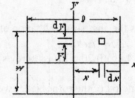

$$I = \int r^2\, dm = \frac{M}{lw} \int\int \left(x^2 + y^2\right) dx\, dy = \frac{M}{lw} \int_{-l/2}^{+l/2} x^2\, dx \int_{-w/2}^{w/2} dy + \int_{-l/2}^{+l/2} dx \int_{-w/2}^{+w/2} y^2\, dy$$

$$I = \frac{M}{lw}\left( \frac{x^3}{3}\bigg|_{-l/2}^{+l/2} \; y\big|_{-w/2}^{+w/2} + x\big|_{-l/2}^{+l/2} \; \frac{y^3}{3}\bigg|_{-w/2}^{+w/2} \right) = \frac{M}{12}\left(l^2 + w^2\right)$$

(*b*) For an axis along the edge parallel to the *y*-axis, we can consider the plate to be an infinite
number of rods of length *l* and width *dy* with a mass (*M*/*w*) *dy*.
The moment of inertia of each rod is
$$dI = (l^2/12)\, dm = (M/w)(l^2/12)\, dx.$$
When we add (integrate), we get
$$I = (M/w)(l^2/12)w = Ml^2/12.$$
Similarly, for the other edge we get $I = Mw^2/12.$

55. (*a*) As the arms are raised some of the person's mass is farther from the axis of rotation, so the
moment of inertia has increased. For the isolated system of platform and person, the
angular momentum is conserved. As the moment of inertia increases, the angular velocity must decrease.
(*b*) If the mass and thus the moment of inertia of the platform can be neglected, for the conservation
of angular momentum, we have
$$L = I_1\omega_1 = I_2\omega_2, \text{ or}$$
$$I_2/I_1 = \omega_1/\omega_2 = (1.30 \text{ rad/s})/(0.80 \text{ rad/s}) = 1.6.$$

59. When the people step onto the merry-go-round, they have no initial angular momentum. For the system of merry-go-
round and people, angular momentum is conserved:
$$I_{\text{merry-go-round}}\omega_0 + I_{\text{people}}\omega_1 = (I_{\text{merry-go-round}} + I_{\text{people}})\omega;$$
$$I_{\text{merry-go-round}}\omega_0 + 4mR^2(0) = (I_{\text{merry-go-round}} + 4mR^2)\omega;$$
(1950 kg · m²)(0.80 rad/s) = [(1950 kg · m²) + 4(65 kg)(2.4 m)²]ω, which gives ω = 0.45 rad/s.
If the people jump off in a radial direction with respect to the merry-go-round, they have the tangential velocity of the
merry-go-round: *v* = *Rω₀*. For the system of merry-go-round and people, angular momentum is conserved:
$$(I_{\text{merry-go-round}} + I_{\text{people}})\omega_0 = I_{\text{merry-go-round}}\omega + I_{\text{people}}\omega_0, \text{ which gives}$$
*ω* = *ω₀* = 0.80 rad/s. The angular speed of the merry-go-round does not change.
Note that the angular momentum of the people will change when contact is made with the ground.

63. Conservation of angular momentum gives us
$I_i\omega_i = I_f\omega_f$,   or   $\omega_f/\omega_i = I_i/I_f$.
The ratio of kinetic energies is
$$K_f/K_i = \tfrac{1}{2}I_f\omega_f^2/\tfrac{1}{2}I_i\omega_i^2 = (I_f/I_i)(\omega_f/\omega_i)^2 = (I_f/I_i)(I_i/I_f)^2 = I_i/I_f = (R_i/R_f)^2$$
$$= [(7 \times 10^5 \text{ km})/(10 \text{ km})]^2 = 5 \times 10^9.$$
The increased kinetic energy came from the loss of gravitational potential energy.

67. (a)  When we use the parallel-axis theorem for
the arms, we get
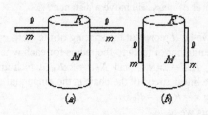
$$I_a = (MR^2/2) + 2[(ml^2/12) + m(R + l/2)^2]$$
$$= [(60 \text{ kg})(0.12 \text{ m})^2/2] +$$
$$2(5.0 \text{ kg})[(0.60)^2/12 + (0.12 \text{ m} + 0.30 \text{ m})^2]$$
$$= 2.5 \text{ kg} \cdot \text{m}^2.$$
(b)  When the arms are at the sides, all of their mass
is the same distance from the axis, so we get
$$I_b = (MR^2/2) + 2mR^2 = [(M/2) + 2m]R^2$$
$$= [(60 \text{ kg})/2 + 2(5.0 \text{ kg})]0.12 \text{ m})^2$$
$$= 0.58 \text{ kg} \cdot \text{m}^2.$$
(c)  From the conservation of angular momentum about the vertical axis, we have
$I_a\omega_a = I_b\omega_b$;
$(2.5 \text{ kg} \cdot \text{m}^2)(1 \text{ rev}/1.5 \text{ s}) = (0.58 \text{ kg} \cdot \text{m}^2)(1 \text{ rev}/t_b)$,  which gives $t_b = 0.35$ s.
(d)  The change in kinetic energy is
$$\Delta K = \tfrac{1}{2}I_a\omega_a^2 - \tfrac{1}{2}I_b\omega_b^2$$
$$= \tfrac{1}{2}(2.5 \text{ kg} \cdot \text{m}^2)(2\pi \text{ rad}/1.5 \text{ s})^2 - \tfrac{1}{2}(0.58 \text{ kg} \cdot \text{m}^2)(2\pi \text{ rad}/0.35 \text{ s})^2 = -72 \text{ J}.$$
(e)  Because the kinetic energy decreases, it will be easier to lift your arms when rotating.
In the rotating system, the arms tend to move away from the center of rotation.

71. The angular speed of the rolling ball is $\omega = v/R$.  The total kinetic energy will have a translational term for the center
of mass and a term for the rotational energy about the center of mass:
$$K_{total} = K_{trans} + K_{rot} = \tfrac{1}{2}Mv^2 + \tfrac{1}{2}I\omega^2 = \tfrac{1}{2}Mv^2 + \tfrac{1}{2}(\tfrac{2}{5}MR^2)(v/R)^2 = 7Mv^2/10$$
$$= 7(7.3 \text{ kg})(5.3 \text{ m/s})^2/10 = 1.4 \times 10^2 \text{ J}.$$

75. If the ball rolls without slipping, the speed of the center of
mass is $v = r_0\omega$.  Because energy is conserved, for the motion from
A to B  we have
$0 = \Delta K + \Delta U$;
$0 = [(\tfrac{1}{2}mv^2 + \tfrac{1}{2}I\omega^2) - 0] + (mgr_0 - mgR_0)$,  or
$\tfrac{1}{2}mv^2 + \tfrac{1}{2}(\tfrac{2}{5}mr_0^2)(v/r_0)^2 = mg(R_0 - r_0)$,  which gives
$v = [10g(R_0 - r_0)/7]^{1/2}$ .

79. (a)  From Example 10–24, we see that while the ball is slipping, the acceleration of the center of mass
is $-\mu_k g$ and is constant.  The ball slips for a time $T = 2v_0/7\mu_k g$, so we find the distance from
$$x_{slip} = v_0 t + \tfrac{1}{2}at^2$$
$$= v_0(2v_0/7\mu_k g) + \tfrac{1}{2}(-\mu_k g)(2v_0/7\mu_k g)^2 = 12v_0^2/49\mu_k g.$$
(b)  Once the ball starts rolling at time $T$, the linear speed is constant
$v = v_0 + at = v_0 + (-\mu_k g)(2v_0/7\mu_k g) = 5v_0/7.$
We can find the angular speed from
$\omega = \omega_0 + \alpha t$,  or from
$\omega = v/R = (5v_0/7)/R = 5v_0/7R.$

83. The subtended angle in radians is the size of the object divided by the distance to the object:
$\theta_{Sun} = 2R_{Sun}/r_{Sun} = 2(6.96 \times 10^5 \text{ km})/(149.6 \times 10^6 \text{ km}) = 9.30 \times 10^{-3} \text{ rad}$  $(0.53°)$;
$\theta_{Moon} = 2R_{Moon}/r_{Moon} = 2(1.74 \times 10^3 \text{ km})/(384 \times 10^3 \text{ km}) = 9.06 \times 10^{-3} \text{ rad}$  $(0.52°)$.
These are almost equal, so eclipses can occur.

87. Because the spool is rolling, $v_{CM} = R\omega$. The velocity of the rope at the top of the spool, which is also the velocity of the person, is

$v = R\omega + v_{CM} = 2R\omega = 2v_{CM}$.

Thus in the time it takes for the person to walk a distance $l$, the center of mass will move a distance $l/2$. Therefore, the length of rope that unwinds is

$l_{rope} = l - (l/2) = l/2$.

The center of mass will move a distance $l/2$.

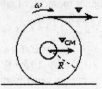

91. (a) If we let $d$ represent the spacing of the teeth, which is the same on both sprockets, we can relate the number of teeth to the radius for each wheel:

$N_F d = 2\pi R_F$, and $N_R d = 2\pi R_R$, which gives $N_F/N_R = R_F/R_R$.

The linear speed of the chain is the tangential speed for each socket:

$v = R_F \omega_F = R_R \omega_R$.

Thus we have

$\omega_R/\omega_F = R_F/R_R = N_F/N_R$.

(b) For the given data we have

$\omega_R/\omega_F = 52/13 = 4.0$.

(c) For the given data we have

$\omega_R/\omega_F = 42/28 = 1.5$.

95. (a) We find the angular acceleration from

$\theta = \omega_0 t + \frac{1}{2}\alpha t^2$;

$(20 \text{ rev})(2\pi \text{ rad/rev}) = 0 + \frac{1}{2}\alpha[(1 \text{ min})(60 \text{ s/min})]^2$, which gives $\alpha = 0.070 \text{ rad/s}^2$.

(b) We find the final angular speed from

$\omega = \omega_0 + \alpha t = 0 + (0.070 \text{ rad/s}^2)(60 \text{ s}) = 4.2 \text{ rad/s} = 40 \text{ rpm}$.

99. We convert the speed: $(90 \text{ km/h})/(3.6 \text{ ks/h}) = 25 \text{ m/s}$.

(a) We assume that the linear kinetic energy that the automobile acquires during each acceleration is not regained when the automobile slows down. For the work-energy principle applied to the 300-km trip we have

$W_{net} = \Delta K + \Delta U$;

$-F_{fr}D = [20(\frac{1}{2}Mv^2) - K_{flywheel}] + 0$, or

$K_{flywheel} = (20)\frac{1}{2}(1400 \text{ kg})(25 \text{ m/s})^2 + (500 \text{ N})(300 \times 10^3 \text{ m}) = 1.6 \times 10^8 \text{ J}$.

(b) We find the angular velocity of the flywheel from

$K_{flywheel} = \frac{1}{2}I\omega^2 = \frac{1}{2}(\frac{1}{2}mR^2)\omega^2$;

$1.6 \times 10^8 \text{ J} = \frac{1}{4}(240 \text{ kg})(0.75 \text{ m})^2\omega^2$, which gives $\omega = 2.2 \times 10^3 \text{ rad/s}$.

(c) We find the time from

$t = K_{flywheel}/P = (1.6 \times 10^8 \text{ J})/(150 \text{ hp})(746 \text{ W/hp}) = 1.43 \times 10^3 \text{ s} = 24 \text{ min}$.

103. (a) By walking to the edge, the moment of inertia of the person changes. Because the system of person and platform is isolated, angular momentum will be conserved:

$L = (I_{platform} + I_{person1})\omega_1 = (I_{platform} + I_{person2})\omega_2$;

$[1000 \text{ kg} \cdot \text{m}^2 + (75 \text{ kg})(0)^2](2.0 \text{ rad/s}) = [1000 \text{ kg} \cdot \text{m}^2 + (75 \text{ kg})(3.0 \text{ m})^2]\omega_2$, which gives

$\omega_2 = 1.2 \text{ rad/s}$.

(b) For the kinetic energies, we have

$K_1 = \frac{1}{2}(I_{platform} + I_{person1})\omega_1^2 = \frac{1}{2}(1000 \text{ kg} \cdot \text{m}^2 + 0)(2.0 \text{ rad/s})^2 = 2.0 \times 10^3 \text{ J}$;

$K_2 = \frac{1}{2}(I_{platform} + I_{person2})\omega_2^2 = \frac{1}{2}[1000 \text{ kg} \cdot \text{m}^2 + (75 \text{ kg})(3.0 \text{ m})^2](1.2 \text{ rad/s})^2 = 1.2 \times 10^3 \text{ J}$.

Thus there is a loss of $8.0 \times 10^2 \text{ J}$, a decrease of 40%.

107. (a) At the top of the loop, if the marble stays on the track, the normal force and the weight provide the radial acceleration:

$F_N + mg = mv^2/R_0$.

The minimum value of the normal force is zero, so we find the minimum speed at the top from

$mg = mv_{min}^2/R_0$, or $v_{min}^2 = gR_0$.

Because the marble is rolling, the corresponding angular velocity at the top is $\omega_{min} = v_{min}/r$, so the minimum kinetic energy at the top is

$K_{min} = \frac{1}{2}mv_{min}^2 + \frac{1}{2}I\omega_{min}^2$

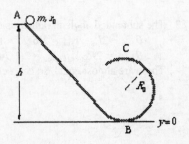

$$= \tfrac{1}{2}mv_{\min}^2 + \tfrac{1}{2}(\tfrac{2}{5}mr^2)(v_{\min}/r)^2 = 7mv_{\min}^2/10 = 7mgR_0/10.$$

If there are no frictional losses, we use energy conservation from the release point to the highest point of the loop:

$$K_i + U_i = K_f + U_f;$$

$$0 + mgh_{\min} = K_{\min} + mg2R_0 = 7mgR_0/10 + 2mgR_0, \text{ which gives } h_{\min} = 2.7R_0.$$

(b) At the top of the loop, if the marble stays on the track, the normal force and the weight provide the radial acceleration:

$$F_N + mg = mv^2/(R_0 - r_0).$$

The minimum value of the normal force is zero, so we find the minimum speed at the top from

$$mg = mv_{\min}^2/(R_0 - r_0), \text{ or } v_{\min}^2 = g(R_0 - r_0).$$

Because the marble is rolling, the corresponding angular velocity at the top is $\omega_{\min} = v_{\min}/r_0$, so the minimum kinetic energy at the top is

$$K_{\min} = \tfrac{1}{2}mv_{\min}^2 + \tfrac{1}{2}I\omega_{\min}^2$$

$$= \tfrac{1}{2}mv_{\min}^2 + \tfrac{1}{2}(\tfrac{2}{5}mr_0^2)(v_{\min}/r_0)^2 = 7mv_{\min}^2/10 = 7mg(R_0 - r_0)/10.$$

If there are no frictional losses, we use energy conservation from the release point to the highest point of the loop:

$$K_i + U_i = K_f + U_f;$$

$$0 + mgh_{\min} = K_{\min} + mg(2R_0 - r_0) = 7mg(R_0 - r_0)/10 + mg(2R_0 - r_0),$$

which gives    $h_{\min} = 2.7R_0 - 1.7r_0$.

# Chapter 11: General Rotation

## Chapter Overview and Objectives

This chapter introduces the vector cross product and applies it to problems of general rotation. The rotational quantities of Chapter 10 are redefined as vector quantities in this chapter.

After completing study of this chapter, you should:

- Know what a vector cross product is and be able to calculate vector cross products.
- Know the definitions of torque and angular momentum as vector quantities.
- Know how Newton's second law for general rotation can be applied to problems.
- Understand the concept of angular momentum and how it can be used to solve problems.
- Understand how Newton's laws do not apply to observers in noninertial reference frames and why pseudoforces are introduced by observers in noninertial reference frames.

## Summary of Equations

Magnitude of vector cross product:
$$|\mathbf{A} \times \mathbf{B}| = AB \sin \theta$$

Vector cross product in terms of Cartesian components:
$$\mathbf{A} \times \mathbf{B} = \begin{vmatrix} \mathbf{i} & \mathbf{j} & \mathbf{k} \\ A_x & A_y & A_z \\ B_x & B_y & B_z \end{vmatrix} = \left(A_y B_z - A_z B_y\right)\mathbf{i} + \left(A_z B_x - A_x B_z\right)\mathbf{j} + \left(A_x B_y - A_y B_x\right)\mathbf{k}$$

Properties of vector cross product:
$$\mathbf{A} \times \mathbf{A} = 0$$
$$\mathbf{A} \times \mathbf{B} = -\mathbf{B} \times \mathbf{A}$$
$$\mathbf{A} \times (\mathbf{B} + \mathbf{C}) = \mathbf{A} \times \mathbf{B} + \mathbf{A} \times \mathbf{C}$$
$$\frac{d}{dt}(\mathbf{A} \times \mathbf{B}) = \frac{d\mathbf{A}}{dt} \times \mathbf{B} + \mathbf{A} \times \frac{d\mathbf{B}}{dt}$$

Definition of torque:
$$\boldsymbol{\tau} = \mathbf{r} \times \mathbf{F}$$

Definition of angular momentum:
$$l = \mathbf{r} \times \mathbf{p}$$

Relationship between angular momentum and angular velocity along symmetry axis:
$$\mathbf{L} = I\boldsymbol{\omega}$$

Precession rate of top:
$$\Omega = \frac{Mgr}{L}$$

Magnitude of Coriolis acceleration:
$$a_{Coriolis} = 2\omega v_{\perp}$$

## Chapter Summary

### Section 11-1 Vector Cross Product

The **vector product** or **cross product** of two vectors is a vector. The cross product of vector **A** and vector **B** is is written as $\mathbf{A} \times \mathbf{B}$. The cross product is a vector with magnitude $AB\sin\theta$, where $\theta$ is the angle between the directions of vector **A** and vector **B** in the plane of the two vectors

$$|\mathbf{A} \times \mathbf{B}| = AB \sin \theta$$

The direction of $\mathbf{A} \times \mathbf{B}$ is perpendicular to the plane of the two vectors and points in the direction given by the right hand rule. To apply the right hand rule to find the direction of the cross product $\mathbf{A} \times \mathbf{B}$, place your fingers of your right hand in the direction of the first factor, $\mathbf{A}$, so that they can curl toward the direction of the second factor, $\mathbf{B}$. Extend your thumb and it will point in the direction of the cross product.

If the vectors $\mathbf{A}$ and $\mathbf{B}$ are written as a linear combination of the Cartesian axes unit vectors as

$$\mathbf{A} = A_x \mathbf{i} + A_y \mathbf{j} + A_z \mathbf{k} \quad \text{and} \quad \mathbf{B} = B_x \mathbf{i} + B_y \mathbf{j} + B_z \mathbf{k}$$

then the cross product of the two vectors can be written

$$\mathbf{A} \times \mathbf{B} = \begin{vmatrix} \mathbf{i} & \mathbf{j} & \mathbf{k} \\ A_x & A_y & A_z \\ B_x & B_y & B_z \end{vmatrix} = \left( A_y B_z - A_z B_y \right)\mathbf{i} + \left( A_z B_x - A_x B_z \right)\mathbf{j} + \left( A_x B_y - A_y B_x \right)\mathbf{k}$$

Some of the properties of the vector cross product are

$$\mathbf{A} \times \mathbf{A} = 0$$
$$\mathbf{A} \times \mathbf{B} = -\mathbf{B} \times \mathbf{A}$$
$$\mathbf{A} \times (\mathbf{B} + \mathbf{C}) = \mathbf{A} \times \mathbf{B} + \mathbf{A} \times \mathbf{C}$$
$$\frac{d}{dt}(\mathbf{A} \times \mathbf{B}) = \frac{d\mathbf{A}}{dt} \times \mathbf{B} + \mathbf{A} \times \frac{d\mathbf{B}}{dt}$$

### Example 11-1-A

Determine the cross product of vector $\mathbf{A} = 3\mathbf{i} + 2\mathbf{j}$ and vector $B = -2\mathbf{i} + \mathbf{j}$.

**Solution:**

$$\mathbf{A} \times \mathbf{B} = \left( A_y B_z - A_z B_y \right)\mathbf{i} + \left( A_z B_x - A_x B_z \right)\mathbf{j} + \left( A_x B_y - A_y B_x \right)\mathbf{k}$$
$$= ((2)(0) - (0)(1))\mathbf{i} + ((0)(-2) - (3)(0))\mathbf{j} + ((3)(1) - (2)(-2))\mathbf{k}$$
$$= 7\mathbf{k}$$

Note that the hand rule is in agreement with this result.

### Section 11-2 The Torque Vector

Torque can be expressed as a vector cross product

$$\boldsymbol{\tau} = \mathbf{r} \times \mathbf{F}$$

where $\boldsymbol{\tau}$ is the torque about a point $O$, $\mathbf{r}$ is the displacement vector of the point of application of the force from the point $O$, and $\mathbf{F}$ is the force acting on the object.

### Example 11-2-A

A force $\mathbf{F} = 3 \, \text{N} \, \mathbf{i} + 1 \, \text{N} \, \mathbf{j} - 2 \, \text{N} \, \mathbf{k}$ acts at a position $\mathbf{r} = -1 \, \text{m} \, \mathbf{i} + 2 \, \text{m} \, \mathbf{j} + 2 \, \text{m} \, \mathbf{k}$ relative to point O. What is the torque of this force about point O?

**Solution:**

$$\boldsymbol{\tau} = \mathbf{r} \times \mathbf{F} = \left( r_y F_z - r_z F_y \right)\mathbf{i} + \left( r_z F_x - r_x F_z \right)\mathbf{j} + \left( r_x F_y - r_y F_x \right)\mathbf{k}$$
$$= ((2\,\text{m})(-2\,\text{N}) - (2\,\text{m})(1\,\text{N}))\mathbf{i} + ((2\,\text{m})(3\,\text{N}) - (-1\,\text{m})(-2\,\text{N}))\mathbf{j} + ((-1\,\text{m})(1\,\text{N}) - (2\,\text{m})(3\,\text{N}))\mathbf{k}$$
$$= -6\,\text{N} \cdot \text{m}\,\mathbf{i} + 4\,\text{N} \cdot \text{m}\,\mathbf{j} - 7\,\text{N} \cdot \text{m}\,\mathbf{k}$$

## Section 11-3 Angular Momentum of a Particle

The angular momentum of a particle about a point is defined as

$$l = \mathbf{r} \times \mathbf{p}$$

where $l$ is the angular momentum of the particle about point $O$, $\mathbf{r}$ is the displacement of the particle from point $O$, and $\mathbf{p}$ is the momentum of the particle. Newton's second law can be written in terms of the torque on a particle about a point and the angular momentum of the particle about the same point

$$\sum \tau = \frac{dl}{dt}$$

### Example 11-3-A

A particle of mass 2 kg travels with constant velocity –3 m/s $\mathbf{i}$ along a path that passes a distance of 2 m from point $O$. What is the angular momentum of the particle about point $O$ as a function of time?

### Solution:

Setting time to zero when the object is at its point of closest approach to point $O$, we can write the position of the object relative to point $O$ as

$$\mathbf{r} = (-3 \text{ m/s})t\, \mathbf{i} + (2 \text{ m})\mathbf{j}$$

The momentum of the particle is

$$\mathbf{p} = m\mathbf{v} = (2 \text{ kg})(-3 \text{ m/s } \mathbf{i}) = -6 \text{ kg} \cdot \text{m/s } \mathbf{i}$$

We determine the angular momentum

$$\mathbf{l} = \mathbf{r} \times \mathbf{p} = \begin{vmatrix} \mathbf{i} & \mathbf{j} & \mathbf{k} \\ r_x & r_y & r_z \\ p_x & p_y & p_z \end{vmatrix} = 12 \text{ kg} \cdot \text{m}^2/\text{s } \mathbf{k}$$

Even though $\mathbf{r}$ is changing with time, the angular momentum is constant as expected, because the force on the particle, and therefore, the torque on the particle is zero.

## Section 11-4 Angular Momentum and Torque for a System of Particles; General Motion

For measurements made in an inertial reference frame, the angular momenta of a system of particles about a given point can be added together. Also, the sum of the torques about the same point acting on the particles can be added. In the sum, torques resulting from forces between particles in the system add to zero because of Newton's third law. The result of the summation is

$$\sum \tau_{ext} = \frac{d}{dt}\sum l = \frac{d\mathbf{L}}{dt}$$

where $\tau_{ext}$ is a torque from an external force and $\mathbf{L}$ is the total angular momentum of the system. If the point about which the torques and angular momenta are calculated is the center of the mass of the system, then

$$\sum \tau_{CM \; ext} = \frac{d}{dt}\sum l_{CM} = \frac{d\mathbf{L}_{CM}}{dt}$$

applies even to non-inertial or accelerating reference frames. The *CM* subscript means the quantities are calculated about the center of mass of the system.

## Section 11-5 Angular Momentum and Torque for a Rigid Body

If the relationship between torque and angular momentum of a system of particles is applied to a rigid extended body, then we can write

$$\mathbf{L} = I\,\omega$$

if the axis of rotation of the body is a reflection symmetry axis of the body. Note that this implies that the angular momentum vector and the angular velocity vector of the body are in the same direction. The angular momentum is calculated about the symmetry axis (or axis of rotation) of the body. If the rotation axis is not a reflection symmetry axis of the body, then the angular momentum vector of the body is not necessarily along the same direction as the rotation axis of the body.

Applying this to the dynamical equation for general rotation, we have

$$\sum \tau_{axis} = \frac{dL}{dt} = \frac{d}{dt}(I\omega) = I\frac{d\omega}{dt}$$

where $\tau_{axis}$ is the component of torque along the rotation axis, $I$ is the moment of inertia of the body about the rotation axis, and $\omega$ is the angular velocity about the rotation axis.

### Example 11-5-A

A uniform cylindrical spool of mass 0.187 kg and radius 2.56 cm has thread wrapped around it. The spool is free to rotate about its cylindrical axis. Attached to the end of the thread is a mass of 0.0542 kg. What is the angular acceleration of the spool?

### Solution:

We will calculate torque and angular momentum about the fixed axis of the spool. The torque acting on the system is the torque caused by the gravitational force on the mass supported by the thread. The torque is

$$\tau = |r \times F| = r_\perp F = (0.0256\,\text{cm})(0.0542\,\text{kg})(9.80\,\text{m/s}^2) = 0.0136\,\text{N}\cdot\text{m}$$

The total angular momentum of the system is the angular momentum of the spool plus the angular momentum of the suspended mass. Using the fact that the velocity (and momentum) of the suspended mass will be vertically downward, we can write the total angular momentum as

$$L = L_{spool} + L_{mass} = \tfrac{1}{2}m_{spool}R_{spool}^2\omega_{spool} + m_{mass}v_{mass}r_{mass}$$

Because the mass is tied to the string, it should be easy to see that the speed of the mass is related to the angular speed of the spool by

$$v_{mass} = \omega_{spool}R_{spool}$$

and that the distance of the mass from axis of rotation is just the radius of the spool, $r_{mass} = R_{spool}$. Writing down Newton's second law for rotation results in

$$\tau = \frac{dL}{dt} = \frac{d}{dt}\left(\tfrac{1}{2}m_{spool}R_{spool}^2\omega_{spool} + m_{mass}R_{spool}^2\omega_{spool}\right)$$

$$= \frac{d}{dt}\left(\tfrac{1}{2}m_{spool} + m_{mass}\right)R_{spool}^2\omega_{spool} = \left(\tfrac{1}{2}m_{spool} + m_{mass}\right)R_{spool}^2\alpha_{spool}$$

Solving for the angular acceleration of the spool gives

$$\alpha_{spool} = \frac{\tau}{\left(\frac{1}{2}m_{spool} + m_{mass}\right)R_{spool}^2} = \frac{0.0136\,\text{N}\cdot\text{m}}{\left(\frac{1}{2}0.0542\,\text{kg} + 0.187\,\text{kg}\right)(0.0256\,\text{m})}$$

$$= 2.48\,\text{s}^{-2}$$

### Section 11-6 Rotational Imbalance

In a case in which the rotation axis of an object is not aligned with a principal axis of the object, torque needs to be applied to the object to keep it rotating at constant angular velocity. This is because the angular momentum vector will rotate around the axis of rotation as the body changes orientation. Thus, there is a $d\mathbf{L}/dt$ not equal to zero. This implies there must be torque acting on the body just to keep it moving with constant angular velocity.

### Section 11-7 Conservation of Angular Momentum

For a system of objects with no net external torque acting on the system, the total angular momentum is conserved. This is a statement of the **law of conservation of angular momentum.**

### Example 11-7-A

A child of mass 32.8 kg stands on a playground merry-go-round that is at rest holding an object with a mass of 1.45 kg. The child is standing 1.34 m from the axis of rotation of the merry-go-round. The moment of inertia of the merry-go-round about its axis is 86.4 kg·m$^2$. The child throws the object to a friend with a speed of 4.65 m/s in a direction that is 66.4° from the direction directly away from the axis of rotation of the merry-go-round as shown. What will be the angular velocity of the merry-go-round after the child releases the object?

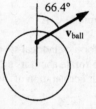

### Solution:

Initially the total angular momentum is zero as everything is at rest. After the ball is released it will have an angular momentum

$$l_{object} = \mathbf{r}\times\mathbf{p} = (1.34\,\text{m}\,\mathbf{j})\times(1.45\,\text{kg})(4.65\,\text{m/s})(\cos 66.5^o\,\mathbf{i} + \sin 66.5^o\,\mathbf{j})$$

$$= -3.60\,\text{kg}\cdot\text{m}^2/\text{s}\,\mathbf{k}$$

The mery-go-round and child rotate together after the object is released. Their angular momentum, **L**, will be the total of their moments of inertia about the merry-go-round axis multiplied by their common angular velocity:

$$\mathbf{L} = \left(I_{merry-go-round} + I_{child}\right)\boldsymbol{\omega}$$

The total angular momentum before the throw must be equal to the total angular momentum after the throw:

$$0 = l_{object} + L = l_{object} + \left(I_{merry-go-round} + I_{child}\right)\boldsymbol{\omega}$$

Solving for ω

$$\omega = -\frac{l_{object}}{\left(I_{merry-go-round} + I_{child}\right)} = -\frac{-3.60\,\text{kg}\cdot\text{m}^2/\text{s}\,\mathbf{k}}{86.4\,\text{kg}\cdot\text{m}^2 + (32.8\,\text{kg})(1.34\,\text{m})^2}$$

$$= 0.0248\,s^{-1}\,\mathbf{k}$$

### Section 11-8 The Spinning Top

The slightly inclined toy top is an example of a system for which the torque on the system is perpendicular to the angular momentum vector of the system at any given instant. The result of this relative orientation of the angular momentum and the net torque is to produce a motion of the angular momentum vector called **precession**. The angular momentum vector does not change magnitude, but the direction precesses around a fixed direction at a constant angular speed. For a toy top, the angular speed of the precession, $\Omega$, is given by

$$\Omega = \frac{Mgr}{L}$$

where $M$ is the mass of the top, $g$ is the acceleration of gravity, $r$ is the distance the center of mass of the top is above the point of the top in contact with the surface it is spinning on, and $L$ is the magnitude of the angular momentum of the top.

## Section 11-9 Rotating Frames of Reference; Inertial Forces

A frame of reference in which Newton's laws of motion do not hold true is called a noninertial reference frame. Typical noninertial reference frames are those that are accelerating relative to inertial reference frames and rotating reference frames. In noninertial reference frames, Newton's second law does not hold true, but we expect it to hold true. We introduce **fictitious forces** or **pseudoforces** to help us understand the motion of objects when observed from noninertial frames of reference.

A common example of this in our everyday experience is what happens to our body while going around a corner in a vehicle at high speed. We "feel a force" that pushes our body away from the inside of the curve that we are traveling around. There is no force. Our bodies our trying to follow a straight line, but our accelerated reference frame (the car is accelerating because it is changing direction) causes us to invent the pseudoforce to explain why our bodies move relative to the reference frame toward the outside of the curve.

## Section 11-10 The Coriolis Effect

A particular pseudoforce that occurs for observers in rotating reference frames is the **Coriolis force**. The acceleration of an object associated with this pseudoforce is called the **Coriolis acceleration**. The magnitude of the Coriolis acceleration, $a_{Coriolis}$ is given by

$$a_{Coriolis} = 2\omega\, v_\perp$$

where $\omega$ is the angular velocity of the noninertial reference frame relative to an inertial reference frame and $v_\perp$ is the component of the object's velocity perpendicular to the angular velocity of the noninertial reference frame.

## Example 11-10-A

A child rides on a playground merry-go-round that is making one rotation every 3.24 s. The child throws a ball with a speed of 8.22 m/s radially outward from the axis of rotation of the merry-go-round. The ball is released a distance 1.16 m from the axis of rotation. What is the magnitude of the initial acceleration of the released ball observed by the child on the merry-go-round? Ignore the relatively small contribution from the Earth being a rotating reference frame.

## Solution:

The child will observe a Coriolis acceleration of the ball and gravitational acceleration of the ball. As the ball is moving initially radially outward it will have a Coriolis acceleration in a direction tangential to the circular path of points on the merry-go-round, but opposite in direction. The magnitude of the Coriolis acceleration will be

$$a_{Coriolis} = 2\omega v_\perp = 2\frac{2\pi}{T}v_\perp = \frac{4\pi}{3.24\,\text{s}}(8.22\,\text{s}) = 31.9\,\text{m/s}$$

The gravitational acceleration will be 9.8 m/s$^2$ and vertically downward. The magnitude of the observed acceleration will be

$$a = \sqrt{a_{Coriolis}^2 + a_{gravity}^2} = \sqrt{(31.9\,\text{m/s}^2)^2 + (9.8\,\text{m/s}^2)^2} = 33.4\,\text{m/s}^2$$

## Practice Quiz

1.    Two vectors have non-zero magnitude.  Under which condition do you know the vectors are perpendicular to each other?

   a)  The vector sum of the two vectors is zero.
   b)  The vector difference of the two vectors is zero.
   c)  The cross product of the two vectors is zero.
   d)  The scalar product of the two vectors is zero.

2.    Which direction does the angular momentum vector of the Earth point?

   a)  Directly upward at your current location
   b)  Directly upward at the North Pole
   c)  Directly upward at the South Pole
   d)  Toward the east

3.    When tightening a screw with right-hand threads (this is the standard type of threads used which require turning clockwise to tighten), which direction is the torque vector required to tighten the screw?

   a)  In the direction the screw will advance
   b)  Opposite the direction the screw will advance
   c)  To the right as you look at the head of the screw
   d)  To the left as you look at the head of the screw

4.    Under what conditions does the equation $\Sigma\tau = d\mathbf{L}/dt$ apply for a system of particles?

   a)  Under all conditions
   b)  Only if the system consists of a single particle
   c)  Only if the center of mass of the system is at rest
   d)  If the quantities are calculated in an inertial reference frame or about the center of mass

5.    In the Southern Hemisphere, which direction is the horizontal component of the Coriolis acceleration for an object moving straight north?

   a)  west
   b)  east
   c)  north
   d)  south

6.    What is the magnitude of the vector cross product of two vectors of magnitude 1 that add to zero?

   a)  1
   b)  2
   c)  Square root of 2
   d)  0

7.    A particle of mass *m* moves along a straight line with an acceleration that is non-zero.  Where can an axis be located such that the angular momentum of the particle is not constant?

   a)  Any point on the path of the particle
   b)  A point that is instantaneously at the location of the particle
   c)  Any point not on the path of the particle
   d)  There are no such points.

8.    An object is spinning freely about a fixed axis when a small piece is ejected from the remainder. The remainder will remain spinning with its initial angular velocity if the eject piece leaves with a velocity

a) that is directly away from the axis of rotation.
b) zero.
c) equal to the velocity of the center of mass of the initial object.
d) equal to its velocity on the object at the moment it is ejected.

9.    What happens to the precession rate of a spinning top as the angular velocity of the top about its axis decreases?

a) The precession rate remains constant.
b) The precession rate decreases.
c) The precession rate increases.
d) It depends on the angle of tilt of the top.

10.    Near the surface of the Earth, where will the Coriolis acceleration be greatest for an object that is allowed to free fall?

a) Near either of the poles
b) Near the equator
c) At a latitude of 45°
d) Same at all latitudes

11.    Calculate the cross product of a vector $\mathbf{A} = 3\mathbf{i} + \mathbf{j} - 4\mathbf{k}$ and a vector $\mathbf{B} = -\mathbf{i} + \mathbf{j} + \mathbf{k}$.

12.    Determine the torque about point $O$ for a force $(2\,\mathbf{i} + 3\,\mathbf{j})$N that acts at a point displaced from $O$ by the vector $\mathbf{r} = (\mathbf{i} - 2\,\mathbf{j})$m.

13.    An asteroid of mass $14 \times 10^6$ kg strikes the earth traveling tangential to the Earth at a latitude of 40° traveling south at a speed of 12 km/s. What will the tilt of the Earth's rotation axis be from its original direction? Treat the earth as a uniform density sphere.

14.    The turbine of a jet engine rotates at 12,000 rev/min. The two bearings that it is mounted on can withstand a 100 N force perpendicular to the axis of rotation. The bearings have a moment arm of 13.8 cm for torque calculated about the center of mass of the turbine. What is the maximum component of angular momentum perpendicular to the axis of rotation that the bearings can support?

15.    Determine the precession rate of a top that is spinning at a rate of 34 rev/s. The top has a 1.6 cm radius of gyration. The center of mass of the top is 3.2 cm above the surface it is spinning on.

## Problem Solutions

3.    The magnitude of the tangential acceleration is $a_T = \alpha r$.
      From the diagram in the text, we see that $\alpha$, $\mathbf{r}$ and $a_{tan}$ are all perpendicular,
      and rotating $\alpha$ into $\mathbf{r}$ gives a vector in the direction of $a_{tan}$.
      Thus we have $a_{tan} = \alpha \times \mathbf{r}$.
      The magnitude of the radial acceleration is $a_R = \omega^2 r = \omega r \omega = \omega v$.
      From the diagram we see that $\omega$, $\mathbf{v}$ and $a_R$ are all perpendicular,
      and rotating $\omega$ into $\mathbf{v}$ gives a vector in the direction of $a_R$.
      Thus we have $a_R = \omega \times \mathbf{v}$.

7.    (a)    For the vectors $\mathbf{A} = 7.0\mathbf{i} - 3.5\mathbf{j}$ and $\mathbf{B} = -8.5\mathbf{i} + 7.0\mathbf{j} - 2.0\mathbf{k}$, we have

$$\mathbf{A} \times \mathbf{B} = \begin{vmatrix} \mathbf{i} & \mathbf{j} & \mathbf{k} \\ 7.0 & -3.5 & 0 \\ -8.5 & 7.0 & 2.0 \end{vmatrix}$$

$$= [(-3.5)(2.0) - (0)(7.0)]\mathbf{i} + [(0)(-8.5) - (7.0)(2.0)]\mathbf{j} + [(7.0)(7.0) - (-3.5)(-8.5)]\mathbf{k}$$

$$= -7.0\mathbf{i} - 14.0\mathbf{j} + 19.3\mathbf{k}$$

(b) The magnitudes of the vectors are
$A = [(7.0)^2 + (-3.5)^2]^{1/2} = 7.83$;
$B = [(-8.5)^2 + (7.0)^2 + (2.0)^2]^{1/2} = 11.2$;
$|\mathbf{A} \times \mathbf{B}| = [(-7.0)^2 + (-14.0)^2 + (19.3)^2]^{1/2} = 24.8$.
We find the angle between **A** and **B** from
$|\mathbf{A} \times \mathbf{B}| = |\mathbf{A}| \, |\mathbf{B}| \sin \theta$;
$24.8 = (7.83)(11.2) \sin \theta$, which gives $\sin \theta = 0.283$,
which gives $\theta = 16°, 164°$. From the diagram, we see that $\theta = 164°$.

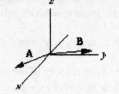

11. We write the force as
$\mathbf{F} = F[(\cos \theta)\mathbf{i} + (\sin \theta)\mathbf{j}] = (188 \text{ N})(\cos 33.0° \, \mathbf{i} + \sin 33.0° \, \mathbf{j})$.
For the torque we have
$\tau = \mathbf{r} \times \mathbf{F} = [(0.220\mathbf{i} + 0.335\mathbf{j}) \text{ m}] \times [(188 \text{ N})(\cos 33.0° \, \mathbf{i} + \sin 33.0°\mathbf{j})]$
$= (188 \text{ N})[(0.220 \text{ m})(\sin 33.0°) - (0.335 \text{ m})(\cos 33.0°)]\mathbf{k}$
$= -(30.0 \text{ m} \cdot \text{N})\mathbf{k}$  (in $-z$-direction).

15. For a particle moving in a circle, the magnitude of the angular momentum is $l = mvr$, and the moment of inertia is $I = mr^2$. The kinetic energy is
$K = mv^2/2 = m(l/mr)^2/2 = l^2/2mr^2 = l^2/2I$.

19. For the angular momentum we have
$l = r \times p = mr \times v$
$= (7.6 \text{ kg}) \left\{ \begin{array}{l} [(2.0\,\text{m})(-3.1\,\text{m/s}) - (3.0\,\text{m})(-4.5\,\text{m/s})]\mathbf{i} + [(3.0\,\text{m})(-5.0\,\text{m/s}) - (1.0\,\text{m})(-3.1\,\text{m/s})]\mathbf{j} \\ + [(1.0\,\text{m})(-4.5\,\text{m/s}) - (2.0\,\text{m})(-5.0\,\text{m/s})]\mathbf{k} \end{array} \right\}$
$= 55\mathbf{i} - 90\mathbf{j} + 42\mathbf{k}$

23. With the positive direction CCW, for the angular momentum about the axis of the pulley we have
$L = R_0 mv + I\omega = R_0 mv + I(v/R_0) = [R_0 m + (I/R_0)]v$.
The net torque is produced by $mg$ and the frictional torque:
$\tau = dL/dt$;
$mgR_0 - \tau_{fr} = [R_0 m + (I/R_0)] \, dv/dt$, which gives
$a = (mgR_0 - \tau_{fr})[R_0 m + (I/R_0)]$
$= [(1.53 \text{ kg})(9.80 \text{ m/s}^2)(0.330 \text{ m}) - 1.10 \text{ m}\cdot\text{N}]/$
$\quad \{(0.330 \text{ m})(1.53 \text{ kg})) + [(0.385 \text{ kg}\cdot\text{m}^2)/(0.330 \text{ m})]\}$
$= 2.30 \text{ m/s}^2$.

27. The impulse changes the linear momentum of the center of mass:
$J = \Delta p = mv_{CM}$;
$8.5 \times 10^{-3} \text{ N}\cdot\text{s} = (0.040 \text{ kg})v_{CM}$, which gives $v_{CM} = 0.213 \text{ m/s}$.
The moment of the impulse about the center of mass changes the angular momentum:
$rJ = \Delta L = I_{CM}\omega_{CM}$;
$[\frac{1}{2}(0.070 \text{ m}) - 0.020 \text{ m}](8.5 \times 10^{-3} \text{ N}\cdot\text{s}) = [(0.040 \text{ kg})(0.070 \text{ m})^2/12]\omega_{CM}$,
which gives $\omega_{CM} = 7.8 \text{ rad/s}$.
The rod rotates at 7.8 rad/s about the center of mass, which moves with constant velocity of 0.21 m/s.

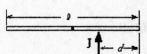

31. If we choose the center of the circle that $m_1$ makes, O′, as origin, the angular momentum will be parallel to $\omega$.
It will have a magnitude
$L = (r \sin \phi)m_1 v_1 = m_1 r^2 \omega \sin^2 \phi$.
**L** is constant in both magnitude and direction, so we have
$\tau_{net} = dL/dt = 0$;
$F_1(d - r \cos \phi) - F_2(d + r \cos \phi) = 0$, or
$F_2 = [(d - r \cos \phi)/(d + r \cos \phi)]F_1$.
The radial acceleration of the mass must be produced by a radial force in the rod:

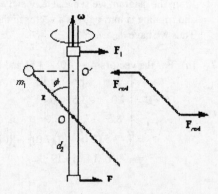

$F_{\text{rod}} = mv^2/(r \sin \phi) = m_1 r \omega^2 \sin \phi.$

Because the rod has no mass, the net force on the rod is zero.

For the net horizontal force on the axle we have

$F_1 + F_2 - F_{\text{rod}} = 0,$ or $F_1 + F_2 = m_1 r \omega^2 \sin \phi.$

When we combine this with the result from the torque equation, we have

$F_1 + [(d - r \cos \phi)/(d + r \cos \phi)]F_1 = m_1 r \omega^2 \sin \phi,$ which gives

$F_1 = [(d + r \cos \phi)/2d]m_1 r \omega^2 \sin \phi,$

$F_2 = [(d - r \cos \phi)/2d]m_1 r \omega^2 \sin \phi.$

35. In the collision, during which we ignore any motion of the rod,
angular momentum about the pivot point will be conserved:

$L_i = L_f;$

$mv(\tfrac{1}{2}l) + 0 = I_{\text{total}}\omega = [\tfrac{1}{3}M l^2 + m(\tfrac{1}{2}l)^2]\omega,$ which gives

$\omega = 6mv/(4M + 3m)l.$

For the rotation about the pivot after the collision, energy will be
conserved. If the center of the rod reaches a height $h$, the bottom
of the rod will swing to a height $H = 2h$, so we have

$K_i + U_i = K_f + U_f;$

$\tfrac{1}{2}I_{\text{total}}\omega^2 + 0 = 0 + (m + M)gh;$

$\tfrac{1}{2}[\tfrac{1}{3}M l^2 + m(\tfrac{1}{2}l)^2][6mv/(4M + 3m) l]^2 = (m + M)gH/2,$

which gives $H = 3m^2v^2/g(3m + 4M)(m + M).$

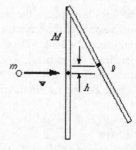

39. (a) We find the speed of the center of mass from the conservation of
linear momentum:

$Mv + 0 = (M + m)v_{\text{CM}};$

$(200 \text{ kg})(18 \text{ m/s}) = (200 \text{ kg} + 50 \text{ kg})v_{\text{CM}},$ which gives $v_{\text{CM}} = 14$ m/s.

(b) During the collision, angular momentum about the center of mass
will be conserved. We find the location of the center of mass relative
to the center of the beam:

$d = m(\tfrac{1}{2}l)/(M + m) = (50 \text{ kg}) \tfrac{1}{2} (2.0 \text{ m})/(200 \text{ kg} + 50 \text{ kg}) = 0.20$ m.

When we use the parallel-axis theorem for the moment of inertia of
the beam, angular momentum conservation gives us

$L_i = L_f;$

$Mvd + 0 = I_{\text{total}}\omega = [(M l^2/12) + Md^2 + m(\tfrac{1}{2}l - d)^2]\omega,$

$(200 \text{ kg})(18 \text{ m/s})(0.20 \text{ m}) = ((200 \text{ kg})\{[(2.0 \text{ m})^2/12]+(0.20 \text{ m})^2\}+(50 \text{ kg})(1.0 \text{ m}-0.20 \text{ m})^2)\omega$

which gives

$\omega = 6.8$ rad/s.

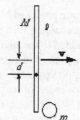

43. We find the rate of precession from

$\Omega = Mgr/L = Mgr/I\omega = Mgr/(\tfrac{1}{2}MR^2)\omega = 2gr/R^2\omega$

$= 2(9.80 \text{ m/s}^2)(0.10 \text{ m})/(0.060 \text{ m})^2(250 \text{ rad/s}) = 2.2$ rad/s  (0.35 rev/s).

47. A convenient dimensionless factor is

$g/r_E\omega^2 = (9.80 \text{ m/s}^2)/(6.38 \times 10^6 \text{ m})[(2\pi \text{ rad})/(86,400 \text{ s})]^2$

$= 290.4.$

(a) At the North Pole there is no radial acceleration,
so the effective acceleration of gravity is
$g$, along a radial line.

(b) At a latitude $\phi$ there will be a pseudoforce $mr\omega^2$ away
from the axis, where $r = r_E \cos \phi$. We use the coordinate
system shown on the diagram, with positive $y$ down
along the radial line. For the components of $g'$ we have

$g_x' = r\omega^2 \sin \phi = r_E\omega^2 \sin \phi \cos \phi,$

$g_y' = g - r\omega^2 \cos \phi = g - r_E\omega^2 \cos^2 \phi.$

We find the angle from

$\tan \theta = g_x'/g_y' = r_E\omega^2 \sin \phi \cos \phi/(g - r_E\omega^2 \cos^2 \phi)$

$= \sin \phi \cos \phi/[(g/r_E\omega^2) - \cos^2 \phi].$

At a latitude of 45°, we get

$\tan \theta = \sin 45° \cos 45°/(290.4 - \cos^2 45°) = 1.725 \times 10^{-3},$ or $\theta = 0.0988°$

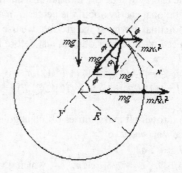

We can find the magnitude from

$g' = g_x'/\sin\theta = g\sin\phi\cos\phi/(g/r_E\omega^2)\sin\theta = g\sin 45°\cos 45°/(290.4)\sin 0.0988° = 0.998g$.

Thus the effective acceleration of gravity is $0.998g$, $0.0988°$ south from a radial line.

(c) At the equator $\phi = 0$, so we have

$\tan\theta = 0$;

as expected the effective acceleration is along a radial line.

We find the magnitude from

$g' = g_y' = g - r_E\omega^2\cos^2\phi = g[1 - 1/(g/r_E\omega^2)] = g[1 - 1/(290.4)] = 0.997g$, along a radial line.

51. (a) In the northern hemisphere, the projectile will be deflected to the right, that is, south.

(b) The initial velocity of the projectile is perpendicular to the rotation axis, so the Coriolis acceleration is $a_{cor} = 2\omega v_0$. The direction of this acceleration is perpendicular to the axis and **v** (thus, parallel to the equator), so it has components $a_{cor,up} = 2\omega v_0\cos\lambda$, and $a_{cor,south} = 2\omega v_0\sin\lambda$. If we ignore the vertical acceleration, which is an effective $g$, the distance the projectile travels east in a time $t$ is $D = v_0 t$. In this time, the projectile will have deflected south a distance

$s_{south} = \frac{1}{2}(2\omega v_0\sin\lambda)t^2$.

Thus we have

$s_{south} = \omega v_0\sin\lambda(D/v_0)^2 = \omega D^2\sin\lambda/v_0$.

(c) For the given data, we have

$s_{south} = (2\pi/86,400\text{ s})(3.0\times10^3\text{ m})^2\sin 45°/(1000\text{ m/s}) = 0.46\text{ m}$.

55. (a) The torque produced by the normal force about the center of mass causes a change $\Delta\mathbf{L}$ in the direction of the tire's motion. When this is added to the angular momentum **L**, we see that the tire will turn in the direction of the lean.

(b) There is no vertical acceleration, so we have $F_N = mg$. The change in the angular momentum about the center of mass is

$\Delta L = \tau\Delta t = mgr\sin\theta\,\Delta t$
$= (9.0\text{ kg})(9.80\text{ m/s}^2)(0.32\text{ m})\sin 10°(0.20\text{ s}) = 0.98\text{ kg·m}^2/\text{s}$.

The original angular momentum is

$L_0 = I\omega = (0.83)[(2.1\text{ m/s})/(0.32\text{ m})] = 5.5\text{ kg·m}^2/\text{s}$.

Thus we have

$\Delta L/L_0 = (0.98\text{ kg·m}^2/\text{s})/(5.5\text{ kg·m}^2/\text{s}) = 0.18$, so $\Delta L = 0.18L_0$.

59. Because there is no friction, the center of mass must fall straight down. The vertical velocity of the right end of the stick must always be zero. If $\omega$ is the angular velocity of the stick just before it hits the table, the velocity of the right end with respect to the center of mass will be $\omega(l/2)$ up. Thus we have

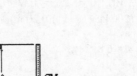

$\omega(l/2) - v_{CM} = 0$, or $\omega = 2v_{CM}/l$.

The kinetic energy will be the translational energy of the center of mass and the rotational energy about the center of mass. With the reference level for potential energy at the ground, we use energy conservation to find the speed of the center of mass just before the stick hits the ground:

$K_i + U_i = K_f + U_f$;
$0 + Mg\frac{1}{2}l = \frac{1}{2}Mv_{CM}^2 + \frac{1}{2}(Ml^2/12)\omega^2 + 0$;
$Mg\frac{1}{2}l = \frac{1}{2}Mv_{CM}^2 + \frac{1}{2}(Ml^2/12)(2v_{CM}/l)^2 = \frac{1}{2}(4Mv_{CM}^2/3)$, which gives $v_{CM} = (3gl/4)^{1/2}$.

63. (a) If the thrown-off mass carries off no angular momentum, from conservation of angular momentum for the star we have

$I_i\omega_i = I_f\omega_f$;
$2/5Mr_s^2\omega_1 = 2/5(\frac{1}{4}M)r_n^2\omega_2$, which gives
$\omega_2 = 4(r_s/r_n)^2\omega_1 = 4(7.0\times10^8\text{ km}/10\text{ km})^2(1\text{ rev}/10\text{ days})/(86,400\text{ s/day}) = 2.3\times10^4\text{ rev/s}$.

(b) If the thrown-off mass carries off ¾ of the initial angular momentum, from conservation of angular momentum for the star we have

$\frac{1}{4}I_i\omega_i = I_f\omega_f$;
$\frac{1}{4}(2/5Mr_s^2)\omega_1 = 2/5(\frac{1}{4}M)r_n^2\omega_2$, which gives
$\omega_2 = (r_s/r_n)^2\omega_1 = (7.0\times10^8\text{ km}/10\text{ km})^2(1\text{ rev}/10\text{ days})/(86,400\text{ s/day}) = 5.7\times10^3\text{ rev/s}$.

# Chapter 12: Static Equilibrium: Elasticity and Fracture

## Chapter Overview and Objectives

This chapter defines equilibrium and stability and discusses the conditions for equilibrium and stability. This chapter also describes the properties of solid materials when forces are applied to those materials. The concepts of stress and strain are introduced. Hooke's law is applied to different types of stress. Several structures are introduced that are used to take advantage of the greater strength of materials under compressive stress.

After completing study of this chapter, you should:

- Know the conditions for static equilibrium.
- Know how to solve statics problems.
- Know the different classifications of stability and what leads to those conditions.
- Know how to relate a strain to a corresponding stress in the Hooke's law limit.
- Know the names of the different types of stresses that can be applied to solids.

## Summary of Equations

Conditions for equilibrium:

$$\sum \mathbf{F} = 0$$
$$\sum \boldsymbol{\tau} = 0$$

General definition of stress:

$$\text{stress} = \frac{\text{force}}{\text{area}} = \frac{F}{A}$$

General definition of strain:

$$\text{strain} = \frac{\text{distortion}}{\text{length}} = \frac{\Delta L}{L_0}$$

Hooke's law for compressive or tensile stress:

$$\frac{F}{A} = E \frac{\Delta L}{L}$$

Hooke's law for shear stress:

$$\frac{F}{A} = G \frac{\Delta L}{L}$$

Hooke's law for isotropic pressure:

$$\Delta P = -B \frac{\Delta V}{V}$$

## Chapter Summary

### Section 12-1 Statics–The Study of Forces in Equilibrium

A body that is not accelerating must have no net force and no net torque acting on it. If there are forces acting on such a body, then the sum of the forces must be zero. A body in this condition is in **equilibrium**. The subject of statics is the study of objects in equilibrium.

### Section 12-2 The Conditions for Equilibrium

For an object to be in equilibrium, the sum of the forces on the object must be zero:

$$\sum \mathbf{F} = 0$$

or, written in terms of the components of the forces in the three Cartesian directions:

$$\sum F_x = 0 \qquad \sum F_y = 0 \qquad \sum F_z = 0$$

Also, the net torque acing on the object must be zero:

$$\sum \tau = 0$$

or, written in terms of the components of the torque in three Cartesian directions:

$$\sum \tau_x = 0 \qquad \sum \tau_y = 0 \qquad \sum \tau_z = 0$$

Most of this chapter is restricted to the case where all the forces are in a single plane. In this case, the equations simplify to

$$\sum F_x = 0 \qquad \sum F_y = 0 \qquad \sum \tau_z = 0$$

where we have chosen the $xy$ plane as the plane the forces are in.

### Example 12-2-A

Two children sit on a seesaw. One child has a mass of 36 kg and sits a distance of 1.3 m from the fulcrum. The second child has a mass of 41 kg. How far from the fulcrum must the second child sit to balance the seesaw? What is the force of the fulcrum on the seesaw?

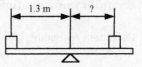

### Solution:

We know that in equilibrium, the children apply a downward force on the seesaw equal to their weight. We draw the free-body diagram for the seesaw:

We recognize that the forces all lie in the same plane. Thus, we have three conditions to satisfy:

$$\sum F_x = 0 \qquad \sum F_y = 0 \qquad \sum \tau_z = 0$$

We use our conventional coordinate system, with $x$ horizontal and positive to the right and $y$ vertical and postive upward (toward the top of the page in the diagram). The condition on the components of forces in the $x$ direction is satisfied by default. The condition on the forces in the y direction is:

$$-F_1 + F_{fulcrum} - F_2 = 0 \quad \Rightarrow \quad F_{fulcrum} = F_1 + F_2 = (m_1 + m_2)g$$

The condition on the sum of the torques can be written in terms of the forces once an axis for calculating the torques is specified. We will use an axis through the point of contact of the fulcrum with the seesaw in a direction perpendicular to the page. Then the condition on the torques is

$$F_1(1.3\,m) + F_{fulcrum}(0\,m) - F_2 x_2 = 0 \quad \Rightarrow \quad x_2 = \frac{F_1(1.3\,m)}{F_2} = \frac{m_1(1.3\,m)}{m_2} = \frac{(36\,kg)(1.3\,m)}{(41\,kg)} = 1.1\,m$$

We have used that clockwise torques are negative and counterclockwise torques are positive.

### Section 12-3 Solving Statics Problems

One procedure to follow in solving statics problems is as follows:

1.  Consider one body in the problem at a time. Draw a free body diagram for the body under consideration, showing a force for each force acting on the body. Draw the force so that it acts at the correct point of application of the force.

2.  Choose a Cartesian coordinate system that is convenient. Often it is helpful to choose a coordinate system that is aligned with the direction of as many of the forces as possible. Resolve each of the forces into components along the Cartesian directions.

3.    Use symbols to represent unknowns and write down the equilibrium conditions.

4.    Choose an axis perpendicular to the plane of the forces to calculate torque. Any axis perpendicular to the plane of the forces is sufficient. The equations to solve are usually simplified by picking the origin at the point of application of one of the unknown forces. Write down the torques being careful to include the proper sign for whether the torque would rotate the object clockwise or counterclockwise.

5.    Solve the system of equations for the unknowns.

**Example 12-3-A**

An irregularly-shaped sign is supported from the ceiling by two cables as shown. The tension in the left cable is 40 N. What is the tension in the right cable? What is the mass of the sign? What can you say about the location of the center of mass of the sign?

**Solution:**

Draw the free body diagram of the sign. The sum of the horizontal components of the forces is

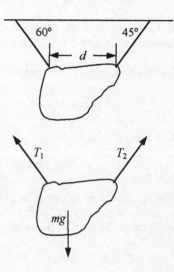

$$\sum F_x = -T_1 \cos 60^\circ + T_2 \cos 45^\circ = 0$$

$$\Rightarrow T_2 = \frac{T_1 \cos 60^\circ}{\cos 45^\circ} = 28.3 \, \text{N}$$

The sum of the vertical components of the forces is

$$\sum F_y = T_1 \sin 60^\circ + T_2 \sin 45^\circ - mg = 0$$

$$\Rightarrow m = \frac{T_1 \sin 60^\circ + T_2 \sin 45^\circ}{g} = \frac{(40 \, \text{N}) \sin 60^\circ + (28.3 \, \text{N}) \sin 45^\circ}{(9.8 \, \text{m}/\text{s}^2)}$$

$$= 5.6 \, \text{kg}$$

To determine the location of the center of mass of the object, consider the torque about the point of intersection of the two lines of force of the two cables, point O. Neither tension in the two cables causes a torque about this point. That means the third force, the force of gravity on the sign must exert zero torque also about this point. The only way that can be possible is if the gravitational force acts on the body at a point along the vertical line that passes through point O. That implies the center of mass of the sign lies on the vertical line passing through point O.

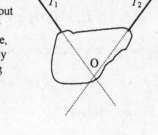

**Example 12-3-B**

Consider the three legged table shown in the diagram. The table top has a uniform density and a mass of 60 kg. What is the force of each leg on the table?

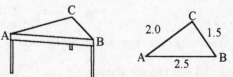

**Solution:**

This is a case where all the forces do not act in the same plane. In these cases, we may need to use that the some of the torques is zero about axes lying in different directions to solve the problem. In this case, there are three unknown forces and the weight force of the table, all in the vertical direction. Because the forces are all vertical, the sums of the forces in the two horizontal directions do not help us in finding the unknown forces acting on the table. We need two sums of different components of torques adding to zero to solve the problem.

Our first torque equation, we pick an axis that lies along the edge of the table from leg C to leg B, as shown. We also use the right-hand coordinate system shown in the diagram. The forces, expressed in terms of the Cartesian unit vectors are

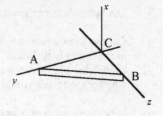

$$\mathbf{F}_A = F_A\,\mathbf{i} \qquad \mathbf{F}_B = F_B\,\mathbf{i} \qquad \mathbf{F}_C = F_C\,\mathbf{i} \qquad \mathbf{F}_W = -mg\,\mathbf{i}$$

The displacements of the positions at which the forces act from the rotation axis are given by

$$\mathbf{r}_A = (2.00\,\text{m})\mathbf{j} \qquad \mathbf{r}_B = 0 \qquad \mathbf{r}_C = 0 \qquad \mathbf{r}_W = (0.67\,\text{m})\mathbf{j}$$

where we have used that the centroid of a triangle is two-thirds of the distance from the vertex to the bisection of the opposite side to locate the center of mass of the table.

The total torque about the rotation axis is given by

$$\sum \tau = \mathbf{r}_A \times \mathbf{F}_A + \mathbf{r}_B \times \mathbf{F}_B + \mathbf{r}_C \times \mathbf{F}_C + \mathbf{r}_W \times \mathbf{F}_W = -(2.00\,\text{m})F_A\,\mathbf{k} + (0.67\,\text{m})mg\,\mathbf{k} = 0$$

If we solve this for $F_A$, we get

$$F_A = (0.333)mg = (0.333)(60\,\text{kg})(9.8\,\text{m/s}) = 196\,\text{N}$$

Now, we choose an axis along the A to C edge of the table, as shown. We use the same coordinate system as before. Now the displacements of the positions at which the forces act from the rotation axis are given by

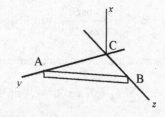

$$\mathbf{r}_A = 0 \qquad \mathbf{r}_B = (1.50\,\text{m})\mathbf{k} \qquad \mathbf{r}_C = 0 \qquad \mathbf{r}_W = (0.50\,\text{m})\mathbf{k}$$

The total torque about the rotation axis is given by

$$\sum \tau = \mathbf{r}_A \times \mathbf{F}_A + \mathbf{r}_B \times \mathbf{F}_B + \mathbf{r}_C \times \mathbf{F}_C + \mathbf{r}_W \times \mathbf{F}_W = (1.50\,\text{m})F_B\,\mathbf{k} - (0.50\,\text{m})mg\,\mathbf{k} = 0$$

If we solve this for $F_B$, we get

$$F_B = (0.333)mg = (0.333)(60\,\text{kg})(9.8\,\text{m/s}) = 196\,\text{N}$$

The sum of the forces in the vertical direction is

$$F_A + F_B + F_C - mg = 0$$

We can solve this for FC:

$$F_C = mg - F_A - F_B = (60\,\text{kg})(9.8\,\text{m/s}^2) - 196\,\text{N} - 196\,\text{N} = 196\,\text{N}$$

The forces on the table from each leg are identical.

### Section 12-4 Stability and Balance

Objects can be in three different conditions of **stability** when in equilibrium. Stability refers to the type of forces and torque that act on the body when it is displaced slightly from its equilibrium position and orientation. If the forces and torque act to return the object to its equilibrium position and orientation, the object is in **stable equilibrium**. If the forces or torque act to push the object further from its equilibrium position or orientation, the object is in **unstable equilibrium**. If there are no forces or torque on the object when it is displaced from its equilibrium position and orientation, the object is in **neutral equilibrium**.

In many cases, we are interested in insuring that objects are in stable equilibrium.

**Section 12-5 Elasticity and Elastic Moduli; Stress and Strain**

Up to this point, bodies have been treated as being rigid. However, all materials change size and shape when forces are applied to them. In this chapter, we look at some simple models of the change in size and shape of materials under applied forces.

In order to separate the properties of a type of material from the size and shape of an object, we introduce the concepts of stress and strain. **Stress** is defined generally as force per area:

$$\text{Stress} = \frac{\text{force}}{\text{area}} = \frac{F}{A}$$

where $F$ is the magnitude of the force applied to the surface of area $A$. In response to a given stress, a piece of material will change size by an amount characterized by a **strain**. Strain is defined as the distance of distortion, $\Delta L$, divided by the length of the object perpendicular to the surface on which force is applied, $L_0$

$$\text{Strain} = \frac{\text{distortion}}{\text{length}} = \frac{\Delta L}{L_0}$$

In the limit of small distortions, the stress is proportional to the strain:

$$\text{stress} \propto \text{strain}$$

The constant of proportionality is called an **elastic modulus**. The elastic modulus is a property of the material. It does not depend on the geometry of the object. Hooke's law for springs is an instance of this general rule. In all cases, once the strain becomes large enough, the relationship between stress and strain changes from proportionality. If the strain is below the **elastic limit**, the object will return to its original undistorted size and shape once the stress is removed. If the material is strained beyond the elastic limit, the material undergoes **plastic deformation**, in which it undergoes internal rearrangement and does not return to its original size or shape when the stress is removed. For even greater stress, the material will reach its **ultimate strength** and break apart.

Stress can be characterized by the direction of the forces applied to a given surface. In each of the following cases, assume that equal magnitude, opposite direction forces are acting on the bottom face of the rectangular prism so that the net force and torque on the rectangular prism is zero.

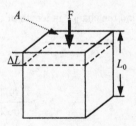

1. **Tension**: The force is directed outward normal to the surface under stress. The strain is an elongation of the rectangular volume element. We write

$$\frac{F}{A} = E\frac{\Delta L}{L}$$

where $E$ is called the elastic or Young's modulus.

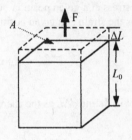

2. **Compression:** The force is directed inward normal to the surface under stress. The strain is a decrease in length of the rectangular volume element. We write

$$\frac{F}{A} = E\frac{\Delta L}{L}$$

where again $E$ is the elastic modulus.

3. **Shear:** The force is directed parallel to the surface under stress. The strain is a relative displacement of the top surface from the bottom surface parallel to the plane of the surfaces. We write

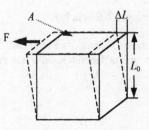

$$\frac{F}{A} = G\frac{\Delta L}{L}$$

where $G$ is the called the shear modulus.

4. **Isotropic pressure:** If a uniform force per area on all surfaces acts perpendicular to the surface then the object is under isotropic pressure. The strain, in this case, is the fractional volume change of the object, $\Delta V/V$. The uniform force per area is called the pressure, $P$. The Hooke's law relationship in this case is written in terms of the change in pressure, $\Delta P$, on the object. This measures the added stress to the object. The relationship between the stress and the strain is

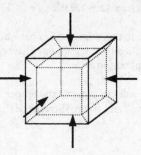

$$\Delta P = -B\frac{\Delta V}{V}$$

where $B$ is called the bulk modulus of the fluid.

**Example 12-5-A**

A 500 m long (with no stress) steel cable is used to suspend a 200 kg mass from a cliff. The diameter of the cable is ½ inch. How much is the cable stretched?

**Solution:**

At a given point along the cable, the stress must support the weight of 200 kg mass at the bottom plus the weight of the cable below that point. If we use a coordinate system with x positive upwards and zero at the bottom of the cable, then the weight of the cable below point x, call it $W(x)$, is given by

$$W(x) = \rho Vg = \rho Axg$$

where $\rho$ is the density of the steel cable, $A$ is the cross-sectional area of the cable, and $g$ is the acceleration of gravity. The total tension force on the cable at a given point is $W(x)$ and the weight of the 200 kg load:

$$F(x) = W(x) + mg = (\rho Ax + m)g$$

The stress at a given position on the cable is the force divided by the cross-sectional area of the cable. Hooke's law then gives the strain at a given position along the cable:

$$\frac{\Delta L}{L} = \frac{1}{E}\left(\rho x + \frac{m}{A}\right)g$$

We will define $d\Delta L$ as the change in length of a segment of length $L = dx$. This is given by

$$d\Delta L = \frac{1}{E}\left(\rho x + \frac{m}{A}\right)g\, dx$$

To find the change in length of the entire cable, we integrate over the length of the cable:

$$\Delta L = \int d\Delta L = \int_0^L \frac{1}{E}\left(\rho x + \frac{m}{A}\right)g\, dx = \frac{1}{2}\frac{\rho gx^2}{E} + \frac{mg}{EA}x\bigg|_0^L = \frac{g}{E}\left(\frac{1}{2}\rho L^2 + \frac{m}{A}L\right)$$

$$= \frac{\left(9.8\,\text{m/s}^2\right)}{\left(2.00\times10^{11}\,\text{N/m}^2\right)}\left[\frac{1}{2}\left(7800\,\text{kg/m}^3\right)\left(500\,\text{m}\right)^2 + \frac{\left(200\,\text{kg}\right)}{\pi\left[\left(0.500\,\text{in}\right)\left(0.0254\,\text{m/in}\right)\right]^2/4}\left(500\,\text{m}\right)\right]$$

$$= 0.057\,\text{m}$$

**Section 12-6 Fracture**

If the stress exceeds a limit called the **ultimate strength**, the material breaks apart. The magnitude of the ultimate strength is dependent on which type of stress is applied to the material.

**Example 12-6-A**

What is the longest steel cable hanging vertically that will support its own weight?

**Solution:**

Assume a cable of cross-sectional area $A$ and length $L$. Then the weight, $W$, of the cable is

$$W = mg = \rho V g = \rho A L g$$

The tensile stress at the top of the cable will be

$$\frac{F}{A} = \frac{\rho A L g}{A} = \rho L g$$

Solving for the length of the cable:

$$L = \frac{F}{A} \frac{1}{\rho g}$$

We want to know the length of the cable when the stress is equal to the ultimate strength of the steel cable. We find this value from Table 12-2 in the text to be $200 \times 10^6 \ \text{N/m}^2$. Solving for $L$:

$$L = \left(200 \times 10^6 \ \text{N/m}^2\right) \frac{1}{(7800 \ \text{kg})(9.8 \ \text{m/s}^2)} = 2.6 \times 10^3 \ \text{m}$$

**Section 12-7 Trusses and Bridges**

A **truss** is a structure that is composed of members that are always connected into triangles. Multiple triangles are used to form arbitrary shape trusses. The points at which beams are joined by pins are called **joints**. The model of a pin that is usually applied to problems is that the forces on a member at a pin cannot provide any net torque to the member about an origin centered on the pin.

**Section 12-8 Arches and Domes**

**Arches** and **domes** are used in construction to take advantage of the greater strength of materials to withstand compressive strain than tensile or shear strain.

## Practice Quiz

1.    Which of the following is an example of an object in neutral equilibrium?

    a) A chair resting on the floor
    b) A pencil standing on its point
    c) A sphere resting atop another sphere
    d) A sphere resting on a flat horizontal surface

2.    What is the ordering from smallest to largest value of the proportional limit, the ultimate strength, and the elastic limit of a given material?

    a) Proportional limit, ultimate strength, elastic limit
    b) Ultimate strength, proportional limit, elastic limit
    c) Ultimate strength, elastic limit, proportional limit
    d) Proportional limit, elastic limit, ultimate strength

3.    The typical behavior of a solid that has reached its proportional limit is that the strain for a given increase in stress is

a) Greater than before the proportional limit is reached
b) Less than before the proportional limit is reached
c) The same as before the proportional limit is reached
d) Unlimited, as this the point at which the material fractures

4.    A block with no applied stress is shown in the figure on the left. The figure on the right is the same block with stress applied. What type of stress has been applied to the block?

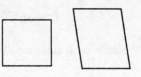

a) Shear and compression
b) Pressure and tension
c) Pressure and shear
d) Shear and tension

5.    What is the purpose of prestressing concrete?

a) To test to see if the concrete is strong enough for the intended load
b) To compress the concrete so that it fits in the space provided
c) To avoid tensile stress on the concrete
d) To stretch the concrete out, so it is more flexible

6.    What structural purpose does an architectural arch serve?

a) None; it only has aesthetic value.
b) To force the loads in the material to be mostly compressive
c) To provide a well-controlled fracture point upon failure
d) To place the structure into a condition of neutral equilibrium

7.    An object is stretched and returns to its original shape when released. The maximum stress that could have been applied for this to be true is

a) The proportional limit
b) The elastic limit
c) The ultimate strength
d) Zero

8.    A pencil standing vertically on its tip is an example of

a) Non-equilibrium
b) Stable equilibrium
c) Unstable equilibrium
d) Neutral equilibrium

9.    What are the predominant types of strains that occur in a bent beam?

a) Shear and pressure
b) Tension and shear
c) Tension and compression
d) Compression and shear

10.   What statement is necessarily true about a system that has no net torque and no net force acting on it?

a) The system is in static equilibrium.
b) The system is in stable static equilibrium.
c) The system is not in equilibrium.
d) The system is not undergoing translational or angular acceleration.

11.    Two children sit on the left side of a seesaw.  One has a mass of 24 kg and sits 1.3 m from the fulcrum.  The second has a mass of 31 kg and sits 1.6 m from the fulcrum.  What is the mass of a person that must sit on the right side of the seesaw a distance 1.8 m from the fulcrum in order to balance the see saw?

12.    In the diagram, a shelf supported by a wall and a wire is shown.  The shelf is 40 cm wide and has a mass of 2.4 kg.  The block sitting on the shelf is 30 cm from the wall and has a mass of 1.8 kg.  The wire is attached at the outer end of the shelf and to the wall a distance 40 cm above the shelf.  Determine the tension in the wire.  The mass of the wire is negligible.

13.    The triangular structure shown has uniform density and a mass of 32 kg.  Determine the maximum mass that can be supported from the upper vertex of the angle, as shown, before the entire structure is unstable.  The base of the triangle is 1.0 m in length, the upper side of the triangle is 2.4 m in length, and the remaining side is 2.0 m in length.

14.    A ladder leans against the wall at a 60° angle as shown.  The ladder is 8.0 ft long and has a mass of 12.5 kg.  The ladder starts to slip when a 88 kg person is three-fourths the way up the ladder.  What is the coefficient of static friction between the ladder and floor?  Ignore friction with the wall.

15.    A concrete column has a uniform cross-section that is 20 in by 32 in and is 10 ft tall.  What is the compression of the column when it carries a load of 300 tons?

## Problem Solutions

3.    We choose the coordinate system shown, with positive torque clockwise.  For the torque from the person's weight about the point B we have
$\tau_B = MgL = (56 \text{ kg})(9.80 \text{ m/s}^2)(3.0 \text{ m}) = 1.6 \times 10^3 \text{ m} \cdot \text{N}$.

7.    We choose the coordinate system shown, with positive torques clockwise.  We write $\Sigma \tau = I\alpha$ about the support point A from the force diagram for the board and people:
$\Sigma \tau_A = - m_1 g(L - d) - Mg(\tfrac{1}{2}L - d) + m_2 gd = 0$;
$-(23.0 \text{ kg})(10 \text{ m} - d) - (12.0 \text{ kg})(5.0 \text{ m} - d) + (67.0 \text{ kg})d = 0$,
which gives $d = 2.84$ m from the adult.

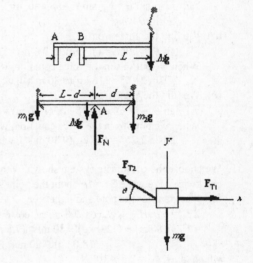

11.    From the force diagram for the mass we can write
$\Sigma F_x = F_{T1} - F_{T2} \cos \theta = 0$,  or
$F_{T1} = F_{T2} \cos 30°$.
$\Sigma F_y = F_{T2} \sin \theta - mg = 0$,  or
$F_{T2} \sin 30° = mg = (200 \text{ kg})(9.80 \text{ m/s}^2)$,
which gives $F_{T2} = 3.9 \times 10^3$ N.
Thus we have
$F_{T1} = F_{T2} \cos 30° = (3.9 \times 10^3 \text{ N}) \cos 30° = 3.4 \times 10^3$ N.

15.    We choose the coordinate system shown, with positive torques clockwise.  We write $\Sigma \tau = I\alpha$ about the lower hinge B from the force diagram for the door:
$\Sigma \tau_B = F_{Ax}(H - 2D) - Mg\tfrac{1}{2}w = 0$;
$F_{Ax}[2.30 \text{ m} - 2(0.40 \text{ m})] - (13.0 \text{ kg})(9.80 \text{ m/s}^2)(1.30 \text{ m})$,
which gives $F_{Ax} = 55.2$ N.
We write $\Sigma \mathbf{F} = m\mathbf{a}$ from the force diagram for the door:
$\Sigma F_x = F_{Ax} + F_{Bx} = 0$;
$55 \text{ N} + F_{Bx} = 0$, which gives $F_{Bx} = -55.2$ N.
The top hinge pulls away from the door, and the bottom hinge pushes on the door.
$\Sigma F_y = F_{Ay} + F_{By} - Mg = 0$.
Because each hinge supports half the weight, we have
$F_{Ay} = F_{By} = \tfrac{1}{2}(13.0 \text{ kg})(9.80 \text{ m/s}^2) = 63.7$ N.
Thus we have:
top hinge: $F_{Ax} = 55.2$ N, $F_{Ay} = 63.7$ N;  bottom hinge: $F_{Bx} = -55.2$ N, $F_{By} = 63.7$ N.

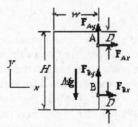

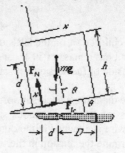

19. Because the person is standing on one foot, the normal force on the
    ball of the foot must support the weight: $F_N = Mg$. We choose the
    coordinate system shown, with positive torques clockwise. We write
    $\Sigma\tau = I\alpha$ about the point A from the force diagram for the foot:
    
    $\Sigma\tau_A = F_T d - F_N D = 0$;
    $F_T d - F_N(2d) = 0$, which gives
    $F_T = 2F_N = 2(70 \text{ kg})(9.80 \text{ m/s}^2) = 1.4 \times 10^3 \text{ N (up)}$.
    We write $\Sigma F_y = ma_y$ from the force diagram:
    $F_T + F_N - F_{bone} = 0$, which gives
    $F_{bone} = F_T + F_N = 3F_N = 3(70 \text{ kg})(9.80 \text{ m/s}^2) = 2.1 \times 10^3 \text{ N (down)}$.

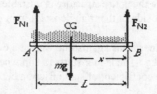

23. We choose the coordinate system shown, with positive torques
    clockwise. For the torques about the center of gravity we have
    $\Sigma\tau_{CG} = F_{N1}(L - x) - F_{N2}x = 0$;
    $(35.1 \text{ kg})g(170 \text{ cm} - x) - (31.6 \text{ kg})gx = 0$,
    which gives $x = 89.5 \text{ cm}$ from the feet.

27. Because the backpack is at the midpoint of the rope, the
    angles are equal. The force exerted by the backpacker is
    the tension in the rope. From the force diagram for the
    backpack and junction we can write
    $\Sigma F_x = F_{T1} \cos\theta - F_{T2} \cos\theta = 0$, or $F_{T1} = F_{T2} = F$;
    $\Sigma F_y = F_{T1} \sin\theta + F_{T2} \sin\theta - mg = 0$, or
    $2F \sin\theta = mg$.
    
    (a) We find the angle from
    $\tan\theta = h/\tfrac{1}{2}L = (1.5 \text{ m})/\tfrac{1}{2}(7.6 \text{ m}) = 0.395$, or $\theta = 21.5°$.
    When we put this in the force equation, we get
    $2F \sin 21.5° = (16 \text{ kg})(9.80 \text{ m/s}^2)$, which gives $F = 2.1 \times 10^2 \text{ N}$.
    
    (b) We find the angle from
    $\tan\theta = h/\tfrac{1}{2}L = (0.15 \text{ m})/\tfrac{1}{2}(7.6 \text{ m}) = 0.0395$, or $\theta = 2.26°$.
    When we put this in the force equation, we get
    $2F \sin 2.26° = (16 \text{ kg})(9.80 \text{ m/s}^2)$, which gives $F = 2.0 \times 10^3 \text{ N}$.

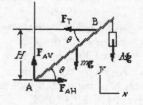

31. We choose the coordinate system shown, with positive torques
    clockwise. We write $\Sigma\tau = I\alpha$ about the point A from the
    force diagram for the pole and light:
    $\Sigma\tau_A = -F_T H + MgL \cos\theta + mg\tfrac{1}{2}L \cos\theta = 0$;
    $-F_T (3.80 \text{ m}) + (12.0 \text{ kg})(9.80 \text{ m/s}^2)(7.5 \text{ m}) \cos 37° +$
    $\qquad (8.0 \text{ kg})(9.80 \text{ m/s}^2)\tfrac{1}{2}(7.5 \text{ m}) \cos 37° = 0$,
    which gives $F_T = 2.5 \times 10^2 \text{ N}$.
    We write $\Sigma\mathbf{F} = m\mathbf{a}$ from the force diagram for the pole and light:
    $\Sigma F_x = F_{AH} - F_T = 0$;
    $F_{AH} - 2.5 \times 10^2 \text{ N} = 0$, which gives $F_{AH} = 2.5 \times 10^2 \text{ N}$.
    $\Sigma F_y = F_{AV} - Mg - mg = 0$;
    $F_{AV} - (12.0 \text{ kg})(9.80 \text{ m/s}^2) - (8.0 \text{ kg})(9.80 \text{ m/s}^2) = 0$, which gives $F_{AV} = 2.0 \times 10^2 \text{ N}$.

35. We choose the coordinate system shown, with positive torques
    clockwise. We assume the normal force acts a distance $x$ from
    the edge of the crate. If the crate does not tip, $\Sigma\tau = 0$, which
    we write about the edge of the crate:
    $(mg \cos\theta)\tfrac{1}{2}h - (mg \sin\theta)d - F_N x = 0$
    We write $\Sigma F_y = ma_y$ from the force diagram:
    $F_N - mg \cos\theta = 0$, which gives $F_N = mg \cos\theta$.
    When we use this in the torque equation, we get
    $x = \tfrac{1}{2}h - d \tan\theta$.

If the crate is not to tip over, $x \geq 0$;

   $\frac{1}{2}h \geq d \tan \theta$, or

   $\tan \theta_{max} = \frac{1}{2}h/d = \frac{1}{2}(2.0 \text{ m})/(1.2 \text{ m}) = 0.833$, $\theta_{max} = 40°$.

If the crate were sliding, the friction force would be kinetic, but in
the limiting case would still act at the corner of the base, so the
angle would be the same.

39.  From the symmetry of the wires, we see that the angle between a
     horizontal line on the ground parallel to the net and the line from
     the base of the pole to the anchoring point is $\theta = 30°$. We find the
     angle between the pole and a wire from

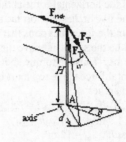

   $\tan \alpha = d/H = (2.0 \text{ m})/(2.6 \text{ m}) = 0.769$, which gives $\alpha = 37.6°$.

   Thus the horizontal component of each tension is $F_T \sin \alpha$.

   We write $\Sigma \tau = I\alpha$ about the horizontal axis through the base A
   perpendicular to the net from the force diagram for the pole:

   $\Sigma \tau_A = F_{net}H - 2(F_T \sin \alpha \cos \theta)H = 0$, or

   $F_{net} = 2F_T \sin \alpha \cos \theta = 2(95 \text{ N}) \sin 37.6° \cos 30° = 1.0 \times 10^2 \text{ N}$.

43.  (a)  We find the stress from

        Stress $= F/A = (2000 \text{ kg})(9.80 \text{ m/s}^2)/(0.15 \text{ m}^2) = 1.3 \times 10^5 \text{ N/m}^2$.

     (b)  We find the strain from

        Strain $= \text{Stress}/E = (1.3 \times 10^5 \text{ N/m}^2)/(200 \times 10^9 \text{ N/m}^2) = 6.5 \times 10^{-7}$.

     (c)  We use the strain to find how much the girder is lengthened:

        Strain $= \Sigma L/L_0$;

        $6.5 \times 10^{-7} = \Sigma L/(9.50 \text{ m})$, which gives $\Sigma L = 6.2 \times 10^{-6} \text{ m} = 0.0062 \text{ mm}$.

47.  The pressure needed is determined by the bulk modulus:

   $\Sigma P = -B \, \Sigma V/V_0 = -(90 \times 10^9 \text{ N/m}^2)(-0.10 \times 10^{-2}) = 9.0 \times 10^7 \text{ N/m}^2$.

   This is $(9.0 \times 10^7 \text{ N/m}^2)/(1.0 \times 10^5 \text{ N/m}^2 \cdot \text{atm}) = 9.0 \times 10^2 \text{ atm}$.

51.  (a)  For the torque from the sign's weight about the
          point A we have

        $\tau_A = Mgd = (5.1 \text{ kg})(9.80 \text{ m/s}^2)(2.2 \text{ m}) = 1.1 \times 10^2 \text{ m} \cdot \text{N}$.

     (b)  The balancing torque must be exerted by the wall, which is the only other
          contact point.  Because the torque from the sign is clockwise, this torque must be
          counterclockwise.

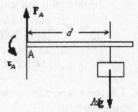

     (c)  If we think of the torque from the wall as a pull on the top of the pole and a push
          on the bottom of the pole, there is tension along the top of the pole and compression
          along the bottom.  There must also be a vertical force at the wall which, in
          combination with the weight of the sign, will create a shear stress in the pole.  Thus all three play a part.

55.  (a)  We want the maximum stress to be (1/7.0) of the tensile strength:

        $\text{Stress}_{max} = F/A_{min} = (\text{Tensile strength})/7.0$;

        $(320 \text{ kg})(9.80 \text{ m/s}^2)/A_{min} = (500 \times 10^6 \text{ N/m}^2)/7.0$, which gives $A_{min} = 4.4 \times 10^{-5} \text{ m}^2$.

     (b)  We find the change in length from

        Strain $= \Sigma L/L_0 = \text{Stress}/E$, or

        $\Sigma L = (\text{Stress})L_0/E = [(500 \times 10^6 \text{ N/m}^2)/7.0](7.5 \text{ m})/(200 \times 10^9 \text{ N/m}^2) = 2.7 \times 10^{-3} \text{ m} = 2.7 \text{ mm}$.

59.  We find the acceleration, and then the force, required to stop:

   $v^2 = v_0^2 + 2as$;

   $0 = (60 \text{ m/s})^2 + 2a(1.0 \text{ m})$; which gives $a = -1.8 \times 10^2 \text{ m/s}^2$.

   The required force is

   $F = ma = (75 \text{ kg})(-1.8 \times 10^2 \text{ m/s}^2) = -1.35 \times 10^5 \text{ N}$.

   This will produce a stress of

   Stress $= F/A = (1.35 \times 10^5 \text{ N})/(0.30 \text{ m}^2) = 4.5 \times 10^5 \text{ N/m}^2$.

   Because this is less than the ultimate strength of $5 \times 10^5 \text{ N/m}^2$, it is possible to escape serious injury.

63.  (a)  Because the tensile strength of steel is the same as the compressive strength, we choose a
          member with the greatest stress, $F_{AB} = Mg/\sqrt{3}$.  The safety condition is

        Stress $= F_{AB}/A \leq (\text{strength})/6.0$, or

        $A \geq (6.0)F_{AB}/\text{strength} = (6.0)(7.0 \times 10^5 \text{ kg})(9.80 \text{ m/s}^2)/(500 \times 10^6 \text{ N/m}^2) \sqrt{3} = 4.8 \times 10^{-2} \text{ m}^2$.

(*b*) If we simplify by putting the additional mass at the center of the truss, and remember that each side of the bridge will support half of the load of the trucks, we have

    $A$   $\Sigma$ (6.0)$F_{AB}$ /strength

    = (6.0)[(7.0 × 10$^5$ kg + ½(50)(1.2 × 10$^4$ kg)](9.80 m/s$^2$)/(500 × 10$^6$ N/m$^2$) $\sqrt{3}$ = 6.8 × 10$^{-2}$ m$^2$.

67. In each arch the horizontal force at the base must equal the horizontal force at the top. Because the two arches support the same load, we see from the force diagrams that the vertical forces will be the same and have the same moment arms. Thus the torque about the base of the horizontal force at the top must be the same for the two arches:

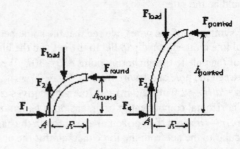

    $\tau = F_{round}h_{round} = F_{round}R = F_{pointed}h_{pointed}$;

    $F_{round}(4.0 \text{ m}) = \frac{1}{3}F_{round}h_{pointed}$ ,

which gives $h_{pointed} = 12$ m.

71. (*a*) The cylinder will roll about the contact point A.

    We write $\Sigma\tau = I\alpha$ about the point A:

    $F_a(2R - h) + F_{N1}[R^2 - (R - h)^2]^{1/2} - Mg[R^2 - (R - h)^2]^{1/2} = I_A\alpha$.

    When the cylinder does roll over the curb, contact with the ground is lost and $F_{N1} = 0$. Thus we get

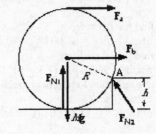

    $F_a = \{I_A\alpha + Mg[R^2 - (R - h)^2]^{1/2}\}/(2R - h)$

    $= [I_A\alpha/(2R - h)] + [Mg(2Rh - h^2)^{1/2}/(2R - h)]$.

    The minimum force occurs when $\alpha = 0$:

    $F_{amin} = Mg[h(2R - h)]^{1/2}/(2R - h) = Mg[h/(2R - h)]^{1/2}$.

(*b*) The cylinder will roll about the contact point A.

    We write $\Sigma\tau = I\alpha$ about the point A:

    $F_b(R - h) + F_{N1}[R^2 - (R - h)^2]^{1/2} - Mg[R^2 - (R - h)^2]^{1/2} = I_A\alpha$.

    When the cylinder does roll over the curb, contact with the ground is lost and $F_{N1} = 0$. Thus we get

    $F_b = \{I_A\alpha + Mg[R^2 - (R - h)^2]^{1/2}\}/(R - h)$

    $= [I_A\alpha/(R - h)] + [Mg(2Rh - h^2)^{1/2}/(R - h)]$.

    The minimum force occurs when $\alpha = 0$:

    $F_{bmin} = Mg[h(2R - h)]^{1/2}/(R - h)$.

75. The force is parallel to the grain. We want the maximum stress to be (1/12) of the compressive strength. For $N$ studs we have

    Stress$_{max}$ = ($Mg/N$)/$A$ = (Compressive strength)/12;

    (12,600 kg)(9.80 m/s$^2$)/$N$(0.040 m)(0.090 m) = (35 × 10$^6$ N/m$^2$)/12, which gives $N = 11.8$.

Thus we need 6 studs on each side.

There are five spaces between the studs, so they will be

    (10.0 m)/5 = 2.0 m apart.

79. From the force diagram for the aircraft we can write

    $\Sigma F_x = F_T - F_D = 0$, or

    $F_D = F_T = 5.0 \times 10^5$ N.

    $\Sigma F_y = F_L - W = 0$, or

    $F_L = W = (77,000 \text{ kg})(9.80 \text{ m/s}^2) = 7.55 \times 10^5$ N.

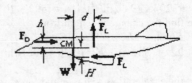

We write $\Sigma\tau = I\alpha$ about the center of mass:

    $\Sigma\tau_{CM} = F_D h_D - F_L d + F_T h_T = 0$.

When we use the previous results, we get

    $F_T(h_D + h_T) = Wd$;

    $(5.0 \times 10^5 \text{ N})(h_D + 1.6 \text{ m}) = (7.55 \times 10^5 \text{ N})(3.2 \text{ m})$, which gives $h_D = 3.2$ m.

83. Because the backpack is at the midpoint of the rope, the angles are equal. From the force diagram for the backpack and junction we can write

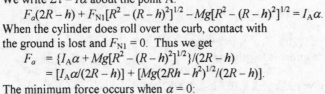

    $\Sigma F_x = F_{T1} \cos\theta - F_{T2} \cos\theta = 0$, or $F_{T1} = F_{T2} = F$;

    $\Sigma F_y = F_{T1} \sin\theta + F_{T2} \sin\theta - mg - F_{bear} = 0$.

When the bear is not pulling, we have

$2F_1 \sin \theta_1 = mg$;

$2F_1 \sin 15° = (23.0 \text{ kg})(9.80 \text{ m/s}^2)$, which gives $F_1 = 435$ N.

When the bear is pulling, we have

$2F_2 \sin \theta_2 = mg + F_{\text{bear}}$;

$2(2)(435 \text{ N}) \sin 30° = (23.0 \text{ kg})(9.80 \text{ m/s}^2) + F_{\text{bear}}$, which gives $F_{\text{bear}} = 6.5 \times 10^2$ N.

87. The ropes can only provide a tension, so the scaffold will be stable if $F_{T1}$ and $F_{T2}$ are greater than zero. The tension will be least in the rope farthest from the painter. To find how far the painter can walk from the right rope toward the right end, we set $F_{T1} = 0$.

We write $\Sigma\tau = 0$ about B from the force diagram:

$\Sigma\tau_B = Mgx_{\text{right}} + F_{T1}(D + 2d) - m_{\text{pail}}g(D + d) - mgD = 0$;

$(60 \text{ kg})(9.80 \text{ m/s}^2)x_{\text{right}} + 0 - (4.0 \text{ kg})(9.80 \text{ m/s}^2)(2.0 \text{ m} + 1.0 \text{ m}) -$

$(25 \text{ kg})(9.80 \text{ m/s}^2)(2.0 \text{ m}) = 0$,

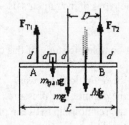

which gives $x_{\text{right}} = 1.03$ m.

Because this is greater than the distance to the end of the plank, 1.0 m, walking to the right end is safe.

To find how far the painter can walk from the right rope toward the left end, we set $F_{T2} = 0$.

We write $\Sigma\tau = 0$ about A from the force diagram:

$\Sigma\tau_A = -Mgx_{\text{left}} - F_{T2}(D + 2d) + m_{\text{pail}}gd + mg2d = 0$;

$-(60 \text{ kg})(9.80 \text{ m/s}^2)x_{\text{left}} - 0 + (4.0 \text{ kg})(9.80 \text{ m/s}^2)(1.0 \text{ m}) + (25 \text{ kg})(9.80 \text{ m/s}^2)2(1.0 \text{ m}) = 0$,

which gives $x_{\text{left}} = 0.90$ m.

Because this is less than the distance to the end of the plank, 1.0 m, walking to the left end is not safe.

The painter can safely walk to within 0.10 m of the left end.

91. From the symmetry we see that the force each beam exerts on the other must be horizontal. We write $\Sigma\mathbf{F} = m\mathbf{a}$ from the force diagram for the left beam:

$\Sigma F_x = F_{\text{fr}} - F = 0$, or $F = F_{\text{fr}}$.

$\Sigma F_y = F_N - Mg = 0$, or $F_N = Mg$.

We write $\Sigma\tau = I\alpha$ about the point B from the force diagram:

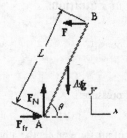

$\Sigma\tau_B = F_N L \cos\theta - Mg\frac{1}{2}L \cos\theta - F_{\text{fr}}L \sin\theta = 0$;

$MgL \cos\theta - Mg\frac{1}{2}L \cos\theta - F_{\text{fr}}L \sin\theta = 0$;

which gives $F_{\text{fr}} = Mg/2 \tan\theta$.

For the friction to remain static, we have

$F_{\text{fr}} = Mg/2 \tan\theta \le \mu_s Mg$, so

$\tan\theta \ge 1/2\mu_s = 1.2(0.60) = 0.833$, $\theta \ge 40°$.

95. We choose the coordinate system shown, with positive torques clockwise. We write $\Sigma\tau = I\alpha$ about the point C:

$\Sigma\tau_C = mgL \sin\theta_A - (F_{TB} \sin\theta_B)L \cos\theta_A + (F_{TB} \cos\theta_B)L \sin\theta_A = 0$,

which gives

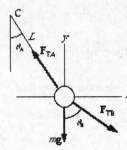

$F_{TB} = mg /[(\sin\theta_B / \tan\theta_A) - \cos\theta_B]$

$= (20 \text{ kg})(9.80 \text{ m/s}^2)/[(\tan 20°/ \sin 50°) - \cos 50°] = 134$ N.

Note that we have found the torque produced by $F_{TB}$ by finding the torques of the components.

We write $\Sigma\mathbf{F} = m\mathbf{a}$ from the force diagram:

$\Sigma F_x = F_{TB} \sin\theta_B - F_{TA} \sin\theta_A = 0$,

which gives

$F_{TA} = F_{TB} \sin\theta_B /\sin\theta_A = (134 \text{ N}) \sin 50°/\sin 20° = 300$ N.

# Chapter 13: Fluids

## Chapter Overview and Objectives

This chapter introduces the physics of fluids. It introduces the concepts of pressure, density, buoyancy, flow rate, viscosity, surface tension, and capillarity. It discusses the relationships between depth and pressure of a fluid at rest. It discusses the types of instruments used to measure pressure. It introduces Archimedes' principle and Bernoulli's equation.

After completing study of this chapter, you should:

- Know the definition of density.
- Know the definition of pressure.
- Know the relationship between pressure and depth within a fluid at rest.
- Know the difference between absolute pressure and gauge pressure.
- Know Pascal's principle and be able to apply it to problems.
- Know how to determine pressure using the height of a column of fluid supported by a pressure difference.
- Know what buoyancy is and how to use Archimedes' principle to calculate buoyant force.
- Know what the equation of continuity is.
- Know Bernoulli's equation and how to apply it to fluids in motion.
- Know what viscosity is.
- Know Poiseuille's equation and how to apply it to problems.
- Know what surface tension is.
- Know what a pump is.

## Summary of Equations

Definition of density:
$$\rho = \frac{m}{V}$$

Definition of pressure:
$$P = \frac{F}{A}$$

Dependence of pressure on depth within fluid:
$$P(h) = P(0) + \rho g h$$

$$P(h) = P(0) + \int_0^h \rho(y) g \, dy$$

Definition of gauge pressure:
$$P_{gauge} = P - P_{atmosphere}$$

Buoyant force on object displacing fluid:
$$F_B = \rho \, g V_{fluid \, displaced}$$

Equation of continuity:
$$\rho_1 A_1 v_1 = \rho_2 A_2 v_2$$

Equation of continuity for a constant density fluid:
$$A_1 v_1 = A_2 v_2$$

Bernoulli's equation:
$$P_1 + \tfrac{1}{2} \rho v_1^2 + \rho g y_1 = P_2 + \tfrac{1}{2} \rho v_2^2 + \rho g y_2$$

Torricelli's theorem:
$$v_1 = \sqrt{2g(y_2 - y_1)}$$

Viscous force:
$$\frac{F}{A} = \eta \frac{dv}{dx}$$

Poiseuille's equation:
$$Q = \frac{\pi R^4 (P_1 - P_2)}{8\eta L}$$

Definition of surface tension:
$$\gamma = \frac{F}{L}$$

## Chapter Summary

### Section 13-1 Density and Specific Gravity

The density, $\rho$, of a material is its mass, $m$, per volume, $V$:

$$\rho = \frac{m}{V}$$

The **specific gravity** of a substance is the ratio of the density of the material to the density of water at 4.00 °C.   The density of water at 4.00 °C is 1000 kg/m$^3$ or 1.000 g/cm$^3$.

### Example 13-1-A

Ice cubes are sold in ten-pound bags.  You need 1000 ice cubes for your party.  Each ice cube is a rectangular prism of dimension 2 cm × 2 cm × 3 cm.  How many bags of ice do you need?

### Solution:

The volume of each ice cube is its length times its width times its height:

$$V_{ice\,cube} = L \times W \times H = 0.02\,\text{m} \times 0.02\,\text{m} \times 0.03\,\text{m} = 1.2 \times 10^{-5}\,\text{m}^3$$

The mass of each ice cube is

$$m_{ice\,cube} = \rho\,V_{ice\,cube} = \left(917\,\text{kg}/\text{m}^3\right)\!\left(1.2 \times 10^{-5}\,\text{m}^3\right) = 1.1 \times 10^{-2}\,\text{kg}$$

Multiplying the number of ice cubes by the mass of an ice cube give the total mass of ice needed:

$$M = N m_{ice\,cube} = 1000\!\left(1.1 \times 10^{-2}\,\text{kg}\right) = 11\,\text{kg}$$

Dividing the total mass needed by the mass per bag gives the number of bags needed:

$$N_{bags} = \frac{M}{M_{bag}} = \frac{11\,\text{kg}}{\left(10\,\text{lbs}\right)\!\left(1\,\text{kg mass}/2.2\,\text{lbs weight}\right)} = 2.42$$

Because you probably can't buy a fraction of a bag, you will have to buy three bags of ice.

### Section 13-2 Pressure in Fluids

Given any area acted on by a force, the pressure, $P$, on that area is the normal force, $F$, to that area divided by the area, $A$:

$$P = \frac{F}{A}$$

The pressure has dimensions of force divided by area so that its SI units are N/m$^2$.  One N/m$^2$ is called a **pascal** and is abbreviated **Pa**.  Another commonly used unit of force is lbs/in$^2$.  Other common ways of expressing pressure will be discussed in later sections.  Pressure in a fluid is isotropic, it is the same in all directions.  In a fluid at rest, the force on a given surface is always perpendicular to the surface.

In a fluid at rest, the pressure is the same at all positions that have the same depth within the fluid. The pressure in a fluid increases with depth in the fluid, because the pressure at a given depth must support the weight of the fluid above it. The pressure at a depth $h$ below the surface of a given fluid at rest is the pressure at the top surface, $P_0$, plus the weight per area of the fluid above the depth $h$:

$$P(h) = P(0) + \rho g h$$

where $\rho$ is the uniform density of the fluid and $g$ is the acceleration of gravity. Most liquids have densities that are only slightly dependent on pressure and so this relationship is valid for most applications involving liquids. However, gases do have a density that depends on pressure. For those cases, a more general relationship between depth and pressure must be used:

$$P(h) = P(0) + \int_0^h \rho(y) g \, dy$$

where $\rho(y)$ is the density as a function of depth $y$ within the fluid.

### Example 13-2-A

A gas has a temperature gradient such that the relationship between pressure and density is given by the relationship:

$$\rho = cP^2$$

In such a gas with a uniform temperature, the pressure at the top of a column of height $H$ is $P_0$ and the density at the top of the column is $\rho_0$. What are the pressure and density at the bottom of this column?

### Solution:

We can solve for the constant $c$ by using the conditions at the top of the column:

$$\rho_0 = cP_0^2 \quad \Rightarrow \quad c = \frac{\rho_0}{P_0^2}$$

Next we can write down the differential form of the pressure-depth relationship:

$$\frac{dP}{dy} = -\rho g = -\frac{\rho_0 P^2}{P_0^2} g$$

Dividing through by $P_2$ and multiplying through by $dy$ on each side and integrating results in

$$\int_{P_{bottom}}^{P_{top}} \frac{dP}{P^2} = -\int_0^H \frac{\rho_0}{P_0^2} g \, dy$$

After completing the two integrals we have

$$-\frac{1}{P_{top}} - \left( -\frac{1}{P_{bottom}} \right) = -\frac{\rho_0}{P_0^2} gH$$

and solving this for the pressure at the bottom of the column gives

$$P_{bottom} = \frac{1}{\dfrac{\rho_0}{P_0^2} gH - \dfrac{1}{P_{top}}}$$

### Section 13-3 Atmospheric Pressure and Gauge Pressure

The pressure of the air at rest at sea level has a mean value of $1.013 \times 10^5$ N/m². This value is defined as 1 **atmosphere** of pressure and is used as a unit of pressure. A **bar** is another unit of pressure convenient for measuring atmospheric pressure. 1 bar is equal to $1.00 \times 10^5$ N/m².

The pressure we have defined previously is sometimes referred to as **absolute pressure**. This is in contrast to gauge pressure. **Gauge pressure** is the difference between absolute pressure and atmospheric pressure:

$$P_{gauge} = P - P_{atmosphere}$$

Most mechanical pressure gauges such as compressed gas gauges measure gauge pressure rather than absolute pressure.

### Example 13-3-A

A child sits on a tire inner tube in such a way that the inner tube supports the child off the floor. The area of contact of the inner tube with the floor is 6.8 square inches. The gauge pressure in the inner tube is 7.2 lbs/in². What is the mass of the child? Assume the weight of the inner tube is negligible.

**Solution:**

We know that the walls of the inner tube themselves are flexible rubber and cannot support the weight of the child, so it must be the air pressure inside the tube that supports the weight. Therefore, the area of contact with the floor times the pressure is the force that the tube pushes on the floor and, by Newton's third law, the force that the floor pushes back upward on the inner tube and child. By Newton's second law, this must be equal to the weight of the child. Now, the question is, should gauge pressure or absolute pressure be used in the calculation of the force? We need to use gauge pressure because the atmospheric pressure outside and inside the tube apply zero net force to the child. Therefore, the weight of the child is

$$W_{child} = m_{child}g = P_{gauge}A = \left(7.2 \text{ lbs/in}^2\right)\left(6.8 \text{ in}^2\right) = 49 \text{ lbs}$$

We know that one kilogram has a weight of 2.2 pounds so

$$m_{child} = 49 \text{ lbs}\left(1 \text{ kg mass} / 2.2 \text{ lbs weight}\right) = 22 \text{ kg}$$

### Section 13-4 Pascal's Principle

**Pascal's principle** states that a pressure change at any given point to a confined fluid at rest is transmitted throughout the fluid so that the same pressure change occurs at all points within the fluid.

### Section 13-5 Measurement of Pressure; Gauges and the Barometer

There are several different instruments that are used for measuring fluid pressure. One common device is a **manometer**. A manometer uses the height of a column of fluid supported by the pressure difference at two surfaces of the liquid to determine the pressure difference. A typical manometer is shown in the diagram. The pressure at the lower surface of the liquid, $P_2$, is related to $P_1$ by

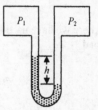

$$P_2 = P_1 + \rho g h$$

where $\rho$ is the density of the liquid in the manometer and $h$ is the difference in vertical position of the two liquid surfaces. Pressures are often expressed by the height of the column of liquid supported, such as mm of mercury. It should be easy to see that to take an expression of the pressure in terms of the height of the supported column of liquid and convert into a measurement of pressure, you need to multiply the height of the column by the density of the fluid and the acceleration of gravity.

A **barometer** is a manometer with one end closed with negligible pressure in the closed end and the other end open to atmospheric pressure. The barometer is used to measure atmospheric pressure.

**Example-13-5-A**

Atmospheric pressure on Earth is 760 mm of mercury. If a mercury barometer were brought to the moon to measure the air pressure inside a spacecraft, what would be the height of the mercury column in the barometer if the air pressure inside the spacecraft were the same as Earth's atmospheric pressure?

**Solution:**

We know that the product of the fluid density, the acceleration of gravity, and the height of the supported column of liquid is the pressure. We can set the expressions for the pressure on the earth and the moon equal to each other:

$$\rho g_{moon} h_{moon} = \rho g_{earth} h_{earth} \quad \Rightarrow \quad h_{moon} = \frac{g_{earth} h_{earth}}{g_{moon}}$$

$$h_{moon} = \frac{9.8 \, \text{m/s}^2}{1.6 \, \text{m/s}^2} \, 0.760 \, \text{m} = 4.7 \, \text{m}$$

A mercury barometer would be somewhat impractical for measuring pressures as large as atmospheric pressure on the moon because of its required height.

### Section 13-6 Buoyancy and Archimedes' Principle

An object either completely or partially immersed in a fluid is acted on by an upward force of a magnitude equal to the weight of the fluid displaced by the object.

$$F_B = \rho g V_{fluid \, displaced}$$

This is statement of Archimedes' principle and the upward force on the object, $F_B$, is called a **buoyant force**. The weight of the displaced fluid is written in terms of the density of the fluid, $\rho$, the acceleration of gravity, $g$, and the volume of the fluid displaced by the object, $V_{fluid \, displaced}$.

**Example 13-6-A**

A raft is made from wood with a density 378 kg/m$^3$. The volume of the logs that make up the raft is 3.87 m$^3$. What is the maximum load that can be placed on the raft before its top surface reaches the surface of the water it is floating in?

**Solution:**

For the raft to float, the buoyant force must be equal in magnitude to the weight of the raft plus the weight of the load:

$$F_B = \left( m_{raft} + m_{load} \right) g \Rightarrow \rho g V_{raft} = \left( \rho V_{raft} + m_{load} \right) g$$

where we have used that the submerged volume is the entire volume of the raft. We can solve this for the mass of the load

$$m_{load} = \left( \rho_{water} - \rho_{raft} \right) V_{raft} = \left( 1000 \, \text{kg/m}^3 - 378 \, \text{kg/m}^3 \right) \left( 3.87 \, \text{m}^3 \right) = 2.41 \times 10^3 \, \text{kg}$$

### Section 13-7 Fluids in Motion; Flow Rate and the Equation of Continuity

Fluid dynamics is the study of moving fluids. Fluid flow can be characterized qualitatively by whether the flow is **laminar** or **turbulent**. Laminar flow exists when the fluid flows in layers. Turbulent flow exists where there are eddies or whirlpools in the flow.

**Steady-state flow** exists when the velocity of the fluid moving past at each fixed point in space remains the same with time. This is not saying that fluid doesn't accelerate, just that any fluid that reaches any given point in space will always have the same velocity at that point in space. In steady-state flow, the amount of mass in a given volume of space must remain constant. This implies that the **equation of continuity** must be satisfied:

$$\rho_1 A_1 v_1 = \rho_2 A_2 v_2$$

where $\rho$ is the density of the fluid, $A$ is the cross-sectional area of the flow tube, and $v$ is the speed of the velocity of the fluid. All of the subscripts 1 apply to one location along a flow tube and the subscripts 2 apply to another location along a flow tube. If the fluid has a constant density, then the equation of continuity can be expressed as

$$A_1 v_1 = A_2 v_2$$

Liquids can be treated as having constant density in most circumstances and gases can be treated as constant density in many situations in which the speeds involved in the problem are small compared to the speed of sound in the gas.

### Section 13-8 Bernoulli's Equation

In the particular case of a fluid that is incompressible under condition of laminar flow that is not changing with time, there is a simple relationship, called **Bernoulli's equation,** that relates the speed of the fluid, the pressure in the fluid, and the height of the fluid within a given flow tube. Bernoulli's equation can be derived from the work energy principle and is written as

$$P_1 + \tfrac{1}{2}\rho v_1^2 + \rho g y_1 = P_2 + \tfrac{1}{2}\rho v_2^2 + \rho g y_2$$

where $P$ is the pressure, $\rho$ is the density of the fluid, $v$ is the speed of the fluid, $g$ is the acceleration of gravity, and $y$ is the vertical position of the fluid (measured positive upward). All of the subscripts 1 apply to one location and the subscripts 2 apply to another location.

### Example 13-8-A

A fire hose sprays water at a rate of 82 gallons per minute at a speed of 17.4 m/s through a nozzle of diameter 2.2 cm. The hose has a diameter of 6.3 cm. What is the pressure in the hose lying on the ground before the water reaches the nozzle held by a fireman at a height 1.4 m above the ground?

### Solution:

We know the water exits the fire hose at atmospheric pressure. The speed of the water leaving the fire hose can be calculated from the flow rate and the area of the nozzle opening:

$$Q = Av \Rightarrow v_{noz} = \frac{Q}{A} = \frac{(82\,\text{gal}/\text{min})(3.78\times10^{-3}\,\text{m}^3/\text{gal})(1\text{min}/60\text{s})}{\pi(0.022\text{m})^2/4}$$

$$v_{noz} = 13.6\,\text{m/s}$$

Using the equation of continuity, we can determine the speed of the fluid in the hose:

$$A_{hose} v_{hose} = A_{noz} v_{noz} \quad \Rightarrow \quad v_{hose} = \frac{A_{noz} v_{noz}}{A_{hose}} = \frac{(0.022\,\text{m})^2(13.6\,\text{m/s})}{(0.063\,\text{m})^2}$$

$$v_{hose} = 1.66\,\text{m/s}$$

We can now use Bernoulli's equation to relate the pressure in the hose to the known quantities:

$$P_{hose} + \tfrac{1}{2}\rho v_{hose}^2 + \rho g y_{hose} = P_{noz} + \tfrac{1}{2}\rho v_{noz}^2 + \rho g y_{noz} \quad \Rightarrow$$

$$P_{hose} = P_{noz} + \rho\left[\tfrac{1}{2}\left(v_{noz}^2 - v_{hose}^2\right) + g\left(y_{noz} - y_{hose}\right)\right] \quad \Rightarrow$$

$$P_{hose} = \left(1.01\times10^5\,\text{N/m}^2\right) + \left(1000\,\text{kg/m}^3\right)\left\{\tfrac{1}{2}\left[(13.6\,\text{m/s})^2 - (1.66\,\text{m/s})^2\right] + \left(9.8\,\text{m/s}^2\right)(1.4\,\text{m})\right\}$$

$$= 2.1\times10^5\,\text{N/m}^2$$

**Section 13-9 Application of Bernoulli's Principle: From Torricelli to Sailboats, Airfoils, and TIA**

For the special case of the pressure being identical at points 1 and 2 of a fluid and the velocity of the fluid being zero at point 2, Bernoulli's equation can be simplified to

$$\tfrac{1}{2}\rho v_1^2 + \rho g y_1 = \rho g y_2$$

If this relationship is solved for the velocity of the fluid at point 1, we obtain

$$v_1 = \sqrt{2g(y_2 - y_1)}$$

This is known as **Torricelli's theorem**. Comparing this to results in Chapter 2, we see that the speed of the fluid is the same as the speed of an object that has freely fallen from rest the same distance, $y_2$-$y_1$. Torricelli's equation applies approximately to the situation where a fluid is in a large container and leaves the container through an opening small compared to the diameter of the container. In that case, the velocity inside the container far from the small exit opening is small compared to the velocity of the fluid moving through the opening.

**Example 13-9-A**

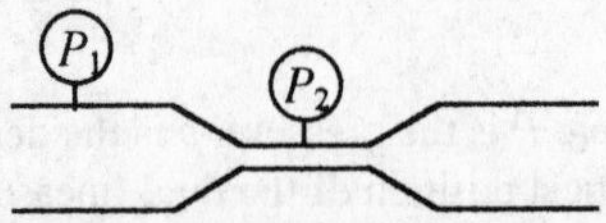

A venturi tube is used to measure rate of flow of gasoline through a line as shown in the diagram. The larger diameter tube has a diameter 1.22 cm and the smaller diameter tube has a diameter 0.26 cm. The gauge pressure in the larger diameter tube is $2.45 \times 10^4\,\text{N/m}^2$ and the pressure in the smaller diameter tube is $1.34 \times 10^4\,\text{N/m}^2$. What is flow rate of gasoline through this tube?

**Solution:**

The flow rate through the tube is equal to $Av$ where we can use either the area and speed in the large tube or the area and speed in the small tube. Because the mean height of the two sections of tubing is identical, we can use Torricelli's simplified form of Bernoulli's equation:

$$P_1 + \tfrac{1}{2}\rho v_1^2 = P_2 + \tfrac{1}{2}\rho v_2^2$$

Because of the equation of continuity, we know $v_2 = (A_1/A_2)v_1$. We substitute this into Torricelli's equation:

$$P_1 + \tfrac{1}{2}\rho v_1^2 = P_2 + \tfrac{1}{2}\rho\left(A_1/A_2\right)^2 v_1^2$$

and solve for $v_1$:

$$v_1 = \sqrt{\frac{2(P_1 - P_2)}{\rho\left[1 + (A_1/A_2)^2\right]}} = \sqrt{\frac{2\left(2.45 \times 10^4\,\text{N/m}^2 - 1.34 \times 10^4\,\text{N/m}^2\right)}{\left(680\,\text{kg/m}^3\right)\left[1 + \left((1.22\,\text{cm})^2/(0.26\,\text{cm})^2\right)^2\right]}} = 0.26\,\text{m/s}$$

Using this to calculate the flow rate:

$$Q = Av = \frac{\pi D^2}{4}v = \frac{\pi\left(1.22 \times 10^{-2}\,\text{m}\right)^2}{4}(0.26\,\text{m/s}) = 3.0 \times 10^{-5}\,\text{m}^3/\text{s}$$

**Section 13-10 Viscosity**

A shear stress, $F/A$, that exists between layers of fluid in relative motion to each other is called **viscosity**.

$$\frac{F}{A} = \eta \frac{dv}{dx}$$

where $\eta$ is the coefficient of viscosity and $dv/dx$ is the spatial derivative of the velocity of the fluid in the direction perpendicular to the fluid velocity.

### Section 13-11 Flow in Tubes: Poiseuille's Equation

The rate of flow, $Q$, of a fluid through a circular cylindrical pipe of radius $R$ and length $L$ is equal to

$$Q = \frac{\pi R^4 (P_1 - P_2)}{8 \eta L}$$

where $P_1 - P_2$ is the pressure difference between the ends of the pipe and $\eta$ is the viscosity of the fluid. This relationship is known as Poiseuille's equation.

### Example 13-11-A

A tank of diameter D is filled with water to a depth H. A pipe with inner radius R much smaller than D and length L is used to drain the water horizontally out of the bottom of the tank. How long does it take to drain the water out of the tank? Assume the flow through the pipe is laminar.

### Solution:

The pressure at the exit to the pipe will be atmospheric pressure the entire time the tank drains. Because the radius of the pipe is small compared to the diameter of the tank, we can approximate that the velocity of the water in the tank is zero and there is no viscous pressure drop in the tank. The pressure at the entrance to the pipe will change as the depth of the water drops in the tank:

$$P_1(h) = P_A + \rho g h$$

The height of the water in the tank depends on the amount of water drained from the tank:

$$h(t) = \frac{V_{\text{in tank}}}{A_{\text{tank}}} = \frac{V_0 - V_{\text{drained}}}{A_{\text{tank}}} = \frac{H A_{\text{tank}} - \int_0^t Q\, dt'}{A_{\text{tank}}}$$

$Q$, the rate at which the water flows out of the tank is given by Poiseuille's equation:

$$Q(t) = \frac{\pi R^4 (P_1 - P_2)}{8 \eta L} = \frac{\pi R^4}{8 \eta L} \left( P_A + \rho g h(t) - P_A \right) = \frac{\pi R^4 \rho g}{8 \eta L} h(t)$$

Putting this into the integrand in the equation above for $h(t)$:

$$h(t) = H - \frac{\pi R^4 \rho g}{8 \eta L A_{\text{tank}}} \int_0^t h(t)\, dt'$$

If we take the derivative of both sides of this equation with respect to $t$, we get

$$\frac{dh(t)}{dt} = -\frac{\pi R^4 \rho g}{8 \eta L A_{\text{tank}}} h(t)$$

We know the solution of this equation is of the form

$$h(t) = He^{-\alpha t}$$

where

$$\alpha = \frac{\pi R^4 \rho g}{8 \eta L A_{\text{tank}}}$$

Because the height of the water in the tank decreases exponentially, there never is a time at which the tank is completely empty. However, for an exponential time dependence, we can characterize the length of time it takes to drain a certain fraction of the amount. It should be easy to see that the height of the tank decreases by a factor $1/e$ during every time interval equal to $1/\alpha$.

### Section 13-12 Surface Tension and Capillarity

To increase the area of the surface of a liquid requires moving molecules of the liquid from the interior of the liquid to the surface. Molecules of the liquid at the surface do not have as many molecules surrounding them as do molecules in the interior of the liquid. Because most liquids have a strong attraction for the other molecules in the liquid, it takes energy to create surface area. This means that work is done on the liquid as the surface is stretched and so a force is needed to stretch the surface. The force per length of surface being stretched is called the surface tension, $\gamma$. This can be expressed as

$$\gamma = \frac{F}{L}$$

where $F$ is the force necessary to stretch a surface of length $L$.

A surface is a boundary between any two materials. If you look up the value of the surface tension of a liquid, it will usually be the surface tension associated with the air-liquid interface unless otherwise specified. There is a surface tension associated with the surface between a liquid and any other material also. The surface tension associated with the boundary between water and glass causes **capillary action**. Capillary action causes water to rise within a small diameter glass tube.

### Section 13-13 Pumps, and the Heart

A pump is a device that is used to maintain a pressure difference in a fluid. The pressure difference is often used to create a flow of the fluid. The human heart is an example of a pump. It creates a pressure difference in the blood to cause the blood to move through the blood vessels. A pump is characterized by the flow rate through the pump as a function of the pressure difference maintained by the pump.

## Practice Quiz

1.  An object floats on water so that of one-half of its volume is above the surface of the water. The object is placed in a container that contains half water and half oil. The oil floats on the water. The object floats at the boundary between the water and the oil. What statement can be made about the fraction of the volume of the object that is now above the surface of the water?

    a) The fraction of the volume above the surface of the water is still 50%.
    b) The fraction of the volume above the surface of the water is greater than 50%.
    c) The fraction of the volume above the surface of the water is less than 50%.
    d) More information about the oil is needed to answer the question.

2.  A pressure gauge on a compressed air tank reads $-6.0$ lbs/in$^2$. How can the pressure be negative?

    a) The gauge must be broken; negative pressure is impossible.
    b) Someone put the gauge on the tank backwards.
    c) Gauge pressure can be negative, if the absolute pressure is less than atmospheric pressure.
    d) The air in the tank is burning.

3.      Which has the greatest mass: air, water, or iron?
        a) Air
        b) Water
        c) Iron
        d) None; mass is not a property of a type of material.

4.      A plunger is often used to clean out blocked drainpipes in homes. The blockage is usually some distance away from there the plunger applies force to the water that is blocked from flowing down the pipe. Which principle explains why the force applied on the water at one position helps to increase the pressure at the location of the blockage to clear the pipe?

        a) Bernoulli's principle
        b) Pascal's principle
        c) Archimedes' principle
        d) Poiseuille's principle

5.      A pump provides a pressure difference to pump water through some pipes of fixed length. Which arrangement of pipes provides the greatest flow rate of water?

        a) One pipe of diameter $D$
        b) Four pipes of diameter $D$
        c) Ten pipes of diameter $D$
        d) One pipe of diameter $2D$

6.      Water is at rest inside a U-shaped tube. The water in the two legs of the U is at identical height. Oil with a density less than the density of water is poured into one leg of the U. What happens to the water level in the other leg?

        a) The water level drops
        b) The water level stays the same
        c) The water level rises to be at the same level as the top of the oil
        d) The water level rises, but will stay below the top of the oil

7.      A compressed air tank has an internal gauge pressure $P_0$ when the atmospheric pressure is $P_A$. The atmospheric pressure then drops to a pressure that is 99% of the original atmospheric pressure. What happens to the gauge pressure on the compressed air tank?

        a) The gauge pressure drops to $0.99\,P_0$
        b) The gauge pressure drops to $P_0 - 0.01P_A$
        c) The gauge pressure rises to $1.01P_0$
        d) The gauge pressure rises to $P_0 + 0.01P_A$

8.      On a freshly cleaned and waxed car, water beads up. On a dirty car the water spreads out and wets the surface. Which statement is consistent with this observation?

        a) The surface tension between the water and the air is smaller than the cohesive forces between the water and the wax.
        b) The cohesive force of the wax for the dirt is larger than the cohesive force for the water.
        c) The cohesive force of the water for the water is stronger than the cohesive force of the water for the wax.
        d) The surface tension of the wax is larger than the surface tension of the dirt.

9.      A small raft of density 500 kg/m$^3$ floats by itself on water. A person of density 980 kg/m$^3$ floats on water. However, if the person stands on the raft, the raft sinks below the surface of the water. Why?

        a) If you add the density of the person to the density of the raft, the result is greater than the density of water.
        b) If you multiply the density of the person by the density of the raft, the result is greater than the density of water.
        c) The volume of the person is smaller than the volume of the raft.
        d) The volume of the person does not displace any water unless the raft sinks.

10.    A pipe carries water up a hill a certain distance. A faucet on the pipe at the top of the hill can control the flow rate or stop the flow all together. The pipe has been under-designed and could burst under certain conditions. When and where is the pipe most likely to burst?

   a) The pipe is most likely to burst at the top when the water is flowing fastest.
   b) The pipe is most likely to burst at the top when the water is stopped.
   c) The pipe is most likely to burst at the bottom when the water is flowing fastest.
   d) The pipe is most likely to burst at the bottom when the water is stopped.

11.    A wooden block floats in water with 32% of its volume above the surface. What is the density of the wood?

12.    A pump transfers water from one container to another through a hose that is 38.4 m long. The pump provides a pressure difference of $5.00 \times 10^4$ N/m² from its intake to its output side. If the hose has a diameter 0.500 in, how long will it take to pump 14,000 gallons of water? Assume the water is at 20° C and the flow is laminar.

13.    The water main to a home is 1.44 m below ground level and has a 1.25 in internal diameter. Water leaves a tap in the home with a ½ in internal diameter at a height of 1.08 m above ground level at a flow rate of 1.54 gallons per minute. What is the absolute pressure in the water main? Ignore viscosity.

14.    A small hydraulic jack is used to jack up a car to change its tire. The piston that pumps fluid into the chamber supporting the car has a diameter of 0.43 cm. The piston that supports the car has a diameter of 5.25 cm. What force is needed to push on the pumping piston to support the weight of a 724 kg car?

15.    Ethyl alcohol is used to make a barometer. What height of column of ethyl alcohol is supported by atmospheric pressure?

## Problem Solutions

3.    When we use the density of gold, we have
   $m = \rho V = \rho LWH = (19.3 \times 10^3$ kg/m³$)(0.60$ m$)(0.25$ m$)(0.15$ m$) = 4.3 \times 10^2$ kg  ($\sim 950$ lb!).

7.    (*a*)  The normal force on the four legs must equal the weight. The pressure of the reaction to the normal force, which is exerted on the floor, is
   $P = F_N/A = mg/A = (60$ kg$)(9.80$ m/s²$)/4(0.05 \times 10^{-4}$ m²$) = 2.9 \times 10^7$ N/m².
   (*b*)  For the elephant standing on one foot, we have
   $P = F_N/A = mg/A = (1500$ kg$)(9.80$ m/s²$)/(800 \times 10^{-4}$ m²$) = 1.8 \times 10^5$ N/m².
   Note that this is a factor of ~100 less than that of the loudspeaker!

11.    Because the force from the pressure on the cylinder supports the automobile, we have
   $PA = mg$;
   $(17.0$ atm$)(1.013 \times 10^5$ N/m² · atm$)\frac{1}{4}\pi(24.5 \times 10^{-2}$ m$)^2 = m(9.80$ m/s²$)$, which gives $m = 8.28 \times 10^3$ kg.
   Note that we use gauge pressure because there is atmospheric pressure on the outside of the cylinder.

15.    Because points a and b are at the same elevation of water, the pressures must be the same. Each pressure is due to the atmospheric pressure at the top of the column and the height of the liquid, so we have
   $P_a = P_b$ , or $P_0 + \rho_{oil}gh_{oil} = P_0 + \rho_{water}gh_{water}$;
   $\rho_{oil}g(27.2$ cm$) = (1.00 \times 10^3$ kg/m³$)g(27.2$ cm $- 9.41$ cm$)$,
   which gives $\rho_{oil} = 6.54 \times 10^2$ kg/m³.

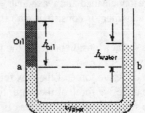

19.    (*a*)  When the height of the mercury in the open tube is greater than the height of the mercury on the tank side of the manometer, the pressure in the tank is
   $P_{tank} = P_{atm} + \rho_{Hg}gh$
   $= (1040$ mbar$)(100$ Pa/mbar$) + (28.0$ cm Hg$)(1330$ Pa/cm Hg$) = 1.41 \times 10^5$ Pa.
   (*b*)  When the height of the mercury in the open tube is less than the height of the mercury on the tank side of the manometer, we use a negative value for *h*:
   $P_{tank} = P_{atm} + \rho_{Hg}gh$
   $= (1040$ mbar$)(100$ Pa/mbar$) + (- 4.2$ cm Hg$)(1330$ Pa/cm Hg$) = 9.8 \times 10^4$ Pa.

$$P_{tank} = P_{atm} + \rho_{Hg}gh$$
$$= (1040 \text{ mbar})(100 \text{ Pa/mbar}) + (- 4.2 \text{ cm Hg})(1330 \text{ Pa/cm Hg}) = 9.8 \times 10^4 \text{ Pa.}$$

23. (a) Because the pressure is a function of the depth, the force on the wall will also be a function of the depth. We choose a coordinate system with $y = 0$ at the bottom of the dam and the water level at height $h$. We find the force on a differential slice $dy$ of length $b$ at height $y$ and then integrate to get the total force:

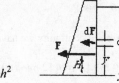

$$F = \int \Delta p_{water}\, dA = \int_0^h \rho g (h - y) b\, dy = \rho g b \left( hy - \tfrac{1}{2} y^2 \right)\Big|_0^h = \tfrac{1}{2}\rho g b h^2$$

   (b) To find the height $H$ of the effective point of action, the torque produced by this force must equal the sum (integral) of the torques produced by all of the differential elements:

$$FH = \int_0^h \rho g (h - y) y b\, dy = \rho g b \left( \tfrac{1}{2} hy^2 - \tfrac{1}{3} y^3 \right)\Big|_0^h = \tfrac{1}{6}\rho g b h^3$$

When we use the result for the force, we have

$$H = \frac{\tfrac{1}{6}\rho g b h^3}{\tfrac{1}{2}\rho g b h^2} = \tfrac{1}{3}h$$

   (c) To prevent overturning, the torque about the base of the dam from the weight of concrete must be greater than the torque from the water pressure:

$$m_{concrete}gt/2 \geq \rho_{water}gbh^3/6;$$
$$\rho_{concrete}(hbt)gt/2 \geq \rho_{water}gbh^3/6, \quad \text{or}$$
$$t^2 \geq (\rho_{water}/\rho_{concrete})h^2/3 = [(1.00 \times 10^3 \text{ kg/m}^3)/(2.3 \times 10^3 \text{ kg/m}^3)]h^2/3, \text{ which gives } t \geq 0.38h.$$

It is not necessary to add atmospheric pressure, because it acts on both sides of the dam.

27. Because the mass of the displaced water is the apparent change in mass of the rock, we have
$$\Delta m = -\rho_{water}V.$$
For the density of the rock we have
$$\rho_{rock} = m_{rock}/V = -(m_{rock}/\Delta m)\rho_{water}$$
$$= [-(7.85 \text{ kg})/(6.18 \text{ kg} - 7.85 \text{ kg})](1.00 \times 10^3 \text{ kg/m}^3) = 4.70 \times 10^3 \text{ kg/m}^3.$$

31. Because the mass of the displaced water is the apparent change in mass of the sample, we have
$$\Delta m = -\rho_{water}V.$$
For the density of the sample we have
$$\rho = m/V = -(m/\Delta m)\rho_{water}$$
$$= [(63.5 \text{ g})/(63.5 \text{ g} - 56.4 \text{ g})](1.00 \times 10^3 \text{ kg/m}^3) = 8.94 \times 10^3 \text{ kg/m}^3.$$

35. (a) The buoyant force is a measure of the net force on the partially submerged object due to the pressure in the fluid. In order to remove the object and have no effect on the fluid, we would have to fill the submerged volume with an equal volume of fluid. As expected, the buoyant force on this fluid is the weight of the fluid. To have no net torque on the fluid, the buoyant force and the weight of the fluid would have to act at the same point, the center of gravity.

   (b) From the diagram we see that, if the center of buoyancy is above the center of gravity, when the ship tilts, the net torque will tend to restore the ship's position. From the diagram we see that, if the center of buoyancy is below the center of gravity, when the ship tilts, the net torque will tend to continue the tilt. If the center of buoyancy is at the center of gravity, there will be no net torque from these forces, so other torques (from the wind and waves) would determine the motion of the ship. Thus stability is achieved when the center of buoyancy is above the center of gravity.

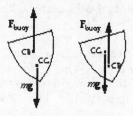

39. From Problem 38, we know that the initial volume out of the water, without the bear on the ice, is
$$V_1 = 0.105V_0 = 0.105(10 \text{ m}^3) = 1.05 \text{ m}^3.$$
Thus we find the submerged volume of ice with the bear on the ice from
$$V_2 = V_0 - \tfrac{1}{2}V_1 = 10 \text{ m}^3 - \tfrac{1}{2}(1.05 \text{ m}^3) = 9.48 \text{ m}^3.$$

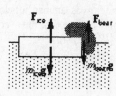

Because the net force is zero, we have

$F_{net} = 0 = F_{bear} + F_{ice} - m_{bear}g - m_{ice}g$,  or

$\rho_{sea\ water}g(0.30)V_{bear} + \rho_{sea\ water}gV_2 = m_{bear}g + \rho_{ice}gV_0$;

$\rho_{sea\ water}g(0.30)(m_{bear}/\rho_{bear}) + \rho_{sea\ water}gV_2 = m_{bear}g + \rho_{ice}gV_0$;

$(1.025)\rho_w(0.30)[m_{bear}/(1.00)\rho_w] + (1.025 \times 10^3\ \text{kg/m}^3)(9.48\ \text{m}^3) = m_{bear} + (0.917)(1.00 \times 10^3\ \text{kg/m}^3)(10\ \text{m}^3)$, which

gives $m_{bear} = 7.9 \times 10^2$ kg.

43. From the equation of continuity we have

Flow rate $= Av$;

$(9.2\ \text{m})(5.0\ \text{m})(4.5\ \text{m})/(12\ \text{min})(60\ \text{s/min}) = \pi(0.15\ \text{cm})^2v$, which gives $v = 4.1$ m/s.

47. If we choose the initial point at the water main where the water is not moving and the final point at the top of the spray, where the water also is not moving, from Bernoulli's equation we have

$P_1 + \frac{1}{2}\rho v_1^2 + \rho g y_1 = P_2 + \frac{1}{2}\rho v_2^2 + \rho g y_2$;

$P_1 + 0 + 0 = P_{atm} + 0 + (1.00 \times 10^3\ \text{kg/m}^3)(9.80\ \text{m/s}^2)(15\ \text{m})$, which gives

$P_1 - P_{atm} = P_{gauge} = 1.5 \times 10^5\ \text{N/m}^2 = 1.5$ atm.

51. If we consider the points at the top and bottom surfaces of the wing compared to the air flow in front of the wing, from Bernoulli's equation we have

$P_0 + \frac{1}{2}\rho v_0^2 + \rho g h_0 = P_1 + \frac{1}{2}\rho v_1^2 + \rho g h_1 = P_2 + \frac{1}{2}\rho v_2^2 + \rho g h_2$;

$P_1 + \frac{1}{2}(1.29\ \text{kg/m}^3)(340\ \text{m/s})^2 + 0 = P_2 + \frac{1}{2}(1.29\ \text{kg/m}^3)(290\ \text{m/s})^2 + 0$,

which gives $P_2 - P_1 = 2.03 \times 10^4\ \text{N/m}^2$.

The net upward force on the wing is

$F = (P_2 - P_1)A = (2.03 \times 10^4\ \text{N/m}^2)(86\ \text{m}^2) = 1.7 \times 10^6$ N.

55. From the equation of continuity we have

Flow rate $= A_1v_1 = A_2v_2$,  or  $v_2 = (A_1/A_2)v_1$.

From Bernoulli's equation we have

$P_1 + \frac{1}{2}\rho v_1^2 + \rho g y_1 = P_2 + \frac{1}{2}\rho v_2^2 + \rho g y_2$;

$P_{atm} + \frac{1}{2}\rho v_1^2 + 0 = P_{atm} + \frac{1}{2}\rho(A_1/A_2)^2 v_1^2 + \rho g h$, which gives

$v_1 = \{2gh/[1 - (A_1/A_2)^2]\}^{1/2}$.

59. (a) We assume $v_2 = 0$. From Bernoulli's equation we have

$\qquad P_1 + \frac{1}{2}\rho v_1^2 + \rho g y_1 = P_2 + \frac{1}{2}\rho v_2^2 + \rho g y_2$;

$\qquad P_{atm} + \frac{1}{2}\rho v_1^2 + \rho g h_1 = P_{atm} + 0 + \rho g h_2$, which gives

$\qquad v_1 = [2g(h_2 - h_1)]^{1/2}$.

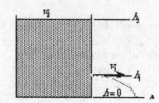

After the fluid leaves the tank, its horizontal velocity is constant. We find the time to reach the ground from

$\qquad y = y_0 + \frac{1}{2}(-g)t^2$;

$\qquad 0 = h_1 - \frac{1}{2}gt^2$, which gives $t = (2h_1/g)^{1/2}$.

In this time the water will have traveled a horizontal distance given by

$\qquad x = v_1t = [2g(h_2 - h_1)]^{1/2}(2h_1/g)^{1/2} = 2[h_1(h_2 - h_1)]^{1/2}$.

(b) To have the same range for a height $h_1'$, we must have

$\qquad h_1'(h_2 - h_1') = h_1(h_2 - h_1)$.

We could solve this quadratic, but we can see by inspection that

$\qquad h_1' = h_2 - h_1$.

63. From Poiseuille's equation we have

$Q = \pi r^4(P_1 - P_2)/8\eta L$;

$(5.6 \times 10^{-6}\ \text{m}^3/\text{min})/(60\ \text{s/min}) = \pi(0.90 \times 10^{-3}\ \text{m})^4(P_1 - P_2)/8(200 \times 10^{-3}\ \text{Pa} \cdot \text{s})(5.5 \times 10^{-2}\ \text{m})$,

which gives $P_1 - P_2 = 4.0 \times 10^3$ Pa.

67. (a) We find the Reynolds number for the blood flow:

$\qquad Re = 2\bar{v}r\rho/\eta = 2(0.30\ \text{m/s})(0.010\ \text{m})(1.05 \times 10^3\ \text{kg/m}^3)/(4.0 \times 10^{-3}\ \text{Pa} \cdot \text{s}) = 1600$.

Thus the flow is laminar but close to turbulent.

(b) If the only change is in the average speed, we have

$\qquad Re_2/Re_1 = \bar{v}_2/\bar{v}_1$;

$\qquad Re_2/1600 = 2$,  or  $Re_2 = 3200$, so the flow is turbulent.

71. For the two surfaces, top and bottom, we have
$F = \gamma 2L = (0.025 \text{ N/m})2(0.182 \text{ m}) = 9.1 \times 10^{-3} \text{ N}.$

75. The liquid pressure is produced from the elevation of the bottle:
$\Delta P = \rho g h.$
(a) $(65 \text{ mm-Hg})(133 \text{ N/m}^2 \cdot \text{mm-Hg}) = (1.00 \times 10^3 \text{ kg/m}^3)(9.80 \text{ m/s}^2)h$, which gives $h = 0.88 \text{ m}.$
(b) $(550 \text{ mm-H}_2\text{O})(9.81 \text{ N/m}^2 \cdot \text{mm-H}_2\text{O}) = (1.00 \times 10^3 \text{ kg/m}^3)(9.80 \text{ m/s}^2)h$, which gives $h = 0.55 \text{ m}.$
(c) If we neglect viscous effects, we must produce a pressure to balance the blood pressure:
$(18 \text{ mm-Hg})(133 \text{ N/m}^2 \cdot \text{mm-Hg}) = (1.00 \times 10^{-3} \text{ kg/m}^3)(9.80 \text{ m/s}^2)h$, which gives $h = 0.24 \text{ m}.$

79. The pressure difference is produced from the elevation:
$\Delta P = \rho g h$
$= (1.29 \text{ kg/m}^3)(9.80 \text{ m/s}^2)(410 \text{ m})/(1.013 \times 10^5 \text{ N/m}^2 \cdot \text{atm}) = 0.051 \text{ atm}.$

83. The net force on the floating continent is zero:
$F_{\text{net}} = 0 = F_{\text{buoy}} - m_{\text{cont}}g = \rho_{\text{rock}}g \, Ah_{\text{rock}} - \rho_{\text{cont}}gAh_{\text{cont}};$
$(3300 \text{ kg/m}^3)gAh_{\text{rock}} = (2800 \text{ kg/m}^3)gA(35 \text{ km}),$
which gives $h_{\text{rock}} = 30 \text{ km}$
Thus the height of the continent above the surrounding rock is $h_{\text{cont}} - h_{\text{rock}} = $

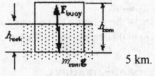

5 km.

87. If we choose the initial point at the pressure head, where the water is not moving, and the final point at the faucet, from Bernoulli's equation we have
$P_1 + \tfrac{1}{2}\rho v_1{}^2 + \rho g h_1 = P_2 + \tfrac{1}{2}\rho v_2{}^2 + \rho g h_2;$
$P_{\text{atm}} + 0 + (1.00 \times 10^3 \text{ kg/m}^3)(9.80 \text{ m/s}^2)h_1 = P_{\text{atm}} + \tfrac{1}{2} (1.00 \times 10^3 \text{ kg/m}^3)(7.2 \text{ m/s})^2 + 0$, which gives
$h_1 = 2.6 \text{ m}.$

91. We find the effective $g$ by considering a volume of the water.
The net force must produce the acceleration:
$F_{\text{buoy1}} - mg = ma;$
$\rho_{\text{water}}g'V_{\text{water}} - \rho_{\text{water}}V_{\text{water}}g = \rho_{\text{water}}V_{\text{water}}a;$
$g' = g + a = g + 2.4g = 3.4g.$

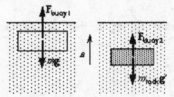

Note that this agrees with adding a pseudoforce $- m\mathbf{a}$ in the accelerating frame of the bucket. The buoyant force on the rock is
$F_{\text{buoy2}} = \rho_{\text{water}}g'V_{\text{rock}} = \rho_{\text{water}}g'(m_{\text{rock}}/\rho_{\text{rock}}) = g'm_{\text{rock}}/\text{SG}_{\text{rock}}$
$= (3.4)(9.80 \text{ m/s}^2)(3.0 \text{ kg})/(2.7)$
$= 37 \text{ N}.$
To see if the rock will float, we find the net force acting on the rock in the accelerated frame:
$F_{\text{net}} = F_{\text{buoy2}} - m_{\text{rock}}g' = 37 \text{ N} - (3.0 \text{ kg})(3.4)(9.80 \text{ m/s}^2) = -63 \text{ N}.$
Because the result is negative, the rock will not float.

95. (a) We choose the reference level at the bottom of the sink. If we apply Bernoulli's equation to the flow from the average depth of water in the sink to the pail, we have
$P_0 + \tfrac{1}{2}\rho v_0{}^2 + \rho g h_0 = P_1 + \tfrac{1}{2}\rho v_1{}^2 + \rho g h_1;$
$P_{\text{atm}} + 0 + \rho g h_0 = P_{\text{atm}} + \tfrac{1}{2}\rho v_1{}^2 + \rho g h_1$, or
$v_1 = [2g(h_0 - h_1)]^{1/2} = \{2(9.80 \text{ m/s}^2)[0.020 \text{ m} - (-0.50 \text{ m})]\}^{1/2} = 3.2 \text{ m/s}.$
(b) We use the flow rate to find the time:
$Q = Av = V/t;$
$\pi(0.010 \text{ m})^2(3.19 \text{ m/s}) = (0.48 \text{ m}^2)(0.040 \text{ m})/t$, which gives $t = 20 \text{ s}.$

# Chapter 14: Oscillations

## Chapter Overview and Objectives

This chapter introduces vibrating or oscillatory motion. It describes in detail a particular type of oscillatory motion called simple harmonic motion. It also introduces damped and driven harmonic oscillators.

After completing study of this chapter, you should:

- Know what periodic motion is and the definitions of frequency, period, and angular frequency.
- Know what simple harmonic motion is and what forces lead to simple harmonic motion.
- Know the parameters which describe simple harmonic motion: amplitude, angular frequency, and phase angle.
- Know how to determine the period or frequency of a mass–spring system.
- Know how to determine the period or frequency of a simple pendulum, a physical pendulum, and a torsion pendulum.
- Know the solution to a damped harmonic oscillator and its characteristics.
- Know the steady-state solution of the driven harmonic oscillator and the characteristics of its solutions.

## Summary of Equations

Relationships between period and frequency:
$$T = \frac{1}{f} \quad \text{and} \quad f = \frac{1}{T}$$

Simple harmonic motion:
$$x(t) = A\cos\left(\sqrt{\frac{k}{m}}t + \phi\right) = A\cos(\omega t + \phi)$$

Relationships between angular frequency, frequency, and period:
$$\omega = \frac{2\pi}{T} = 2\pi f$$

Total energy of mass–spring system in simple harmonic motion:
$$E = \tfrac{1}{2}kA^2$$

Period of motion of simple pendulum:
$$T = 2\pi\sqrt{\frac{L}{g}}$$

Period of motion of physical pendulum:
$$T = 2\pi\sqrt{\frac{I}{mgh}}$$

Angular frequency of torsion pendulum:
$$\omega = \sqrt{\frac{K}{I}}$$

Motion of damped harmonic oscillator:
$$x(t) = Ae^{-\alpha t}\cos(\omega' t + \phi)$$

Steady-state solution of driven harmonic oscillator:
$$x(t) = \frac{F_0}{m\sqrt{(\omega^2 - \omega_0^2) + b^2\omega^2/m^2}}\cos(\omega t + \phi_0)$$

Quality factor or $Q$ of driven harmonic oscillator:
$$Q = \frac{m\omega_0}{b}$$

Full width at half-maximum power:
$$\Delta\omega \approx \frac{\omega_0}{Q}$$

# Chapter Summary

### Section 14-1 Oscillations of a Spring

When a motion repeats itself over and over, it is called **periodic motion**. The time for one repetition of the motion is called the **period** of the motion. The number of repetitions per unit time is called the **frequency** of the motion. The period, $T$, and the frequency, $f$, are related by

$$T = \frac{1}{f} \qquad \text{and} \qquad f = \frac{1}{T}$$

In the particular case of a mass attached to a spring, the mass vibrates back and forth about **the equilibrium position** of the mass on the spring. The equilibrium position of the mass is the position at which the mass has no net force acting on it (see Chapter 11). The difference in the position of the mass and its equilibrium position is called the **displacement** of the mass. The magnitude of the maximum displacement is called the **amplitude** of the periodic motion.

### Section 14-2 Simple Harmonic Motion

For a force that is a restoring force proportional to the displacement of the object such as Hooke's law, $F = -kx$, the resulting periodic motion is called **simple harmonic motion**. Writing down Newton's second law for a system with a Hooke's law force:

$$F = ma \qquad \Rightarrow \qquad -kx = m\frac{d^2x}{dt^2}$$

If we look for solutions to this equation, we find the most general solution to be

$$x(t) = A\cos\left(\sqrt{\frac{k}{m}}\,t + \phi\right) = A\cos(\omega t + \phi)$$

This type of dependence of position on time is simple harmonic motion. The constant $\omega$ is the angular frequency of the motion and is determined by the properties of the mass and the spring. The constant $A$ is the **amplitude** of the motion and the constant $\phi$ is called the **phase angle**. The amplitude and the phase are determined by the **initial conditions** of the motion.

The relationships between angular frequency, frequency, and period are

$$\omega = \frac{2\pi}{T} = 2\pi f$$

### Example 14-2-A

A mass of 2.44 kg is attached to a spring. The motion of the mass is

$$x(t) = (3.32\,\text{cm})\cos\left(2.29\,\text{s}^{-1}t + 1.19\right)$$

What is the spring constant of the spring? What is the amplitude of the motion? What is the position of the mass at time $t = 1.45$ s?

**Solution:**

By inspection, we know $\omega$ is equal to 2.29 s$^{-1}$. The relationship between $\omega$, the mass, $m$, and the spring constant, $k$, is

$$\omega = \sqrt{\frac{k}{m}} \qquad \Rightarrow \qquad k = m\omega^2 = (2.44\,\text{kg})(2.29\,\text{s}^{-1})^2 = 12.8\,\text{N}/\text{m}$$

By inspection, the amplitude of the motion is 3.32 cm.

Determining the position at time $t = 1.45$ s:

$$x(t) = (3.32\,\text{cm})\cos(2.29\,\text{s}^{-1}(1.45\,\text{s}) + 1.19) = -0.666\,\text{cm}$$

### Example 14-2-B

For the system in Example 14-2-A, determine the velocity and the acceleration of the mass as a function of time. Also determine the maximum speed and the maximum magnitude of the acceleration.

**Solution:**

$$v(t) = \frac{dx(t)}{dt} = -(3.32\,\text{cm})(2.29\,\text{s}^{-1})\sin(2.29\,\text{s}^{-1}(1.45\,\text{s}) + 1.19) = -(7.60\,\text{cm/s})\sin(2.29\,\text{s}^{-1}(1.45\,\text{s}) + 1.19)$$

$$a(t) = \frac{dv(t)}{dt} = -(7.60\,\text{cm/s})(2.29\,\text{s}^{-1})\cos(2.29\,\text{s}^{-1}(1.45\,\text{s}) + 1.19) = -(17.4\,\text{cm/s}^2)\cos(2.29\,\text{s}^{-1}(1.45\,\text{s}) + 1.19)$$

The maximum speed is the amplitude of the velocity, 7.60 cm/s. The maximum magnitude of the acceleration is the amplitude of the acceleration, 17.4 cm/s$^2$.

### Section 14-3 Energy in the Simple Harmonic Oscillator

The total energy of a mass and spring in simple harmonic motion is given by the sum of the kinetic and potential energies:

$$E = K + U = \tfrac{1}{2}mv^2 + \tfrac{1}{2}kx^2$$

When the object is at its maximum displacement from equilibrium, its speed must be zero. This allows us to write the total energy in terms of the amplitude, $A$, of the motion of the mass:

$$E = \tfrac{1}{2}kA^2$$

### Example 14-3-A

Determine the amplitude of a mass–spring system that has a speed of 2.34 m/s when the displacement of the mass is 1.45 cm. The spring constant of the spring is 98.2 N/m and the mass is 287.2 g.

**Solution:**

We can determine the total energy of the system:
$$E = K + U = \tfrac{1}{2}mv^2 + \tfrac{1}{2}kx^2 = \tfrac{1}{2}(0.2872\,\text{kg})(2.34\,\text{m/s})^2 + \tfrac{1}{2}(98.2\,\text{N/m})(0.0145\,m)^2 = 0.797\,\text{J}$$

We also know the total energy is equal to ½$kA^2$. We can use this to solve for the amplitude:

$$E = \tfrac{1}{2}kA^2 \quad \Rightarrow \quad A = \sqrt{\frac{2E}{k}} = \sqrt{\frac{2(0.797\,\text{J})}{98.2\,\text{N/m}}} = 0.127\,\text{m}$$

### Section 14-4 Simple Harmonic Motion Related to Uniform Circular Motion

The Cartesian components on the plane of the motion of an object in circular motion undergo simple harmonic motion. The motions in the two Cartesian directions differ by only a phase angle difference of $\pi/2$. The motions have identical amplitude and frequency.

## Section 14-5 The Simple Pendulum

A simple pendulum is a small mass hanging from the end of a string of length $L$ and of negligible mass. The mass is free to swing back and forth with only the force of gravity and the string acting on it. Newton's second law written for the motion along the tangential direction of the circular motion is

$$F = ma \quad \Rightarrow \quad mL\frac{d^2\theta}{dt^2} = -mg\sin\theta$$

This is a difficult equation to solve with the calculus we are expected to know at this point. However, under the circumstances that $\theta$ is a small angle, we can use the approximation

$$\sin\theta \approx \theta$$

where $\theta$ is in radians. Then Newton's second law becomes

$$mL\frac{d^2\theta}{dt^2} = -mg\theta \quad \Rightarrow \quad \frac{d^2\theta}{dt^2} = -\frac{g}{L}\theta$$

which we can see is the same form of equation of motion as the mass on the spring where the angular frequency of the motion, in this case, is

$$\omega = \sqrt{\frac{g}{L}}$$

The frequency is given by

$$f = \frac{\omega}{2\pi} = \frac{1}{2\pi}\sqrt{\frac{g}{L}}$$

and the period is given by

$$T = \frac{1}{f} = 2\pi\sqrt{\frac{L}{g}}$$

## Example 14-5-A

A planet has a radius $4.75 \times 10^3$ km. A pendulum with a length 1.00 m has a period of 2.1 s on the planet's surface. What is the mass of the planet?

## Solution:

Form the period, $T$, of the pendulum and the length, $L$, of the pendulum we can determine the surface gravitational acceleration on the planet:

$$T = 2\pi\sqrt{\frac{L}{g}} \quad \Rightarrow \quad g = \frac{4\pi^2 L}{T^2} = \frac{4\pi^2(1.00\,\text{m})}{(2.11\,\text{s})} = 8.87\,\text{m/s}^2$$

The acceleration of gravity at the surface is related to the mass and radius of the planet:

$$g = \frac{GM_{planet}}{R_{planet}^2} \Rightarrow M_{planet} = \frac{gR_{planet}^2}{G} = \frac{(8.87\,\text{m/s}^2)(4.75\times10^6\,\text{m})^2}{6.67\times10^{-11}\,\text{N}\cdot\text{m}^2/\text{kg}^2} = 3.00\times10^{24}\,\text{kg}$$

### Section 14-6 The Physical Pendulum and the Torsion Pendulum

Any body that is suspended about a fixed axis that does not pass through its center of mass will swing back and forth like the pendulum. Such a system is called a **physical pendulum**. The period, $T$, of such a pendulum is given by

$$T = 2\pi \sqrt{\frac{I}{mgh}}$$

where $I$ is the moment of inertia of the object about the axis of rotation, $m$ is the mass of the object, and $h$ is the distance of the axis of rotation from the center of mass of the object.

An object suspended from a thin rod or wire will also oscillate back and forth in simple harmonic rotational motion if the rod or wire is initially twisted and released. This configuration is called a **torsion pendulum**. A rod or wire will twist by an angle proportional to the applied torque on the angle. The proportionality constant is called the torsional constant, $K$, of the rod or wire. The angular frequency of the torsion pendulum is given by

$$\omega = \sqrt{\frac{K}{I}}$$

where $I$ is the moment of inertia of the object about the twist axis of the system.

### Example 14-7-A

A uniform rod of mass 1.3 kg and length 59 cm is suspended on a pivot at one end of the rod. At the other end of the rod is a small object with a mass of 1.4 kg. What is the period of the physical pendulum?

### Solution:

This is a physical pendulum so we need to know the moment of inertia about the pivot and the distance of the center of mass from the pivot. The moment of inertia of this system, $I$, is the moment of inertia of the rod about its end, $I_{rod}$, plus the moment of inertia of the mass at the end of the rod, $I_{mass}$.

$$I = I_{rod} + I_{mass} = \tfrac{1}{2} m_{rod} L_{rod}^2 + m_{mass} L_{rod}^2$$
$$= \tfrac{1}{2}(1.3\,\text{kg})(0.59\,\text{m})^2 + (1.4\,\text{kg})(0.59\,\text{m})^2 = 0.714\,\text{kg} \cdot \text{m}^2$$

The center of mass is located at a distance from the pendulum given by

$$x_{CM} = \frac{\sum m_i x_i}{\sum m_i} = \frac{m_{rod} L_{rod}/2 + m_{mass} L_{rod}}{m_{rod} + m_{mass}}$$
$$= \frac{(1.3\,\text{kg})(0.59\,\text{m})/2 + (1.4\,\text{kg})(0.59\,\text{m})}{(1.3\,\text{kg} + 1.4\,\text{kg})} = 0.448\,\text{m}$$

The period of the physical pendulum is given by

$$T = 2\pi \sqrt{\frac{I}{mgh}} = 2\pi \sqrt{\frac{0.714\,\text{kg} \cdot \text{m}^2}{(2.7\,\text{kg})(9.8\,\text{m/s}^2)(0.448\,\text{m})}} = 1.5\,\text{s}$$

### Example 14-6-B

A square plate that is 12 cm on a side has a mass of 1.05 kg. The plate is suspended from a wire that requires a torque of 0.21 N·m to twist it through an angle of 45°. What is the period of the torsion pendulum?

**Solution:**

To determine the period of the torsion pendulum, we must know the moment of inertia of the system about the suspending wire and the torsion constant of the wire. The torsion constant of the wire is determined from

$$\tau = K\theta \quad \Rightarrow \quad K = \frac{\tau}{\theta} = \frac{0.21\,\text{N}\cdot\text{m}}{\pi/4} = 0.267\,\text{N}\cdot\text{m}\,(\text{per radian})$$

The moment of inertia of a square plate about an axis passing through its center perpendicular to the plate is

$$I = \tfrac{1}{6}mL^2 = \tfrac{1}{6}(1.05\,\text{kg})(0.12\,\text{m})^2 = 2.52\times10^{-3}\,\text{kg}\cdot\text{m}^2$$

The period of the torsion pendulum is

$$T = 2\pi\sqrt{\frac{I}{K}} = 2\pi\sqrt{\frac{2.52\times10^{-3}\,\text{kg}\cdot\text{m}^2}{0.267\,\text{N}\cdot\text{m}}} = 0.61\,\text{s}$$

**Section 14-7 Damped Harmonic Motion**

There is always some friction present in any macroscopic sized system. As the frictional force will do negative work on the system, the total energy of the system will decrease with time. For a system in simple harmonic motion, this means that the amplitude of the motion will decrease with time. Such a system is in **damped[1] harmonic motion**.

In general, the friction or damping force can have a complicated mathematical dependence, which can make Newton's second law difficult to solve for the motion as a function of time. One particular force for which the solution is relatively simple is a frictional force that is proportional to the velocity of the object

$$F_{damping} = -bv$$

This results in Newton's second law for the damped mass on a spring

$$F = ma \quad \Rightarrow \quad -kx - bv = ma \quad \Rightarrow \quad -kx - b\frac{dx}{dt} = m\frac{dx^2}{dt^2}$$

In that case, the solution to Newton's second law is

$$x(t) = Ae^{-\alpha t}\cos(\omega' t + \phi)$$

where

$$\alpha = \frac{b}{2m} \quad \text{and} \quad \omega' = \sqrt{\frac{k}{m} - \frac{b^2}{4m^2}}$$

and again, as in the undamped case, $A$ is the amplitude and $\phi$ is the phase angle. Notice that the frequency is not a real number if $b^2/4m^2 > k/m$. In these cases, the system does not oscillate, but returns to the equilibrium position without oscillating. In this situation, the system is said to be **overdamped**. The system is said to be **underdamped** if $b^2/4m^2 < k/m$ and the system is **critically damped** if $b^2/4m^2 = k/m$.

---

[1]A note on vocabulary is appropriate here. Many students confuse damping and damped with dampening and dampened. The latter two words refer to making moist. Use the correct words!

**Example 14-7-A**

How long does it take a damped harmonic oscillator amplitude to reach one-half its original amplitude?

**Solution:**

We want to determine when

$$A_0 e^{-at} = \tfrac{1}{2} A_0 \quad \Rightarrow \quad e^{-at} = \tfrac{1}{2} \quad \Rightarrow \quad t = -\frac{1}{a}\ln \tfrac{1}{2} = \frac{1}{a}\ln 2$$

**Section 14-8 Forced Vibrations; Resonance**

Sometimes an external force is applied to an oscillatory system. If the time dependence of this external force has a time dependence of the form

$$F_{external} = F_0 \cos \omega t$$

then Newton's second law for damped mass–spring system becomes

$$F_{external} - kx - bv = ma \quad \Rightarrow \quad F_0 \cos \omega t - kx - b\frac{dx}{dt} = m\frac{d^2 x}{dt^2}$$

A solution of this equation is

$$x(t) = \frac{F_0}{m\sqrt{\left(\omega^2 - \omega_0^2\right)^2 + b^2\omega^2 / m^2}} \cos\left(\omega t + \phi_0\right)$$

where

$$\omega_0 = \sqrt{\frac{k}{m}} \quad \text{and} \quad \phi_0 = \tan^{-1}\frac{\omega_0^2 - \omega^2}{\omega(b/m)}$$

This is called the **steady-state** solution to the problem. In addition to this solution, a term that is the same as the solution to the non-driven damped harmonic oscillator must be added to complete the solution. That additional term is called the **transient** solution because it eventually becomes negligible in size because of the decreasing exponential factor.

Resonance occurs when the $\omega$ is equal to $\omega_0$. For situations when the damping force is relatively small, the amplitude of the steady-state solution becomes large near resonance, because the denominator in the amplitude becomes small. The **quality factor** or $Q$ of the system is a measurement of the ratio of the amplitude of oscillation of the system at resonance to the amplitude at low frequency. The $Q$ value is defined to be

$$Q = \frac{m\omega_0}{b}$$

The width of the resonance peak in the graph of amplitude versus frequency is usually measured by the full width at half maximum power. The power is proportional to the square of the amplitude of the oscillator. That implies the half maximum power is at amplitudes that are the maximum amplitude divided by the square root of 2. By this definition, the full width at half maximum power, $\Delta\omega$, is approximately

$$\Delta\omega \approx \frac{\omega_0}{Q}$$

when the damping is relatively small.

## Practice Quiz

1.    Which of the following changes in a mass–spring system does not increase the frequency of the system?

    a) Increase the spring constant of the spring.
    b) Decrease the mass.
    c) Increase the stiffness of the spring.
    d) Increase the amplitude of the motion.

2.    When a simple harmonic oscillator is at maximum displacement, which statement is true about another kinematical quantity describing the motion?

    a) The harmonic oscillator is at its maximum speed.
    b) The harmonic oscillator is at zero acceleration.
    c) The harmonic oscillator is at maximum acceleration in the direction of the displacement.
    d) The harmonic oscillator is at maximum acceleration in the direction opposite the displacement.

3.    If the amplitude of a simple harmonic oscillator is doubled, what happens to the total eenrgy of the system?

    a) The total energy doubles.
    b) The total energy triples.
    c) The total energy quadruples.
    d) The total energy becomes half as great.

4.    Which set of quantities is sufficient for specifying the motion of a simple harmonic oscillator?

    a) Amplitude, period, and frequency
    b) Phase angle, period, and frequency
    c) Amplitude, phase angle, and period
    d) Spring constant, mass, and phase angle

5.    A mass-spring system has a frequency $f$ when the mass slides horizontally across a frictionless surface. If instead of sliding horizontally, the mass is suspended from the spring vertically, what will be the frequency of the mass-spring system?

    a) Depends on the gravitational acceleration
    b) Depends on how far the spring stretches to support the weight of the mass at equilibrium
    c) $2f$
    d) $f$

6.    We have seen how a pendulum is approximately a system that undergoes simple harmonic motion for small amplitudes. What happens to the period of the pendulum as the amplitude increases to large angles?

    a) Period remains the same.
    b) Period increases.
    c) Period decreases.
    d) It depends on the length of the pendulum.

7.    A damped harmonic oscillator decreases to half its initial amplitude in time $t$. How much longer does it take for the amplitude to reduce to one fourth of its original amplitude?

    a) $t/2$
    b) $t/4$
    c) $t \ln 2$
    d) $t$

8.    The acceleration of gravity at the surface of the moon is approximately one sixth the acceleration of gravity at the surface of the earth. A pendulum at the surface of the earth has length $L$. On the surface of the moon, a pendulum with the same period will have a length

    a) $6L$
    b) $L/6$
    c) $36L$
    d) $L/\sqrt{6}$

9.    Which of these systems would benefit most by having the system critically damped?

    a) A pendulum on a clock
    b) The suspension of an automobile
    c) A tuning fork
    d) A child on a swing

10.    For a driven harmonic oscillator, the amplitude of the system is $A$. Which of the following will definitely reduce the amplitude of the system?

    a) Reduce the frequency of the driving force.
    b) Increase the frequency of the driving force.
    c) Increase the quality factor of the system.
    d) Decrease the quality factor of the system.

11.    A mass–spring system has a total energy of 18.4 J, oscillates with a period of 1.46 s, and has an amplitude of 4.6 cm. What are the spring constant and the mass of the system?

12.    A torsion pendulum is a wire with a torsion constant of 24.6 N·m/rad that supports a uniform rod at its center of mass. The rod has a length of 10.8 cm and a mass of 0.422 kg. What is the period of oscillation of the torsion pendulum?

13.    For a damped harmonic oscillator, the amplitude is initially 3.64 cm. After 24.5 s the amplitude is 2.99 cm. How long does it take the amplitude to reach 1.00 cm?

14.    For a driven harmonic oscillator, determine the limit of the phase angle as the driving frequency goes to zero, the driving frequency goes to infinity, and the driving frequency goes to the natural frequency.

15.    A driven harmonic oscillator consists of a spring with a spring constant 145 N/m attached to a 0.388 kg mass. The driving force has an amplitude of 3.65 N and a frequency 2.87 Hz. The damping force is given by $F_{\text{damping}} = -8.45$ N·s/m $v$. What is the amplitude of the oscillation of the mass?

## Problem Solutions

3.    We find the spring constant from the compression caused by the increased weight:
    $k = mg/x = (80 \text{ kg})(9.80 \text{ m/s}^2)/(0.0140 \text{ m}) = 5.60 \times 10^4$ N/m.
    The frequency of vibration will be
    $f = (k/m)^{1/2}/2\pi = [(5.60 \times 10^4 \text{ N/m})/(1000 \text{ kg})]^{1/2}/2\pi = 1.19$ Hz.

7.    Because the mass is released at the maximum displacement, we have
    $x = x_{\text{max}} \cos(\omega t); \quad v = -v_{\text{max}} \sin(\omega t); \quad a = -a_{\text{max}} \cos(\omega t).$
    ($a$) We find $\omega t$ from
        $v = -\tfrac{1}{2}v_{\text{max}} = -v_{\text{max}} \sin(\omega t)$, which gives $\omega t = 30°$.
    Thus the distance is
        $x = x_{\text{max}} \cos(\omega t) = x_{\text{max}} \cos 30° = 0.866 \, x_{\text{max}}.$
    ($b$) We find $\omega t$ from
        $a = -\tfrac{1}{2}a_{\text{max}} = -a_{\text{max}} \cos(\omega t)$, which gives $\omega t = 60°$.
    Thus the distance is
        $x = x_{\text{max}} \cos(\omega t) = x_{\text{max}} \cos 60° = 0.500 \, x_{\text{max}}.$

11. In the equilibrium position, the net force is zero. When the mass is pulled down a distance $x$, the net restoring force is the sum of the additional forces from the springs, so we have

$F_{net} = \Delta F_2 + \Delta F_1 = - k_2 x - k_1 x = - (k_1 + k_2)x,$

which gives an effective force constant of $k_1 + k_2$.
We find the frequency of vibration from

$f = (k_{eff}/m)^{1/2}/2\pi = [(k_1 + k_2)/m]^{1/2}/2\pi.$

15. The dependence of the frequency on the mass is

$f = (k/m)^{1/2}/2\pi.$

Because the spring constant does not change, we have

$f_2/f_1 = (m_1/m_2)^{1/2};$

$f_2/(3.0 \text{ Hz}) = [(0.50 \text{ kg})/(0.35 \text{ kg})]^{1/2},$ which gives $f_2 = 3.6$ Hz.

19. (a) We find the frequency from

$f = (k/m)^{1/2}/2\pi = [(345 \text{ N/m})/(0.250 \text{ kg})]^{1/2}/2\pi = 5.91$ Hz, so

$\omega = 2\pi f = 2\pi(5.91 \text{ Hz}) = 37.1 \text{ s}^{-1}.$

Because the mass starts at the equilibrium position moving in the negative (downward) direction, we have a sine function:

$y = - A \sin(\omega t) = - (0.220 \text{ m}) \sin [(37.1 \text{ s}^{-1})t].$

(b) The period of the motion is

$T = 1/f = 1/(5.91 \text{ Hz}) = 0.169$ s.

It will take one-quarter period to reach the maximum extension, so the spring will have maximum extensions at 0.0423 s, 0.211 s, 0.381 s, … .
It will take three-quarters period to reach the minimum extension, so the spring will have minimum extensions at 0.127 s, 0.296 s, 0.465 s, … .

23. (a) If we apply a force $F$ to stretch the springs, the total displacement $\Delta x$ is the sum of the displacements of the two springs: $\Delta x = \Delta x_1 + \Delta x_2.$
The effective spring constant is defined from $F = - k_{eff} \Delta x.$
Because they are in series, the force must be the same in each spring:

$F_1 = F_2 = F = - k_1 \Delta x_1 = - k_2 \Delta x_2.$

Then $\Delta x = \Delta x_1 + \Delta x_2$ becomes

$- F/k_{eff} = - (F/k_1) - (F/k_2),$  or  $1/k_{eff} = (1/k_1) + (1/k_2).$

For the period we have

$T = 2\pi (m/k_{eff})^{1/2} = 2\pi \{m[(1/k_1) + (1/k_2)]\}^{1/2}.$

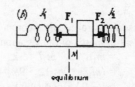

(b) In the equilibrium position, we have

$F_{net} = F_{20} - F_{10} = 0,$  or  $F_{10} = F_{20}.$

When the object is moved to the right a distance $x$, we have

$F_{net} = F_{20} - k_2 x - (F_{10} + k_1 x) = - (k_1 + k_2)x.$

The effective spring constant is $k_{eff} = k_1 + k_2$, so the period is

$T = 2\pi (m/k_{eff})^{1/2} = 2\pi [m/(k_1 + k_2)]^{1/2}.$

27. (a) The amplitude is the maximum value of $x$: 0.650 m.
(b) We find the frequency from the coefficient of $t$:

$2\pi f = 8.40 \text{ s}^{-1},$ which gives $f = 1.34$ Hz.

(c) The maximum speed is

$v_0 = \omega A = (8.40 \text{ s}^{-1})(0.650 \text{ m}) = 5.46$ m/s.

We find the total energy from the maximum kinetic energy:

$E = K_{max} = \tfrac{1}{2}mv_0^2 = \tfrac{1}{2}(2.00 \text{ kg})(5.46 \text{ m/s})^2 = 29.8$ J.

(d) We find the velocity at the position from

$v = v_0[1 - (x^2/A^2)]^{1/2}$

$= (5.46 \text{ m/s})\{1 - [(0.260 \text{ m})^2/(0.650 \text{ m})^2]\}^{1/2} = 5.00$ m/s.

The kinetic energy is

$K = \tfrac{1}{2}mv^2 = \tfrac{1}{2}(2.00 \text{ kg})(5.00 \text{ m/s})^2 = 25.0$ J.

The potential energy is

$U = E - K = 29.8 \text{ J} - 25.0 \text{ J} = 4.8$ J.

31. We can compare the two maximum potential energies:
$U_1/U_2 = \frac{1}{2}k_1A_1{}^2/\frac{1}{2}k_2A_2{}^2 = (k_1/k_2)(A_1/A_2)^2$;
$10 = 2(A_1/A_2)^2$,  or  $A_1 = 2.24A_2$.

35. For the collision of the bullet and block, momentum is conserved:
$mv = (m + M)V$,  so  $V = mv/(m + M)$.
The kinetic energy of the bullet and block immediately after the collision is stored in the potential energy of the spring when the spring is fully compressed:
$\frac{1}{2}(m + M)V^2 = \frac{1}{2}kA^2$;
$(m + M)[mv/(m + M)]^2 = kA^2$;
$[(0.007870 \text{ kg})v]^2/(0.007870 \text{ kg} + 6.023 \text{ kg}) = (142.7 \text{ N/m})(0.09460 \text{ m})^2$, which gives $v = 352.6$ m/s.

39. We find the length from
$T = 2\pi (L/g)^{1/2}$;
$2.000 \text{ s} = 2\pi [L/(9.80 \text{ m/s}^2)]^{1/2}$, which gives $L = 0.9929$ m.

43. We assume that 14° is small enough that we can consider this a simple pendulum, with a period
$T = 2\Delta(L/g)^{1/2} = 2\Delta[(0.30 \text{ m})/(9.80 \text{ m/s}^2)]^{1/2} = 1.10 \text{ s}$, and $\omega = 2\Delta/T = 2\Delta/(1.10 \text{ s}) = 5.71 \text{ s}^{-1}$.
Because the pendulum is released at the maximum angle, the angle will oscillate as a cosine function:
$\theta = \theta_0 \cos (\omega t) = (14°) \cos [(5.71 \text{ s}^{-1})t]$.
   (a)  $\theta = (14°) \cos [(5.71 \text{ s}^{-1})(0.65 \text{ s})] = -12°$.
   This is reasonable, because the time is slightly over half a period.
   (b)  $\theta = (14°) \cos [(5.71 \text{ s}^{-1})(1.95 \text{ s})] = +2.1°$.
   (c)  $\theta = (14°) \cos [(5.71 \text{ s}^{-1})(5.00 \text{ s})] = -13°$.
   This is reasonable; the time is 4 periods and 0.60 s, so it should be close to the answer for part (a).

47. We use the parallel-axis theorem to find the moment of
inertia about the pin A:
$I_A = I_{CM} + Mh^2 = \frac{1}{2}MR^2 + Mh^2 = M(\frac{1}{2}R^2 + h^2)$.
The period for small oscillations is
$\begin{aligned} T &= 2\pi (I_A/Mgh)^{1/2} = 2\pi [(\frac{1}{2}R^2 + h^2)/gh]^{1/2} \\ &= 2\pi \{[\frac{1}{2}(0.200 \text{ m})^2 + (0.18 \text{ m})^2]/(9.80 \text{ m/s}^2)(0.18 \text{ m})\}^{1/2} \\ &= 1.08 \text{ s}. \end{aligned}$

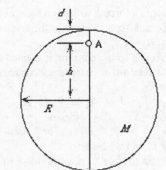

51. (a) We use the parallel-axis theorem for the lower mass to find the
moment of inertia of the leg about the hip:
$\begin{aligned} I &= m_1L_1{}^2/3 + m_2L_2{}^2/12 + m_2(L_1 + \frac{1}{2}L_2)^2 \\ &= (m_1/3 + m_2/12 + 9m_2/4)L^2 \\ &= [(7.0 \text{ kg})/3 + (4.0 \text{ kg})/12 + 9(4.0 \text{ kg})/4](0.50 \text{ m})^2 = 2.92 \text{ kg} \cdot \text{m}^2. \end{aligned}$
The distance of the center of mass of the leg from the hip is
$\begin{aligned} h &= [m_1\frac{1}{2}L_1 + m_2(L_1 + \frac{1}{2}L_2)]/(m_1 + m_2) = (\frac{1}{2}m_1 + 3/2m_2)L/(m_1 + m_2) \\ &= [\frac{1}{2}(7.0 \text{ kg}) + 3/2 (4.0 \text{ kg})](0.50 \text{ m})/(7.0 \text{ kg} + 4.0 \text{ kg}) = 0.432 \text{ m}. \end{aligned}$
The natural period is
$\begin{aligned} T &= 2\pi [I/(m_1 + m_2)gh]^{1/2} \\ &= 2\pi [(2.92 \text{ kg} \cdot \text{m}^2)/(7.0 \text{ kg} + 4.0 \text{ kg})(9.80 \text{ m/s}^2)(0.432 \text{ m})]^{1/2} \\ &= 1.6 \text{ s}. \end{aligned}$

55. (a) The angular frequency for the damped motion is
$\begin{aligned} \omega' &= [(k/m) - (b^2/4m^2)]^{1/2} \\ &= \{[(56.0 \text{ N/m})/(0.750 \text{ kg})] - [(0.162 \text{ N} \cdot \text{s/m})^2/4(0.750 \text{ kg})^2]\}^{1/2} = 8.64 \text{ s}^{-1}. \end{aligned}$
The period is
$T = 2\pi /\omega' = 2\Delta/(8.64 \text{ s}^{-1}) = 0.727 \text{ s}$.
   (b) We evaluate the factor
$b/2m = (0.162 \text{ N} \cdot \text{s/m})/2(0.750 \text{ kg}) = 0.108 \text{ /s}$.
The amplitude is proportional to $e^{-bt/2m}$. The fractional decrease over one period is
$\begin{aligned} \text{fractional decrease} &= [e^{-bt/2m} - e^{-b(t + T)/2m}]/e^{-bt/2m} = 1 - e^{-bT/2m} \\ &= 1 - e^{-(0.108 \text{ /s})(0.727 \text{ s})} = 0.0755. \end{aligned}$

(c) The general expression for the displacement is
$$x = Ae^{-bt/2m} \cos(\omega' t + \phi).$$
We use the given data to evaluate the constants:
$0 = Ae^{-0} \cos\phi$, which gives $\phi = -\pi/2$, or we can change to a sine function.
$0.120 \text{ m} = Ae^{-(0.108 /s)(1.00 s)} \sin[(8.64 \text{ s}^{-1})(1.00 \text{ s})]$, which gives $A = 0.189$ m.
Thus the displacement is
$$x = (0.189 \text{ m})e^{-(0.108 /s)t} \sin[(8.64 \text{ s}^{-1})t].$$

59. (a) In the equilibrium position, we have
$$F_{net} = F_{20} - F_{10} = 0, \quad \text{or} \quad F_{10} = F_{20}.$$
When the object is moved to the right a distance $x$,
we have
$$F_{net} = F_{20} - kx - (F_{10} + kx) = -2kx.$$
The effective spring constant is $k_{eff} = 2k$, so the frequency is
$$f = (k_{eff}/m)^{1/2}/2\pi = (2k/m)^{1/2}/2\pi$$
$$= [2(100 \text{ N/m})/(0.200 \text{ kg})]^{1/2}/2\pi = 5.03 \text{ Hz}.$$

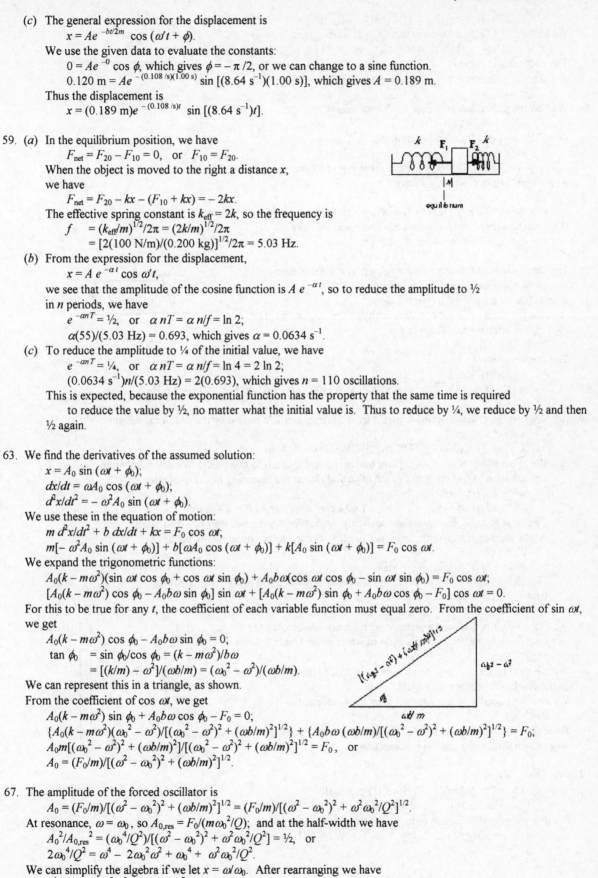

(b) From the expression for the displacement,
$$x = A e^{-\alpha t} \cos \omega' t,$$
we see that the amplitude of the cosine function is $A e^{-\alpha t}$, so to reduce the amplitude to ½
in $n$ periods, we have
$$e^{-\alpha nT} = \tfrac{1}{2}, \quad \text{or} \quad \alpha nT = \alpha n/f = \ln 2;$$
$\alpha(55)/(5.03 \text{ Hz}) = 0.693$, which gives $\alpha = 0.0634 \text{ s}^{-1}$.

(c) To reduce the amplitude to ¼ of the initial value, we have
$$e^{-\alpha nT} = \tfrac{1}{4}, \quad \text{or} \quad \alpha nT = \alpha n/f = \ln 4 = 2 \ln 2;$$
$(0.0634 \text{ s}^{-1})n/(5.03 \text{ Hz}) = 2(0.693)$, which gives $n = 110$ oscillations.
This is expected, because the exponential function has the property that the same time is required
to reduce the value by ½, no matter what the initial value is. Thus to reduce by ¼, we reduce by ½ and then
½ again.

63. We find the derivatives of the assumed solution:
$$x = A_0 \sin(\omega t + \phi_0);$$
$$dx/dt = \omega A_0 \cos(\omega t + \phi_0);$$
$$d^2x/dt^2 = -\omega^2 A_0 \sin(\omega t + \phi_0).$$
We use these in the equation of motion:
$$m\, d^2x/dt^2 + b\, dx/dt + kx = F_0 \cos \omega t;$$
$$m[-\omega^2 A_0 \sin(\omega t + \phi_0)] + b[\omega A_0 \cos(\omega t + \phi_0)] + k[A_0 \sin(\omega t + \phi_0)] = F_0 \cos \omega t.$$
We expand the trigonometric functions:
$$A_0(k - m\omega^2)(\sin \omega t \cos \phi_0 + \cos \omega t \sin \phi_0) + A_0 b\omega(\cos \omega t \cos \phi_0 - \sin \omega t \sin \phi_0) = F_0 \cos \omega t;$$
$$[A_0(k - m\omega^2) \cos \phi_0 - A_0 b\omega \sin \phi_0] \sin \omega t + [A_0(k - m\omega^2) \sin \phi_0 + A_0 b\omega \cos \phi_0 - F_0] \cos \omega t = 0.$$
For this to be true for any $t$, the coefficient of each variable function must equal zero. From the coefficient of $\sin \omega t$,
we get
$$A_0(k - m\omega^2) \cos \phi_0 - A_0 b\omega \sin \phi_0 = 0;$$
$$\tan \phi_0 = \sin \phi_0/\cos \phi_0 = (k - m\omega^2)/b\omega$$
$$= [(k/m) - \omega^2]/(\omega b/m) = (\omega_0^2 - \omega^2)/(\omega b/m).$$
We can represent this in a triangle, as shown.
From the coefficient of $\cos \omega t$, we get
$$A_0(k - m\omega^2) \sin \phi_0 + A_0 b\omega \cos \phi_0 - F_0 = 0;$$
$$\{A_0(k - m\omega^2)(\omega_0^2 - \omega^2)/[(\omega_0^2 - \omega^2)^2 + (\omega b/m)^2]^{1/2}\} + \{A_0 b\omega (\omega b/m)/[(\omega_0^2 - \omega^2)^2 + (\omega b/m)^2]^{1/2}\} = F_0;$$
$$A_0 m[(\omega_0^2 - \omega^2)^2 + (\omega b/m)^2]/[(\omega_0^2 - \omega^2)^2 + (\omega b/m)^2]^{1/2} = F_0, \quad \text{or}$$
$$A_0 = (F_0/m)/[(\omega^2 - \omega_0^2)^2 + (\omega b/m)^2]^{1/2}.$$

67. The amplitude of the forced oscillator is
$$A_0 = (F_0/m)/[(\omega^2 - \omega_0^2)^2 + (\omega b/m)^2]^{1/2} = (F_0/m)/[(\omega^2 - \omega_0^2)^2 + \omega^2 \omega_0^2/Q^2]^{1/2}.$$
At resonance, $\omega = \omega_0$, so $A_{0,res} = F_0/(m\omega_0^2/Q)$; and at the half-width we have
$$A_0^2/A_{0,res}^2 = (\omega_0^4/Q^2)/[(\omega^2 - \omega_0^2)^2 + \omega^2 \omega_0^2/Q^2] = \tfrac{1}{2}, \quad \text{or}$$
$$2\omega_0^4/Q^2 = \omega^4 - 2\omega_0^2\omega^2 + \omega_0^4 + \omega^2 \omega_0^2/Q^2.$$
We can simplify the algebra if we let $x = \omega/\omega_0$. After rearranging we have
$$x^4 - (2 - 1/Q^2)x^2 + (1 - 2/Q^2) = 0.$$
This is a quadratic for $x^2$ with the solutions

$x^2 = \frac{1}{2}\{(2 - 1/Q^2) \pm [(2 - 1/Q^2)^2 - 4(1 - 2/Q^2)]^{1/2}\} = 1 - 1/2Q^2 \pm \frac{1}{2}(4/Q^2 + 1/Q^4)^{1/2}$.

We know that $Q > 1$ and assume that $Q^2 \gg 1$, which gives

$x^2 \approx 1 - 1/2Q^2 + (1/Q^2)^{1/2} \approx 1 \pm 1/Q$,   or   $x = (1 \pm 1/Q)^{1/2} \approx 1 \pm 1/2Q$.

The width of the curve is

$\Delta\omega/\omega_0 = \Delta x = x_+ - x_- = (1 + 1/2Q) - (1 - 1/2Q) = 1/Q$.

71. We find the period from the time for $N$ oscillations:

$T = t/N = (34.7 \text{ s})/8 = 4.34 \text{ s}$.

From this we can get the spring constant:

$T = 2\pi(m/k)^{1/2}$;

$4.34 \text{ s} = 2\pi[(72.0 \text{ kg})/k]^{1/2}$, which gives $k = 151$ N/m.

At the equilibrium position, we have

$mg = kx_0$;

$(72.0 \text{ kg})(9.80 \text{ m/s}^2) = (151 \text{ N/m})x_0$, which gives $x_0 = 4.68$ m.

Because this is how much the cord has stretched, we have

$L = D - x_0 = 25.0 \text{ m} - 4.68 \text{ m} = 20.3$ m.

75. The shear stress and shear strain are related by the shear modulus:

$G = (F/A)/(\Delta x/h)$,   or   $F = (GA/h)\,\Delta x$,

where $A$ is the area of the top of the block.

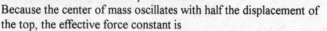

Because the center of mass oscillates with half the displacement of
the top, the effective force constant is

$k_{\text{eff}} = F/(\Delta x/2) = 2GA/h$.

The frequency is

$f = (k_{\text{eff}}/m)^{1/2}/2\pi$

$= [(2GA/h)/\rho Ah]^{1/2}/2\pi = [2(520 \text{ N/m}^2)/(1300 \text{ kg/m}^3)(0.040 \text{ m})^2]^{1/2}/2\pi = 3.6$ Hz.

79. (a) We find the effective spring constant from the displacement caused by the additional weight:

$k\,\Delta y = mg$,   or

$k = mg/\Delta y = (0.55 \text{ kg})(9.80 \text{ m/s}^2)/(0.060 \text{ m}) = 90$ N/m.

(b) We assume the collision takes place before the springs start to compress. We use momentum
conservation to find the speed of the table and clay at the beginning of the oscillation:

$mv + 0 = (m + M)V$;

$(0.55 \text{ kg})(1.65 \text{ m/s}) = (0.55 \text{ kg} + 1.60 \text{ kg})V$, which gives $V = 0.422$ m/s.

The table and clay are 6.0 cm above the equilibrium position when they have this speed.
At the maximum amplitude, the speed is zero. From energy conservation, we have

$K_i + U_i = K_f + U_f$;

$\frac{1}{2}(m + M)V^2 + \frac{1}{2}kx^2 = 0 + \frac{1}{2}kA^2$;

$\frac{1}{2}(0.55 \text{ kg} + 1.60 \text{ kg})(0.422 \text{ m/s})^2 + \frac{1}{2}(89.8 \text{ N/m})(0.060 \text{ m})^2 = \frac{1}{2}(89.8 \text{ N/m})A^2$,

which gives $A = 8.87 \times 10^{-2}$ m $= 8.9$ cm.

83. When the water is displaced a distance $\Delta x$ from equilibrium, the net
restoring force is the unbalanced weight of water in the height $2\,\Delta x$:

$F_{\text{net}} = -2\rho gA\,\Delta x$.

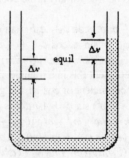

We see that the net restoring force is proportional to the displacement,
so the block will oscillate with SHM.
We find the effective spring constant from the coefficient of $\Delta x$:

$k = 2\rho gA$.

From the formula for $k$, we see that the effective spring constant
depends on the density and the cross section.

87. (a) The period is

$T = 2\pi\,(I/mgh)^{1/2} = 2\pi[(ml^2/3)/mg(l/2)]^{1/2}$

$= 2\pi(2l/3g)^{1/2} = 2\pi[2(1.00 \text{ m})/3(9.80 \text{ m/s}^2)]^{1/2} = 1.64$ s.

(b) For the simple pendulum to have the same period, we have

$T = 2\pi(2l/3g)^{1/2} = 2\pi(l_0/g)^{1/2}$,   or   $l_0 = \frac{2}{3}l = \frac{2}{3}(1.00 \text{ m}) = 0.67$ m.

# Chapter 15: Wave Motion

## Chapter Overview and Objectives

This chapter describes the phenomena of mechanical waves. It introduces properties of waves and looks at the behavior of pulse and sinusoidal waves.

After completing study of this chapter, you should:

- Know what a wave is.
- Know the difference between longitudinal and transverse waves.
- Know the parameters that are necessary to define a sinusoidal wave and the relationship between them.
- Know how wave speed depends on properties of the material carrying the wave.
- Know the linear wave equation and the nature of solutions of that equation.
- Know how mechanical waves carry energy.
- Know what the wave phenomena of reflection, transmission, refraction, diffraction, and resonance are.
- Know how to calculate the resonant frequencies of a string anchored at both ends.

## Summary of Equations

Relationship between period and frequency of a sinusoidal wave:

$$T = \frac{1}{f}$$

Relationship between wave speed, frequency, wavelength, and period:

$$v = \lambda f = \frac{\lambda}{T}$$

General dependence of wave speed on properties of materials:

$$v = \sqrt{\frac{elastic\ force\ factor}{inertial\ factor}}$$

Dependence of transverse wave speed on a string on tension and linear density of string:

$$v = \sqrt{\frac{F_T}{\mu}}$$

Dependence of longitudinal acoustic wave speed in a solid on elastic modulus and density of material:

$$v = \sqrt{\frac{E}{\rho}}$$

Dependence of longitudinal acoustic wave speed in a fluid on bulk modulus and density of material:

$$v = \sqrt{\frac{B}{\rho}}$$

Work done on string as one period of sinusoidal wave passes a given point:

$$W = 2\pi^2 \mu f^2 D_M^2 \lambda$$

Average power transmitted on string carrying transverse sinusoidal wave:

$$\overline{P} = \frac{W}{T} = \frac{2\pi^2 \mu f^2 D_M^2 \lambda}{T} = 2\pi^2 \mu f^2 D_M^2 v$$

Intensity of three-dimensional mechanical wave:    $I = \frac{\overline{P}}{A} = 2\pi^2 \rho f^2 D_M^2 v$

Dependence of intensity of spherical wave on radius of wavefront:

$$\frac{I_2}{I_1} = \frac{r_1^2}{r_2^2}$$

Dependence of amplitude of spherical wave on radius of wavefront:

$$\frac{D_{M2}}{D_{M1}} = \frac{r_1}{r_2}$$

Definition of wave number:    $k = \frac{2\pi}{\lambda}$

Linear wave equation:    $\frac{d^2 D}{dt^2} = \frac{1}{v^2} \frac{d^2 D}{dx^2}$

Resonant frequencies of string tied down at both ends:

$$f_n = \frac{nv}{2L}, \text{ where } n = 1, 2, 3, \ldots$$

Dependence of direction of incident and refracted wave velocities on wave speed:

$$\frac{\sin \theta_2}{\sin \theta_1} = \frac{v_2}{v_1}$$

## Chapter Summary

### Section 15-1 Characteristics of Wave Motion

Waves can be divided into two categories, those that last for a limited time, called **pulse** waves, and those that last for extended lengths of time, called **continuous** waves. A special category of continuous waves is **periodic** waves. Periodic waves have a time dependence that is repetitive. A sinusoidal wave is a particular type of periodic wave in which the motion of the wave at a particular point in space undergoes simple harmonic motion.

A sinusoidal wave can be characterized by quantities similar to those used to describe simple harmonic motion. The **amplitude** of the wave is the maximum displacement from equilibrium of the material carrying the wave. The **period**, $T$, of the wave is the amount of time for one cycle of the sinusoidal motion to take place at a given point in space. The **frequency**, $f$, of the wave is the number of cycles of sinusoidal motion that take place per unit time. The period and frequency are related by

$$T = \frac{1}{f}$$

Because the wave moves through space at a constant velocity, there is also a spatial sinusoidal dependence of the displacement of the material at a given time. The spatial extent of one sinusoidal cycle of the wave is called the **wavelength**, $\lambda$, of the wave. The speed of the wave is related to the wavelength and the period or frequency by

$$v = \lambda f = \frac{\lambda}{T}$$

## Section 15-2 Wave Types

We can classify waves by the direction of the displacement of the material relative to the direction of the velocity of the wave. If the direction of the displacement of the material is parallel to the direction of the velocity of the wave, the wave is called a **longitudinal** wave. If the direction of the displacement of the material is perpendicular to the direction of the velocity of the wave, the wave is called a **transverse** wave.

The speed of wave motion is determined by properties of the material in which the wave travels. The general form of the wave velocity dependence for mechanical waves is

$$v = \sqrt{\frac{elastic\ force\ factor}{inertial\ factor}}$$

The velocity of waves on string with a mass per unit length $\mu$ and stretched with a tension $F_T$ is

$$v = \sqrt{\frac{F_T}{\mu}}$$

The velocity of longitudinal acoustic waves in a solid with density $\rho$ and elastic modulus $E$ is

$$v = \sqrt{\frac{E}{\rho}}$$

The velocity of longitudinal acoustic waves in a gas with density $\rho$ and bulk modulus $B$ is

$$v = \sqrt{\frac{B}{\rho}}$$

## Example 15-2-A

The velocity of sound in air is about 340 m/s. The density of air is 1.29 kg/m$^3$. What is the change in pressure of air that is compressed by a 1% change in volume?

**Solution:**

The change in pressure is related to the fractional change in volume (Chapter 12) by

$$\Delta P = -B \frac{\Delta V}{V}$$

We are given $\Delta V/V$ is 0.01 and we can determine the bulk modulus, $B$, from

$$v = \sqrt{\frac{B}{\rho}} \Rightarrow B = \rho v^2 = \left(1.29\,\text{kg/m}^3\right)\left(343\,\text{m/s}\right)^2 = 1.52 \times 10^5\,\text{kg/m} \cdot \text{s}^2$$

The change in pressure is

$$\Delta P = \left(1.52 \times 10^5\,\text{kg/m} \cdot \text{s}^2\right)\left(0.01\right) = 1.52 \times 10^3\,\text{N/m}^2$$

**Example 15-2-B**

A string is stretched to a tension of 247 N. Waves traveling on the string with a wavelength of 26.4 cm have a frequency of 489 Hz. What is the linear density of the string?

**Solution:**

The information about wavelength and frequency allows us to calculate the speed of waves on the string:

$$v = \lambda f = (0.264\,\text{m})(489\,\text{Hz}) = 65.7\,\text{m/s}$$

The wave speed is related to the tension and the linear density by

$$v = \sqrt{\frac{F_T}{\mu}} \Rightarrow \mu = \frac{F_T}{v^2} = \frac{247\,\text{N}}{(65.7\,\text{m/s})^2} = 0.0572\,\text{kg/m}$$

**Section 15-3 Energy Transported by Waves**

Energy is transported through a medium carrying a wave. A mechanical wave traveling from one adjacent part of the medium to the next causes the preceding section of the medium to work on the succeeding section of the medium. Consider a wave traveling on a rope as shown in the diagram. The sinusoidal wave is traveling from left to right. The rope is moving downward as the wave travels further to the right. The tension of the rope to the left of point $x$ is pulling downward on the rope to the right of point $x$. This means the tension in the rope to the left of point $x$ will be doing work on the part of the rope to the right of point $x$. The work will be done at a rate equal to the downward velocity of the rope multiplied by the downward component of the tension in the rope. For a wave traveling to the right, the left part of the rope is almost always doing positive work on the right part of the rope. Conversely, it should be easy to see that the right part of the rope is almost always doing negative work on the left part of the rope as a right-moving wave passes a given point on the rope. If the direction of the wave is to the left, then the positive work is done by the right part of the rope on the left part of the rope.

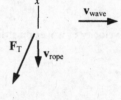

For a sinusoidal wave on a strong, we can calculate the work, $W$, done by one part of a rope on the other for a complete period. That work is

$$W = 2\pi^2 \mu f^2 D_M^2 \lambda$$

where $\mu$ is the mass per length of the string, $f$ is the frequency of the sinusoidal wave, $D_M$ is the amplitude of the wave, and $\lambda$ is the wavelength of waves on the string. The average power into the succeeding part of the rope is given by

$$\overline{P} = \frac{W}{T} = \frac{2\pi^2 \mu f^2 D_M^2 \lambda}{T} = 2\pi^2 \mu f^2 D_M^2 v$$

where $v$ is the speed of the waves on the string.

For three-dmensional waves, it is necessary to introduce **intensity**. Intensity is the average power per area of wavefront carried by a wave. The intensity, $I$, of a sinusoidal wave traveling through an elastic material is given by

$$I = \frac{\overline{P}}{A} = 2\pi^2 \rho f^2 D_M^2 v$$

where $\rho$ is the volume density of the material.

For spherically-shaped wavefronts, the area of the wavefront grows as the square of the radius of the spherical wavefront. Conservation of the energy in the wave implies that the intensity of the wave drops off proportional to the square of the radius of the wavefront:

$$\frac{I_2}{I_1} = \frac{r_1^2}{r_2^2}$$

where $I_1$ and $I_2$ are the intensities at radii $r_1$ and $r_2$ respectively. Knowing that the intensity is proportional to the square of the amplitude of the wave, we can use this to conclude that

$$\frac{D_{M2}}{D_{M1}} = \frac{r_1}{r_2}$$

**Example 15-3-A**

The intensity of sound waves that is barely audible is $1 \times 10^{-12}$ W/m². What is the displacement amplitude of these sound waves when the speed of sound in air is 343 m/s and the frequency of the sound waves is 1000 Hz? The density of air is 1.29 kg/m³.

**Solution:**

The intensity of the waves is related to the frequency of the waves, density of the material, and the speed of the waves by

$$I = \frac{\overline{P}}{A} = 2\pi^2 \rho f^2 D_M^2 v \quad \Rightarrow \quad D_M = \frac{1}{\pi f}\sqrt{\frac{I}{2\rho v}}$$

Using the values given we find

$$D_M = \frac{1}{\pi(1000\,\text{Hz})}\sqrt{\frac{1\times10^{-12}\,\text{W/m}^2}{2(1.29\,\text{kg/m}^3)(343\,\text{m/s})}} = 1.07\times10^{-11}\,\text{m}$$

**Section 15-4 Mathematical Representation of a Traveling Wave**

A wave can be represented by a wavefunction. A wavefunction is a function of position and time with a value equal to the **displacement** of the wave at each value of position and time. In general, we write such a function as $f(x,t)$. A traveling sinusoidal wave can be represented by the function

$$D(x,t) = D_M \sin(kx - \omega t + \phi)$$

where $D(x,t)$ is the displacement of the wave as a function of position and time, $D_M$ is the amplitude of the sinusoidal wave, $k$ is the **wave number** of the wave given by

$$k = \frac{2\pi}{\lambda}$$

where $\lambda$ is the wavelength of the wave, $\omega$ is the angular frequency of the wave, and $\phi$ is the **phase** of the wave. Alternate expressions for the same wave are

$$D(x,t) = D_M \sin\left(\frac{2\pi}{\lambda}x - \frac{2\pi}{T}t + \phi\right) = D_M \sin\left(\frac{2\pi}{\lambda}(x - vt) + \phi\right)$$

where $T$ is the period of the wave.

**Example 15-4-A**

A wave is described by the function $D(x,t) = (1.6 \text{ cm}) \sin [1.2 \text{ cm}^{-1} (x + 6.8 \text{ cm/s } t)]$.   What are the amplitude, wavelength, wave number, frequency, period, and speed of the wave?  In which direction is the wave traveling?

**Solution:**

We can immediately recognize the coefficient of the sine term as the amplitude.  The amplitude of the wave is 1.6 cm.

We can also immediately recognize the coefficient of the $x$ term in the argument of the sine function as the wave number $k$.  So $k = 1.2 \text{ cm}^{-1}$.

We can calculate the wavelength from the wave number:

$$\lambda = \frac{2\pi}{k} = \frac{2\pi}{1.2 \text{ cm}^{-1}} = 5.3 \text{ cm}$$

The coefficient of the time, $t$, is the wave speed.  The wave speed is 6.8 cm/s.

We can calculate the frequency from the wavelength and the wave speed:

$$f = \frac{v}{\lambda} = \frac{6.8 \text{ cm/s}}{5.3 \text{ cm}} = 1.3 \text{ Hz}$$

The period can be calculated from the frequency:

$$T = \frac{1}{f} = \frac{1}{1.3 \text{ Hz}} = 0.77 \text{ Hz}$$

Finally, the direction of the wave is in the negative $x$ direction because the sign of the $x$ term and the $t$ term are the same.

**Section 15-5 The Wave Equation**

Applying Newton's second law to infinitesimal mass elements along a string allows us to derive the equation of motion of the string when it is carrying a wave.  The equation that describes the transverse displacement of the string is

$$\frac{d^2 D}{dt^2} = \frac{F_T}{\mu} \frac{d^2 D}{dx^2} = \frac{1}{v^2} \frac{d^2 D}{dx^2}$$

where $D$ is the transverse displacement of the string, $F_T$ is the tension in the string, $\mu$ is the mass per length of the string, and $v$ is the speed of the waves on the string.  This equation is called the one-dimensional **linear wave equation**.

**Example 15-5-A**

Show that $D(x,t) = D_M \cos (Ax) \sin (Bt)$ is a solution to the wave equation and determine the speed of waves in this medium.

**Solution:**

We find the position and time derivatives of the wavefunction:

$$\frac{d}{dt} D(x,t) = D_M B \cos(Ax) \cos(Bt) \qquad \frac{d}{dx} D(x,t) = -D_M A \sin(Ax) \sin(Bt)$$

$$\frac{d^2}{dt^2} D(x,t) = -D_M B^2 \cos(Ax) \sin(Bt) \qquad \frac{d^2}{dx^2} D(x,t) = -D_M A^2 \cos(Ax) \sin(Bt)$$

It should be easy to see that

$$\frac{d^2}{dt^2}D(x,t) = \frac{B^2}{A^2}\frac{d^2}{dx^2}D(x,t)$$

This implies $D(x,t)$ is a solution to the wave equation with the wave speed given by

$$v = \frac{A}{B}$$

## Section 15-6 The Principle of Superposition

Because the linear wave equation only has terms with the first power of the wavefunction or its derivatives, any linear combination of solutions is also a solution to the wave equation. That is, if $f(x,t)$ is a solution to the wave equation and $g(x,t)$ is a solution to the wave equation, then $Af(x,t) + Bg(x,t)$ is also a solution to the linear wave equation where $A$ and $B$ are any numbers. This is called the **principle of superposition**.

## Example 15-6-A

Show that the principle of superposition holds for the linear wave equation.

## Solution:

Assume $f(x,t)$ and $g(x,t)$ are solutions to the linear wave equation. This means

$$\frac{d^2 f(x,t)}{dt^2} = \frac{1}{v^2}\frac{d^2 f(x,t)}{dx^2} \quad \text{and} \quad \frac{d^2 g(x,t)}{dt^2} = \frac{1}{v^2}\frac{d^2 g(x,t)}{dx^2}$$

Consider $Af(x,t) + Bg(x,t)$. Its second derivative with respect to time is

$$\frac{d^2}{dt^2}\left[Af(x,t) + Bg(x,t)\right] = A\frac{d^2 f(x,t)}{dt^2} + B\frac{d^2 g(x,t)}{dt^2} = \frac{A}{v^2}\frac{d^2 f(x,t)}{dx^2} + \frac{B}{v^2}\frac{d^2 g(x,t)}{dx^2}$$

$$= \frac{1}{v^2}\frac{d^2}{dx^2}\left[Af(x,t) + Bg(x,t)\right]$$

which shows the superposition of the two solutions is also a solution to the wave equation.

## Section 15-7 Reflection and Transmission

When waves reach a boundary or discontinuity in the medium in which they are propagating, a wave will usually be created traveling back into the initial medium. This is called a **reflected wave**. A wave will also be created in the medium past the boundary. This is called the **transmitted wave**.

In one dimension, the reflected wave will travel back into the medium with the velocity opposite the initial velocity. The amplitude and phase of a reflected sinusoidal wave may be different than the initial incident wave. In two or three dimensions, the reflected wave has its component of velocity perpendicular to the boundary reversed. The other components of the velocity remain the same. This is summarized in the **law of reflection**: The angle of incidence is equal to the angle of reflection.

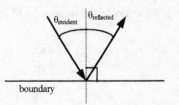

### Section 15-8 Interference

When multiple waves are traveling through the same media, the displacement of the media at a given point in space is the sum of the displacements of the individual waves at that point in space. If the displacement is less than the displacement that would result from the individual waves, the condition is called **destructive interference**. If the resulting wave displacement is greater than the displacement that would result from the individual waves, the condition is called **constructive interference**. Interference may be a misleading word in the sense that the waves *do not* interfere with the motion of each other, only in the displacement of the medium in which they travel.

### Section 15-9 Standing Waves; Resonance

When two sinusoidal waves of identical frequency and amplitude travel in opposite direction through a medium, the displacement of the media is described by a **standing wave**. The name "standing wave" results from the fact that under these conditions, even though there are waves traveling in two directions, no motion along the direction of wave travel is apparent.

For a system with fixed dimensions and perfectly reflective boundaries, standing waves are formed between the reflections from the boundaries of the system only for particular frequencies. These frequencies are called the **natural frequencies** or **resonant frequencies** of the system.

The lowest resonant frequency of a system is called the **fundamental frequency**. The higher frequency resonances are called **overtones**. Any frequency that is an integer multiple of the fundamental frequency is called a **harmonic** of the fundamental frequency. In some systems, the overtones are harmonics of the fundamental frequency, in others they are not harmonic.

For waves on a string that is tied down at both ends, the resonant frequencies are given by the expression

$$f_n = \frac{nv}{2L}, \quad \text{where} \quad n = 1, 2, 3, \ldots$$

$L$ is the length of the string, and $v$ is the speed of the waves traveling on the string.

### Example 15-9-A

A string on a guitar has a fundamental frequency of 467 Hz when under a tension of 204 N. Assuming its mass per length and length remain constant when tuning, what should the tension in the string be so that the fundamental frequency of the string is 440 Hz?

**Solution:**

The fundamental frequency of a string is

$$f_1 = \frac{v}{2L} = \frac{\sqrt{F_T/\mu}}{2L}$$

If we let the final frequency be $f_1'$ and the final tension $F_T'$, we get

$$\frac{f_1'}{f_1} = \sqrt{\frac{F_T'}{F_T}} \quad \Rightarrow \quad F_T' = \left(\frac{f_1'}{f_1}\right)^2 F_T = \left(\frac{440\,\text{Hz}}{467\,\text{Hz}}\right)^2 204\,\text{N} = 181\,\text{N}$$

## Section 15-10 Refraction

When waves pass through a boundary between two materials such that the wave speed is different in the two materials, the direction of the wave velocity changes upon the wave passing through the boundary. If we measure the direction of the velocity by an angle away from the direction perpendicular to the boundary, then the angle in the two materials are related to the two velocities by

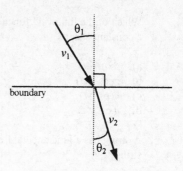

$$\frac{\sin \theta_2}{\sin \theta_1} = \frac{v_2}{v_1}$$

### Example 15-10-A

Sound waves in the water of a swimming pool strike the wall of the swimming pool at an angle of 12° and are partially transmitted into the wall of the swimming pool where they travel at an angle of 33°. The speed of sound waves in the water is 1450 m/s. What is the speed of the sound waves in the wall of the pool?

**Solution:**

The angle of incidence and the refracted angle are related to the velocities by

$$\frac{v_{wall}}{v_{water}} = \frac{\sin \theta_{wall}}{\sin \theta_{water}} \Rightarrow v_{wall} = \frac{\sin \theta_{wall}}{\sin \theta_{water}} v_{water}$$

$$v_{wall} = \frac{\sin 33^o}{\sin 12^o} 1450 \, \text{m/s} = 3.8 \times 10^3 \, \text{m/s}$$

## Section 15-11 Diffraction

Waves do not cast geometrically sharp shadows. Waves tend to bend around objects placed in their path by a phenomenon called diffraction. An angle characteristic of the amount of bending of the direction of the waves passing an obstacle is

$$\theta \approx \frac{\lambda}{L}$$

where $\lambda$ is the wavelength of the wave and $L$ is a characteristic size of the obstacle.

## Practice Quiz

1.    You watch a wave on string travel from point $A$ to point $B$. What has physically moved from point $A$ to point $B$?

    a)  Some of the mass of the string
    b)  Energy
    c)  Air surrounding the string
    d)  Nothing

2.    A picture of a string is shown at the right. What can you conclude about any transverse waves traveling on the string?

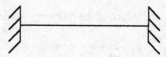

    a)  There is no possibility of any waves traveling on the string at this time.
    b)  The waves traveling on the string are destructively interfering everywhere and always.
    c)  The displacements of a left-traveling and a right-traveling wave might be momentarily canceling each other.
    d)  The string is not in motion.

3.    Which of the following functions describes a wave moving in the positive $x$ direction with a speed $v$ and keeping a constant shape?

a)  $f(x/v - t)$
b)  $g(x^2 - vt)$
c)  $h(x - vt)$
d)  $i(x - v^2 t)$

4.    A string has a fundamental frequency of 324 Hz.  Which of the following is not one of the harmonics of the string?

a)  648 Hz
b)  972 Hz
c)  162 Hz
d)  324,000,000 Hz

5.    Harmonic waves travel along a string with a wavelength $\lambda$ and with an amplitude such that the average power carried by the waves is $P_0$.  What is the average power carried by waves of identical amplitude, but wavelength $2\lambda$?

a)  $2P_0$
b)  $\frac{1}{2}P_0$
c)  $4P_0$
d)  $\frac{1}{4}P_0$

6.    Waves are incident on a boundary at an angle of 42° to the perpendicular to the boundary.  What angle does the velocity of the reflected wave make with the perpendicular to the boundary?

a)  42°
b)  21°
c)  132°
d)  Need to know wave speeds in the two media to answer question

7.    In the diagram to the right, the velocity, $v_1$, of a wave incident on the boundary and the velocity, $v_2$, of the wave transmitted through the boundary are shown.  What does this picture allow you to conclude about the speed of the waves in the two media on either side of the boundary?

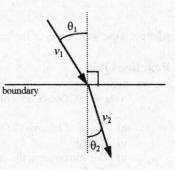

a)  The speed of the waves in the upper medium is greater than the speed of the waves in the lower medium.
b)  The speed of the waves in the upper medium is the same as the speed of the waves in the lower medium.
c)  The speed of the waves in the upper medium is less than the speed of the waves in the lower medium.
d)  No conclusion can be reached about the wave speeds from the drawing.

8.    Which of the following would raise the fundamental resonant frequency of a string?

a)  Lower the tension in the string.
b)  Shorten the string.
c)  Make the string more massive.
d)  Increase the linear density of the string.

9.    If you are sitting in a small boat on a lake, you observe that the short wavelength ripples on the surface of the water do not appear on the side of the boat opposite the direction the ripples are traveling on the surface. You do, however, observe the long wavelength swells appearing on the both sides of the boat. This can be explained by

    a)   Reflection
    b)   Refraction
    c)   Resonance
    d)   Diffraction

10.   A violin player tunes her instrument when it is cold. As the violin warms up, the tension in the strings drop, but the strings get slightly longer. What happens to the fundamental resonant frequencies of the strings?

    a)   All fundamental resonant frequencies increase.
    b)   All fundamental resonant frequencies decrease.
    c)   Some fundamental resonant frequencies decrease, others increase.
    d)   It depends on which of the two changes, length or tension, is larger.

11.   Determine the elastic modulus for a solid material in which longitudinal waves travel at a speed of 4.2 km/s. The density of the material is 5.32 g/cm$^3$.

12.   Spherical waves have an intensity of 3.24 W/m$^2$ at a distance of 13.4 m from the source. How much energy does the source emit in these waves in a time of 32.5 s?

13.   The displacement of a wave is represented by the function $D(x,t) = 3.2$ cm sin $(1.2$ cm$^{-1}$ $x + 6.8$ s$^{-1}$ $t)$. Determine the amplitude, wavelength, frequency, period, and velocity of the wave.

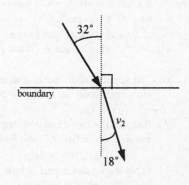

14.   A string has a tension of 436 N, a length 68.4 cm, and a mass of 13.3 g. Determine the fundamental frequency of the string.

15.   A wave traveling in the first material reaches the boundary between the two materials traveling in the direction shown in the diagram. It is transmitted past the boundary into the second material and leaves the boundary in the direction shown. If the speed of the wave in the first material is 16.4 m/s, what is the speed of the wave in the second material?

## Problem Solutions

3.   We find the wavelength from
    $v = f\lambda$;
    330 m/s = (262 Hz)$\lambda$, which gives $\lambda = 1.26$ m.

7.   We find the tension from the speed of the wave:
    $v = [F_T/(m/L)]^{1/2}$;
    (4.8 m)/(0.85 s) = $\{F_T/[(0.40$ kg)/(4.8 m)$]\}^{1/2}$ , which gives $F_T = 2.7$ N.

11.   (a)   Because both waves travel the same distance, we have
    $\Delta t = (d/v_S) - (d/v_P) = d[(1/v_S) - (1/v_P)]$;
    94 s = $d\{[1/(5.5$ km/s)] $- [1/(9.0$ km/s)]$\}$, which gives $d = 1.3 \times 10^3$ km.
    (b)   The direction of the waves is not known, thus the position of the epicenter cannot be determined. Because two circles have two intersections, it would take at least two more stations.

15.   If we consider two concentric circles around the spot where the waves are generated, the same energy must go past each circle in the same time. The intensity of a wave depends on $D_M{}^2$, so for the energy passing through a circle of radius $r$, we have
    $E = I(2\pi r) = kD_M{}^2 2\pi r$ = a constant.
Thus $D_M$ must vary with $r$; in particular, we have $D_M \propto 1/\sqrt{r}$.

19. To represent a wave traveling to the left, we replace $x$ by $x + vt$:

$$D = D_M \sin (2\Delta x/\lambda + \phi) = D_M \sin [(2\Delta/\lambda)(x + vt) + \phi] = D_M \sin [2\Delta(x/\lambda + t/T) + \phi].$$

23. (a) At $t = 0$ the traveling wave is

$$D = (0.45 \text{ m}) \sin [(3.0 \text{ m}^{-1})x + 1.2],$$

with $\lambda = 2\pi/((3.0 \text{ m}^{-1})) = 2.09$ m.

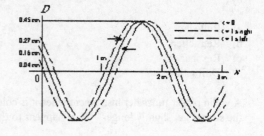

(b) For a wave traveling to the right $(+ x)$,
we replace $x$ with $x - vt$:

$$\begin{aligned} D &= (0.45 \text{ m}) \sin [(3.0 \text{ m}^{-1})(x - vt) + 1.2] \\ &= (0.45 \text{ m}) \sin \{(3.0 \text{ m}^{-1})[x - (2.0 \text{ m/s})t] + 1.2\} \\ &= (0.45 \text{ m}) \sin [(3.0 \text{ m}^{-1})x - (6.0 \text{ s}^{-1})t + 1.2]. \end{aligned}$$

(d) For a wave traveling to the left $(- x)$,
we replace $x$ with $x + vt$:

$$\begin{aligned} D &= (0.45 \text{ m}) \sin [(3.0 \text{ m}^{-1})(x + vt) + 1.2] \\ &= (0.45 \text{ m}) \sin \{(3.0 \text{ m}^{-1})[x + (2.0 \text{ m/s})t] + 1.2\} \\ &= (0.45 \text{ m}) \sin [(3.0 \text{ m}^{-1})x + (6.0 \text{ s}^{-1})t + 1.2]. \end{aligned}$$

27. We find the various derivatives for the given function:

$$D = D_M \sin kx \cos \omega t;$$
$$dD/dx = kD_M \cos kx \cos \omega t;$$
$$d^2D/dx^2 = - k^2 D_M \sin kx \cos \omega t;$$
$$dD/dt = - \omega D_M \sin kx \sin \omega t;$$
$$d^2D/dt^2 = - \omega^2 D_M \sin kx \cos \omega t.$$

From $v = f\lambda = (\omega /2\pi)(2\pi /k)$, we see that $k = \omega /v$. If we use this in $d^2D/ dx^2$, we get

$$d^2D/ dx^2 = - k^2 D_M \sin kx \cos \omega t = - (\omega^2/v^2) \sin kx \cos \omega t = (1/v^2)\, d^2D/ dt^2,$$

which is the wave equation.   Thus the function is a solution.

31. (a) The speed of the wave in a string is $v = [F_T/\mu]^{1/2}$.  Because the tensions must be the same
anywhere along the string, for the ratio of velocities we have

$$v_2/v_1 = (\mu_1/\mu_2)^{1/2}.$$

(b) Because the motion of one string is creating the motion of the other, the frequencies must be the
same.  For the ratio of wavelengths we have

$$\lambda_2/\lambda_1 = v_2/v_1 = (\mu_1/\mu_2)^{1/2}.$$

(c) From the result for part $(b)$ we see that if $\mu_2 > \mu_1$, we have $\lambda_2 < \lambda_1$, so the lighter cord will
have the greater wavelength.

35. (a)  For the sum of the two waves we have

$$\begin{aligned} D &= D_1 + D_2 = D_M \sin (kx - \omega t) + D_M \sin (kx - \omega t + \phi) \\ &= 2D_M \sin \tfrac{1}{2}(kx - \omega t + kx - \omega t + \phi) \cos \tfrac{1}{2}(kx - \omega t - kx + \omega t - \phi) \\ &= 2D_M \cos \tfrac{1}{2}(- \phi) \sin \tfrac{1}{2}(2kx - 2\omega t + \phi) = 2D_M \cos (\tfrac{1}{2}\phi) \sin (kx - \omega t + \tfrac{1}{2}\phi). \end{aligned}$$

(b) The amplitude is the coefficient of the sine function:    $2D_M \cos (\tfrac{1}{2}\phi)$.
The variation in $x$ and $t$ is purely sinusoidal.

(c) If $\phi = 0, 2\pi, 4\pi, \dots$ ; $\tfrac{1}{2}\phi = 0, \pi, 2\pi, \dots$ ; so $\cos (\tfrac{1}{2}\phi) = \pm 1$.
Thus the amplitude is maximum and we have complete constructive interference.
If $\phi = \pi, 3\pi, 5\pi, \dots$ ; $\tfrac{1}{2}\phi = \pi/2, 3\pi/2, 5\pi/2, \dots$ ; so $\cos (\tfrac{1}{2}\phi) = 0$.
Thus the amplitude is zero and we have destructive interference.

(d) If $\phi = \pi/2$, we have

$$D = 2D_M \cos (\pi/4) \sin (kx - \omega t + \pi/4) = \sqrt{2} D_M \sin (kx - \omega t + \pi/4).$$

The resultant wave has amplitude $D_M \sqrt{2}$, travels toward $+ x$, and at $x = 0$, $t = 0$,
the displacement is $+ D_M$.

39. The oscillation corresponds to the fundamental with a frequency:

$$f_1 = 1/T = 1/(2.0 \text{ s}) = 0.50 \text{ Hz}.$$

This is similar to the vibrating string, so all harmonics are present:

$$f_n = nf_1 = n(0.50 \text{ Hz}), n = 1, 2, 3, \dots .$$

We find the corresponding periods from

$$T_n = 1/f_n = 1/nf_1 = T/n = (2.0 \text{ s})/n, n = 1, 2, 3, \dots .$$

43. The speed of the wave depends on the tension and the mass density:
$$v = (F_T/\mu)^{1/2}.$$
The wavelength of the fundamental for a string is $\lambda_1 = 2L$. We find the fundamental frequency from
$$f_1 = v/\lambda_1 = (1/2L)(F_T/\mu)^{1/2}.$$
All harmonics are present in a vibrating string, so we have
$$f_n = nf_1 = (n/2L)(F_T/\mu)^{1/2}, n = 1, 2, 3, \dots.$$

47. The traveling wave is
$$D = (4.2 \text{ cm}) \sin [(0.71 \text{ cm}^{-1})x - (47 \text{ s}^{-1})t + 2.1].$$
(a) We get a wave traveling in the opposite direction with the same amplitude, frequency and speed by changing the relative sign of the $x$ and $t$ terms:
$$D_2 = (4.2 \text{ cm}) \sin [(0.71 \text{ cm}^{-1})x + (47 \text{ s}^{-1})t + 2.1].$$
(b) When we add the two sine functions and use a trigonometric identity, we get
$$
\begin{aligned}
D_{\text{resultant}} &= D + D_2 \\
&= (4.2 \text{ cm}) \sin [(0.71 \text{ cm}^{-1})x - (47 \text{ s}^{-1})t + 2.1] + (4.2 \text{ cm}) \sin [(0.71 \text{ cm}^{-1})x + (47 \text{ s}^{-1})t + 2.1] \\
&= 2(4.2 \text{ cm}) \sin [(0.71 \text{ cm}^{-1})x + 2.1] \cos [- (47 \text{ s}^{-1})t] \\
&= (8.4 \text{ cm}) \sin [(0.71 \text{ cm}^{-1})x + 2.1] \cos [(47 \text{ s}^{-1})t].
\end{aligned}
$$

51. (a)                                              (b)

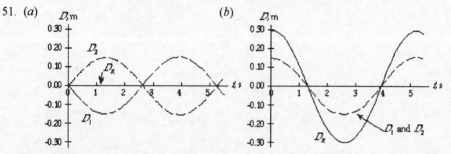

From the plots we see that $D_R = 0$ at $x = 0$ for any $t$, so it is a node. At $x = \lambda/4$, the displacement varies with the maximum amplitude, so it is an antinode.

55. The speed of the longitudinal (compression) wave for the solid rock depends on the modulus and the density: $v = (E/\rho)^{1/2}$. The modulus does not change, so we have $v \propto (1/\rho)^{1/2}$.
For the refraction of the waves we have
$$v_2/v_1 = (\rho_1/\rho_2)^{1/2} = (SG_1/SG_2)^{1/2} = (\sin \theta_2)/(\sin \theta_1);$$
$$[(3.7)/(2.8)]^{1/2} = (\sin \theta_2)/(\sin 25°), \text{ which gives } \theta_2 = 29°.$$

59. The speed of the longitudinal wave in a solid is given by
$$v = (B/\rho)^{1/2}.$$
If we form the ratio for two rods with the same bulk modulus, we get
$$v_2/v_1 = (\rho_1/\rho_2)^{1/2} = (1/2)^{1/2}, \quad \text{or} \quad v_1 = \sqrt{2}v_1.$$
The speed will be greater in the less dense rod by a factor of $\sqrt{2}$.

63. An object will leave the ground when the maximum acceleration of the ground during the SHM as the wave passes is greater than the acceleration due to gravity:
$$a_{\text{max}} = D_M \omega^2 > g, \quad \text{or}$$
$$D_M > g/\omega^2 = g/4\pi^2 f^2 = (9.80 \text{ m/s}^2)/4\pi^2(0.50 \text{ Hz})^2 = 0.99 \text{ m}.$$

67. (a) All harmonics are present in a vibrating string: $f_n = nf_1$, $n = 1, 2, 3, \dots$.
The first overtone is $f_2$ and the second overtone is $f_3$.
For G we have
$$f_2 = 2(392 \text{ Hz}) = 784 \text{ Hz}; \qquad f_3 = 3(392 \text{ Hz}) = 1176 \text{ Hz}.$$
For A we have
$$f_2 = 2(440 \text{ Hz}) = 880 \text{ Hz}; \qquad f_3 = 3(440 \text{ Hz}) = 1320 \text{ Hz}.$$
(b) The speed of the wave in a string is $v = [F_T/(M/L)]^{1/2}$. Because the lengths are the same, the wavelengths of the fundamentals must be the same. For the ratio of frequencies we have
$$f_A/f_G = v_A/v_G = (M_G/M_A)^{1/2};$$
$$(440 \text{ Hz})/(392 \text{ Hz}) = (M_G/M_A)^{1/2}, \text{ which gives } M_G/M_A = 1.26.$$

(c) Because the mass densities and the tensions are the same, the speeds must be the same. The wavelengths are proportional to the lengths, so for the ratio of frequencies we have

$f_A/f_G = \lambda_G/\lambda_A = L_G/L_A$;

$(440 \text{ Hz})/(392 \text{ Hz}) = L_G/L_A$ , which gives $L_G/L_A = 1.12$.

(d) The speed of the wave in a string is $v = [F_T/(M/L)]^{1/2}$. Because the lengths are the same, the wavelengths of the fundamentals must be the same. For the ratio of frequencies we have

$f_A/f_G = v_A/v_G = (F_{TA}/F_{TG})^{1/2}$;

$(440 \text{ Hz})/(392 \text{ Hz}) = (F_{TA}/F_{TG})^{1/2}$, which gives $F_{TG}/F_{TA} = 0.794$.

71. From the figure we see that

$D_M = 3.5 \text{ cm}$,   and   $\lambda = 6.0 \text{ cm}$, so $k = 2\pi/\lambda = 2\pi/(6.0 \text{ cm}) = 1.05 \text{ cm}^{-1}$.

The problem states the wave is moving to the right.

The displacement $y = 0$ is at $x = 1.5 \text{ cm}$ at $t = 0$, and at $x = 5.5 \text{ cm}$ at $t = 3.0 \text{ s}$.

Thus the velocity of the wave is

$v = (5.5 \text{ cm} - 1.5 \text{ cm})/(3.0 \text{ s}) = 1.33 \text{ cm/s}$,   and   $\omega = vk = (1.33 \text{ cm/s})(1.05 \text{ cm}^{-1}) = 1.39 \text{ s}^{-1}$.

We also see that $y$ is maximum at $x = 0$ and $t = 0$, so we can use a cosine function without a phase angle:

$y = D_M \cos (kx - \omega t) = (3.5 \text{ cm}) \cos[(1.05 \text{ cm}^{-1})x - (1.39 \text{ s}^{-1})t]$.

# Chapter 16: Sound

## Chapter Overview and Objectives

This chapter describes the phenomenon of sound and the physics associated with it. It shows how sound can be described by the displacement of molecules by a sound wave or by the change in pressure caused by the sound wave. This chapter introduces the concept of intensity and sound intensity level. This chapter discusses the vibration frequencies of strings and air columns. Also, some other characteristics of sound waves are discussed.

After completing study of this chapter, you should:

- Know what sound is.
- Know that sound can be described by a displacement function or a pressure function.
- Know how the displacement amplitude is related to the pressure amplitude.
- Know what intensity is.
- Know what sound intensity level is and how to calculate it.
- Know what the inverse square law is.
- Know how to determine vibrational frequencies of strings and air columns.
- Know how to determine the beat frequency of two sound sources.
- Know how to calculate frequency changes that are the result of the Doppler effect.
- Know what a shock wave is, how it's produced, and how to calculate the apex angle of a shock wave.

## Summary of Equations

Approximate speed of sound in air:

$$v = (331 + 0.60T)\,\text{m/s}$$

Expression for harmonic displacement wave:

$$D(x,t) = D_M \sin(kx - \omega t)$$

Expression for associated harmonic pressure wave:

$$\Delta P(x,t) = -(BD_M k)\cos(kx - \omega t)$$

Relationships between pressure and displacement amplitudes:

$$\Delta P_M = BD_M k$$

$$\Delta P_M = 2\pi \rho v D_M f$$

Definition of intensity:

$$I = \frac{E}{A \cdot t} = \frac{P}{A}$$

Definition of sound intensity level:

$$\beta = 10 \log_{10} \frac{I}{I_0}$$

Inverse square law:

$$\frac{I_2}{I_1} = \frac{r_1^2}{r_2^2}$$

Vibration frequencies of a string or air column with identical end conditions:

$$f_n = \frac{nv}{2L} \quad n \in 1, 2, 3, 4, \ldots$$

Vibration frequencies of a string in terms of the tension, mass of string, and length of string:

$$f_n = \frac{n}{2L}\sqrt{\frac{T}{m/L}} \quad n \in 1, 2, 3, 4, \ldots$$

Vibration frequencies of an air column with one open end and one closed end:

$$f_n = \frac{(2n-1)v}{2L} \quad n \in 1,2,3,4,\ldots$$

Beat frequency of two sources:             $f_b = |f_1 - f_2|$

Doppler shifted frequency:                 $f' = \frac{v \pm v_o}{v \mp v_s} f$

Apex angle of shock wave:                  $\sin\theta = \dfrac{v_{wave}}{v_{source}}$

## Chapter Summary

### Section 16-1 Characteristics of Sound

Sound is a mechanical wave that travels through materials.   Its velocity in air is a function of temperature that is approximately given by the expression

$$v = (331 + 0.60T)\,\text{m/s}$$

where T is the temperature of the air on the Celsius temperature scale.

### Section 16-2 Mathematical Representation of Longitudinal Waves

Sound can be described mathematically by a function that describes the displacement of the atoms or molecules of the medium that the sound wave moves through as a function of position and time.   For example, a harmonic sound wave with wave number k and angular frequency $\omega$ has a displacement $D$ that is a function of position $x$ and time $t$:

$$D(x,t) = D_M \sin(kx - \omega t)$$

where $D_M$ is the amplitude of the displacement.   Sound waves in air are longitudinal waves.   The displacement is parallel to the direction of travel of the sound wave.

Rather than considering the displacement of the molecules, we could consider the change in pressure in the air as the sound wave moves through the medium.   The pressure as a function of position and time for the above displacement is given by

$$\Delta P(x,t) = -(BD_M k)\cos(kx - \omega t)$$

where $B$ is the bulk modulus of the medium.   We see that the pressure amplitude, $\Delta P_M$, is related to the displacement amplitude by

$$\Delta P_M = BD_M k$$

We can rewrite this in terms of the wave speed, $v$, and wave frequency, $f$, as

$$\Delta P_M = 2\pi\rho\, v D_M f$$

where $\rho$ is the density of the medium.

### Example 16-2-A

The pressure amplitude in a loud sound is $1.00\ \text{N/m}^2$.  If this sound has a frequency of 1000 Hz, what is the displacement amplitude of this sound?  Assume the speed of sound in air is 340 m/s.

**Solution:**

We use the relationship between pressure and displacement amplitudes above:

$$\Delta P_M = 2\pi\rho v D_M f \quad \Rightarrow \quad D_M = \frac{\Delta P_M}{2\pi\rho \; vf} = \frac{\left(1.00\,\text{N}/\text{m}^2\right)}{2\pi\left(1.29\,\text{kg}/\text{m}^3\right)\left(340\,\text{m}/\text{s}\right)\left(1000\,\text{s}^{-1}\right)}$$

$$= 3.6\times10^{-7}\,\text{m}$$

### Section 16-3 Intensity of Sound; Decibels

As with all mechanical waves, sound waves carry energy. The energy being carried by a sound wave is spread out over the area of the wavefront of the wave. The amount of energy into a receiver of waves is proportional to the area of the receiver. The appropriate measurement to characterize the energy of the wave itself is the **intensity** of the wave. The intensity, I, is the energy per area per time or power per area:

$$I = \frac{E}{A \cdot t} = \frac{P}{A}$$

The range of sound intensity that is detected by the human ear without pain is $10^{-12}\,\text{W}/\text{m}^2$ to $1\,\text{W}/\text{m}^2$. To avoid this large range in magnitudes in numbers, the logarithmic scale called **sound intensity level** is often used to specify the intensity. The sound loudness level is defined to be

$$\beta = 10\log_{10}\frac{I}{I_0}$$

The label given to the value of the sound intensity level is decibels or dB.

Sound waves that are free to expand in all directions away from a small source have an intensity, $I$, that depends on distance as the inverse square of the distance, $r$, from the source:

$$\frac{I_2}{I_1} = \frac{r_1^2}{r_2^2}$$

This is known as the **inverse square law**.

### Example 16-3-A

At a distance of 3.8 m from a source of sound, you measure the sound intensity level to be 79 dB. At a distance of 42.6 m from the same source, you measure the sound intensity level to be 62 dB. Does the sound field behave according to the inverse square law?

**Solution:**

The distances to the two source from the two observation points are given. We must determine the intensity at those two points to answer the question. We can use the definition of sound intensity level to solve for the intensity as a function of sound intensity level:

$$\beta = 10\log_{10}\frac{I}{I_0} \quad \Rightarrow \quad I = 10^{\frac{\beta}{10}} \cdot I_0$$

Dividing $I_2$ by $I_1$ results in:

$$\frac{I_2}{I_1} = \frac{10^{\frac{\beta_2}{10}} \cdot I_0}{10^{\frac{\beta_1}{10}} \cdot I_0} = 10^{\frac{\beta_2 - \beta_1}{10}} = 10^{\frac{62-79}{10}} = 0.020$$

The ratio of the inverse distances squared is

$$\frac{r_1^2}{r_2^2} = \frac{(3.8\,\text{m})^2}{(42.6\,\text{m})^2} = 0.080$$

So this sound field is not obeying the inverse square law. Some sound may be reflecting off of surfaces in the vicinity, increasing the sound intensity above the inverse square law intensity.

### Section 16-4 Sources of Sound: Vibrating Strings and Air Columns

A vibrating string vibrates with particular frequencies that depend on the string's mass, $m$, the string's length, $L$, and tension in the string, $T$. The set of frequencies, $f_n$, of vibration is given by

$$f_n = \frac{nv}{2L} \qquad n \in 1, 2, 3, 4, \ldots$$

where $v$ is the speed of waves traveling on the string. This can be written in terms of the strings mass and the tension in the string as

$$f_n = \frac{n}{2L}\sqrt{\frac{T}{m/L}} \qquad n \in 1, 2, 3, 4, \ldots$$

The frequency $f_1$ is called the fundamental frequency of the string. The other frequencies are called the **overtones**. For a string, the overtones are harmonics of the fundamental frequency. The $n^{\text{th}}$ harmonic of the fundamental frequency has a frequency $nf_1$.

An air column that has both ends of the column closed off by a rigid solid or both ends open to the atmosphere has frequencies of vibration identical to those given for a string in terms of the wave speed, $v$. An air column that is closed at one end and open at the other has a set of frequencies of vibration given by

$$f_n = \frac{(2n-1)v}{2L} \qquad n \in 1, 2, 3, 4, \ldots$$

where $v$ is the speed of sound in air and $L$ is the length of the air column.

### Example 16-4-A

What should the length of a string be that has a $3^{\text{rd}}$ overtone frequency of 940 Hz and a tension of 240 N? A piece of the string that is 2.8 m long has a mass of 12 g.

### Solution:

We know the frequencies of the string are given by the expression

$$f_n = \frac{n}{2L}\sqrt{\frac{T}{m/L}} \qquad n \in 1, 2, 3, 4, \ldots$$

The mass per length, $m/L$, is the same regardless of the length of the string. We can calculate m/L from the information given in the problem:

$$m/L = \frac{0.012\,\text{kg}}{2.8\,\text{m}} = 4.29 \times 10^{-3}\,\text{kg/m}$$

Be careful in determining the value of $n$ to use! The fundamental frequency has $n = 1$. The first overtone frequency has $n = 2$. This implies that the third overtone frequency has $n = 4$. We can solve the equation above for $L$, the length of the string needed:

$$L = \frac{n}{2f_n}\sqrt{\frac{T}{m/L}} = \frac{4}{2(940\,\text{s}^{-1})}\sqrt{\frac{240\,\text{N}}{4.29 \times 10^{-3}\,\text{kg/m}}} = 0.50\,\text{m}$$

## Section 16-5 Quality of Sound, and Noise

Most sounds are not harmonic waves of a single frequency. However, many sounds are approximately periodic for a finite length of time. Periodic waves can be constructed as a sum of harmonic waves that have frequencies that are multiples of the fundamental frequency of the wave. The fundamental frequency is the frequency that corresponds to the period of the wave. The relative amplitudes of the different components and their phases determine the waveform of the resulting wave.

## Section 16-6 Interference of Sound Waves; Beats

When there are waves traveling through a medium from different sources, the principle of superposition tells us that the waves can interfere with each other. In general, application of the principle of superposition will tell us the displacement of the medium at a given time and place for any number of sources acting to create waves. However, there are some special cases of interest for which we can determine the resulting wave field in general.

Consider two sources of sound that are emitting identical frequency and amplitude waves. Constructive interference will occur at points whose distance from one source differs from the distance to the other source by a whole number of wavelengths. Destructive interference will occur at points whose distance from one source differs from the distance to the other source by an odd number of half wavelengths. We will investigate this more thoroughly when we study optics.

A second situation for which it is easy to make a statement about the results of the superposition principle is the case of two sources with comparable amplitude but with slightly different frequencies, $f_1$ and $f_2$. The result of summing the displacements of two such waves can be described as a wave with a frequency equal to the average of the two frequencies, but an amplitude that varies with time. The frequency of the amplitude variation is called the beat frequency, $f_b$. The beat frequency is related to the frequencies of the sources by

$$f_b = |f_1 - f_2|$$

## Example 16-6-A

Two sounds are reaching an observer. The observer hears a beat frequency of 5.8 Hz. The observer knows that one of the frequencies is 984.6 Hz. What is the other frequency?

## Solution:

The beat frequency is related to the two frequencies of the sounds by

$$f_b = |f_1 - f_2|$$

This implies either

$$f_b = f_1 - f_2 \quad \Rightarrow \quad f_2 = f_1 - f_b = 984.6\,\text{Hz} - 5.8\,\text{Hz} = 978.8\,\text{Hz}$$

or

$$f_b = f_2 - f_1 \quad \Rightarrow \quad f_2 = f_1 + f_b = 984.6\,\text{Hz} + 5.8\,\text{Hz} = 990.4\,\text{Hz}$$

## Section 16-7 Doppler Effect

When a source and observer of sound waves are in relative motion to each other, the frequency of sound waves emitted by the source is different than the frequency of sound waves received by the observer. The relationship between the frequency of the sound waves received by the observer and the frequency of sound waves emitted by the source is given by

$$f' = \frac{v \pm v_o}{v \mp v_s} f$$

where $v$ is the speed of the sound waves in the medium, $v_o$ is the speed of the observer relative to the medium, and $v_s$ is the speed of the source relative to the medium.

## Example 16-7-A

A car is traveling north at a speed of 24 m/s. The car is catching up to a truck that is also traveling north at a speed of 18 m/s. A 5 m/s wind is blowing toward the south. The car driver honks the horn of the car which emits sound at a frequency of 400 Hz. What is the frequency of sound observed by the truck driver?

## Solution:

All of the velocities in the Doppler shift expression are relative to the medium. It should be easy to see that the speed of velocity of the car is 29 m/s toward the north relative to the air and the velocity of the truck is 23 m/s toward the north relative to the air. We also need to determine the correct signs to use in the equation. The car, the source, is moving toward the position of the truck. This means that the negative sign should be used in the denominator of the expression. The truck, the observer, is moving away from the position of the source. This means that the negative sign should be used in the numerator of the expression also. Solving for the frequency of the observer:

$$f' = \frac{v \pm v_o}{v \mp v_s} f = \frac{(340\,\text{m/s}) - (23\,\text{m/s})}{(340\,\text{m/s}) - (29\,\text{m/s})} (400\,\text{Hz}) = 408\,\text{Hz}$$

## Section 16-8 Shock Waves and the Sonic Boom

A **shock wave** is generated whenever a sound source is moving relative to the medium at a speed greater than the speed of sound in that medium. The shock moves outward from the source with a conical wavefront. The apex angle, $\theta$, of the conical wavefront depends on the ratio of the wave speed to the source speed as

$$\sin \theta = \frac{v_{wave}}{v_{source}}$$

## Section 16-9 Applications; Sonar, Ultrasound, and Ultrasound Imaging

Sound waves can be used as an exploratory and diagnostic tool. **Sonar** and **diagnostic ultrasound** are two of the applications of sound waves of this sort.

# Practice Quiz

1.    What happens to the wavelength of a particular frequency of sound in air as the temperature rises?

   a) Wavelength remains the same.
   b) Wavelength becomes longer.
   c) Wavelength becomes shorter.
   d) Wavelength is unpredictable.

2.    Which perceptual judgment of a sound is primarily determined by the intensity of a sound wave?

   a) Pitch
   b) Loudness
   c) Timbre
   d) tonal quality

3.    In which of the following materials is the speed of sound the greatest?

   a) Air
   b) Helium
   c) Water
   d) Iron

4.    Sound in air is which type of wave?

   a) Longitudinal wave
   b) Transverse wave
   c) Shear wave
   d) Torsional wave

5.    The string of a guitar has a frequency that is 2% low.  What can be done to correct the frequency of the string?

   a) Increase the tension in the string by 2%.
   b) Increase the tension in the string by 4%.
   c) Decrease the tension in the string by 2%.
   d) Decrease the tension in the string by 1%.

6.    Sound B has a sound intensity level that is 10 dB higher than the sound intensity level of sound A.  The ratio of the intensity of sound B to the intensity of sound A is 10.  Sound C has a sound intensity level that is 20 dB higher than the sound intensity level of sound A.  What is the ratio of the intensity of sound C to the intensity of sound A?

   a) 20
   b) 40
   c) 100
   d) 200

7.    A column of air that is open at one end and closed at the other has a fundamental resonant frequency of 300 Hz. What is the frequency of the first overtone?

   a) 150 Hz
   b) 300 Hz
   c) 600 Hz
   d) 900 Hz

8.    You know that a source that is stationary relative to you emits a frequency 400 Hz. When it is moving, you hear a frequency of 404 Hz. What can you say about the motion of the source?

a)  The source is not moving.
b)  The source is moving directly toward you.
c)  The source is moving with a component of its velocity toward you.
d)  The source is moving with a component of its velocity away from you.

9.    Velocities are relative to a reference frame. What must be the reference frame for the velocities in the Doppler shift equation?

a)  Any frame of reference
b)  The frame of reference at rest relative to the source of the sound
c)  The frame of reference at rest relative to the observer of the sound
d)  The frame of reference at rest relative to the medium carrying the sound

10.    You detect the following frequencies being emitted from an air column: 100 Hz, 300 Hz, 500 Hz, 700 Hz. What can you conclude about the air column?

a)  It is open at one end and closed at the other.
b)  It is closed at both ends.
c)  It is open at both ends.
d)  You cannot conclude any of the above about the air column.

11.    Determine the lowest three frequencies of vibration of a string that is 69 cm long, has a mass of 12.5 g, and has a tension of 232 N.

12.    The sound field of a source follows the inverse square law. At a distance of 12.4 m from the source, the sound intensity level is 98 dB. What is the sound intensity level at a distance 29.6 m from the source?

13.    A person is driving a car toward a large building. When the person honks the horn of the car, which emits a 466 Hz tone, the driver hears a 6 Hz beat frequency. What is the speed of the car toward the building? Assume the speed of sound in air is 341m/s.

14.    The apex angle of a shock wave produced by a source moving through the air is 35.6°. Assume the speed of sound in air is 340 m/s. What is the speed of the sound source relative to the air?

15.    A trombone player tunes their instrument when the trombone is at 20°C. By what percentage will the trombone be out of tune when it and the air inside warm up to 32°C?

## Problem Solutions

3.    The speed in the concrete is determined by the elastic modulus:
$$v_{concrete} = (E/\rho)^{1/2} = [(20 \times 10^9 \text{ N/m}^2)/(2.3 \times 10^3 \text{ kg/m}^3)]^{1/2} = 2.95 \times 10^3 \text{ m/s}.$$
For the time interval we have
$$\Delta t = (d/v_{air}) - (d/v_{concrete});$$
$$1.4 \text{ s} = d\{[1/(343 \text{ m/s})] - [1/(2.95 \times 10^3 \text{ m/s})]\}, \text{ which gives } d = 5.4 \times 10^2 \text{ m}.$$

7.    We find the displacement amplitude from
$$\Delta P_M = 2\pi\rho v D_M f.$$
(a)  For the frequency of 100 Hz, we have
$$3.0 \times 10^{-3} \text{ Pa} = 2\pi(1.29 \text{ kg/m}^3)(331 \text{ m/s})D_M(100 \text{ Hz}), \text{ which gives } D_M = 1.1 \times 10^{-8} \text{ m}.$$
(b)  For the frequency of 10 kHz, we have
$$3.0 \times 10^{-3} \text{ Pa} = 2\pi(1.29 \text{ kg/m}^3)(331 \text{ m/s})D_M(10 \times 10^3 \text{ Hz}), \text{ which gives } D_M = 1.1 \times 10^{-10} \text{ m}.$$

11. (a) We find the intensity level from
$$\beta = 10 \log_{10}(I/I_0) = 10 \log_{10}[(8.5 \times 10^{-8} \text{ W/m}^2)/(10^{-12} \text{ W/m}^2)] = 49 \text{ dB}.$$
(b) We find the intensity from
$$\beta = 10 \log_{10}(I/I_0);$$
$$25 \text{ dB} = 10 \log(I/10^{-12} \text{ W/m}^2), \text{ which gives } I = 3.2 \times 10^{-10} \text{ W/m}^2.$$

15. (a) The intensity of the sound wave is given by
$$I = 2\pi^2 \rho f^2 D_M^2 v.$$
Because the frequency, density, and velocity are the same, for the ratio we have
$$I_2/I_1 = (D_{M2}/D_{M1})^2 = 3^2 = 9.$$
(b) We find the change in intensity level from
$$\Delta\beta = 10 \log_{10}(I_2/I_1) = 10 \log_{10}(9) = 9.5 \text{ dB}.$$

19. (a) We find the intensity of the sound from
$$\beta = 10 \log_{10}(I/I_0);$$
$$40 \text{ dB} = 10 \log(I/10^{-12} \text{ W/m}^2), \text{ which gives } I = 1.0 \times 10^{-8} \text{ W/m}^2.$$
The rate at which energy is absorbed is the power of the sound wave:
$$P = IA = (1.0 \times 10^{-8} \text{ W/m}^2)(5.0 \times 10^{-5} \text{ m}^2) = 5.0 \times 10^{-13} \text{ W}.$$
(b) We find the time from
$$t = E/P = (1.0 \text{ J})/(5.0 \times 10^{-13} \text{ W}) = 2.0 \times 10^{12} \text{ s} = 6.3 \times 10^4 \text{ yr}.$$

23. (a) We find the intensity of the sound from
$$\beta = 10 \log_{10}(I/I_0);$$
$$120 \text{ dB} = 10 \log(I/10^{-12} \text{ W/m}^2), \text{ which gives } I = 1.00 \text{ W/m}^2.$$
We find the maximum displacement from
$$I = 2\pi^2 \rho f^2 D_M^2 v$$
$$1.00 \text{ W/m}^2 = 2\pi^2(1.29 \text{ kg/m}^3)(210 \text{ Hz})^2 D_M^2(343 \text{ m/s}), \text{ which gives } D_M = 5.10 \times 10^{-5} \text{ m}.$$
(b) We find the pressure amplitude from
$$I = (\Delta P_M)^2/2\rho v;$$
$$1.00 \text{ W/m}^2 = (\Delta P_M)^2/2(1.29 \text{ kg/m}^3)(343 \text{ m/s}), \text{ which gives } \Delta P_M = 29.8 \text{ Pa}.$$

27. (a) Because the sounds are in the same medium, the variation in intensity is due only to the change in pressure amplitude. Thus we have
$$\beta = 10 \log_{10}(I/I_0) = 10 \log_{10}[(\Delta P_M)^2/(\Delta P_{M0})^2] = 10 \log_{10}(\Delta P_M/\Delta P_{M0})^2 = 20 \log_{10}(\Delta P_M/\Delta P_{M0}).$$
(b) For the given data we have
$$\beta = 20 \log_{10}(\Delta P_M/\Delta P_{M0}) = 20 \log_{10}[(1.013 \times 10^5 \text{ Pa})/(3.0 \times 10^{-5} \text{ Pa})] = 190 \text{ dB}.$$

31. For an open pipe the wavelength of the fundamental frequency is $\lambda = 2L$. We find the required lengths from
$$v = \lambda f = 2Lf;$$
$$343 \text{ m/s} = 2L_{lowest}(20 \text{ Hz}), \text{ which gives } L_{lowest} = 8.6 \text{ m};$$
$$343 \text{ m/s} = 2L_{highest}(20{,}000 \text{ Hz}), \text{ which gives } L_{highest} = 8.6 \times 10^{-3} \text{ m} = 8.6 \text{ mm}.$$
Thus the range of lengths is $8.6 \text{ mm} < L < 8.6 \text{ m}$.

35. (a) We find the speed of sound at 21°C:
$$v = (331 + 0.60T) \text{ m/s} = [331 + (0.60)(21°C)] \text{ m/s} = 344 \text{ m/s}.$$
The fundamental wavelength for an open pipe has an antinode at each end, so the wavelength is $\lambda = 2L$. We find the length from
$$v = \lambda f;$$
$$344 \text{ m/s} = 2L(262 \text{ Hz}), \text{ which gives } L = 0.66 \text{ m}.$$
(b) The frequency must be the same: 262 Hz.
The wave length will be
$$\lambda = 2L = 2(0.66 \text{ m}) = 1.32 \text{ m}.$$
(c) The wavelength and frequency in the outside air will be the same as in the air in the organ pipe:
1.32 m, 262 Hz.

39. (a) We find the speed of sound at 15°C:

$v = (331 + 0.60T)$ m/s $= [331 + (0.60)(15°C)]$ m/s $= 340$ m/s.

For an open pipe, the wavelength of the fundamental frequency is $\lambda_1 = 2L$. We find the length from

$v = \lambda_1 f_1 = 2Lf_1$ ;

340 m/s $= 2L(294$ Hz), which gives $L = 0.578$ m.

(b) For helium we have

$v = \lambda_1 f_1 = 2Lf_1$ ;

1005 m/s $= 2(0.578$ m)$f_1$ , which gives $f_1 = 869$ Hz.

Note that we have no correction for the 5 C° temperature change.

43. Because the intensity is proportional to the square of both the amplitude and the frequency, we have

$I_2/I_1 = (A_2/A_1)^2(f_2/f_1)^2 = (0.4)^2(2)^2 = 0.64$;

$I_3/I_1 = (A_3/A_1)^2(f_3/f_1)^2 = (0.15)^2(3)^2 = 0.20$.

We find the relative intensity levels from

$\beta_2 - \beta_1 = 10 \log_{10}(I_2/I_1) = 10 \log_{10}(0.64) = -2$ dB;

$\beta_3 - \beta_1 = 10 \log_{10}(I_3/I_1) = 10 \log_{10}(0.20) = -7$ dB.

47. (a) The beat frequency will be the difference in frequencies. Because the second frequency could be higher or lower, we have

$f_{beat} = \Delta f = f_2 - f_1$;

$\pm (3$ beats$)/(2.0$ s$) = f_2 - 132$ Hz, which gives $f_2 = 130.5$ Hz, or 133.5 Hz.

(b) We assume that the change in tension does not change the mass density, so the velocity variation depends only on the tension. Because the wavelength does not change, we treat the small changes in $f$ and $F_T$ as differentials to get

$\lambda\, df = \frac{1}{2}(1/F_T\mu)^{1/2}\, dF_T$ .

When we divide both sides by $v$, we get

$df/f = (\pm 1.5$ Hz$)/(132$ Hz$) = \frac{1}{2}\, dF_T/F_T$ , which gives $dF_T/F_T = \pm 0.023 = \pm 2.3\%$.

51. The two sound waves travel in the same medium, so the same wavelength means the same frequency. We take the initial phase of each wave to be zero. If the sources radiate uniformly in all directions, the intensity decreases as $1/r^2$, so the amplitude will decrease as $1/r$. If we add the displacements of the two waves, we have

$D = D_1 + D_2 = (D_M/r_A) \sin (kr_A - \omega t) + (D_M/r_B) \sin (kr_B - \omega t)$.

Because $r_A \approx r_B$, the two coefficients of the sine functions are approximately equal, so we have

$D \approx (D_M/r_A)[\sin (kr_A - \omega t) + \sin (kr_B - \omega t)] = (D_M/r_A)2 \sin [\frac{1}{2}k(r_A + r_B) - \omega t] \cos [\frac{1}{2}k(r_A - r_B)]$.

Because $r_A \approx r_B$, we have $r_A + r_B \approx 2r_A$, and we get

$D \approx (2D_M/r_A) \sin (kr_A - \omega t) \cos [\frac{1}{2}k(r_A - r_B)]$, or $D \approx (2D_M/r_A) \cos [(\pi/\lambda)(r_A - r_B)] \sin (kr_A - \omega t)$.

This is a traveling wave with an amplitude of $(2D_M/r_A) \cos [(\pi/\lambda)(r_A - r_B)]$ .

55. We see that there is no path difference for a listener on the bisector of the two speakers. The maximum path difference occurs for a listener on the line of the speakers. If this path difference is less than $\lambda/2$, there will be no location where destructive interference can occur. Thus we have

path difference $= d_B - d_A = d \geq \lambda/2$.

59. The wavelength in front of a moving object decreases, so the wavelength traveling toward the wall is

$\lambda_1 = (v - v_{bat})/f_0$ .

Because the wall is stationary, this is also the wavelength of the reflected sound. This wavelength approaches the bat at a relative speed of $v + v_{bat}$, so the frequency received by the bat is

$f \quad = (v + v_{bat})/\lambda_1 = [(v + v_{bat})/(v - v_{bat})]f_0$

$= [(343$ m/s $+ 5.0$ m/s$)/(343$ m/s $- 5.0$ m/s$)](30{,}000$ Hz$) = 30{,}890$ Hz.

63. Because the source is at rest, the wavelength traveling toward the blood is

$\lambda_1 = v/f_0$ .

This wavelength approaches the blood at a relative speed of $v - v_{blood}$. The ultrasound strikes and reflects from the blood with a frequency

$f_1 = (v - v_{blood})/\lambda_1 = (v - v_{blood})/(v/f_0) = (v - v_{blood})f_0/v$.

This frequency can be considered emitted by the blood, which is moving away from the source. Because the wavelength behind the moving blood increases, the wavelength approaching the source is

$\lambda_2 = (v + v_{blood})/f_1 = (v + v_{blood})/[(v - v_{blood})f_0/v] = [1 + (v_{blood}/v)]v/f_0[1 - (v_{blood}/v)]$.

This wavelength approaches the source at a relative speed of $v$, so the frequency received by the source is

$f_2 = v/\lambda_2 = v/\{[1 + (v_{blood}/v)]v/f_0[1 - (v_{blood}/v)]\} = [1 - (v_{blood}/v)]f_0/[1 + (v_{blood}/v)]$.

Because $v_{blood} \ll v$, we use $1/[1 + (v_{blood}/v)] \approx 1 - (v_{blood}/v)$, so we have

$f_2 \approx f_0[1 - (v_{blood}/v)]^2 \approx f_0[1 - 2(v_{blood}/v)]$.

For the beat frequency we have

$f_{beat} = f_0 - f_2 = 2(v_{blood}/v)f_0 = 2[(0.020 \text{ m/s})/(1540 \text{ m/s})](3.5 \times 10^6 \text{ Hz}) = 91 \text{ Hz}$.

67. Because the wind velocity is a movement of the medium, it adds or subtracts from the speed of sound in the medium.

(a) Because the wind is blowing away from the observer, the effective speed of sound is $v - v_{wind}$.

Therefore the wavelength traveling toward the observer is

$\lambda_a = (v - v_{wind})/f_0$ .

This wavelength approaches the observer at a relative speed of $v - v_{wind}$ . The observer will hear a frequency

$f_a = (v - v_{wind})/\lambda_a = (v - v_{wind})/[(v - v_{wind})/f_0] = f_0 = 570 \text{ Hz}$.

(b) Because the wind is blowing toward the observer, the effective speed of sound is $v + v_{wind}$.

From the analysis in part (a), we see that there will be no change in the frequency: 570 Hz.

(c) Because the wind is blowing perpendicular to the line toward the observer, the effective speed of sound is $v$. Because there is no relative motion of the whistle and the observer, there will be no change in the frequency: 570 Hz.

(d) Because the wind is blowing perpendicular to the line toward the observer, the effective speed of sound is $v$. Because there is no relative motion of the whistle and the observer, there will be no change in the frequency: 570 Hz.

(e) Because the wind is blowing toward the cyclist, the effective speed of sound is $v + v_{wind}$.

Therefore the wavelength traveling toward the cyclist is

$\lambda_e = (v + v_{wind})/f_0$ .

This wavelength approaches the cyclist at a relative speed of $v + v_{wind} + v_{cycle}$ . The cyclist will hear a frequency

$\begin{aligned} f_e &= (v + v_{wind} + v_{cycle})/\lambda_e = (v + v_{wind} + v_{cycle})/[(v + v_{wind})/f_0] \\ &= (v + v_{wind} + v_{cycle})f_0/(v + v_{wind}) \\ &= (343 \text{ m/s} + 12.0 \text{ m/s} + 15.0 \text{ m/s})(570 \text{ Hz})/(343 \text{ m/s} + 12.0 \text{ m/s}) = 594 \text{ Hz}. \end{aligned}$

(f) Because the wind is blowing perpendicular to the line toward the cyclist, the effective speed of sound is $v$. Therefore the wavelength traveling toward the cyclist is

$\lambda_f = v/f_0$ .

This wavelength approaches the cyclist at a relative speed of $v + v_{cycle}$ . The cyclist will hear a frequency

$\begin{aligned} f_f &= (v + v_{cycle})/\lambda_f = (v + v_{cycle})/(v/f_0) \\ &= (v + v_{cycle})f_0/v = (343 \text{ m/s} + 15.0 \text{ m/s})(570 \text{ Hz})/(343 \text{ m/s}) = 595 \text{ Hz}. \end{aligned}$

71. (a) From the definition of the Mach number, we have

$v = (\text{Mach number})v_{sound}$ ;

$(15,000 \text{ km/h})/(3.6 \text{ ks/h}) = (\text{Mach number})(35 \text{ m/s})$, which gives Mach number = 120.

(b) We find the angle of the shock wave from

$\sin \theta = v_{sound}/v_{object} = (35 \text{ m/s})/[(15,000 \text{ km/h})/(3.6 \text{ ks/h})] = 0.0084$, so $\theta = 0.48°$.

Thus the apex angle is $2\theta = 0.96°$.

75. Because the reflected pulse must be received before the emission of the next pulse, the minimum time between pulses is

$t_{min} = 2d/v_{sound} = 2(200 \text{ m})/(1440 \text{ m/s}) = 0.278 \text{ s}$.

79. Because the wavelength in front of a moving source decreases, the wavelength from the approaching car is

$\lambda_1 = (v - v_{car})/f_0$ .

This wavelength approaches the stationary listener at a relative speed of $v$, so the frequency heard by the listener is

$f_1 = v/\lambda_1 = v/[(v - v_{car})/f_0] = vf_0/(v - v_{car})$.

Because the wavelength behind a moving source increases, the wavelength from the receding car is

$\lambda_2 = (v + v_{car})/f_0$ .

This wavelength approaches the stationary listener at a relative speed of $v$, so the frequency heard by the listener is

$f_2 = v/\lambda_2 = v/[(v + v_{car})/f_0] = vf_0/(v + v_{car})$.

If the frequency drops by one octave, we have

$f_1/f_2 = [vf_0/(v - v_{car})]/[vf_0/(v + v_{car})] = (v + v_{car})/(v - v_{car}) = 2$,   or

$2(v - v_{car}) = v + v_{car}$ , which gives $v_{car} = \frac{1}{3}v = \frac{1}{3} (343 \text{ m/s}) = 114 \text{ m/s} = 410 \text{ km/h} (257 \text{ mi/h})$.

83. We find the intensity of the sound from
    $\beta = 10 \log_{10}(I/I_0)$;
    $100 \text{ dB} = 10 \log(I/10^{-12} \text{ W/m}^2)$, which gives $I = 1.00 \times 10^{-2} \text{ W/m}^2$.
    If the speaker radiates equally in all directions, the power output of the speaker is
    $P = IA = I4\pi r^2 = (1.00 \times 10^{-2} \text{ W/m}^2)4\pi(12.0 \text{ m})^2 = 18.1 \text{ W}$.

87. Because the speakers are at rest, the wavelength traveling toward the person is
    $\lambda = v/f_0$.
    This wavelength from the speaker in front of the person approaches the person at a relative speed
    of $v + v_{\text{person}}$. The frequency heard is
    $f_1 = (v + v_{\text{person}})/\lambda$.
    This wavelength from the speaker behind the person approaches the person at a relative speed
    of $v - v_{\text{person}}$. The frequency heard is
    $f_2 = (v + v_{\text{person}})/\lambda$.
    Thus the beat frequency is
    $\Delta f = f_2 - f_1 = [(v + v_{\text{person}})/\lambda] - [(v + v_{\text{person}})/\lambda]$
    $= 2v_{\text{person}}/\lambda = 2v_{\text{person}}f_0/v = 2(1.4 \text{ m/s})(280 \text{ Hz})/(343 \text{ m/s}) = 2.3 \text{ Hz}$.

91. We find the gain from
    $\beta = 10 \log_{10}(P_2/P_1) = 10 \log_{10}[(100 \text{ W})/(1 \times 10^{-3} \text{ W})] = 50 \text{ dB}$.

95. Because the source is at rest, the wavelength traveling toward the blood is
    $\lambda_1 = v/f_0$.
    This wavelength approaches the blood at a relative speed of $v - v_{\text{blood}}$. The ultrasound strikes and reflects from the blood
    with a frequency
    $f_1 = (v - v_{\text{blood}})/\lambda_1 = (v - v_{\text{blood}})/(v/f_0) = (v - v_{\text{blood}})f_0/v$.
    This frequency can be considered emitted by the blood, which is moving away from the source. Because the wavelength
    behind the moving blood increases, the wavelength approaching the source is
    $\lambda_2 = (v + v_{\text{blood}})/f_1 = (v + v_{\text{blood}})/[(v - v_{\text{blood}})f_0/v] = [1 + (v_{\text{blood}}/v)]v/f_0[1 - (v_{\text{blood}}/v)]$.
    This wavelength approaches the source at a relative speed of $v$, so the frequency received by the source is
    $f_2 = v/\lambda_2 = v/\{[1 + (v_{\text{blood}}/v)]v/f_0[1 - (v_{\text{blood}}/v)]\} = [1 - (v_{\text{blood}}/v)]f_0/[1 + (v_{\text{blood}}/v)]$.
    Because $v_{\text{blood}} \ll v$, we use $1/[1 + (v_{\text{blood}}/v)] \approx 1 - (v_{\text{blood}}/v)$, so we have
    $f_2 \Delta f_0[1 - (v_{\text{blood}}/v)]^2 \approx f_0[1 - 2(v_{\text{blood}}/v)]$.
    For the beat frequency we have
    $f_{\text{beat}} = f_0 - f_2 = 2(v_{\text{blood}}/v)f_0 = 2[(0.32 \text{ m/s})/(1.54 \times 10^3 \text{ m/s})](5.50 \times 10^6 \text{ Hz}) = 2.29 \times 10^3 \text{ Hz} = 2.29 \text{ kHz}$.

99. (a) The rod will have a node at the middle and antinodes at the ends. The wavelength for
    the fundamental frequency is
    $\lambda_1 = 2L$.
    Thus the fundamental frequency is
    $f_1 = v/\lambda_1 = v/2L = (5100 \text{ m/s})/2(0.90 \text{ m}) = 2.8 \times 10^3 \text{ Hz}$.
    (b) The wavelength in the rod is
    $\lambda_{\text{rod}} = \lambda_1 = 2L = 2(0.90 \text{ m}) = 1.80 \text{ m}$.
    (c) The frequency in the air is the frequency in the rod, so the wavelength in the air is
    $\lambda_{\text{air}} = v/f_1 = (343 \text{ m/s})/(2.8 \times 10^3 \text{ Hz}) = 0.12 \text{ m}$.

# Chapter 17: Temperature, Thermal Expansion, and the Ideal Gas Law

## Chapter Overview and Objectives

This chapter introduces the atomic theory of matter, thermal equilibrium, temperature, thermal expansion, and gas laws.

After completing study of this chapter, you should:

- Know what the atomic theory of matter is.
- Know the conditions for thermal equilibrium.
- Know what temperature measures.
- Know how to calculate the amount of thermal expansion of a material.
- Know how to use the gas laws and under what conditions they are valid.
- Know what an absolute temperature scale is.

## Summary of Equations

Atomic Mass Units:
$$1u = 1.66 \times 10^{-27} \text{ kg}$$

Conversion, Celsius–Fahrenheit:
$$T\left(^\circ C\right) = \tfrac{5}{9}\left[T\left(^\circ F\right) - 32\right]$$

$$T\left(^\circ F\right) = \tfrac{9}{5}T\left(^\circ C\right) + 32$$

Linear Thermal Expansion:
$$L = L_0\left(1 + \alpha\,\Delta T\right)$$

Volume Thermal Expansion:
$$V = V_0\left(1 + \beta\,\Delta T\right)$$

Thermal Stress:
$$\frac{F}{A} = \alpha\,E\Delta T$$

Ideal Gas Law:
$$PV = nRT = NkT$$

## Chapter Summary

### Section 17-1 Atomic Theory of Matter

The atomic theory of matter models matter as a collection of indivisible units called **atoms**. A unit of mass that is commonly used for atoms and molecules is the unified atomic mass unit (u):

$$1u = 1.66 \times 10^{-27} \text{ kg}$$

In 1905, Albert Einstein estimated the mass of molecules from measurements made on the random motion of dust particles suspended in air. This motion is called **Brownian Motion**.

### Example 17-1-A

The atomic mass of silicon is 28.1 u. How many atoms of silicon are in a 17.4 g piece of silicon?

**Solution:**

The average mass of a silicon atom is

$$m_{Si} = 28.1\text{u} = (28.1)\left(1.66\times10^{-27}\,\text{kg}\right) = 4.665\times10^{-26}\,\text{kg}$$

The number of atoms in a 17.4 g piece of silicon will be

$$N = \frac{m}{m_{Si}} = \frac{17.4\times10^{-3}\,\text{kg}}{4.665\times10^{-26}\,\text{kg}} = 3.73\times10^{23}$$

## Section 17-2 Temperature and Thermometers

In everyday life, temperature is used to measure how hot or cold something is. It turns out that our sense of hot or cold isn't exactly a temperature-measuring system, but in many cases our judgment of how hot or cold something is will be related to its temperature.

Thermometers are used to measure temperature. All thermometers rely on the temperature dependence of some property of a material to measure temperature. Some of the properties of materials that change with temperature that are used to make thermometers are length, density, and electrical resistivity.

There are two temperature scales in use in everyday life. These are the Fahrenheit scale (°F) and the Celsius scale (°C). The relationship between temperatures on the two scales is given by the following two relationships:

$$T\left(^{\circ}\text{C}\right) = \tfrac{5}{9}\left[T\left(^{\circ}\text{F}\right) - 32\right]$$

$$T\left(^{\circ}\text{F}\right) = \tfrac{9}{5}T\left(^{\circ}\text{C}\right) + 32$$

## Section 17-3 Thermal Equilibrium and the Zeroth Law of Thermodynamics

Temperature is a means of determining whether two systems are in thermal equilibrium. Two systems at the same temperature will not exchange energy through the process of heat transfer when in thermal contact. When one system at a higher temperature is in thermal contact with another system at lower temperature, energy will flow by heat transfer from the higher temperature system to the lower temperature system.

The zeroth law of thermodynamics is a way of stating that assigning a temperature to a system is meaningful. It states that thermal equilibrium is a transitive relationship. If system A is in thermal equilibrium with system B and system B is in thermal equilibrium with system C, then system A is in thermal equilibrium with system C. If this were not true, then the concept of temperature would not be a very useful idea.

## Section 17-4 Thermal Expansion

Most materials expand when their temperature is increased and contract when their temperature is decreased. The amount of the expansion or contraction depends on the type of material. The change in a linear dimension, $\Delta L$, of a material that undergoes a temperature change, $\Delta T$, can be approximated by a linear relationship:

$$\Delta L = \alpha L_0 \Delta T$$

where $L_0$ is the original linear dimension and $\alpha$ is the coefficient of linear expansion. The values of $\alpha$ for various materials are given in Table 17-1 of the text. Typical values of $\alpha$ for solids have a magnitude of $10^{-5}/^{\circ}\text{C}$. This relationship can be written in terms of the new length, $L$, at the new temperature:

$$L = L_0\left(1 + \alpha\,\Delta T\right)$$

Similar relationships hold for the volume change of a material with temperature change $\Delta T$:

$$\Delta V = V_o \beta\,\Delta T \qquad\qquad V = V_0\left(1 + \beta\,\Delta T\right)$$

The initial volume of the material is $V_0$, the change in volume is $\Delta V$, $\beta$ is the coefficient of volume expansion, and $V$ is the volume of the material at the final temperature. The coefficient of volume expansion is typically three times the coefficient of linear expansion.

Increasing volume with increasing temperature implies that the potential energy of atoms or molecules as a function of separation distance is asymmetrical.

Water in the temperature range of 0 °C to 4 °C behaves differently than most other materials. In that range of temperatures, water decreases in volume with increasing temperature.

**Example 17-4-A**

A copper ring of diameter 3.47 cm fits snugly around an iron rod of the same diameter at a temperature 20 °C. To remove the ring easily from this rod, it is necessary for the ring to have a diameter that is 0.05 mm larger than that of the rod. Determine the temperature the rod and ring must be heated or cooled to so that the ring slides easily off of the rod.

**Solution:**

We write down the change in length for each of the objects:

$$\Delta L_{ring} = \alpha_{Cu} L_0 \Delta T \qquad \Delta L_{rod} = \alpha_{Fe} L_0 \Delta T$$

The difference in the change in diameters of the two objects must be the given clearance:

$$\Delta L_{ring} - \Delta L_{rod} = 0.05\,\text{mm} \Rightarrow (\alpha_{Cu} - \alpha_{Fe}) L_0 \Delta T = 0.05\,\text{mm}$$

$$\Rightarrow \Delta T = \frac{0.05\,\text{mm}}{(\alpha_{Cu} - \alpha_{Fe}) L_0}$$

$$\Rightarrow T_f = 20\,°\text{C} + \frac{0.005\,\text{cm}}{(17-12) \times 10^{-6} \left(\text{C}°\right)^{-1} (3.47\,\text{cm})}$$

$$\Rightarrow T_f = 310\,°\text{C}$$

**Example 17-4-B**

A bulb type thermometer holds 0.132 cm³ of ethyl alcohol. The bore of the capillary tube in which the level of the alcohol rises with temperature has a diameter of 0.061 cm. Assuming the thermal expansion of the quartz bulb and capillary are negligible, what distance should be marked as one degree Celsius on the thermometer?

**Solution:**

First we calculate the change in volume of the liquid for a one degree change in Celsius temperature:

$$\Delta V = \beta V_o \Delta T = \left(1.10 \times 10^{-3}\,\text{C}°^{-1}\right)\left(0.132\,\text{cm}^3\right)\left(1°\text{C}\right) = 1.45 \times 10^{-4}\,\text{cm}$$

This additional volume of fluid goes into a cylinder with a cross-sectional area the same as the bore and a height equal to the rise in the level of the fluid in the bore:

$$\Delta V = hA = h\frac{\pi D^2}{4} \quad \Rightarrow \quad h = \frac{4\Delta V}{\pi D^2} = \frac{4\left(1.45 \times 10^{-4}\,\text{cm}\right)}{\pi (0.061\ \text{cm})^2} = 0.05\,\text{cm}$$

**Section 17-5 Thermal Stresses**

If materials are not allowed to freely expand, the strain of thermal expansion will induce a stress called thermal stress. As an example, consider a piece of material with elastic constant $E$ that is constrained from increasing its length. The piece of material would have a change in length given by

$$\Delta L = \alpha \, L_0 \, \Delta T$$

if its ends were free. The ends being fixed, there must be an elastic strain of equal size but opposite direction to keep the ends fixed. This means there must be an elastic stress equal to

$$\frac{F}{A} = E \frac{\Delta L}{L} = E \alpha \, \Delta T$$

### Example 17-5-A

A steel beam of length 10.6 m with a cross-sectional area of 0.183 $m^2$ is wedged between two bedrock outcroppings when it is at a temperature of 14.6 °C. On a hot day the temperature of the steel beam rises to 46.2 °C. What is the compressive force on the beam if the spacing between the bedrock outcroppings does not change?

**Solution:**

If the beam were free to expand, the change in length of the beam would be

$$\Delta L = \alpha L \Delta T = \left(12 \times 10^{-6} \, C^{\circ -1}\right)(10.6 \, m)\left(46.2 \, ^\circ C - 14.6 \, ^\circ C\right) = 4.020 \times 10^{-3} \, m$$

Because the ends of beam are not allowed to move, an elastic compression equal to this thermal expansion must exist in the beam. This elastic compression is related to the stress in the beam by Hooke's Law:

$$\frac{F}{A} = E \frac{\Delta L}{L}$$

Young's modulus, $E$, for steel is $200 \times 10^9$ N/$m^2$ (Chapter 12 of text). The cross-sectional area, $A$, of the beam is known, so this can be solved for the compressive force

$$F = EA \frac{\Delta L}{L} = \left(200 \times 10^9 \, N/m^2\right)\left(0.183 \, m^2\right)\frac{4.020 \times 10^{-3} \, m}{10.6 \, m} = 1.39 \times 10^7 \, N$$

### Section 17-6 The Gas Laws and Absolute Temperature

An equation of state of a system relates the volume, pressure, temperature, and the amount of material in the system to each other when the system is in equilibrium. Gases have relatively simple equations of state, particularly when the volume of the molecules is small compared to the volume of the gas and the temperature of the gas is much greater than the temperature at which the gas would condense.

For a fixed quantity of ideal gas, the volume of the gas is inversely proportional to the pressure of the gas at fixed temperature:

$$PV = \text{Constant}$$

This relationship is known as Boyle's law. Also, for a fixed quantity of ideal gas, the volume of a gas is directly proportional to the absolute temperature of the gas when it is held at a fixed pressure:

$$V = \text{Constant} \times T$$

This relationship is known as Charles's law. The absolute temperature scale arises from the discovery that when the pressure of a gas is not too high, the volume of gas is linearly related to temperature when far above its liquefaction temperature. If the linear relationship is solved for the temperature at which the volume of the gas would be zero, it is the common temperature –273.15 °C, independent of the type of gas. The Kelvin temperature scale uses this temperature as zero and has an increment of temperature of one Kelvin equal to 1°C.

A third relationship between gas quantities relates the pressure of a gas to its absolute temperature. Gay-Lussac's Law states that the pressure of a gas is directly proportional to the absolute temperature of the gas when the volume is held constant:

$$P = \text{Constant} \times T$$

### Section 17-7 The Ideal Gas Law

A **mole** of a substance is an amount of the substance that contains as many molecules of the substance as there are carbon atoms in 12.00 g of carbon. That is the same as a number of grams of the substance equal to the molecular weight of the substance in atomic mass units. The ideal gas law relates the pressure, volume, temperature, and amount of a gas to each other:

$$PV = nRT$$

where $P$ is the absolute pressure of the gas, $V$ is the volume of the gas, $n$ is the number of moles of gas, and $T$ is the absolute pressure of the gas. The constant $R$ is called the universal gas constant and has a value of 8.315 J/mole·K. The relationship is also called the equation of state of an ideal gas.

### Section 17-8 Problem Solving with the Ideal Gas Law

Always remember that the temperature must be the absolute temperature and the pressure must be the absolute pressure in applying the ideal gas law.

Standard temperature and pressure (STP) refer to a temperature of 273 K and a pressure of $1.013 \times 10^5$ N/m$^2$. The volume of 1mole of ideal gas at STP is 22.4L or $2.24 \times 10^{-2}$ m$^3$.

### Example 17-8-A

A hot air balloon has a volume of 134 m$^3$ when inflated and the payload has negligible volume. It needs to lift a total of its mass and payload of 23.8 kg. The temperature of the ambient air is 26 °C. What does the temperature of the air inside the balloon need to be to give enough buoyant force to lift the balloon and payload float in the air?

### Solution:

We know the buoyant force must be equal to the weight of the system for the system to float in the air. The mass of the atmospheric air displaced is the volume of the balloon times the density of the outside air:

$$m_{air\,displaced} = \rho V_{balloon} = \frac{n m_{mol}}{V_{balloon}} V_{balloon}$$

where $n$ is the number of moles of air in the balloon and $m_{mol}$ is the mass of one mole of air, 0.029 kg. We can use the ideal gas law to write

$$m_{air\,displaced} = \frac{P}{RT_{outside}} m_{mol} V_{balloon}$$

It easily follows that the mass of the warm air inside the balloon will be the same expression, but with the inside temperature substituted for the outside temperature:

$$m_{air\,inside} = \frac{P}{RT_{inside}} m_{mol} V_{balloon}$$

Using Archimedes' principle the mass of the displaced air must be equal to the mass of the system in order for it to float:

$$m_{air\,displaced} = m_{air\,inside} + m_{balloon\,and\,payload}$$

$$\Rightarrow \frac{P}{RT_{outside}} m_{mol} V_{balloon} = \frac{P}{RT_{inside}} m_{mol} V_{balloon} + m_{balloon\,and\,payload}$$

$$\Rightarrow T_{inside} = \frac{1}{\dfrac{1}{T_{outside}} - \dfrac{Rmp_{balloon\,and\,payload}}{Pm_{mol} V_{balloon}}}$$

$$= \frac{1}{\dfrac{1}{(26+273)\,\text{K}} - \dfrac{(8.315\,\text{J/K})(23.8\,\text{kg})}{(1.01 \times 10^5\,\text{N/m}^2)(0.029\,\text{kg})(134\,\text{m}^3)}} = 352\,\text{K} = 79^\circ\text{C}$$

## Example 17-8-B

A container of volume of $0.180\,\text{m}^3$ contains $1.22\,\text{kg}$ of liquid nitrogen and nitrogen gas in the remainder of the volume. The gas is at atmospheric pressure and the temperature is 78 K.   The container is sealed and the temperature rises to 300 K.   At this temperature, all of the liquid has become a gas.   What is the pressure inside the container now?   The density of liquid nitrogen is $810\,\text{kg/m}^3$ and the molecular weight of nitrogen gas is $0.028\,\text{kg/mol}$.

## Solution:

We know the volume and the temperature of the final gas.   If we can determine the number of moles of gas in the container, we can determine the pressure using the ideal gas law.   We can easily determine the number of moles in the liquid by dividing the mass of liquid by the mass per mole:

$$n_{liquid} = \frac{m}{m_{mol}} = \frac{1.22\,\text{kg}}{0.028\,\text{kg}} = 43.6$$

We can use the ideal gas law to determine the number of moles of nitrogen gas under the initial conditions.   Initially the temperature and pressure of the gas are known.   The initial volume of gas will be the volume of the container less the volume of liquid in the container.   Using the ideal gas law,

$$n_{gas} = \frac{PV}{RT} = \frac{P(V_{container} - V_{liquid})}{RT} = \frac{P(V_{container} - m_{liquid}/\rho)}{RT}$$

$$= \frac{(1.01 \times 10^5\,\text{N/m}^2)(0.180\,\text{m}^3 - 1.22\,\text{kg}/810\,\text{kg/m}^3)}{(8.315\,\text{J/K})(78\,\text{K})}$$

$$= 27.8$$

The total number of moles of gas in the final conditions of the container will be the sum of the moles in the initial liquid and gas:

$$n = n_{liquid} + n_{gas} = 43.6 + 27.8 = 71.4$$

We can now use the ideal gas law to determine the final pressure:

$$P = \frac{nRT}{V} = \frac{(71.4)(8.315\,\text{J/K})(300\,\text{K})}{0.180\,\text{m}^3} = 9.9 \times 10^5\,\text{N/m}^2$$

## Section 17-9 Ideal Gas Law in Terms of Molecules: Avogadro's Number

The number of molecules in a mole of substance is called Avogadro's number, $N_A$:

$$N_A = 6.022 \times 10^{23}$$

The ideal gas law can be written in the form

$$PV = NkT$$

where $N$ is the number of molecules of a substance and $k = R/N_A$ and is called Boltzmann's constant:

$$k = R/N_A = 1.381 \times 10^{-23} \text{ J/K}$$

**Example 17-9-A**

At atmospheric pressure and a temperature of 20 °C, how many air molecules are in a room that is 20 m × 15 m × 3 m?

**Solution:**

There are several approaches to this problem that we can take, but we will use the ideal gas law written in terms of the number of molecules:

$$PV = NkT \Rightarrow N = \frac{PV}{kT} = \frac{\left(1.01 \times 10^5 \text{ N/m}^2\right)\left(20\,\text{m}\right)\left(15\,\text{m}\right)\left(3.0\,\text{m}\right)}{\left(1.38 \times 10^{-23} \text{ J/K}\right)\left(20 + 273\,\text{K}\right)}$$

$$= 2.2 \times 10^{28}$$

**Section 17-10 Ideal Gas Temperature Scale – a Standard**

The many methods of measuring temperature can be made precisely equivalent at two fixed points, but the assumption of approximate temperature dependencies of material properties causes the different scales based on these approximations to disagree away from the fixed points. In order to create a standard of temperature, one particular method of temperature method must be agreed upon as a standard and all other methods may be calibrated against that standard by placing those thermometers in equilibrium with the standard thermometer. The Ideal Gas Temperature Scale has characteristics of universality and repeatability that make it a good choice as a temperature standard. The Kelvin temperature on the ideal gas temperature scale is defined to be:

$$T = 273.16 \text{ K}(P/P_{tp})$$

for an ideal gas at constant volume where $T$ is the Kelvin temperature of the gas, $P$ is the absolute pressure of the gas, and $P_{tp}$ is the absolute pressure of the gas at the triple point of water (that temperature at which ice, liquid water, and water vapor are in thermal equilibrium). No real gas behaves exactly as an ideal gas, but as the pressure of the gas decreases at temperatures far from the liquefaction temperature, the behavior approaches ideal gas behavior. In practice, the ideal gas temperature scale is defined in the limit of the pressure of a real gas approaching zero pressure.

## Practice Quiz

1.     The random motion of dust particles suspended in a gas or liquid is

    a) Brownian motion
    b) Impossible
    c) The same regardless of the temperature
    d) Stopped at a temperature of 0 °C

2.     When are two systems in thermal equilibrium?

    a) When they have the same thermal expansion
    b) When one system has the negative of the thermal stress of the other system
    c) When the two systems have the same pressure
    d) When the two systems have the same temperature

3.    How does the coefficient of linear expansion in units of $(F°)^{-1}$ compare to the coefficient of linear expansion in units of $(C°)^{-1}$?

a) They are the same.
b) The coefficient of linear expansion in $(F°)^{-1}$ is 5/9 the coefficient of linear expansion in $(C°)^{-1}$.
c) The coefficient of linear expansion in $(F°)^{-1}$ is 9/5 the coefficient of linear expansion in $(C°)^{-1}$.
d) The coefficient of linear expansion in $(F°)^{-1}$ is 5/9 the coefficient of linear expansion in $(C°)^{-1}$ + 32.

4.    Can thermal stress ever be tensile rather than compressive?

a) No; objects always expand on heating and it takes a compressive stress to prevent the expansion.
b) Yes, but only if the material has a negative coefficient of expansion.
c) Yes, when objects are cooled to a lower temperature.
d) No, because it is impossible to cool an object below absolute zero temperature.

5.    What is meant by the anomalous behavior of water in regards to thermal expansion?

a) Water has a zero coefficient of linear thermal expansion at all temperatures.
b) Water has a negative coefficient of linear thermal expansion at all temperatures.
c) Water has a negative coefficient of thermal expansion at temperatures below 4° C.
d) Water has an unusually large coefficient of thermal expansion.

6.    For most materials, how is the coefficient of volume expansion related to the coefficient of linear expansion?

a) The coefficient of volume expansion is equal to the coefficient of linear expansion.
b) The coefficient of volume expansion is one third the coefficient of linear expansion.
c) The coefficient of volume expansion is three times the coefficient of linear expansion.
d) The coefficient of volume expansion is twice the coefficient of linear expansion.

7.    When applying the ideal gas law to a system that changes temperature, the temperature can be measured on which temperature scale?

a) Fahrenheit
b) Celsius
c) Kelvin
d) Any scale

8.    Which of the gas laws relates the volume of gas to its pressure at a fixed temperature and fixed amount of gas?

a) Boyle's law
b) Charles's law
c) Gay-Lussac's law
d) Zeroth law of thermodynamics

9.    Which of the gas laws relates the temperature of gas to its pressure at a fixed volume and fixed amount of gas?

e) Boyle's law
f) Charles's law
g) Gay-Lussac's law
h) Zeroth law of thermodynamics

10.    What is special about the ideal gas temperature scale?

a) It is the best temperature scale to use for finding the temperature of gases.
b) It defines the temperature scale independently from the properties of a particular gas.
c) It is the most practical scale for determining the fixed points of the scale at home.
d) It uses the nicest set of numbers for working with everyday temperatures.

11. The floors of a building are to be made level. As the building is being built, the 12 foot long posts that support the second floor on the south side of the building are exposed to the sun and warm to 120° F. The posts on the north side remain at the 70° F ambient temperature. The second floor is leveled under these conditions. How far out of level will the floor be when the building all cools down to 60° F at night? Do you think builders bother to consider this when constructing a building?

12. Determine the temperature on the Celsius scale at which the Kelvin temperature scale and the Fahrenheit temperature scale have the same value.

13. A 134 m long concrete sidewalk is made without expansion joints when the temperature is 45° F. The temperature of the concrete rises to 110° F on a sunny day. What is the thermal stress in the sidewalk? Assume the ends of the sidewalk are fixed in position.

14. A container of nitrogen gas contains 128 moles of nitrogen at a pressure of $2.4 \times 10^5$ N/m$^2$ at a temperature of 290 K. As nitrogen is added to the container, the pressure rises to $6.9 \times 10^5$ N/m$^2$ and the temperature rises to 354 K. How many moles of nitrogen were added to the container?

15. The unit of pressure **torr** is often used in measuring pressure in a vacuum system. One torr is equal to the pressure that supports a column of mercury one millimeter high. Determine the number of molecules per cubic centimeter in a vacuum of $10^{-10}$ torr at a temperature of 300 K.

## Problem Solutions

3. (a) $T(°C) = (5/9)[T(°F) - 32] = (5/9)(68°F - 32) = 20°C$.
   (b) $T(°F) = (9/5)T(°C) + 32 = (9/5)(1800°C) + 32 = 3272°F \approx 3300°F$.

7. We set $T(°F) = T(°C) = T$ in the conversion between the temperature scales:
   $T(°F) = (9/5)T(°C) + 32$
   $T = (9/5)T + 32$, which gives $T = -40°F = -40°C$.

11. We can treat the change in diameter as a simple change in length, so we have
   $\Delta L = \alpha L_0 \Delta T$;
   $1.869$ cm $- 1.871$ cm $= [12 \times 10^{-6} (C°)^{-1}](1.871$ cm$)(T - 20°C)$, which gives $T = -69°C$.

15. We find the change in volume from
   $\Delta V = V_0 \beta \Delta T = 4\pi r^3 \beta \Delta T /3$
   $= 4\pi (4.375$ cm$)^3[1 \times 10^{-6} (C°)^{-1}](200°C - 30°C) = 0.060$ cm$^3$.

19. The increase in temperature will cause the length of the brass rod to increase. The period of the pendulum depends on the length,
   $T = 2\pi(L/g)^{1/2}$,
   so the period will be greater. This means the pendulum will make fewer swings in a day, so the clock will be slow and the clock will lose time.
   We use $T_C$ for the temperature to distinguish it from the period.
   For the length of the brass rod, we have
   $L = L_0(1 + \alpha \Delta T_C)$.
   Thus the ratio of periods is
   $T/T_0 = (L/L_0)^{1/2} = (1 + \alpha \Delta T_C)^{1/2}$.
   Because $\alpha \Delta T_C$ is much less than 1, we have
   $T/T_0 \approx 1 + \frac{1}{2}\alpha \Delta T_C$, or $\Delta T/T_0 = \frac{1}{2}\alpha \Delta T_C$.
   The number of swings in a time $t$ is $N = t/T$. For the same time $t$, the change in period will cause a change in the number of swings:
   $\nabla N = (t/T) - (t/T_0) = t(T_0 - T)/TT_0 \approx -t(\Delta T/T_0)/T_0$,
   because $T \approx T_0$. The time difference in one year is
   $\Delta t = T_0 \Delta N = -t(\Delta T/T_0) = -t(\frac{1}{2}\alpha \Delta T_C)$
   $= -(1$ yr$)(3.16 \times 10^7$ s/yr$)\frac{1}{2}[19 \times 10^{-6} (C°)^{-1}](25°C - 17°C) = -2.4 \times 10^3$ s $= -40$ min.

23. The compressive strain must compensate for the thermal expansion. From the relation between stress and strain, we have

$$\text{Stress} = E(\text{Strain}) = E\alpha\,\Delta T;$$
$$F/A = (70 \times 10^9 \text{ N/m}^2)[25 \times 10^{-6}\,(\text{C}°)^{-1}](35°C - 15°C) = 3.5 \times 10^7 \text{ N/m}^2.$$

27. On the Celsius scale, absolute zero is

$$T(°C) = T(K) - 273.15 = 0\text{ K} - 273.15 = -273.15°C.$$

On the Fahrenheit scale, we have

$$T(°F) = (9/5)T(°C) + 32 = (9/5)(-273.15°C) + 32 = -459.7°F.$$

31. If we assume oxygen is an ideal gas, we have

$$PV = nRT = (m/M)RT;$$
$$(1.013 \times 10^5 \text{ Pa})V = [m/(32 \text{ g/mol})(10^3 \text{ kg/g})](8.315 \text{ J/mol} \cdot \text{K})(273 \text{ K}), \text{ which gives}$$
$$m/V = 1.43 \text{ kg/m}^3.$$

35. If we assume argon is an ideal gas, we have

$$PV = nRT = (m/M)RT;$$
$$P(35.0 \times 10^{-3} \text{ m}^3) = [(105.0 \text{ kg})(10^3 \text{ g/kg})/(40 \text{ g/mol})](8.315 \text{ J/mol} \cdot \text{K})(293 \text{ K}),$$

which gives $P = 1.83 \times 10^8 \text{ Pa} = 1.80 \times 10^3 \text{ atm}.$

39. For the two states of the gas we can write

$$P_1V_1 = nRT_1 \quad \text{and} \quad P_2V_2 = nRT_2, \text{ which can be combined to give}$$
$$(P_2/P_1)(V_2/V_1) = T_2/T_1;$$
$$(P_2/2.45 \text{ atm})(48.8 \text{ L}/61.5 \text{ L}) = (323.2 \text{ K}/291.2 \text{ K}), \text{ which gives } P_2 = 3.43 \text{ atm}.$$

43. If we write the ideal gas law as $PV = NkT$, we have

$$N/V = P/kT = (1.013 \times 10^5 \text{ Pa})/(1.38 \times 10^{-23} \text{ J/K})(273 \text{ K}) = 2.69 \times 10^{25} \text{ molecules/m}^3.$$

47. The volume and mass are constant. For the two states of the gas we can write

$$P_1V_1 = nRT_1, \quad \text{and} \quad P_2V_2 = nRT_2, \text{ which can be combined to give}$$
$$P_2/P_1 = T_2/T_1;$$
$$(P_2/1.00 \text{ atm}) = (455 \text{ K}/293 \text{ K}), \text{ which gives } P_2 = 1.55 \text{ atm}.$$

We find the length of a side of the box from

$$V = L^3;$$
$$5.1 \times 10^{-2} \text{ m}^3 = L^3, \text{ which gives } L = 0.371 \text{ m}.$$

The net force is the same on each side of the box. Because there is atmospheric pressure outside the box, the net force is

$$F = A\,\Delta P = L^2(P_2 - P_1) = (0.371 \text{ m})^2(1.55 \text{ atm} - 1.00 \text{ atm})(1.013 \times 10^5 \text{ Pa/atm}) = 7.7 \times 10^3 \text{ N}.$$

Note that we have assumed no change in dimensions from the increased pressure.

51. (a) From Fig. 17–16, we read a temperature of 373.34 K from the oxygen curve at a pressure of 268 torr. Thus the inaccuracy is

$$\Delta T = 373.34 \text{ K} - (273.15 + 100.00)\text{ K} = 0.19 \text{ K}.$$

(b) As a percentage, this is

$$(\Delta T/T)100 = (0.19 \text{ K}/373.15)100 = 0.051\%.$$

55. For the two conditions of the gas in the cylinder, we can write

$$P_1V = n_1RT, \quad \text{and} \quad P_2V = n_2RT, \text{ which can be combined to give}$$
$$P_2/P_1 = n_2/n_1;$$
$$(5 \text{ atm} + 1 \text{ atm})/(35 \text{ atm} + 1 \text{ atm}) = n_2/n_1, \text{ which gives } n_2/n_1 = 1/6.$$

59. The pressure at a depth $h$ is

$$P = P_0 + \rho gh = 1.013 \times 10^5 \text{ Pa} + (1000 \text{ kg/m}^3)(9.80 \text{ m/s}^2)(10 \text{ m}) = 1.99 \times 10^5 \text{ Pa}.$$

For the two states of the gas we can write

$$PV = nRT, \quad \text{and} \quad P_0V_0 = nRT, \text{ which can be combined to give}$$
$$P/P_0 = V_0/V;$$
$$(1.99 \times 10^5 \text{ Pa}/1.013 \times 10^5 \text{ Pa}) = (V_0/5.5 \text{ L}), \text{ which gives } V_0 = 11 \text{ L}.$$

This doubling of the volume is definitely not advisable.

63. The pressure on a small area of the surface can be considered to be due to the weight of the air column above the area:

$P = Mg/A$.

When we consider the total surface of the Earth, we have

$$M_{\text{total}} = PA_{\text{total}}/g = P4\pi R^2/g$$
$$= (1.013 \times 10^5 \text{ Pa})4\pi(6.37 \times 10^6 \text{ m})^2/(9.80 \text{ m/s}^2) = 5.27 \times 10^{18} \text{ kg}.$$

If we use the average mass of an air molecule, we find the number of molecules from

$N = M_{\text{total}}/m = (5.27 \times 10^{18} \text{ kg})/(28.8 \text{ u})(1.66 \times 10^{-27} \text{ kg}) = 1.1 \times 10^{44}$ molecules.

67. We treat the circumference of the band as a length, which will expand according to

$2\pi R = 2\pi R_0(1 + \alpha\Delta T)$,   or

$R - R_0 = R_0\alpha\Delta T = (6.38 \times 10^6 \text{ m})[12 \times 10^{-6} \text{ (C°)}^{-1}](35°C - 20°C) = 1.1 \times 10^3 \text{ m}$.

71. We consider a fixed mass of the substance:

$m = \rho_0 V_0 = \rho V = \rho V_0(1 + \beta\Delta T)$;

$\rho_0 = \rho(1 + \beta\Delta T)$;

$0.68 \times 10^3 \text{ kg/m}^3 = \rho\{1 + [950 \times 10^{-6} \text{ (C°)}^{-1}](32°C - 0°C)\}$, which gives $\rho = 0.66 \times 10^3 \text{ kg/m}^3$.

75. (a) If $P$ refers to the pressure in the atmosphere, for the two states of the gas we can write

$1.05P_0V_0 = nRT_0$   and   $1.05PV = nRT_1$, which can be combined to give

$(P/P_0)(V/V_0) = T_1/T_0$.

When we use the dependence of the pressure in the atmosphere on the altitude,

$P = P_0e^{-cy}$, where $c = \rho_0 g/P_0$, we get

$(P_0 e^{-cy}/P_0)(V/V_0) = T_1/T_0$,   or   $V = V_0(T_1/T_0) e^{+cy}$.

(b) The density of the air is

$\rho = m/V = mP/nRT_1$, so

$\rho/\rho_0 = (P/P_0)(T_0/T_1)$.

The buoyant force at an altitude $y$ is

$F_{\text{buoy}} = \rho gV = \rho_0(P/P_0)(T_0/T_1)gV_0(T_1/T_0)e^{+cy} = \rho_0 e^{-cy}gV_0 e^{+cy} = \rho_0 gV_0 = F_{\text{buoy0}}$.

# Chapter 18: Kinetic Theory of Gases

## Chapter Overview and Objectives

This chapter introduces the kinetic theory of gases. It relates the ideal gas law to a model of a gas as a collection of particles. It introduces the concepts of saturated vapor pressure, relative humidity, and partial pressure of gases. The chapter discusses non-ideal equations of state for gases. The chapter introduces the concepts of mean free path and diffusion.

After completing study of this chapter, you should:

- Understand the microscopic model of an ideal gas as a collection of particles.
- Know how the mean kinetic energy of the particles in a gas is related to the absolute temperature.
- Know how the speeds of an ideal gas are distributed according to Maxwell's distribution of speeds.
- Know the difference in behavior of real gases and ideal gases.
- Understand the concepts of saturated vapor pressure, partial pressure, and relative humidity.
- Be familiar with the Clausius and van der Waals equations of state of a gas.
- Understand the concepts of mean free path and diffusion.

## Summary of Equations

Mean kinetic energy of particles in an ideal gas:
$$\bar{K} = \left[ \frac{1}{2} m \left( v^2 \right) \right]_{av} = \frac{3}{2} kT$$

Root-mean-square speed of particles in ideal gas:
$$v_{RMS} = \sqrt{\frac{3kT}{m}}$$

Maxwell distribution of speeds in an ideal gas:
$$f(v) = 4\pi N \left( \frac{m}{2\pi kT} \right)^{\frac{3}{2}} v^2 e^{-\frac{1}{2}\frac{mv^2}{kT}}$$

Definition of relative humidity:
$$\text{Relative humidity} = \frac{\text{Partial pressure } H_2O}{\text{Saturated Vapor pressure of } H_2O} \times 100\%$$

Clausius equation of state:
$$P(V - nb) = nRT$$

Van der Waals equation of state:
$$\left( P + \frac{a}{(V/n)^2} \right) \left( \frac{V}{n} - b \right) = RT$$

Mean free path in gas of spheres:
$$l_M = \frac{1}{4\pi\sqrt{2}\, r^2 (N/V)}$$

Diffusion equation or Fick's law:
$$J = DA \frac{dC}{dx}$$

# Chapter Summary

### Section 18-1 The Ideal Gas Law and the Molecular Interpretation of Temperature

An ideal gas is a model of the gas phase of matter. The properties of the ideal gas model are:

1. The particles in the gas have negligible volume.
2. The number of particles in the gas is large.
3. Particles of the gas move independently and interact only during collisions.
4. All collisions with walls of containers and other gas particles are elastic collisions.

One result of the ideal gas model is that the average kinetic energy, $\overline{K}$, of the particles in the gas must be proportional to the absolute temperature, $T$, of the gas:

$$\overline{K} = \left[\tfrac{1}{2}m(v^2)\right]_{av} = \tfrac{3}{2}kT$$

where $m$ is the mass of the gas particles, $v$ is their speed, and $k$ is Boltzmann's constant.

We can use this result to determine the root-mean-square speed, $v_{RMS}$, of the gas particles:

$$v_{RMS} = \sqrt{\frac{3kT}{m}}$$

Root-mean-square (RMS) means the square root of the mean (average) of the square of the quantity. Root-mean-square speed means the square root of the average of the square of the speed of each particle. This is different from the mean speed, $\overline{v}$, which is just the average of the speeds of each particle.

### Example 18-1-A

Suppose you have a "gas" of spherical dust particles with each dust particle having a radius of 0.1 μm and a density of 1800 kg/m³. What is the RMS speed of these particles at a temperature 300 K?

**Solution:**

To find the RMS speed of the gas particles, we need to know the mass of the particles and the temperature. We are given the density of the particles, so we can determine their mass:

$$m = \rho V = \rho \tfrac{4}{3}\pi r^3 = \left(1800\,\text{kg}/\text{m}^3\right)\tfrac{4}{3}\pi\left(0.1\times10^{-6}\,\text{m}\right)^3 = 7.5\times10^{-18}\,\text{kg}$$

The RMS speed is then:

$$v_{RMS} = \sqrt{\frac{3kT}{m}} = \sqrt{\frac{3\left(1.38\times10^{-23}\,\text{J}/\text{K}\right)\left(300\,\text{K}\right)}{7.5\times10^{-18}\,\text{kg}}} = 0.041\,\text{m}/\text{s} = 4.1\,\text{cm}/\text{s}$$

### Section 18-2 Distribution of Molecular Speeds

The speeds of the particles in an ideal gas are not all the same. The expected number of particles to be found with a speed between $v$ and $v + dv$ in an ideal gas of $N$ particles is $f(v)dv$ with $f(v)$ given by the Maxwell distribution of speeds:

$$f(v) = 4\pi N\left(\frac{m}{2\pi kT}\right)^{\tfrac{3}{2}} v^2 e^{-\tfrac{1}{2}\tfrac{mv^2}{kT}}$$

where $m$ is the mass of the gas particles, $k$ is Boltzmann's constant, $T$ is the absolute temperature, and $N$ is the number of particles in the gas.

**Example 18-2-A**

For a fixed speed $v$, determine the ratio of the number of molecules at temperature $T_2$ to the number of molecules at temperature $T_1$ within the interval $v$ to $v + dv$.

**Solution:**

The ratio of the number of molecules within the given velocity interval will be the ratio of the Maxwell distribution at those two temperatures:

$$\frac{f_{T_2}(v)}{f_{T_1}(v)} = \frac{4\pi \ N\left(\dfrac{m}{2\pi \ kT_2}\right)^{\frac{3}{2}} v^2 e^{-\frac{1}{2}\frac{mv^2}{kT_2}}}{4\pi \ N\left(\dfrac{m}{2\pi \ kT_1}\right)^{\frac{3}{2}} v^2 e^{-\frac{1}{2}\frac{mv^2}{kT_1}}} = \left(\frac{T_1}{T_2}\right)^{\frac{3}{2}} e^{-\frac{1}{2}\frac{mv^2}{k}\left(\frac{1}{T_1}-\frac{1}{T_2}\right)}$$

## Section 18-3 Real Gases and Changes of Phase

Real gases have a behavior that under conditions of high temperature and low density is very close to that predicted by the ideal gas model. However, when the temperature becomes low or the density becomes high, the behavior of real gases can depart significantly from that predicted by the ideal gas model. It is quite easy to see why this is. The ideal gas model assumes that the particle volume is negligible. Under conditions of high density, the finite size of atoms and molecules can become a significant fraction of the volume of the gas. This increases the probability of collisions which causes the pressure in a real gas to be greater than that predicted by the ideal gas law. Also, all real gas particles interact with each other, resulting in a slightly attractive force acting between them. When the temperature drops far enough so that the size of the average kinetic energy of the gas particles becomes small enough to be comparable to the size of the potential energy associated with the interaction between the gas molecules, this interaction becomes important and the gas particles no longer move independently.

Under appropriate conditions of temperature and pressure, the molecular interaction is great enough for the molecules to stick together to form liquids or solids. There is a maximum temperature at which a liquid will exist. This temperature is called the **critical temperature**. The conditions of density and pressure at the critical temperature determine the **critical point** of the material.

A **phase diagram** is a pressure–temperature plot that shows the equilibrium phases for a given pressure and temperature. The **triple point** of a material is point on the pressure–temperature plot of a material at which there exist solid, liquid, and gas phase of the material in a state of equilibrium.

The common phases discussed are not the only phases of matter. For example, several different **liquid crystal** phases of matter exist for some materials.

## Section 18-4 Vapor Pressure and Humidity

The molecules in a liquid can escape from the surface if their kinetic energy exceeds the potential energy that binds them to the remaining liquid. At any finite temperature, the speeds of the molecules have a distribution that ensures at any given instant that there is a finite probability that a molecule has enough energy to escape from the surface of the liquid. The escape of molecules from the surface of a liquid is called **evaporation**.

The molecules that escape from the surface of the liquid undergo collisions with the gas molecules above the liquid. There is some probability that the molecules will return to the liquid. This is called **condensation**. The rate of condensation increases with an increased number density of molecules in the gas above the liquid. In a closed container, with both liquid and vapor molecules, the number density of vapor molecules will eventually reach an equilibrium such that the condensation rate will equal the evaporation rate. The pressure in the gas caused by the molecules under conditions of equilibrium is called the **saturated vapor pressure**. The saturated vapor pressure for a given liquid depends on the temperature.

In the ideal gas model, the total pressure in a gas is the sum of the pressure for each of the constituents of the gas. For example, consider a gas containing two different types of molecules, A and B. If only the A molecules were present, the pressure would be $P_A$. If only the B molecules were present, the pressure would be $P_B$. Each of these pressures is called the **partial pressure** of the corresponding molecule. The pressure in the gas composed of both types of molecules will be the total of all the partial pressures, in this case

$$P = P_A + P_B.$$

The relative humidity is the ratio of the partial pressure of water in the air to the saturated vapor pressure at the temperature of the air:

$$\text{Relative humidity} = \frac{\text{Partial pressure } H_2O}{\text{Saturated Vapor pressure of } H_2O} \times 100\%$$

It is possible for the air to have a partial pressure of water vapor greater than the saturated vapor pressure. In this condition, the air is said to be **supersaturated** with water vapor.

The **dew point** is the temperature at which the existing partial pressure of water in the air would be equal to the saturated vapor pressure of water.

**Example 18-4-A**

If 0.500 kg of water evaporates into a room of dry air at a temperature of 20° C, what will the relative humidity be? The room is 3.00 m × 4.00 m × 2.5 m. Treat the water vapor as an ideal gas.

**Solution:**

First determine the number of moles of water that will be in the air after it evaporates (or alternatively, the number of molecules):

$$n = \frac{M}{m} = \frac{0.500 \, \text{kg}}{0.018 \, \text{kg}} = 27.8$$

Next, we can determine the partial pressure of the water vapor, treating it as an ideal gas:

$$P_{H_2O} = \frac{n_{H_2O} RT}{V} = \frac{(27.8)(8.31 \, \text{J/K})(293.15 \, \text{K})}{3.00 \, \text{m} \times 4.00 \, \text{m} \times 2.5 \, \text{m}} = 2.26 \times 10^3 \, \text{N/m}^2$$

The relative humidity is then calculated using this partial pressure and the saturated vapor pressure of water at this temperature:

$$\text{RH} = \frac{\text{Partial pressure } H_2O}{\text{Saturated Vapor pressure of } H_2O} \times 100\% = \frac{2.26 \times 10^3 \, \text{N/m}}{2.33 \times 10^3 \, \text{N/m}} \times 100\% = 97\%$$

About one pint of water is enough to almost saturate the air of a small room with water at 20° C!

**Section 18-5 Van der Waals Equation of State**

Corrections to the ideal gas model can be made to overcome the shortcomings of the ideal gas model in describing real gases. If the particles of the gas have a volume that is not negligible, a correction can be made to the ideal gas law by reducing the actual volume by the volume of the gas particles:

$$P(V - nb) = nRT$$

where $nb$ is the total volume of all the particles in the gas. This is called the **Clausius equation of state**. Notice that this is consistent with the earlier statements and assumptions of the ideal gas model. If the volume of the particles, $nb$, is negligible compared to the volume of the gas, $V$, then this reduces to the ideal gas law. In the other limit, when $V$

becomes equal to the volume of the gas particles, we see that the pressure will climb without limit. That is what would be expected for a gas made of rigid particles of finite size.

A correction can also be made to correct for the attractive force between the gas particles. The **van der Waals equation of state** makes such a correction to the ideal gas law:

$$\left( P + \frac{a}{(V/n)^2} \right)\left( \frac{V}{n} - b \right) = RT$$

where $a$ is constant proportional to the magnitude of the force of interaction between the gas particles.

### Example 18-5-A

The fit of van der Waals equation of state to a particular gas results in a constant $a = 2.6 \times 10^{-2} \, \text{Nm}^4/\text{mol}^2$ and a constant $b = 6.3 \times 10^{-5} \, \text{m}^3/\text{mol}$. At what temperature are the pressures given by the van der Waals equation of state and the ideal gas law identical if the molar density of the gas is 76 moles/m$^3$?

### Solution:

We will denote the pressure from the van der Waals equation of state as $P_{vdw}$ and the pressure from the ideal gas law as $P_i$. The van der Waals equation of state and the ideal gas law are, respectively,

$$\left( P_{vdw} + \frac{a}{(V/n)^2} \right)\left( \frac{V}{n} - b \right) = RT$$

$$P_i V = nRT$$

Solving each of these for the respective pressure:

$$P_{vdw} = \frac{RT}{\left( \dfrac{V}{n} - b \right)} - \frac{a}{(V/n)^2}$$

$$P_i = \frac{nRT}{V}$$

Setting the two pressures equal:

$$\frac{RT}{\left( \dfrac{V}{n} - b \right)} - \frac{a}{(V/n)^2} = \frac{nRT}{V}$$

Solving this for the temperature:

$$T = \frac{a}{R(V/n)^2}\left( \frac{1}{\left( \dfrac{V}{n} - b \right)} - \frac{n}{V} \right)$$

$$= \frac{\left( 2.6 \times 10^{-2} \, \text{Nm}^4/\text{mol}^2 \right)}{(8.31 \text{J/K})\left( 1/76 \, \text{mol/m}^3 \right)^2}\left( \frac{1}{\left( 1/76 \, \text{mol/m}^3 - 6.3 \times 10^{-5} \, \text{m}^3/\text{mol} \right)} - 76 \, \text{mol/m}^3 \right)$$

$$= 6.6 \, \text{K}$$

## Section 18-6 Mean Free Path

Because of the finite size of the particles in a real gas, the particles undergo collisions with each other. A quantity that is useful for understanding some phenomena associated with a gas is the **mean free path** of the particles in the gas. The mean free path is the average distance a gas particle travels after colliding with another gas particle before colliding with a second gas particle. It is easy to see that this depends on the size of the gas particles and the average spacing between them. For spherical gas particles that have a distribution of speeds given by the Maxwell distribution of speeds, the mean free path, $l_M$, is given by

$$l_M = \frac{1}{4\pi\sqrt{2}\,r^2(N/V)}$$

where $r$ is the radius of the spherical gas particles, $N$ is the number of gas particles, and $V$ is the volume of the gas.

## Example 18-6-A

Many processes are carried out in a vacuum so that the mean free path of some particle is large enough so that it can move from one part of the system to another with a small probability of colliding with a gas molecule. This condition can be met by decreasing the number density of the gas in the system so that the mean free path is less than the necessary path length. What should the pressure be in a system that needs a mean free path for air molecules of 10.0 cm at a temperature 300 K? Use a diameter of $3 \times 10^{-10}$ m for air molecules.

## Solution:

The mean free path is calculated from the number density of the molecules. We can rewrite the number density in terms of the pressure and temperature using the ideal gas law:

$$PV = NkT \quad \Rightarrow \quad \frac{N}{V} = \frac{P}{kT}$$

Replacing $N/V$ in the expression for the mean free path, we get

$$l_M = \frac{1}{4\pi\sqrt{2}\,r^2(N/V)} = \frac{kT}{4\pi\sqrt{2}\,r^2 P}$$

If we solve this for the pressure:

$$P = \frac{kT}{4\pi\sqrt{2}\,r^2 l_M} = \frac{\left(1.38\times10^{-23}\,\text{J/K}\right)\left(300\,\text{K}\right)}{4\pi\sqrt{2}\left(3\times10^{-10}\,\text{m}\right)^2\left(0.10\,\text{m}\right)} = 0.026\,\text{N/m}^2$$

This is less than one millionth of atmospheric pressure, but is easily obtainable with vacuum systems used for the processes that require this size of mean free path.

## Section 18-7 Diffusion

For gas particles to move from one place to another that is significantly more than a mean free path length away, they will follow a path which includes many collisions with other particles. As the number of particles and number of collisions per time in a gas becomes large, the only practical method of treating the motion of particles over distances of many mean free path lengths is statistically. The motion of gas particles over distances of many mean free path lengths is related to a mathematical problem called the random walk problem.

The rate of diffusion, $J$, of a given species of gas particle across a surface of area $A$ is related to the gradient of the concentration of that species of particle perpendicular to that area by the **diffusion equation**:

$$J = DA\frac{dC}{dx}$$

where $D$ is called the diffusion constant, $A$ is the area of the surface through which the species is diffusing. The concentration of the species is $C$ and the derivative of the concentration of the species with respect to position along a line perpendicular to the surface is $dC/dx$. The diffusion constant depends on the species that is diffusing and the material it is diffusing through. The diffusion constant also has dependence on temperature and density. The direction of the diffusion is from high concentration to low concentration.

### Example 18-7-A

Consider a room of volume $V$ connected to the outside air by a narrow passage way with a cross-sectional area $A$ and length $L$. In the room, there are perfume molecules uniformly distributed with a concentration $C_0$. Estimate the concentration of the perfume molecules in the room as a function of time. Assume the dimensions of the room are small compared to the length of the passage. The diffusion constant of the molecules is $D$.

### Solution:

The concentration in the room as a function of time will be C(t). The concentration at the outside end of the passage way will remain close to zero. This means that the gradient of the concentration through the passage will be approximately $dC/dx = C(t)/L$. The rate of diffusion out of the room will be

$$J = \frac{DAC(t)}{L}$$

But the rate of change of the number of perfume molecules in the room will be equal to the negative of the diffusion rate:

$$\frac{dN}{dt} = -J = -\frac{DAC(t)}{L}$$

The concentration in the room depends on $N$, the number of molecules in the room:

$$C(t) = \frac{N(t)}{V}$$

This implies

$$\frac{dC(t)}{dt} = \frac{1}{V}\frac{dN}{dt} = -\frac{DAC(t)}{L}$$

Separating the time and concentration variables on the two sides of the equation gives

$$\frac{dC(t)}{C(t)} = -\frac{DA}{L}dt$$

Both sides of this equation can be integrated:

$$\int_{C(0)}^{C(t)}\frac{dC(t)}{C(t)} = -\frac{DA}{L}\int_0^t dt \Rightarrow \ln C(t) - \ln C(0) = -\frac{DA}{L}t$$

Taking the exponential function of each side of the equation gives

$$C(t) = C(0)e^{-\frac{DAt}{L}}$$

The reason this is an approximate solution is that we have assumed that the concentration within the room is constant and the concentration near the outside opening is zero. In general, these are not exactly true.

## Practice Quiz

1.      Consider the three gas models given by the Clausius equation of state and the van der Waals equation of state. Which statement about pressure predictions of the models is definitely true for a given density and temperature? Assume the *a* and *b* constants of the Clausius and van der Waals models to be positive.

        a) Ideal gas pressure is greater than Clausius pressure.
        b) Van der Waals pressure is greater than Clausius pressure.
        c) Clausius pressure is greater than ideal gas pressure.
        d) Van der Waals pressure is greater than ideal gas pressure.

2.      Why is one of the assumptions kinetic theory of the ideal gas model that there are a large number of molecules?

        a) Many of the quantities derived require averages over the molecules.
        b) If you lose some molecules by leakage, it won't matter as much.
        c) So that the pressure will be large.
        d) So that there are molecules traveling in every direction.

3.      You place a 1 m tall beaker filled with water on a hot plate and a 10 cm beaker filled with water on the hot plate. Of course, it takes longer for the water in the 1 m tall beaker to boil because there is more water in the tall beaker to heat, but the temperature at the bottom of the tall beaker will be higher than the temperature at the bottom of the short beaker when the water is boiling. Why?

        a) The top of the tall beaker is at a lower atmospheric pressure.
        b) There is more water in the tall beaker.
        c) The pressure is higher at the bottom of the tall beaker.
        d) The volume of the large beaker is greater.

4.      Why is a humidifier necessary in many heated homes during the winter?

        a) The pressure of the air in the house drops as it is heated and the partial pressure of the added water is increased to help the indoor pressure match the outdoor pressure.
        b) As the air is heated, the average speed of the molecules in the air increases. Water molecules move slower than other air molecules and help to slow the other molecules down.
        c) Water molecules are less massive than most air molecules and therefore move faster than the other air molecules and therefore make it easier to warm up the air.
        d) As the air is heated, its relative humidity drops. To keep the relative humidity high enough for comfort, water is added to the air.

5.      By what factor does the RMS speed of an ideal gas change when the absolute temperature of the gas is doubled?

        a) 2
        b) 4
        c) ½
        d) √2

6.      Why is the diffusion constant for hydrogen molecules larger than the diffusion constant for oxygen molecules?

        a) Hydrogen is more reactive than oxygen.
        b) Hydrogen molecules are less massive and smaller than oxygen molecules.
        c) Hydrogen molecules have fewer protons than oxygen molecules.
        d) Hydrogen molecules are larger than oxygen molecules.

7.      Which if the following changes would increase the mean free path in a gas that is at constant volume (assume it remains above the critical temperature)?

        a) Increase the number of molecules per volume.
        b) Increase the temperature of the gas.
        c) Decrease the temperature of the gas.
        d) Decrease the radius of the particles.

8.    What condition exists when the partial pressure of a vapor exceeds the saturated vapor pressure of the vapor?

    a) Saturation
    b) Evaporation
    c) Boiling
    d) Supersaturation

9.    Why does it take longer to cook food in boiling water on top of mountain than at sea level?

    a) The water's saturated vapor pressure reaches the atmospheric pressure at a lower temperature at the top of a mountain than at sea level.
    b) The food has more potential energy stored in it at the top of the mountain.
    c) It is usually colder on top of a mountain.
    d) The relative humidity is low at the top of a mountain.

10.   A small drop of dye is placed in the center of a beaker of water that has a diameter $D$. It takes a time $T$ for the dye molecules to diffuse to the edge of the beaker. If the beaker diameter is $2D$, then, roughly, how much time does it take the dye molecules to diffuse to the edge of the beaker?

    *a) T*
    b) $2T$
    c) $3T$
    d) $4T$

11.   At what temperature would the RMS speed of hydrogen molecules be 1.00 cm/s?

12.   Using the Maxwell distribution of speeds, calculate the expected number of gas particles that have a kinetic energy between $E$ and $E + dE$ in a gas of $N$ particles at temperature $T$.

13.   Determine the total mass of water in a cubic meter of air at a relative humidity of 85% at a temperature 25° C.

14.   What is the number of molecules per cubic meter in a gas with a mean free path of 10 cm? Assume the effective diameter of the molecules is $3.0 \times 10^{-10}$ m.

15.   Determine the $a$ and $b$ constants of the van der Waals equation of state for 4 moles of a gas that has a pressure $1.000 \times 10^5$ N/m$^2$ at a temperature of 300.0 K and volume 0.0900 m$^3$ and a pressure $0.600 \times 10^5$ N/m$^2$ at a temperature of 240.0 K and the same volume.

## Problem Solutions

3.    The average kinetic energy depends on the temperature:
    $\frac{1}{2}mv_{rms}^2 = \frac{3}{2}kT$.
    If we form the ratio for the two temperatures, we have
    $(v_{rms2}/v_{rms1})^2 = T_2/T_1 = 373$ K/273 K, which gives $v_{rms2}/v_{rms1} = 1.17$.

7.    The average kinetic energy depends on the temperature:
    $\frac{1}{2}mv_{rms}^2 = \frac{3}{2}kT$.
    We form the ratio at the two temperatures, and use the ideal gas law:
    $(v_{rms2}/v_{rms1})^2 = T_2/T_1 = P_2V_2/P_1V_1 = P_2/P_1$ ;
    $(v_{rms2}/v_{rms1})^2 = 2$, which gives $v_{rms2}/v_{rms1} = \sqrt{2}$.

11.   (*a*)  The average kinetic energy depends on the temperature:
    $\frac{1}{2}mv_{rms}^2 = \frac{3}{2}kT$, which gives
    $v_{rms} = (3kT/m)^{1/2} = [3(1.38 \times 10^{-23}$ J/K$)(273$ K$)/(32$ u$)(1.66 \times 10^{-27}$ kg/u$)]^{1/2} = 461$ m/s.
    (*b*)  The molecule, on the average, will have a component in one direction less than the average speed. If we take the RMS speed as the average speed, from the analysis of the molecular motion, we know that $(v_x^2)_{av} = \frac{1}{3}v_{rms}^2$. Thus the time to go back and forth is
    $t = 2L/(v_x)_{av} = 2L\sqrt{3}/v_{rms}$.
    The frequency of collisions with one wall is
    $N = 1/t = v_{rms}/2L\sqrt{3} = (461$ m/s$)/2(7.0$ m$)\sqrt{3} = 19$ s$^{-1}$.

15. (a) The total number of molecules is represented by the area under the Maxwell distribution, which we find by integration:

$$\int f(v) \ dv = \int_0^\infty 4\pi N\left(\frac{m}{2\pi kT}\right)^{3/2} v^2 e^{-mv^2/2kT} \ dv$$

This can be simplified by a change in variable:

$$u = \left(\frac{m}{2\pi kT}\right)^{3/2} v \qquad du = \left(\frac{m}{2\pi kT}\right)^{3/2} dv$$

The integral becomes

$$\int f(v) \ dv = \int_0^\infty 4\pi N\left(\frac{1}{\pi}\right)^{3/2} u^2 e^{-u^2} \ du = 4\pi N\left(\frac{1}{\pi}\right)^{3/2}\left(\frac{\pi^{1/2}}{4}\right) = N$$

where we have used the result from integral tables.

(b) We make the same change in variable for the integral to find the RMS speed:

$$\frac{1}{N}\int f(v) \ dv = \int_0^\infty 4\pi\left(\frac{m}{2\pi kT}\right)^{3/2} v^4 e^{-mv^2/2kT} \ dv$$

$$= 4\pi\left(\frac{1}{\pi}\right)^{3/2}\left(\frac{2kT}{m}\right)\int_0^\infty u^4 e^{-u^2} \ du = 4\pi\left(\frac{1}{\pi}\right)^{3/2}\left(\frac{2kT}{m}\right)\left(\frac{3\pi^{1/2}}{8}\right) = \frac{3kT}{m}$$

19. (a) From Figure 18–6 we see that water is a gas at 220 atm and 100°C. As we lower the pressure, it becomes a liquid at 218 atm, and a vapor at 1.0 atm.

(b) Water is a gas at 220 atm and 0°C. As we lower the pressure, it becomes a liquid at 218 atm, a solid at 1.0 atm, and a vapor at a pressure < 0.006 atm.

23. We find the saturated vapor pressure from
$P = (RH)P_s$ ;
530 Pa = $0.40P_s$ , which gives $P_s$ = 1325 Pa.
This saturated vapor pressure corresponds to a temperature between 10°C and 15°C. We use the values at 10°C and 15°C to find the temperature;
$T = 10°C + [(15°C - 10°C)(1.325 \times 10^3 \text{ Pa} - 1.23 \times 10^3 \text{ Pa})/(1.71 \times 10^3 \text{ Pa} - 1.23 \times 10^3 \text{ Pa})] = 11°C$.

27. Because there is only steam in the autoclave, the saturated vapor pressure is the gauge pressure plus atmospheric pressure:
1.0 atm + 1.0 atm = 2.0 atm = $2.03 \times 10^5$ Pa.
From Table 18–2 we see this saturated vapor pressure occurs at 120°C.

31. (a) We write the Van der Waals equation as
$P = -(an^2/V^2) + nRT/(V - nb)$.
When we differentiate, with the temperature constant, we get
$dP/dV = (2an^2/V^3) - nRT/(V - nb)^2$.
At the critical point, we have
$dP/dV = 0 = (2an^2/V_{cr}^3) - nRT_{cr}/(V_{cr} - nb)^2 = [2an^2(V_{cr} - nb)^2 - nRT_{cr}V_{cr}^3]/V_{cr}^3(V_{cr} - nb)^2$, so
$2an(V_{cr} - nb)^2 = RT_{cr}V_{cr}^3$.
For the second derivative we get
$d^2P/dV^2 = -(6an^2/V^4) + 2nRT/(V - nb)^3$.
Because the critical point is an inflection point, we have
$d^2P/dV^2 = 0 = -(6an^2/V_{cr}^4) + 2nRT_{cr}/(V_{cr} - nb)^3 = [6an^2(V_{cr} - nb)^3 - 2nRT_{cr}V_{cr}^4]/V_{cr}^4(V_{cr} - nb)^3$, so
$3an(V_{cr} - nb)^3 = RT_{cr}V_{cr}^4$.
When we divide the two equations, we get
$3(V_{cr} - nb) = 2V_{cr}$, or $V_{cr} = 3nb$.
If w use this result in one of the equations, we get
$2an(3nb - nb)^2 = RT_{cr}(3nb)^3$, which gives $T_{cr} = 8a/27bR$.
For the pressure at the critical point, we have
$P_{cr} = -(an^2/V_{cr}^2) + nRT_{cr}/(V_{cr} - nb) = -[an^2/(3nb)^2] + nR(8a/27bR)/(3nb - nb) = a/27b^2$.

(*b*)  We find *b* from

$T_{cr}/P_{cr} = (8a/27bR)/(a/27b^2) = 8b/R$;

$(304 \text{ K})/(72.8 \text{ atm})(1.013 \times 10^5 \text{ N/m}^2 \cdot \text{atm}) = 8b/(8.315 \text{ J/mol} \cdot \text{K})$,

which gives $b = 4.28 \times 10^{-5} \text{ m}^3/\text{mol}$.

We find *a* from

$T_{cr}^2/P_{cr} = (8a/27bR)^2/(a/27b^2) = 64a/27R^2$;

$(304 \text{ K})^2/(72.8 \text{ atm})(1.013 \times 10^5 \text{ N/m}^2 \cdot \text{atm}) = 64a/27(8.315 \text{ J/mol} \cdot \text{K})^2$,

which gives $a = 0.365 \text{ N} \cdot \text{m}^4/\text{mol}^2$.

35.  The length of a side of the cube is

$L = V^{1/3} = (4.4 \times 10^{-3} \text{ m}^3)^{1/3} = 0.164 \text{ m}$.

(*a*)  We assume that vigorous shaking will disperse the marbles throughout the box.  Because the diameter of a marble $\approx 0.1L$ and there are on average about four marbles along a side,  we use the approximation of an ideal gas:

$l_M = 1/4\pi\sqrt{2}r^2(N/V) = 1/4\pi\sqrt{2}(0.75 \times 10^{-2} \text{ m})^2(70/4.4 \times 10^{-3} \text{ m}^3) = 6.29 \times 10^{-2} \text{ m} = 6.3 \text{ cm}$.

(*b*)  If the box is slightly shaken , all marbles will be on the floor of the box.  There are on average about eight marbles along a side. Because $8(1.5 \text{ cm}) = 12 \text{ cm} < L$, there will be space between marbles.  The volume occupied by the marbles is

$V_b = A(2r) = 2L^2r$.

We assume we can still use the expression for the mean free path:

$l_{Mb} = 1/4\pi\sqrt{2}r^2(N/V_b) = 2L^2r/4\pi\sqrt{2}r^2N = \sqrt{2}L^2/4\pi rN$

$= \sqrt{2}(0.164 \text{ m})^2/4\pi(0.75 \times 10^{-2} \text{ m})(70) = 5.8 \times 10^{-3} \text{ m} = 0.58 \text{ cm}$.

39.  (*a*)  If the average speed of a molecule is $\mu$, the average time between collisions is

$\Delta t = l_M/\mu$,

so the frequency of collisions is

$f = 1/\Delta t = \mu/l_M$.

When we use the expression for the mean free path, we get

$f = \mu/[1/4\pi\sqrt{2}r^2(N/V)] = 4\sqrt{2}\pi r^2\mu N/V$.

(*b*)  We make use of the ideal gas law and the Maxwell distribution to get

$f = 4\sqrt{2}\pi r^2\mu N/V = 4\sqrt{2}\pi r^2(8kT/\pi m)^{1/2}P/kT = 16r^2P(\pi/kTm)^{1/2}$

$= 16(1.5 \times 10^{-10} \text{ m})^2(1.0 \times 10^{-2} \text{ atm})(1.013 \times 10^5 \text{ Pa}) \times$

$[\pi/(1.38 \times 10^{-23} \text{ J/K})(293 \text{ K})(28 \text{ u})(1.66 \times 10^{-27} \text{ kg/u})]^{1/2} = 4.7 \times 10^7 \text{ s}^{-1}$.

43.  From the result of Example 18–8, we have

$$t = \left(\frac{\overline{C}}{\Delta C}\right)\frac{(\Delta x)^2}{D} = \frac{1}{2}\frac{(1.5 \text{ m})^2}{4.0 \times 10^{-5} \text{ m}^2/\text{s}} = 2.8 \times 10^4 \text{ s} = 7.8 \text{ h}$$

Our experience is that the odor is detected much sooner than this, which means that convection is much more important than diffusion.

47.  The average kinetic energy depends on the temperature:

$\tfrac{1}{2}mv_{rms}^2 = \tfrac{3}{2}kT$, which gives

$v_{rms} = (3kT/m)^{1/2} = [3(1.38 \times 10^{-23} \text{ J/K})(2.7 \text{ K})/(1 \text{ u})(1.66 \times 10^{-27} \text{ kg/u})]^{1/2} = 2.6 \times 10^2 \text{ m/s}$.

We find the pressure from

$PV = NkT$,  or

$P = (N/V)kT = (1 \text{ atom/cm}^3)(10^6 \text{ cm}^3/\text{m}^3)/(1.38 \times 10^{-23} \text{ J/K})(2.7 \text{ K})$

$= 4 \times 10^{-17} \text{ N/m}^2 \approx 4 \times 10^{-22} \text{ atm}$.

51. We find the average volume occupied by a molecule from the ideal gas law:

$PV = NkT$,   or

$V/N = kT/P = (1.38 \times 10^{-23} \text{ J/K})(273 \text{ K})/(1.013 \times 10^5 \text{ Pa}) = 3.7 \times 10^{-26}$ m$^3$/molecule.

The volume of a molecule is

$V_{molecule} = \frac{4}{3}\pi r^3 = \frac{4}{3}\pi (0.1 \times 10^{-9} \text{ m})^3 = 4 \times 10^{-30}$ m$^3 \approx 0.01\%$ of average volume occupied by a molecule.

Thus the assumption is reasonable.

If we consider the average volume occupied by a molecule to be a cube, the side of the cube (which we can use as the average distance between molecules) is

$l = (3.7 \times 10^{-26} \text{ m}^3)^{1/3} = 3.3 \times 10^{-9}$ m.

If we scale the diameter of a molecule to the size of a ping-pong ball, for the average distance between molecules we have

$D = [(4 \text{ cm})/(0.2 \times 10^{-9} \text{ m})](3.3 \times 10^{-9} \text{ m}) \approx 50$ cm.

55. For the two conditions of the gas in the cylinder, we can write

$P_1 V = nRT_1$,   and   $P_2 V = nRT_2$, which can be combined to give

$P_2/P_1 = T_2/T_1$ ;

$P_2/P_1 = (563 \text{ K}/393 \text{ K}) = 1.43$.

The average kinetic energy depends on the temperature:

$\frac{1}{2}mv_{rms}^2 = \frac{3}{2}kT$.

We form the ratio for the two temperatures:

$(v_{rms2}/v_{rms1})^2 = T_2/T_1 = (563 \text{ K}/393 \text{ K})$, which gives $v_{rms2}/v_{rms1} = 1.20$.

59. (a) From Table 18–2 the saturated vapor pressure at 30°C is $4.24 \times 10^3$ Pa.

At 40% humidity the water vapor pressure is

$P = 0.40P_s = 0.40(4.24 \times 10^3 \text{ Pa}) = 1.7 \times 10^3$ Pa.

(b) From Table 18–2 the saturated vapor pressure at 5°C is $8.72 \times 10^2$ Pa.

At 80% humidity the water vapor pressure is

$P = 0.80P_s = 0.80(8.72 \times 10^2 \text{ Pa}) = 7.0 \times 10^2$ Pa.

The ratio of summer to winter is 2.4.

# Chapter 19: Heat and the First Law of Thermodynamics

## Chapter Overview and Objectives

This chapter defines heat as the flow of energy between bodies at different temperatures and presents the first law of thermodynamics. It also introduces specific heat and the latent heat of phase change. It shows how to calculate the work done by a system in general and for several specific thermodynamic processes. This chapter discusses adiabatic processes in ideal gases. The mechanisms of heat transfer are also introduced in this chapter.

After completing study of this chapter, you should:

- Know what heat is.
- Know the first law of thermodynamics.
- Know the relationship between heat, specific heat, and change of temperature.
- Know what Calories, calories, and kilocalories are and how they are related to the Joule.
- Know how the internal energy of a monatomic ideal gas is related to temperature.
- Know how to calculate the work done by a system during a process.
- Know the definitions of isothermal, adiabatic, isochoric, and isobaric processes.
- Know what is meant by equipartition of energy.
- Know what the three mechanisms of heat transfer are and how conductivity and radiation rates depend on temperature.

## Summary of Equations

Mechanical equivalent of heat:
$$4.186\,J = 1\,calorie$$
$$4186\,J = 1\,kilocalorie$$

Internal energy of ideal monatomic gas:
$$U = \tfrac{3}{2}nRT$$

Relationship between heat added and temperature change:
$$dQ = mc(T)\,dT$$
$$Q = mc\Delta T$$

Latent heat of phase change:
$$Q = mL$$

First law of thermodynamics:
$$\Delta U = Q - W$$

Differential form of first law of thermodynamics:
$$dU = dQ - dW$$

Work done by a system during a process:
$$W = \int_{V_i}^{V_f} P\,dV$$

Work done by ideal gas during isothermal process:
$$W_{\substack{isothermal \\ ideal\,gas}} = nRT \ln \frac{V_b}{V_a}$$

Work done during isobaric process:
$$W_{isobaric} = P(V_B - V_A)$$

Heat added during adiabatic process:
$$Q_{adiabatic} = 0$$

Work done during isochoric process:
$$W_{isochoric} = 0$$

Relationship between $C_P$ and $C_V$:
$$C_P - C_V = R$$

Definition of adiabatic gas constant:    $\gamma = \dfrac{C_P}{C_V}$

Relationship of pressure and volume during adiabatic process:

$$PV^\gamma = \text{constant}$$

Rate of heat conduction:

$$\frac{dQ}{dt} = kA\frac{dT}{dx}$$

$$\frac{\Delta Q}{\Delta t} = kA\frac{\Delta T}{l}$$

Rate of radiation from a body:

$$\frac{dQ}{dt} = e\sigma AT^4$$

Net rate of radiative heat transfer to a body at temperature $T_1$ in surroundings at temperature $T_2$:

$$\frac{dQ}{dt} = e\sigma A\left(T_2^4 - T_1^4\right)$$

## Chapter Summary

### Section 19-1 Heat as Energy Transfer

**Heat** is a measure of the spontaneous flow of energy from one material to another driven by temperature difference. Heat flows from materials at higher temperature to materials at lower temperature. Heat can be measure in Joules, but is often measured in **calories** or **kilocalories**. A calorie is the amount of heat needed to raise the temperature of one gram of water from 14.5 C to 15.5 C. A kilocalorie is 1000 calories and is sometimes written as **Calorie**. Care must be taken to notice whether the heat unit is calorie or Calorie, it is often difficult to determine because handwritten uppercase C's are similar to lowercase c's.

It can be determined experimentally that the **mechanical equivalent of heat** is that 4.186 J of mechanical work will transfer the same amount of energy to an object as 1 calorie of heat. Writing this in equation form:

$$4.186\,J = 1\,\text{calorie}$$

or, in terms of kilocalories:

$$4186\,J = 1\,\text{kilocalorie}$$

### Section 19-2 Internal Energy

**Internal energy** is the kinetic and potential energy associated with the random motion of atoms and molecules. Sometimes internal energy is also called **thermal energy**. The internal energy, $U$, of a monatomic ideal gas is given by:

$$U = \tfrac{3}{2}nRT$$

where $n$ is the number of moles of gas, $R$ is the gas constant, 8.315 J/K, and $T$ is the absolute temperature. For polyatomic ideal gases, the internal energy at a given temperature is higher than for a monatomic ideal gas because vibrational and rotational energies are present in addition to translational kinetic energy.

The internal energy of liquids and solids is difficult to determine because of the large contribution of potential energy due to the interatomic forces making contributions of the same order of magnitude as the kinetic energy terms.

### Example 19-2-A

A monatomic ideal gas is at a pressure of $1.00 \times 10^5$ N/m$^2$ and has a volume 1.65 m$^3$. The gas has energy added to it so its new pressure is $2.00 \times 10^5$ N/m2 and its new volume is 2.44 m$^3$. How much energy was added to the gas?

**Solution:**

We know that the internal energy $U$ is related to the temperature:

$$U = \tfrac{3}{2}nRT$$

but we can relate the temperature to the pressure and volume using the ideal gas law:

$$PV = nRT \quad \Rightarrow \quad U = \tfrac{3}{2}PV$$

The change in energy of the gas will be the final energy minus the initial energy:

$$\begin{aligned}
\Delta U = U_2 - U_1 &= \tfrac{3}{2}\left(P_2 V_2 - P_1 V_1\right) \\
&= \tfrac{3}{2}\left[\left(2.00 \times 10^5 \text{ N/m}^2\right)\left(2.44 \text{ m}^3\right) - \left(1.00 \times 10^5 \text{ N/m}^2\right)\left(1.65 \text{ m}^3\right)\right] \\
&= 4.85 \times 10^5 \text{ N/m}^2
\end{aligned}$$

## Section 19-3 Specific Heat

At most temperatures, the amount of heat added to an object causes an approximately proportional change in the temperature. The amount of heat added for a particular temperature change is also proportional to the mass of the material. We can write this relationship as

$$dQ = mc(T)dT$$

where $dQ$ is the amount of heat added to the object, $m$ is the mass of the object, $c(T)$ is called the **specific heat** (it depends on the material and the temperature and pressure), and $dT$ is the change in temperature of the object.

If a finite amount of heat, $Q$, is added to a material, the change in temperature can be related to the heat by integrating the above relationship:

$$Q = \int dQ = \int mc(T)dT$$

In many circumstances, the specific heat is approximately constant over the range of temperatures encountered. In this case, we can factor the specific heat out of the integral:

$$Q = mc\int dT = mc\left(T_f - T_i\right)$$

or

$$Q = mc\Delta T$$

where $\Delta T$ is the change in temperature of the object.

## Example 19-3-A

A typical water heater for a home might hold 40 gallons of water. If water enters the water heater at 55° F and is warmed by the water heater to 105° F, how much energy is required?

**Solution:**

We know the specific heat of water is 4186 J/kg·C°. We know the amount of water is 40 gallons. The mass of the water is

$$m = \rho V = \left(1000 \text{ kg/m}^3\right)\left(40 \text{ gal}\right)\left(0.00378 \text{ m}^3/\text{gal}\right) = 151 \text{ kg}$$

The temperature difference in Celsius degrees is 5/9 of its temperature difference in Fahrenheit degrees:

$$\Delta T_C = \tfrac{5}{9}\Delta T_F = \tfrac{5}{9}\left(105^\circ F - 55^\circ F\right) = 27.8^\circ C$$

The heat added to the water is

$$Q = mc\Delta T = \left(151\,kg\right)\left(4186\,J/kg\cdot C^\circ\right)\left(27.8^\circ C\right) = 1.8\times10^7\,J$$

### Section 19-4 Calorimetry—Solving Problems

A common problem of interest in thermodynamics is placing two systems together in thermal contact. Thermal contact exists when heat is able to flow from one system to another. Systems in thermal contact proceed toward thermal equilibrium with each other. Remember, thermal equilibrium exists when two systems are at the same temperature. Often what we are interested in finding is the equilibrium temperature of the two systems in contact. Many times, we can treat the two systems in thermal contact as isolated, that is, the two systems are not in thermal contact with any other system and no mechanical work is done on the system. Applying the law of conservation of energy implies that whatever heat is lost by one system must be gained by the other system. Writing this down in an equation:

$$Q_1 + Q_2 = 0$$

Where $Q_1$ is the heat added to the first system and $Q_2$ is the heat added to the second system. Remember that $Q$ always means the heat added to a system. Any heat lost will result in a negative $Q$.

### Example 19-4-A

A mass of 0.349 kg of a solid at 55° C are added to 0.894 kg of water at 22° C. The final temperature of the mixture is 28° C. What is the specific heat of the solid? Assume no heat is exchanged with the environment.

### Solution:

No heat is exchanged with the environment so the heat added to the water plus the heat added to the solid is zero:

$$Q_w + Q_s = 0 \quad\Rightarrow\quad m_w c_w \Delta T_w + m_s c_s \Delta T_s = 0$$

where the $w$ subscripts refer to the water and the $s$ subscripts refer to the solid. The final temperature of the two substances is the same, $T_f$. Rewriting the above equation in terms of initial and final temperatures:

$$m_w c_w\left(T_f - T_{iw}\right) + m_s c_s\left(T_f - T_{is}\right) = 0$$

Solving this for the specific heat of the solid:

$$c_s = \frac{-m_w c_w\left(T_f - T_{iw}\right)}{m_s\left(T_f - T_{is}\right)} = -\frac{\left(0.894\,kg\right)\left(1.00\,kcal/kg\cdot C^\circ\right)\left(28^\circ C - 22^\circ C\right)}{\left(0.349\,kg\right)\left(28^\circ C - 55^\circ C\right)} = 0.57\,kcal/kg\cdot C^\circ$$

### Section 19-5 Latent Heat

When a material changes from a solid to a liquid, liquid to a solid, gas to a liquid, or liquid to a gas we say the material undergoes a **change of phase**. There are also other types of phase changes that take place that are not as obvious to us. There are several liquid crystal phases. There are superconducting and superfluid phases. There are several other different types of phases of matter. All phase changes are accompanied by some type of discontinuity in a property of the material.

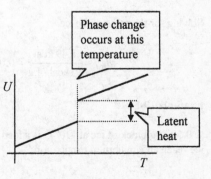

Many of the phase changes, but not all, have a discontinuity in their internal energy as a function of temperature at a phase transition. This means when a material makes a change of phase of this type, energy must be added to or removed from one phase, without temperature changing, to create the other phase before the temperature can change. This energy difference per mass of material between the two phases is called the **latent heat** of phase change. We can write the amount of heat, $Q$, needed to change the phase of a mass $m$ of material as

$$Q = mL$$

$L$ is the latent heat of phase change.

The **heat of fusion** is the latent heat required to melt a solid. The **heat of vaporization** is the latent heat required to vaporize a liquid.

When two systems are in thermal contact and proceeding toward equilibrium, latent heats must be accounted for if any phase change is encountered as the temperatures head toward the equilibrium temperature.

### Example 19-5-A

How much ice at $-12°$ C must be added to 0.566 kg of water at 20° C to cool the water down to 5° C?

**Solution:**
As the ice has heat added to it from the water, it first warms to the freezing point. During this process, the heat added to the ice will be

$$Q_1 = m_{ice}c_{ice}\Delta T = m_{ice}\left(0.50\,\text{kcal}/\text{kg}\cdot\text{C}°\right)\!\left(0°\,\text{C}-\left(-12°\,\text{C}\right)\right)$$
$$= m_{ice}\left(6.0\,\text{kcal}/\text{kg}\right)$$

The ice will then melt at 0° C. The amount of heat added during this process will be

$$Q_2 = m_{ice}L_{fusion} = m_{ice}\left(80\,\text{kcal}/\text{kg}\right)$$

The melted ice will then increase in temperature to the final temperature. The heat added during this process will be:

$$Q_3 = m_{ice}c_{water}\Delta T = m_{ice}\left(1.0\,\text{kcal}/\text{kg}\cdot\text{C}°\right)\!\left(5°\,\text{C}-0°\,\text{C}\right) = m_{ice}\left(5.0\,\text{kcal}/\text{kg}\right)$$

The total heat added to the ice will be

$$Q_{ice} = Q_1 + Q_2 + Q_3 = m_{ice}\left(91\,\text{kcal}/\text{kg}\right)$$

The heat to cool the water from its initial temperature to its final temperature will be

$$Q_{water} = mc\Delta T = \left(0.566\,\text{kg}\right)\!\left(1.0\,\text{kcal}/\text{kg}\cdot\text{C}°\right)\!\left(5°\,\text{C}-20°\text{C}\right) = -8.49\,\text{kcal}$$

The total heat added to the mixture of ice and water is zero:

$$Q = Q_{ice} + Q_{water} = 0 \quad\Rightarrow\quad m_{ice}\left(91\,\text{kcal}/\text{kg}\right)+\left(-8.49\,\text{kcal}\right)=0$$

Solving for the mass of the ice:

$$m_{ice} = \frac{8.49\,\text{kcal}}{91\,\text{kcal}/\text{kg}} = 0.093\,\text{kg}$$

### Example 19-5-B

A 0.500 kg block of ice at $-10°$ C is added to 0.500 kg of water at 20° C. How much ice will there be when the water and ice reach an equilibrium temperature?

**Solution:**

The ice will warm to a temperature of 0° C before any ice melts. The heat to warm the ice to 0° C will be

$$Q_1 = m_{ice} c_{ice} \Delta T = (0.500 \text{ kg})(0.50 \text{ kcal} / \text{kg} \cdot \text{C}°)(0° \text{C} - (-10° \text{C}))$$
$$= 2.5 \text{ kcal}$$

The heat to cool the water to 0° C will be

$$Q_2 = m_{water} c_{water} \Delta T = (0.500 \text{ kg})(1.00 \text{ kcal} / \text{kg})(0° \text{C} - 20° \text{C})$$
$$= -10.0 \text{ kcal}$$

The total of all the heat transferred will be the heat to warm the ice to 0° C, the heat to cool the water to 0° C, and the heat to melt some ice. The total of the heat transferred must be zero:

$$Q_1 + Q_2 + Q_{melt} = 0 \quad \Rightarrow \quad Q_{melt} = -Q_1 - Q_2 = -2.5 \text{ kcal} - (-10.0 \text{ kcal}) = 7.5 \text{ kcal}$$

The amount of ice melted by this much heat is determined from the latent heat of fusion of water:

$$Q_{melt} = m_{melted} L_{fusion} \quad \Rightarrow \quad m_{melted} = \frac{Q_{melt}}{L_{fusion}} = \frac{7.5 \text{ kcal}}{80 \text{ kcal/kg}} = 0.094 \text{ kg}$$

This will leave 0.500 kg – 0.094 kg = 0.406 kg of ice not melted when the water and ice reach equilibrium.

### Section 19-6 The First Law of Thermodynamics

The first law of thermodynamics is a statement of the law of conservation of energy applied to the internal energy of a system. Here a **system** is whatever you identify as being considered in the group of things for which you are calculating the total energy. A **closed system** is a system that does not lose or gain matter. An **open system** can exchange matter with the rest of the universe. A system can exchange energy with the remainder of the universe either by exchange of heat because of a temperature difference or by work being done, either by the universe on the system or by the system on the universe. We will adopt the convention that we will use the quantity of *work done by the system on the universe, W*, in our equations. We will also use the convention that we will use the *heat added to the system from the universe, Q*, in our equations. We can write that the change in internal energy, $\Delta U$, of a system is related to the work done by the system on the universe and the heat added to the system from the universe as

$$\Delta U = Q - W$$

This statement is the first law of thermodynamics. Sometimes we write the first law of thermodynamics in differential form:

$$dU = dQ - dW$$

### Example 19-6-A

While the temperature of a gas remains constant, 536 J of heat are added to the gas. How much work is done by the gas?

**Solution:**

We must recognize that a gas at constant temperature has no change in internal energy, $\Delta U = 0$. Then we can use the first law of thermodynamics to calculate the work done by the gas:

$$\Delta U = Q - W \quad \Rightarrow \quad W = Q - \Delta U = 536 \text{ J} - 0 = 536 \text{ J}$$

### Section 19-7 Applying the First Law of Thermodynamics; Calculating the Work

It is very useful to consider processes that occur in the system under consideration by following the process on a pressure–volume diagram because it helps us to determine the work done by the system during any process. The work done on the universe, $W$, when a system changes volume from $V_i$ to $V_f$ is

$$W = \int_{V_i}^{V_f} P \ dV$$

where $P$ is the pressure in the system. The pressure–volume diagram is useful because the area under the locus points the system follows during the process is the work done on the universe during that process.

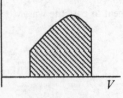

We next consider the work done during some specific types of processes.

**Isothermal Processes** are processes that occur while the temperature remains constant. During an isothermal process in a closed ideal gas system, the product of the pressure and volume remain constant. On the pressure–volume diagram, the locus of points is a hyperbola. We can solve for the work done during an isothermal process on an ideal gas:

$$W_{\substack{isothermal \\ ideal\ gas}} = \int_{V_A}^{V_B} P\ dV = \int_{V_A}^{V_B} \frac{nRT}{V}\ dV = nRT \ln \frac{V_b}{V_a}$$

where in the second step above we have used the ideal gas law. Because the internal energy, U, depends on the temperature, it remains constant during an isothermal process. Using the first law of thermodynamics

$$\Delta U = Q - W \quad \Rightarrow \quad 0 = Q - W \quad \Rightarrow \quad Q = W$$

**Adiabatic processes** are processes in which no heat enters or leaves the system, $Q = 0$.

**Isochoric processes** are processes in which the volume remains constant. If the volume remains constant, no work is done by the system, $W = 0$.

Isobaric processes are processes in which the pressure remains constant. It is relatively simple to calculate the work done during an isobaric process:

$$W_{isobaric} = \int_{V_A}^{V_B} P\ dV = P(V_B - V_A)$$

### Example 19-7-A

A hand pump has a volume of 740 cm$^3$ when the handle is pulled back and the chamber filled with an ideal gas at atmospheric pressure, $1.01 \times 10^5$ N/m$^2$, and a temperature of 300 K. The handle is pushed down and the air compressed to a volume of 148 cm$^3$. The compression is done slowly so that the process is isothermal. How much work will have been done in compressing the gas and what will the final pressure of the gas be?

### Solution:

The process is isothermal so the work done is

$$W = nRT \ln \frac{V_b}{V_a}$$

We are not given the number of moles of gas, but we can calculate it using the ideal gas law:

$$P_a V_a = nRT \quad \Rightarrow \quad n = \frac{P_a V_a}{RT}$$

We can use this to write the work done *by the gas* as

$$W = P_a V_a \ln\frac{V_b}{V_a} = \left(1.01\times10^5\ \text{N/m}^2\right)\left(7.40\times10^{-4}\ \text{m}^3\right)\ln\frac{1.48\times10^{-4}\ \text{m}^3}{7.40\times10^{-4}\ \text{m}^3}$$

$$= -120\,\text{J}$$

This is the work done by the gas, so the work done to compress the gas by the outside force must be +120 J.

To calculate the final pressure, we can use Boyle's law because the final temperature is the same as the initial temperature:

$$P_b V_b = P_a V_a \Rightarrow P_b = \frac{P_a V_a}{V_b} = \frac{\left(1.01\times10^5\ \text{N/m}^2\right)\left(740\,\text{cm}^3\right)}{148\,\text{cm}^3} = 5.05\times10^5\ \text{N/m}^2$$

## Section 19-8 Molar Specific Heats for Gases, and the Equipartition of Energy

The molar specific heat of a substance is the heat that must be added to one mol of the substance to raise the temperature 1 C°. The molar specific heat at constant volume, $C_V$, applies to processes for which the volume of the substance remains constant:

$$Q = nC_V\Delta T \qquad \text{Constant volume}$$

The molar specific heat at constant pressure applies to processes in which the pressure remains constant:

$$Q = nC_P\Delta T \qquad \text{Constant pressure}$$

There is a simple relationship between $C_P$ and $C_V$:

$$C_P - C_V = R$$

where $R$ is the gas constant, 8.315 J/mol·K. For a monatomic ideal gas, $C_V = {}^3/_2R$ and $C_P = {}^5/_2R$.

In general, each **degree of freedom** of a system in thermal equilibrium will have an average energy $^1/_2kT$. This is called the **principle of equipartition of energy**.

## Section 19-9 Adiabatic Expansion of a Gas

During an adiabatic expansion of an ideal gas, the pressure and volume of the gas will be related by

$$PV^\gamma = \text{constant}$$

where $\gamma$ is the ratio of the specific heat at constant pressure to the specific heat at constant volume, $C_P/C_V$.

## Example 19-9-A

A hand pump has a volume of 740 cm³ when the handle is pulled back and the chamber filled with an ideal gas at atmospheric pressure, $1.01\times10^5$ N/m², and a temperature of 300 K. The handle is pushed down and the air compressed to a volume of 148 cm³. The compression is done rapidly so that the process is adiabatic. How much work will have been done in compressing the gas and what will the final pressure and temperature of the gas be?

**Solution:**

We know that during an adiabatic compression,

$$PV^\gamma = \text{constant} = P_a V_a^\gamma \qquad \Rightarrow \qquad P = \frac{P_a V_a^\gamma}{V^\gamma}$$

We can use this to calculate the work done during an adiabatic compression:

$$W = \int_{V_a}^{V_b} P\, dV = \int_{V_a}^{V_b} \frac{P_a V_a^{\gamma}}{V^{\gamma}}\, dV = P_a V_a^{\gamma} \frac{V^{1-\gamma}}{(1-\gamma)}\Big|_{V_a}^{V_b} = \frac{P_a V_a^{\gamma}}{(1-\gamma)}\left(V_b^{1-\gamma} - V_a^{1-\gamma}\right)$$

For a monatomic ideal gas, the adiabatic gas constant is 5/3:

$$W = \frac{P_a V_a^{5/3}}{(1-5/3)}\left(V_b^{1-5/3} - V_a^{1-5/3}\right) = -\tfrac{3}{2}\left(1.01\times10^5 \text{ N}/\text{m}^2\right)\left(7.40\times10^{-4} \text{ m}^3\right)^{5/3}$$

$$\times\left[\left(1.48\times10^{-4} \text{ m}^3\right)^{-2/3} - \left(7.40\times10^{-4} \text{ m}^3\right)^{-2/3}\right]$$

$$W = -216 \quad \text{J}$$

This is the work done by the gas, so the work done on the gas is +216 J. Compare this to the work done in an isothermal process between the same two states of the gas in Example 19-7-A. Why is the work more during an adiabatic compression than during an isothermal compression?

We use the relationship between pressure and volume in adiabatic processes to determine the final pressure:

$$P_b V_b^{\gamma} = P_a V_a^{\gamma} \quad \Rightarrow \quad P_b = \frac{P_a V_a^{\gamma}}{V_b^{\gamma}} = \frac{\left(1.01\times10^5 \text{ N}/\text{m}^2\right)\left(740 \text{ cm}^3\right)^{5/3}}{\left(148 \text{ cm}^3\right)^{5/3}} = 1.48\times10^6 \text{ N}/\text{m}^2$$

We can use the ideal gas law to find the final temperature:

$$\frac{P_b V_b}{T_b} = \frac{P_a V_a}{T_a} \Rightarrow T_b = \frac{P_b V_b}{P_a V_a} T_a = \frac{\left(1.48\times10^6 \text{ N}/\text{m}^2\right)\left(148 \text{ cm}^3\right)}{\left(1.01\times10^5 \text{ N}/\text{m}^2\right)\left(740 \text{ cm}^3\right)}(300\,\text{K}) = 879\,\text{K}$$

## Section 19-10 Heat Transfer: Conduction, Convection, Radiation

There are three mechanisms of heat transfer.

**Conduction** is transfer of heat by the direct contact of atoms from a higher temperature substance with those of a lower temperature substance. The rate at which heat flows through a given cross-section of material by conduction, $dQ/dt$, depends on the temperature gradient, $dT/dx$, perpendicular to the area, the cross-sectional area, $A$, through which the heat is transferred, and a property of the material:

$$\frac{dQ}{dt} = kA\frac{dT}{dx}$$

where $k$ is a property of the material through which the heat is flowing called the **thermal conductivity**. If the gradient of the temperature through a uniform cross-section piece of material is approximately constant, we can write this as

$$\frac{dQ}{dt} = kA\frac{\Delta T}{l}$$

where $\Delta T$ is the temperature difference between two points separated by distance $l$ in the material.

Materials with a relatively high value of thermal conductivity are called **conductors** and those with a relatively low conductivity are called **insulators**.

**Example 19-10-A**

An arrangement of two layers of materials separates a system at temperature $T_1$ and a system at temperature $T_2$ as shown in the diagram. The two layers have equal thickness $d$ and area $A$. The first layer has a conductivity $k_1$ and the second a conductivity $k_2$. What is the rate of heat transfer from the system at temperature $T_1$ to the system at temperature $T_2$?

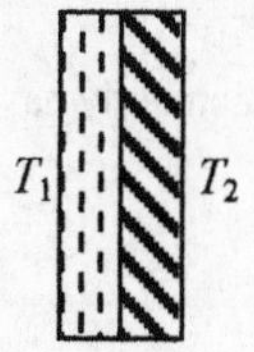

**Solution:**

The key to this problem is that the rate of heat transfer through each of the two layers must be identical, otherwise energy would accumulate at the boundary between the two layers. Let $T_m$ be the temperature at the interface between the two layers. Then the rate of heat transfer from the system at temperature $T_1$ to the interface through the first layer will be

$$\frac{dQ_1}{dt} = k_1 A \frac{T_1 - T_m}{d}$$

and the rate of heat transfer from interface to the body at temperature $T_2$ through the second layer will be

$$\frac{dQ_2}{dt} = k_1 A \frac{T_m - T_2}{d}$$

Solving the two equations above to eliminate the temperature at the interface and using that the two rates of heat transfer must be identical:

$$\frac{dQ}{dt} = \frac{k_1 k_2}{k_1 + k_2} A(T_1 - T_2)$$

**Convection** is the transfer of heat in fluids by the macroscopic motion of the fluid. If the motion of the fluid is driven by the density difference due to a temperature difference within the fluid, the convection is called **natural convection**. This is the phenomenon that most people are referring to when they say "heat rises". What they should say is that higher temperature air has a lower density than lower temperature air resulting in a buoyant force on the higher temperature air, forcing it upward. If a fan or pump is used to circulate the fluid, the convection is called **forced convection**.

**Radiation** is the transfer of energy by emission of electromagnetic radiation. All surfaces radiate electromagnetic radiation. The rate at which an object radiates energy depends on the absolute temperature, $T$, of the surface of the body, the surface area, $A$, and a property of the surface object, $e$, called the emissivity of the object. The rate at which energy leaves the surface, $dQ/dt$, of an object by radiation is

$$\frac{dQ}{dt} = e\sigma A T^4$$

where $\sigma$ is the **Stefan-Boltzmann constant** and is equal to $5.67 \times 10^{-8}$ W/m$^2$·K$^4$. The relationship above is called the **Stefan-Boltzmann equation**.

All objects absorb electromagnetic radiation from the surroundings. If an object at temperature $T_1$ is in an environment at temperature $T_2$, then the net rate of heat flow into the object is

$$\frac{dQ}{dt} = e\sigma A(T_2^4 - T_1^4)$$

**Example 19-10-B**

Assume a human body has a surface temperature of 98.6° F and an emissivity of 0.8. Estimate the rate that a bare human body loses heat by radiation.

**Solution:**

A human body has approximately $1 m^2$ of surface to emit radiation. The rate at which energy will be radiated is

$$\frac{dQ}{dt} = e\sigma AT^4 = (0.8)\left(5.67\times10^{-8} \text{ W/m}^2\cdot\text{K}^4\right)\left(1 m^2\right)\left[\tfrac{5}{9}(98.6-32)+273\right]^4$$

$$= 400 \text{ W}$$

## Practice Quiz

1.  An amount of work $W$ is done by an ideal gas on its environment. During this process, the temperature of the ideal gas remains constant. Home much heat was added to the ideal gas during this process?

    a)  $Q = 0$
    b)  $Q = W$
    c)  $Q = -W$
    d)  Need more information to determine the amount of heat added

2.  What happens to the adiabatic gas constant as the value of the molar specific heat at constant volume is increased without bound?

    a)  The adiabatic gas constant remains unchanged.
    b)  The adiabatic gas constant increases unbounded.
    c)  The adiabatic gas constant decreases to zero.
    d)  The adiabatic gas constant decreases to one.

3.  Usually, in home construction, the insulation in the ceiling of a home is made much thicker than the insulation in the walls of a home. Why?

    a)  Ceilings have more area than walls.
    b)  Insulation has higher conductive when laid horizontally than standing vertically.
    c)  The ceiling is usually warmer than the walls because of convection in the house.
    d)  Ceiling insulation is easier to install.

4.  During an isochoric process on an ideal gas, what doesn't change?

    a)  The temperature of the gas
    b)  The pressure of the gas
    c)  The internal energy of the gas
    d)  The volume of the gas

5.  During which type of process on an ideal gas will the pressure change by the greatest factor for a given change in the volume?

    a)  Isothermic
    b)  Adiabatic
    c)  Isobaric
    d)  All processes will have the same pressure change.

6.  A system consisting of a fixed amount of gas starts at pressure $P_1$ and volume $V_1$ and ends up at pressure $P_2$ and volume $V_2$ after some thermodynamic process. Which of the following quantities do not depend on the path taken on a pressure–volume diagram during the process?

    a)  Work done on the environment by the system during the process
    b)  Work done by the environment on the system during the process
    c)  The heat added to the system during the process
    d)  The final temperature of the system

7.    By what mechanism is energy transferred from the Sun to the Earth?

a)    Conduction
b)    Convection
c)    Radiation
d)    Injection

8.    You are in a room at normal room temperature. You touch two different objects that have been in the room for quite some time. One object is made of wood, the other is made of metal. The wood feels warmer than the metal. Why is that?

a)    Metal is always at a lower temperature than wood.
b)    The metal reflects the radiation from the surroundings more than wood.
c)    Metal has a higher thermal conductivity than wood.
d)    More work is done on the wood in touching it because it is softer than the metal and compresses further, raising its internal energy more than the metal.

9.    An object is under conditions in which it is cooling mostly by conduction. When the object is at 50° C and is being cooled by conduction to a 20° C body, it takes 2 minutes to cool 1C°. Approximately how long would it take the body to cool 1C° if it started at 35° C and was cooled by conduction to the 20° C body?

a)    0.5 minutes
b)    1 minute
c)    2 minutes
d)    4 minutes

10.    A gas goes from an initial state with pressure $P_1$ and volume $V_1$ and reaches a final state with pressure $P_2$ and volume $V_2$. Which of the processes shown in the pressure–volume diagrams requires the most heat transfer to the gas during this change in pressure and volume?

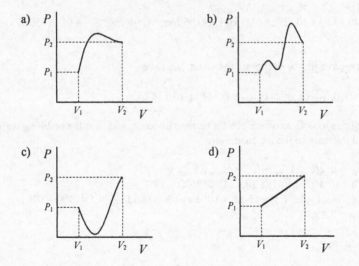

11.    Warm water with a mass of 1.67 kg and a temperature of 85.2° C is added to cold water with a mass of 1.17 kg and a temperature 12.4° C. What is the temperature of the water when it reaches thermal equilibrium?

12.    If heat is added at a rate of 8.56 kJ/s to a mixture of 0.500 kg of ice at -20° C and 0.678 kg of water at +30° C, how long does it take to boil way all of the water? Assume no heat is lost to the environment.

13.    A process occurs such that the pressure is related to the volume by

$$P = 1.65 \times 10^5 \text{ N}/\text{m}^8 V^2$$

while the volume changes from 2.00 m³ to 1.00 m³. How much work is done by the gas during this process?

14.    Assume the Earth is at a uniform surface temperature of 50° F and the emissivity of the Earth is 1. What amount of energy would the earth radiate into space in one year? (The emissivity of the Earth is not 1, and the presence of green house gases in the atmosphere decreases the emissivity.)

15.    Five moles of a monatomic ideal gas are in a container of fixed size. The pressure is initially $1.00 \times 10^5$ N/m² and the temperature is 300 K. 4.87 kJ of heat is added to the gas. What are the final pressure and temperature of the gas?

## Problem Solutions

3.    The heat flow generated must equal the kinetic energy loss:
$\Delta Q = -(\frac{1}{2}mv_f^2 - \frac{1}{2}mv_i^2) = \frac{1}{2}m(v_i^2 - v_f^2)$
$= \frac{1}{2}(3.0 \times 10^{-3} \text{ kg})[(400 \text{ m/s})^2 - (200 \text{ m/s})^2]^2 = 1.8 \times 10^2 \text{ J}.$

7.    The heat flow generated must equal the kinetic energy loss:
$\Delta Q = \frac{1}{2}mv^2 = \frac{1}{2}(1000 \text{ kg})[(95 \text{ km/h})/(3.6 \text{ ks/h})]^2(1 \text{ kcal}/4186 \text{ J}) = 83 \text{ kcal}.$

11.    If all the kinetic energy in the hammer blows is absorbed by the nail, we have
$K = 10(\frac{1}{2}mv^2) = mc\,\Delta T$;
$10[\frac{1}{2}(1.20 \text{ kg})(6.5 \text{ m/s})^2] = (0.014 \text{ kg})(450 \text{ J/kg} \cdot \text{C}°)\,\Delta T$, which gives $\Delta T = 40 \text{ C}°$.

15.    The water must be heated to the boiling temperature, 100°C. We find the time from
$t\quad = (\text{heat gained})/P = [(m_{\text{Al}}c_{\text{Al}} + m_{\text{water}}c_{\text{water}})\,\Delta T_{\text{water}}]/P$
$= [(0.360 \text{ kg})(900 \text{ J/kg} \cdot \text{C}°) + (0.75 \text{ L})(1.00 \text{ kg/L})(4186 \text{ J/kg} \cdot \text{C}°)](100°\text{C} - 8.0\text{C})/(750 \text{ W})$
$= 425 \text{ s} = 7.1 \text{ min}.$

19.    If we assume that the heat is required just to evaporate the water, we have
$Q = m_{\text{water}}L_{\text{water}}$;
$180 \text{ kcal} = m_{\text{water}}(539 \text{ kcal/kg})$, which gives $m_{\text{water}} = 0.334 \text{ kg}$ (0.334 L).

23.    The temperature of the ice will rise to 0°C, at which point melting will occur, and then the resulting water will rise to the final temperature. We find the mass of the ice cube from
heat lost = heat gained;
$(m_{\text{Al}}c_{\text{Al}} + m_{\text{water}}c_{\text{water}})\,\Delta T_{\text{Al}} = m_{\text{ice}}(c_{\text{ice}}\,\Delta T_{\text{ice}} + L_{\text{ice}} + c_{\text{water}}\,\Delta T_{\text{water}})$;
$[(0.075 \text{ kg})(900 \text{ J/kg} \cdot \text{C}°) + (0.300 \text{ kg})(4186 \text{ J/kg} \cdot \text{C}°)](20°\text{C} - 17°\text{C}) =$
$\qquad m_{\text{ice}}\{(2100 \text{ J/kg} \cdot \text{C}°)[0°\text{C} - (-8.5°\text{C})] + (3.33 \times 10^5 \text{ J/kg}) + (4186 \text{ J/kg} \cdot \text{C}°)(17°\text{C} - 0°\text{C})\}$,
which gives $m_{\text{ice}} = 9.4 \times 10^3 \text{ kg} = 9.4 \text{ g}.$

27.    We find the latent heat of fusion from
heat lost = heat gained;
$(m_{\text{Al}}c_{\text{Al}} + m_{\text{water}}c_{\text{water}})\,\Delta T_{\text{water}} = m_{\text{Hg}}(L_{\text{Hg}} + c_{\text{Hg}}\,\Delta T_{\text{Hg}})$
$[(0.620 \text{ kg})(900 \text{ J/kg} \cdot \text{C}°) + (0.430 \text{ kg})(4186 \text{ J/kg} \cdot \text{C}°)](12.80°\text{C} - 5.06°\text{C}) =$
$\qquad\qquad\qquad (1.00 \text{ kg})\{L_{\text{Hg}} + (138 \text{ J/kg} \cdot \text{C}°)[5.06°\text{C} - (-39.0°\text{C})]\}$,
which gives $L_{\text{Hg}} = 1.22 \times 10^4 \text{ J/kg}.$

31.    (a)  The internal energy of an ideal gas depends only on the temperature, $U = \frac{3}{2}nRT$, so we have
$\Delta U = \frac{3}{2}nR\,\Delta T = 0.$
    (b)  We use the first law of thermodynamics to find the heat absorbed:
$\Delta U = Q - W$;
$0 = Q - 5.00 \times 10^3 \text{ J}$, which gives $Q = 5.00 \times 10^3 \text{ J}.$

35.    (a)  Because there is no change in volume, we have

$W = 0$.

(b) We use the first law of thermodynamics to find the change in internal energy:
$$\Delta U = Q - W$$
$$= -1300 \text{ kJ} - 0 = -1300 \text{ kJ}.$$

39. The work done during an isothermal process is
$$W = nRT \ln(V_2/V_1)$$
$$= (2.00 \text{ mol})(8.315 \text{ J/mol} \cdot \text{K})(300 \text{ K}) \ln(7.00 \text{ m}^3/3.50 \text{ m}^3)$$
$$= 3.46 \times 10^3 \text{ J}.$$

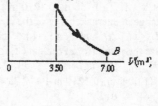

The internal energy of an ideal gas depends only on the temperature, so
$$\Delta U = 0.$$
We use the first law of thermodynamics to find the heat flow for the process:
$$\Delta U = Q - W;$$
$0 = Q - (+3.46 \times 10^3 \text{ J})$, which gives $Q = +3.46 \times 10^3 \text{ J}$ (into the gas).

43. We consider a differential area $dA$ on the surface of the arbitrary volume. Because the force from the outside pressure is perpendicular to the surface, the work done when the surface expands a distance $dl$ is
$$dW = P \, dA \, dl.$$
The work done in a finite expansion is
$$W = \int dW = \int \Delta P \, dA \, dl.$$
The pressure is the same over the surface of the volume, so we can integrate over the area:
$$W = \int P \left[ \int dA \right] dl = \int PA \, dl = \int P \, dV.$$

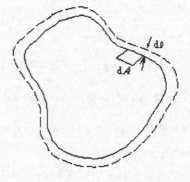

47. The pressure is a function of volume for a van der Waals gas:
$$[P + a/(V/n)^2][(V/n) - b)] = RT.$$
With $n = 1$, we can write this as
$$P = [RT/(V - b)] - a/V^2.$$
We find the work by integrating:
$$W = \int_{V_1}^{V_2} P \ dV = \int_{V_1}^{V_2} \left[ \frac{RT}{V-b} - \frac{a}{V^2} \right] dV = \left[ RT \ln(V-b) + \frac{a}{V} \right]\Bigg|_{V_1}^{V_2}$$
$$= RT \ln \frac{V_2 - b}{V_1 - b} + a \left( \frac{1}{V_2} - \frac{1}{V_1} \right)$$

51. Because hydrogen is diatomic, at room temperature there are 5 degrees of freedom. Thus we have
$C_V = 5R/2 = 5(1.99 \text{ cal/mol} \cdot \text{K})/2 = 4.98 \text{ cal/mol} \cdot \text{K};$
$c_V = C_V/M = (4.98 \text{ cal/mol} \cdot \text{K})/(2 \text{ g/mol}) = 2.49 \text{ cal/g} \cdot \text{K} = 2.49 \text{ kcal/kg} \cdot \text{K};$
$C_P = C_V + R = 4.98 \text{ cal/mol} \cdot \text{K} + 1.99 \text{ cal/mol} \cdot \text{K} = 6.97 \text{ cal/mol} \cdot \text{K};$
$c_P = C_P/M = (6.97 \text{ cal/mol} \cdot \text{K})/(2 \text{ g/mol}) = 3.48 \text{ cal/g} \cdot \text{K} = 3.48 \text{ kcal/kg} \cdot \text{K}.$

55. If there are no heat losses, we have
$Q = mc_V \Delta T = \rho V c_V \Delta T,$ so the rate is
$(2500)(70 \text{ W})(2.0 \text{ h})(3600 \text{ s/h}) = (1.29 \text{ kg/m}^3)(30,000 \text{ m}^3)(0.17 \text{ kcal/kg} \cdot \text{C}°) \Delta T,$
which gives $\Delta T = 46 \text{ C}°.$

59. For an adiabatic process we have
$P_2 V_2{}^\gamma = P_1 V_1{}^\gamma,$ or $P_2/P_1 = (V_1/V_2)^\gamma;$
$P_2/1.00 \text{ atm} = (\frac{1}{2})^{7/5},$ which gives $P_2 = 0.379 \text{ atm}.$
We find the temperature from the ideal gas equation:
$P_2 V_2/P_1 V_1 = T_2/T_1 ;$
$(0.379 \text{ atm}/1.00 \text{ atm})(2) = T_2/293 \text{ K},$ which gives $T_2 = 222 \text{ K} = -51°\text{C}.$

63. (a) We find the initial temperature from the ideal gas equation:

$P_1 V_1 = nRT_1$ ;

$(1.013 \times 10^5 \text{ N/m}^2)(0.1210 \text{ m}^3) = (4.65 \text{ mol})(8.315 \text{ J/mol} \cdot \text{K})T_1$, which gives $T_1 = 317 \text{ K}$.

For an adiabatic process we have

$P_2 V_2^\gamma = P_1 V_1^\gamma$, or $P_2/P_1 = (V_1/V_2)^\gamma = (0.1210 \text{ m}^3/0.750 \text{ m}^3)^{7/5} = 7.78 \times 10^{-2}$.

We find the final temperature from the ideal gas equation:

$P_2 V_2 / P_1 V_1 = T_2/T_1$ ;

$(7.78 \times 10^{-2})(0.750 \text{ m}^3/0.1210 \text{ m}^3) = T_2/317 \text{ K}$, which gives $T_2 = 153 \text{ K}$.

(b) The change in internal energy of the ideal gas is

$\Delta U = nC_V \Delta T = n\tfrac{5}{2}R \Delta T = (4.65 \text{ mol}) \tfrac{5}{2} (8.315 \text{ J/mol} \cdot \text{K})(153 \text{ K} - 317 \text{ K}) = -1.59 \times 10^4 \text{ J}$.

(c) If we use the result from Problem 61, for the work done by the gas we have

$W = (P_1 V_1 - P_2 V_2)/(\gamma - 1)$

$= [(1.00 \text{ atm})(0.1210 \text{ m}^3) - (7.78 \times 10^{-2} \text{ atm})(0.750 \text{ m}^3)](1.013 \times 10^5 \text{ N/m}^2 \cdot \text{atm})/[(7/5) - 1]$

$= 1.59 \times 10^4 \text{ J}$.

Thus the work done on the gas is

$W_{\text{on}} = -W = -1.59 \times 10^4 \text{ J}$.

(d) For an adiabatic process, there is no heat flow, so $Q = 0$.

67. (a) We find the radiated power from

$\Delta Q/\Delta t = e\sigma A T^4$

$= (0.35)(5.67 \times 10^{-8} \text{ W/m}^2 \cdot \text{K}^4)4\pi(0.180 \text{ m})^2(298 \text{ K})^4 = 64 \text{ W}$.

(b) We find the net flow rate from

$\Delta Q/\Delta t = e\sigma A(T_2^4 - T_1^4)$

$= (0.35)(5.67 \times 10^{-8} \text{ W/m}^2 \cdot \text{K}^4)4\pi(0.180 \text{ m})^2[(298 \text{ K})^4 - (268 \text{ K})^4] = 22 \text{ W}$.

71. The cross-sectional area of the beam that falls on an area $A$ is $A \cos \theta$. Thus the rate at which energy is absorbed is

$P = IeA \cos \theta$.

There is no change in temperature of the ice or melted water. We find the time to provide the energy to melt the ice from

$Q = mL = \rho AhL = Pt = (IeA \cos \theta)t$ ;

$(917 \text{ kg/m}^3)(0.016 \text{ m})(3.33 \times 10^5 \text{ J/kg}) = (1000 \text{ W/m}^2)(0.050)(\cos 30°)t$, which gives

$t = 1.13 \times 10^5 \text{ s} = 31 \text{ h}$.

Note that the result is independent of the area.

75. (a) We call the temperatures at the interfaces $T_a$ and $T_b$, as shown. In the steady state, the rate of heat flow is the same for each layer:

$\Delta Q/\Delta t = k_1 A(T_2 - T_a)/l_1 = k_2 A(T_a - T_b)/l_2 = k_3 A(T_b - T_1)/l_3$ .

We treat this as three equations:

$T_2 - T_a = [(\Delta Q/\Delta t)/A]l_1/k_1$ ;

$T_a - T_b = [(\Delta Q/\Delta t)/A]l_2/k_2$ ;

$T_b - T_1 = [(\Delta Q/\Delta t)/A]l_3/k_3$ .

If we add these equations, we get

$T_2 - T_1 = [(\Delta Q/\Delta t)/A][(l_1/k_1) + (l_2/k_2) + (l_3/k_3)]$, which gives

$\Delta Q/\Delta t = A(T_2 - T_1)/[(l_1/k_1) + (l_2/k_2) + (l_3/k_3)]$.

(b) We can generalize this by recognizing that more layers will mean more equations, similar to the three that we had. When we eliminate the intermediate temperatures by adding all the equations, we get

$\Delta Q/\Delta t = A(T_2 - T_1)/\Sigma(l_i/k_i)$.

79. The liberated heat is

$\Delta Q = mc \Delta T = \rho Vc \Delta T$

$= (1.00 \times 10^3 \text{ kg/m}^3)(1 \times 10^3 \text{ m})^3(4186 \text{ J/kg} \cdot \text{C}°)(1 \text{ C}°) = 4 \times 10^{15} \text{ J}$.

83. We use the first law of thermodynamics to find the rate at which the internal energy is changing:

$\Delta U/\Delta t = Q/\Delta t - W/\Delta t = (- 1.5 \text{ kW}) - (- 10 \text{ kW}) = 8.5 \text{ kW}.$

The time for the compression is the time for half a revolution:

$\Delta t = \frac{1}{2}(1/400 \text{ rpm})(60 \text{ s/min}) = 0.075 \text{ s}.$

If we assume an ideal gas, we have

$\Delta U/\Delta t = nC_V \Delta T/\Delta t;$

$8.5 \text{ kW} = (1.00 \text{ mol})(5.0 \text{ cal/mol} \cdot \text{C}°)(4.186 \text{ J/cal}) \Delta T/(0.075 \text{ s}),$ which gives $\Delta T = 31 \text{ C}°.$

87. We find the temperature rise from

$Q = mc \Delta T;$

$(0.80)(200 \text{ kcal/h})(1.00 \text{ h}) = (70 \text{ kg})(0.83 \text{ kcal/kg} \cdot \text{C}°)\Delta T,$ which gives $\Delta T = 2.8 \text{ C}°.$

91. If we assume that all the energy evaporates the water, with the latent heat at 20°C given in the text, we have

$Q = m_{water}L_{water}$

$(1000 \text{ kcal/h})(2.5 \text{ h}) = m_{water}(585 \text{ kcal/kg}),$ which gives $m_{water} = 4.3 \text{ kg}.$

95. (a) If we assume that all of the radiation is absorbed to raise the temperature of the leaf, we have

$P = IeA = m_{leaf}c_{leaf} (\Delta T /\Delta t);$

$(1000 \text{ W/m}^2)(0.85)(40 \times 10^{-4} \text{ m}^2) = (4.5 \times 10^{-4} \text{ kg})(0.80 \text{ kcal/kg} \cdot \text{C}°)(4186 \text{ J/kcal})(\Delta T/\Delta t),$

which gives $\Delta T/\Delta t = 2.3 \text{ C}°/\text{s}.$

(b) When the leaf reaches the temperature at which the absorbed energy is re-radiated to the surroundings from both sides of the leaf, we have

$IeA = e\sigma 2A(T_2^4 - T_1^4),$ or $I = 2\sigma (T_2^4 - T_1^4);$

$(1000 \text{ W/m}^2) = 2(5.67 \times 10^{-8} \text{ W/m}^2 \cdot \text{K}^4)[T_2^4 - (293 \text{ K})^4],$ which gives $T_2 = 365 \text{ K} = 84°\text{C}.$

(c) The major ways that heat can be dissipated are by convection from conduction to the air in contact with the leaf and evaporation.

99. (a) Because the pressure is constant, we find the work from

$W = p(V_2 - V_1)$

$= (1.013 \times 10^5 \text{ N/m}^2)(4.1 \text{ m}^3 - 2.2 \text{ m}^3) = 1.9 \times 10^5 \text{ J}.$

(b) We use the first law of thermodynamics to find the change in internal energy:

$\Delta U = Q - W$

$= + 5.30 \times 10^4 \text{ J} - 1.9 \times 10^5 \text{ J} = -1.4 \times 10^5 \text{ J}.$

(c)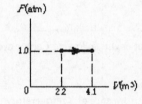

103. The temperature difference from conduction through the glass:

$\Delta Q/\Delta t = kA(\Delta T/L);$

$95 \text{ W} = (0.84 \text{ J/s·m·C}°)4\pi m(0.030 \text{ m})^2\Delta T/(1.0 \times 10^{-3}),$ which gives $\Delta T = 10 \text{ C}°$

# Chapter 20: Second Law of Thermodynamics

## Chapter Overview and Objectives

This chapter introduces the second law of thermodynamics and its applications to physical systems. It shows the equivalence of several different statements of the second law of thermodynamics. This chapter introduces heat engines, refrigerators, and heat pumps and measures of their efficiency. This chapter introduces the concepts of both macroscopic and microscopic entropy and shows their equivalence.

After completing study of this chapter, you should:

- Know the various equivalent forms of the second law of thermodynamics.
- Know what heat engines, refrigerators, and heat pumps are.
- Know the definition of efficiency of heat engines.
- Know the definition of the coefficient of performance of refrigerators and heat pumps.
- Know that a Carnot cycle heat engine has the maximum possible theoretical efficiency.
- Know how to calculate the efficiency of a Carnot cycle heat engine.
- Know the definition of a change in macroscopic entropy and how to calculate the change.
- Know that entropy measures the disorder of a system.
- Know the distinction between a macrostate and a microstate.
- Know how entropy can be related to the distribution of microstates belonging to macrostates.

## Summary of Equations

Efficiency of a heat engine:
$$e = \frac{|W|}{|Q_H|}$$

Efficiency of heat engine following Carnot cycle:
$$e = 1 - \frac{T_C}{T_H}$$

Coefficient of performance of a refrigerator:
$$CP = \frac{|Q_C|}{W}$$

Coefficient of performance of a Carnot cycle refrigerator:
$$CP_{Carnot} = \frac{T_L}{T_H - T_C}$$

Coefficient of performance of a heat pump:
$$CP = \frac{|Q_H|}{W}$$

Coefficient of performance of a Carnot cycle heat pump:
$$CP_{Carnot} = \frac{T_H}{T_H - T_C}$$

Definition of change in entropy of a system:
$$dS = \frac{dQ}{T}$$

Change in entropy for finite amount of heat:
$$\Delta S = \int_a^b \frac{dQ}{T}$$

Statistical definition of entropy:
$$S = k \ln W$$

Definition of thermodynamic temperature scale:
$$T = (273.16 K)\left(\frac{|Q|}{|Q_{TP}|}\right)$$

# Chapter Summary

## Section 20-1 The Second Law of Thermodynamics—Introduction

The **second law of thermodynamics** states that heat does not flow spontaneously from lower temperature materials to higher temperature materials, but only from higher temperature materials to lower temperature materials.

A **heat engine** is a mechanical device used to transform internal energy into mechanical energy.

## Section 20-2 Heat Engines

A simple model of a heat engine is a device that allows heat to be transferred from a high temperature system partially to a low temperature system while transforming some of the heat to work. The heat engine satisfies the first law of thermodynamics in that the heat that leaves the high temperature system is equal to the heat that enters the low temperature system added to the work done by the heat engine. The temperatures of the two systems involved in the heat transfer are called the operating temperatures of the heat engine.

The efficiency, $e$, of a heat engine is the ratio of the work done by the engine divided by the amount of heat taken from the high temperature system:

$$e = \frac{|W|}{|Q_H|}$$

where $|W|$ is the work done by the engine and $|Q_H|$ is the heat extracted from the high temperature system.

Another way of stating the second law of thermodynamics is to say that:

**No heat engine can have an efficiency of 1.**

or

**No device can exist that transforms a given amount of heat completely into work.**

## Example 20-2-A

A heat engine is needed that will output 240 horsepower. At what rate does heat need to be put into the engine if the engine efficiency is 0.323?

**Solution:**

The efficiency is given in terms of the heat in and work done per cycle:

$$e = \frac{|W|}{|Q_H|}$$

The numerator and denominator can each be divided by the time to express the efficiency in terms of the output power and the rate of heat addition to the engine:

$$e = \frac{|W|}{|Q_H|} = \frac{|W|/t}{|Q_H|/t} = \frac{P}{Q_H/t}$$

We can solve this for the rate of heat addition to the engine:

$$|Q_H/t| = \frac{P}{e} = \frac{(240\,\text{hp})(746\,\text{W/hp})}{0.323} = 5.54 \times 10^5\,\text{W}$$

**Section 20-3 Reversible and Irreversible Processes; Carnot Engine**

A **reversible process** is one in which the system is always in thermodynamic equilibrium during the process. This means that the temperature must be constant throughout the system. It is called reversible because there is never a flow of heat from higher temperature to lower temperature, so that we could reverse the process without requiring the heat to flow from lower temperature to higher temperature. As any real process occurs in a finite time, any real process will create finite temperature differences in a system and not be reversible. Any process that is not reversible is said to be an **irreversible process**. If we were to reverse the direction of an irreversible process, we would observe heat flowing from lower temperature to higher temperature.

A **Carnot cycle** is a thermodynamic cycle that is reversible. In a Carnot cycle, the system undergoes an isothermal change in volume at a temperature $T_H$. This is followed by an adiabatic expansion. The adiabatic expansion is followed by an isothermal compression at temperature $T_C$. Finally, an adiabatic compression brings the system back to its original state. The efficiency of a Carnot cycle is

$$e = 1 - \frac{T_C}{T_H}$$

This is maximum possible efficiency of a heat engine. The efficiency of any irreversible engine is less than the efficiency of a Carnot cycle.

**Example 20-3-A**

A heat engine loses its waste heat at a temperature of 30° C. What is the minimal possible temperature that heat must be added to the engine so that the efficiency of the engine is 82%?

**Solution:**

We know the maximum efficiency is the Carnot efficiency,

$$e = 1 - \frac{T_L}{T_H}$$

where the temperatures are absolute temperatures. The $T_L$ temperature is 30° C + 273.15 = 303.15 K. Solving for $T_H$:

$$T_H = \frac{T_L}{1-e} = \frac{303.15 \, \text{K}}{1 - 0.82} = 1.7 \times 10^3 \, \text{K}$$

**Example 20-3-B**

You know that the gas leaving the engine of a car is still very hot (and still may be combusting). It is possible to use this hot exhaust gas to add heat to an auxiliary engine to get additional work out of the energy put into the car. Suppose the exhaust temperature is 600 K and we could construct an auxiliary engine that can exhaust its waste heat at 400 K. Compare the efficiency of a single Carnot engine acting between temperature 4200 K and 400 K with the total efficiency of two Carnot engines with the first acting between 4200 K and 600 K and the second acting between 600 K and 400 K utilizing the waste heat from the first Carnot engine. The total efficiency of the pair of engines will be the total work output of each divided by the heat into the first engine.

**Solution:**

The efficiency of the single Carnot engine will be

$$e = 1 - \frac{T_L}{T_H} = 1 - \frac{400 \, \text{K}}{4200 \, \text{K}} = 0.905 = 90.5 \%$$

The first part of the two engine system will have an efficiency of

$$e = 1 - \frac{T_L}{T_H} = 1 - \frac{600\,\text{K}}{4200\,\text{K}} = 0.857 = 85.7\%$$

This means that the work from the first of the two engines will be $0.857\,Q_H$. The waste heat will be $(1 - 0.857)Q_H = 0.143Q_H$. This is the heat into the second engine in the two engine system. The efficiency of the second of the two-engine systems will be

$$e = 1 - \frac{T_L}{T_H} = 1 - \frac{400\,\text{K}}{600\,\text{K}} = 0.333 = 33.3\%$$

The work done by the second engine will be $0.333\,(0.143\,Q_H) = 0.048\,Q_H$. The total work done by the first and second engines in the two engine system is $0.857Q_H + 0.048Q_H = 0.905Q_H$. This is an efficiency of 90.5% just like the single engine acting between the initial and final temperatures. Of course, we can show this to be true in general. (Show it! Follow through the example but just leave the temperatures as symbols rather than putting in numbers.) We should also have expected it from knowing that any two reversible engines acting between the same two temperatures have the same efficiency. Whether there are one or two component engines to the system doesn't affect its reversibility.

### Section 20-4 Refrigerators, Air Conditioners, and Heat Pumps

Rather than making use of the work output of a thermodynamic cycle, it is also useful to run a thermodynamic cycle in the opposite direction so that there is a net work into the system upon completing a cycle. In this case the useful output of the thermodynamic cycle will be heat removed from the low temperature system and added to the high temperature system. These devices are called **refrigerators** if the system is used for its ability to remove heat from the low temperature system. The devices are called heat **pumps** if the system is used for its ability to add heat to the high temperature system.

A measure of how well a refrigerator is performing is the ratio of the heat removed from the cold system divided by the amount of work done by external forces on the system during one cycle. This is called the **coefficient of performance**, $CP$, of the refrigerator:

$$CP = \frac{|Q_C|}{W}$$

where $Q_C$ is the heat removed from the low temperature side of the system and $W$ is the work done on the system by external forces during a thermodynamic cycle. The coefficient of performance of a Carnot cycle refrigerator is

$$CP_{Carnot} = \frac{T_L}{T_H - T_C}$$

where $T_L$ is the temperature of the low temperature isotherm of the Carnot cycle and $T_H$ is the temperature of the high temperature isotherm of the Carnot cycle.

The coefficient of performance of a heat pump is the heat that is added to the high temperature side of the system, $Q_H$, divided by the work done, $W$, by external forces on the system during each thermodynamic cycle:

$$CP = \frac{|Q_H|}{|W|}$$

The coefficient of performance of a Carnot cycle heat pump is the maximum coefficient of performance for a given high and low temperature and is given by

$$CP_{Carnot} = \frac{T_H}{T_H - T_C}$$

**Example 20-4-A**

A heat pump is used to warm a house in winter. The desired temperature inside the house is 68° F, which can be taken to be high temperature of the heat pump. (In practice, the high temperature will be somewhat greater than the desired temperature.) Calculate the amount of work that must be done on a Carnot cycle heat pump to add $1.00 \times 10^5$ J to the house if the outside air temperature is 40° F and if it is –20° F.

**Solution:**

The coefficent of performance of a heat pump is the ratio of the output heat to the work input. A Carnot cycle heat pump has a coefficient of performance

$$CP = \frac{T_H}{T_H - T_L}$$

where the temperatures are absolute temperatures. Because all the temperatures are Fahrenheit temperatures, we will use the Rankin temperature scale for temperature. The Rankin scale is the absolute temperature scale that corresponds to the Fahrenheit scale. To convert from Fahrenheit to Rankin, add 459.7 to the Fahrenheit temperature. (Verify that absolute zero on the Fahrenheit scale is –459.7 by using the Celsius temperature for absolute zero and converting to Fahrenheit.) For the higher outdoor temperature:

$$CP = \frac{68 + 459.7}{68 + 459.7 - (40 + 459.7)} = 18.8$$

The definition of the coefficient of performance allows us to calculate the work needed:

$$CP = \frac{|Q_H|}{|W|} \Rightarrow |W| = \frac{|Q_H|}{CP} = \frac{1.00 \times 10^5 \text{ J}}{18.8} = 5.3 \times 10^3 \text{ J}$$

For the lower outdoor temperature:

$$CP = \frac{68 + 459.7}{68 + 459.7 - (-20 + 459.7)} = 6.0$$

and calculating the work needed:

$$|W| = \frac{|Q_H|}{CP} = \frac{1.00 \times 10^5 \text{ J}}{6.0} = 1.7 \times 10^4 \text{ J}$$

**Section 20-5 Entropy**

The change in entropy, $dS$, of a system that has a infinitesimal amount of heat, $dQ$, added to it when the system is at temperature $T$ is

$$dS = \frac{dQ}{T}$$

The change in entropy, $\Delta S$, that occurs when a finite amount of heat is added to a system while it undergoes a process that takes it from a state $a$ to a state $b$ is given by

$$\Delta S = \int_a^b \frac{dQ}{T}$$

where we note that $T$ is not constant unless the process is isothermal.

**Example 20-5-A**

Calculate the entropy change when 1 kg of water at 100° C is changed to steam at 100° C.

**Solution:**

We need to find

$$\Delta S = \int \frac{dQ}{T}$$

but because the temperature is a constant 100° C = 373.15 K, we need to find the integral of $dQ$ which is just the latent heat added to vaporize the water. We find this from

$$Q = mL = (1.0 \, \text{kg})(22.6 \times 10^5 \, \text{J} / \text{kg}) = 22.6 \times 10^5 \, \text{J}$$

We can then calculate the change in entropy:

$$\Delta S = \int \frac{dQ}{T} = \frac{1}{T} \int dQ = \frac{Q}{T} = \frac{22.6 \times 10^5 \, \text{J}}{373.15 \, \text{K}} = 6.1 \times 10^3 \, \text{J/K}$$

**Section 20-6 Entropy and the Second Law of Thermodynamics**

We can restate the second law of thermodynamics in terms of entropy:

> **The entropy of an isolated system never decreases. It remains constant during reversible processes and it increases during irreversible processes.**

**Example 20-6-A**

A pot containing 10 kg of water at 20° C is placed on an electric element that is at a temperature 200° C. What is the change in entropy of the system as the water is warmed to 100° C? Ignore the change in entropy of the pot. Assume the range element remains at 200° C.

**Solution:**

The change in entropy of each component is calculated from $dS = dQ/T$. For the change in the entropy of the range element, the temperature is constant, so we only need to know the total heat transferred to the water. We can calculate this using the heat capacity of water:

$$Q = mC\Delta T = (10 \, kg)(4186 \, \text{J} / \text{kg} \cdot \text{C}°)(100° \, \text{C} - 20° \, \text{C}) = 3.35 \times 10^6 \, \text{J}$$

The heat is leaving the range element so the change in entropy of the range element is

$$\Delta S_{range} = \int \frac{dQ}{T} = -\frac{3.35 \times 10^6 \, \text{J}}{473.15 \, \text{K}} = -7.08 \times 10^3 \, \text{J/K}$$

The change in the entropy of the water is a little more difficult to calculate because the water changes temperature as the heat is added. But we can also relate the infinitesimal addition of heat to an infinitesimal change in temperature:

$$dQ = mC[(T + dT) - T] = mC \, dT$$

We can use this to write the change in entropy as

$$\Delta S_{water} = \int \frac{dQ}{T} = \int_{T_i}^{T_f} \frac{mC\,dT}{T} = mC \ln \frac{T_f}{T_i}$$

$$= (10\,\text{kg})(4186\,\text{J}/\text{kg} \cdot \text{C}^\circ) \ln \frac{373.15\,\text{K}}{293.15\,\text{K}} = 1.01 \times 10^4\,\text{J}/\text{K}$$

The total change in entropy of the system will be:

$$\Delta S = \Delta S_{range} + \Delta S_{water} = -7.08 \times 10^3\,\text{J}/\text{K} + 1.01 \times 10^4\,\text{J}/\text{K} = +3.0 \times 10^3\,\text{J}/\text{K}$$

### Section 20-7 Order to Disorder

Entropy is a measure of the disorder of the system. A state of low entropy is a state with a great amount of ordering of a system. A state of high entropy is a state with a great amount of disorder. We can write the second law of thermodynamics as a statement about the disorder of a system:

**Natural processes tend to move systems from more ordered states to more disordered states.**

### Section 20-8 Energy Availability; Heat Death

All real processes are not reversible thermodynamic processes. This means that the change in entropy is greater than zero for all real processes. This implies that some energy that was available for transformation into some other form of energy has become internal energy at a low temperature. This internal energy at low temperature is unavailable to produce mechanical work unless a subsystem of even lower temperature can be found to exchange heat with. Eventually, in any finite universe, all available energy will be transformed into internal energy at a uniform low temperature and entropy will have reached its maximum. No further processes can occur. This is called the **heat death** of the universe.

### Section 20-9 Statistical Interpretation of Entropy and the Second Law

The laws of classical physics are deterministic. That means if we completely know the exact position and velocity of all particles within a system and the forces that act between those particles, we can, in principle, determine the position and velocity of each particle at any future time. There are several reasons we are unable to apply this in practice. One reason is the number of particles (atoms) in any macroscopic system is too large to keep track of, let alone solve each of the equations of motion for each particle. A second reason is because the values of position and velocity at any given time are measurements. Those measurements have limited precision and accuracy. Small errors at one time can result in very large errors in the knowledge of the position and velocity at later times, predicted using earlier inaccurate values.

Fortunately, there are some rules that can be applied to sums or averages over many particles in a system. Specifying all the detailed information about a system, such as the position and velocity of each particle, determines a **microstate** of the system. Specifying a sum or average of some quantity for many particles in the system determines a **macrostate** of the system. As an example, in the kinetic theory of gases, we know each gas particle has a position and velocity at any given moment. If we know the position and velocity of each particle at any given moment, we know the microstate of the gas. That is impossible . However, it is possible to determine the temperature of the gas. We saw in Chapter 18 that knowing the temperature of a gas is equivalent to knowing the average kinetic energy of the gas particles. The temperature, along with pressure and density, determines a macrostate of the gas.

If we only have information about macrostates of a system, we don't have the detailed information necessary to utilize Newton's laws to solve for the evolution of the system with time. However, it is possible to determine many things about the system. We turn to statistical physics to help us answer the questions we have any chance of answering about the system with our lack of detailed knowledge. The basis of statistical physics lies in making the assumption that for any macrostate, it is equally probable to find the system in any of the microstates consistent with the given macrostate. The probability of finding a system in a given macrostate is then equal to the number of microstates corresponding to that particular macrostate divided by the total number of microstates of the system consistent with any physical constraints on the system.

The example of coin tosses is often used as an illustrative example. Let's look at what happens if we toss three coins and observe whether they land as heads or tails. We make a table of the possible results of the coin toss. The first column represents the face showing on the first coin, the second column represents the face showing on the second coin, and third column represents the face showing on the third coin.

HHH
HHT
HTH
HTT
THH
THT
TTH
TTT

This list exhausts the possibilities. These are the microstates of the system. They have the maximum amount of information about the system — the face showing on each of the coins. Often we ask questions about macrostates. One type of question we can ask about a macrostate is the probability that we have a certain number of heads. The macrostate that has two heads has the three microstates HHT, HTH, THH that belong to that macrostate. We can see that having two heads is a macrostate because it lacks the detailed information about which particular two coins have the heads. The probability of two heads is 3/8. Three microstates belong to this particular macrostate out of a total of eight possible microstates of the system.

It can be shown that the entropy change $dS = dQ/T$ is consistent with a definition of entropy of a given macrostate based on counting the number of microstates corresponding to that macrostate. A statistical definition of entropy of a given macrostate is

$$S = k \ln W$$

where $k$ is Boltzmann's constant and $W$ is the number of microstates corresponding to the macrostate.

The second law of thermodynamics becomes a probabilistic statement:

**The most likely state to find a system in is that with the greatest entropy.**

This means the second law of thermodynamics is a different sort of law of nature than Newton's laws. It does not make a definite statement about what will happen, but just states what is the most likely thing to happen. For any macroscopic sized system, the probability that we would witness a system decrease its entropy spontaneously is so exceedingly small that we can ignore the possibility with a great deal of confidence.

## Example 20-9-A

Calculate the entropies for the macrostates of the results of tossing three coins as discussed above.

## Solution:

The number of microstates corresponding to the macrostate with three heads is 1, so

$$S_{3\ heads} = k \ln W = \left(1.38 \times 10^{-23} \text{ J/K}\right) \ln 1 = 0$$

The number of microstates corresponding to the macrostate with two heads is 3, so

$$S_{2\ heads} = k \ln W = \left(1.38 \times 10^{-23} \text{ J/K}\right) \ln 3 = 1.52 \times 10^{-23} \text{ J/K}$$

The number of microstates corresponding to the macrostate with one heads is 3, so

$$S_{1\ head} = k \ln W = \left(1.38 \times 10^{-23} \text{ J/K}\right) \ln 3 = 1.52 \times 10^{-23} \text{ J/K}$$

The number of microstates corresponding to the macrostate with no heads is 1, so

$$S_{no\ heads} = k \ln W = \left(1.38 \times 10^{-23} \text{ J/K}\right) \ln 1 = 0$$

**Section 20-10 Thermodynamic Temperature Scale; Absolute Zero, and the Third Law of Thermodynamics**

Temperature can be defined in terms of the amount of heat exchanged during the two isothermal processes of a Carnot cycle. The definition of the Kelvin or thermodynamic temperature scale is

$$T = (273.16\,\text{K})\left(\frac{|Q|}{|Q_{TP}|}\right)$$

where $|Q|$ is the heat exchanged along the isotherm at temperature $T$ and $|Q_{TP}|$ is the heat exchanged along the isotherm a the triple point temperature of water. The triple point temperature of water is define to be 273.16 K. This definition is in agreement with the definition using a constant volume ideal gas thermometer. It does not depend on the properties of a particular substance.

It is impossible to bring a system to absolute zero from any finite temperature above absolute zero in any finite number of cooling processes. This is a statement of the **third law of thermodynamics**.

## Practice Quiz

1.  You measure the efficiency of an engine to be different than a Carnot engine acting between the same two temperatures. What statement can you make about the efficiency and the engine cycle?

    a) The efficiency can be less than or greater than the Carnot efficiency and the cycle is a reversible cycle.
    b) The efficiency of the engine is less than the Carnot efficiency and the cycle can be reversible or irreversible.
    c) The efficiency of the engine is greater than the Carnot efficiency and the cycle is reversible.
    d) The efficiency of the engine is less than the Carnot efficiency and the cycle is irreversible.

2.  The work done in proceeding around a thermodynamic cycle is the area enclosed by a cycle. The largest area for a given perimeter is that of a circle. Why isn't the highest efficiency cycle shaped like a circle on a pressure volume graph?

    a) A circular cycle is impossible.
    b) It isn't the same shape as a Carnot cycle.
    c) It could be if the cycle were followed reversibly.
    d) It requires negative pressure, which is impossible.

3.  Heat pumps are used more often where winters are moderate than where winters are very cold. Why?

    a) The coefficient of performance decreases the further the outdoor temperature drops.
    b) The coefficient of performance increases the further the outdoor temperature drops.
    c) Fossil fuel for heating is cheap where the winters are cold.
    d) Heat pumps can't operate where the temperature is very cold.

4.  Can you heat your home using the internal energy contained in ice?

    a) No, heat flows from hot to cold only.
    b) Yes, but only to heat your home from below the freezing temperature of water up to the freezing temperature of water.
    c) No, there is not enough internal energy in ice to heat anything.
    d) Yes, but it requires work.

5.  Which of the following is closest to being a reversible process?

    a) A vase falls off a shelf and breaks upon hitting the floor.
    b) A cup of water at room temperature evaporates.
    c) An ice cube is melted by dropping it into a cup of very hot water.
    d) A car engine which operates with an efficiency of half that of a Carnot engine.

6.      Which of the following changes would decrease the input work needed to remove a given amount of heat from the inside of a refrigerator using a Carnot cycle?

a) Raise the exterior temperature.
b) Lower the interior temperature of the refrigerator.
c) Raise the interior temperature of the refrigerator.
d) Change to an irreversible cycle refrigerator.

7.      What is the heat death of the universe?

a) Eventually, the universe will become so hot from waste heat that it will vaporize.
b) Eventually, the entropy of the universe will maximize and no further irreversible thermodynamic changes will occur.
c) One day the Sun will become a giant star and burn up everything.
d) My brain will overheat from trying to learn thermodynamics and the universe will cease to exist as far as I am concerned.

8.      Which of these statements is not a statement of the second law of thermodynamics?

a) The net result of any physical process cannot be the conversion of heat completely into work.
b) Heat cannot flow spontaneously from a low temperature system to a higher temperature system.
c) Internal energy cannot be transferred from a lower temperature system to a higher temperature system.
d) In any physical process, the change in entropy of the universe is never negative.

9.      What is the probability of flipping a fair coin twice and getting identical results on both flips?

a) 1
b) 1/2
c) 1/3
d) 1/4

10.     Flipping 1000 coins at one time, which of the following is the most probable outcome?

a) All heads
b) All tails
c) 400 heads, 600 tails
d) 472 heads, 528 tails

11.     Determine the efficiency of a heat engine that lifts 200 kg a distance of 14.6 m when $5.67 \times 10^4$ J of heat are added to the engine.

12.     Determine the rate at which heat must flow into a Carnot cycle engine acting between 2400 K and 860 K if the output power of the engine is 47.2 hp.

13.     Calculate the total change in entropy of the water in a 50 g ice cube at $-10°$ C when it is warmed to the melting temperature, melted and then warmed to $+20°$ C.

14.     Calculate the total change in entropy of the universe if the ice cube in Question 13 received all of its heat from a source at 20° C.

15.     Two dice are rolled simultaneously. Calculate the probability for each possible total of the spots (macrostate) showing on the upper surface. List the microstates that correspond to each macrostate.

## Problem Solutions

3.  We find the rate of heat input from the efficiency:

$e = W/Q_H = (W/t)/(Q_H/t)$;

$0.38 = (500 \text{ MW})/(Q_H/t)$, which gives $Q_H/t = 1316$ MW.

We find the rate of heat discharge from

$Q_H/t = (Q_L/t) + (W/t)$;

1316 MW = $(Q_L/t)$ + 500 MW, which gives $Q_L/t = 816$ MW.

7.  We find the heat input rate from

$e = W/Q_H$;

$0.38 = (810 \text{ MW})/(Q_H/t)$, which gives $Q_H/t = 2132$ MW.

We find the discharge heat flow from

$Q_L/t = (Q_H/t) - (W/t) = 2132$ MW − 810 MW = 1322 MW.

If this heat flow warms the air, we have

$Q_L/t = (n/t)c\Delta T$;

$(1322 \times 10^6 \text{ W})(3600 \text{ s/h})(24 \text{ h/day}) = (n/t)(7.0 \text{ cal/mol} \cdot \text{C}°)(7.5 \text{ C}°)(4.186 \text{ J/cal})$,

which gives $n/t = 5.20 \times 10^{11}$ mol/day.

To find the volume rate, we use the ideal gas law:

$P(V/t) = (n/t)RT$;

$(1.013 \times 10^5 \text{ Pa})(V/t) = (5.20 \times 10^{11} \text{ mol/day})(8.315 \text{ J/mol} \cdot \text{K})(293 \text{ K})$,

which gives $V/t = 1.25 \times 10^{10}$ m³/day = 13 km³/day.

Depending on the dispersal by the winds, the local climate could be heated significantly.

We find the area from

$A = (V/t)t/h = (12.5 \text{ km}^3/\text{day})(1 \text{ day})/(0.200 \text{ km}) = 63 \text{ km}^2$.

11. (*a*)  The net work is

$$W = W_{ab} + W_{bc} + W_{cd} + W_{da}$$

$$= \int_a^b P\,dV + \int_b^c P\,dV + \int_c^d P\,dV + \int_d^a P\,dV$$

$$= \int_a^b P\,dV + \int_b^c P\,dV - \left( \int_d^c P\,dV + \int_a^d P\,dV \right)$$

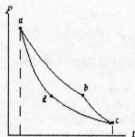

The sum of the first two terms is the area under the abc path, and the sum of the last two terms is the area under the adc path. Thus the net work done is the area enclosed by the cycle.

(*b*)  For any reversible cycle, we can consider it a large number of paths with $W = \Sigma W_i$. We select the two points with the maximum and minimum volumes. We could then apply the above reasoning to the upper and lower paths to arrive at the same result.

15. If we assume the efficiency of a reversible engine, we have

$e_1 = 1 - (T_L/T_H) = 1 - (293 \text{ K}/310 \text{ K}) = 0.055$.

We find the rate at which work can be done from

$e = W/Q_H$;

$0.055 = W/(4000 \text{ kcal/day})$, which gives $W = 219$ kcal/day.

We can estimate the maximum height by assuming all this work increases the potential energy:

$W = mgh_{max}$;

$(219 \text{ kcal/day})(4186 \text{ J/kcal}) = (65 \text{ kg})(9.80 \text{ m/s}^2)h_{max}$, which gives $h_{max} = 1.4 \times 10^3$ m/day.

19. We find the discharge heat flow for the plant from

$Q_{L2}/t = (Q_{H1}/t) - (W/t) = 3.29 \times 10^9 \text{ J/s} - 900 \times 10^6 \text{ J/s} = 2.39 \times 10^9 \text{ J/s}$.

We find the rate at which water must pass through the plant from

$Q_{L2}/t = (m/t)c\Delta T$;

$(2.39 \times 10^9 \text{ J/s})(3600 \text{ s/h}) = (m/t)(4186 \text{ J/kg} \cdot \text{C}°)(5.5 \text{ C}°)$, which gives $m/t = 3.7 \times 10^8$ kg/h.

23. The maximum coefficient of performance for the cooling coil is

$CP = Q_L/W = T_L/(T_H - T_L) = (258 \text{ K})/(303 \text{ K} - 258 \text{ K}) = 5.7$.

27. The efficiency of the engine is
$$e = 1 - (T_L/T_H).$$
For the coefficient of performance, we have
$$CP = T_H/(T_H - T_L) = 1/[1 - (T_L/T_H)] = 1/e = 1/0.35 = 2.9. \quad P = (2.07 \times 10^3 \text{ W}) /(746 \text{ W/hp}) = 2.8 \text{ hp}.$$

31. The power input is
$$P = W/t = Q_L/(CP)t = mL /(CP)t = \rho V/(CP)t;$$
$$1000 \text{ W} = (1.00 \times 10^3 \text{ kg/m}^3)V(3.33 \times 10^5 \text{ J/kg})/(7.0)(3600 \text{ s}),$$
which gives $V = 7.6 \times 10^{-2} \text{ m}^3 = 76 \text{ L}.$

35. We assume when the ice is formed at 0°C it is removed, so its entropy change will be the same as in Problem 34. The heat flow from the water went into the ice at –10°C. Because there is a great deal of ice, its temperature will not change. We find the entropy change of the block of ice from
$$\Delta S_{ice} = Q/T_{ice} = + mL/T = (1.00 \text{ m}^3)(1000 \text{ kg/m}^3)(79.7 \text{ kcal/kg})/(263 \text{ K}) = + 303 \text{ kcal/K}.$$
Thus the total entropy change is
$$\Delta S_{total} = \Delta S_{ice} + \Delta S = + 303 \text{ kcal/K} + (- 292 \text{ kcal/K}) = + 11 \text{ kcal/K}.$$
If the new ice were not removed, we would include an additional heat term which would be negative for cooling the new ice and positive for the great deal of ice. The net additional entropy change would be small.

39. The heat flow to freeze the water is
$$Q = m_{water}L = (2.5 \text{ kg})(3.33 \times 10^5 \text{ J/kg} ) = 8.33 \times 10^5 \text{ J}.$$
We find the final temperature of the ice from
$$Q = m_{ice}c_{ice} \Delta T_{ice} ;$$
$$8.33 \times 10^5 \text{ J} = (450 \text{ kg})(2100 \text{ J/kg} \cdot \text{C}°)[T - (-15°C)], \text{ which gives } T = -14.12°C.$$
The heating of the ice does not occur at constant temperature. Because the temperature change is small, to estimate the entropy change, we will use the average temperature:
$$T_{ice,av} = \frac{1}{2}(T_{ice} + T) = \frac{1}{2} [(-15°C) + (-14.12°C)] = -14.56°C;$$
The total entropy change is
$$\Delta S = \Delta S_{water} + \Delta S_{ice} = (-Q/T_{water}) + (+Q/T_{ice,av})$$
$$= [(-8.33 \times 10^5 \text{ J})/(273.2 \text{ K})] + [(+8.33 \times 10^5 \text{ J})/(258.6 \text{ K})] = +172 \text{ J/K}.$$

43. (a) The heating does not occur at constant temperature. We add (integrate) the differential changes in entropy:
$$\Delta S_{water} = \int (m_{water}c_{water} \, dT)/T = m_{water}c_{water} \ln(T_f/T_i)$$
$$= (1.00 \text{ kg})(1.00 \text{ kcal/kg} \cdot \text{C}°) \ln(373 \text{ K}/273 \text{ K}) = 0.312 \text{ kcal/K}.$$
(b) If the process were reversible, the energy change of the universe would be zero, so
$$\Delta S_{surr} = - \Delta S_{water} = - 0.312 \text{ kcal/K}.$$
Because the process is not reversible, $\Delta S_{surr}$ will be greater, that is, less negative:
$$\Delta S_{surr} > - 0.312 \text{ kcal/K}.$$

47. (a) Because entropy is a state function and the system returns to the initial state in a cycle, the change in entropy for the system is zero. Because all processes are reversible, the change in entropy for the universe is zero, so the change in entropy for the environment is zero.
(b) For the adiabatic processes
$$\Delta Q = 0, \text{ so } \Delta S_{ad} = \Delta dQ/T = 0.$$
We let $V_2$ represent the higher volume for the isothermal processes, so the entropy change is
$$\Delta S = \Delta S_{hot} + \Delta S_{cold} = nR \ln(V_{2H}/V_{1H}) + nR \ln(V_{1L}/V_{2L}).$$
For a Carnot cycle, $V_{2H}/V_{1H} = V_{2L}/V_{1L}$, so we get
$$\Delta S = nR [\ln(V_{2H}/V_{1H}) - \ln(V_{2L}/V_{1L})] = nR [\ln(V_{2H}/V_{1H}) - \ln(V_{2H}/V_{1H})] = 0.$$

51. We use H for head, and T for tail.  For the microstates we construct the following table:

| Macrostate | Microstates | Number |
|---|---|---|
| 6 heads | H H H H H H | 1 |
| 5 heads, 1 tail | H H H H H T, H H H H T H, H H H T H H, H H T H H H, H T H H H H, | 6 |
| | T H H H H H | |
| 4 heads, 2 tails | H H H H T T, H H H T H T, H H H T T H, H H T H H T, H H T H T H, H H T T H H, | 15 |
| | T H H T H H, T H T H H H, T T H H H H | |
| 3 heads, 3 tails | H H H T T T, H H T H T T, H H T T H T, H H T T T H, H T H H T T, H T H T H T, | 20 |
| | H T H T T H, H T T H H T, H T T H T H, H T T T H H, T H H H T T, T H H T H T, | |
| | T H H T T H, T H T H H T, H T H T H, T H T T H H, T T H H H T, T T H H T H, | |
| | T T H T H H, T T T H H H | |
| 2 heads, 4 tails | H H T T T T, H T H T T T, H T T H T T, H T T T H T, H T T T T H, T H H T T T, | 15 |
| | T H T H T T, T H T T H T, T H T T T H, T T H H T T, T T H T H T, T T H T T H, | |
| | T T T H H T, T T T H T H, T T T T H H | |
| 1 head, 5 tails | H T T T T T, T H T T T T, T T H T T T, T T T H T T, T T T T H T, | 6 |
| | T T T T T H | |
| 6 tails | T T T T T T | 1 |

There are a total of 64 microstates.

(a)  The probability of obtaining three heads and three tails is

$P_{33}$ = (20 microstates)/(64 microstates) = 5/16.

(b)  The probability of obtaining six heads is

$P_{60}$ = (1 microstate)/(64 microstates) = 1/64.

55. The maximum possible efficiency is the efficiency of the Carnot cycle:

$e = 1 - (T_L/T_H) = 1 - [(90 \text{ K})/(293 \text{ K})] = 0.69 = 69\%$.

59. (a)  The coefficient of performance for the heat pump is

$CP = T_H/(T_H - T_L) = (297 \text{ K})/(297 \text{ K} - 279 \text{ K}) = 17$.

(b)  We find the heat delivered at the high temperature from

$CP = Q_H/W$ ;

$17 = Q_H/(1000 \text{ J/s})(3600 \text{ s/h})$ , which gives $Q_H = 6.1 \times 10^7$ J/h.

63. The kinetic energy the rock loses when it hits the ground becomes a heat flow to the ground.
We assume the temperature of the ground, $T$, does not change appreciably.  The entropy change is

$\Delta S = Q/T = K/T$.

67. Because process $ab$ is isothermal, $\Delta U_{ab} = 0$, and

$Q_{ab} = W_{ab} = nRT_H \ln(V_b/V_a)$.

Because process $bc$ is constant volume, we have

$Q_{bc} = \Delta U_{bc} = nC_V(T_L - T_H) = n\tfrac{3}{2}R(T_L - T_H)$;

$W_{bc} = 0$.

Because process $cd$ is isothermal, $\Delta U_{cd} = 0$, and

$Q_{cd} = W_{cd} = nRT_L \ln(V_d/V_c)$

$= nRT_L \ln(V_a/V_b) = - nRT_L \ln(V_b/V_a)$.

Because process $da$ is constant volume, we have

$Q_{da} = \Delta U_{da} = nC_V(T_H - T_L) = n\tfrac{3}{2}R(T_H - T_L)$;

$W_{da} = 0$.

The net work done by the cycle is

$W = W_{ab} + W_{cd} = nRT_H \ln(V_b/V_a) - nRT_L \ln(V_b/V_a)$

$= nR(T_H - T_L)\ln(V_b/V_a)$.

The heat added to the system is

$Q_{in} = Q_{ab} + Q_{da} = nRT_H \ln(V_b/V_a) + n\tfrac{3}{2}R(T_H - T_L)$.

The efficiency is

$e_{Stirling} = W/Q_{in} = nR(T_H - T_L)\ln(V_b/V_a)/[nRT_H \ln(V_b/V_a) + n\tfrac{3}{2}R(T_H - T_L)]$

$= (T_H - T_L)\ln(V_b/V_a)/[T_H \ln(V_b/V_a) + \tfrac{3}{2}(T_H - T_L)]$.

The efficiency of a Carnot cycle is

$e_{Carnot} = (T_H - T_L)/T_H$.

We rearrange the expression for the Stirling cycle:

$e_{Stirling} = [(T_H - T_L)/T_H]\{1/[1 + \tfrac{3}{2}(T_H - T_L)/T_H \ln(V_b/V_a)]\}$.

Because the denominator of the second factor is greater that 1, we see that

$e_{Stirling} < e_{Carnot}$.

71. (*a*) We assume that our hand is the first dealt and we receive the aces in the first four cards.  The probability of the first card being an ace is 4/52.  For the next three cards dealt to the other players, there must be no aces, so the probabilities are 48/51, 47/50, 46/49.  The next card is ours; the probability of it being an ace is 3/48.  For the next three cards, there must be no aces, so the probabilities are 45/47, 44/46, 43/45.  For the next cards, we have

2/44, 42/43, 41/42, 40/41, 1/40, 39/39, 38/38, ... .

For the product of all these, we have

$P_1$ = ( 4/52)(48/51)(47/50)(46/49)(3/48)(45/47)(44/46)(43/45)(2/44)(42/43)(41/42)
        (40/41)(1/40)

= (4)(3)(2)(1)/(52)(51)(50)(49).

Because we do not have to receive the aces as the first four cards, we multiply this probability by the number of ways 4 cards can be drawn from a total of 13, which is

13!/4!9! = (13)(12)(11)(10)/(4)(3)(2)(1).

Thus the probability of being dealt four aces is

$P$   = [(4)(3)(2)(1)/(52)(51)(50)(49)][(13)(12)(11)(10)/(4)(3)(2)(1)]

= (13)(12)(11)(10)/(52)(51)(50)(49) = 11/4165 ≈ 1/379

(*b*) We assume that our hand is the first dealt.  Because the suit is not specified, any first card is acceptable, so the probability is (52/52) = 1.  For the next three cards dealt to the other players, there must be no cards in the suit we were dealt, so the probabilities are

39/51, 38/50, 37/49.

The next card is ours; the probability of it being in the same suit is 12/48.

For the next three cards, there must be no cards in the suit we were dealt, so the probabilities are

36/47, 35/46, 34/45.

For the next cards, we have

11/44, 33/43, 32/42, 31/41, 10/40, ... .

For the product of all these, we have

$P_1$   = (1)(39/51)(38/50)(37/49)(12/48)(36/47)(35/46)
                               ×(34/45)(11/44))(33/43)(32/42)(31/41)(10/40)...

= 12! 39!/51! = 6.3 × 10⁻¹² = 1/1.59 × 10¹¹.

# Chapter 21: Electric Charge and Electric Field

## Chapter Overview and Objectives

This chapter introduces electric charges, Coulomb's law, and electric fields. It describes how electrostatic forces can be calculated for discreet and continuous charge distributions.

After completing study of this chapter, you should:

- Know that there are two signs of electric charge and charge is conserved in physical processes.
- Know that like charges repel and unlike charges attract.
- Know the difference between conductors, insulators, and semiconductors.
- Know what induced charge is.
- Know Coulomb's law and how to calculate the force on charged particles caused by other charged particles.
- Know the definition of electric field.
- Know how to calculate the force on a charged particle using the electric field.
- Know how to calculate electric fields for distributions of point charges and continuous distributions of charge.
- Know what electric field lines are and what they represent.
- Know that, in static conditions, the electric field inside a conductor is zero and that the electric field is perpendicular to the surface of the conductor external to the conductor at its surface.
- Know what an electric dipole is and how to determine its electric dipole moment.

## Summary of Equations

Scalar form of Coulomb's law:
$$F = k \frac{Q_1 Q_2}{r^2}$$

Vector form of Coulomb's law:
$$\mathbf{F}_{12} = k \frac{Q_1 Q_2}{r_{21}^2} \hat{\mathbf{r}}_{21}$$

Definition of electric field:
$$\mathbf{E} = \frac{\mathbf{F}}{q}$$

Electric field of a point charge:
$$\mathbf{E} = k \frac{Q}{r^2} \hat{\mathbf{r}}$$

Electric field from a continuous distribution of charge: $\mathbf{E} = \int d\mathbf{E} = \int k \frac{\hat{\mathbf{r}}}{r^2} dq$

Force on a charged particle in an electric field:
$$\mathbf{F} = q\mathbf{E}$$

Electric dipole moment:
$$\mathbf{p} = Q\mathbf{d}$$

Torque on an electric dipole in an electric field:
$$\boldsymbol{\tau} = \mathbf{p} \times \mathbf{E}$$

Potential energy of a dipole in an electric field:
$$U = -\mathbf{p} \cdot \mathbf{E}$$

## Chapter Summary

### Section 21-1 Static Electricity; Electric Charge and Its Conservation

A property of matter is **electric charge**. There are two types of electric charges, positive and negative. Positive charge interacts with other positive charges to create a repulsive force on each charge, as does the interaction between a negative charge with other negative charges.  Positive charge interacts with negative charge to create an attractive force between the two charges.  This is usually stated as **like charges repel and unlike charges attract**.

In all observed processes, the total charge of the universe remains unchanged.  If an amount +$Q$ of charge is created in a process, then an amount –$Q$ also is created somewhere in the process.   This is called **the law of conservation of electric charge**.

### Section 21-2 Electric Charge in the Atom

The current accepted model of ordinary matter is that matter consists of atoms.  Atoms have a structure in which they have a massive nucleus that has most of the mass of the atom and is positively charged.  The nucleus consists of positively charged protons and uncharged neutrons.  The diameter of the nucleus is about one hundred thousandth of the diameter of the atom.   Negative charged particles called electrons orbit the nucleus.  Neutral atoms have the same number of protons and electrons.  Electrons have the same magnitude, but opposite sign of electric charge as protons, so a neutral atom has a zero total electric charge.

Atoms can lose or gain electrons and have a net electric charge.  We call atoms that have a net electric charge **ions**.

Molecules may have no net charge, but will still interact with other charges.  This can happen if the charge in the molecule is distributed asymmetrically.   If the charges in the molecule are distributed asymmetrically, we say the molecule is a **polar** molecule.

### Section 21-3 Insulators and Conductors

Materials through which electric charge is easily transported are called **conductors**.  Materials through which it is very difficult for charge to transport through are called **insulators**.  An intermediate class of materials are called **semiconductors**.

### Section 21-4 Induced Charge; the Electroscope

Charge can be transferred from one electrically charged object to another by direct contact.  Conductors can also be charged without direct contact.  Following the pictures to the right in sequence, first a pair of conductive bodies attached by a thin conductive wire has no net electric charge.  In the second picture, a positively charged body is brought near one end of the pair of conductive bodies.  The positive charge attracts negative charges to the end it is near and repels positive charges toward the far end.  In the third picture, the thin conductive wire between the two bodies is cut, leaving a net negative charge on the body on the left and a net positive charge on the body on the right.  In the fourth picture, when the other bodies are removed, the charge redistributes itself because the negative charges repel each other.  This process of charging bodies is called charging by induction.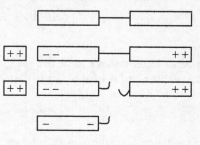

**Electroscopes** and **electrometers** are instruments that can be used to measure electric charge.

### Section 21-5 Coulomb's Law

**Coulomb's law** is a quantitative relationship of the size of the force between charges to the amount of charge and the distance separating the charges.  The electric force is proportional to the product of the two interacting charges divided by the square of the distance between them.  We write this as

$$F = k\frac{Q_1 Q_2}{r^2}$$

where $F$ is the magnitude of the force between the two charges, $Q_1$ and $Q_2$ are the magnitudes of the two interacting charges, and $r$ is the distance between the charges.  The charge is measured in units called coulombs (C). The charge of an electron is $1.602 \times 10^{-19}$ C. The constant $k$ is called Coulomb's constant and has a value of

$$k = 8.988 \times 10^9 \ \text{N} \cdot \text{m}^2/\text{C}^2$$

Sometimes $k$ is written in terms of another constant:

$$k = \frac{1}{4\pi\varepsilon_0}$$

where $\varepsilon_0$ is called the permittivity of free space and has a value

$$\varepsilon_0 = 8.85 \times 10^{-12} \ \text{C}^2/\text{N}\cdot\text{m}^2$$

We can rewrite Coulomb's law in a vector expression that includes the information about the direction of the force:

$$\mathbf{F}_{12} = k \frac{Q_1 Q_2}{r_{21}^2} \hat{\mathbf{r}}_{21}$$

where $\mathbf{F}_{12}$ is the force of charge $Q_2$ on charge $Q_1$, $r_{21}$ is the distance from charge $Q_2$ to charge $Q_1$, and $\hat{\mathbf{r}}_{21}$ is the unit vector that points in the direction from charge $Q_2$ to charge $Q_1$. Note that here $Q_1$ and $Q_2$ carry the sign of the charge in contrast to the non-vector form of Coulomb's law for which the signs are dropped.

### Example 21-5-A

A charge of $-3.6 \times 10^{-4}$ C is located 0.45 m from a $-8.3 \times 10^{-4}$ C charge. At what position along the line between the two charges is the total force on any third charge equal to zero?

### Solution:

Along the line segment joining the two charges, the force from each charge points in a direction opposite the other. That means to have zero net force, the magnitudes of the force from each charge must be equal:

$$k \frac{Q_1 Q_3}{r_1^2} = k \frac{Q_2 Q_3}{r_2^2}$$

where $r_1$ is the distance to the charge $Q_1$, $r_2$ is the distance to the charge $Q_2$, and $Q_3$ is the charge of the third charge. We know that $r_1$ and $r_2$ must add up to the total distance, $d = 0.45$ m, between Q1 and Q2. We can write

$$r_2 = d - r_1$$

Substituting this expression for $r_2$ we get

$$k \frac{Q_1 Q_3}{r_1^2} = k \frac{Q_2 Q_3}{(d - r_1)^2}$$

Solving for $r_1$, we get

$$r_1 = \frac{Q_1 \pm \sqrt{Q_1 Q_2}}{Q_1 - Q_2} d$$

Substituting the values given in the problem, we get

$$r_1 = \frac{-3.6 \times 10^{-4} \ \text{C} \pm \sqrt{\left(-3.6 \times 10^{-4} \ \text{C}\right)\left(-8.3 \times 10^{-4} \ \text{C}\right)}}{\left(-3.6 \times 10^{-4} \ \text{C}\right) - \left(-8.3 \times 10^{-4} \ \text{C}\right)} (0.45 \ \text{m})$$

$$= 0.18 \ \text{m} \quad or \quad -0.87 \ \text{m}$$

Because we know the position at which the charge must be lies between the other two charges, the 0.18 m solution is the one that answers the question. The other solution also corresponds to a position at which the magnitudes of the two forces are equal, but the directions of the forces are the same and so the vector sum is not zero.

### Example 21-5-B

A +3.4 μC charge is located at position (0.0, 0.0, 0.0). A –4.5 μC charge is located at position (0.0, 2.0 m, 3.0 m). A –1.3 μC charge is located at (1.0 m, 1.0 m, 0.0 m). Determine the force on the –1.3 μC charge.

**Solution:**

The displacement of the position of the –1.3 μC charge from the +3.4 μC charge, $\mathbf{r}_{13}$, is

$$\mathbf{r}_{13} = (1.0\,\text{m}\,\mathbf{i} + 1.0\,\text{m}\,\mathbf{j}) - \vec{0} = 1.0\,\text{m}\,\mathbf{i} + 1.0\,\text{m}\,\mathbf{j}$$

The distance between the –1.3 μC charge and the +3.4 μC charge is the magnitude of this vector:

$$r_{13} = \sqrt{(1.0\,\text{m})^2 + (1.0\,\text{m})^2} = 1.41\,\text{m}$$

The unit vector pointing in the direction from the –1.3 μC charge to the +3.4 μC charge is

$$\hat{\mathbf{r}}_{13} = \frac{\mathbf{r}_{13}}{r_{13}} = \frac{1.0\,\text{m}\,\mathbf{i} + 1.0\,\text{m}\,\mathbf{j}}{1.41\,\text{m}} = 0.707\,\text{m}\,\mathbf{i} + 0.707\,\text{m}\,\mathbf{j}$$

Calculating the force of the +3.4 μC charge on the –1.3 μC charge, $\mathbf{F}_{13}$, using Coulomb's law, we get

$$\mathbf{F}_{13} = k\frac{Q_1 Q_3}{r_{13}^2}\hat{\mathbf{r}}_{13} = \left(9.0 \times 10^9\,\text{N}\cdot\text{m}^2/\text{C}^2\right)\frac{\left(3.4 \times 10^{-6}\,\text{C}\right)\left(-1.3 \times 10^{-6}\,\text{C}\right)}{(1.41\,\text{m})^2}(0.707\,\mathbf{i} + 0.707\,\mathbf{j})$$

$$= -1.41 \times 10^{-2}\,\text{N}\,\mathbf{i} - 1.41 \times 10^{-2}\,\text{N}\,\mathbf{j}$$

Similarly, we calculate the force of the –4.5 μC charge on the –1.3μC charge:

$$\mathbf{r}_{23} = (1.0\,\text{m}\,\mathbf{i} + 1.0\,\text{m}\,\mathbf{j}) - (2.0\,\text{m}\,\mathbf{j} + 3.0\,\text{m}\,\mathbf{k}) = 1.0\,\text{m}\,\mathbf{i} - 1.0\,\text{m}\,\mathbf{j} - 3.0\,\text{m}\,\mathbf{k}$$

$$r_{23} = \sqrt{(1.0\,\text{m})^2 + (-1.0\,\text{m})^2 + (-3.0\,\text{m})^2} = 3.32\,\text{m}$$

$$\hat{\mathbf{r}}_{23} = \frac{\mathbf{r}_{23}}{r_{23}} = \frac{1.0\,\text{m}\,\mathbf{i} - 1.0\,\text{m}\,\mathbf{j} + 3.0\,\text{m}\,\mathbf{k}}{3.32\,\text{m}} = 0.301\,\text{m}\,\mathbf{i} - 0.301\,\text{m}\,\mathbf{j} - 0.904\,\text{m}\,\mathbf{k}$$

$$\mathbf{F}_{23} = k\frac{Q_2 Q_3}{r_{23}^2}\hat{\mathbf{r}}_{23} = \left(9.0 \times 10^9\,\text{N}\cdot\text{m}^2/\text{C}^2\right)\frac{\left(-4.5 \times 10^{-6}\,\text{C}\right)\left(-1.3 \times 10^{-6}\,\text{C}\right)}{(3.32\,\text{m})^2}(0.301\,\mathbf{i} - 0.301\,\mathbf{j} - 0.904\,\mathbf{k})$$

$$= 1.44 \times 10^{-3}\,\text{N}\,\mathbf{i} - 1.44 \times 10^{-3}\,\text{N}\,\mathbf{j} - 4.31 \times 10^{-3}\,\text{N}\,\mathbf{k}$$

The net force on the –1.3 μC charge is the sum of the two forces:

$$\mathbf{F} = \mathbf{F}_{13} + \mathbf{F}_{23} = \left(-1.41 \times 10^{-2}\,\text{N}\,\mathbf{i} - 1.41 \times 10^{-2}\,\text{N}\,\mathbf{j}\right) + \left(1.44 \times 10^{-3}\,\text{N}\,\mathbf{i} - 1.44 \times 10^{-3}\,\text{N}\,\mathbf{j} - 4.31 \times 10^{-3}\,\text{N}\,\mathbf{k}\right)$$

$$= -1.27 \times 10^{-2}\,\text{N}\,\mathbf{i} - 1.55 \times 10^{-2}\,\text{N}\,\mathbf{j} - 4.31 \times 10^{-3}\,\text{N}\,\mathbf{k}$$

**Section 21-6 The Electric Field**

The electric field, **E**, at a given point in space is defined as the electric force, **F**, on a small test charge, $q$, divided by the charge:

$$\mathbf{E} = \frac{\mathbf{F}}{q}$$

The meaning of a small test charge is a charge small enough to not change the distribution of charge that cause the force on the test charge.

The electric field created by a point charge, $Q$, is

$$\mathbf{E} = \frac{\mathbf{F}}{q} = k\frac{Qq}{qr^2}\hat{\mathbf{r}} = k\frac{Q}{r^2}\hat{\mathbf{r}}$$

where $r$ is the distance from the charge $Q$ to the point at which the electric field is being evaluated and $\hat{\mathbf{r}}$ is the unit vector that points from the charge $Q$ to the point at which the electric field is being evaluated.

Just as forces from several electric charges add together as vectors, so do electric fields:

$$\mathbf{E} = \mathbf{E}_1 + \mathbf{E}_2 + \mathbf{E}_3 + \cdots$$

where **E** is the electric field and $\mathbf{E}_1$, $\mathbf{E}_2$, $\mathbf{E}_3$, ... are the electric fields due to individual point charges. This is called the **principle of superposition** for electric fields.

**Example 21-6-A**

A 2.17 μC charge is located on the x-axis a distance 0.345 m from the origin and a –4.21μC charge is located on the y-axis a distance 0.576 m from the origin. What is the electric field at the origin?

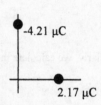

**Solution:**

It should be easy to see that the direction of the electric field at the origin due to the +2.17 μC charge is in the –**i** direction and the direction of the electric field at the origin due to the –4.21 μC charge is in the +**j** direction. The net electric field is

$$\mathbf{E} = k\frac{q_1}{r_1^2}\hat{\mathbf{r}}_1 + k\frac{q_2}{r_2^2}\hat{\mathbf{r}}_2 = \left(9.0\times10^9 \text{ N}\cdot\text{m}^2/\text{C}^2\right)\frac{\left(2.17\times10^{-6}\text{ C}\right)}{\left(0.345\,\text{m}\right)^2}(-\mathbf{i}) + \left(9.0\times10^9 \text{ N}\cdot\text{m}^2/\text{C}^2\right)\frac{\left(-4.21\times10^{-6}\text{ C}\right)}{\left(0.576\,\text{m}\right)^2}(+\mathbf{j})$$

$$= -1.64\times10^5 \text{ N/C}\,\mathbf{i} - 1.14\times10^5 \text{ N/C}\,\mathbf{j}$$

**Example 21-6-B**

An equilateral triangle with sides of length $a$ has charges at each of its vertices. One vertex has a charge $-q$ and the two other vertices have a charge $+q$. What is the electric field at the centroid of the equilateral triangle?

**Solution:**

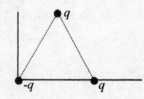

To help in completing the problem, we draw the arrangement of charges with a coordinate system. We choose the origin to be at the location of the $-q$ charge and we let one side of the triangle lie on the +x-axis as shown. We can write the location of each of the charges as

$$-q \quad at \quad \mathbf{r}_1 = \vec{0}$$
$$+q \quad at \quad \mathbf{r}_2 = a\mathbf{i}$$
$$+q \quad at \quad \mathbf{r}_3 = a\cos 60°\,\mathbf{i} + a\sin 60°\,\mathbf{j}$$

We know from elementary geometry that the centroid of a triangle lies two-thirds of the distance from any vertex to the bisector of the opposite side.  It should be easy to see that the position of the centroid is

$$\mathbf{r}_c = \frac{2}{3}\frac{\sqrt{3}}{2}a\left(\cos 30^\circ\,\mathbf{i} + \sin 30^\circ\,\mathbf{j}\right) = \frac{1}{2}a\,\mathbf{i} + \frac{\sqrt{3}}{6}a\,\mathbf{j}$$

We calculate the vector displacement of the centroid from each vertex:

$$\mathbf{r}_{1c} = \mathbf{r}_c - \mathbf{r}_1 = \frac{1}{2}a\,\mathbf{i} + \frac{\sqrt{3}}{6}a\,\mathbf{j} - \vec{0} = \frac{1}{2}a\,\mathbf{i} + \frac{\sqrt{3}}{6}a\,\mathbf{j}$$

$$\mathbf{r}_{2c} = \mathbf{r}_c - \mathbf{r}_2 = \frac{1}{2}a\,\mathbf{i} + \frac{\sqrt{3}}{6}a\,\mathbf{j} - a\,\mathbf{i} = -\frac{1}{2}a\,\mathbf{i} + \frac{\sqrt{3}}{6}a\,\mathbf{j}$$

$$\mathbf{r}_{3c} = \mathbf{r}_c - \mathbf{r}_3 = \frac{1}{2}a\,\mathbf{i} + \frac{\sqrt{3}}{6}a\,\mathbf{j} - \frac{1}{2}a\,\mathbf{i} - \frac{\sqrt{3}}{2}a\,\mathbf{j} = -\frac{\sqrt{3}}{3}a\,\mathbf{j}$$

The magnitudes of these displacement vectors are all the same:

$$r_{1c} = r_{2c} = r_{3c} = \frac{\sqrt{3}}{3}a$$

We can now calculate the unit vectors in the directions of the displacements:

$$\hat{\mathbf{r}}_{1c} = \frac{\mathbf{r}_{1c}}{r_{1c}} = \frac{\frac{1}{2}a\,\mathbf{i} + \frac{\sqrt{3}}{6}a\,\mathbf{j}}{\frac{\sqrt{3}}{3}a} = \frac{\sqrt{3}}{2}\mathbf{i} + \frac{1}{2}\mathbf{j}$$

$$\hat{\mathbf{r}}_{2c} = \frac{\mathbf{r}_{2c}}{r_{2c}} = \frac{-\frac{1}{2}a\,\mathbf{i} + \frac{\sqrt{3}}{6}a\,\mathbf{j}}{\frac{\sqrt{3}}{3}a} = -\frac{\sqrt{3}}{2}\mathbf{i} + \frac{1}{2}\mathbf{j}$$

$$\hat{\mathbf{r}}_{1c} = \frac{\mathbf{r}_{1c}}{r_{1c}} = \frac{-\frac{\sqrt{3}}{3}a\,\mathbf{j}}{\frac{\sqrt{3}}{3}a} = -\mathbf{j}$$

We can now calculate the electric field at the centroid due to each charge:

$$\mathbf{E}_{1c} = k\frac{-q}{r_{1c}^2}\hat{\mathbf{r}}_{1c} = \frac{-kq}{\left(\frac{\sqrt{3}}{3}a\right)^2}\left(\frac{\sqrt{3}}{2}\mathbf{i} + \frac{1}{2}\mathbf{j}\right) = -\frac{3kq}{a^2}\left(\frac{\sqrt{3}}{2}\mathbf{i} + \frac{1}{2}\mathbf{j}\right)$$

$$\mathbf{E}_{2c} = k\frac{q}{r_{2c}^2}\hat{\mathbf{r}}_{2c} = \frac{kq}{\left(\frac{\sqrt{3}}{3}a\right)^2}\left(-\frac{\sqrt{3}}{2}\mathbf{i} + \frac{1}{2}\mathbf{j}\right) = \frac{3kq}{a^2}\left(-\frac{\sqrt{3}}{2}\mathbf{i} + \frac{1}{2}\mathbf{j}\right)$$

$$\mathbf{E}_{3c} = k\frac{q}{r_{3c}^2}\hat{\mathbf{r}}_{3c} = \frac{kq}{\left(\frac{\sqrt{3}}{3}a\right)^2}(-\mathbf{j}) = -\frac{3kq}{a^2}\mathbf{j}$$

The net electric field is the sum of the electric fields from each point charge:

$$\mathbf{E} = \mathbf{E}_{1c} + \mathbf{E}_{2c} + \mathbf{E}_{3c} = \left[-\frac{3kq}{a^2}\left(\frac{\sqrt{3}}{2}\mathbf{i} + \frac{1}{2}\mathbf{j}\right)\right] + \left[\frac{3kq}{a^2}\left(-\frac{\sqrt{3}}{2}\mathbf{i} + \frac{1}{2}\mathbf{j}\right)\right] + \left[-\frac{3kq}{a^2}\mathbf{j}\right]$$

$$= -\frac{3kq}{a^2}\left(\sqrt{3}\,\mathbf{i} + \mathbf{j}\right)$$

### Section 21-7 Electric Field Calculations for Continuous Charge Distributions

In many cases, there are so many charges present that it is impossible to sum the electric field from each of the individual charges, but the charge distribution can be treated by a continuous charge density.  For each infinitesimal charge element $dq$ in the distribution of charge, there is an infinitesimal contribution to the net electric field, $d\mathbf{E}$, where

$$d\mathbf{E} = k\frac{\hat{\mathbf{r}}}{r^2}dq$$

The distance from the infinitesimal charge element to the point at which the field is being evaluated is $r$, and the unit vector in the direction from the infinitesimal charge element to the point the field is being evaluated is $\hat{\mathbf{r}}$. To determine the electric field at a given point, we integrate over all of the infinitesimal charge elements:

$$\mathbf{E} = \int d\mathbf{E} = \int k\frac{\hat{\mathbf{r}}}{r^2}dq$$

Because the charge distribution is often a known function of spatial location, we almost always change variables in the integral and write the charge density as a function of position. For a linear (or one-dimensional) density of charge, we write

$$dq = \lambda\, dl$$

where $\lambda$ is the linear charge density (charge per unit length) and $dl$ is an infinitesimal line element along the charge distribution. If the charge distribution is a surface charge density, then we write

$$dq = \sigma\, dA$$

where $\sigma$ is the surface charge density (charge per unit area) and $dA$ is an infinitesimal area element of the charge distribution. If the charge distribution is three-dimensional, then we write

$$dq = \rho\, dV$$

where $\rho$ is the volume charge density (charge per unit volume), and $dV$ is an infinitesimal volume element.

### Example 21-7-A

A linear charge density lies along the $+x$-axis from $x = 0$ to $x = L$. The charge density as a function of position along the this segment is given by

$$\lambda(x) = \lambda_0\frac{x^2}{L^2}$$

Determine the electric field due to this charge distribution along the $x$-axis for all $x$ greater than $L$.

**Solution:**

This is a linear continuous charge distribution. We calculate the electric field by finding the integral

$$E = \int k\frac{\hat{\mathbf{r}}}{r^2}\lambda\, dl$$

In this case, the charge distribution lies on the x-axis and an infinitesimal line element of the charge is the same as an infinitesimal increment of $x$ so $dl = dx$. We are calculating the electric field on the $x$ axis at $x > L$ so that for all points in the charge distribution, the direction to the point at which the electric field is being evaluated is the $\mathbf{i}$ direction. This means that $\hat{\mathbf{r}} = \mathbf{i}$. Let the point at which we are evaluating the field be called $x'$. Then the distance, $r$, from the charge density at $x$ to the point at which the field is being evaluated is $x' - x$. We can now write the integral as

$$\mathbf{E} = \int_{x=0}^{L} k\frac{\mathbf{i}}{(x'-x)^2}\lambda_0\frac{x}{L}dx$$

This integral can easily be completed by using the substitution $u = x' - x$. The result is

$$\mathbf{E} = k \frac{\lambda_0 \mathbf{i}}{L} \left[ \ln(x' - x) + \frac{x'}{x' - x} \right]_{x=0}^{L}$$

$$= k \frac{\lambda_0 \mathbf{i}}{L} \left[ \ln\left( \frac{x' - L}{x'} \right) + \frac{L}{x' - L} \right]$$

### Section 21-8 Field Lines

A type of diagram that is used to help visualize the electric field is a diagram of **electric field lines**. An electric field line is a path such that the electric field at each point along the path is tangent to the path at that point. An electric field line carries no information about the magnitude of the electric field, only about its direction. If a diagram of electric field lines is made, it is created according to the following rules:

1.  Electric field lines only begin on positive charges and end on negative charges. The number of lines starting or ending on a given charge is proportional to the magnitude of the charge.

2.  A given line has its tangent at a given point in the direction of the electric field at that point.

3.  The magnitude of the electric field is proportional to the density of the electric field lines passing through a plane perpendicular to the direction of the electric field.

### Section 21-9 Electric Fields and Conductors

Because charges move freely inside a conductor, there will be no electric field inside a conductor under static conditions. If there were an electric field, charges would continue to experience a force and move within the conductor. The only condition consistent with static conditions is that there be no electric field inside the conductor. The electric field at the surface just external to the conductor must be perpendicular to the conductor under static conditions.

### Section 21-10 Motion of a Charged Particle in an Electric Field

The force on an object with charge $q$ in an electric field $\mathbf{E}$ is given by

$$\mathbf{F} = q\mathbf{E}$$

### Example 21-10-A

An electron has a velocity of $3.99 \times 10^4$ m/s $\mathbf{i}$. It moves in an electric field $2.76 \times 10^2$ N/C $\mathbf{i}$. How far does the particle travel before its velocity reaches zero?

### Solution:

We can calculate the force on the electron:

$$\mathbf{F} = q\mathbf{E}$$

We can then use Newton's second law to calculate the acceleration of the electron:

$$\mathbf{a} = \frac{\mathbf{F}}{m} = \frac{q\mathbf{E}}{m}$$

We can use the kinematic equation

$$v_f^2 - v_i^2 = 2a(x_f - x_i)$$

to relate the initial and final velocities to the acceleration and the displacement because the acceleration is constant. Solving this kinematic equation for the displacement, we get

$$x_f - x_i = \frac{v_f^2 - v_i^2}{2a} = \frac{\left(v_f^2 - v_i^2\right)m}{2qE} = \frac{\left[0 - \left(3.99 \times 10^4 \text{ m/s}\right)^2\right]\left(9.11 \times 10^{-31} \text{ kg}\right)}{2\left(-1.60 \times 10^{-19} \text{ C}\right)\left(2.76 \times 10^2 \text{ N/C}\right)}$$

$$= 1.64 \times 10^{-5} \text{ m}$$

### Section 21-11 Electric Dipoles

A particular distribution of charge that is often a good approximation to many physical situations is the **electric dipole**. An electric dipole is two charges of equal magnitude but opposite sign, $+Q$ and $-Q$, separated by a displacement **d**. This displacement is the displacement of the positive charge from the negative charge, i.e., it points from the location of the negative charge to the location of the positive charge. The **electric dipole moment, p**, is a vector given by

$$\mathbf{p} = Q\mathbf{d}$$

If $l$ is the length of the displacement vector **d**, we can write the magnitude of the dipole moment, $p = Ql$.

In an electric field, a dipole experiences a torque:

$$\boldsymbol{\tau} = \mathbf{p} \times \mathbf{E}$$

As the dipole is rotated by the electric field, work is done on the dipole by the torque. The potential energy associated with the orientation of the dipole moment in the electric field is given by:

$$U = -\mathbf{p} \cdot \mathbf{E}$$

A dipole in a uniform electric field feels no net electric force. In a uniform electric field, the positive charge and the negative charge of the dipole experience equal magnitude and opposite direction forces. An electric dipole in a non-uniform electric field can experience a net electric force because the electric field can be different at the locations of the positive and negative charges.

### Example 21-11-A

A rigid electric dipole consists of a +6.32 μC charge separated from a –6.32 μC charge by a distance of 3.45 cm. The dipole moment points is in the direction of the unit vector **–i**. The dipole has a moment of inertia about the **k** direction of $1.31 \times 10^{-4}$ kg·m². The dipole is in an electric field given by $(3.11 \times 10^1 \text{ N/C})(2\mathbf{i} + \mathbf{j})$. The torque on the dipole caused by the electric field will cause it to rotate. Calculate the torque on the dipole in its current orientation and calculate the rotational speed of the dipole when its dipole moment is pointing in the +**i** direction.

### Solution:

First, we calculate the magnitude of the dipole moment:

$$p = Qd = \left(6.32 \times 10^{-6} \text{ C}\right)\left(3.45 \times 10^{-2} \text{ m}\right) = 2.18 \times 10^{-7} \text{ C} \cdot \text{m}$$

We know from the problem statement that the current direction of the dipole moment is –**i**, so

$$\mathbf{p} = 2.18 \times 10^{-7} \text{ C} \cdot \text{m}\,\mathbf{i}$$

We can now calculate the torque

$$\boldsymbol{\tau} = \mathbf{p} \times \mathbf{E} = \left(-2.18 \times 10^{-7} \text{ C} \cdot \text{m}\,\mathbf{i}\right) \times \left(3.11 \times 10^1 \text{ N/C}\right)\left(2\mathbf{i} + \mathbf{j}\right)$$

$$= -6.78 \times 10^{-6} \text{ N} \cdot \text{m}\,\mathbf{k}$$

We use the principle of conservation of energy to determine the angular speed when the dipole moment points in the $+\mathbf{i}$ direction:

$$\Delta K + \Delta U = 0$$

Writing the kinetic energy in terms of the angular speed and the potential energy in terms of the dipole moment–electric field interaction we have

$$\tfrac{1}{2}I\left(\omega_f^2 - \omega_i^2\right) + \left[\left(-\mathbf{p}_f \cdot \mathbf{E}\right) - \left(-\mathbf{p}_i \cdot \mathbf{E}\right)\right] = 0$$

where the $f$ subscripts indicate the final values of the quantity and the $i$ subscripts indicate the initial values and $I$ is the moment of inertia of the dipole about an axis through its center of mass perpendicular to the dipole moment. The dipole is initially at rest, so the initial angular speed is zero. Solving the above equation for the final angular speed gives

$$\omega_f = \sqrt{\frac{2\left[\left(\mathbf{p}_f - \mathbf{p}_i\right) \cdot \mathbf{E}\right]}{I}} = \sqrt{\frac{2\left(2.18\times10^{-7}\ \mathrm{C}\cdot\mathrm{m}\right)\left[\mathbf{i} - (-\mathbf{i})\right] \cdot 3.11\times10^{1}\ \mathrm{N/C}}{1.31\times10^{-4}\ \mathrm{kg}\cdot\mathrm{m}^2}}$$

$$= 0.207\,\mathrm{s}^{-1}$$

## Practice Quiz

1.    A charged object A repels a charged object B. Charged object A also repels charged object C. What type of force exists between charged object B and charged object C?

   a)   Object B will repel object C.
   b)   Object B will attract object C.
   c)   Object B will apply no force on object C.
   d)   There is not enough information to determine the type of force of B on C.

2.    Given two electrical charges separated by a distance $d$, will there always be a position at which a third charge will experience no electrical force?

   a)   There will always be a position at which a third charge will have no electrical force on it.
   b)   There will be a position at which a third charge experiences no force only if the first two charges are of opposite sign.
   c)   There will be a position at which a third charge experiences no force only if the first two charges are of the same sign.
   d)   There will be cases in which no such position exists.

3.    Electric field lines are:

   a)   The direction of the force on a charged particle.
   b)   The direction a charged particle follows in the electric field.
   c)   Lines of constant magnitude of the electric field.
   d)   Lines tangent to the direction of the electric field at each point.

4.    Two surfaces, each of which are uncharged, are rubbed together. After rubbing the surfaces together, one has a charge of 3.4 μC and the other as a charge of –4.6 μC. The two surfaces only exchanged charged particles and no charges from other sources were lost or gained. Is this possible?

   a)   No, it violates conservation of charge
   b)   No, materials do not have that great of amount of charge to share
   c)   No, the materials must each have the same charge as they started with
   d)   Yes, the total charge can change in a process

5.    At a distance of $d$ from a point charge, the magnitude of the electric force on a charge $q$ is $F$. What is the magnitude of the electric force if the charge is moved to a distance $3d$ from the other charge?

      a) $3F$
      b) $9F$
      c) $F/3$
      d) $F/9$

6.    Suppose the convention on the sign of charges was reversed. That is, positive charges are called negative and negative charges are called positive. If Coulomb's law were written the same as it is now, what would happen to the prediction of electric forces?

      a) The direction of the predicted force would be opposite in direction to the actual force.
      b) The direction would be correct for like sign charges, but wrong for opposite sign charges.
      c) The direction would be correct for opposite sign charges, but wrong for like sign charges.
      d) The direction of the predicted force would be correct.

7.    The gravitational constant in Newton's law of gravity is 20 orders of magnitude smaller than Coulomb's constant in SI units, but gravitational forces seem to dominate over electrical forces in most of everyday life. Why is that?

      a) Most objects have a relatively small net electrical charge.
      b) The distance between the masses is much smaller than the distance between the charges.
      c) Gravity is actually an electric force.
      d) Air is electrically conductive and the electric field is zero inside conductive media.

8.    Which of the electric field line pictures is correct?

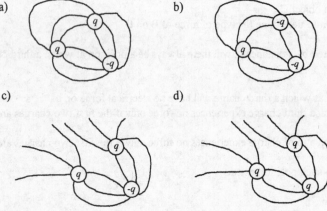

9.    Why is it difficult to accelerate a charged, thin flat object along a conductive surface using electrostatic fields?

      a) Flat objects are difficult to keep charged.
      b) Electrostatic fields near a conductive surface are always zero in magnitude.
      c) Electrostatic fields always increase the frictional force between the object and the surface.
      d) Electrostatic fields are perpendicular to conductive surfaces at the surface.

10.   Can an electric field cause a translational acceleration of an electric dipole?

      a) No, it can only cause a torque on the electric dipole.
      b) Yes, all electric dipoles accelerate in the direction of the electric field.
      c) No, an electric field has no effect on an electric dipole because of its total zero charge.
      d) Yes, if the electric field changes with position.

11.    Three charges lie on a straight line.  A 32 µC charge is 1.62 m to the east of a –6.73 µC charge and 3.64 m to the east of a +24.5 µC.  What is the net electric force on the 32 µC charge?

12.    A 5.6 µC charge is at the origin.  A 3.5 µC charge is located at a position ($x = 1.00$ m, $y = 0.00$).  A –2.3 µC charge is located at the position ($x = 0.00$, $y = 0.78$ m).  What is the net electric force on the 5.6 µC charge?

13.    What is the electric field at the position ($x = 0.00$ m, $y = 0.00$ m, $z = 1.00$ m) for the arrangement of three charges given in quiz problem 12?

14.    A particle has an initial velocity 245 m/s $\mathbf{i}$. The particle has a charge $4.76 \times 10^{-4}$ C and a mass $3.78 \times 10^{-3}$ kg. The particle is moving in a uniform electric field 365 N/C $\mathbf{i}$.   How far will the particle move in 3.87 s?

15.    In a particular electric field, a 14.6 µC charge feels a force 34 N $\mathbf{i}$ + 28 N $\mathbf{j}$.  What is the torque on an electric dipole with electric dipole moment 0.24 uC·m $\mathbf{i}$ + 0.16 µC·m $\mathbf{k}$ in this field?

## Problem Solutions

3.    The magnitude of the Coulomb force is
    $F$   $= kQ_1Q_2/r^2$
        $= (9.0 \times 10^9$ N · m²/C²)(26)(1.60 × 10⁻¹⁹ C)(1.60 × 10⁻¹⁹ C)/(1.5 × 10⁻¹² m)² = 2.7 × 10⁻³ N.

7.    The magnitude of the Coulomb force is
    $F = kQ_1Q_2/r^2$.
    If we divide the expressions for the two forces, we have
    $F_2/F_1 = (r_1/r_2)^2$;
    $3 = [(15.0$ cm)/$r_2]^2$, which gives $r_2 = 8.66$ cm.

11.  Because all the charges and their separations are equal,
    we find the magnitude of the individual forces:
    $F_1$   $= kQQ/L^2 = kQ^2/L^2$
        $= (9.0 \times 10^9$ N · m²/C²)(11.0 × 10⁻⁶ C)²/(0.150 m)²
        $= 48.4$ N.

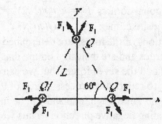

The directions of the forces are determined from the signs
of the charges and are indicated on the diagram.
For the forces on the top charge, we see that the horizontal
components will cancel.  For the net force, we have
    $F$   $= F_1 \cos 30° + F_1 \cos 30° = 2F_1 \cos 30°$
        $= 2(48.4$ N) cos 30°
        $= 83.8$ N up, or away from the center of the triangle.
From the symmetry each of the other forces will have the same magnitude and a direction away from the center:  The net force on each charge is 83.8 N away from the center of the triangle.
Note that the sum for the three charges is zero.

15.  The magnitudes of the individual forces on the charges are
    $F_{12} = kQ2Q/l^2 = 2kQ^2/l^2$;
    $F_{13} = kQ3Q/(l\sqrt{2})^2 = 3kQ^2/2l^2$;
    $F_{14} = kQ4Q/l^2 = 4kQ^2/l^2$;
    $F_{23} = k2Q3Q/l^2 = 6kQ^2/l^2$;
    $F_{24} = k2Q4Q/(l\sqrt{2})^2 = 4kQ^2/l^2$;
    $F_{34} = k3Q4Q/l^2 = 12kQ^2/l^2$.

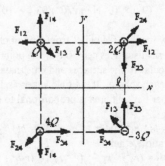

The directions of the forces are determined from the signs
of the charges and are indicated on the diagram.
For charge $Q$ we have
    $\mathbf{F}_1$   $= (-F_{12} + F_{13} \cos 45°)\mathbf{i} + (-F_{13} \sin 45° + F_{14})\mathbf{j}$
        $= [-(2kQ^2/l^2) + (3kQ^2/2l^2)(\sqrt{2}/2)]\mathbf{i} +$
            $[-(3kQ^2/2l^2)(\sqrt{2}/2) + (4kQ^2/l^2)]\mathbf{j}$
        $=   (kQ^2/l^2)[(-2 + 3\sqrt{2}/4)\mathbf{i} + (4 - 3\sqrt{2}/4)\mathbf{j}]$.
For charge $2Q$ we have
    $\mathbf{F}_2$   $= (F_{12} + F_{24} \cos 45°)\mathbf{i} + (F_{24} \sin 45° - F_{23})\mathbf{j}$

$$= [(2kQ^2/l^2) + (4kQ^2/l^2)(\sqrt{2}/2)]\mathbf{i} + [(4kQ^2/l^2)(\sqrt{2}/2) - (6kQ^2/l^2)]\mathbf{j}$$
$$= (kQ^2/l^2)[(2 + 2\sqrt{2})\mathbf{i} + (-6 + 2\sqrt{2})\mathbf{j}].$$

For charge $-3Q$ we have

$$\mathbf{F}_3 = (-F_{34} - F_{13}\cos 45°)\mathbf{i} + (F_{13}\sin 45° + F_{23})\mathbf{j}$$
$$= [(-12kQ^2/l^2) - (3kQ^2/2l^2)(\sqrt{2}/2)]\mathbf{i} + [(3kQ^2/2l^2)(\sqrt{2}/2) + (6kQ^2/l^2)]\mathbf{j}$$
$$= (kQ^2/l^2)[(-12 - 3\sqrt{2}/4)\mathbf{i} + (6 + 3\sqrt{2}/4)\mathbf{j}].$$

For charge $4Q$ we have

$$\mathbf{F}_4 = (F_{34} - F_{24}\cos 45°)\mathbf{i} + (-F_{24}\sin 45° - F_{14})\mathbf{j}$$
$$= [(12kQ^2/l^2) - (4kQ^2/l^2)(\sqrt{2}/2)]\mathbf{i} + [-(4kQ^2/l^2)(\sqrt{2}/2) - (4kQ^2/l^2)]\mathbf{j}$$
$$= (kQ^2/l^2)[(12 - 2\sqrt{2})\mathbf{i} + (-4 - 2\sqrt{2})\mathbf{j}].$$

19. Because the charges have the same sign, they repel each other. The force from the third charge must balance the repulsive force for each charge, so the third charge must be positive and between the two negative charges. For each of the negative charges, we  have

$$Q_0: kQ_0Q/x^2 = kQ_0(3Q_0)/l^2, \quad \text{or} \quad l^2Q = 3x^2Q_0;$$
$$3Q_0: \quad k3Q_0Q/(l-x)^2 = kQ_0(3Q_0)/l^2, \quad \text{or} \quad l^2Q = (l-x)^2Q_0.$$

Thus we have

$$3x^2 = (l-x)^2, \quad \text{which gives } x = -1.37l, +0.366l.$$

Because the positive charge must be between the charges, it must be $0.366l$ from $Q_0$. When we use this value in one of the force equations, we get

$$Q = 3(0.366l)^2Q_0/l^2 = 0.402Q_0.$$

Thus we place a charge of $0.402Q_0$, $0.366l$ from $Q_0$.

Note that the force on the middle charge is also zero.

23. From the symmetry, we see that there are only three magnitudes for the seven forces from the other charges: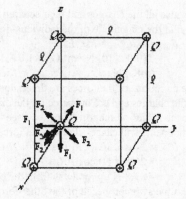

3 adjacent corners: $F_1 = kQQ/l^2 = kQ^2/l^2$

3 diagonal corners: $F_2 = kQQ/(l\sqrt{2})^2 = kQ^2/2l^2$;

1 opposite corner: $F_3 = kQQ/(l\sqrt{3})^2 = kQ^2/3l^2$.

The directions of the forces are determined from the signs of the charges and are indicated on the diagram. We could add the seven forces, but we can use the symmetry to reduce the process. Each of the components will have the same magnitude and will be in the corresponding negative direction. Thus we find one of them:

$$F_x = -F_1 - F_2/\sqrt{2} - F_2/\sqrt{2} - F_3/\sqrt{3}$$
$$= -[(kQ^2/l^2) + 2(kQ^2/2l^2)/\sqrt{2} + (kQ^2/3l^2)/\sqrt{3}]$$
$$= -(kQ^2/l^2)[1 + (1/\sqrt{2}) + (1/3\sqrt{3})] = -1.90kQ^2/l^2.$$

Thus the resultant force is

$$\mathbf{F} = -(1.90kQ^2/l^2)(\mathbf{i} + \mathbf{j} + \mathbf{k}).$$

27. The electric field above a positive charge will be away from the charge, or up. We find the magnitude from

$$E = kQ/r^2$$
$$= (9.0 \times 10^9 \text{ N} \cdot \text{m}^2/\text{C}^2)(33.0 \times 10^{-6} \text{ C})/(0.200 \text{ m})^2 = 7.43 \times 10^6 \text{ N/C (up)}.$$

31. We know that the field lines from a point charge are radial. If we choose two spherical surfaces centered at the point charge, the field lines will be perpendicular to the surfaces and must have the same number passing through each sphere. If the field varies as $1/r^{2+x}$, and we assume that the number per unit area is proportional to the electric field magnitude, we have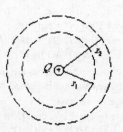

$$N_1 = b(1/r_1^{2+x})4\pi r_1^2, \quad \text{and} \quad N_2 = b(1/r_2^{2+x})4\pi r_2^2.$$

For the ratio we get

$$N_1/N_2 = (r_2^{2+x}/r_1^{2+x})(r_1^2/r_2^2) = (r_2/r_1)^x \neq 1.$$

Thus we could not draw the field lines proportional to the magnitude of the field.

35. The acceleration is produced by the force from the electric field:

$$F = qE = ma;$$

$(-1.60 \times 10^{-19} \text{ C})E = (9.11 \times 10^{-31} \text{ kg})(145 \text{ m/s}^2)$, which gives $E = -8.26 \times 10^{-10}$ N/C.

Because the charge on the electron is negative, the direction of the force, and thus the acceleration, is opposite to the direction of the electric field, so the electric field is $8.26 \times 10^{-10}$ N/C (south).

39. The directions of the individual fields are shown in the figure.  We find the magnitudes of the individual fields:

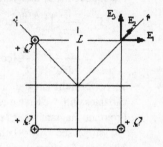

$E_1 = E_3 = kQ/L^2$
$= (9.0 \times 10^9 \text{ N} \cdot \text{m}^2/\text{C}^2)(3.25 \times 10^{-6} \text{ C})/(1.00 \text{ m})^2$
$= 2.93 \times 10^4$ N/C.
$E_2 = kQ/(L\sqrt{2})^2 = \frac{1}{2}kQ_2/L^2$
$= \frac{1}{2}(9.0 \times 10^9 \text{ N} \cdot \text{m}^2/\text{C}^2)(3.25 \times 10^{-6} \text{ C})/(1.00 \text{ m})^2$
$= 1.46 \times 10^4$ N/C.

From the symmetry, we see that the resultant field will be along the diagonal shown as the $x$-axis.  For the net field, we have
$E = 2E_1 \cos 45° + E_2 = 2(2.93 \times 10^4 \text{ N/C}) \cos 45° + 1.46 \times 10^4$ N/C
$= 5.61 \times 10^4$ N/C.

Thus the field at the unoccupied corner is
$5.61 \times 10^4$ N/C away from the opposite corner.

43. (a) From the symmetry of the charges, we see that the electric field points along the $y$-axis.  Thus we have
$\mathbf{E} = 2(Q/4\pi\varepsilon_0) \sin \theta/(y^2 + l^2) \, \mathbf{j} = 2Qy/4\pi\varepsilon_0(y^2 + l^2)^{3/2} \, \mathbf{j}$.

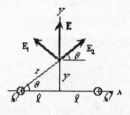

(b) To find the position where the magnitude is maximum, we differentiate and set the first derivative equal to zero:

$$\frac{dE}{dy} = \frac{2Q}{4\pi\varepsilon_0}\left[\frac{1}{\left(y^2 + l^2\right)^{3/2}} - \frac{(3/2)y(2y)}{\left(y^2 + l^2\right)^{5/2}}\right] = \frac{2Q}{4\pi\varepsilon_0}\left[\frac{y^2 + l^2 - 3y^2}{\left(y^2 + l^2\right)^{5/2}}\right] = 0,$$

which gives
$y = \pm \, l/\sqrt{2}$.

47. We choose a differential element of the rod $dy$ a distance $y$ from the center of the rod, as shown in the diagram.  The charge of this element is $dq = (Q/L) \, dy$.  We find the field, which has both $x$- and $y$-components, by integrating along the rod:

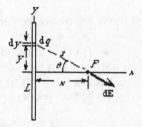

$$\mathbf{E} = \frac{1}{4\pi\varepsilon_0}\int_{y=-L/2}^{+L/2}\frac{(\cos\theta \, \mathbf{i} - \sin\theta \, \mathbf{j})dq}{r^2}$$

$$= \frac{\lambda}{4\pi\varepsilon_0}\int_{y=-L/2}^{+L/2}\frac{(\cos\theta \, \mathbf{i} - \sin\theta \, \mathbf{j})dy}{r^2}$$

From the symmetry we see that there will be no $y$-component.
To perform the integration, we must eliminate variables until we have one, for which we choose $\theta$.
From the diagram we see that $r = x/\cos \theta$, and $y = x \tan \theta$.  This gives
$dy = x \sec^2 \theta \, d\theta = (x \, d\theta)/\cos^2 \theta$.
The limits for $\theta$ are $\pm \, \theta_0 = \pm \sin^{-1} \{\frac{1}{2}L/[x^2 + (\frac{1}{2}L)^2]^{1/2}\} = \pm \sin^{-1} [L/(4x^2 + L^2)^{1/2}]$.
When we make these substitutions, we have

$$\mathbf{E}(x,0) = \frac{\lambda}{4\pi\varepsilon_0}\int_{-\theta_0}^{\theta_0}\frac{(x \, d\theta)/\cos^2\theta}{(x/\cos\theta)^2}(\cos\theta \, \mathbf{i}) = \frac{\lambda}{4\pi\varepsilon_0 x}\int_{-\theta_0}^{\theta_0}(\cos\theta \, \mathbf{i})d\theta$$

$$= \frac{\lambda \, \mathbf{i}}{4\pi\varepsilon_0 x}\sin\theta\Big|_{-\theta_0}^{\theta_0} = \frac{2\lambda \sin\theta_0 \, \mathbf{i}}{4\pi\varepsilon_0 x}$$

Thus for the magnitude we have
$E = \lambda L/2\pi\varepsilon_0 x(L^2 + 4x^2)^{1/2}$.

51. (a) We choose a differential element of the rod $dy$ a
distance $y$ from the end of the rod, as shown in the
diagram. The charge of this element is $dq = \lambda \, dy$.
We find the field, which has both $x$- and
$y$-components, by integrating along the rod:

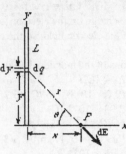

$$\mathbf{E} = \frac{1}{4\pi\varepsilon_0} \int_{y=0}^{y=L} \frac{dq}{r^2}(\cos\theta \, \mathbf{i} - \sin\theta \, \mathbf{j})$$

$$= \frac{\lambda}{4\pi\varepsilon_0} \int_{y=0}^{y=L} \frac{dy}{r^2}(\cos\theta \, \mathbf{i} - \sin\theta \, \mathbf{j})$$

To perform the integration, we must eliminate
variables until we have one, for which we choose $\theta$.
From the diagram we see that $r = x/\cos\theta$, and $y = x\tan\theta$. This gives
$$dy = x\sec^2\theta \, d\theta = (x \, d\theta)/\cos^2\theta.$$
The limits for $\theta$ are 0 and $\theta_0 = \sin^{-1}[L/(x^2 + L^2)^{1/2}]$.
When we make these substitutions, we have

$$\mathbf{E} = \frac{\lambda}{4\pi\varepsilon_0} \int_0^{\theta_0} \frac{(x \, d\theta)/\cos^2\theta}{(x/\cos\theta)^2}(\cos\theta \, \mathbf{i} - \sin\theta \, \mathbf{j}) = \frac{\lambda}{4\pi\varepsilon_0 x} \int_0^{\theta_0} (\cos\theta \, \mathbf{i} - \sin\theta \, \mathbf{j})d\theta$$

$$= \frac{\lambda}{4\pi\varepsilon_0 x}\left(\sin\theta \, \mathbf{i} + \cos\theta \, \mathbf{j}\right)\Big|_0^{\theta_0} = \frac{\lambda}{4\pi\varepsilon_0 x}[\sin\theta_0 \, \mathbf{i} + (\cos\theta_0 - 1)\mathbf{j}]$$

$$= \frac{\lambda}{4\pi\varepsilon_0 x}\left[\frac{L}{(x^2+L^2)^{1/2}}\,\mathbf{i} + \left(\frac{L}{(x^2+L^2)^{1/2}} - 1\right)\mathbf{j}\right] = \frac{\lambda}{4\pi\varepsilon_0 x (x^2+L^2)^{1/2}}\left\{L\,\mathbf{i} + \left[x - \left(x^2+L^2\right)^{1/2}\right]\mathbf{j}\right\}$$

(b) The angle the field makes with the $x$-axis is found from
$$\tan\alpha = E_y/E_x = [x - (x^2 + L^2)^{1/2}]/L.$$
When $L \to \infty$, we have
$$\tan\alpha \to (x - L)/L = -1,$$ so the angle is 45° below the $x$-axis, independent of $x$.

55. (a) We find the acceleration produced by the electric field:
$$q\mathbf{E} = m\mathbf{a};$$
$$(-1.60 \times 10^{-19} \text{ C})[(2.0 \times 10^4 \text{ N/C})\,\mathbf{i} + (8.0 \times 10^4 \text{ N/C})\,\mathbf{j}] = (9.11 \times 10^{-31} \text{ kg})\mathbf{a},$$
which gives $\mathbf{a} = -(3.5 \times 10^{15} \text{ m/s}^2)\,\mathbf{i} - (1.41 \times 10^{16} \text{ m/s}^2)\,\mathbf{j}$.
Because the field is constant, the acceleration is constant.

(b) We find the velocity from
$$\mathbf{v} = \mathbf{v}_0 + \mathbf{a}t$$
$$= (8.0 \times 10^4 \text{ m/s})\,\mathbf{i} + [-(3.5 \times 10^{15} \text{ m/s}^2)\,\mathbf{i} - (1.41 \times 10^{16} \text{ m/s}^2)\,\mathbf{j}](1.0 \times 10^{-9} \text{ s})$$
$$= (-3.43 \times 10^6 \text{ m/s})\,\mathbf{i} - (1.41 \times 10^7 \text{ m/s})\,\mathbf{j}.$$
The direction of the electron is the direction of its velocity:
$$\tan\theta = v_y/v_x = (-1.41 \times 10^7 \text{ m/s})/(-3.43 \times 10^6 \text{ m/s}) = 4.11, \text{ or } \theta = -104°.$$

59. (a) The field along the axis of the ring is
$$\mathbf{E} = (-Q/4\pi\varepsilon_0)[x/[(x^2 + R^2)^{3/2}]\,\mathbf{i},$$
so the force on the charge is
$$\mathbf{F} = q\mathbf{E} = (-qQ/4\pi\varepsilon_0)[x/[(x^2 + R^2)^{3/2}]\,\mathbf{i} = (-qQx/4\pi\varepsilon_0 R^3)/[1 + (x/R)^2]^{3/2}\,\mathbf{i}.$$
If $x \ll R$, we can use the approximation $(1 + u)^{-n} \approx 1 - nu$:
$$\mathbf{F} \approx (-qQx/4\pi\varepsilon_0 R^3)[1 - \tfrac{3}{2}(x/R)^2] \approx -qQx/4\pi\varepsilon_0 R^3.$$

We see that the force is a restoring force proportional to the displacement, so the motion will be
simple harmonic.

(b) The effective spring constant is
$$k = qQ/4\pi\varepsilon_0 R^3,$$
so the period is
$$T = 2\pi(m/k)^{1/2} = 2\pi(4\pi\varepsilon_0 m R^3/qQ)^{1/2}.$$

63. (*a*) The torque on the dipole, which is in a direction to decrease $\theta$, produces an angular acceleration:

$\tau = I\alpha$;

$-pE \sin \theta = I\, d^2\theta/dt^2$.

If  $\theta \ll 1$, $\sin \theta \approx \theta$, so we get

$-pE\theta = I\, d^2\theta/dt^2$,

which produces simple harmonic motion, with the effective force constant $k = pE$.

(*b*) The period is

$T = 2\pi(I/k)^{1/2} = 2\pi(I/pE)^{1/2}$, so the frequency is $f = 1/T = (pE/I)^{1/2}/2\pi$.

67. Because the charge on the Earth can be considered to be at the center, we can use the expression for the force between two point charges. For the Coulomb force to be equal to the weight, we have

$kQ^2/R^2 = mg$;

$(9.0 \times 10^9\ \text{N} \cdot \text{m}^2/\text{C}^2)Q^2/(6.38 \times 10^6\ \text{m})^2 = (1050\ \text{kg})(9.80\ \text{m/s}^2)$, which gives $Q = 6.8 \times 10^3$ C.

71. We find the magnitude of the forces between the pairs:

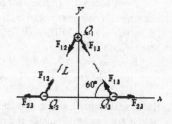

$F_{12} = kQ_1Q_2/L^2$

$= (9.0 \times 10^9\ \text{N} \cdot \text{m}^2/\text{C}^2)(4.0 \times 10^{-6}\ \text{C})(8.0 \times 10^{-6}\ \text{C})/(1.20\ \text{m})^2$

$= 0.20$ N;

$F_{13} = kQ_1Q_3/L^2$

$= (9.0 \times 10^9\ \text{N} \cdot \text{m}^2/\text{C}^2)(4.0 \times 10^{-6}\ \text{C})(6.0 \times 10^{-6}\ \text{C})/(1.20\ \text{m})^2$

$= 0.15$ N;

$F_{23} = kQ_2Q_3/L^2$

$= (9.0 \times 10^9\ \text{N} \cdot \text{m}^2/\text{C}^2)(8.0 \times 10^{-6}\ \text{C})(6.0 \times 10^{-6}\ \text{C})/(1.20\ \text{m})^2$

$= 0.30$ N.

The directions of the forces are determined from the signs of the charges and are indicated on the diagram. For the resultant forces, we have

$\mathbf{F}_1 \quad = (F_{13} \sin 30° - F_{12} \sin 30°)\ \mathbf{i} - (F_{13} \cos 30° + F_{12} \cos\ 30°)\ \mathbf{j}$

$= (0.15\ \text{N} \sin 30° - 0.20\ \text{N} \sin 30°)\ \mathbf{i} - (0.15\ \text{N} \cos 30° + 0.20\ \text{N} \cos\ 30°)\ \mathbf{j}$

$= (-\,0.025\ \text{N})\ \mathbf{i} - (0.30\ \text{N})\ \mathbf{j}$, with magnitude $F_1 = [(0.025\ \text{N})^2 + (0.30\ \text{N})^2]^{1/2} = 0.30$ N.

For the direction we have

$\tan \phi_1 = (-\,0.30\ \text{N})/(-\,0.025\ \text{N}) = 12.1$, $\phi_1 = 265°$, so

$\mathbf{F}_1 = 0.30$ N, 265° from *x*-axis.

$\mathbf{F}_2 \quad = (F_{12} \cos 60° - F_{23})\ \mathbf{i} + (F_{12} \sin 60°)\ \mathbf{j}$

$= (0.20\ \text{N} \cos 60° - 0.30\ \text{N})\ \mathbf{i} + (0.20\ \text{N} \sin 60°)\ \mathbf{j}$

$= (-\,0.20\ \text{N})\ \mathbf{i} + (0.17\ \text{N})\ \mathbf{j}$, with magnitude $F_2 = [(0.20\ \text{N})^2 + (0.17\ \text{N})^2]^{1/2} = 0.26$ N.

For the direction we have

$\tan \phi_2 = (0.17\ \text{N})/(-\,0.20\ \text{N}) = -\,0.866$, $\phi_2 = 139°$, so

$\mathbf{F}_2 = 0.26$ N, 139° from *x*-axis.

$\mathbf{F}_3 \quad = (F_{23} - F_{13} \cos 60°)\ \mathbf{i} + (F_{13} \sin 60°)\ \mathbf{j}$

$= (0.30\ \text{N} - 0.15\ \text{N} \cos 60°)\ \mathbf{i} + (0.15\ \text{N} \sin 60°)\ \mathbf{j}$

$= (0.225\ \text{N})\ \mathbf{i} + (0.13\ \text{N})\ \mathbf{j}$, with magnitude $F_1 = [(0.225\ \text{N})^2 + (0.13\ \text{N})^2]^{1/2} = 0.26$ N.

For the direction we have

$\tan \phi_3 = (0.13\ \text{N})/(0.225\ \text{N}) = 0.577$, $\phi_3 = 30°$, so

$\mathbf{F}_3 = 0.26$ N, 30° from *x*-axis.

75. Because the charges have the same sign, they repel each other.

The force from the third charge must balance the repulsive force for each charge, so the third charge must be positive and between the two negative charges. For each of the negative charges, we have

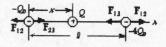

$Q_0 : kQ_0Q/x^2 = kQ_0(4Q_0)/l^2$,  or  $l^2Q = 4x^2Q_0$;

$4Q_0 : \quad k4Q_0Q/(l-x)^2 = kQ_0(4Q_0)/l^2$,  or  $l^2Q = (l-x)^2Q_0$.

Thus we have

$4x^2 = (l-x)^2$, which gives $x = -l,\ +\,0.333l$.

Because the positive charge must be between the charges, it must be $0.333l$ from $Q_0$. When we use this value in one of the force equations, we get

$Q = 4(0.333l)^2Q_0/l^2 = 0.444Q_0$.

Thus we place a charge of $0.444Q_0$, $0.333l$ from $Q_0$.

Note that the force on the middle charge is also zero.

79. (a) The acceleration of the electron, and thus the force produced by the electric field, must be opposite its velocity. Because the electron has a negative charge, the direction of the electric field will be opposite that of the force, so the direction of the electric field is in the direction of the velocity, to the right.

(b) Because the field is constant, the acceleration is constant, so we find the required acceleration from
$$v^2 = v_0^2 + 2ax;$$
$$0 = (2.0 \times 10^6 \text{ m/s})^2 + 2a(0.054 \text{ m}), \text{ which gives } a = -3.70 \times 10^{13} \text{ m/s}^2.$$
We find the electric field from
$$F = qE = ma;$$
$$(-1.60 \times 10^{-19} \text{ C})E = (9.11 \times 10^{-31} \text{ kg})(-3.70 \times 10^{13} \text{ m/s}^2), \text{ which gives } E = 2.1 \times 10^2 \text{ N/C}.$$

83. The angular frequency of the SHM is
$$\omega = (k/m)^{1/2} = [(126 \text{ N/m})/(0.800 \text{ kg})]^{1/2} = 12.5 \text{ s}^{-1}.$$
If we take down as positive with respect to the equilibrium position, the ball will start at maximum displacement, so the position as a function of time is
$$x = A \cos(\omega t) = (0.0500 \text{ m}) \cos [(12.5 \text{ s}^{-1})t].$$
Because the charge is negative, the electric field at the table will be up and the distance from the table is
$$r = H - x = 0.150 \text{ m} - (0.0500 \text{ m}) \cos [(12.5 \text{ s}^{-1})t].$$
The electric field is
$$E = kQ/r^2 = (9.0 \times 10^9 \text{ N} \cdot \text{m}^2/\text{C}^2)(3.00 \times 10^{-6} \text{ C})/\{0.150 \text{ m} - (0.0500 \text{ m}) \cos [(12.5 \text{ s}^{-1})t]\}^2$$
$$= (1.08 \times 10^7 \text{ N/C})/\{3.00 - \cos [(12.5 \text{ s}^{-1})t]\}^2 \text{ up}.$$

87. Because the charges have opposite signs, the location where the electric field is zero must be outside the two charges, as shown.
The fields from the two charges must balance:
$$kQ_1/(d + x)^2 = kQ_2/x^2;$$
$$Q/(d + x)^2 = Q/2x^2, \text{ or } d + x = \pm x\sqrt{2}.$$
which gives $x = d/(\sqrt{2} - 1), -d/(\sqrt{2} + 1), \text{ or } x = d(1 + \sqrt{2}), d(1 - \sqrt{2}),$
Because $d(1 - \sqrt{2})$ is between the charges, the location is
$$d(1 + \sqrt{2}) \text{ from the negative charge, and } d(2 + \sqrt{2}) \text{ from the positive charge.}$$
Other than at infinity, there is no place on the $x$-axis, where the vectors sum to zero.

# Chapter 22: Gauss's Law

## Chapter Overview and Objectives

This chapter introduces the concept of electric flux. It also introduces Gauss's Law which is equivalent to Coulomb's Law for situations involving static charge distributions.

After completing study of this chapter, you should:

- Know the definition of electric flux.
- Be able to calculate the electric flux through a surface.
- Know Gauss's Law.
- Be able to calculate electric fields using Gauss's Law for certain symmetric charge distributions.

## Summary of Equations

Electric flux through a planar area element in a uniform electric field:

$$\Phi_E = EA_\perp = EA\cos\theta$$

Electric flux through a general area in an electric field: $\Phi_E = \int E \cdot dA$

Gauss's Law:

$$\oint E \cdot dA = \frac{Q_{encl}}{\varepsilon_0}$$

## Chapter Summary

### Section 22-1 Electric Flux

The **electric flux**, $\Phi_E$, passing through a planar surface of area $A$ that lies in a plane perpendicular to a uniform electric field of magnitude $E$ is defined to be

$$\Phi_E = EA$$

If the plane of the area $A$ is not perpendicular to the electric field, then the electric flux is given by

$$\Phi_E = EA_\perp = EA\cos\theta$$

where $A_\perp$ is the area of projection of the area $A$ onto a plane perpendicular to the electric field and $\theta$ is the angle between the electric field and the direction of the perpendicular to the area. If we define a vector **A** that has magnitude $A$ and a direction of the perpendicular to the area, we can write the electric flux as

$$\Phi_E = EA$$

In the case of a non-uniform electric field and a general surface, we can consider the flux through infinitesimal areas $dA$ and then integrate the flux over the entire surface:

$$\Phi_E = \int E \cdot dA$$

When the surface is a closed surface, a surface that separates space into an interior and an exterior, we write the integral as

$$\Phi_E = \oint E \cdot dA$$

A closed surface has a definite inside and outside. We adopt the convention that the area vector, $dA$, always points outward for a closed volume. Thus electric field entering a closed surface contributes to negative flux and electric field leaving a closed surface contributes to positive flux.

**Example 22-1-A**

The electric field as a function of position along the $yz$-plane is given by

$$\mathbf{E}(y, z) = E_0 yz / L^2 \mathbf{i} + E_1 y^2 z^2 / L^4 \mathbf{j}$$

Determine the flux through a square with vertices $(0,0,0)$, $(0,L,0)$, $(0,L,L)$, and $(0,0,L)$.

**Solution:**

We need to determine the direction perpendicular to the plane of the square. In this case, it should be easy to see that the square is in the $yz$ plane and the perpendicular is parallel to the $\mathbf{i}$ direction. What if the result were not obvious, how would we determine the perpendicular direction then? We know that the vector cross product is perpendicular to the two factors of the cross product. If we can find two vectors that lie in the plane of the surface, we can take their cross product to determine the direction perpendicular to the surface. In this case we can take vectors along two of the edges of the square, $\mathbf{j}$ and $\mathbf{k}$. We know the cross product $\mathbf{j} \times \mathbf{k} = \mathbf{i}$. We calculate the flux through the surface:

$$\phi = \int \mathbf{E} \cdot d\mathbf{A} = \int_{z=0}^{L} \int_{y=0}^{L} \left( E_0 yz / L^2 \, \mathbf{i} + E_1 y^2 z^2 / L^4 \, \mathbf{j} \right)(\mathbf{i}\, dy\, dz)$$

$$= \int_{z=0}^{L} \int_{y=0}^{L} E_0 yz / L^2 \, dy\, dz = \frac{E_0}{L^2} \int_{z=0}^{L} y\, dy \int_{y=0}^{L} z\, dz$$

$$= \frac{E_0 L^2}{4}$$

**Section 22-2 Gauss's Law**

The electric flux through a closed surface is proportional to the charge within the closed surface:

$$\oint E \cdot dA = \frac{Q_{encl}}{\varepsilon_0}$$

This relationship is called **Gauss's Law**. Gauss's Law is equivalent to Coulomb's Law for static charge distributions. Gauss's Law is more general than Coulomb's Law in that it applies to situations when there is an electric field that is not produced by a static charge distribution.

**Example 22-2-A**

A spherical conductive shell of inner radius $R_1$ and outer radius $R_2$ has a charge $Q$ at its center. Determine the surface charge density on the inner and outer surfaces of the conductive shell and the electric field at the outer surface of the conductive shell.

**Solution:**

We know there must be zero electric field inside the material of the conductive shell. That implies that the total charge interior to the inner surface of the shell must be zero. Because there is a charge $Q$ at the center of the shell, there must be a charge $-Q$ on its inner surface. To determine the surface charge density, we divide the charge by the area of the inner surface of the shell:

$$\sigma_{in} = \frac{-Q}{4\pi R_1^2}$$

Because the net charge of the shell is zero and the electric field within the shell is zero, the $-Q$ charge must be balanced by a charge $+Q$ on the outer surface of the shell. The charge density on the outer surface of the shell is this charge divided by the area of the outer surface:

$$\sigma_{out} = \frac{+Q}{4\pi R_2^2}$$

## Section 22-3 Applications of Gauss's Law

To use Gauss's Law to determine an unknown electric field from a given charge distribution, it is necessary to choose a surface that has the correct symmetry.

## Example 22-3-A

A very long cylinder of charge has charge uniformly distributed along its length, but has a radial variation of charge density given by

$$\rho(r) = \begin{matrix} \rho_0(A-r)/A & if & r < A \\ 0 & if & r > A \end{matrix}$$

Determine the electric field for $r < A$ and for $r > A$.

## Solution:

We take a surface that is a cylinder of radius $R$ and length $L$ concentric with the cylindrical axis of the charge distribution including the bases of the cylinder, as shown in the diagram. By symmetry, the electric field must point radially from the axis of the charge distribution. (Suppose the flux through one of the bases of the cylinder was to the left in the diagram. If we flipped the charge distribution end for end about the 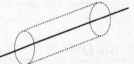 center point of the base of the cylinder, it would look identical, but now the flux would point to the right, which is inconsistent. So we conclude that the electric field must point radially.) This means that the flux through the ends of the cylinder must be zero. Rotational symmetry implies that the electric field pointing out through the cylinder is equal in magnitude everywhere on the cylinder's surface. As the field must point radially, the flux through the cylindrical surface must be the area of the cylinder multiplied by the magnitude of the electric field at the surface:

$$\phi = 2\pi RLE(R)$$

To find the charge inside the cylinder, we integrate the charge density over the volume of the cylinder. If $R$ is less than $A$, the charge inside the cylinder will be

$$Q_{inside} = \int \rho \, dV = \int_{z=0}^{L} \int_{r=0}^{R} \rho_0 \frac{A-r}{A} 2\pi r \, dr \, dz$$

$$= 2\pi \rho_0 L \left( \frac{r^2}{2} - \frac{r^3}{3A} \right) \Bigg|_{r=0}^{R} = 2\pi \rho_0 L \left( \frac{R^2}{2} - \frac{R^3}{3A} \right)$$

We now apply Gauss's Law to determine the electric field:

$$\phi = \frac{Q_{inside}}{\varepsilon_0} \quad \Rightarrow \quad 2\pi RLE(R) = \frac{2\pi \rho_0 L}{\varepsilon_0} \left( \frac{R^2}{2} - \frac{R^3}{3A} \right)$$

$$E(R) = \frac{2\pi \rho_0}{\varepsilon_0} \left( \frac{R}{2} - \frac{R^2}{3A} \right)$$

Once $R$ becomes greater than the radius of the charge distribution, $A$, the charge inside the cylinder remains the same as for $R = A$:

$$Q_{inside} = 2\pi \rho_0 L \left( \frac{A^2}{2} - \frac{A^3}{3A} \right) = 2\pi \rho_0 L \left( \frac{A^2}{2} - \frac{A^2}{3} \right) = 2\pi \rho_0 L \left( \frac{A^2}{6} \right)$$

Again, applying Gauss's Law:

$$\phi = \frac{Q_{inside}}{\varepsilon_0} \quad \Rightarrow \quad 2\pi RLE(R) = \frac{2\pi \rho_0 L}{6\varepsilon_0} A^2$$

$$E(R) = \frac{\rho_0 A^2}{6\varepsilon_0 R}$$

### Section 22-4 Experimental Basis of Gauss's and Coulomb's Law

Experimental tests of Gauss's Law and Coulomb's Law are in agreement with the laws to a very high precision. If we express any measured deviation from Coulomb's Law as

$$F = k \frac{Q_1 Q_2}{r^{2+\delta}}$$

the experimental limits on $\delta$ are $(2.7 \pm 3.1) \times 10^{-16}$.

## Practice Quiz

This is a short chapter, so we have a short quiz!

1. A charge of $+5 \mu C$ is at the origin. Through which of these closed surfaces is the electric flux the least?

   a) A cube with sides of 1.0 m that is centered on the origin
   b) A sphere with a radius 1.0 m that is centered on the origin
   c) A sphere with a radius of 0.8 m and a center that is 0.5 m from the origin
   d) A sphere with a radius of 0.8 m and a center that is 1.0 m from the origin

2. Suppose Coulomb's Law was $F = kQ_1Q_2/r^{2-\varepsilon}$ rather than $F = kQ_1Q_2/r^2$ where $\varepsilon$ is a small positive number. If you take a gaussian surface that is a sphere around a positive point charge, how does the flux through the sphere depend on the radius of the sphere?

   a) The flux does not depend on the radius of the sphere.
   b) The flux increases as the radius of the sphere increases.
   c) The flux decreases as the radius of the sphere increases.
   d) There is insufficient information given to answer the question.

3. A cube has sides $L$. On one face of the cube, the electric field is uniform with magnitude $E$ and has a direction pointing directly into the cube. The total charge on and within the cube is zero. Which statement is necessarily true?

   a) The electric field on the opposite face of the box has magnitude $E$ and points directly out of the face.
   b) The total flux through the remaining five faces is out of the box and equal to $EL^2$.
   c) At least one other face of the cube must have an inward flux.
   d) None of the previous statements must be true.

4.     Suppose our universe were squashed into an almost zero thickness two-dimensional surface and suppose Gauss's Law still held true.  What would the distance dependence of Coulomb's law be in such a universe?

a) $F \propto r$
b) $F \propto 1/r$
c) $F \propto 1/r^2$
d) $F \propto 1/r^3$

5.     The electric flux through a spherical surface is zero.  Which condition is necessarily true?

a) There are no charges inside the sphere.
b) The electric field is zero everywhere within the sphere.
c) The electric field is zero everywhere on the sphere.
d) None of the above statements is necessarily true.

6.     The electric flux through a given open surface is zero.  What condition is certain to be true?

a)  The electric field is zero everywhere along the surface.
b)  There are no electric charges near the surface.
c)  If the electric flux is non-zero for some part of the surface, there is some other part of the surface that the electric flux is non-zero.
d)  The electric field is zero everywhere.

7.     In the diagram to the right is an arrangement of charges and the projection of the outline of some closed surfaces that are right cylinders with the bottom surface below the page and the top surface above the page. Through which of the closed surfaces is the flux zero?

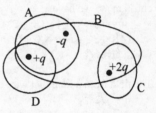

a) A
b) B
c) C
d) D

8.     A uniform electric field of magnitude 30.4 µN/C passes through an equilateral triangle with sides that are 13.8 cm.  The electric field is in a direction that makes a 37.5° angle with the perpendicular to the plane of the triangle.  What is the electric flux through the triangle?

9.     A cubical box with edges 3.2 cm long has an electric field of magnitude 0.46 N/C on one of its faces pointing directly into the cube.  The box contains an arrangement of charge such that the electric field is zero on all other faces.  What is the total charge within the box?

10.    A distribution of charge is spherically symmetric.  The charge density as a function of distance from the center of the distribution is given by

$$\rho(r) = \begin{matrix} \rho_0 (R-r)^2 / R^2 & \text{if} & r < R \\ 0 & \text{if} & r > R \end{matrix}$$

Determine the electric field for $r < R$ and for $r > R$.

11.    An infinite slab of charge lies in the $yz$ plane from $-L/2 < x < L/2$. On either side of the slab, the electric field is zero.  Within the slab is an electric field given by

$$E(x) = E_0 \left( 1 - x^2 \right) \mathbf{i}$$

What is the charge density in the slab as a function of $x$?

12.     The electric field in a region of space is given by $\mathbf{E} = E_0 \cos(\pi y/2L) \cos(\pi z/2L) \, \mathbf{i}$.  What is the electric flux through a square with corners $(0,0,0)$, $(0,L,0)$, $(0,L,L)$, $(0,0,L)$?

## Problem Solutions

3.     All field lines enter and leave the cube, so the net flux is
       $\Phi_{net} = 0$.
       We find the flux through a face from
       $\Phi = \int \mathbf{E} \cdot d\mathbf{A}$.
       There are two faces with the field lines perpendicular to the face, say one at $x = 0$ and one at $x = l$.
       Thus for these two faces we have
       $\Phi_{x=0} = -EA = -(6.50 \times 10^3 \text{ N/C}) \, l^2 = -(6.50 \times 10^3 \text{ N/C}) \, l^2$;
       $\Phi_{x=l} = +EA = +(6.50 \times 10^3 \text{ N/C}) \, l^2 = +(6.50 \times 10^3 \text{ N/C}) \, l^2$.
       For all other faces, the field is parallel to the face, so we have
       $\Phi_{all\ others} = 0$.

7.     The total electric flux through the surface depends
       only on the enclosed charge:
       $\Phi = \int \mathbf{E} \cdot d\mathbf{A} = Q/\varepsilon_0$.
       The only contributions to the integral are from the
       faces perpendicular to the electric field.  Over each
       of these two surfaces, the magnitude of the field is
       constant, so we have
       $\Phi = E_l A - E_0 A = (E_l - E_0)A = Q/\varepsilon_0$;
       $(410 \text{ N/C} - 560 \text{ N/C})(30 \text{ m})^2 = Q/(8.85 \times 10^{-12} \text{ C}^2/\text{N} \cdot \text{m}^2)$,
       which gives $Q = -1.2 \times 10^{-6} \text{ C} = -1.2 \, \mu\text{C}$.

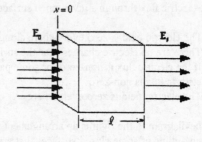

11.    The field from a long thin wire is radial with a magnitude given by
       $E = \lambda/2\pi\varepsilon_0 r$.
       (a)  At a distance of 5.0 m the field is
            $E = (-2.8 \times 10^{-6} \text{ C/m})/2\pi(8.85 \times 10^{-12} \text{ C}^2/\text{N} \cdot \text{m}^2)(5.0 \text{ m}) = -1.0 \times 10^4 \text{ N/C}$ (toward the wire).
       (b)  At a distance of 2.0 m the field is
            $E = (-2.8 \times 10^{-6} \text{ C/m})/2\pi(8.85 \times 10^{-12} \text{ C}^2/\text{N} \cdot \text{m}^2)(2.0 \text{ m}) = -2.5 \times 10^4 \text{ N/C}$ (toward the wire).

15.    From the symmetry of the charge distribution, we know that
       the electric field must be radial, with a magnitude independent
       of the direction.
       (a)  For a spherical gaussian surface within the spherical
            cavity, we have
            $\int \mathbf{E} \cdot d\mathbf{A} = E 4\pi r^2 = Q_{enclosed}/\varepsilon_0 = Q/\varepsilon_0$, so we have
            $E = (1/4\pi\varepsilon_0)Q/r^2$
              $= (9.0 \times 10^9 \text{ N} \cdot \text{m}^2/\text{C}^2)(5.50 \times 10^{-6} \text{ C})/(0.030 \text{ m})^2$
              $= 5.5 \times 10^7 \text{ N/C}$ (away from center).

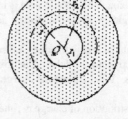

       (b)  The point 6.0 cm from the center is inside the conductor,
            thus the electric field is 0.
       Note that there must be a negative charge of $-5.50 \, \mu\text{C}$ on the surface of the cavity and a positive charge of $+5.50 \, \mu\text{C}$
       on the outer surface of the sphere.

19.    Each positive plate produces an electric field directed
       away from the plate with a magnitude
       $E_+ = \sigma/2\varepsilon_0$.
       (a)  Between the plates, the two fields are in opposite
            directions, so we have
            $E_{between} = E_+ - E_+ = (\sigma/2\varepsilon_0) - (\sigma/2\varepsilon_0) = 0$.

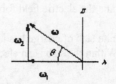

(b)  Outside the plates, the two fields are in the same
     direction, so we have
     $E_{\text{outside}} = E_+ + E_+ = (\sigma/2\varepsilon_0) + (\sigma/2\varepsilon_0) = \sigma/\varepsilon_0$ .

(c)  If the plates were nonconductors, the fields from each plate would be the same, so the results will
     be unaffected.  Note that the field inside the plates would change.

23.  From the symmetry of the charge distribution, we know that the
     electric field must be radial, with a magnitude independent of
     the direction.  The charge density of the sphere is

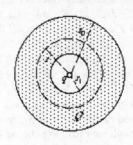

$\rho = Q/[\frac{4}{3}\pi(r_0^3 - r_1^3)]$ .

We can add the field of the point charge at the center to the
fields found in Problem 22:

(a)  For the region where $r < r_1$,  the electric field is that of
     the point charge at the center:
     $E = q/4\pi\varepsilon_0 r^2$; $r < r_1$.

(b)  For the region where $r_1 < r < r_0$, we add the two fields:
     $\begin{aligned} E &= Q(r^3 - r_1^3)/4\pi\varepsilon_0(r_0^3 - r_1^3)r^2 + q/4\pi\varepsilon_0 r^2 \\ &= (1/4\pi\varepsilon_0)[Q(r^3 - r_1^3) + q(r_0^3 - r_1^3)]/(r_0^3 - r_1^3)r^2; \ r_1 < r < r_0 \, . \end{aligned}$

(c)  For the region where $r > r_0$, the electric field is that of a point charge equal to the total charge:
     $E = (q + Q)/4\pi\varepsilon_0 r^2$; $r > r_0$.

27.  From the symmetry of the charge distribution, for points far
     from the ends and not too far from the shell, we know that the
     electric field must be radial, away from the axis of the cylinder,
     with a magnitude independent of the direction.  For a gaussian
     surface we choose a cylinder of length $l$ and radius $r$, centered on
     the axis.  On the ends of this surface, the electric field is not
     constant but $\mathbf{E}$ and $d\mathbf{A}$ are perpendicular, so we have $\mathbf{E} \cdot d\mathbf{A} = 0$.
     On the curved side, the field has a constant magnitude and
     $\mathbf{E}$ and $d\mathbf{A}$ are parallel, so we have $\mathbf{E} \cdot d\mathbf{A} = E\,dA$.

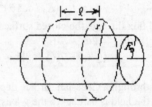

(a)  For the region where $r > R_0$, the charge inside the
     gaussian surface is $Q = \sigma 2\pi R_0\, l$.
     For Gauss's law we have
     $\int \mathbf{E} \cdot d\mathbf{A} = \int_{\text{ends}} \mathbf{E} \cdot d\mathbf{A} + \int_{\text{side}} \mathbf{E} \cdot d\mathbf{A} = Q_{\text{enclosed}}/\varepsilon_0$ ;
     $0 + E2\pi r\, l = \sigma 2\pi R_0\, l/\varepsilon_0$, or $E = \sigma R_0/\varepsilon_0 r$; $r > R_0$ .

(b)  For the region where $r < R_0$, the charge inside the
     gaussian surface is $Q = 0$, so we have
     $E = 0$ for $r < R_0$.

(c)  We find the equivalent linear charge density from
     $Q = \sigma 2\pi R_0 L = \lambda_{\text{eq}} L$, which gives $\lambda_{\text{eq}} = \sigma 2\pi R_0$.
     If we treat the cylinder as a line of charge, the field is
     $E = 2\lambda_{\text{eq}}/4\pi\varepsilon_0 r = 2(\sigma 2\pi R_0)/4\pi\varepsilon_0 r = \sigma R_0/\varepsilon_0 r$; which is the same as the result for part (a).

31.  From the symmetry of the charge distribution, for points far
     from the ends and not too far from the outer shell, we know
     that the electric field must be radial, away from the axis of
     the cylinders, with a magnitude independent of the direction.
     For a gaussian surface we choose a cylinder of length $l$ and
     radius $r$, centered on the axis.  On the ends of this surface, the
     electric field is not constant but $\mathbf{E}$ and $d\mathbf{A}$ are perpendicular,
     so we have $\mathbf{E} \cdot d\mathbf{A} = 0$.  On the curved side, the field has a
     constant magnitude and $\mathbf{E}$ and $d\mathbf{A}$ are parallel, so we have
     $\mathbf{E} \cdot d\mathbf{A} = E\,dA$.

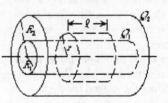

(a)  A point 3.0 cm from the axis is inside the inner shell.
     For the region where $r < R_1$,  there is no charge inside the gaussian surface, so we have
     $E = 0$; $r = 3.0$ cm.

(b)  A point 6.0 cm from the axis is between the shells.  For the region where $R_1 < r < R_2$, the charge
     inside the gaussian surface is $Q_{\text{enclosed}} = (Q_1/L)\, l$.  For Gauss's law we have
     $\int \mathbf{E} \cdot d\mathbf{A} = \int_{\text{ends}} \mathbf{E} \cdot d\mathbf{A} + \int_{\text{side}} \mathbf{E} \cdot d\mathbf{A} = Q_{\text{enclosed}}/\varepsilon_0$ ;

$0 + E2\pi r\, l = (Q_1/L)\, l\, /\varepsilon_0$, so

$E = Q_1/2\pi\varepsilon_0 Lr$

$= (-3.8 \times 10^{-6}\ \text{C})/2\pi(8.85 \times 10^{-12}\ \text{C}^2/\text{N} \cdot \text{m}^2)(5.0\ \text{m})(0.060\ \text{m})$

$= -2.3 \times 10^5\ \text{N/C}$ (toward the axis), $r = 6.0$ cm.

(c) A point 12.0 cm from the axis is outside the shells. For the region where $r > R_2$, the charge inside the Gaussian surface is $Q_{enclosed} = [(Q_1/L) + (Q_2/L)]\, l$. For Gauss's law we have

$\int \mathbf{E} \cdot d\mathbf{A} = \int_{ends} \mathbf{E} \cdot d\mathbf{A} + \int_{side} \mathbf{E} \cdot d\mathbf{A} = Q_{enclosed}/\varepsilon_0$;

$0 + E2\pi r\, l = [(Q_1/L) + (Q_2/L)]\, l\, /\varepsilon_0$, so

$E = (Q_1 + Q_2)/2\pi\varepsilon_0 Lr$

$= [(-3.8 \times 10^{-6}\ \text{C}) + (+3.2 \times 10^{-6}\ \text{C})]/2\pi(8.85 \times 10^{-12}\ \text{C}^2/\text{N} \cdot \text{m}^2)(5.0\ \text{m})(0.120\ \text{m})$

$= -1.8 \times 10^4\ \text{N/C}$ (toward the axis), $r = 12.0$ cm.

35. On the ends of the cylinder the electric field will vary in magnitude and direction. Thus we must integrate to find the flux through the ends. We choose a circular ring of radius $y$ and thickness $dy$. From the diagram we see that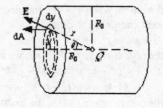

$R_0 = r \cos\theta$,

$y = R_0 \tan\theta$,

$dy = R_0 \sec^2\theta\, d\theta = (R_0/\cos^2\theta)\, d\theta$.

The flux through one end is

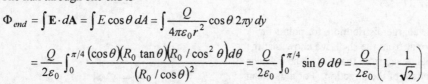

$$\Phi_{end} = \int \mathbf{E} \cdot d\mathbf{A} = \int E\cos\theta\, dA = \int \frac{Q}{4\pi\varepsilon_0 r^2}\cos\theta\, 2\pi y\, dy$$

$$= \frac{Q}{2\varepsilon_0}\int_0^{\pi/4} \frac{(\cos\theta)(R_0\tan\theta)(R_0/\cos^2\theta)\,d\theta}{(R_0/\cos\theta)^2} = \frac{Q}{2\varepsilon_0}\int_0^{\pi/4}\sin\theta\, d\theta = \frac{Q}{2\varepsilon_0}\left(1 - \frac{1}{\sqrt{2}}\right)$$

The total flux through the closed surface is $Q/\varepsilon_0$, so the flux through the curved sides is

$\Phi_{sides} = \Phi_{total} - 2\Phi_{end}$

$= (Q/\varepsilon_0) - 2(Q/2\varepsilon_0)[1 - 1/\sqrt{2}] = Q/\varepsilon_0\sqrt{2}$.

39. The flux through a Gaussian surface depends on the enclosed charge. Because the field is parallel to the $y$-axis, the only faces that will have flux through them are the ones perpendicular to the $y$-axis. Thus we have

$\int \mathbf{E} \cdot d\mathbf{A} = \int_{y=l} \mathbf{E} \cdot d\mathbf{A} + \int_{y=0} \mathbf{E} \cdot d\mathbf{A} = Q_{enclosed}/\varepsilon_0$;

$(a + bl)l^2 - (a + 0)\, l^2 = Q_{enclosed}/\varepsilon_0$, which gives $Q_{enclosed} = \varepsilon_0 bl^3$.

43. (a) Because there is no charge within the sphere, every field line will enter and leave the sphere, so the net flux through the sphere is 0.

(b) The maximum electric field will be at the point on the sphere closest to the point charge, the top of the sphere:

$E_{max} = Q/4\pi\varepsilon_0(\tfrac{1}{2}r_0)^2 = Q/\pi\varepsilon_0 r_0^2$.

The minimum electric field will be at the point on the sphere farthest from the point charge, the bottom of the sphere:

$E_{min} = Q/4\pi\varepsilon_0(r_0)^2 = Q/25\pi\varepsilon_0 r_0^2$.

Thus the range of values is $Q/25\pi\varepsilon_0 r_0^2 \leq E \leq Q/\pi\varepsilon_0 r_0^2$.

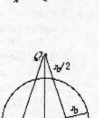

(c) The electric field is not perpendicular at all points. It is perpendicular only at the top and bottom of the sphere. At some points it is almost parallel to the surface of the sphere.

(d) The electric field is not perpendicular or constant over the surface of the sphere. Gauss's law is not useful for obtaining $E$ because a gaussian surface cannot be chosen that simplifies the integral for the flux.

47. From symmetry, we know that the field inside a uniformly charged sphere must be radial and depends only on the distance from the center. At a distance $r$ from the center, $r \leq r_0$, only the charge inside a spherical surface with a radius $r$ will provide the field:

$$E(r \leq r_0) = Q_{\text{enclosed}}/4\pi\varepsilon_0 r^2 = \tfrac{4}{3}\pi r^3 \rho_E/4\pi\varepsilon_0 r^2 = \rho_E r/3\varepsilon_0 .$$

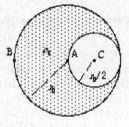

We create the cavity by adding to the original sphere, with charge density $\rho_E$, a sphere with charge density $-\rho_E$ and radius $\tfrac{1}{2} r_0$, centered at C. At any point the total field will be the sum of the fields from the two spheres, which we label $E_+$ and $E_-$.

(a) At the point A we have

$$E_A = E_+ + E_- = 0 - [-\rho_E(\tfrac{1}{2} r_0)/3\varepsilon_0] = \rho_E r_0/6\varepsilon_0 \text{ (right)}.$$

(b) At the point B the cavity can be treated as a point charge, so we have

$$E_B = E_+ + E_- = -\rho_E r_0/3\varepsilon_0 - [-\rho_E \tfrac{4}{3}\pi(\tfrac{1}{2} r_0^3)/4\pi\varepsilon_0(\tfrac{3}{2} r_0)^2] = -17\rho_E r_0/54\varepsilon_0 \text{ (left)}.$$

By considering other points in the cavity, it can be shown that the field inside the cavity is uniform.

# Chapter 23: Electric Potential

## Chapter Overview and Objectives

This chapter introduces the concept of electric potential. It discusses the relationship of electric potential to work done in moving charged particles in electric fields, the relationship between electric fields and electric potential, and the electric potential of point charges and continuous distributions of charge. This chapter introduces the concept of electric dipoles and derives an approximate potential energy function for electric dipoles.

After completing study of this chapter, you should:

- Know the definition of electric potential and be able to relate work done by electrical fields to electric potential.
- Know the electric potential of a point charge.
- Be able to calculate the electric potential difference between different points in space for a collection of point charges.
- Be able to calculate the electric potential of a continuous charge distribution.
- Know both the integral and differential relationships between electric field and electric potential and be able to calculate electric field from electric potential and electric potential from electric field.
- Know what an electric dipole is and how to calculate its dipole moment.
- Know how to calculate the approximate electric potential of an electric dipole.

## Summary of Equations

Definition of electric potential:
$$V_a = \frac{U_a}{q}$$

Relationship between electric potential difference and work done by electric field:

$$V_{ba} = V_b - V_a = -\frac{W_{ba}}{q}$$

Integral relationship between electric potential and electric field:

$$V_b - V_a = -\int_a^b E \cdot dl$$

Electric potential of a point charge:
$$V(r) = \frac{kQ}{r}$$

Electric potential due to a continuous distribution of point charge:

$$V_a = \frac{1}{4\pi\varepsilon_0} \int \frac{dq}{r} = \frac{1}{4\pi\varepsilon_0} \int \frac{\rho}{r} dV$$

Definition of magnitude of dipole moment:
$$p = Ql$$

Approximate electric potential of a dipole:
$$V(x,y,z) \approx \frac{1}{4\pi\varepsilon_0} \frac{p\cos\theta}{r^2}$$

Differential relationship between electric potential and components of electric field:

$$E_x = \frac{\partial V}{\partial x} \qquad E_y = \frac{\partial V}{\partial y} \qquad E_z = \frac{\partial V}{\partial z}$$

# Chapter Summary

### Section 23-1 Electric Potential and Potential Difference

As a charge is moved through an electric field, the electric force due to a static charge distribution does work on the charge. The static electric force is a conservative force. The work done by the force does not depend on the path taken but only on the initial and final position. Because the static electric force is conservative, we can write a potential energy function of position for the static electric force, $U_a$, where $a$ labels the position at which the potential energy is evaluated. In a manner similar to how we defined electric field from electric force, we define **electric potential** at point $a$, $V_a$, as the potential energy at point $a$ per unit charge:

$$V_a = \frac{U_a}{q}$$

Just as only the change in potential energy is meaningful, only the change in electric potential has physical meaning. This means that any potential energy function that is used can have a constant added to it at all points in space and still be a correct potential energy function.

The work done by the electric field in moving a charge $q$ from point $a$ to point $b$, $W_{ab}$, is equal to the negative of the change in potential energy. We can write this in terms of the electric potential at point $a$, $V_a$, and at point $b$, $V_b$:

$$W_{ab} = -\Delta U_{ab} = -q(V_b - V_a)$$

This can be rewritten as the change in electric potential, $V_{ba}$, in moving from point $a$ to point $b$, called the **potential difference**, is equal to the negative of the work done divided by the charge:

$$V_{ba} = V_b - V_a = -\frac{W_{ba}}{q}$$

It should be easy to see that the dimensions of electric potential are energy/charge. The corresponding SI units are J/C. This combination of units is defined to be a **volt** (abbreviated V):

$$1\,\text{V} = 1\,\text{J/C}$$

### Example 23-1-A

The electric potential at a given position in space is 256 V. A particle with a mass of $3.46 \times 10^{-5}$ kg and a charge $-4.63$ μC starts at this location from rest and moves to another location where it has a speed of 453 m/s. What is the electric potential at the new location?

### Solution:

Only conservative forces are involved in this problem, so we can use the law of conservation of energy in the form

$$\Delta K + \Delta U = 0 \quad \Rightarrow \quad \tfrac{1}{2}mv_2^2 - \tfrac{1}{2}mv_1^2 + q(V_2 - V_1) = 0$$

We solve this for the unknown electric potential:

$$V_2 = -\frac{\tfrac{1}{2}mv_2^2 - \tfrac{1}{2}mv_1^2}{q} + V_1 = -\frac{\tfrac{1}{2}\left(3.46 \times 10^{-8}\ \text{kg}\right)\left[(453\ \text{m/s})^2 - 0\right]}{-4.63 \times 10^{-6}\ \text{C}} + 256\ \text{V} = 1.02 \times 10^3\ \text{V}$$

## Section 23-2 Relation Between Electric Potential and Electric Field

The relationship between potential energy change and force is derived from the definition of work done by a force:

$$U_b - U_a = -\int_a^b F \cdot dl$$

If we divide both sides of this equation by the charge, $q$, we get a relationship between the change in electric potential and electric field:

$$\frac{U_b - U_a}{q} = -\frac{\int_a^b F \cdot dl}{q} \quad \Rightarrow \quad V_b - V_a = -\int_a^b E \cdot dl$$

In the case of a uniform electric field, this simplifies to

$$V_b - V_a = Ed$$

where $E$ is the magnitude of the electric field and $d$ is the distance between the two points in the direction of the electric field.

## Example 23-2-A

The electric field in a given region of space is given by $\mathbf{E}(x,y) = Axy\,\mathbf{i} + Bx^2 y\,\mathbf{j}$. Determine the difference in electric potential in going from (0,0) to (1,1).

## Solution:

We know the electric potential difference between two points in space is related to the electric field by

$$V_b - V_a = -\int_a^b E \cdot dl$$

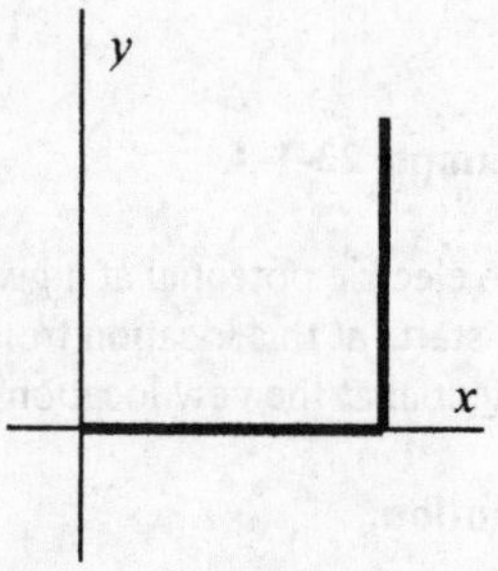

This is a path integral. We need to choose a path from the initial point to the final point along which we calculate the integral. We try to choose a path over which the integral becomes easy to determine. Let's try a path that first goes along the $x$-axis from $x = 0$ to $x = 1$ and then follows a line at $x = 1$ that goes from $y = 0$ to $y = 1$ as shown in the diagram. Along the first segment of the path, $dl$ is in the $\mathbf{i}$ direction, so we write $dl = \mathbf{i}\,dx$. Along the second segment of the path, $dl$ is in the $\mathbf{j}$ direction, so we write $dl = \mathbf{j}\,dy$. The path integral becomes

$$V_b - V_a = -\int_a^b \mathbf{E} \cdot d\mathbf{l} = -\int_a^c \mathbf{E} \cdot d\mathbf{l} - \int_c^b \mathbf{E} \cdot d\mathbf{l}$$

$$= -\left[ \int_{x=0}^1 \left( Axy\,\mathbf{i} + Bx^2 y\,\mathbf{j} \right) \cdot \mathbf{i}\,dx \Big|_{y=0} + \int_{x=0}^1 \left( Axy\,\mathbf{i} + Bx^2 y\,\mathbf{j} \right) \cdot \mathbf{j}\,dy \Big|_{x=1} \right]$$

Performing the vector dot products in the integrands leaves

$$V_b - V_a = -\left[ \int_{x=0}^1 Axy\,dx \Big|_{y=0} + \int_{y=0}^1 Bx^2 y\,dy \Big|_{x=1} \right]$$

$$= -\left\{ \left[ \tfrac{1}{2} Ax^2 y \right]_{x=0}^1 \Big|_{y=0} + \left[ \tfrac{1}{2} Bx^2 y^2 \right]_{y=0}^1 \Big|_{x=1} \right\} = -\tfrac{1}{2} \left[ Ay \Big|_{y=0} + Bx^2 \Big|_{x=1} \right]$$

$$= -\tfrac{1}{2} B$$

## Section 23-3 Electric Potential Due to Point Charges

It is relatively easy to derive the potential energy for a point charge. For a point charge $Q$, the electric field is given by

$$E = \frac{kQ}{r^2}\hat{\mathbf{r}}$$

where $r$ is the distance from the charge and $\hat{\mathbf{r}}$ is the unit vector that points from the charge to the point at which the electric field is being evaluated. The change in electric potential in going from position $\mathbf{r}_a$ to position $\mathbf{r}_b$ is

$$V_b - V_a = \int_{r_a}^{r_b} E \cdot dl = \int_{r_a}^{r_b} \frac{kQ}{r^2}\hat{r} \cdot dl = \int_{r_a}^{r_b} \frac{kQ}{r^2} dr = \frac{kQ}{r_b} - \frac{kQ}{r_a}$$

where $dr$ is the change in distance from the charge when moving along the path element $dl$. It is easy to see that an electric potential function of the form

$$V(r) = \frac{kQ}{r} + C$$

where $C$ is an arbitrary constant potential, is consistent with the difference in potential energy derived above. If we choose the constant potential so that the electric potential goes to zero as the distance from the charge becomes infinitely large, then

$$V(r) = \frac{kQ}{r}$$

## Example 23-1-A

Two charges are located a distance 1.00 m apart as shown in the diagram. If a 10.0 μC charge is moved from infinity to the position ( 1.00 m, −1.00 m ), what will be its change in electric potential and its change in electric potential energy?

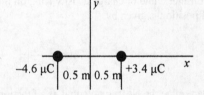

## Solution:

The electric potential at an infinite distance from the charges will be zero. The electric potential at the position ( 1.00 m, −1.00 m ) will be the sum of the electric potentials due to each charge. The distance, $r_1$, to the point ( 1.00 m, −1.00 m ) from the location of the −4.6 μC charge is determined using the Pythagorean theorem:

$$r_1 = \sqrt{(x - x_1)^2 + (y - y_1{}^2)} = \sqrt{[1.00\,\text{m} - (-0.5\,\text{m})]^2 + [-1.00\,\text{m} - 0.0\,\text{m}]^2} = 1.80\,\text{m}$$

and the distance, $r^2$, to the point ( 1.00 m, −1.00 m ) from the location of the +3.4 μC charge is

$$r_2 = \sqrt{(x - x_2)^2 + (y - y_2{}^2)} = \sqrt{[1.00\,\text{m} - (0.5\,\text{m})]^2 + [-1.00\,\text{m} - 0.0\,\text{m}]^2} = 1.12\,\text{m}$$

We can calculate the potential at the point ( 1.00 m, −1.00 m ):

$$V = \frac{kQ_1}{r_1} + \frac{kQ_2}{r_2} = \frac{(9.0 \times 10^9\ \text{N}\cdot\text{m}^2/\text{C}^2)(-4.6 \times 10^{-6}\ \text{C})}{1.80\,\text{m}} + \frac{(9.0 \times 10^9\ \text{N}\cdot\text{m}^2/\text{C}^2)(3.4 \times 10^{-6}\ \text{C})}{1.12\,\text{m}}$$

$$= 4.3 \times 10^3\ \text{J/C}$$

The potential energy of the charge at this position is

$$U = QV = \left(10.0 \times 10^{-6} \text{ C}\right)\left(4.3 \times 10^3 \text{ J/C}\right) = 4.3 \times 10^{-2} \text{ J}$$

As the potential at an infinite distance is zero, the potential difference and the electric potential energy difference are the same as the values calculated for these quantities at the final location.

### Section 23-4 Potential Due to Any Charge Distribution

Because the electric fields of individual point charges are superpose or add to determine the net electric field, the electric potentials of individual point charges add to determine the total electric potential at a point in space. We write the potential at point $a$, $V_a$, due to $n$ other charges as

$$V_a = \sum_{i-1}^{n} V_{ia} = \frac{1}{4\pi\varepsilon_0} \frac{kQ_i}{r_{ia}}$$

where $i$ is an index that labels each of the $n$ charges, $V_{ia}$ is the potential at position $a$ due to the $i$th charge, $Q_i$ is the value of the $i$th charge, and $r_i$ is the distance from the $i$th charge to point $a$. If the charge distribution is continuous or can be approximated as continuous, then the sum becomes an integral:

$$V_a = \frac{1}{4\pi\varepsilon_0} \int \frac{dq}{r} = \frac{1}{4\pi\varepsilon_0} \int \frac{\rho}{r} dV$$

where $r$ is the distance of a charge $dq$ from point $a$ and $\rho$ is the density of charge in the volume $dV$.

### Example 23-4-A

Consider a line of charge density lying on the x-axis between x = –L and x = +L. The linear charge density as a function of position is given by

$$\lambda(x) = \frac{\lambda_0}{L} |x|$$

Determine the electric potential as a function of position along the y-axis.

### Solution:

Here there is a linear, 1-dimensional distribution of charge. In such a case, the infinitesimal quantity of charge, $dq$, is equal to the linear charge density multiplied by an infinitesimal length, $\lambda dx$. We write the electric potential as

$$V = \frac{1}{4\pi\varepsilon_0} \int \frac{dq}{r} = \frac{1}{4\pi\varepsilon_0} \int \frac{\lambda}{r} dx$$

In this case, since we are interested in the electric potential along the y-axis, the distance from a given $dx$ element of the charge density to a point $y$ along the y-axis is given by

$$r = \sqrt{x^2 + y^2}$$

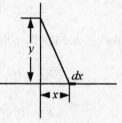

We substitute this expression and the given expression for the charge density into the integral:

$$V = \frac{1}{4\pi\varepsilon_0} \int_{x=-L}^{+L} \frac{\lambda_0 |x|}{L\sqrt{x^2+y^2}} dx$$

$$= \frac{1}{4\pi\varepsilon_0} \int_{x=-L}^{0} \frac{-\lambda_0 x}{L\sqrt{x^2+y^2}} dx + \frac{1}{4\pi\varepsilon_0} \int_{x=0}^{+L} \frac{\lambda_0 x}{L\sqrt{x^2+y^2}} dx$$

$$= \frac{-1}{4\pi\varepsilon_0} \frac{\lambda_0}{L} \sqrt{x^2+y^2} \Big|_{x=-L}^{0} + \frac{1}{4\pi\varepsilon_0} \frac{\lambda_0}{L} \sqrt{x^2+y^2} \Big|_{x=0}^{L}$$

$$= \frac{1}{4\pi\varepsilon_0} \frac{\lambda_0}{L} \left[ \sqrt{L^2+y^2} + \sqrt{L^2+y^2} \right] = \frac{1}{2\pi\varepsilon_0} \frac{\lambda_0}{L} \sqrt{L^2+y^2}$$

## Section 23-5 Equipotential Surfaces

Surfaces at the same potential are called **equipotential surfaces**. It is often useful to draw equipotential surfaces for a given arrangement of charge. The electric field is always perpendicular to any equipotential surface. If a contour map of equipotential surfaces is drawn with equal potential differences between the different contour lines, the electric field magnitude is proportional to the density of equipotential lines in a given region of space.

The surface of a conductor is an equipotential surface and the electric field always points perpendicular to the surface of a conductor when there is a static charge distribution.

## Section 23-6 Electric Dipoles

A particular arrangement of charge that is often encountered in systems of interest or is often a good approximation to systems of interest is the **electric dipole**. An electric dipole is defined to be two charges of equal magnitude $Q$ but opposite sign separated by a distance $l$. The magnitude of the **dipole moment**, $p$, of a pair of charges arranged in this manner is

$$p = Ql$$

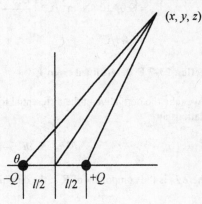

The dipole moment is a vector that points in the direction from the position of the negative charge to the positive charge.
The exact electric potential due to a dipole is easily calculated. Assume the dipole lies on the x-axis as shown in the diagram. The distance to a point $(x,y,z)$ to the charge $Q$ located at $(l/2,0,0)$ is

$$r_1 = \sqrt{(x-l/2)^2 + y^2 + z^2}$$

and the distance to the charge $-Q$ located at $(-l/2,0,0)$ is

$$r_2 = \sqrt{(x+l/2)^2 + y^2 + z^2}$$

The electric potential at the point $(x,y,z)$ is

$$V(x,y,z) = \frac{1}{4\pi\varepsilon_0} \left( \frac{Q}{\sqrt{(x-l/2)^2+y^2+z^2}} + \frac{-Q}{\sqrt{(x-l/2)^2+y^2+z^2}} \right)$$

In the situation that any of $x$, $y$, or $z$ is considerably larger than $l$, we can make a simple approximation. Using the approximation $1/\sqrt{1+p} \approx 1 - p/2$, we get

$$V(x, y, z) = \frac{1}{4\pi\varepsilon_0} \frac{1}{\sqrt{x^2 + y^2 + z^2}} \left( \frac{Q(xl + l^2)/2}{x^2 + y^2 + z^2} + \frac{-Q(-xl + l^2)/2}{x^2 + y^2 + z^2} \right)$$

$$= \frac{1}{4\pi\varepsilon_0} \frac{1}{\sqrt{x^2 + y^2 + z^2}} \frac{Qxl}{x^2 + y^2 + z^2}$$

From the diagram, it should be easy to see that $x/\sqrt{x^2 + y^2 + z^2} = \cos\theta$ and $x^2 + y^2 + z^2 = r^2$, thus

$$V(x, y, z) \approx \frac{1}{4\pi\varepsilon_0} \frac{p\cos\theta}{r^2}$$

Note that $\theta$ is the angle between the direction of the dipole moment vector and the direction to the point at which the electric potential is being evaluated.

**Example 23-6-A**

An electric dipole with a dipole moment of magnitude $3.8 \times 10^{-10}$ C·m is at the origin. Its dipole moment points in the +**i** direction. Determine the difference in electric potential between a position that is on the x-axis at $x = 1.0$ m and a position on the x-axis at $x = -1.0$ m.

**Solution:**

The direction of the dipole moment vector is in the +**i** direction. This implies that the $\theta = 0$ direction is in the +**i** direction (+x-axis) and the $\theta = 180°$ direction is in the −**i** direction (−x-axis). The electric potential difference is then

$$\Delta V = V_b - V_a = \frac{1}{4\pi\varepsilon_0} \left[ \frac{p\cos\theta_b}{r_b^2} - \frac{p\cos\theta_a}{r_a^2} \right]$$

$$= \left(9.0 \times 10^9 \text{ N·m}^2/\text{C}^2\right) \left[ \frac{\left(3.8 \times 10^{-10} \text{ C·m}\right)\cos 180°}{(1.0 \text{ m})^2} - \frac{\left(3.8 \times 10^{-10} \text{ C·m}\right)\cos 0°}{(1.0 \text{ m})^2} \right]$$

$$= 6.8 \text{ J/C}$$

## Section 23-7 E Determined from $V$

If we take the derivative of the differential form of the definition of work and divide by the charge, we have the relationship

$$dV = -\mathbf{E} \cdot d\mathbf{l} = -E_l dl$$

where $E_l$ is the component of **E** in the direction of the displacement $d\mathbf{l}$. We can rearrange this to

$$E_l = -\frac{dV}{dl}$$

If we let $d\mathbf{l}$ be along the Cartesian directions, we see that

$$E_x = -\frac{\partial V}{\partial x} \qquad E_y = -\frac{\partial V}{\partial y} \qquad E_z = -\frac{\partial V}{\partial z}$$

where $\partial V/\partial x$ is the partial derivative of $V$ with respect to $x$.

**Example 23-7-A**

An electric potential is given by $V(x,y,z) = V_0 x^2 y z^3 / L^6$. Determine the electric field.

**Solution:**

We find the electric field's Cartesian components by taking the partial derivative of the electric potential function along the Cartesian axes directions:

$$E_x = \frac{\partial V}{\partial x} = \frac{\partial}{\partial x}\left(\frac{V_0}{L^6}x^2 yz^3\right) = 2\frac{V_0}{L^6}xyz^3$$

$$E_y = \frac{\partial V}{\partial y} = \frac{\partial}{\partial y}\left(\frac{V_0}{L^6}x^2 yz^3\right) = \frac{V_0}{L^6}x^2 z^3$$

$$E_z = \frac{\partial V}{\partial z} = \frac{\partial}{\partial z}\left(\frac{V_0}{L^6}x^2 yz^3\right) = 3\frac{V_0}{L^6}x^2 yz^2$$

Expressing the electric field as a vector,

$$\mathbf{E}(x,y,z) = \frac{V_0}{L^6}xz^2\left(2yz\,\mathbf{i} + xz\,\mathbf{j} + 3xy\,\mathbf{k}\right)$$

### Section 23-8 Electrostatic Potential Energy; the Electron Volt

A convenient unit of energy for dealing with electrons, atoms, and molecules is the **electron volt** (eV). The electron volt is defined as the increase in electrostatic energy of the charge equal to that of the charge of an electron when it changes electric potential by –1 V:

$$1\,\text{eV} = \left(1.602\times10^{-19}\,\text{C}\right)\left(1\,\text{V}\right) = 1.602\times10^{-19}\,\text{CV} = 1.602\times10^{-19}\,\text{J}$$

### Section 23-9 Cathode Ray Tube: TV and Computer Monitors, Oscilloscope

The **cathrode ray tube (CRT)** is a devise that accelerates electrons through a potential difference and allows the electrons to impinge on a screen coated with a fluorescent material that glows with visible light when the kinetic energy of the electrons is absorbed by the material. Thus, by moving the position that the electrons strike the screen with other electric fields (or magnetic fields) the electron beam can "write" on the screen to display information. Devices using this type of display device include television, computer monitors, and oscilloscopes.

## Practice Quiz

1.    A negatively charged particle is located at a position where the electric potential is 124 V and is traveling at a speed $v$. The particle moves to a position where the electric potential is 248 V. What can you say about the speed of the particle at this location?

   a) The speed will be $v/2$.
   b) The speed will be $2v$.
   c) The speed will be less than $v$.
   d) The speed will be greater than $v$.

2.    The electric field is constant in magnitude and points toward the east. In which direction does the electric potential increase?

   a) East
   b) North
   c) West
   d) South

3.    A conductor has no charge initially. Electrons are added one at a time to the conductor. The electrons are initially infinitely far from the conductor and at rest. How does the work done to add the second electron to the conductor compare to the work done to add the tenth electron to the conductor?

a) The work done to add the second electron is less than the work done to add the tenth electron.
b) The work done to add the second electron is the same as the work done to add the tenth electron.
c) The work done to add the second electron is more than the work done to add the tenth electron.
d) Not enough information is given to compare the work done to add the two electrons to the conductor.

4.    A surface is an equipotential surface at a potential of zero. What is true about the electric field on the surface?

a) The electric field on the surface is zero.
b) The electric field on the surface is perpendicular to the surface.
c) The electric field on the surface points parallel to the surface.
d) None of the above statements is necessarily true about the electric field.

5.    A particle has a kinetic energy of 346 eV. The particle moves to a position where the electric potential is 346 V higher. What is the kinetic energy of the particle at the new position?

a) 0 eV
b) 346 eV
c) 692 eV
d) Not enough information is given to determine the kinetic energy.

6.    An electric dipole is placed in a uniform electric field. What orientation of the electric dipole minimizes its electric potential energy?

a) Any orientation such that the line joining the charges is perpendicular to the electric field
b) An orientation such that the displacement of the positive charge from the negative charge is in the direction of the electric field
c) An orientation such that the displacement of the positive charge from the negative charge is in the opposite direction of the electric field
d) The electric field is uniform, so all orientations of the dipole have the same electric potential energy.

7.    Point *a* and point *b* are at the same electric potential. What can you conclude about the electric field between point *a* and point *b*?

a) The electric field between point *a* and point *b* is zero everywhere.
b) The electric field always points perpendicular to the line between point *a* and point *b*.
c) Given a path from *a* to *b*, if the field has a component in the direction of the path in some region, it must have a component opposite to the path somewhere else.
d) Nothing can be concluded about the electric field.

8.    An electron is moved through a potential difference of +10 V. What is its change in electrical potential energy?

a) +10eV
b) −10 eV
c) −1.602 × 10$^{-18}$ eV
d) −10 J

9.    The electric field is uniform and has a magnitude of 100 V/m and a direction east. What is the difference in electric potential at any given position and a position 5.0 m to the north?

a) +500 V
b) −500 V
c) 0 V
d) 20 V

10.    The electric field is uniform and has a magnitude of 100 V/m and a direction east. What is the difference in electric potential at any given position and a position 5.0 m to the west?

 a)  +500 V
 b)  −500 V
 c)  0 V
 d)  −20 V

11.    A charge of 24.6 μC is located at position 1.56 m from the origin on the +x-axis. A second charge of +32.7 μC is located 2.21 m from the origin on the +y-axis. How much work does it take to bring a charge of +4.11 μC from a position 5.63 m on the z-axis to the origin?

12.    Determine the amount of work that is required to assemble an arrangement of charge that is four 1.22 μC charges each located on a different corner of a square that is 9.44 cm on a side if the charges are initially separated by an infinite distance.

13.    An electron is accelerated from rest by a potential difference of +480 V. What are the kinetic energy and speed of the electron after being accelerated?

14.    A dipole consists of charges $+Q$ and $-Q$ separated along the x-axis by a distance $l$ as shown in the diagram. Determine the percent difference between the approximate dipole electric potential and the exact electric potential for a dipole at a positions $x = l$ and $x = 10l$.

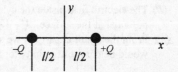

15.    The electric potential in a region of space is given by $V(x, y, z) = V_0(xy/z^2 + y/x)$. Determine the components of the electric field in this region of space.

## Problem Solutions

3.    Because the total energy of the electron is conserved, we have
$$\Delta K + \Delta U = 0, \text{ or}$$
$$\Delta K = -q(V_b - V_a) = -(-1.60 \times 10^{-19} \text{ C})(21,000 \text{ V}) = 3.4 \times 10^{-15} \text{ J}.$$

7.    For the uniform electric field between two large, parallel plates, we have
$$E = \Delta V/d;$$
$$640 \text{ V/m} = \Delta V/(11.0 \times 10^{-3} \text{ m}), \text{ which gives } \Delta V = 7.04 \text{ V}.$$

11.    The potential difference between two points in an electric field is found from
$$\Delta V = -\int \mathbf{E} \cdot d\mathbf{l}.$$
 (a)  For $V_{BA}$ we have
$$V_{BA} = -\int_A^B \mathbf{E} \cdot d\mathbf{l} = -\int_A^B (-300 \text{ N/C}) \mathbf{i} \cdot dy \, \mathbf{j} = 0.$$

 (b)  For $V_{CB}$ we have
$$V_{CB} = -\int_B^C \mathbf{E} \cdot d\mathbf{l} = -\int_B^C (-300 \text{ N/C}) \mathbf{i} \cdot dx \, \mathbf{i} = \int_{4\text{m}}^{-3\text{m}} (300 \text{ N/C}) dx$$
$$= (300 \text{ N/C})(-3 \text{ m} - 4 \text{ m}) = -2100 \text{ V}$$

 (c)  For $V_{CA}$ we have

$$V_{CA} = -\int_A^C \mathbf{E} \cdot d\mathbf{l} = -\int_A^C (-300 \text{ N/C}) \mathbf{i} \cdot (dx \, \mathbf{i} + dy \, \mathbf{j}) = \int_{4\text{m}}^{-3\text{m}} (300 \text{ N/C}) dx$$
$$= (300 \text{ N/C})(-3 \text{ m} - 4 \text{ m}) = -2100 \text{ V}$$

 Note that $V_{CA} = V_{CB} + V_{BA}$.

15. (*a*) After the connection, if the two spheres were at different potentials, there would be a flow of charge in the wire.  Thus the potentials must be the same.

(*b*) We assume the spheres are so far apart that the potential of one sphere at the other sphere is essentially zero.  The initial potentials are

$V_{01} = Q/4\pi\varepsilon_0 r_1$, $V_{02} = 0$.

After the connection, $Q_2$ is transferred to the second sphere, so we have

$V_1 = (Q - Q_2)/4\pi\varepsilon_0 r_1 = V_2 = Q_2/4\pi\varepsilon_0 r_2$,  or

$r_2(Q - Q_2) = r_1 Q_2$, which gives $Q_2 = r_2 Q/(r_1 + r_2)$.

19. The field outside the cylinder is the same as that of a long wire.  We find the equivalent linear charge density from the charge on the length $L$:

$Q = \sigma 2\pi R_0 L = \lambda L$, which gives $\lambda = \sigma 2\pi R_0$.

(*a*) The radial electric field outside the cylinder is

$E = \lambda/2\pi\varepsilon_0 r = \sigma 2\pi R_0/2\pi\varepsilon_0 r = \sigma R_0/\varepsilon_0 r$.

We find the potential difference from

$$V - V_0 = -\int_{R_0}^{r} \mathbf{E} \cdot d\mathbf{l} = -\int_{R_0}^{r} \frac{\sigma R_0}{\varepsilon_0 r} dr = -\frac{\sigma R_0}{\varepsilon_0} \ln\left(\frac{r}{R_0}\right), \text{ or}$$

$V = V_0 + (\sigma R_0/\varepsilon_0) \ln(R_0/r)$, $r > R_0$.

(*b*) The electric field inside the cylinder is zero, so the potential inside is constant and equal to the potential at the surface:  $V = V_0$, $r < R_0$.

(*c*) From the result in part (*a*) we see that the potential at $r = \infty$ is undefined.  $V \neq 0$ because there would be charge at infinity for an infinite cylinder.

23. We find the electric potentials of the stationary charges at the initial and final points:

$V_a = (1/4\pi\varepsilon_0)[(Q_1/r_{1a}) + (Q_2/r_{2a})]$
    $= (9.0 \times 10^9 \text{ N} \cdot \text{m}^2/\text{C}^2)\{[(25 \times 10^{-6} \text{ C})/(0.030 \text{ m})] + [(25 \times 10^{-6} \text{ C})/(0.030 \text{ m})]\} = 1.50 \times 10^7 \text{V}$

$V_b = (1/4\pi\varepsilon_0)[(Q_1/r_{1b}) + (Q_2/r_{2b})]$
    $= (9.0 \times 10^9 \text{ N} \cdot \text{m}^2/\text{C}^2)\{[(25 \times 10^{-6} \text{ C})/(0.040 \text{ m})] + [(25 \times 10^{-6} \text{ C})/(0.020 \text{ m})]\} = 1.69 \times 10^7 \text{V}$

Because there is no change in kinetic energy, we have

$W_{a \oslash b} = \Delta K + \Delta U = 0 + q(V_b - V_a)$
    $= (0.10 \times 10^{-6} \text{ C})(1.69 \times 10^7 \text{ V} - 1.50 \times 10^7 \text{ V}) = + 0.19 \text{ J}$.

27. When the electron is far away, the potential from the fixed charge is zero.
Because energy is conserved, we have

$\Delta K + \Delta U = 0$;

$\tfrac{1}{2}mv^2 - 0 + (- e)(0 - V) = 0$,  or

$\tfrac{1}{2}mv^2 = - e(kQ/r)$

$\tfrac{1}{2}(9.11 \times 10^{-31} \text{ kg})v^2 = - (1.60 \times 10^{-19} \text{ C})(9.0 \times 10^9 \text{ N} \cdot \text{m}^2/\text{C}^2)(- 0.125 \times 10^{-6} \text{ C})/(0.725 \text{ m})$,

which gives $v = 2.33 \times 10^7$ m/s.

31. We choose a ring of radius $r$ and width $dr$ for a differential element, with charge $dq = \sigma 2\pi r\, dr$.  The potential of this element on the axis a distance $x$ from the ring is

$dV = dq/4\pi\varepsilon_0(x^2 + r^2)^{1/2}$
    $= \sigma 2\pi r\, dr/4\pi\varepsilon_0(x^2 + r^2)^{1/2} = \sigma r\, dr/2\varepsilon_0(x^2 + r^2)^{1/2}$.

We integrate to get the potential:

$$V = \int_{R_1}^{R_2} \frac{\sigma r\, dr}{2\varepsilon_0 \left(x^2 + r^2\right)^{1/2}} = \frac{\sigma}{2\varepsilon_0} \left(x^2 + r^2\right)^{1/2} \Big|_{R_1}^{R_2}$$

$$= \frac{\sigma}{2\varepsilon_0} \left[ \left(x^2 + R_2^2\right)^{1/2} - \left(x^2 + R_1^2\right)^{1/2} \right]$$

35. We choose a ring of radius $r$ and width $dr$ for a differential element, with charge $dq = \sigma\, 2\pi r\, dr$. The potential of this element on the axis a distance $x$ from the ring is

$$dV = dq/4\pi\varepsilon_0(x^2 + r^2)^{1/2}$$
$$= \sigma\, 2\pi r dr/4\pi\varepsilon_0(x^2 + r^2)^{1/2} = ar^3\, dr/2\varepsilon_0(x^2 + r^2)^{1/2}.$$

We integrate to get the potential:

$$V = \frac{a}{2\varepsilon_0}\int_0^R \frac{r^3}{\left(x^2 + r^2\right)^{1/2}}\, dr = \frac{a}{2\varepsilon_0}\left[r^2\left(x^2 + r^2\right)^{3/2} - \tfrac{2}{3}\left(x^2 + r^2\right)^{3/2}\right]_0^R$$

$$= \frac{a}{6\varepsilon_0}\left[\left(x^2 + R^2\right)^{1/2}\left(R^2 - 2x^2\right) + 2x^3\right]$$

39. (a) We find the dipole moment from
$$p = eL = (1.60 \times 10^{-19}\ \text{C})(0.53 \times 10^{-10}\ \text{m}) = 8.5 \times 10^{-30}\ \text{C} \cdot \text{m}.$$

(b) The dipole moment will point from the electron toward the proton. As the electron revolves about the proton, the dipole moment will spend equal times pointing in any direction. Thus the average over time will be zero.

43. (a) Because $p_1 = p_2$, from the vector addition we have
$$p = 2p_1 \cos(\tfrac{1}{2}\phi) = 2qL \cos(\tfrac{1}{2}\phi);$$
$$6.1 \times 10^{-30}\ \text{C} \cdot \text{m} = 2q(0.96 \times 10^{-10}\ \text{m}) \cos[\tfrac{1}{2}(104°)],$$
which gives $q = 5.2 \times 10^{-20}$ C.

(b) We find the potential by adding the potentials from the two dipoles:
$$V = V_1 + V_2 = p_1 \cos(\theta - \tfrac{1}{2}\phi)/4\pi\varepsilon_0 r^2 + p_2 \cos(\theta + \tfrac{1}{2}\phi)/4\pi\varepsilon_0 r^2$$
$$= (p_1/4\pi\varepsilon_0 r^2)[\cos(\theta - \tfrac{1}{2}\phi) + \cos(\theta + \tfrac{1}{2}\phi)].$$
When we use a trigonometric identity and the above result for $p$ we get
$$V = (p_1/4\pi\varepsilon_0 r^2)(2 \cos\theta \cos\tfrac{1}{2}\phi) = p \cos\theta/4\pi\varepsilon_0 r^2.$$

47. From the spatial dependence of the electric potential, $V(x, y, z) = y^2 + 2xy - 4xyz$, we find the components of the electric field from the partial derivatives of $V$:
$$E_x = -\partial V/\partial x = -(2y - 4yz);$$
$$E_y = -\partial V/\partial y = -(2y + 2x - 4xz);$$
$$E_z = -\partial V/\partial z = -(-4xy).$$
We can write the electric field: $\mathbf{E} = 2y(2z - 1)\mathbf{i} - 2(y + x - 2xz)\mathbf{j} + (4xy)\mathbf{k}.$

51. Because the total energy of the helium nucleus is conserved, we have
$$\Delta K + \Delta U = 0;$$
$$\Delta K + q\Delta V = 0;$$
$$48 \times 10^3\ \text{eV} + [(3.2 \times 10^{-19}\ \text{C})/(1.60 \times 10^{-19}\ \text{C/e})]\Delta V; \text{ which gives } \Delta V = -2.4 \times 10^4\ \text{V}.$$
The negative sign means the helium nucleus gains kinetic energy by going to a lower potential.

55. (a) The kinetic energy of the electron $(q = -e)$ is
$$K_e = -qV_{BA} = -(-e)V_{BA} = eV_{BA}.$$
The kinetic energy of the proton $(q = +e)$ is
$$K_p = -qV_{AB} = -(+e)(-V_{BA}) = eV_{BA} = 2.0\ \text{keV}.$$

(b) We find the ratio of their speeds, starting from rest, from
$$\tfrac{1}{2}m_e v_e^2 = \tfrac{1}{2}m_p v_p^2, \text{ or } v_e/v_p = (m_p/m_e)^{1/2} = [(1.67 \times 10^{-27}\ \text{kg})/(9.11 \times 10^{-31}\ \text{kg})]^{1/2} = 42.8.$$

59. The charge density of the sphere is $\rho_E = Q/\tfrac{4}{3}\pi r_0^3$. To find the total potential energy of the sphere, we consider it to be made up of differential shells and add (integrate) the work required to bring each shell in from infinity. If a sphere of radius $r < r_0$ with charge $q$ has been formed, the potential at the surface is
$$V = (1/4\pi\varepsilon_0)(q/r) = (1/4\pi\varepsilon_0)(\rho_E \tfrac{4}{3}\pi r^3/r) = \rho_E r^2/3\varepsilon_0.$$
The work to bring the charge of the next shell, $dq = \rho_E 4\pi r^2\, dr$, in from infinity is
$$dW = dq\, V = (\rho_E 4\pi r^2\, dr)(\rho_E r^2/3\varepsilon_0) = \rho_E^2 4\pi r^4\, dr/3\varepsilon_0.$$
The total work and thus the total potential energy stored is
$$W = \int_0^{r_0} \frac{\rho_E^2 4\pi r^4\, dr}{3\varepsilon_0} = \frac{4\pi\rho_E^2}{3\varepsilon_0}\int_0^{r_0} r^4\, dr = \frac{4\pi\rho_E^2 r_0^5}{15\varepsilon_0} = \frac{3Q^2}{20\pi\varepsilon_0 r_0}$$

63. The potential at the surface of a charged sphere is

$V = Q/4\pi\varepsilon_0 r = (9.0 \times 10^9 \text{ N} \cdot \text{m}^2/\text{C}^2)(10^{-8} \text{ C})/(0.10 \text{ m}) = 9 \times 10^2 \text{ V}.$

67. When the proton is accelerated by a potential difference, it acquires a kinetic energy:

$K = Q_p V_{accel}$ .

If it is far from the iron nucleus, the potential is zero. The proton will slow as it approaches the positive charge of the nucleus, because the potential produced by the iron nucleus is increasing. At the proton's closest point the kinetic energy will be zero. We find the required accelerating potential from

$\Delta K + \Delta U = 0;$
$0 - K + Q_p(V_{Si} - 0) = 0,$ or
$Q_p V_{accel} = Q_p Q_{Si}/4\pi\varepsilon_0(r_p + r_{Si});$
$V_{accel} = (9.0 \times 10^9 \text{ N} \cdot \text{m}^2/\text{C}^2)(26)(1.60 \times 10^{-19} \text{ C})/(1.2 \times 10^{-15} \text{ m} + 4.0 \times 10^{-15} \text{ m})$
$\quad\quad = 7.2 \times 10^6 \text{ V} = 7.2 \text{ MV}.$

71. The acceleration produced by a potential difference of 1000 V over a distance of 1 cm is

$a = eE/m = eV/md = (1.60 \times 10^{-19} \text{ C})(1000 \text{ V})/(9.11 \times 10^{-31} \text{ kg})(0.01 \text{ m}) = 2 \times 10^{16} \text{ m/s}^2.$

Because this is so much greater than $g$, yes, the electron can easily move upward.

To find the potential difference to hold the electron stationary, we have

$mg = eE = eV/d;$
$(9.11 \times 10^{-31} \text{ kg})(9.80 \text{ m/s}^2) = (1.60 \times 10^{-19} \text{ C})V/(0.030 \text{ m}),$ which gives $V = 1.7 \times 10^{-12} \text{ V}.$

75. The distances from the midpoint of a side to the three charges
are $L/2$, $L/2$, and $L \cos 30°$.

At point a, we have

$V_a = (1/4\pi\varepsilon_0)\{[(-Q)/(L/2)] + [(+Q)/(L/2)] + [(-3Q)/(L\cos 30°)]\}$
$\quad = (Q/4\pi\varepsilon_0 L)[(-2) + (+2) + (-3/\cos 30°)] = -3.5\ Q/4\pi\varepsilon_0 L.$

At point b, we have

$V_b = (1/4\pi\varepsilon_0)\{[(+Q)/(L/2)] + [(-3Q)/(L/2)] + [(-Q)/(L\cos 30°)]\}$
$\quad = (Q/4\pi\varepsilon_0 L)[(+2) + (-6) + (-1/\cos 30°)] = -5.2\ Q/4\pi\varepsilon_0 L.$

At point c, we have

$V_c = (1/4\pi\varepsilon_0)\{[(-3Q)/(L/2)] + [(-Q)/(L/2)] + [(+Q)/(L\cos 30°)]\}$
$\quad = (Q/4\pi\varepsilon_0 L)[(-6) + (-2) + (+1/\cos 30°)] = -6.8\ Q/4\pi\varepsilon_0 L.$

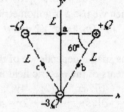

79. In the region between the wire and cylinder, the radial electric field will be produced by the central wire:

$E = \lambda/2\pi\varepsilon_0 r, R_a < r < R_b$ .

We find the potential difference from

$$V_b - V_a = -\int_a^b \mathbf{E} \cdot d\mathbf{l} = -\int_{R_a}^{R_b} \frac{\lambda}{2\pi\varepsilon_0 r}\,dr = -\frac{\lambda}{2\pi\varepsilon_0}\ln\left(\frac{R_b}{R_a}\right),\ \text{or}$$

$V_a - V_b = (\lambda/2\pi\varepsilon_0)\ln(R_b/R_a).$

83. (a) The reference level for the potential is $V = 0$ at $r = \infty$. At points outside the spherical shell,
it is equivalent to a point charge. Thus the potential when $r > r_2$ is

$V = Q/4\pi\varepsilon_0 r = \rho_E \frac{4}{3}\pi(r_2^3 - r_1^3)/4\pi\varepsilon_0 r = \rho_E(r_2^3 - r_1^3)/3\varepsilon_0 r,\ r > r_2$ .

(b) The electric field within the spherical conductor, $r_1 < r < r_2$, is due to the charge within a radius $r$:

$E = \rho_E \frac{4}{3}\pi(r^3 - r_1^3)/4\pi\varepsilon_0 r^2 = (\rho_E/3\varepsilon_0)[r - (r_1^3/r^2)].$

We find the potential by integrating along a radial line from $r$ to $r_2$:

$$\int_{r_1}^{r_2} dV = -\int_r^{r_2} \mathbf{E} \cdot d\mathbf{l} = -\int_r^{r_2} E\,dr$$

$$\frac{\rho_E\left(r_2^3 - r_1^3\right)}{3\varepsilon_0 r_2} - V = -\frac{\rho_E}{3\varepsilon_0}\int_r^{r_2}\left(r - \frac{r_1^3}{r^2}\right)dr = -\frac{\rho_E}{3\varepsilon_0}\left(\frac{r_2^2 - r^2}{2} + \frac{r_1^3}{r_2} - \frac{r_1^3}{r}\right),\ \text{which gives}$$

$V = (\rho_E/6\varepsilon_0)[3r_2^2 - r^2 - (2r_1^3/r)],\ r_1 < r < r_2$ .

(c) Inside the cavity the electric field is zero, so the potential is constant and equal to the potential at the inner surface of the shell:

$$V = V(r_1) = (\rho_E/6\varepsilon_0)[3r_2^2 - r_1^2 - (2r_1^3/r_1)] = (\rho_E/2\varepsilon_0)(r_2^2 - r_1^2), r < r_1.$$

The potential is continuous at $r_1$ and $r_2$.

# Chapter 24: Capacitance, Dielectrics, Electric Energy Storage

## Chapter Overview and Objectives

This chapter introduces the concept of capacitance. It discusses how to calculate the capacitance of arrangements of conductors and the effect of dielectrics on the capacitance. It also discusses electrical potential energy storage in capacitors.

After completing study of this chapter, you should:

- Know what capacitance is.
- Know how to calculate the capacitance of parallel plates, concentric cylinders, and concentric spherical shells with and without dielectric material between the conductors.
- Know how to determine the amount of electrical potential energy stored in a capacitor.
- Know how to calculate equivalent capacitances of serial and parallel networks of capacitances.

## Summary of Equations

Definition of capacitance:
$$C = \frac{Q}{V}$$

Capacitance of parallel plate capacitor:
$$C = \frac{\varepsilon_0 A}{d}$$

Capacitance of concentric cylinders:
$$C = \frac{2\pi\varepsilon_0 L}{\ln(R_a / R_b)}$$

Capacitance of concentric spheres:
$$C = 4\pi\varepsilon_0 \frac{r_a r_b}{r_b - r_a}$$

Capacitance of isolated sphere:
$$C = 4\pi\varepsilon_0 r$$

Equivalent capacitance of parallel network:
$$C_{eq} = \sum_{i=1}^{n} C_i$$

Equivalent capacitance of series network:
$$\frac{1}{C_{eq}} = \sum_{i=1}^{n} \frac{1}{C_i}$$

Electric potential energy of charged capacitor:
$$U = \tfrac{1}{2} QV = \tfrac{1}{2} CV^2 = \tfrac{1}{2} \frac{Q^2}{C}$$

Energy density in electric field:
$$u = \tfrac{1}{2} \varepsilon_0 E^2$$

Capacitance of capacitor with dielectric:
$$C = KC_0$$

## Chapter Summary

### Section 24-1 Capacitors

For a particular geometry of two conductors, a particular charge of $+Q$ on one of the conductors and a charge of $-Q$ on the other conductor will cause a potential difference $V$ between the two conductors. The ratio of the charge $Q$ to the potential difference $V$ is a constant called the **capacitance**, $C$, of the system of conductors:

$$C = \frac{Q}{V}$$

The dimensions of capacitance are charge per volt and the SI unit of capacitance is the farad. One **farad** is one coulomb/volt. The abbreviation for farad is F.

A device constructed to provide capacitance to an electrical circuit is called a **capacitor**. The symbol for a capacitor in an electrical circuit is

**Example 24-1-A**

An initially uncharged capacitor has a current of 3.64 mA flow from one plate to another for a time of 23.9s. The resulting voltage across the capacitor is 167 V. What is the capacitance of the capacitor?

**Solution:**

The charge is the time integral of the current:

$$Q = \int I\, dt = I \int dt = I\Delta t = \left(3.64 \times 10^{-3}\ \text{A}\right)\!\left(23.9\ \text{s}\right) = 8.70 \times 10^{-2}\ \text{C}$$

where we have used the fact that the current is constant with time to complete the integral. We can now calculate the capacitance:

$$C = \frac{Q}{V} = \frac{8.70 \times 10^{-2}\ \text{C}}{167\ \text{V}} = 5.21 \times 10^{-4}\ \text{F} = 521\,\mu\text{F}$$

**Section 24-2 Determination of Capacitance**

For some very symmetric arrangements of conductors, we can determine the capacitance by using our knowledge of electric fields near certain arrangements of charge. For two conductive parallel flat plats of equal area, $A$, separated by a distance $d$ perpendicular to the plane of the areas, the capacitance is

$$C = \frac{\varepsilon_0 A}{d}$$

For two concentric conductive circular cylinders of length $L$, the capacitance is

$$C = \frac{2\pi\varepsilon_0 L}{\ln(R_a / R_b)}$$

where $R_a$ is the radius of the outer cylinder and $R_b$ is the radius of the inner cylinder. For two concentric conductive spherical shells, the capacitance is

$$C = 4\pi\varepsilon_0 \frac{r_a r_b}{r_b - r_a}$$

where $r_a$ is the inner radius of the outer shell and $r_b$ is the outer radius of the inner shell.

A single conductor has a capacitance defined as the ratio of the charge on the conductor to the potential difference of conductor to the potential an infinite distance away. It is easy to show that the capacitance of an isolated conductive sphere or spherical shell of outer radius $r$ is

$$C = 4\pi\varepsilon_0 r$$

**Example 24-2-A**

A coaxial cable has an inner wire of diameter 0.67 mm and an outer conductor of inside diameter 4.65 mm. What is the capacitance per length of cable?

**Solution:**

The capacitance per length of concentric cylindrical conductors is

$$\frac{C}{L} = \frac{2\pi\varepsilon_0}{\ln(R_a / R_b)} = \frac{2\pi(8.85\times10^{-12}\ \text{F/m})}{\ln(4.65\ \text{mm}/0.67\ \text{mm})} = 1.31\times10^{-11}\ \text{F/m}$$

**Example 24-2-B**

What is the minimum plate area of a parallel plate capacitor needs to have a capacitance of 2.85 μF? The capacitor must be able to be charged to a voltage of 1000 V. The maximum electric field between the plates allowed is $1.00 \times 10^4$ V/cm or the air in between the plates will break down and become conductive.

**Solution:**

The maximum of the electric field allows us to calculate the minimum separation between the capacitor plates:

$$E = \frac{V}{d} \quad \Rightarrow \quad d_{min} = \frac{V}{E_{max}} = \frac{1000\ \text{V}}{1.00\times10^6\ \text{V/m}} = 1.00\times10^{-3}\ \text{m}$$

We can now use the expression for the capacitance of parallel plates to determine their area:

$$C = \frac{\varepsilon_0 A}{d} \quad \Rightarrow \quad A = \frac{dC}{\varepsilon_0} = \frac{(1.00\times10^{-3}\ \text{m})(2.85\times10^{-6}\ \text{F})}{8.85\times10^{-12}\ \text{F/m}} = 322\ \text{m}^2$$

This is quite a large area!

**Section 24-3 Capacitors in Series and Parallel**

Capacitors are connected in parallel if they are connected in such a way that the potential difference across each is identical. An example of three capacitors connected in parallel is shown in the diagram. The conductors connecting the upper plates ensure that all of the upper plates are at the same potential as each other. The conductors connecting the lower plates also ensure that all of the lower plates are all at the same potential. When capacitors are connected in this manner, the capacitance of the circuit between the top plates and bottom plates is the sum of the capacitance of each capacitor:

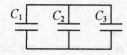

$$C_{eq} = C_1 + C_2 + C_3$$

in the case of three capacitors connected in parallel. This generalizes to the case of $n$ capacitors connected in parallel:

$$C_{eq} = \sum_{i=1}^{n} C_i$$

where $C_i$ is the capacitance of the $i$th capacitor.

Capacitors are connected in series if the charge on each capacitance is forced to always be identical. Three capacitors connected in series are shown in the diagram. The equivalent capacitance between the first and final terminals is the reciprocal of the sum of the reciprocals of the individual capacitances:

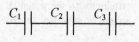

$$\frac{1}{C_{eq}} = \frac{1}{C_1} + \frac{1}{C_2} + \frac{1}{C_3}$$

This can be generalized to the case of $n$ capacitors in series:

$$\frac{1}{C_{eq}} = \sum_{i=1}^{n} \frac{1}{C_i}$$

where $C_i$ is the capacitance of the $i$th capacitor.

**Example 24-3-A**

Determine the equivalent capacitance of the capacitor network shown in the diagram.

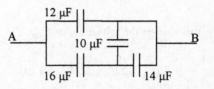

**Solution:**

First, we recognize that the 10 µF and 14 µF capacitor are in parallel with each other. The equivalent cpacitance is

$$C_{eq1} = 10\,\mu F + 14\,\mu F = 24\,\mu F$$

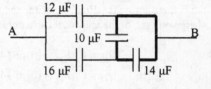

We redraw the network with the equivalent capacitance in place of the 10 µF and 12 µF capacitors:

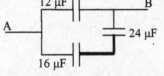

We then recognize that the 16 µF and the 26 µF capacitances are in series. We calculate the equivalent capacitance of this series:

$$\frac{1}{C_{eq2}} = \frac{1}{16\,\mu F} + \frac{1}{24\,\mu F} \quad \Rightarrow \quad C_{eq2} = 9.6\,\mu F$$

We again redraw the circuit with this equivalent capacitance replacing the individual capacitances.

We recognize that the 12 µF and the 9.6 µF capacitors are in parallel and calculate their equivalent capacitance:

$$C_{eq3} = 12\,\mu F + 9.6\,\mu F = 21.6\,\mu F$$

We have reduced the capacitor network to a single equivalent 22 µF capacitance.

**Example 24-3-B**

Determine the charge on each capacitor in the network in Example 24-3-A and the voltage across each capacitor if point A is at a potential 12 V higher than point B.

**Solution:**

To determine the charges on and voltages across the capacitors in the network, we work backwards through the steps of reducing the circuit. First, if point A is 12 V higher in potential than point B, the potential difference across the 12 µF capacitor is 12 V and the charge on the 12 µF capacitor is

$$Q_{12} = C_{12}V_{12} = (12\,\mu F)(12\,V) = 144\,\mu C$$

$C_{eq2}$ also has a 12 V potential difference across it. Remember, capacitors in parallel have the same potential difference. The charge across $C_{eq2}$ then is

$$Q_{eq2} = C_{eq2}V_{eq2} = (9.6\,\mu\text{F})(12\,\text{V}) = 115\,\mu\text{C}$$

Next we know that $C_{eq2}$ is the equivalent of the 16 μF capacitor in series with $C_{eq1}$. Capacitors in series have the same charge as their equivalent capacitance. Therefore, we know the charge on the 16 μF capacitor is 112 μC, so we can calculate the potential difference across it:

$$V_{16} = \frac{Q_{16}}{C_{16}} = \frac{115\,\mu\text{C}}{16\,\mu\text{F}} = 7.2\,\text{V}$$

Similarly, we calculate the voltage across $C_{eq1}$:

$$V_{eq1} = \frac{Q_{eq1}}{C_{eq1}} = \frac{115\,\mu\text{C}}{24\,\mu\text{F}} = 4.8\,\text{V}$$

Again, the voltage across $C_{eq1}$ is the same as that across each of the capacitors that make up that parallel equivalent. So the voltage across the 10 μF capacitor and the voltage across the 14 μF capacitor are 4.8 V. We can then calculate the charge on each of the individual capacitors:

$$Q_{10} = C_{10}V_{10} = (10\,\mu\text{F})(4.8\,\text{V}) = 48\,\mu\text{C}$$
$$Q_{14} = C_{14}V_{14} = (14\,\mu\text{F})(4.8\,\text{V}) = 67\,\mu\text{C}$$

## Section 24-4 Electric Energy Storage

The electric potential energy is stored in a capacitor when it is charged. This potential energy, $U$, can be written several ways:

$$U = \tfrac{1}{2}QV = \tfrac{1}{2}CV^2 = \tfrac{1}{2}\frac{Q^2}{C}$$

where $Q$ is the charge on the capacitor, $C$ is the capacitance, and $V$ is the potential difference across the capacitor.

Because the electric field between the plates of the capacitor is a well-defined function of the charge on the plates of the capacitor, we can alternatively write the potential energy stored in the capacitor in terms of the electric field between the plates of the capacitor. The potential energy can be written as

$$U = \tfrac{1}{2}\varepsilon_0 E^2 Ad$$

where $E$ is the magnitude of the electric field between the plates of the capacitor, $A$ is the area of the plates, and $d$ is the separation of the plates. As $Ad$ is the volume between the plates of the capacitor, we can divide through by the volume to obtain an energy density, $u$, the energy per volume in the electric field:

$$u = \tfrac{1}{2}\varepsilon_0 E^2$$

This expression can be shown to be general and is useful in situations other than for parallel plate capacitors. It is often applied to situations for which the electric field is known, but the charge distribution that created the electric field is unknown.

## Section 24-5 Dielectrics

When an insulating material, also called a **dielectric**, is placed between the plates of a capacitor, the capacitance is altered from the capacitance without the insulator in place. We write the capacitance with the dielectric in place, $C$, as

$$C = KC_0$$

where $C_0$ is the capacitance without any material between the plates and $K$ is called the dielectric constant of the material. If we combine this expression for the capacitance with the expression for a parallel plate capacitor, we get

$$C = \frac{K\varepsilon_0 A}{d}$$

Often, the dielectric constant and the free space permittivity are combined and called the **permittivity**, $\varepsilon$, of the dielectric material:

$$\varepsilon = K\varepsilon_0$$

The parallel plate capacitance can then be written

$$C = \frac{\varepsilon A}{d}$$

### Example 24-5-A

Consider a system of a charged isolated parallel plate capacitors with plates of length $L$, width $W$, and no dielectric, of capacitance $C_0$ charged to a potential difference $V$. Consider a dielectric material of the correct thickness to fill the space between the plates of the capacitor if inserted. Considering the electrical potential energy of this system as a function of the distance, $x$, that the dielectric is inserted between the plates, determine the force on the dielectric as it is inserted between the plates.

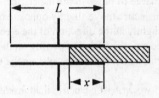

### Solution:

As the capacitor is isolated during the insertion of the dielectric, the charge on the capacitor remains constant and equal to

$$Q = CV$$

If the dielectric is inserted into the gap a distance $x$, then the capacitance looks like a capacitor with plate of length $L - x$ by width $W$ with no dielectric of capacitance

$$C_1 = \varepsilon_0 \frac{(L-x)W}{d}$$

where $d$ is the separation distance of the two plates. This is in parallel with a capacitor with plates of length $x$ and width $W$ with a dielectric with a dielectric constant $K$ between the plates with capacitance

$$C_2 = K\varepsilon_0 \frac{x}{d}$$

The total capacitance of these in parallel is

$$C = C_1 + C_2 = \varepsilon_0 \frac{(L-x)W}{d} + K\varepsilon_0 \frac{xW}{d} = \varepsilon_0 \frac{WL}{d}\left(1 - \frac{1-K}{L}x\right) = C_0\left(1 - \frac{1-K}{L}x\right)$$

Note that this expression is only valid for $0 < x < L$. We calculate the potential energy of the system as a function of $x$:

$$U = \frac{1}{2}\frac{Q^2}{C} = \frac{1}{2}\frac{Q^2}{C_0\left(1 - \dfrac{1-K}{L}x\right)}$$

To determine the force, we know that the force is the negative of the derivative with respect to position:

$$F = -\frac{\partial U}{\partial x} = -\frac{1}{2}\frac{Q^2}{C_0\left(1-\frac{1-K}{L}x\right)^2}\left(\frac{1-K}{L}\right)$$

Now, rewriting this in terms of the original voltage and capacitance, we get

$$F = \frac{1}{2}\frac{C_0V^2}{\left(1+\frac{K-1}{L}x\right)^2}\left(\frac{K-1}{L}\right)$$

As $K$ is a number larger than one, this force is positive. The dielectric is pulled into the gap with this force. Think carefully about why we wrote the energy in terms of the constant charge and the capacitance rather than writing it in terms of the voltage and capacitance before we took the derivative.

### Section 24-6 Molecular Description of Dielectrics

The reason a dielectric reduces the electric field in the interior of the dielectric is because the applied electric field causes charge to move in the dielectric. The charge does not move freely in the dielectric, but the electrons in the atoms of the dielectric move slightly opposite to the direction of the electric field and the nuclei of the atoms of the dielectric move slightly in the direction of the electric field. If this occurs for all the atoms in the material, the net effect is that there will be a positive charge density on the surface of the material that the electric field is pointing toward and a negative surface charge density on the surface the electric field is pointing away from.

If we write the electric field inside the dielectric, $E_D$, it is reduced by a factor of the dielectric constant, $K$, from the applied electric field, $E_0$:

$$E_D = \frac{E_0}{K}$$

We can determine the electric field due to the polarization of the molecules within the material:

$$E_D = E_0 + E_{induced} \quad \Rightarrow \quad E_{induced} = E_D - E_0 = -E_0\left(1-\frac{1}{K}\right)$$

The minus sign means that the induced field points in the direction opposite to the applied field. The induced field is due to the induced surface charge density. We can use the induced electric field to determine the induced surface charge density:

$$E_{induced} = \frac{\sigma_{induced}}{\varepsilon_0}$$

## Practice Quiz

1.    If electrostatic potential energy is given by $QV$ as stated in Chapter 23, why is the potential energy stored in a capacitor $\frac{1}{2}QV$?

    a)  Half of the energy is lost in charging up the capacitor.
    b)  We divide by two because half of the charge is on each plate of the capacitor.
    c)  The average voltage difference during the charge transfer is $\frac{1}{2}V$.
    d)  The other half of the energy is stored in the electric field between the plates of the capacitor.

2.      You have three capacitors of equal capacitance. How should they be connected together to reach the minimum capacitance of the combination of capacitors?

        a) Connect the three capacitors in parallel.
        b) Connect the three capacitors in series.
        c) Connect the first two capacitors in parallel and then in series with the third.
        d) Connect the first two capacitors in series and then in parallel with the third.

3.      What direction is the electrical force on the plates of a charged capacitor?

        a) One plate is attracted toward the other.
        b) One plate is repelled from the other.
        c) The force is parallel to the surface of the plates.
        d) There is no electric force on the plates.

4.      The dielectric strength of dry air at room temperature and pressure is about 30,000 V/cm. What is the breakdown voltage of a capacitor with air as a dielectric and a gap of 1.00 mm?

        a) 300 V
        b) 3,000 V
        c) 30,000 V
        d) 300,000 V

5.      You have a parallel plate capacitor with air between the plates. You have enough dielectric material with a relatively high dielectric constant to fill only half of the volume between the plates. To create the greatest capacitance, how should you fill the volume between the plates?

        a) Fill the gap between the plates completely over one half the area of the capacitor.
        b) Fill half the gap between the plates over the entire area of the capacitor.
        c) Don't fill the gap at all.
        d) It doesn't matter how you fill the gap, as long as half the volume is filled.

6.      Suppose you have a capacitor with capacitance $C$ charged to voltage $V$ and a second capacitor with capacitance $C$ charged to voltage $2V$. The two capacitors are connected in parallel. What will be the voltage across the two capacitors after being connected in parallel?

        a) The first capacitor will have voltage $V$ and the second capacitor will have voltage $2V$.
        b) Both capacitors will be charged to voltage $3V$.
        c) Both capacitors will be charged to voltage $1.5V$.
        d) The first capacitor will have voltage $2V$ and the second capacitor will have voltage $V$.

7.      Make an estimate of the capacitance of the human body based on the expression for the capacitance of an isolated sphere.

        a) 10 pF
        b) 10 nF
        c) 10 μF
        d) 10 mF

8.      In order to double the capacitance of a parallel plate capacitor you can

        a) Double the area of its plates.
        b) Double the distance between its plates.
        c) Double the length of each of the edges of its plates.
        d) Any of the above.

9.    In order to increase the capacitance of a concentric cylinder capacitor by a factor of 10 you can

   a) Make the outer conductor's inner radius a factor 10 larger.
   b) Make the inner conductor's outer radius a factor 10 larger.
   c) Make the outer conductor's inner radius a factor $e^{10}$ larger.
   d) Make the ratio of the outer conductor's radius to the inner conductor's radius the tenth root of its current ratio.

10.   Why does water have a relatively high dielectric constant?

   a)  Its molecules have a large permanent dipole moment and are free to rotate.
   b)  Its molecules are lower in mass than most diatomic molecules.
   c)  It has a high specific heat.
   d)  It has very low conductivity.

11.   Determine the capacitance of two rectangular plates 24 cm × 36 cm separated by a distance of 0.02 cm.

12.   Determine the equivalent capacitance between points A and B of the capacitor network shown.

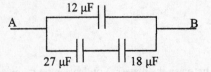

13.   Determine the charge in each capacitor and the energy stored in each capacitor when the capacitor network in quiz Problem 12 is connected to a potential difference of 12 V.

14.   A capacitor with no dielectric material between its plates has a capacitance $C$. If half of the volume between its plates is filled with a dielectric material with dielectric constant $K_1$ and half is filled with a dielectric material with dielectric constant $K_2$ as shown, what will the new capacitance be?

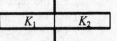

15.   A capacitor with capacitance $C$ is charged to a potential difference $V$ and then disconnected from the potential difference source. A double layered dielectric with dielectric constant $K_1$ on one layer and dielectric $K_2$ on the other layer is placed in between the plates of the capacitor, as shown. How much work is done to insert the dielectric?

## Problem Solutions

3.   From $Q = CV$, we have
        75 pC = $C$(12.0 V), which gives $C$ = 6.3 pF.

7.   We assume that the charge transferred is small compared to the initial charge on the plates so the potential difference between the plates is constant. The energy required to move the charge is
        $W = qV$.
   Thus the charge on each plate is
        $Q = CV = C(W/q) = (16 \times 10^{-6}$ F$)(25$ J$)/(0.20 \times 10^{-3}$ C$) = 2.0$ C.
   Because this is much greater than the charge moved, our assumption is justified.

11.  The potential at the surface of a spherical conductor is
        $V = Q/4\pi\varepsilon_0 r_E$, so we have
        $C = Q/V = 4\pi\varepsilon_0 r_E = 4\pi(8.85 \times 10^{-12}$ F/m$)(6.38 \times 10^6$ m$) = 7.1 \times 10^{-4}$ F.

15.  When the two cylinders are separated by $d$, we have $R_a = R_b + d$. For a cylindrical capacitor, we have
        $C = L2\pi\varepsilon_0/\ln(R_a/R_b) = L2\pi\varepsilon_0/\ln[(R_b + d)/R_b] = L2\pi\varepsilon_0/\ln[1 + (d/R_b)]$.
   If $d \ll R_b$, we have
        $C \approx L2\pi\varepsilon_0/(d/R_b) = L2\pi R_b\varepsilon_0/d = \varepsilon_0 A/d$,
   which is the expression for a parallel-plate capacitor.

19. (a) When the uncharged plate is placed between the two charged plates, charges will separate so
    that there is a charge $+Q$ on the side facing the negative plate and a charge $-Q$ on the side facing
    the positive plate.  Thus we have the same uniform electric field in each gap:
    $$E = \sigma/\varepsilon_0 = Q/A\varepsilon_0 \,.$$
    If $x$ is the separation on one side of the sheet, the potentials across the gaps are
    $$V_1 = Ex, \quad V_2 = E(d - l - x).$$
    Thus the potential across the capacitor is
    $$V_1 + V_2 = Ex + E(d - l - x) = (Q/A\varepsilon_0)(d - l).$$
    The capacitance is
    $$C = Q/(V_1 + V_2) = \varepsilon_0 A/(d - l).$$
    (b) If $l = \tfrac{2}{3}d$, we have
    $$C/C_0 = d/(d - l) = d/(d - \tfrac{2}{3}d) = 3.$$

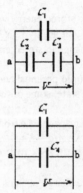

23. (a) From the circuit, we see that $C_2$ and $C_3$ are in series
    and find their equivalent capacitance from
    $$1/C_4 = (1/C_2) + (1/C_3), \text{ which gives } C_4 = C_2 C_3/(C_2 + C_3).$$
    From the new circuit, we see that $C_1$ and $C_4$ are in parallel,
    with an equivalent capacitance
    $$\begin{aligned}C_{\text{eq}} &= C_1 + C_4 = C_1 + [C_2 C_3/(C_2 + C_3)]\\ &= (C_1 C_2 + C_1 C_3 + C_2 C_3)/(C_2 + C_3).\end{aligned}$$
    (b) Because $V$ is across $C_1$ , we have
    $$Q_1 = C_1 V = (14.0\ \mu\text{F})(25.0\ \text{V}) = 350\ \mu\text{C}.$$
    Because $C_2$ and $C_3$ are in series, the charge on each is the
    charge on their equivalent capacitance:
    $$\begin{aligned}Q_2 &= Q_3 = C_4 V = [C_2 C_3/(C_2 + C_3)]V\\ &= [(14.0\ \mu\text{F})(7.00\ \mu\text{F})/(14.0\ \mu\text{F} + 7.00\ \mu\text{F})](25.0\ \text{V}) = 117\ \mu\text{C}.\end{aligned}$$

27. The capacitance increases with a parallel connection, so the maximum capacitance is
    $$\begin{aligned}C_{\text{max}} &= C_1 + C_2 + C_3\\ &= 3000\ \text{pF} + 5000\ \text{pF} + 0.010\ \mu\text{F} = 3.0\ \text{nF} + 5.0\ \text{nF} + 10\ \text{nF} = 18\ \text{nF (parallel)}.\end{aligned}$$
    The capacitance decreases with a series connection, so we find the minimum capacitance from
    $$\begin{aligned}1/C_{\text{min}} &= (1/C_1) + (1/C_2) + (1/C_3) = [1/(3000\ \text{pF})] + [1/(5000\ \text{pF})] + [1/(0.010\ \mu\text{F})]\\ &= [1/(3.0\ \text{nF})] + [1/(5.0\ \text{nF})] + [1/(10\ \text{nF})], \text{ which gives } C_{\text{min}} = 1.6\ \text{nF (series)}.\end{aligned}$$

31. When the switch is down, the initial charge on $C_2$ is
    $$Q_2 = C_2 V_0.$$
    When the switch is connected upward, some charge will flow from $C_2$ to $C_1$
    until the potential difference across the two capacitors is the same:
    $$V_1 = V_2 = V.$$
    Because charge is conserved, we have
    $$Q = Q_1' + Q_2' = Q_2, \text{ or}$$
    $$C_1 V + C_2 V = C_2 V_0, \text{ which gives } V = C_2 V_0/(C_1 + C_2).$$
    For the charges we have
    $$Q_1' = C_1 V = C_1 C_2 V_0/(C_1 + C_2); \quad Q_2' = C_2 V = C_2{}^2 V_0/(C_1 + C_2).$$

35. When there is no reading on the voltmeter, we have $V_{ab} = 0$, so
    $$V_1 = V_2, \quad \text{or} \quad Q_1/C_1 = Q_2/C_2; \quad \text{and}$$
    $$V_x = V_3, \quad \text{or} \quad Q_x/C_x = Q_3/C_3.$$
    If we divide the two equations, we get
    $$(Q_1/Q_x)(C_x/C_1) = (Q_2/Q_3)(C_3/C_2).$$
    Because $V_{ab} = 0$, we could remove the connection between
    $a$ and $b$ without affecting the circuit.  This means that
    $$Q_1 = Q_x, \text{ and } Q_2 = Q_3, \text{ so we have}$$
    $$C_x/C_1 = C_3/C_2, \quad \text{or}$$
    $$C_x = (C_3/C_2)C_1 = [(6.0\ \mu\text{F})/(18.0\ \mu\text{F})](8.9\ \mu\text{F}) = 3.0\ \mu\text{F}.$$

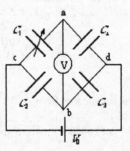

39. The energy stored in the capacitor is
    $$U = \tfrac{1}{2}CV^2 = \tfrac{1}{2}(2800 \times 10^{-12}\ \text{F})(1200\ \text{V})^2 = 2.0 \times 10^{-3}\ \text{J}.$$

43. From Problem 23 we know that the equivalent capacitance is
$$C_{eq} = (C_1C_2 + C_1C_3 + C_2C_3)/(C_2 + C_3) = 3C_1/2 = 3(2200 \text{ pF})/2 = 3300 \text{ pF}.$$
Because this capacitance is equivalent to the three capacitors, the energy stored in it is the energy stored in the network:
$$U = \tfrac{1}{2}C_{eq}V^2 = \tfrac{1}{2}(3300 \times 10^{-12} \text{ F})(10.0 \text{ V})^2 = 1.65 \times 10^{-7} \text{ J}.$$

47. (a) Because there is no stored energy on the uncharged 5.0 μF capacitor, the total stored energy is
$$U_a = \tfrac{1}{2}C_1V_0^2 = \tfrac{1}{2}(3.0 \times 10^{-6} \text{ F})(12 \text{ V})^2 = 2.2 \times 10^{-4} \text{ J}.$$
  (b) We find the initial charge on the 3.0 μF capacitor when it is connected to the battery;
$$Q = C_1V_0 = (3.0 \text{ μF})(12 \text{ V}) = 36 \text{ μC}.$$
   When the capacitors are connected, some charge will flow from $C_1$ to $C_2$ until the potential difference across the two capacitors is the same; so the two capacitors are connected in parallel:
$$C_{eq} = C_1 + C_2 = 3.0 \text{ μF} + 5.0 \text{ μF} = 8.0 \text{ μF}.$$
   For the stored energy we have
$$U_b = \tfrac{1}{2}C_{eq}V^2 = \tfrac{1}{2}Q^2/C_{eq} = \tfrac{1}{2}(36 \text{ μC})^2/(8.0 \text{ μF}) = 81 \text{ μJ} = 8.1 \times 10^{-5} \text{ J}.$$
  (c) The change in stored energy is
$$\Delta U = U_b - U_a = 8.1 \times 10^{-5} \text{ J} - 2.2 \times 10^{-4} \text{ J} = -1.4 \times 10^{-4} \text{ J}.$$
  (d) The stored potential energy is not conserved. During the flow of charge before the final steady state, some of the stored energy is dissipated as thermal and radiant energy.

51. We find the capacitance from
$$C \quad = K\varepsilon_0 A/d = K\varepsilon_0 \pi r^2/d$$
$$= (7)(8.85 \times 10^{-12} \text{ C}^2/\text{N} \cdot \text{m}^2)\pi(0.050 \text{ m})^2/(3.2 \times 10^{-3} \text{ m}) = 1.5 \times 10^{-10} \text{ F}.$$

55. Because the charge remains constant, we express the energy as
$$U = \tfrac{1}{2}CV^2 = \tfrac{1}{2}Q^2/C.$$
If we form the ratio for the two conditions, we have
$$U/U_0 = C_0/C = 1/K, \text{ so}$$
$$U = U_0/K = (2.33 \times 10^3 \text{ J})/7 = 3.3 \times 10^2 \text{ J}.$$

59. (a) Because each capacitor acquires $Q_0$, the (equal) initial capacitance of each is
$$C = Q_0/V_0.$$
   When the dielectric is inserted, the capacitors are still in parallel and charge will flow to make the potential the same. Charge is conserved so, if we call $Q_1$ the charge on the capacitor without the dielectric, we have
$$Q_1/C = Q_2/KC = (2Q_0 - Q_1)/KC, \text{ which gives}$$
$$Q_1 = 2Q_0/(1 + K) = 2Q_0/(1 + 4.0) = 0.40Q_0.$$
   For $Q_2$ we have
$$Q_2 = 2Q_0 - Q_1 = 1.60Q_0.$$
  (b) We find the common voltage from
$$V = Q_1/C = 0.40Q_0/C = 0.40V_0.$$

63. We find the energy in each region from the energy density and the volume:
$$U_{\text{dielectric}} = u_{\text{dielectric}}V_{\text{dielectric}} = \tfrac{1}{2}K\varepsilon_0 E_{\text{dielectric}}{}^2 Al;$$
$$U_{\text{gap}} = u_{\text{gap}}V_{\text{gap}} = \tfrac{1}{2}\varepsilon_0 E_{\text{gap}}{}^2 A(d - l).$$
When we use $E_{\text{gap}} = KE_{\text{dielectric}}$ and cancel common factors, we have
$$\begin{aligned} U_{\text{dielectric}}/(U_{\text{dielectric}} + U_{\text{gap}}) \quad &= KE_{\text{dielectric}}{}^2 l/[KE_{\text{dielectric}}{}^2 l + E_{\text{gap}}{}^2(d - l)] \\ &= KE_{\text{dielectric}}{}^2 l/[KE_{\text{dielectric}}{}^2 l + K^2 E_{\text{dielectric}}{}^2(d - l)] \\ &= l/[l + K(d - l)] \\ &= (1.00 \text{ mm})/[1.00 \text{ mm} + (3.50)(2.00 \text{ mm} - 1.00 \text{ mm})] \\ &= 0.22 = 22\%. \end{aligned}$$

67. We find the capacitance from
$$U = \tfrac{1}{2}CV^2;$$
$$200 \text{ J} = \tfrac{1}{2}C(6000 \text{ V})^2, \text{ which gives } C = 1.1 \times 10^{-5} \text{ F} = 11 \text{ μF}.$$

71. Because the charged capacitor is disconnected from the plates, the charge must be constant.  The paraffin will change the capacitance, so we have
$Q = C_1V_1 = C_2V_2 = KC_1V_2$ ;
24.0 V = (2.2)$V_2$ , which gives $V_2$ = 10.9 V.

75. Because the capacitor is isolated, the charge will not change.  The initial stored energy is
$U_1 = \tfrac{1}{2}C_1V_1^2 = \tfrac{1}{2}Q^2/C_1$ , with $C_1 = \varepsilon_0A/d_1$ .
The changes will change the capacitance:
$C_2 = K\varepsilon_0A/d_2$ .
For the ratio of stored energies, we have
$U_2/U_1 = C_1/C_2 = (\varepsilon_0A/d_1)/(K\varepsilon_0A/d_2) = d_2/Kd_1 = \tfrac{1}{2}/K = 1/2K$.
The stored energy decreases from two factors.  Because the plates attract each other, when the separation is halved, work is done by the field, so the energy decreases.  When the dielectric is inserted, the induced charges on the dielectric are attracted to the plates; again work is done by the field and the energy decreases.
The uniform electric field between the plates is related to the potential difference across the plates:
$E = V/d$.
For a parallel-plate capacitor, we have
$Q = C_1V_1 = C_1E_1d_1 = C_2E_2d_2$ ,  or
$E_2/E_1 = C_1d_1/C_2d_2 = \varepsilon_0A/K\varepsilon_0A = 1/K$.

79. (a) Because the capacitor is disconnected from the power supply, the charge is constant.  We find the new voltage from
$Q = C_1V_1 = C_2V_2$ ;
(10 pF)(10,000 V) = (1 pF)$V_2$ , which gives $V_2 = 1.0 \times 10^5$ V = 0.10 MV.
(b) A major disadvantage is that, when the stored energy is used, the voltage will decrease exponentially, so it can be used for only short bursts.

83. When the switch is connected left, the initial charge on $C_1$ is
$Q_0 = C_1V_{ab} = (1.0~\mu F)(24~V) = 24~\mu C$.
When the switch is connected right, some charge will flow from $C_1$ to $C_2$ and $C_3$ until the potential difference across $C_1$ is the potential difference across the series combination of $C_2$ and $C_3$:
$V_1 = V_{23} = V$.
Because $C_2$ and $C_3$ are in series, we know that $Q_2 = Q_3$, and we find their equivalent capacitance from
$1/C_{23} = 1/C_2 + 1/C_3 = 1/2.0~\mu F + 1/3.0~\mu F$, which gives $C_{23} = 1.2~\mu F$.
Because charge is conserved, we have
$Q_0 = Q_1 + Q_2$ .
$Q_0 = C_1V + C_{23}V$, which gives
$V = Q_0/(C_1 + C_{23}) = (24~\mu C)/(1.0~\mu F + 1.2~\mu F) = 11~V = V_1$ .
For the charges we have
$Q_1 = C_1V = (1.0~\mu F)(11~V) = 11~\mu C$;
$Q_2 = Q_3 = C_{23}V = (1.2~\mu F)(11~V) = 13~\mu C$.
We find the potential differences from
$V_2 = Q_2/C_2 = (13~\mu C)/(2.0~\mu F) = 6.5~V$;
$V_3 = Q_3/C_3 = (13~\mu C)/(3.0~\mu F) = 4.4~V$.

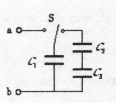

87. (a) The capacitance is
$C_0 = \varepsilon_0A/d = (8.85 \times 10^{-12}~C^2/N \cdot m^2)(2.5~m^2)/(3.0 \times 10^{-3}~m) = 7.38 \times 10^{-9}$ F = 7.4 nF.
The charge on the capacitor is
$Q_0 = C_0V = (7.38 \times 10^{-9}~F)(45~V) = 3.32 \times 10^{-7}$ C = 0.33 μC.
The electric field is
$E_0 = V/d = (45~V)/(3.0 \times 10^{-3}~m) = 1.5 \times 10^4$ V/m.
The stored energy is
$U_0 = \tfrac{1}{2}C_0V^2 = \tfrac{1}{2}(7.38 \times 10^{-9}~F)(45~V)^2 = 7.5 \times 10^{-6}$ J.
(b) With the addition of the dielectric we have
$C = KC_0 = (3.6)(7.38 \times 10^{-9}~F) = 2.66 \times 10^{-8}$ F = 27 nF;
$Q = CV = (2.66 \times 10^{-8}~F)(45~V) = 1.20 \times 10^{-7}$ C = 1.2 μC;
$E = V/d = (45~V)/(3.0 \times 10^{-3}~m) = 1.5 \times 10^4$ V/m;
$U = \tfrac{1}{2}CV^2 = \tfrac{1}{2}KC_0V^2 = KU_0 = (3.6)(7.5 \times 10^{-6}~J) = 2.7 \times 10^{-5}$ J.

# Chapter 25: Electric Currents and Resistance

## Chapter Overview and Objectives

This chapter introduces electric current, electric circuits, and Ohm's Law. This chapter discusses topics related to circuits such as power, resistivity, rms quantities, current density, and drift velocity.

After completing study of this chapter, you should:

- Know that electric current is the rate of flow of electric charge.
- Know Ohm's Law.
- Be able to determine the resistance of a conductor given its cross-sectional area, length and resistivity.
- Know how to relate resistivities and resistances at one temperature to those at another temperature.
- Know how to calculate power in both dc and ac resistive electrical circuits.
- Know what conventional current is.
- Know what rms quantities are and how to relate them to peak values in sinusoidal functions.
- Know what current density is and how to calculate current from a known current density.
- Know what drift velocity is.

## Summary of Equations

Definition of average current:
$$\bar{I} = \frac{\Delta Q}{\Delta t}$$

Definition of instantaneous current:
$$I = \frac{dQ}{dt}$$

Definition of an ampere:   $1\,A = 1\,C/S$

Ohm's Law:   $V = IR$

Definition of an ohm:   $1\,\Omega = 1\,V/A$

Dependence of resistance on resistivity and geometry of conductor:

$$R = \rho \frac{L}{A}$$

Relationship between resistivity and conductivity:   $\sigma = \frac{1}{\rho}$

Linear temperature dependence of resistivity:   $\rho(T) = \rho(T_0)[1 + \alpha(T - T_0)]$

Dependence of power on voltage, current, and resistance in dc circuit:

$$P = IV = \frac{V^2}{I} = I^2 R$$

Instantaneous power in resistive sinusoidal ac circuit:   $P = I_0 V_0 \sin^2 \omega t$

Time average power in resistive sinusoidal ac circuit:   $\bar{P} = \frac{1}{2}\frac{V_0^2}{R} = \frac{1}{2}I_0^2 R = \frac{1}{2}I_0 V_0$

Definition of rms value of a periodic function:

$$f_{RMS} = \sqrt{\frac{1}{T}\int_0^T [f(t)]^2\, dt}$$

Relationship between peak and rms values for a sinusoidal voltage and current:

$$V_{RMS} = \frac{V_0}{\sqrt{2}}$$

$$I_{RMS} = \frac{I_0}{\sqrt{2}}$$

Relationship between average power and rms voltage and current for a sinusoidal ac voltage and current in a resistive circuit:

$$\overline{P} = I_{RMS} V_{RMS} = I_{RMS}^2 R = \frac{V_{RMS}^2}{R}$$

Definition of current density:

$$j = \frac{I}{A}$$

Relationship of current density to drift velocity:    $j = nq v_d$

Relationship between current density and electric field:

$$j = \sigma E \qquad \text{or} \qquad E = \rho j$$

## Chapter Summary

### Section 25-1 The Electric Battery

When two dissimilar metal plates are immersed in an electrolyte, a potential difference develops from plate to the other. Such an arrangement is called an **electric cell** and the metal plates are called the **electrodes** of the cell. The potential difference between the electrodes depends on which metals are used for the electrodes. A **battery** is either a single electric cell or a collection of electric cells connected together. The parts of the metal plates that lie outside the electrolyte to which external conductors can be connected to the electrodes are called the **terminals** of the electric cell.

### Section 25-2 Electric Current

If a conductive path is connected between the terminals of a battery, electric charge moves along the conductive path. The flow of charge is called an **electric current**. Electric current measures the amount of charge per unit time that passes a given cross-section of the conductive path. The time average current is given by

$$\overline{I} = \frac{\Delta Q}{\Delta t}$$

where $\Delta Q$ is the amount of charge that passes through a given cross-section of the conductive path during a time $\Delta t$. If the limit as the time interval $\Delta t$ goes to zero is taken, we get the instantaneous current:

$$I = \frac{dQ}{dt}$$

An **ampere** (A) of current is defined as one Coulomb of charge per second passing through a given cross-section of the conductive path

$$1\,A = 1\,C/s$$

When we consider the direction that current flows in a circuit, there is a possible ambiguity. If the moving charges are positive charges, the charges move from higher electrical potential to lower electrical potential, as that is the direction of the electrical force on a positive charge. If the moving charges are negative charges, the charges move from lower electrical potential to higher electrical potential, as that is the direction of the electrical force on a negative charge. However, it is not necessary to know the direction of motion and sign of the charges. The net transfer of charge *in a circuit* of an amount of charge Q from a higher to a lower potential is identical to an amount of charge –Q from the lower to the higher potential. We adopt the **conventional current** as the direction of current flow to avoid ambiguity. Conventional current flows from higher potential to lower potential.

**Example 25-2-A**

The current in a particular circuit as a function of time is given as

$$I(t) = 1.32\,\text{A} + 0.64t\,\text{A/s} + 0.24t^2\,\text{A/s}^2$$

What is the total amount of charge transferred from time $t = 0$ to time $t = 10$ s?

**Solution:**

The definition of current is

$$I = \frac{dQ}{dt}$$

To determine the charge transferred, $\Delta Q$, we need to multiply each side by $dt$ and integrate:

$$dQ = I\,dt \quad \Rightarrow \quad \int_{Q_i}^{Q_f} dQ = \int_{t=0}^{t=10\,\text{s}} I\,dt$$

The integral on the left side of this equation evaluates to

$$\int_{Q_i}^{Q_f} dQ = Q_f - Q_i = \Delta Q$$

The integral on the right side of the equation evaluates to

$$\int_{t=0}^{t=10\,\text{s}} I\,dt = \int_{t=0}^{t=10\,\text{s}} 1.32\,\text{A} + 0.64t\,\text{A/s} + 0.24t^2\,\text{A/s}^2\,dt$$

$$= \left[1.32\,\text{A}\,t + 0.32t^2\,\text{A/s} + 0.080t^3\,\text{A/s}^2\right]_{t=0\,\text{s}}^{t=10\,\text{s}}$$

$$= 13.2\,\text{As} + 32\,\text{As} + 80\,\text{As} = 125\,\text{C}$$

We find that 125 C of charge is transferred by the given current over the specified time interval.

**Section 25-3 Ohm's Law: Resistance and Resistors**

The amount of current that flows through a given piece of conductor is proportional to the potential difference across the conductor. This relationship is mathematically expressed as Ohm's Law:

$$V = IR$$

where $V$ is the potential difference across the conductor, $I$ is the current flowing through the conductor and $R$ is a property of the piece of conductive material called **resistance**. Ohm's Law is not universal behavior, but is true for a wide range of conductive materials under a particular range of conditions. Resistance has dimensions of electric potential divided by current. The SI unit of resistance is the ohm ($\Omega$), defined as one volt/ampere:

$$1\Omega = 1\,\text{V/A}$$

Resistance in the schematic diagram of a circuit is shown by the symbol

**Example 25-3-A**

A circuit powered by a 9.0 V battery has a current of 17.0 mA flowing through it. What is the resistance of the circuit?

**Solution:**

We use Ohm's law to determine the resistance of the circuit:

$$V = IR \quad \Rightarrow \quad R = \frac{V}{I} = \frac{9.0\,\text{V}}{17.0 \times 10^{-3}\,\text{A}} = 5.3 \times 10^2\,\Omega$$

**Section 25-4 Resistivity**

For a given type of material, the resistance of a conductor is proportional to the length of the conductor and inversely proportional to the cross-sectional area of the conductor:

$$R = \rho \frac{L}{A}$$

The constant of proportionality, $\rho$, is called the **resistivity** of the material. The resistivity depends on the type of material, its temperature, and pressure.

Sometimes it is more useful to work with a quantity called **conductivity**. Conductivity, $\sigma$, is the inverse of resistivity:

$$\sigma = \frac{1}{\rho}$$

Resistivity depends on temperature. For metals, the resistivity can often be approximated with a linear dependence on temperature:

$$\rho(T) = \rho(T_0)\left[1 + \alpha(T - T_0)\right]$$

where $\rho(T)$ is the resistivity of the material at temperature $T$, $\rho(T_0)$ is the resistivity at temperature $T_0$, and $\alpha$ is the **linear temperature coefficient of resistivity**.

**Example 25-4-A**

Two round wires are made of the same material. The first wire has a diameter of 0.450 mm and a length of 23.4 cm. It has a resistance of 25.8 $\Omega$ at a temperature of 64.2 °C. The second wire has a diameter of 0.750 mm and a length of 184 cm. It has a resistance of 64.8 $\Omega$ at a temperature of 33.2 °C. What is the temperature coefficient of resistivity of the material? What is the resistivity of this material at 20 °C?

**Solution:**

We use the relationship between resistance, resistivity, and geometry to calculate the resistivity of two samples at their given temperatures:

$$R_1 = \rho_1 \frac{L_1}{A_1} \quad \Rightarrow \quad \rho_1 = \frac{R_1 A_1}{L_1} = \frac{(25.8\,\Omega)\pi(0.450 \times 10^{-3}\,\text{m})^2 / 4}{0.234\,\text{m}} = 1.75 \times 10^{-5}\,\Omega \cdot \text{m}$$

$$R_2 = \rho_2 \frac{L_2}{A_2} \quad \Rightarrow \quad \rho_2 = \frac{R_2 A_2}{L_2} = \frac{(64.8\,\Omega)\pi(0.750 \times 10^{-3}\,\text{m})^2 / 4}{1.84\,\text{m}} = 1.56 \times 10^{-5}\,\Omega \cdot \text{m}$$

To determine the temperature coefficient of resistivity, we relate the resistivity at the given temperatures in terms of the temperature coefficient of resistivity:

$$\rho_{64.2} = \rho_{33.2}\left[1 + \alpha(64.2\,^\circ\text{C} - 33.2\,^\circ\text{C})\right]$$

Solving this for the linear temperature coefficient of resistivity:

$$\alpha = \frac{\dfrac{\rho_{64.2}}{\rho_{33.2}} - 1}{64.2\ ^\circ C - 33.2\ ^\circ C} = \frac{\dfrac{1.75 \times 10^{-5}\ \Omega \cdot m}{1.56 \times 10^{-5}\ \Omega \cdot m} - 1}{64.2\ ^\circ C - 33.2\ ^\circ C} = 3.9 \times 10^{-3}\ ^\circ C^{-1}$$

The resistivity at 20 °C is easily determined:

$$\begin{aligned}
\rho_{20} &= \rho_{33.2}\left[1 + \alpha\left(20\ ^\circ C - 33.2\ ^\circ C\right)\right] \\
&= \left(1.56 \times 10^{-5}\ \Omega \cdot m\right)\left[1 + \left(3.9 \times 10^{-3}\ ^\circ C^{-1}\right)\left(20\ ^\circ C - 33.2\ ^\circ C\right)\right] \\
&= 1.48 \times 10^{-5}\ \Omega \cdot m
\end{aligned}$$

## Section 25-5 Electric Power

As electric charge moves from one electric potential to another in a circuit, its electrical potential energy changes. If we apply the conservation of energy or the work–energy theorem to a circuit, when a charge moves through a resistive circuit element it must have negative work done on it by the resistance. The change in the potential energy, $dU$, of a charge $dq$ moving through a potential difference $V$ will be

$$dU = V\, dq$$

If we divide through by $dt$, we get

$$\frac{dU}{dt} = V\frac{dq}{dt} = VI$$

But $dU/dt$ is power, $P$:

$$P = VI$$

If we consider the current as conventional current, the current flows from higher potential to a lower electrical potential. This implies that the potential energy decreases in an electrical circuit. The work done by the resistance is always negative! Resistance is identical to the kinetic friction force in mechanical motion in this sense. The change in potential energy of the electric charge moving through the resistance of the circuit ends up as internal energy in the material of the resistance.

Using Ohm's Law, we can rewrite the power in terms of either current or voltage and resistance:

$$P = IV = \frac{V^2}{I} = I^2 R$$

A convenient unit of electrical energy supplied from electrical utility companies is the kilowatt·hour. One kilowatt·hour is equal to $3.600 \times 10^6$ J.

## Section 25-6 Power in Household Circuits

Electric power is supplied to customers at standardized voltages. The voltage supplied is such that an alternating current (AC) is created in the circuit. This means that the supplied voltage has a simple harmonic dependence on time. Loads or devices are connected to the supplied voltage in *parallel*. This means that each device connected to the supply has the standardized voltage across its terminals. The total current flowing from the energy source is the sum of the currents through each device.

To protect from the possibility of fire from overheated wires, circuits include a fuse or circuit breaker that disconnects the circuit from the energy source when the current in the circuit exceeds a certain value.

**Example 25-6-A**

A motor that draws a maximum power of 2.6 horsepower is to be connected to a 120 V household circuit. What is the minimum current capacity of the circuit needed to supply the motor without tripping a circuit breaker? (We must assume that the motor acts as a resistance under these circumstances.)

**Solution:**

The power is the product of the voltage and the current:

$$P = IV \quad \Rightarrow \quad I = \frac{P}{V} = \frac{(2.6\,\text{hp})(746\,\text{W / hp})}{120\,\text{V}} = 16\,\text{A}$$

Standard household circuits have capacities of 15 A, 20 A, and 30 A. This motor could operate on a 20 A or greater household circuit.

**Section 25-7 Alternating Current**

A current that flows with a constant magnitude in a given direction is called a **direct current (dc)**. A current that varies with time, such as current that has a simple harmonic dependence on time, is called an **alternating current (ac)**. In linear circuits (circuits such that the current in the circuit is linearly related to the voltage) an ac current implies that an ac voltage is present. We write an ac voltage as

$$V = V_0 \sin 2\pi ft = V_0 \sin \omega t$$

where $V_0$, the amplitude of the simple harmonic function, is called the **peak voltage** and $f$ is the frequency of the simple harmonic function.

In a resistive circuit, Ohm's Law is satisfied at each instant in time. This implies that the current $I$ as a function of time is given by

$$I = \frac{V}{R} = \frac{V_0}{R} \sin 2\pi ft = I_0 \sin \omega t$$

where $I_0$ is called the peak current and is equal to the peak voltage divided by the resistance, $R$.

The instantaneous power into the resistance of an ac circuit is equal to the instantaneous voltage multiplied by the instantaneous current:

$$P = IV = I_0 \sin \omega t \times V_0 \sin 2\pi ft = I_0 V_0 \sin^2 \omega t$$

This can alternatively be written as

$$P = I_0^2 R \sin^2 \omega t = \frac{V_0^2}{R} \sin^2 \omega t$$

In many situations, it is the time average power that is useful information. If we average the power over one period of the simple harmonic function, we get

$$\overline{P} = \frac{1}{T}\int_0^T I_0^2 R \sin^2 \omega t\, dt = \frac{1}{T}\int_0^T I_0^2 R \sin^2 \frac{2\pi}{T}t\, dt = \tfrac{1}{2} I_0^2 R$$

where $T$ is the period of the simple harmonic function. We can also write this as

$$\overline{P} = \tfrac{1}{2}\frac{V_0^2}{R}$$

Commonly, the quantities **root-mean-square (rms) voltage** and **root-mean-square current** are used to describe the magnitude of ac voltages and currents rather than the peak values of voltage and current. The definition of the root-mean-square value, $f_{RMS}$, of a periodic function of time, $f(t)$, is the square root of the average value of the square of the function over one period, $T$:

$$f_{RMS} = \sqrt{\frac{1}{T} \int_0^T [f(t)]^2 \, dt}$$

For a voltage that is a simple harmonic function with amplitude $V_0$, the root-mean-square voltage is

$$V_{RMS} = \sqrt{\frac{1}{T} \int_0^T V_0^2 \sin^2 \omega t \, dt} = \frac{V_0}{\sqrt{2}}$$

Similarly, the rms current for a current that is a simple harmonic function of time is

$$I_{RMS} = \sqrt{\frac{1}{T} \int_0^T I_0^2 \sin^2 \omega t \, dt} = \frac{I_0}{\sqrt{2}}$$

We can write the average power in terms of the rms voltage and current as

$$\overline{P} = I_{RMS} V_{RMS} = I_{RMS}^2 R = \frac{V_{RMS}^2}{R}$$

There are some advantages to using rms values of currents and voltages in ac circuits. The above relationship between average power and rms current and voltage is identical to the relationship between power, current, and voltage in a dc circuit. You don't need to include the factor of one half as when dealing with peak currents and voltages. Also, the factor of one half in the relationship between average power and peak current and voltage only applies to voltages and currents that are simple harmonic functions of time. The relationship between average power and rms currents and voltages applies to voltages and currents that are arbitrary periodic functions of time in resistive circuits.

**Example 25-7-A**

A resistive circuit with a resistance of 150 Ω is connected to a sinusoidal ac potential difference with an rms voltage of 120 V. Determine the average power in the circuit, the peak power, the rms current, and the peak current.

**Solution:**

The average power in the circuit can be determined from

$$\overline{P} = \frac{V_{rms}^2}{R} = \frac{(120\,\text{V})^2}{150\,\Omega} = 96\,\text{W}$$

The peak power is given by the square of the peak voltage divided by the resistance. Because this is a sinusoidal voltage, the peak voltage is the square root of two times the rms voltage:

$$P_P = \frac{V_P^2}{R} = \frac{\left(\sqrt{2}\,V_{rms}\right)^2}{R} = \frac{2(120\,\text{V})^2}{150\,\Omega} = 192\,\text{W}$$

The current is determined using Ohm's Law:

$$I_{rms} = \frac{V_{rms}}{R} = \frac{120\,\text{V}}{150\,\Omega} = 0.80\,\text{A}$$

$$I_P = \sqrt{2}\,I_{rms} = \sqrt{2}\,(0.80\,\text{A}) = 1.13\,\text{A}$$

**Section 25-8 Microscopic View of Electric Current: Current Density and Drift Velocity**

All conductors have a finite cross-sectional area, $A$. The **current density**, $j$, in a conductor is the current, $I$, per cross-sectional area of the conductor,

$$j = \frac{I}{A}$$

The current density is actually a function of position, which implies that we need to write

$$j = \frac{dI}{dA}$$

where $dI$ is the infinitesimal current flowing through the infinitesimal cross-section $dA$. Given the current density as a function of position with the cross-section of the wire, we can calculate the current flowing through the total cross-section:

$$I = \int j \cdot dA$$

The vector $\mathbf{j}$ is in the direction of current flow and the direction of $d\mathbf{A}$ is perpendicular to the infinitesimal cross-section area element.

Although an exact description of microscopic current flow requires a quantum mechanical description, we can describe a classical model of microscopic current flow that will allow us to understand many of the behaviors of currents in conductors. When an electric field is applied to a conductor, the charges initially accelerate, but are eventually scattered into a random direction by what we can picture as collisions with atoms or other moving charged particles. After a short time, equilibrium is reached between the acceleration and scattering processes such that the average velocity of the charged particles carrying the current reaches a time-independent value. This average velocity is called the **drift velocity**, $\mathbf{v_d}$. The magnitude of the drift velocity is much smaller than the average speed of the charged particles, similar to how wind speeds are much less than the average speed of molecules in a gas. The current density, $\mathbf{j}$, is related to the drift velocity by

$$j = nqv_d$$

where $n$ is the number of charged particles per unit volume and $q$ is the amount of charge on one charged particle.

We can use the definition of current density and Ohm's law to rewrite Ohm's law in terms of local quantities[1]:

$$j = \sigma E \qquad or \qquad E = \rho j$$

**Example 25-8-A**

A particular material has electrons as conductive charge carriers and a charge carrier density of $1.96 \times 10^{19}/cm^3$. If the drift velocity is 3.84 cm/s when an electric field of 1.45 V/m exists within the conductor, what is the resistivity of the material?

**Solution:**

We know the relationship between resistivity, current density, and electric field is

$$E = \rho j$$

We also know the relationship between current density and drift velocity

$$j = nqv_d$$

---

[1] This assumes that the local current is in the direction of the local electric field. This is typical of many materials, but there are some materials in which the current density is not parallel to the electric field.

If we eliminate the current density from these two equations and solve for the resistivity, we have

$$\rho = \frac{E}{nqv_d} = \frac{1.45\,\text{V}/\text{m}}{\left(1.96\times10^{25}\,\text{m}^{-3}\right)\left(1.602\times10^{-19}\,\text{C}\right)\left(3.84\times10^{-2}\,\text{m}/\text{s}\right)}$$
$$= 1.20\times10^{-5}\,\Omega\cdot\text{m}$$

### Section 25-9 Superconductivity

Some materials, when their temperature drops below a certain temperature, have immeasurably low resistivity.  These materials are called **superconductors** and this condition of extremely low conductivity is called **superconductivity**. Superconductors have another property in addition to the lack of resistivity.  This is a magnetic property that we will learn about in later chapters.

### Section 25-10 Electric Hazards; Leakage Currents

Electric currents passing through the human body present dangers through two different mechanisms.  First, large currents passing through tissue deposit energy into the tissue, causing the tissue's internal energy and temperature to rise.  If the temperature rise is great enough, a burn is the result.  Second, and more dangerously, currents flowing through the body cause potential differences across neurons that cause the neurons to fire uncontrollably.  If the neurons of the heart are caused to fire uncontrollably from a current passing through the body, the heart fibrillates.  A heart that is fibrillating is unable to pump blood.

It is impossible to maintain an exactly zero potential and infinite resistance between people and equipment that uses electrical power.  This means that in most circumstance, making contact with electrically powered equipment does cause a current to flow through the human body.  Safety design of electrically powered equipment ensures that this leakage current from the equipment stays at magnitudes far below that which would cause injury or death in properly used equipment.

## Practice Quiz

1.    The resistance of a piece of a wire depends on

   a) the length of the piece of material.
   b) the cross-sectional area of the piece of material.
   c) the resistivity of the piece of material.
   d) all of the above.

2.    If the voltage across a fixed resistance is doubled, what happens to the power into the resistor?

   a) Power becomes twice as large.
   b) Power becomes four times as large.
   c) Power becomes half as large.
   d) Power becomes one fourth as large.

3.    A resistor is connected across a fixed potential difference.  The current is 2.00 amps and the resistance is 4.00 $\Omega$. If the resistance of the resistor is doubled, what will the power be?

   a) 32.0 W
   b) 16.0 W
   c) 8.00 W
   d) 4.00 W

4.    If you take a 120 V 100 W light bulb and measure its resistance with an ohm meter, you might measure a resistance of 40 Ω. If you calculate the power from a 120 V rms source with a 40 Ω load using $P=V^2/R$ you get a power of 360 W. Why is the light bulb only rated 100 W?

   a) The bulb rating only gives the power that goes into producing visible light, not into heat.
   b) The power calculation is incorrect because the rms voltage value was used.
   c) The resistance of the bulb is considerably higher at the operating temperature.
   d) The calculated power of 360 W is the peak power, the 100 W rating is the average power.

5.    For a given amplitude sinusoidal voltage, what happens to the rms voltage as the frequency is increased?

   a) When the frequency is increased, the rms value increases.
   b) When the frequency is increased, the rms value decreases.
   c) When the frequency is increased, the rms value remains the same.
   d) When the frequency is increased, the rms value could increase or decrease.

6.    Two copper wires, each of the same length but different cross-sectional area, are connected across a potential difference. Which statement is true?

   a) The current through each of the wires is the same.
   b) The current density in each wire is inversely proportional to its cross-sectional area.
   c) The current density in each of the wires is the same.
   d) The current in each of the wires is inversely proportional to its cross-sectional area.

7.    Two wires of identical resistance, but made from different materials, are connected across a potential difference at room temperature. The wires begin to warm up due to the electrical power added to them. After current flows for some time, one of the wires is at a higher temperature than the other. Which wire is at the higher temperature?

   a) The wire with the greater temperature coefficient of resistivity.
   b) The wire with the lesser temperature coefficient of resistivity.
   c) The wire with the greater length.
   d) The wire with the greater cross-sectional area.

8.    Four identically shaped cylinders are made from four different metals. Which of these four materials will have the greatest resistance cylinder?

   a) Copper
   b) Platinum
   c) Gold
   d) Iron

9.    A particular electrical circuit related quantity is measured in units of $kg \cdot m^2/(C^2 \cdot s)$. Which electrical circuit quantity is being measured?

   a) Voltage
   b) Current
   c) Resistance
   d) Power

10.   The rms of a sinusoidal current in an ac circuit is 20.0 A. What is the peak current in the circuit?

   a) 14.1 A
   b) 28.2 A
   c) 20.0 A
   d) 40.0 A

11.   A dc electrical source supplies a potential difference of 12.0 V. Loads with resistances 120 Ω, 240 Ω, and 300 Ω are connected in parallel to the source. What is the total current and power supplied by the source?

12.   A lightning strike occurs between a potential difference of $3.8 \times 10^7$ V and an average current of 88 A flows for a time of 124 ms. If a power company charges 10 cents per kilowatt·hr, what would the amount of energy in the lightening strike be worth to the power company?

13.   A time-varying voltage is a periodic triangular function as shown in the diagram. Determine the rms voltage of this voltage. The positive peaks are at $V_0$ and the negative peaks are at $-V_0$.

14.   The change in resistance of a conductor with temperature can be used to construct a temperature measuring device. For a thermometer that needs to resolve a $0.5°$ C temperature difference, what must be the precision in measuring the resistance of a $1000\ \Omega$ resistor. Assume the conductor is made of copper.

15.   Calculate the drift velocity of electrons in a copper wire with a cross-sectional area of $1.67 \times 10^{-5}$ m$^2$ when it is carrying a current of 11.6 A.

## Problem Solutions

3.   We find the current from
$$I = \Delta Q / \Delta t = (1000\ \text{ions})(1.60 \times 10^{-19}\ \text{C/ion})/(7.5 \times 10^{-6}\ \text{s}) = 2.1 \times 10^{-11}\ \text{A}.$$

7.   The rate at which electrons leave the battery is the current:
$$I = V/R = [(9.0\ \text{V})/(1.6\ \Omega)](60\ \text{s/min})/(1.60 \times 10^{-19}\ \text{C/electron}) = 2.1 \times 10^{21}\ \text{electron/min}.$$

11.   We find the radius from
$$R = \rho L/A = \rho L/\pi r^2;$$
$$0.22\ \Omega = (5.6 \times 10^{-8}\ \Omega \cdot \text{m})(1.00\ \text{m})/\ \pi r^2,\ \text{which gives}\ r = 2.85 \times 10^{-4}\ \text{m},$$
so the diameter is $5.7 \times 10^{-4}$ m $= 0.57$ mm.

15.   Because the material and area of the two pieces are the same, from the expression for the resistance,
$R = \rho L/A$, we see that the resistance is proportional to the length:
$$R_1/R_2 = L_1/L_2 = 5.0.$$
Because $L_1 + L_2 = L$, we have
$$5.0L_2 + L_2 = L,\ \text{or}\ L_2 = L/6.0,\ \text{and}\ L_1 = 5.0L/6.0,\ \text{so the wire should be cut at 1/6 the length.}$$
We find the resistance of each piece from
$$R_1 = (L_1/L)R = (5.0/6.0)(10.0\ \Omega) = 8.3\ \Omega;$$
$$R_2 = (L_2/L)R = (1/6.0)(10.0\ \Omega) = 1.7\ \Omega.$$

19.   We find the temperature from
$$R = R_0(1 + \alpha_{\text{Cu}}\ \Delta T);$$
$$140\ \Omega = (12\ \Omega)\{1 + [0.0060\ (\text{C}°)^{-1}](T - 20.0°\text{C})\},\ \text{which gives}\ T = 1.8 \times 10^3\ °\text{C}.$$

23.   (*a*) For a length $x$ of the uniform wire we have
$$V = IR = Ix/\sigma A.$$
If we find the potential gradient by differentiating, we get
$$dV/dx = I/\sigma A.$$
Because the current is defined to flow from higher to lower potential, that is, opposite to the potential gradient, we must introduce a negative sign:
$$I = dq/dt = -\ \sigma A\ dV/dx.$$
(*b*) The expression for heat conduction through an area $A$ is
$$dQ/dt = -\ kA\ dT/dx,$$
where the negative sign indicates heat flow from higher to lower temperature.
We expect $\sigma$ and $k$ to be related, because the free electrons can easily acquire thermal energy and transmit it along the wire.

27.   We find the power from
$$P = IV = (0.350\ \text{A})(9.0\ \text{V}) = 3.2\ \text{W}.$$

31. (a) We find the resistance from
$$P_1 = V^2/R_1 ;$$
$60 \text{ W} = (120 \text{ V})^2/R_1$, which gives $R_1 = 240 \ \Omega$.
The current is
$$I_1 = V/R_1 = (120 \text{ V})/(240 \ \Omega) = 0.50 \text{ A}.$$
(b) We find the resistance from
$$P_2 = V^2/R_2 ;$$
$150 \text{ W} = (120 \text{ V})^2/R_2$, which gives $R_2 = 96 \ \Omega$.
The current is
$$I_2 = V/R_2 = (120 \text{ V})/(96 \ \Omega) = 1.25 \text{ A}.$$

35. $90 \text{ A} \cdot \text{h}$ is the total charge that passed through the battery when it was charged.
We find the energy from
$$\text{Energy} = Pt = VIt = VQ = (12 \text{ V})(90 \text{ A} \cdot \text{h})(10^{-3} \text{ kW/W}) = 1.1 \text{ kWh} = 3.9 \times 10^6 \text{ J}.$$

39. The required current to deliver the power is $I = P/V$, and the wasted power (thermal losses in the wires) is $P_{\text{loss}} = I^2 R$.
For the two conditions we have
$$I_1 = (520 \text{ kW})/(12 \text{ kV}) = 43.3 \text{ A}; \ P_{\text{loss1}} = (43.3 \text{ A})^2(3.0 \ \Omega)(10^{-3} \text{ kW/W}) = 5.63 \text{ kW};$$
$$I_2 = (520 \text{ kW})/(50 \text{ kV}) = 10.4 \text{ A}; \ P_{\text{loss2}} = (10.4 \text{ A})^2(3.0 \ \Omega)(10^{-3} \text{ kW/W}) = 0.324 \text{ kW}.$$
Thus the decrease in power loss is
$$\Delta P_{\text{loss}} = P_{\text{loss1}} - P_{\text{loss2}} = 5.63 \text{ kW} - 0.324 \text{ kW} = 5.3 \text{ kW}.$$

43. We find the peak current from the peak voltage:
$$V_0 = \sqrt{2} \ V_{\text{rms}} = I_0 R;$$
$\sqrt{2}(120 \text{ V}) = I_0(1.8 \times 10^3 \ \Omega)$, which gives $I_0 = 0.094 \text{ A}.$

47. The peak voltage is
$$V_0 = \sqrt{2} \ V_{\text{rms}} = \sqrt{2}(450 \text{ V}) = 636 \text{ V}.$$
We find the peak current from
$$P = I_{\text{rms}} V_{\text{rms}} = (I_0/\sqrt{2})V_{\text{rms}} ;$$
$1800 \text{ W} = (I_0/\sqrt{2})(450 \text{ V})$, which gives $I_0 = 5.66 \text{ A}.$

51. From Example 25-12 we know that the density of free electrons in copper is
$$n = 8.4 \times 10^{28} \text{ m}^{-3}.$$
(a) We find the drift speed from
$$I = neAv_d = ne(\tfrac{1}{4}\pi D^2)v_d ;$$
$2.5 \times 10^{-6} \text{ A} = (8.4 \times 10^{28} \text{ m}^{-3})(1.60 \times 10^{-19} \text{ C})[\ \tfrac{1}{4}\pi \ (0.55 \times 10^{-3} \text{ m})^2]v_d ,$
which gives $v_d = 7.8 \times 10^{-10} \text{ m/s}.$
(b) The current density is
$$j = I/A = (2.5 \times 10^{-6} \text{ A})/ \ \tfrac{1}{4}\pi \ (0.55 \times 10^{-3} \text{ m})^2 = 10.5 \text{ A/m}^2 \text{ along the wire.}$$
(c) We find the electric field from
$$j = E/\rho,$$
$10.5 \text{ A/m}^2 = E/(1.68 \times 10^{-8} \ \Omega \cdot \text{m})$, which gives $E = 1.8 \times 10^{-7} \text{ V/m}.$

55. We find the current when the lights are on from
$$P = IV;$$
$92 \text{ W} = I(12 \text{ V})$, which gives $I = 7.67 \text{ A}.$
$90 \text{ A} \cdot \text{h}$ is the total charge that passes through the battery when it is completely discharged. Thus the time for complete discharge is
$$t = Q/I = (90 \text{ A} \cdot \text{h})/(7.67 \text{ A}) = 12 \text{ h}.$$

59. (a) We find the resistance from
$$V = IR;$$
$2(1.5 \text{ V}) = (0.350 \text{ A})R$, which gives $R = 8.6 \ \Omega.$
The power dissipated is
$$P = IV = (0.350 \text{ A})(3.0 \text{ V}) = 1.1 \text{ W}.$$
(b) We assume that the resistance does not change, so we have
$$P_2/P_1 = (V_2/V_1)^2 = (6.0 \text{ V}/3.0 \text{ V})^2 = 4\times.$$
The increased power would last for a short time, until the increased temperature of the filament would burn out the bulb.

63. (a) The dependence of the power output on the voltage is $P = V^2/R$. When we form the ratio for the two conditions, we get
$P_2/P_1 = (V_2/V_1)^2$.
For the percentage change we have
$[(P_2 - P_1)/P_1](100) = [(V_2/V_1)^2 - 1](100) = [(105\ \text{V}/117\ \text{V})^2 - 1](100) = -19.5\%$.

(b) The decreased power output would cause a decrease in the temperature, so the resistance would decrease. This means that for the reduced voltage, the percentage decrease in the power output would be less than calculated.

67. (a) We find the input power from
$P_{output} = (\text{efficiency})P_{input}$ ;
$900\ \text{W} = (0.60)P_{input}$ , which gives $P_{input} = 1500\ \text{W} = 1.5\ \text{kW}$.

(b) We find the current from
$P = I_{rms}V_{rms}$ ;
$1500\ \text{W} = I_{rms}(120\ \text{V})$, which gives $I_{rms} = 12.5\ \text{A}$.

71. For the resistance, we have
$R = \rho L/A = \rho L/\tfrac{1}{4}\pi\, d^2$ ;
$6.50\ \Omega = 4(1.68 \times 10^{-8}\ \Omega \cdot \text{m})L/\pi d^2$.
The mass of the wire is
$m = \rho_m AL$ ;
$0.0180\ \text{kg} = (8.9 \times 10^3\ \text{kg/m}^3)\ \tfrac{1}{4}\pi\, d^2 L$.
This gives us two equations with two unknowns, $L$ and $d$. When we solve them, we get
$d = 3.03 \times 10^{-4}\ \text{m} = 0.303\ \text{mm}$, and $L = 28.0\ \text{m}$.

75. The time for a proton to travel completely around the accelerator is
$t = L/v$.
In this time all the protons stored in the beam will pass a point, so the current is
$I = Ne/t = Nev/L$ ;
$11 \times 10^{-3}\ \text{A} = N(1.60 \times 10^{-19}\ \text{C})(3.0 \times 10^8\ \text{m/s})/(6300\ \text{m})$, which gives $N = 1.4 \times 10^{12}$ protons.

# Chapter 26: DC Circuits

## Chapter Overview and Objectives

This chapter applies the laws of electricity learned thus far to DC circuits. It introduces Kirchoff's circuit rules and how to apply those rules to DC circuits. It discusses RC circuits with DC sources. It describes how voltage and current measuring instruments are constructed from simple galvanometers.

After completing study of this chapter, you should:

- Know what the terminal voltage of a battery is and how to calculate it from the emf and the internal resistance of the source.
- Know how to recognize when resistances are connected in series and in parallel.
- Know how to calculate equivalent resistances.
- Know Kirchoff's current rule and Kirchoff's voltage rule and how to apply these rules to solve for unknown currents and voltages in DC circuits.
- Know the time dependence of charge and current in an RC circuit with DC sources.
- Know how to calculate the time constant of an RC circuit.
- Know how to determine the resistances necessary to create ammeters and voltmeters from galvanometers.

## Summary of Equations

Terminal voltage of a battery:

$$V_{ab} = \mathcal{E} - Ir$$

Equivalent resistance to resistances in series:

$$R_{eq} = \sum R_i$$

Equivalent resistance to resistances in parallel:

$$\frac{1}{R_{eq}} = \sum \frac{1}{R_i}$$

Charge on a capacitor as function of time in RC circuit:

$$Q = C\mathcal{E} + (Q_0 - C\mathcal{E})e^{-t/RC}$$

Current in RC circuit as function of time:

$$I = -\frac{Q_0 - C\mathcal{E}}{RC}e^{-t/RC}$$

Time constant of RC circuit:

$$\tau = RC$$

Series resistance to make voltmeter from galvanometer:

$$R_s = \frac{V_{fs}}{I_g} - R_g$$

Parallel or shunt resistance to make current meter from galvanometer:

$$R_p = \frac{I_g R_g}{I_{fs} - I_g}$$

## Chapter Summary

### Section 26-1 EMF and Terminal Voltage

The **electromotive force** or **emf** of a device that creates an electric potential difference is the electric potential difference across the two terminals that exists when no current flows from the source. When current does flow through the device, the device may have some **internal resistance** that causes a voltage drop internal to the device, so that the potential difference across the terminals of the device falls below the emf of the source. The potential difference between the two terminals of the device is called the **terminal voltage**. The terminal voltage of the device is related to the emf, $\mathcal{E}$, the internal resistance, $r$, and the current, $I$, that flows from the terminals of the source by:

$$V_{ab} = \mathcal{E} - Ir$$

### Example 26-1-A

Determine the electrical maximum power that can flow from a battery with an emf $\mathcal{E}$ and an internal resistance $r$ and also determine the resistance of the external circuit that causes this maximum power to flow from the battery.

### Solution:

The power out of the battery will be the terminal voltage multiplied by the current:

$$P = I^2 R$$

The current will be the emf divided by the series equivalent of the internal resistance, $r$, and the external resistance, $R$:

$$I = \frac{\mathcal{E}}{r + R}$$

We can then write the power as

$$P = \left(\frac{\mathcal{E}}{r + R}\right)^2 R$$

To find the maximum power into the external resistance, we take the derivative of the power with respect to the external resistance and set it equal to zero:

$$\frac{dP}{dR} = \frac{d}{dR}\left(\frac{\mathcal{E}}{r + R}\right)^2 R = -2\left(\frac{\mathcal{E}^2}{r + R}\right)\left(\frac{1}{r + R}\right)^2 R + \left(\frac{\mathcal{E}}{r + R}\right)^2 = 0$$

which, on solving for $R$, gives

$$R = r$$

Putting this value of the external resistance into the expression for power above gives

$$P = \left(\frac{\mathcal{E}}{r + r}\right)^2 r = \frac{\mathcal{E}^2}{4r}$$

### Section 26-2 Resistors in Series and in Parallel

In Chapter 24, we learned that capacitors could be connected in series or parallel. Resistors can also be connected in series or parallel. Resistors are connected in series if identical currents must flow through each resistor. Resistors that are connected in series have an equivalent resistance that is the sum of the resistances in the series network. The diagram shows three resistors connected in series. The equivalent resistance of this series network is

$$R_{eq} = R_1 + R_2 + R_3$$

In general, for an arbitrary number of resistors connected in series, the equivalent resistance, $R_{eq}$, is

$$R_{eq} = \sum R_i$$

where $R_i$ is the $i$th resistor in the network. Resistors that are connected in such a way that the potential difference across each of the resistors must be identical are connected in parallel. The three resistors shown in the diagram are connected in parallel. The equivalent resistance to the three resistors is given by

$$\frac{1}{R_{eq}} = \frac{1}{R_1} + \frac{1}{R_2} + \frac{1}{R_3}$$

In general, the equivalent resistance of an arbitrary number of resistors connected in parallel is given by

$$\frac{1}{R_{eq}} = \sum \frac{1}{R_i}$$

### Example 26-2-A

Determine the current through each resistor in the circuit shown in the diagram.

### Solution:

We recognize that the 2.0 Ω and the 4.0 Ω resistors are in parallel. Calculating their equivalent resistance we get

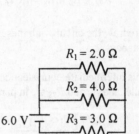

$$\frac{1}{R_A} = \frac{1}{2.0\,\Omega} + \frac{1}{4.0\,\Omega} \quad \Rightarrow \quad R_A = 1.33\,\Omega$$

We redraw the circuit with the equivalent resistance in place of the 2 Ω and the 4 Ω resistors. We then recognize that the 1.33 Ω and the 3.0 Ω in the equivalent circuit are in series. Calculating the equivalent series resistance, we get

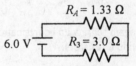

$$R_B = 1.33\,\Omega + 3.0\,\Omega = 4.33\,\Omega$$

We can draw the equivalent circuit as:

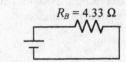

6.0 V

We can now apply Ohm's law to this circuit to find the current flowing from the battery through the equivalent resistance $R_B$:

$$I = \frac{6.0\,\text{V}}{4.33\,\Omega} = 1.38\,\text{A}$$

This will be the current through both series resistors that the equivalent resistance $R_B$ replaces. So the current through $R_3$ and the resistance $R_A$ are both 1.38 A. A current of 1.38 A through $R_A$ allows us to calculate the voltage drop across $R_A$:

$$V_{R_B} = IR_B = (1.38\,\text{A})(1.33\,\Omega) = 1.85\,\text{V}$$

This is the voltage drop across both of the parallel resistors that are equivalent to RA. We can use Ohm's law to calculate the current through each of them:

$$I_{R_1} = \frac{V_{R_B}}{R_1} = \frac{1.85\,\text{V}}{2\,\Omega} = 0.92\,\text{A}$$

$$I_{R_2} = \frac{V_{R_B}}{R_2} = \frac{1.85\,\text{V}}{4\,\Omega} = 0.46\,\text{A}$$

**Example 26-2-B**

Determine the equivalent resistance between points A and B of the resistance network shown in the diagram.

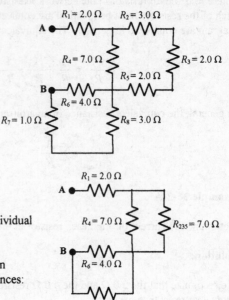

**Solution:**

We look for resistors that are in series or parallel with each other. We recognize that $R_2$, $R_3$, and $R_5$ are in series with each other and $R_7$ and $R_8$ are in series with each other. Calculating the equivalent resistances:

$$R_{235} = R_2 + R_3 + R_5 = 3.0\,\Omega + 2.0\,\Omega + 2.0\,\Omega = 7.0\,\Omega$$
$$R_{78} = R_7 + R_8 = 1.0\,\Omega + 3.0\,\Omega = 4.0\,\Omega$$

We redraw the circuit with these equivalent resistances in place of the individual resistances:

In this intermediate equivalent circuit, we recognize that $R_4$ and $R_{235}$ are in parallel and $R_6$ and $R_{78}$ are in parallel. We calculate the equivalent resistances:

$$\frac{1}{R_{4235}} = \frac{1}{R_4} + \frac{1}{R_{235}} = \frac{1}{7.0\,\Omega} + \frac{1}{7.0\,\Omega} \quad \Rightarrow \quad R_{4235} = 3.5\,\Omega$$

$$\frac{1}{R_{678}} = \frac{1}{R_6} + \frac{1}{R_{78}} = \frac{1}{4.0\,\Omega} + \frac{1}{4.0\,\Omega} \quad \Rightarrow \quad R_{678} = 2.0\,\Omega$$

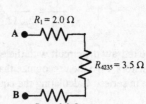

We again redraw an equivalent circuit with these equivalent resistors replacing the individual resistors:

We then recognize that the remaining three resistances are in series and we calculate their equivalent resistance:

$$R_{eq} = R_1 + R_{4235} + R_{678} = 2.0\,\Omega + 3.5\,\Omega + 2.0\,\Omega = 7.5\,\Omega$$

This is the equivalent resistance of the entire network between points A and B.

### Section 26-3 Kirchhoff's Rules

We already know that electrical charge is conserved in any physical process. That fact, together with the fact that charge cannot accumulate at a given point in a conductor can be written as Kirchhoff's current rule or Kirchhoff's node rule:

**The sum of the currents into any given point (or node) in a circuit is zero.**

We also know the electric potential is a well-defined function of position. If we move around through space and return to the same point that we started at, the electric potential will be the same at a given time. If this is applied to a circuit, the sum of the potential differences across circuit elements must add up to zero. This is Kirchhoff's voltage rule or Kirchhoff's loop rule:

**The sum of the changes in potential around any closed path in an electric circuit is zero.**

Kirchhoff's rules can be used to solve for currents and potential differences in circuits when there are interconnections of circuit elements that are not series or parallel. To make use of Kirchhoff's rules, there are some bookkeeping details that require careful attention.

1. Each unknown current must have a direction of positive current flow defined. It is arbitrary which direction is defined as the positive direction, but consistency must be maintained throughout the solution of the problem. Put an arrow on the circuit diagram defining the direction of positive current for each current.

2.  When Kirchhoff's current node rules are used, currents are treated as positive if they flow into the node and negative if they flow out of the node, as defined by the direction defined as positive current flow for each current.

3.  When Kirchhoff's voltage rules are written down, make sure a complete closed path through the circuit is used and components are traversed in the same direction around a loop.  If a resistor is traversed when following the path, the potential change is $-IR$ when traversing it in the direction defined as positive current flow for current $I$ and is $+IR$ when traversing it in the opposite direction.  If a potential difference source is traversed along the path, it is a positive change in potential if the source is traversed in the direction going from its negative terminal to its positive terminal and it is a negative change in potential if it is traversed in a direction from its positive terminal to its negative terminal.

4.  To obtain a complete set of equations that can be solved for the unknown currents, each unknown current must appear in at least one Kirchhoff's current rule equation and at least one Kirchhoff's voltage rule equation.

These issues will be pointed out in the following example.

**Example 26-3-A**

Determine the currents through and the voltages across each of the resistors in the circuit shown in the diagram.

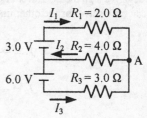

**Solution:**

First, we should be able to see that there will be three independent unknown currents in this circuit, one through each of the resistances.  Each of the currents from the batteries are identical to the currents through one of the resistances so that additional independent currents don't need to be defined for the batteries.  The current from the 3.0 V battery is the same as the current through the 2.0 Ω resistor and the current from the 6.0 V battery is the same as the current through the 3.0 Ω resistor.

We draw an arrow on the circuit diagram for each of the unknown currents and label them appropriately:

Next we write down enough Kirchhoff's current rule equations, so that each unknown current appears in at least one of the equations.  We start with the node marked A in the diagram to the right.  We see that currents $I_1$ and $I_3$ are defined such that positive current flow is into node A and current $I_2$ is defined such that positive current flow is out of node A.  We write Kirchhoff's current rule as

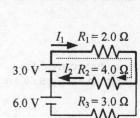

$$I_1 - I_2 + I_3 = 0$$

We see that each of the unknown currents appears in this equation, so this will be a sufficient set of Kirchhoff's current rule equations.

Next we need to write down enough Kirchhoff's voltage rules so that each unknown current appears in at least one of these equations.  We start with the closed path indicated by the heavy line in the diagram to the right.  We traverse the path clockwise as indicated by the dashed arrow.  The first component we encounter is resistor $R_1$, which we travel across in the direction of the current $I_1$, so there is a potential change of $-I_1R_1$ as we cross it.  The next component we reach is $R_2$, which we travel across in the direction of current $I_2$, so there is a potential change of $-I_2R_2$ as we cross it.  Finally, we cross the 3.0 V potential source, which we cross in the direction from its negative terminal to its positive terminal, so there is a potential change of $+3.0$ V as we cross it. We have now completed the closed path.  Adding up the potential differences along the closed path, we have

$$-I_1R_1 - I_2R_2 + 3.0V = 0$$

We see that the unknown current $I_3$ is missing from this equation, so we need one more Kirchhoff's voltage rule equation.  We take the closed path indicated by the bold line in the diagram to the right.  We traverse the path clockwise as indicated by the dashed arrow. The first component we encounter is resistor $R_2$, which we travel across in the direction opposite to the current $I_2$, so there is a potential change of $+I_2R_2$ as we cross

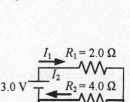

it. The next component we reach is $R_3$, which we travel across in the direction opposite to current $I_3$, so there is a potential change of $+I_3R_3$ as we cross it. Finally, we cross the 6.0 V potential source, which we cross in the direction from its negative terminal to its positive terminal, so there is a potential change of +6.0 V as we cross it. We have now completed the closed path. Adding up the potential differences along the closed path, we have

$$+I_2R_2 + I_3R_3 + 6.0V = 0$$

We now have three independent linear equations for the three unknown currents that can be solved by elementary methods. One method of solution is to solve the two Kirchhoff's voltage rule equations for $I_1$ and $I_3$ and then substitute the expressions for these currents into the Kirchhoff's current rule equation:

$$I_1 = \frac{3.0\,V - I_2R_2}{R_1}$$

$$I_3 = \frac{-6.0\,V - I_2R_2}{R_3}$$

$$I_1 - I_2 + I_3 = \frac{3.0\,V - I_2R_2}{R_1} - I_2 + \frac{-6.0\,V - I_2R_2}{R_3} = 0$$

This final equation is a single equation in a single unknown, $I_2$, which can be solved for $I_2$:

$$I_2 = \frac{3.0\,V/R_1 - 6.0\,V/R_3}{1 + R_2/R_1 + R_2/R_3} = \frac{3.0\,V/2.0\,\Omega - 6.0\,V/3.0\,\Omega}{1 + 4.0\,\Omega/2.0\,\Omega + 4.0\,\Omega/3.0\,\Omega} = -0.115\,A$$

The negative sign means current $I_2$ flows in the direction opposite to what we defined as the positive current direction. We can now solve for the other unknown currents using this result in the two Kirchhoff's voltage rule equations:

$$I_1 = \frac{3.0\,V - I_2R_2}{R_1} = \frac{3.0\,V - (-0.115\,A)(4.0\,\Omega)}{2.0\,\Omega} = 1.73\,A$$

$$I_3 = \frac{-6.0\,V - I_2R_2}{R_3} = \frac{-6.0\,V - (-0.115\,A)(4.0\,\Omega)}{3.0\,\Omega} = -1.85\,A$$

### Section 26-4 Circuits Containing Resistor and Capacitor (*RC* Circuits)

If we apply Kirchhoff's voltage rule to a circuit consisting of a potential difference source, a resistor, and a capacitor in series as show in the diagram, we get an equation of the form

$$\mathcal{E} - IR - \frac{Q}{C} = 0$$

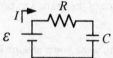

if we adopt the current direction shown in the diagram as positive and the charge $Q$ is the charge on the upper plate of the capacitor. In this equation, we recognize that $I$ and $Q$ are functions of time and we can write the relationship

$$I = \frac{dQ}{dt}$$

between the current and the charge on the capacitor. Substituting this into Kirchhoff's voltage rule equation for this circuit gives:

$$\mathcal{E} - \frac{dQ}{dt}R - \frac{Q}{C} = 0$$

The most general solution to this equation is

$$Q = C\mathcal{E} + Ae^{-t/RC} = C\mathcal{E} + Ae^{-t/\tau}$$

where $\tau = RC$ is called the time constant of the circuit. The constant A is determined by the value of the charge on the capacitor at the time $t = 0$. If $Q_0$ is the charge on the capacitor at time $t = 0$, then

$$Q_0 = C\mathcal{E} + A \quad \Rightarrow \quad A = Q_0 - C\mathcal{E}$$

So we can write the solution as

$$Q = C\mathcal{E} + (Q_0 - C\mathcal{E})e^{-t/RC}$$

The current flowing in the circuit is the time derivative of the charge on the capacitor:

$$I = \frac{dQ}{dt} = \frac{d}{dt}\left[C\mathcal{E} + (Q_0 - C\mathcal{E})e^{-t/RC}\right] = -\frac{Q_0 - C\mathcal{E}}{RC}e^{-t/RC}$$

The voltage on the capacitor is the charge on the capacitor divided by the capacitance:

$$V_C = \mathcal{E} + (Q_0/C - \mathcal{E})e^{-t/RC}$$

### Example 26-4-A

For the RC circuit shown to the right, determine the charge on the capacitor as a function of time if the switch is closed at time $t = 0$. The capacitor has an initial voltage across it of 12.0 V before the switch is closed with the upper plate positive. Determine the time at which the voltage across the capacitor is equal to 5.0 V. Determine the current in the circuit at time $t = 0$.

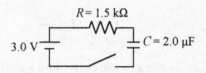

### Solution:

We know the solution for the voltage across the capacitor as a function of time is given by

$$V_C = \mathcal{E} + (Q_0/C - \mathcal{E})e^{-t/RC}$$

We recognize $Q_0/C$ as the voltage across the capacitor at time $t = 0$. The time constant of the circuit is

$$\tau = RC = (1.5\times10^3\,\Omega)(2.0\times10^{-6}\,\text{F}) = 3.0\times10^{-3}\,\text{s}$$

We can now write the voltage across the capacitor as a function of time as

$$V_C = 3.0\,\text{V} + (12.0V - 3.0\,\text{V})e^{-t/3.0\times10^{-3}\,\text{s}} = 3.0\,\text{V} + (9.0\,\text{V})e^{-t/3.0\times10^{-3}\,\text{s}}$$

To find the time at which the voltage across the capacitor is 5.0 V, we use this value for the voltage across the capacitor in the above equation and solve for $t$:

$$5.0\,\text{V} = 3.0\,\text{V} + (9.0\,\text{V})e^{-t/3.0\times10^{-3}\,\text{s}} \quad \Rightarrow \quad t = -(3.0\times10^{-3}\,\text{s})\ln\left(\frac{5.0\,\text{V} - 3.0\,\text{V}}{9.0\,\text{V}}\right) = 4.5\times10^{-3}\,\text{s}$$

An easy way to determine the current at time $t = 0$ is to use Kirchhoff's voltage loop rule around the circuit:

$$\varepsilon - IR - V_C = 0$$

where we have defined positive current flow across the resistor as positive from left to right in the diagram. Solving this for the current, we get

$$I = \frac{-V_C + \varepsilon}{R} = \frac{-12.0\,\text{V} + 3.0\,\text{V}}{1.5\times10^3\,\Omega} = -6.0\times10^{-3}\,\text{A}$$

The minus sign tells us that the current flows from right to left at time $t = 0$.

## Section 26-5 DC Ammeters and Voltmeters

Devices for measuring current are called **ammeters**.  Devices for measuring voltage are called **voltmeters**.  Devices for measuring resistance are called **ohmmeters**.  A single instrument that can operate as more than one of these devices is called a **multimeter**.  All of these devices can be constructed from a basic instrument that can measure either current or voltage.  There are two basic devices in common use today.  A **galvanometer** is a meter that has an indicator with deflection that is proportional to the current passing through it.  A digital voltmeter is a device that displays a numerical voltage reading on a display that is proportional to the voltage across its terminals.

A galvanometer has an indicator deflection that is proportional to the current through it.  The current required to deflect the indicator to the maximum deflection of the indicator is called the full scale current, $I_g$, of the galvanometer. The galvanometer has a finite resistance we will call $R_g$.  If we want to construct a voltmeter with a full-scale reading, $V_{fs}$, from a galvanometer, we must add a series resistance, $R_s$, to the galvanometer so that when $V_{fs}$ is applied across the series network of the galvanometer and series resistance, the current will be $I_g$.  The equivalent resistance of the galvanometer and the series resistance is $R = R_g + R_s$.  Using Ohm's law, the resistance $R$ must also be equal to the desired full-scale voltage divided by the galvanometer current required for full deflection of the indicator:

$$R = \frac{V_{fs}}{I_g} \quad \Rightarrow \quad R_s = \frac{V_{fs}}{I_g} - R_g$$

To construct a current meter with a greater full scale current reading, $I_{fs}$, then the current required to give maximum deflection of the galvanometer by itself, a resistance, $R_p$, must be added in parallel to the galvanometer resistance.  This resistance is often called a **shunt resistance** because it shunts a fraction of the current around the meter.  To determine the resistance needed we must ensure that the desired full scale current through the network of the galvanometer and shunt resistance causes the current through the galvanometer to be that which causes full-scale deflection.  The easiest way to do this analysis is by using the fact that the voltage across a parallel network is the same across each component resistance.  The voltage across the galvanometer when its deflection is maximum is

$$V_g = I_g R_g$$

This must be the voltage across the shunt resistance when the remainder of the current, $I = I_{fs} - I_g$ flows through it:

$$V_g = I_p R_p \quad \Rightarrow \quad R_p = \frac{V_g}{I_p} = \frac{I_g R_g}{I_{fs} - I_g}$$

## Example 26-5-A

A galvanometer reaches full-scale when a current of 50 μA passes through it.  The galvonometer has a resistance of 100 Ω.  How can a voltmeter with a full-scale deflection of 1.00 V be made?

## Solution:

A voltmeter is to be constructed, so a series resistance must be added to the galvanometer.  The value of the series resistance is

$$R_s = \frac{V_{fs}}{I_g} - R_g = \frac{1.00 \text{ V}}{50 \times 10^{-6} \text{ A}} - 100 \, \Omega = 1.99 \times 10^4 \, \Omega$$

**Section 26-6 Transducers and the Thermocouple**

Transducers are devices that accept one type of input signal and generate a corresponding output signal of some other type. Some examples are thermocouples, microphones, loudspeakers, and strain gauges.

## Practice Quiz

1.  Consider Example 26-1-A in this study guide, where the maximum electrical power out of the battery was calculated to occur when the external resistance of the circuit is equal to the internal resistance of the battery. Why doesn't decreasing the external resistance further, causing an increase in current, increase the electrical power out of the battery?

    a) Decreasing the resistance causes a decrease in the current drawn from the battery.
    b) Decreasing the resistance causes chemical energy to be transformed into electrical energy at a greater rate, but more of the electrical energy goes into heating the internal resistance of the battery.
    c) Decreasing the resistance decreases the voltage drop across the internal resistance of the battery.
    d) The electrical power out of the battery remains the same regardless of the external resistance.

2.  Given three equal resistances with resistance $R$, how can they be interconnected so their equivalent resistance is $\frac{2}{3}R$ between the right and left ends of the network?

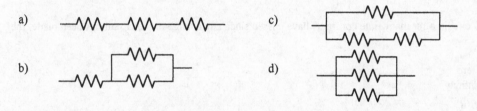

3.  A network of resistors, all with resistance $R$, is connected such that it can be reduced using equivalent parallel and/or series resistance to a single resistor. The equivalent resistance is greater than $R$. What must be true about the network of resistors?

    a) There must be at least two resistors in series.
    b) There must be at least two resistors in parallel.
    c) A series equivalent must be used at some point in reducing the network to its equivalent resistance.
    d) A parallel equivalent must be used at some point in reducing the network to its equivalent resistance.

4.  If in an arbitrary network of batteries and resistances, the voltage of each battery is doubled, what is true about the currents in the circuit?

    a) They will all increase by a factor $2^N$, where $N$ is the number of batteries.
    b) They will all increase by a factor of two.
    c) How the currents change will depend on the details of the circuit.
    d) The currents will not change.

5.  A parallel network of resistances has an equivalent resistance that

    a) is always greater than any of the component resistances.
    b) is always less than any of the component resistances.
    c) is always less than the greatest component resistance, but greater than the least component resistance.
    d) can be less than, equal to, or greater than any of the component resistances.

6.  A series network of resistances has an equivalent resistance that

    a) is always greater than any of the component resistances.
    b) is always less than any of the component resistances.
    c) is always less than the greatest component resistance, but greater than the least component resistance.
    d) can be less than, equal to, or greater than any of the component resistances.

7.    The time constant of an *RC* circuit is the amount of time for the charge on the capacitor

    a)  to go from its initial to its final value.
    b)  to go a fraction $1/e$ of the charge difference from its initial to its final value.
    c)  to go to zero.
    d)  to not change.

8.    How can the time constant of an RC circuit be doubled?

    a)  Double the voltage.
    b)  Double the capacitance.
    c)  Double both the resistance and the capacitance simultaneously.
    d)  All of the above.

9.    What is the minimum full-scale voltage possible for a voltmeter made from a galvanometer with a full deflection current $I_g$ and a resistance $R$?

    a)  There is no minimum full-scale voltage.
    b)  $I_g/R$
    c)  $I_g R$
    d)  $R/I_g$

10.   In a series RC circuit, after many time constants have passed since any change in the circuit has been made, the current will

    a)  approach zero.
    b)  approach infinity.
    c)  approach $\mathcal{E}/R$.
    d)  become opposite in direction to its initial direction.

11.   A minimum of 7.4 V is needed to power a radio. If the power is supplied by a battery with a 9.0 V emf and the circuit draws 0.18 A when the terminal voltage is 7.4 V, what can be the maximum internal resistance of the battery in order to power the radio?

12.   Determine the equivalent resistance between points A and B of the resistance network shown in the diagram to the right.

13.   Determine the current through each of the resistors in the circuit shown in the diagram below.

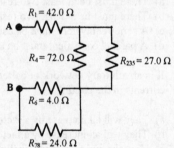

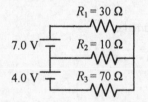

14.   The capacitor in the circuit shown in the diagram to the right is initially uncharged. The switch is closed at time $t = 0$. At what time is the voltage across the capacitor equal to 8.0 V?

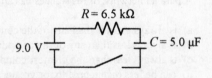

15.   Determine the shunt resistance necessary to make an ammeter with a full-scale deflection of 1.00 A from a galvanometer with a full-scale deflection of 100 μA and a resistance of 250 Ω.

## Problem Solutions

3.    If we can ignore the resistance of the ammeter, for the single loop we have

$I = \mathcal{E}/r$;

25 A = (1.5 V)/$r$, which gives $r$ = 0.060 Ω.

7.  If we use them as single resistors, we have

$R_1$ = 25 Ω; $R_2$ = 70 Ω.

When the resistors are connected in series, the equivalent resistance is

$R_{series} = \Sigma R_i = R_1 + R_2$ = 25 Ω + 70 Ω = 95 Ω.

When the resistors are connected in parallel, we find the equivalent resistance from

$1/R_{parallel} = \Sigma(1/R_i) = (1/R_1) + (1/R_2)$ = [1/(25 Ω)] + [1/(70 Ω)], which gives $R_{parallel}$ = 18 Ω.

11. We can reduce the circuit to a single loop by successively combining parallel and series combinations.

We combine $R_1$ and $R_2$, which are in series:

$R_7 = R_1 + R_2$ = 2.8 kΩ + 2.8 kΩ = 5.6 kΩ.

We combine $R_3$ and $R_7$, which are in parallel:

$1/R_8 = (1/R_3) + (1/R_7)$ = [1/(2.8 kΩ)] + [1/(5.6 kΩ)],

which gives $R_8$ = 1.87 kΩ.

We combine $R_4$ and $R_8$, which are in series:

$R_9 = R_4 + R_8$ = 2.8 kΩ + 1.87 kΩ = 4.67 kΩ.

We combine $R_5$ and $R_9$, which are in parallel:

$1/R_{10} = (1/R_5) + (1/R_9)$ = [1/(2.8 kΩ)] + [1/(4.67 kΩ)],

which gives $R_{10}$ = 1.75 kΩ.

We combine $R_{10}$ and $R_6$, which are in series:

$R_{eq} = R_{10} + R_6$ = 1.75 kΩ + 2.8 kΩ = 4.6 kΩ.

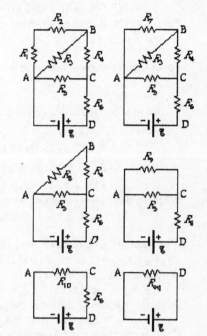

15. Fortunately the required resistance is less. We can reduce the resistance by adding a parallel resistor, which does not require breaking the circuit. We find the necessary resistance from

$1/R = (1/R_1) + (1/R_2)$;

1/(320 Ω) = [1/(480 Ω)] + (1/$R_2$), which gives $R_2$ = 960 Ω in parallel.

19. (a)  When the switch is closed the addition of $R_2$ to the parallel set will decrease the equivalent

resistance, so the current from the battery will increase. This causes an increase in the voltage across $R_1$, and a corresponding decrease across $R_3$ and $R_4$. The voltage across $R_2$ increases from zero. Thus we have $V_1$ and $V_2$ increase; $V_3$ and $V_4$ decrease.

(b)  The current through $R_1$ has increased. This current is now split into three, so currents through $R_3$ and $R_4$ decrease. Thus we have

$I_1$ (= $I$) and $I_2$ increase; $I_3$ and $I_4$ decrease.

(c)  The current through the battery has increased, so the power output of the battery increases.

(d)  Before the switch is closed, $I_2$ = 0. We find the resistance for $R_3$ and $R_4$ in parallel from

$1/R_A = \Sigma(1/R_i) = 2/R_3 = 2/(100 \text{ Ω})$,

which gives $R_A$ = 50 Ω.

For the single loop, we have

$I = I_1 = V/(R_1 + R_A)$

= (45.0 V)/(100 Ω + 50 Ω) = 0.300 A.

This current will split evenly through $R_3$ and $R_4$:

$I_3 = I_4 = \frac{1}{2}I = \frac{1}{2}(0.300 \text{ A})$ = 0.150 A.

After the switch is closed, we find the resistance for $R_2$, $R_3$, and $R_4$ in parallel from

$1/R_B = \Sigma(1/R_i) = 3/R_3 = 3/(100 \text{ Ω})$,

which gives $R_B$ = 33.3 Ω.

For the single loop, we have

$I = I_1 = V/(R_1 + R_B)$

= (45.0 V)/(100 Ω + 33.3 Ω) = 0.338 A.

This current will split evenly through $R_2$, $R_3$, and $R_4$:

$I_2 = I_3 = I_4 = \frac{1}{3}I = \frac{1}{3}(0.338 \text{ A})$ = 0.113 A.

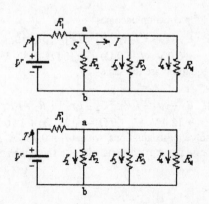

23. For the current in the single loop, we have

$I \quad = V/(R_1 + R_2 + r)$

$\quad = (9.0 \text{ V})/(8.0 \text{ }\Omega + 12.0 \text{ }\Omega + 2.0 \text{ }\Omega) = 0.41 \text{ A}.$

For the terminal voltage of the battery, we have

$V_{ab} = \mathcal{E} - Ir = 9.0 \text{ V} - (0.41 \text{ A})(2.0 \text{ }\Omega) = 8.18 \text{ V}.$

The current in a resistor goes from high to low potential.

For the voltage changes across the resistors, we have

$V_{bc} = - IR_2 = - (0.41 \text{ A})(12.0 \text{ }\Omega) = - 4.91 \text{ V};$

$V_{ca} = - IR_1 = - (0.41 \text{ A})(8.0 \text{ }\Omega) = - 3.27 \text{ V}.$

For the sum of the voltage changes, we have

$V_{ab} + V_{bc} + V_{ca} = 8.18 \text{ V} - 4.91 \text{ V} - 3.27 \text{ V} = 0.$

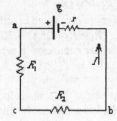

27. For the conservation of current at point b, we have

$I_{in} = I_{out} ;$

$I_1 + I_3 = I_2 .$

For the two loops indicated on the diagram, we have

loop 1:  $\mathcal{E}_1 - I_1R_1 - I_2R_2 = 0;$

$\quad\quad\quad + 9.0 \text{ V} - I_1(15 \text{ }\Omega) - I_2(20 \text{ }\Omega) = 0;$

loop 2:  $- \mathcal{E}_2 + I_2R_2 + I_3R_3 = 0;$

$\quad\quad\quad - 12.0 \text{ V} + I_2(20 \text{ }\Omega) + I_3(40 \text{ }\Omega) = 0.$

When we solve these equations, we get

$I_1 = 0.18 \text{ A right}, I_2 = 0.32 \text{ A left}, I_3 = 0.14 \text{ A up}.$

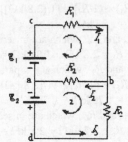

31. If we assume the current in $R_4$ is to the right, we have

$V_{cd} = I_4R_4 = (3.50 \text{ mA})(4.0 \text{ k}\Omega) = 14.0 \text{ V}.$

We can now find the current in $R_8$ :

$I_8 = V_{cd}/R_8 = (14.0 \text{ V})/(8.0 \text{ k}\Omega) = 1.75 \text{ mA}.$

From conservation of current at the junction c, we have

$I = I_4 + I_8 = 3.50 \text{ mA} + 1.75 \text{ mA} = 5.25 \text{ mA}.$

If we go clockwise around the outer loop, starting at a, we have

$V_{ba} - IR_5 - I_4R_4 - \mathcal{E} - Ir = 0, \quad \text{or}$

$V_{ba} = (5.25 \text{ mA})(5.0 \text{ k}\Omega) + (3.50 \text{ mA})(4.0 \text{ k}\Omega) + 12.0 \text{ V} + (5.25 \text{ mA})(1.0 \times 10^{-3} \text{ k}\Omega) = \quad 52 \text{ V}.$

If we assume the current in $R_4$ is to the left, all currents are reversed, so we have

$V_{dc} = 14.0 \text{ V}; \quad I_8 = 1.75 \text{ mA}, \quad \text{and} \quad I = 5.25 \text{ mA}.$

If we go counterclockwise around the outer loop, starting at a, we have

$- Ir + \mathcal{E} - I_4R_4 - IR_5 + V_{ab} = 0, \quad \text{or} \quad V_{ba} = - V_{ab} = - Ir + \mathcal{E} - I_4R_4 - IR_5 ;$

$V_{ba} = - (5.25 \text{ mA})(1.0 \times 10^{-3} \text{ k}\Omega) + 12.0 \text{ V} - (3.50 \text{ mA})(4.0 \text{ k}\Omega) - (5.25 \text{ mA})(5.0 \text{ k}\Omega) = - 28 \text{ V}.$

The negative value means the battery is facing the other direction.

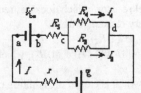

35. The lower loop equation becomes

loop 2:  $\mathcal{E}_3 - I_3r_3 + I_2R_2 - \mathcal{E}_2 + I_2r_2 - I_3R_4 = 0;$

$\quad\quad\quad + 6.0 \text{ V} - I_3(1.0 \text{ }\Omega) + I_2(10 \text{ }\Omega) - 12.0 \text{ V} + I_2(1.0 \text{ }\Omega) - I_3(15 \text{ }\Omega) = 0.$

The other equations are the same:

$I_2 + I_3 = I_1 .$

loop 1:  $\mathcal{E}_1 - I_1r_1 - I_1R_3 + \mathcal{E}_2 - I_2r_2 - I_2R_2 - I_1R_1 = 0;$

$\quad\quad\quad + 12.0 \text{ V} - I_1(1.0 \text{ }\Omega) - I_1(8.0 \text{ }\Omega) + 12.0 \text{ V} - I_2(1.0 \text{ }\Omega) - I_2(10 \text{ }\Omega) - I_1(12 \text{ }\Omega) = 0.$

When we solve these equations, we get $I_1 = 0.783 \text{ A}, I_2 = 0.686 \text{ A}, I_3 = 0.097 \text{ A}.$

We have carried an extra decimal place to show the agreement with the junction equations.

39. (*a*) On the diagram, we show the potential difference applied
between points a and c and the six currents.
For the conservation of current we have
    at point a:    $I = I_1 + I_2 + I_3$ ;                      (1)
    at point b:    $I_3 = I_4 + I_5$ ;
    at point d:    $I_2 + I_4 = I_6$ .

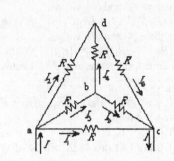

For a loop CW around the right triangle, we have
    $-I_4R - I_6R' + I_5R = 0$;
    $-I_4R - (I_2 + I_4)R' + (I_3 - I_4)R = 0$,   or
    $(2R + R')I_4 = I_3R - I_2R'$.                         (2)
For a loop CW around the left triangle, we have
    $-I_2R + I_4R + I_3R = 0$,   or   $I_4 = I_2 - I_3$.   (3)
For a loop CW around the bottom triangle, we have
    $-I_3R - I_5R + I_1R = 0$;
    $-I_3R - (I_3 - I_4)R + I_1R = 0$,   or   $I_4 = 2I_3 - I_1$.   (4)
When we combine (3) and (4), we get
    $3I_3 = I_1 + I_2$ .                                        (5)
When we combine (2) and (3), we get
    $(2R + R')I_2 - (2R + R')I_3 = I_3R - I_2R'$,   or
    $I_3 = 2(R + R')I_2/(3R + R')$.                       (6)
When we combine (5) and (6), we get
    $I_2 = (3R + R')I_1/(3R + 5R')$.                     (7)
When we combine (6) and (7), we get
    $I_3 = 2(R + R')I_1/(3R + 5R')$.                     (8)
When we use (7) and (8) in (1), we get
    $I = 8(R + R')I_1/(3R + 5R')$.
Because $I_1 = V_{ac}/R$, for the equivalent resistance we have
    $R_{eq} = V_{ac}/I = I_1R/I = R(3R + 5R')/8(R + R')$.

(*b*) If we apply a potential difference between points a and b,
all currents leaving a must enter b. From the symmetry
$I_1 = I_2$ and $I_4 = I_5$, so $I_2 = -I_4$ and $I_1 = -I_5$. Consequently there
is no current in $R'$, which can be removed. When we redraw
the circuit, we see that we have three parallel branches
between points a and b. We find the equivalent resistance from
    $1/R_{ab} = (1/R) + (1/2R) + (1/2R)$,
which gives $R_{ab} = R/2$.

43. The charge on the capacitor increases with time to a final charge $Q_0$ :
    $Q = Q_0(1 - e^{-t/\tau})$.
When we express the stored energy in terms of charge we have
    $U = \frac{1}{2}CV^2 = \frac{1}{2}Q^2/C = \frac{1}{2}(Q_0^2/C)(1 - e^{-t/\tau})^2 = U_{max}(1 - e^{-t/\tau})^2$.
We find the time to reach half the maximum from
    $\frac{1}{2} = (1 - e^{-t/\tau})^2$,   or   $e^{-t/\tau} = 1 - 1/\sqrt{2}$, which gives $t = 1.23\,\tau$.

47. Because we have no simple series or parallel connections, we analyze the circuit. On the diagram, we show the
potential difference applied between points c and d and the four currents.
For the conservation of current at points c and d, we have
    $I = I_1 + I_3 = I_2 + I_4$ .                          (1)

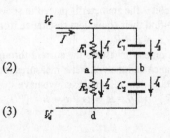

The current $I_3$ is charging the capacitor $C_1$ : $I_3 = dQ_1/dt$.
The current $I_4$ is charging the capacitor $C_2$ : $I_4 = dQ_2/dt$.
For a CW loop acba, we have
    $I_1R_1 - Q_1/C_1 = 0$, or
    $Q_1/C_1 = (I - I_3)R_1 = (I - dQ_1/dt)R_1$ .      (2)
For a CW loop dabd, we have
    $I_2R_2 - Q_2/C_2 = 0$, or
    $Q_2/C_2 = (I - I_4)R_2 = (I - dQ_2/dt)R_2$ .      (3)
For the path cbd, we have
    $V_{cd} = (Q_1/C_1) + (Q_2/C_2)$.                    (4)
If we differentiate this, we get
    $dV_{cd}/dt = 0 = (1/C_1)(dQ_1/dt) + (1/C_2)(dQ_2/dt)$,   or

$dQ_2/dt = -(C_2/C_1) \, dQ_1/dt.$    (5)

When we combine (3) and (5), we get

$Q_2/C_2 = R_2[I + (C_2/C_1) \, dQ_1/dt].$    (6)

When we combine (2) and (6), we get

$Q_2/C_2 = (R_2/R_1C_1)Q_1 + [R_2(C_1 + C_2)/C_1] \, dQ_1/dt.$    (7)

When we combine (4) and (7), we get

$V_{cd} = [(R_1 + R_2)/R_1C_1]Q_1 + [R_2(C_1 + C_2)/C_1] \, dQ_1/dt.$

This has the same form as the simple $RC$ circuit:

$\mathcal{E} = R \, dQ/dt + Q/C,$

with $R = R_2(C_1 + C_2)/C_1$, and $C = R_1C_1/(R_1 + R_2)$.

Thus the time constant is

$\tau = R_1R_2(C_1 + C_2)/(R_1 + R_2)$

$= (8.8 \ \Omega)(4.4 \ \Omega)(0.48 \ \mu\text{F} + 0.24 \ \mu\text{F})/(8.8 \ \Omega + 4.4 \ \Omega) = 2.1 \ \mu\text{s}.$

51. (*a*)  The current for full-scale deflection of the galvanometer is

$I = 1/(\text{sensitivity}) = 1/(35 \ \text{k}\Omega/\text{V}) = 2.86 \times 10^{-2} \ \text{mA}.$

We make an ammeter by putting a resistor in parallel with the galvanometer.  For full-scale deflection, we have

$V_{meter} = I_G r = I_s R_s \, ;$

$(2.86 \times 10^{-5} \ \text{A})(20.0 \ \Omega) = (2.0 \ \text{A} - 2.86 \times 10^{-5} \ \text{A})R_s \, ,$

which gives $R_s = 2.9 \times 10^{-4} \ \Omega$ in parallel.

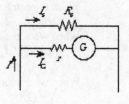

(*b*)  We make a voltmeter by putting a resistor in series with the galvanometer.  For full-scale deflection, we have

$V_{meter} = I(R_x + r) = I_G(R_x + r);$

$1.00 \ \text{V} = (2.86 \times 10^{-5} \ \text{A})(R_x + 20 \ \Omega),$

which gives $R_x = 3.5 \times 10^4 \ \Omega = 35 \ \text{k}\Omega$ in series.

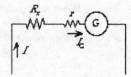

55.  We find the equivalent resistance of the voltmeter in parallel with one of the resistors:

$1/R = (1/R_1) + (1/R_V) = (1/9.0 \ \text{k}\Omega) + (1/11.5 \ \text{k}\Omega),$

which gives $R = 5.05 \ \text{k}\Omega.$

The current in the circuit, which is read by the ammeter, is

$I = \mathcal{E}/(r + R_A + R + R_2)$

$= (12.0 \ \text{V})/(1.0 \ \Omega + 0.50 \ \Omega + 5.05 \ \text{k}\Omega + 9.0 \ \text{k}\Omega)$

$= 0.85 \ \text{mA}.$

The reading on the voltmeter is

$V_{ab} = IR = (0.85 \ \text{mA})(5.05 \ \text{k}\Omega) = 4.3 \ \text{V}.$

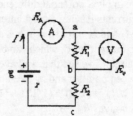

59. (*a*)  In circuit 1 the voltmeter is placed in parallel with the resistor, so we find their equivalent resistance from

$1/R_{eq1} = (1/R) + (1/R_V).$

The ammeter measures the current through this equivalent resistance and the voltmeter measures the voltage across this equivalent resistance, so we have

$R_{eq1} = V/I.$

Thus we have

$I/V = (1/R) + (1/R_V), \quad \text{or} \quad 1/R = (I/V) - (1/R_V).$

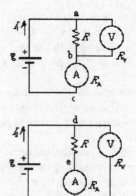

(*b*)  In circuit 2 the ammeter is placed in series with the resistor, so we find their equivalent resistance from

$R_{eq2} = R + R_A \, .$

The ammeter measures the current through this equivalent resistance and the voltmeter measures the voltage across this equivalent resistance, so we have

$R_{eq2} = V/I = R + R_A, \quad \text{or} \quad R = (V/I) - R_A \, .$

63. The voltage is the same across resistors in parallel, but is less across a resistor in a series connection. We connect two resistors in series as shown in the diagram. Each resistor has the same current:

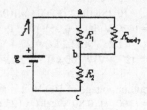

$$I = V/(R_1 + R_2) = (6.0 \text{ V})/(R_1 + R_2).$$

If the desired voltage is across $R_1$, we have

$$V_{ab} = IR_1 = (6.0 \text{ V})R_1/(R_1 + R_2);$$
$$0.25 \text{ V} = (6.0 \text{ V})R_1/(R_1 + R_2) = (6.0 \text{ V})/[1 + (R_2/R_1)],$$

which gives $R_2/R_1 = 23$.

When the body is connected across ab, we want very negligible current through the body, so the potential difference does not change. This requires $R_{body} = 2000 \text{ } \Omega \gg R_1$. If we also do not want a large current from the battery, a possible combination is

$$R_1 = 4 \text{ } \Omega, R_2 = 92 \text{ } \Omega.$$

67. The time between firings is

$$t = (60 \text{ s})/(72 \text{ beats}) = 0.833 \text{ s}.$$

In this time the capacitor reaches 45% of maximum, so we have

$$V = V_0(1 - e^{-t/\tau}) = 0.45V_0, \text{ which gives } e^{-t/\tau} = 0.55, \text{ or } \tau = t/\ln(1.82) = (0.833 \text{ s})/\ln(1.82) = 1.39 \text{ s}.$$

We find the required resistance from

$$\tau = RC;$$
$$1.39 \text{ s} = R(7.5 \text{ } \mu\text{F}), \text{ which gives } R = 0.19 \text{ M}\Omega.$$

71. For the conservation of current at point b, we have

$$I = I_1 + I_2.$$

For the two loops indicated on the diagram, we have

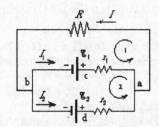

loop 1:  $\mathcal{E}_1 - I_1r_1 - IR = 0;$
$$+ 2.0 \text{ V} - I_1(0.10 \text{ } \Omega) - I(4.0 \text{ } \Omega) = 0;$$

loop 2:  $\mathcal{E}_2 - I_2r_2 - IR = 0;$
$$+ 3.0 \text{ V} - I_2(0.10 \text{ } \Omega) - I(4.0 \text{ } \Omega) = 0.$$

When we solve these equations, we get

$$I_1 = -4.69 \text{ A}, I_2 = 5.31 \text{ A}, I = 0.62 \text{ A}.$$

For the voltage across $R$ we have

$$V_{ab} = IR = (0.62 \text{ A})(4.0 \text{ } \Omega) = 2.5 \text{ V}.$$

Note that one battery is charging the other with a significant current.

75. The resistance along the potentiometer is proportional to the length, so we find the equivalent resistance between points b and c:

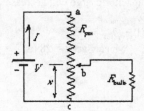

$$1/R_{eq} = (1/xR_{pot}) + (1/R_{bulb}), \text{ or }$$
$$R_{eq} = xR_{pot}R_{bulb}/(xR_{pot} + R_{bulb}).$$

We find the current in the loop from

$$I = V/[(1 - x)R_{pot} + R_{eq}].$$

The potential difference across the bulb is

$$V_{bc} = IR_{eq}, \text{ so the power expended in the bulb is}$$
$$P = V_{bc}{}^2/R_{bulb}.$$

(a) For $x = 1.00$ we have

$$R_{eq} = (1.00)(100 \text{ } \Omega)(200 \text{ } \Omega)/[(1.00)(100 \text{ } \Omega) + 200 \text{ } \Omega] = 66.7 \text{ } \Omega.$$
$$I = (120 \text{ V})/[(1 - 1.00)(100 \text{ } \Omega) + 66.7 \text{ } \Omega] = 1.80 \text{ A}.$$
$$V_{bc} = (1.80 \text{ A})(66.7 \text{ } \Omega) = 120 \text{ V}.$$
$$P = (120 \text{ V})^2/(200 \text{ } \Omega) = 72.0 \text{ W}.$$

(b) For $x = 0.50$ we have

$$R_{eq} = (0.50)(100 \text{ } \Omega)(200 \text{ } \Omega)/[(0.50)(100 \text{ } \Omega) + 200 \text{ } \Omega] = 40.0 \text{ } \Omega.$$
$$I = (120 \text{ V})/[(1 - 0.50)(100 \text{ } \Omega) + 40.0 \text{ } \Omega] = 1.33 \text{ A}.$$
$$V_{bc} = (1.33 \text{ A})(40.0 \text{ } \Omega) = 53.3 \text{ V}.$$
$$P = (53.3 \text{ V})^2/(200 \text{ } \Omega) = 14.2 \text{ W}.$$

(c) For $x = 0.25$ we have

$R_{eq} = (0.25)(100\ \Omega)(200\ \Omega)/[(0.25)(100\ \Omega) + 200\ \Omega] = 22.2\ \Omega$.

$I = (120\ V)/[(1 - 0.25)(100\ \Omega) + 22.2\ \Omega] = 1.23\ A$.

$V_{bc} = (1.23\ A)(22.2\ \Omega) = 27.4\ V$.

$P = (27.4\ V)^2/(200\ \Omega) = 3.76\ W$.

79. To get an output voltage of 100 V, it is necessary to put solar cells in series. The number of series cells required is

$N_{series} = (100\ V)/(0.80\ V/cell) = 125$ cells.

To get an output current of 1.0 A, it is necessary to put solar cells in parallel. The number of parallel cells required is

$N_{parallel} = (1.0\ A)/(0.350\ A) = 2.8 = 3$ cells.

We connect 125 cells in series and then 3 of these units in parallel. Thus the total number of cells is

$N = N_{series}N_{parallel} = (125)(3) = 375$ cells.

The size of the panel is

$(125)(0.030\ m) \times (3)(0.030\ m) = 3.8\ m \times 0.090\ m$.

Of course it is possible to adjust the dimensions by changing the wiring of the cells. To optimize the output it is necessary to have the panel move so that it is always perpendicular to the sunlight.

83. (a) After a long time there will be a steady state; there will be no current in the capacitor branch:

$I_5 = 0;\quad I_1 = I_3,\quad$ and $\quad I_2 = I_4$.

For the two resistor branches we have

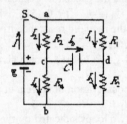

$V_a - V_b = \mathcal{E} = I_2(R_2 + R_4)$;

$12.0\ V = I_2(1.0\ \Omega + 9.0\ \Omega)$, which gives $I_2 = 1.20\ A$;

$V_a - V_b = \mathcal{E} = I_1(R_1 + R_3)$;

$12.0\ V = I_1(10.0\ \Omega + 5.0\ \Omega)$, which gives $I_1 = 0.80\ A$.

We can find the potential difference across the capacitor from

$V_c - V_d = (V_c - V_b) - (V_d - V_b) = I_4R_4 - I_3R_3$

$\qquad\qquad = (1.20\ A)(9.0\ \Omega) - (0.80\ A)(5.0\ \Omega) = +6.8\ V$.

The charge on the capacitor is

$Q = C(V_c - V_d) = (1.5\ \mu F)(6.8\ V) = 10.2\ \mu C$.

(b) When the switch is opened, we find the equivalent resistance between c and d:

$1/R_{eq} = 1/(R_1 + R_2) + 1/(R_3 + R_4)$;

$1/R_{eq} = [1/(10.0\ \Omega + 1.0\ \Omega) + 1/(5.0\ \Omega + 9.0\ \Omega)]$, which gives $R_{eq} = 6.16\ \Omega$.

The time constant for this circuit is

$\tau = R_{eq}C = (6.16\ \Omega)(1.5\ \mu F) = 9.24\ \mu s$.

The capacitor charge will decrease exponentially:

$Q = Q_{max}e^{-t/\tau}$;

$0.050Q_{max} = Q_{max}e^{-t/(9.24\ \mu s)}$,   or   $t/(9.24\ \mu s) = \ln(20.0) = 3.00$,

which gives $t = 28\ \mu s$.

# Chapter 27: Magnetism

## Chapter Overview and Objectives

This chapter introduces magnetic fields, magnetic forces on currents, and magnetic forces on charged particles. In addition, the magnetic force is applied to determine the torque on a current loop, the trajectory of a charged particle in a magnetic field, and the Hall effect. The concept of a magnetic dipole moment is introduced and the torque on and potential energy of a magnetic dipole in a magnetic field are determined.

After completing study of this chapter, you should:

- Know how to calculate the magnetic force on a current.
- Know how to calculate the magnetic charge on a charged particle.
- Know the right-hand rules for calculating forces on currents and charged particles.
- Know how to determine the cyclotron period and frequency of a particle in a magnetic field.
- Know how to calculate the torque due to a magnetic field on a current loop or magnetic dipole.
- Know how to calculate the magnetic dipole moment of a current loop.
- Know how to calculate the potential energy of a magnetic dipole moment in a magnetic field.
- Know what the Hall emf and Hall field are and be able to calculate these quantities.

## Summary of Equations

Force on a current in a magnetic field:
$$\mathbf{F} = I\mathbf{l} \times \mathbf{B}$$

Force on an infinitesimal length of current in a magnetic field:

$$d\mathbf{F} = I\,d\mathbf{l} \times \mathbf{B}$$

Magnitude of the force on a current in a magnetic field:

$$F = IlB \sin \theta$$

Force on a charged particle in a magnetic field:
$$\mathbf{F} = q\mathbf{v} \times \mathbf{B}$$

Magnitude of force on a charged particle in a magnetic field:

$$F = qvB \sin \theta$$

Cyclotron period:
$$T = \frac{2\pi m}{qB}$$

Cyclotron frequency:
$$f = \frac{qB}{2\pi m}$$

Magnitude of torque on coil:
$$\tau = NIAB \sin \theta$$

Torque on a coil:
$$\boldsymbol{\tau} = NI\mathbf{A} \times \mathbf{B} = \boldsymbol{\mu} \times \mathbf{B}$$

Magnetic dipole moment of a coil:
$$\mu = NIA$$

Potential energy of magnetic dipole in a magnetic field:
$$U = -\boldsymbol{\mu} \cdot \mathbf{B}$$

Hall emf: $\qquad\qquad\qquad\qquad\qquad\qquad\qquad \varepsilon_H = v_d Bl$

Determination of mass of particle from mass spectrometer parameters:

$$m = \frac{qBr}{v}$$

## Chapter Summary

### Section 27-1 Magnets and Magnetic Fields

A magnet has regions on it called **poles**. These regions come in two different types. One type of region is called a **north pole** and the other type of region is called a **south pole**. The north pole regions of a magnet are attracted to the south pole regions of other magnets. The north pole regions of a magnet are repelled by the north pole regions of other magnets and the south pole regions of one magnet are repelled by the south pole magnets of other magnets. The north pole magnet is named this because the north pole end of a suspended magnet points toward the north geographic pole of the earth. The earth acts like a magnet with its south type magnetic pole pointing toward the north geographic pole, so the north pole end of a suspended magnet is attracted toward the north geographic pole of the earth.

As with electric field for the case of the electric force, it is useful to introduce the concept of **magnetic field** in the case of magnetic forces. We can also draw **magnetic field lines** that are everywhere parallel to the magnetic field in analogy to electric field lines. A major difference between magnetic field lines and electric field lines is that magnetic field lines are continuous closed curves without a beginning or an end. Remember that electric field lines begin on positive charges and end on negative charges.

### Section 27-2 Electric Currents Produce Magnetism

Electric currents create magnetic fields. The direction of the magnetic field created by a current is given by the **right-hand rule**. If you grasp a wire carrying a current with you right hand with your thumb extended in the direction the current flows through the wire, your fingers wrap around the wire in the same direction that the magnetic field lines wrap around the wire.

### Section 27-3 Force on an Electric Current in a Magnetic Field; Definition of B

A magnetic field exerts a force on a current-carrying wire. The force exerted on the wire by the magnetic field is always in a direction that is perpendicular to both the direction of the current flowing through the wire and the direction of the magnetic field. The direction of the force is given by a second **right-hand rule**. We can write the relationship between the force on the wire, **F**, the current $I$, and the uniform magnetic field, **B**, in terms of the vector cross product:

$$\mathbf{F} = I\mathbf{l} \times \mathbf{B}$$

where **l** is a vector with a magnitude equal to the length of the wire and pointing in the direction of the current flow. If the magnetic field is not uniform, we can look at the force on a small length $dl$ of the current:

$$d\mathbf{F} = I\,d\mathbf{l} \times \mathbf{B}$$

where $d\mathbf{F}$ is the infinitesimal force on the segment of current of length $dl$ for which the magnetic field at the location of $dl$ is **B**. The vector $d\mathbf{l}$ has a magnitude equal to $dl$ and points in the direction of the current.

We can write an equation for the magnitude of the magnetic force on a wire, $F$, the current, $I$, the length of the wire, $l$, and the magnitude of the magnetic field, $B$:

$$F = IlB \sin\theta$$

where $\theta$ is the angle between the direction of the current and the direction of the magnetic field.

The dimensions of magnetic field are force/(current × length). The SI units of magnetic field are tesla (T). One tesla is equal to 1 N/A·m. In cgs units, the unit of magnetic field is called a gauss (G). One tesla is ten thousand gauss.

**Example 27-3-A**

The circuit shown to the right carries a current of 5.22 A that flows clockwise around the circuit. A magnetic field points downward into the page with a magnitude of 3.55 T. Each longer segment is 1.000 m in length and each shorter segment is 0.500 m in length. What is the magnetic force on each segment of the circuit and what is the total magnetic force on the circuit?

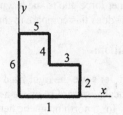

**Solution:**

We have numbered the segments 1 through 6 in the diagram. Starting with segment 1, we calculate the force on each segment. For segment 1, the vector $l = 1.000$ m $(-\mathbf{i})$ because the current will flow in the $-\mathbf{i}$ direction in segment 1 if it flows clockwise around the circuit. The force on segment 1 is

$$\mathbf{F}_1 = I l_1 \times \mathbf{B} = (5.22 \text{ A})(-1.000\,\mathbf{i}) \times (-3.55 \text{ T}\,\mathbf{k}) = -18.5 \text{ N}\,\mathbf{j}$$

Similarly, we calculate the forces on each of the other segments:

$$\mathbf{F}_2 = I l_2 \times \mathbf{B} = (5.22 \text{ A})(1.000\,\mathbf{j}) \times (-3.55 \text{ T}\,\mathbf{k}) = -18.5 \text{ N}\,\mathbf{i}$$
$$\mathbf{F}_3 = I l_3 \times \mathbf{B} = (5.22 \text{ A})(0.500\,\text{m}\,\mathbf{i}) \times (-3.55 \text{ T}\,\mathbf{k}) = 9.27 \text{ N}\,\mathbf{j}$$
$$\mathbf{F}_4 = I l_4 \times \mathbf{B} = (5.22 \text{ A})(-0.500\,\text{m}\,\mathbf{j}) \times (-3.55 \text{ T}\,\mathbf{k}) = 9.27 \text{ N}\,\mathbf{i}$$
$$\mathbf{F}_5 = I l_5 \times \mathbf{B} = (5.22 \text{ A})(0.500\,\text{m}\,\mathbf{i}) \times (-3.55 \text{ T}\,\mathbf{k}) = 9.27 \text{ N}\,\mathbf{j}$$
$$\mathbf{F}_6 = I l_6 \times \mathbf{B} = (5.22 \text{ A})(-0.500\,\text{m}\,\mathbf{j}) \times (-3.55 \text{ T}\,\mathbf{k}) = 9.27 \text{ N}\,\mathbf{i}$$

If we add all these together, we get a total force of zero (to within the expected number of significant figures).

**Section 27-4 Force on an Electric Charge Moving in a Magnetic Field**

The force, $\mathbf{F}$, on a particle of charge $q$ moving with a velocity $\mathbf{v}$ in a magnetic field $\mathbf{B}$ is given by

$$\mathbf{F} = q\mathbf{v} \times \mathbf{B}$$

Notice the direction of the force is perpendicular to the direction of $\mathbf{v}$ and the direction of $\mathbf{B}$ and is given by the right hand rule by curling your fingers from the direction of $\mathbf{v}$ to the direction of $\mathbf{B}$ and your thumb extends in the direction of the force for a positive charge $q$. The force is in the opposite direction if $q$ is negative. The magnitude of the magnetic force on the charged particle can be written as

$$F = qvB \sin \theta$$

where $\theta$ is the angle between the direction of the velocity and the direction of the magnetic field.

A particle moving in a uniform magnetic field with its velocity perpendicular to the direction of the magnetic field moves in a circular path. The period of the circular motion, $T$, is independent of the speed of the particle and is given by

$$T = \frac{2\pi m}{qB}$$

where $m$ is the mass of the particle, $q$ is the magnitude of the charge of the particle, and $B$ is the magnitude of the magnetic field. The frequency of the circular motion is called the **cyclotron frequency** and is given by

$$f = \frac{1}{T} = \frac{qB}{2\pi m}$$

**Example 27-4-A**

Approximate the magnetic field of the earth as pointing horizontally toward the north with a magnitude of 0.500 T. For a particle moving with a speed of 300 m/s, what would the minimum charge-to-mass ratio of the particle need to be so that

352    Giancoli, *Physics for Scientists & Engineers*: Study Guide

the net force due to the magnetic field and gravity is zero? What direction does the velocity need to be for this to be true? How does this compare to the charge-to-mass ratio of an electron?

**Solution:**

First, let's use the right-hand rule to determine the direction the velocity must be in. We know the magnetic force must be vertically upward to give a net force of zero because the gravitational force is vertically downward. Using the right-hand rule, for a north magnetic field and upward force, the velocity must have a component toward the east if the particle is positively charged, and it must have a component toward the west if it is negatively charged. Taking the vertical components of forces to be positive upward, we have:

$$F_B - F_g = 0 \quad \Rightarrow \quad qvB\sin\theta = mg$$

Solving for the charge-to-mass ratio of the particle, we get

$$\frac{q}{m} = \frac{g}{vB\sin\theta}$$

The minimum charge to mass ratio occurs when $\sin\theta$ is maximum, i.e., 1. Therefore,

$$\left(\frac{q}{m}\right)_{min} = \frac{g}{vB} = \frac{9.81\,\text{m/s}^2}{(300\,\text{m/s})(0.500\,\text{T})} = 0.0654\,\text{C/kg}$$

The charge-to-mass ratio of the electron is

$$\left(\frac{q}{m}\right)_{electron} = \frac{1.602\times10^{-19}\,\text{C}}{9.11\times10^{-31}\,\text{kg}} = 1.76\times10^{11}\,\text{C/kg}$$

This is much higher than our minimum necessary charge-to-mass ratio, so it should be possible to assemble something with the minimum charge-to-mass ratio for the conditions of the problem.

**Example 27-4-B**

What is the magnetic force on a particle with a charge 3.77 μC with a velocity 3.04 m/s $\mathbf{i}$ + 4.76 m/s $\mathbf{j}$ + 3.11 m/s $\mathbf{k}$ moving in a magnetic field 1.23 T $\mathbf{i}$ –2.11 T $\mathbf{k}$?

**Solution:**

We can calculate the magnetic force directly:

$$\mathbf{F} = q\mathbf{v}\times\mathbf{B} = \left(-3.77\times10^{-6}\,\text{C}\right)\left(3.04\,\text{m/s}\,\mathbf{i}+4.76\,\text{m/s}\,\mathbf{j}+3.11\,\text{m/s}\,\mathbf{k}\right)\left(1.23\,\text{T}\,\mathbf{i}-2.11\,\text{T}\,\mathbf{k}\right)$$

$$= \left(-3.77\times10^{-6}\,\text{C}\right)\left[\begin{array}{l}(3.04\,\text{m/s})(1.23\,\text{T})\mathbf{i}\times\mathbf{i}+(4.76\,\text{m/s})(1.23\,\text{T})\mathbf{j}\times\mathbf{i}+(3.11\,\text{m/s})(1.23\,\text{T})\mathbf{k}\times\mathbf{i}\\ +(3.04\,\text{m/s})(-2.11\,\text{T})\mathbf{i}\times\mathbf{k}+(4.76\,\text{m/s})(-2.11\,\text{T})\mathbf{j}\times\mathbf{k}+(3.11\,\text{m/s})(-2.11\,T)\mathbf{k}\times\mathbf{k}\end{array}\right]$$

$$= 3.79\times10^{-5}\,\text{N}\,\mathbf{i}-3.86\times10^{-5}\,\text{N}\,\mathbf{j}+2.21\times10^{-5}\,\text{N}\,\mathbf{k}$$

**Section 27-5 Torque on a Current Loop; Magnetic Dipole Moment**

In a complete electrical circuit, if the current flows with a component in a certain direction in one part of the circuit, it must necessarily flow with that component in the opposite direction in some other part of the circuit. This means that an electrical circuit can experience a torque in a magnetic field. For a planar circuit loop, the magnitude of the torque, $\tau$, on the loop is given by

$$\tau = NIAB\sin\theta$$

where $N$ is the number of turns in the coil, $I$ is the current flowing in the coil, $A$ is the area of the coil, and $B$ is the magnitude of the magnetic field. This can be written in vector form to help clarify the direction of the torque:

$$\tau = NI\mathbf{A} \times \mathbf{B}$$

where the vector **A** is a vector with a magnitude equal to the area of the loops of the coil and a direction given by a right hand rule. The right hand rule to determine the direction of the vector **A** is to place your hand over the coil so that your fingers curl around in the direction that the current flows around the loops of the coil. Then your extended thumb points in the direction of the **A** vector.

We can define the magnetic dipole moment, **μ**, of a coil. The magnitude of the magnetic moment, $\mu$, is

$$\mu = NIA$$

The magnetic dipole moment vector has the same direction as the area vector, **A**.

The torque on a magnetic dipole is easily seen to be

$$\tau = \mu \times \mathbf{B}$$

Work is done when a magnetic dipole is rotated in a magnetic field. We can write a potential energy associated with the torque on the magnetic dipole:

$$U = -\mu \cdot \mathbf{B}$$

**Example 27-5-A**

A current of 3.76 A flows through a circular coil with a diameter of 1.34 cm and 24 turns of wire. The wire lies in the *xy*-plane and the current flows counterclockwise around the coil when viewed as shown. There is a magnetic field of magnitude 1.78 T ( 3 **i** + 4 **k** ). What is the torque on the coil?

**Solution:**

Because the coil lies in the *xy*-plane, the area vector points either in the positive **k** direction or the negative **k** direction. Using the right hand rule to obtain the direction of the area vector, we should be able to see that the area vector points in the positive **k** direction. The magnitude of the area vector, $A$, is the area of the circular coil:

$$A = \pi \frac{D^2}{4} = \pi \frac{(0.0134\,\text{m})^2}{4} = 1.41 \times 10^{-4}\,\text{m}^2$$

So that the vector **A** is

$$\mathbf{A} = 1.41 \times 10^{-4}\,\text{m}^2\,\mathbf{k}$$

The torque on the coil is

$$\tau = NI\mathbf{A} \times \mathbf{B} = (24)(3.76\,\text{A})(1.41 \times 10^{-4}\,\text{m}^2\,\mathbf{k})[1.78\,\text{T}(3\mathbf{i}+4\mathbf{k})]$$
$$= 6.79 \times 10^{-2}\,\text{N}\cdot\text{m}\,\mathbf{j}$$

**Section 27-6 Applications: Galvanometers, Motors, Loudspeakers**

There are several practical applications of the magnetic torque on a current loop. **Galvanometers** are electric meters that balance the magnetic torque on a current loop against the torque of a torsional spring. As the magnetic torque is proportional to the current and the spring torque is proportional to the rotation of the coil, the rotation of the coil is proportional to the current. An indicator attached to the coil rotates through an angle proportional to the current, giving a indication on a scale of the magnitude of the current. An **electric motor** can cause a shaft to rotate by passing a current through a coil of wore that is within the magnetic field. Some means of switching current directions through the coils of a motor must be provided to keep the torque acting in the same direction for any orientation of the coils.

## Section 27-7 Discovery and Properties of the Electron

By measuring the deflection of the path of electrons in a magnetic field, J. J. Thompson was able to measure the ratio of the charge of the electron to the mass of the electron. Robert A. Millikan was able to determine the charge on the electron using his oil-drop experiment.

## Section 27-8 The Hall Effect

When a current-carrying conductor is held statically in a magnetic field so that the current flows with some component of its direction perpendicular to the magnetic field, a voltage difference develops across the conductor due to the deflection of the charge carriers by the magnetic field. A dynamic equilibrium is reached such that the electric field from this potential difference creates an electrostatic force that balances the magnetic force on the charge carriers so that there is no net force on the charge carriers perpendicular to the current. This is called the Hall effect. The potential difference across the wire is called the **Hall emf** or **Hall voltage**, $\mathcal{E}_H$. When the direction of the current is perpendicular to the direction of the magnetic field, the Hall emf is given by

$$\mathcal{E}_H = v_d B l$$

where $v_d$ is the drift velocity of the charge carriers in the conductor and $l$ is the dimension of the wire transverse to the direction of the current.

The Hall effect can be used to measure magnetic fields using a device called a Hall probe. The Hall effect is also used to determine the drift velocity or the density of charged carriers in a conductor.

### Example 27-8-A

What is the Hall emf across a 2.00 mm diameter copper wire that carries a current of 5.78 A perpendicular to a magnetic field of magnitude 6.61 T?

### Solution:

To determine the Hall emf, we need to determine the drift velocity of the charge carriers. To do this we need the current, the cross-sectional area of the wire, the charge of the carriers, and the carrier concentration within the wire. From Chapter 25 we know that electrons are the charge carriers in metal conductors. The cross-sectional area of the wire is

$$A = \pi \frac{D^2}{4} = \pi \frac{\left(2.00 \times 10^{-3} \text{ m}\right)^2}{4} = 3.14 \times 10^{-6} \text{ m}^2$$

To determine the carrier concentration we use the atomic mass of copper atoms, $m$, the density of copper, $\rho$, and the fact that each atom contributes one electron to the carrier concentration. The average atomic mass, $m$, of copper is, from the periodic table, 63.546 u ($= 1.06 \times 10^{-25}$ kg). The density, $\rho$, of copper is $8.9 \times 10^3$ kg/m$^3$. The concentration of charge carriers will be

$$n = \frac{N}{V} = \frac{M}{Vm}(\text{\# of electrons per atom}) = \frac{\rho}{m}(1)$$

$$= \frac{8.9 \times 10^3 \text{ kg/m}^3}{1.06 \times 10^{-23} \text{ kg}}(1) = 8.4 \times 10^{28} \text{ m}^{-3}$$

We can now calculate the drift velocity, $v_d$:

$$v_d = \frac{I}{Ane} = \frac{5.78 \text{ A}}{\left(3.14 \times 10^{-6} \text{ m}^2\right)\left(8.4 \times 10^{28} \text{ m}^{-3}\right)\left(1.60 \times 10^{-19} \text{ C}\right)} = 1.37 \times 10^{-4} \text{ m/s}$$

Using the drift velocity we can determine the Hall emf:

$$\mathcal{E}_H = v_d B l = \left(1.37 \times 10^{-4} \text{ m/s}\right)\left(6.61 \text{ T}\right)\left(2.00 \times 10^{-3} \text{ m}\right) = 1.81 \mu\text{V}$$

## Section 27-9 Mass Spectrometer

A mass spectrometer is a device that measures the mass of atoms, molecules, or other small particles. One type of mass spectrometer can be made using the trajectory of charged particles in a magnetic field. From Example 27-4 in the text, we know that charged particles follow a circular path in the plane perpendicular to the magnetic field with a radius, $r$, given by

$$r = \frac{mB}{qv}$$

where $m$ is the mass of the particle, $B$ is the magnitude of the magnetic field, $q$ is the charge of the particle, and $v$ is the magnitude of the velocity of the particle perpendicular to the magnetic field. We can use this to determine the mass of the particle when the other quantities are known:

$$m = \frac{qBr}{v}$$

### Example 27-9-A

A particular ion has had two electrons removed. The ion is accelerated from rest by a 300 V potential difference. It then enters a magnetic field perpendicular to its velocity with a magnitude 2.24 T in which it travels along a trajectory with a radius of curvature equal to 12.7 cm. What is the mass of the ion?

**Solution:**

We know the charge of the ion, the magnetic field, and the radius of curvature of the trajectory in the magnetic field. We can determine the velocity using the conservation of energy:

$$\Delta K + \Delta U = 0 \quad \Rightarrow \quad \tfrac{1}{2}m\left(v_f^2 - v_i^2\right) + q\Delta V = 0$$

Solving for $v_f$, we get

$$v_f = \sqrt{\frac{-2q\Delta V}{m} + v_i^2} = \sqrt{\frac{-2q\Delta V}{m}}$$

Inserting this velocity into the expression for the mass of the particle we get

$$m = \frac{qBr}{\sqrt{\dfrac{-2q\Delta V}{m}}} \quad \Rightarrow \quad m = -\frac{qB^2 r^2}{2\Delta V}$$

$$m = -\frac{\left(1.60\times10^{-19}\,\text{C}\right)\left(2.24\,\text{T}\right)^2\left(0.127\,\text{m}\right)^2}{2\left(300\,\text{V}\right)} = 2.16\times10^{-23}\,\text{kg}$$

## Practice Quiz

1.  A negative charge is moving into this page and a magnetic field points to the right parallel to this page. Which direction is the magnetic force on the charge?

    a) toward the left
    b) toward the top of the page
    c) toward the bottom of the page
    d) out of the page

2.    A loop is partially inserted into a region with a magnetic field pointing into the page as shown in the diagram. Current flows around the loop in a counter-clockwise direction. What is the direction of the magnetic force on the loop?

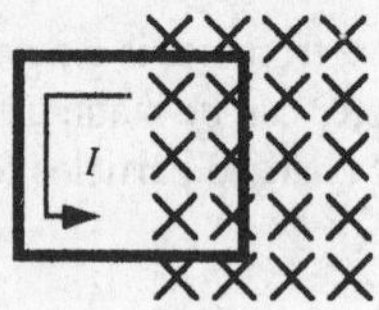

a) toward the top of the page
b) into the page
c) toward the right
d) toward the left

3.    If you know the charge and velocity of a particle and the magnetic force acting on the particle, is this enough information to determine the magnetic field?

a) yes
b) No, you also need to know the mass of the object.
c) Yes, unless the velocity is zero.
d) no, only the component of magnetic field perpendicular to the velocity can be determined.

4.    A particle moves in a uniform magnetic field from point A to C as shown in the picture. The path length from A to C is $L$. How much work is done by the magnetic field shown as the particle moves from A to C?

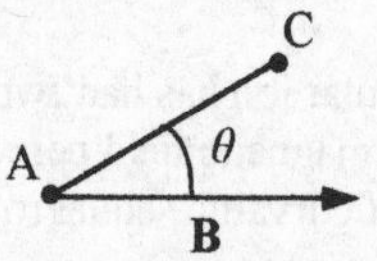

*a)* $qvBL \sin \theta$
*b)* $qvBL \cos \theta$
c) $-qvBL \cos \theta$
d) zero

5.    A loop has a constant current $I$ flowing through it while it is in a uniform magnetic field. What is a description of the most general type of motion the loop will undergo due to the interaction with the magnetic field if it is released from rest?

a) The loop will spin with constant angular speed.
b) The loop will spin with constant angular velocity.
c) The loop will rotate back and forth.
d) The loop will translationally accelerate with constant acceleration.

6.    For a given length of wire, what shape of loop will have the greatest torque for a given current and magnetic field?

a) square
b) rectangle with the long sides twice as long as the short sides
c) rectangle with the long sides the square root of two longer than the short sides
d) circle

7.    A power line runs east-west in a place where there is no magnetic dip. It carries a DC current to the west. What direction is the power line deflected from the direction it would hang with no current flowing through it?

a) north
b) south
c) up
d) down

8.    In the situation shown in the diagram, which direction is the Hall field in the conductor if electrons are the charge carriers of the current?

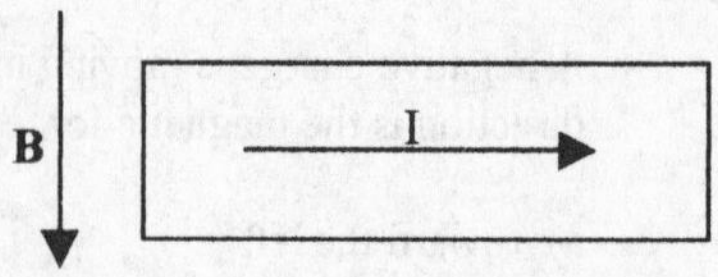

a) toward the top of the page
b) toward the bottom of the page
c) out of the page
d) into the page

9.    Two conductors are made of identical materials.  They are in the same uniform magnetic field and have identical currents passing through them.  The length of the first conductor in the direction perpendicular to the current and the magnetic field is twice the length of the second conductor in the direction perpendicular to the current and the magnetic field.  How do the Hall fields compare in the two conductors?

    a)  The Hall field of the first conductor is twice the Hall field in the second conductor.
    b)  The Hall field of the second conductor is twice the Hall field in the first conductor.
    c)  The Hall field in the first conductor is the same as the Hall field in the second conductor.
    d)  There is insufficient information given to compare the Hall fields.

10.    Two conductors have identical geometry.  They are in the same uniform magnetic field and have identical currents passing through them.  The density of charge carriers in the first conductor is twice the density of charge carriers in the second conductor.  How do the Hall emfs compare in the two conductors?

    a)  The Hall emf of the first conductor is twice the Hall emf in the second conductor.
    b)  The Hall emf of the second conductor is twice the Hall emf in the first conductor.
    c)  The Hall emf in the first conductor is the same as the Hall emf in the second conductor.
    d)  There is insufficient information given to compare the Hall emf's.

11.    A current-carrying wire lies along the $\mathbf{i}$ direction.  What is the force per length of wire on the wire if it carries 3.56 A of current in the $\mathbf{i}$ direction in a magnetic field $\mathbf{B} = 1.65$ T ( $\mathbf{i} + 1.34\,\mathbf{j}$ )?

12.    Determine the acceleration of an electron with a velocity given by $\mathbf{v} = 7.72$ m/s $(3.42\,\mathbf{i} + 4.34\,\mathbf{j}$ ) when moving through a magnetic field $\mathbf{B} = 2.75$ T $(-2.21\,\mathbf{i} + 3.67\,\mathbf{k})$.

13.    An electron moves through an electric field $\mathbf{E} = E_0\,\mathbf{j}$ and a magnetic field $\mathbf{B} = B_0\,\mathbf{k}$.  With what velocity will the electrons move through these fields with no acceleration?

14.    A mass spectrometer is designed with a magnetic field of 1.43 T and particles are injected with a velocity of 380 m/s.  Determine the difference in radii of curvature of the trajectories for singly ionized carbon-12 and carbon-14 atoms.

15.    What is the concentration of charge carriers in a piece of material if a 2 mm wide ribbon wire with a thickness of 0.5 mm that carries a 1.69 A current in the direction perpendicular to a 4.76 T magnetic field causes a 0.761 V Hall emf.

## Problem Solutions

3.    For the maximum force, the wire is perpendicular to the field, so we find the current from
    $F = ILB$;
    0.900 N $= I(4.20$ m$)(0.0800$ T$)$, which gives $I = 2.68$ A.

7.    (a)  We see from the diagram that the magnetic field is up, so the top pole face is a south pole.
    (b)  We find the current from the length of wire in the field:
        $F = ILB$;
        5.30 N $= I(0.10$ m$)(0.15$ T$)$, which gives $I = 3.5 \times 10^2$ A.
    (c)  The new force is
        $F' = ILB \sin \theta = F \sin \theta = (5.30$ N$) \sin 80° = 5.22$ N.
    Note that the wire could be tipped either way.

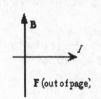

11.    We choose the coordinate system shown in the diagram.
We select a differential element of the curved wire,
    $d\mathbf{l} = dx\,\mathbf{i} + dy\,\mathbf{j}$,
on which the force is
    $d\mathbf{F} = I\,d\mathbf{l} \times \mathbf{B} = I(dx\,\mathbf{i} + dy\,\mathbf{j}) \times (-B\mathbf{k}) = IB(dx\,\mathbf{j} - dy\,\mathbf{i})$.
We find the resultant force by integration:
    $\mathbf{F} = \int d\mathbf{F} = IB \int (dx\,\mathbf{j} - dy\,\mathbf{i}) = IB(\Delta x\,\mathbf{j} - \Delta y\,\mathbf{i})$,
where $\Delta x = x_b - x_a$, and $\Delta y = y_b - y_a$.
If we have the same current in the straight wire, the

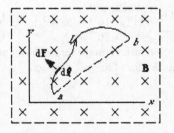

resultant force is
$$\mathbf{F} = I(\Delta x\ \mathbf{i} + \Delta y\ \mathbf{j})\ \times\ (-B\mathbf{k}) = IB(\Delta x\ \mathbf{j} - \Delta y\ \mathbf{i}),\ \text{which is the same result.}$$

15.  We assume that we want the direction of **B** that produces the maximum force, i. e., perpendicular to **v**.  Because the charge is positive, we point our thumb in the direction of **F** and our fingers in the direction of **v**.  To find the direction of **B**, we note which way we should curl our fingers, which will be the direction of the magnetic field **B**.
  (*a*)  Thumb out, fingers left, curl down.
  (*b*)  Thumb up, fingers right, curl in.
  (*c*)  Thumb down, fingers in, curl right.

19.  (*a*)  The magnetic force provides the centripetal acceleration:
$$qvB = mv^2/r,\quad \text{or} \quad mv = qBr.$$
  The kinetic energy of the electron is
$$K = \tfrac{1}{2}mv^2 = \tfrac{1}{2}(qBr)^2/m = (q^2B^2/2m)r^2.$$
  (*b*)  The magnetic force provides the centripetal acceleration:
$$qvB = mv^2/r,\quad \text{or} \quad mv = p = qBr.$$
  The angular momentum is
$$L = mvr = qBr^2.$$

23.  (*a*)  We find the speed acquired from the accelerating voltage from energy conservation:
$$0 = \Delta K + \Delta U;$$
$$0 = \tfrac{1}{2}mv^2 - 0 + q(-V),\ \text{which gives}$$
$$v = (2qV/m)^{1/2} = [2(2)(1.60 \times 10^{-19}\ \text{C})(2100\ \text{V})/(6.6 \times 10^{-27}\ \text{kg})]^{1/2} = 4.51 \times 10^5\ \text{m/s.}$$
  For the radius of the path, we have
$$\begin{aligned} r &= mv/qB = (6.6 \times 10^{-27}\ \text{kg})(4.51 \times 10^5\ \text{m/s})/(2)(1.60 \times 10^{-19}\ \text{C})(0.340\ \text{T}) \\ &= 2.7 \times 10^{-2}\ \text{m} = 2.7\ \text{cm.} \end{aligned}$$
  (*b*)  The period of revolution is
$$T = 2\pi r/v = 2\pi m/qB = 2\pi(6.6 \times 10^{-27}\ \text{kg})/(2)(1.60 \times 10^{-19}\ \text{C})(0.340\ \text{T}) = 3.8 \times 10^{-7}\ \text{s.}$$

27.  The total force on the proton is
$$\begin{aligned} \mathbf{F} &= e(\mathbf{E} + \mathbf{v}\ \times\ \mathbf{B}) \\ &= e\{(3.0\mathbf{i} - 4.2\ \mathbf{j})\ \times\ 10^3\ \text{V/m} + [(6.0\mathbf{i} + 3.0\mathbf{j}\ - 5.0\mathbf{k})\ \times\ 10^3\ \text{m/s}\ \times\ (0.45\mathbf{i} + 0.20\ \mathbf{j})\text{T}]\} \\ &= (1.60 \times 10^{-19}\ \text{C})[(3.0\mathbf{i} - 4.2\ \mathbf{j}) + (1.0\mathbf{i} - 2.25\mathbf{j}\ - 0.15\mathbf{k})]\ \times\ 10^3\ \text{V/m} \\ &= (6.4\mathbf{i} - 10.3\mathbf{j}\ - 0.24\mathbf{k})\ \times\ 10^{-16}\ \text{N.} \end{aligned}$$

31.  (*a*)  Because the velocity is perpendicular to the magnetic field, the proton will travel in a circular arc.  From the symmetry of the motion we see that the upper half is a mirror image of the lower half, so the exit angle is the same as the incident angle: 45°.

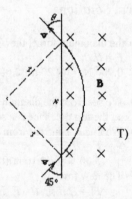

  (*b*)  The magnetic force provides the radial acceleration, so we have
$$\begin{aligned} F &= evB = mv^2/r,\ \text{so} \\ r &= mv/eB = (1.67 \times 10^{-27}\ \text{kg})(2.0 \times 10^5\ \text{m/s})/(1.60 \times 10^{-19}\ \text{C})(0.850\ \text{T}) \\ &= 2.46 \times 10^{-3}\ \text{m.} \end{aligned}$$
  Thus the distance *x* is
$$x = r\sqrt{2} = 3.5 \times 10^{-3}\ \text{m.}$$

35.  (*a*)  The angle between the normal to the coil and the field is 24.0°, so the torque is
$$\begin{aligned} \tau &= NIAB \sin \theta \\ &= (12)(7.10\ \text{A})\pi(0.0850\ \text{m})^2(5.50 \times 10^{-5}\ \text{T}) \sin 24.0° \\ &= 4.33 \times 10^{-5}\ \text{m} \cdot \text{N.} \end{aligned}$$
  (*b*)  From the directions of the forces shown on the diagram, the north edge of the coil will rise.

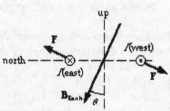

39. When the coil comes to rest, the magnetic torque is balanced by the restoring torque:
$$NIAB = k\phi.$$
Because the full-scale deflection is the same, we have
$$I_1/k_1 = I_2/k_2 \ ;$$
$$(36 \ \mu A)/k_1 = I_2/(0.80 k_1), \text{ which gives } I_2 = 29 \ \mu A.$$

43. (a) The Hall emf is across the width of the sample, so the Hall field is
$$E_H = å_H/w = (6.5 \times 10^{-6} \ V)/(0.030 \ m) = 2.2 \times 10^{-4} \ V/m.$$
   (b) The forces from the electric field and the magnetic field balance. We find the drift speed from
$$E_H = v_d B;$$
$$2.2 \times 10^{-4} \ V/m = v_d(0.80 \ T), \text{ which gives } v_d = 2.7 \times 10^{-4} \ m/s.$$
   (c) We find the density from
$$I = neAv_d \ ;$$
$$42 \ A = n(1.60 \times 10^{-19} \ C)(0.030 \ m)(500 \times 10^{-6} \ m)(2.7 \times 10^{-4} \ m/s),$$
   which gives $n = 6.4 \times 10^{28}$ electrons/m$^3$.

47. We find the velocity of the velocity selector from
$$v = E/B = (2.48 \times 10^4 \ V/m)/(0.58 \ T) = 4.28 \times 10^4 \ m/s.$$
For the radius of the path, we have
$$r = mv/qB' = [(4.28 \times 10^4 \ m/s)/(1.60 \times 10^{-19} \ C)(0.58 \ T)]m = (4.61 \times 10^{23} \ m/kg)m.$$
If we let $A$ represent the mass number, we can write this as
$$r = (4.61 \times 10^{23} \ m/kg)(1.66 \times 10^{-27} \ kg)A = (7.65 \times 10^{-4} \ m)A = (0.765 \ mm)A.$$
The separation of the lines is the difference in the diameter, or
$$\Delta D = 2 \ \Delta r = 2(0.765 \ mm) \ \Delta A = (1.53 \ mm)(1) = 1.53 \ mm.$$
If the ions were doubly charged, all radii would be reduced by one-half, so the separation would be 0.76 mm.

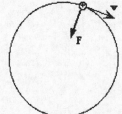

51. The magnetic force must be toward the center of the circular path,
so the magnetic field must be up.
The magnetic force provides the centripetal acceleration:
$$qvB = mv^2/r, \quad \text{or} \quad mv = qBr;$$
$$4.8 \times 10^{-16} \ kg \cdot m/s = (1.60 \times 10^{-19} \ C)B(1.0 \times 10^3 \ m),$$
which gives $B = 3.0$ T up.

55. Even though the Earth's field dips, the current and the field are perpendicular. The direction of the force will be perpendicular to both the cable and the Earth's field, so it will be 68° above the horizontal toward the north. For the magnitude, we have
$$F = ILB$$
$$= (330 \ A)(10 \ m)(5.0 \times 10^{-5} \ T) = 0.17 \ N, \ 68° \text{ above the horizontal toward the north.}$$

59. (a) The radius of the circular orbit is
$$r = mv/qB.$$
   The time to complete a circle is
$$T = 2\pi r/v = 2\pi m/qB, \text{ so the frequency is}$$
$$f = 1/T = qB/2\pi m.$$
   Note that this is independent of $r$.
   Because we want the AC voltage to be maximum when the proton reaches the gap and minimum (reversed) when the proton has made half a circle, the frequency of the AC voltage must be the same: $f = 1/T = qB/2\pi m.$
   (b) In a full circle, the proton crosses the gap twice. If the gap is small, the AC voltage will not change significantly from its maximum magnitude while the proton is in the gap.
   The energy gain from the two crossings is
$$\Delta K = 2qV_0 .$$
   (c) From $r = mv/qB$, we see that the maximum speed, and thus the maximum kinetic energy, occurs at the maximum radius of the path. The maximum kinetic energy is
$$K_{max} = \tfrac{1}{2}mv_{max}^2 = \tfrac{1}{2}m(qBr_{max}/m)^2 = (qBr_{max})^2/2m$$
$$= [(1.60 \times 10^{-19} \ C)(0.50 \ T)(2.0 \ m)]^2/2(1.67 \times 10^{-27} \ kg)$$
$$= 7.66 \times 10^{-12} \ J = (7.66 \times 10^{-12} \ J)/(1.60 \times 10^{-13} \ J/MeV) = 48 \ MeV.$$

63. We find the required acceleration from
    $v^2 = v_0^2 + 2ax$;
    $(30 \text{ m/s})^2 = 0 + 2a(1.0 \text{ m})$, which gives $a = 450 \text{ m/s}^2$.
    This acceleration is provided by the force from the magnetic field:
    $F = ILB = ma$;
    $I(0.20 \text{ m})(1.7 \text{ T}) = (1.5 \times 10^{-3} \text{ kg})(450 \text{ m/s}^2)$, which gives $I = 2.0 \text{ A}$.
    The force is away from the battery, so fingers in the direction of $I$ would have to curl down; thus the field points down.

67. The magnetic force provides the centripetal acceleration of the circular motion:
    $evB = mv^2/r$,   or   $mv = eBr$.
    The kinetic energy of the proton is
    $K = \frac{1}{2}mv^2 = \frac{1}{2}(eBr)^2/m = (e^2B^2/2m)r^2$.
    Thus the change in the kinetic energy is
    $\Delta K = (e^2B^2/2m)(r_Q^2 - r_P^2)$
    $= [(1.60 \times 10^{-19} \text{ C})^2(0.010 \text{ T})^2/2(1.67 \times 10^{-27} \text{ kg})][(8.5 \times 10^{-3} \text{ m})^2 - (10.0 \times 10^{-3} \text{ m})^2]$
    $= -2.1 \times 10^{-20} \text{ J}$.

# Chapter 28: Sources of Magnetic Field

## Chapter Overview and Objectives

This chapter introduces the different sources of magnetic fields and how to calculate the fields for simple arrangements of currents. It also introduces the magnetic properties of materials.

After completing study of this chapter, you should:

- Know how to calculate the magnetic field in the vicinity of a long straight wire.
- Know how to calculate the force between two long straight current-carrying wires.
- Know Ampere's law and how to use it to calculate magnetic fields for symmetrical current distributions.
- Know how to calculate the magnetic field inside a solenoid.
- Know how to calculate the magnetic field inside a toroidal coil.
- Know the Biot-Savart law and how to calculate magnetic fields using the Biot-Savart law.
- Know what ferromagnetic materials and their properties are.
- Know what paramagnetic and diamagnetic materials are.

## Summary of Equations

Magnitude of magnetic field around a straight wire:
$$B = \frac{\mu_0}{2\pi} \frac{I}{r}$$

Magnetic field around a straight wire:
$$\mathbf{B} = \frac{\mu_0}{2\pi} \frac{\mathbf{I} \times \hat{\mathbf{r}}}{r}$$

Force per length between two parallel current-carrying wires:
$$\frac{F}{l} = \frac{\mu_0}{2\pi} \frac{I_1 I_2}{d}$$

Ampere's law:
$$\oint \mathbf{B} \cdot d\mathbf{l} = \mu_0 I_{encl}$$

Magnetic field in interior of a solenoid:
$$B = \mu_0 n I$$

Magnetic field inside of toroidal coil:
$$B = \frac{\mu_0 N I}{2\pi r}$$

Biot-Savart law:
$$d\mathbf{B} = \frac{\mu_0 I}{4\pi} \frac{d\mathbf{l} \times \hat{\mathbf{r}}}{r^2}$$

$$\mathbf{B} = \int \frac{\mu_0}{4\pi} \frac{I\, d\mathbf{l} \times \hat{\mathbf{r}}}{r^2}$$

## Chapter Summary

### Section 28-1 Magnetic Field Due to a Straight Wire

The magnetic field created by a long straight wire carrying a current $I$ has a magnitude

$$B = \frac{\mu_0}{2\pi} \frac{I}{r}$$

where $r$ is the distance from the axis of the wire to the point the magnetic field is being evaluated at and $\mu_0$ has the value $4\pi \times 10^{-7}$ T·m/A and is called the **permeability of free space**. The direction of the magnetic field is always perpendicular to the long axis of the wire and the displacement vector from the wire to the point at which the magnetic field is being evaluated. This is along a tangent to the circle concentric with the wire, as shown in the diagram. A right-hand rule specifies the direction the magnetic field points along the circle's tangent. Grasp the wire with your right hand with your thumb pointing in the direction of the current. The magnetic field points in the same direction around the circle as your fingers point around the wire. We can write the magnetic field around the wire in a vector form as

$$\mathbf{B} = \frac{\mu_0}{2\pi} \frac{\mathbf{I} \times \hat{\mathbf{r}}}{r}$$

where $\mathbf{I}$ is a vector with its magnitude equal to the current and its direction in the direction of the current. The vector $\hat{\mathbf{r}}$ is a unit vector in the direction pointing from the wire to the point the magnetic field is being evaluated.

### Example 28-1-A

Two long wires run parallel to each other and are separated by a distance 1.5 cm. One wire carries a current of 3.9 A in one direction and the other wire carries a current of 5.6 A in the opposite direction. What is the magnetic field at the point indicated in the diagram, which is located a distance 1.5 cm above the plane of the two wires and directly above the wire carrying the 3.9 A current. The diagram is an end view of the wires.

### Solution:

Let's set up a coordinate system with $\mathbf{i}$ horizontal to the right, $\mathbf{j}$ toward the top of the page, and $\mathbf{k}$ out of the page with the origin located at the 5.6 A wire in the plane of the diagram. Let the vector $\mathbf{r}_{5.6}$ be the displacement vector from the wire carrying 5.6 A current to the point P. It should be easy to see that this vector is given by

$$\mathbf{r}_{5.6} = (0.015\,\text{m})\mathbf{i} + (0.015\,\text{m})\mathbf{j}$$
$$r_{5.6} = \sqrt{(0.015\,\text{m})^2 + (0.015\,\text{m})^2} = 0.0212\,\text{m}$$
$$\hat{\mathbf{r}}_{5.6} = \frac{\mathbf{r}_{5.6}}{r} = \frac{(0.015\,\text{m})\mathbf{i} + (0.015\,\text{m})\mathbf{j}}{0.0212\,\text{m}} = (0.707)\mathbf{i} + (0.707)\mathbf{j}$$

Let the vector $\mathbf{r}_{3.9}$ be the displacement vector from the wire carrying the 3.9 A current to the point P. It should be easy to see that this vector is given by

$$\mathbf{r}_{3.9} = (0.015\,\text{m})\mathbf{j}$$
$$r_{3.9} = 0.015\,\text{m}$$
$$\hat{\mathbf{r}}_{3.9} = \mathbf{j}$$

Adding the magnetic fields from the two wires together gives:

$$\mathbf{B} = \frac{\mu_0}{2\pi} \frac{\mathbf{I}_{5.6} \times \hat{\mathbf{r}}_{5.6}}{r_{5.6}} + \frac{\mu_0}{2\pi} \frac{\mathbf{I}_{3.9} \times \hat{\mathbf{r}}_{3.9}}{r_{3.9}}$$
$$= \frac{\mu_0}{2\pi} \left[ \frac{(5.6\,\text{A})\mathbf{k} \times (0.707\,\mathbf{i} + 0.707\,\mathbf{j})}{0.0212\,\text{m}} + \frac{(-3.9\,\text{A})\mathbf{k} \times \mathbf{j}}{0.015\,\text{m}} \right]$$
$$= \frac{4\pi \times 10^{-7}\,\text{T} \cdot \text{m/A}}{2\pi} \left[ (187\,\text{A/m}\,\mathbf{j} - 187\,\text{A/m}\,\mathbf{i}) + 260\,\text{A/m}\,\mathbf{i} \right]$$
$$= 1.46 \times 10^{-5}\,\text{T}\,\mathbf{i} + 3.74 \times 10^{-5}\,\text{T}\,\mathbf{j}$$

**Section 28-2 Force Between Two Parallel Wires**

The magnitude of the magnetic force per length of wire between two long straight parallel current-carrying wires is given by

$$\frac{F}{l} = \frac{\mu_0}{2\pi} \frac{I_1 I_2}{d}$$

where $I_1$ is the current in one wire and $I_2$ is the current in the other wire and $d$ is the distance between the two wires. This assumes that the wires are circular in cross-section, have a uniform current density, and the radius of the cross-section of the wire is small compared to the distance between the wires.

If the current is in the same direction in the two wires, the force is attractive. If the current in one wire is opposite in direction to the current in the other wire, the force is repulsive.

**Example 28-2-A**

Determine the net force per length on the wire with the 3.9 A current in Example 28-1-A.

**Solution:**

The currents in the two wires are 3.9 A and 5.6 A and the separation of the wires is 0.015 m. We can directly calculate the magnitude of the force per length:

$$\frac{F}{l} = \frac{\mu_0}{2\pi} \frac{I_1 I_2}{d} = \frac{4\pi \times 10^{-7} \text{ T} \cdot \text{m/A}}{2\pi} \frac{(3.9 \text{ A})(5.6 \text{ A})}{(0.015 \text{ m})} = 2.9 \times 10^{-4} \text{ N}$$

**Section 28-3 Operational Definitions of the Ampere and the Coulomb**

The modern definition of an ampere is in terms of the magnitude of the magnetic force between two parallel current-carrying wires. The definition of the **ampere** is the current that must flow through each of two parallel wires one meter apart that the magnitude of the magnetic force per length of the wires acting between the wires is $2\pi \times 10^{-7}$ N/m. The modern definition of the **coulomb** is that one coulomb is one ampere·second.

**Section 28-4 Ampère's Law**

Consider any smooth closed path in space. We can move around the loop with infinitesimal displacements $dl$. The $dl$ is a vector quantity of infinitesimal length $dl$ and points in the direction tangent to the curve in the sense of the direction we move around the path. **Ampere's law** states that the integral around the path of the scalar product of the magnetic field with $dl$ is equal to $\mu_0$ times the current that passes through the loop. We write this as

$$\oint \mathbf{B} \cdot dl = \mu_0 I_{encl}$$

Ampere's law, like Gauss's law for electric fields in Chapter 22, can be used to determine the magnetic field, but only for a few very symmetric current distributions.

**Example 28-4-A**

A long straight wire has a circular cross-section of radius $R$. The current density (current per area), $j$, is a function of distance from the circular cross-section, $r$, and is equal to

$$j(r) = j_0 r$$

Determine the total current flowing through the wire and the magnetic field within the wire.

**Solution:**

To determine the total current in the wire we integrate the current density over the cross-section of the wire:

$$I = \int j\, dA$$

Because the current density only depends on $r$, we can take our differential area elements as circular rings of thickness $dr$. A ring of radius $r$ and thickness $dr$ has an area $dA = 2\pi\, dr$. The integral becomes

$$I = \int_0^R j_0 r\, 2\pi r\, dr = 2\pi j_0 \int_0^R r^2\, dr = \frac{2\pi j_0 r^3}{3}$$

To find the magnetic field with the wire, we apply Ampere's law. We know from the symmetry of the current that the magnitude of the magnetic field is constant at a given value of $r$ and it points tangent to the circle of radius $r$. The line integral of the magnetic field easily evaluates to

$$\oint \mathbf{B}(r)\cdot d\mathbf{l} = 2\pi B(r)$$

To use Ampere's law we must also know the current through a circle of radius $r$. This is the same integral for which we evaluated the total current in the wire, but with an upper limit $r$ rather than $R$:

$$I_{insider} = 2\pi j_0 \int_0^r r'^2\, dr' = \frac{2\pi j_0 r}{3}$$

Using Ampere's law, we get

$$\oint \mathbf{B}\cdot dl = \mu_0 I_{encl} \quad \Rightarrow \quad 2\pi B(r) = \frac{\mu_0\, 2\pi\, j_0 r}{3}$$

$$B(r) = \frac{\mu_0\, j_0 r}{3}$$

### Section 28-5 Magnetic Field of a Solenoid and a Toroid

A **solenoid** is a relatively long coil with many loops of wire, usually uniformly spaced down the length of the solenoid and wrapped around the cylindrical axis of the coil. The magnetic field inside the solenoid, far from the ends, is uniform and has a magnitude

$$B = \mu_0 n I$$

where $n$ is the number of turns or loops per length of the solenoid and $I$ is the current flowing through the wire of the solenoid. The direction of the magnetic field inside the solenoid is parallel to the cylindrical axis of the solenoid and toward which end its direction is can be determined from the right hand rule.

A toroidal coil is shaped like a solenoid bent around in a circle so that the two ends of the solenoid are located next to each other, making a donut shape. The magnetic field in a tightly and uniformly round toroidal coil will be zero outside the toroid. Inside the toroid, the magnetic field will be of magnitude

$$B = \frac{\mu_0 N I}{2\pi r}$$

where $N$ is the number of turns of wire in the toroidal coil, $I$ is the current in the coil, and $r$ is the distance to the point the field is being evaluated from the center of toroid (the center of the donut hole).

### Example 28-5-A

A solenoid is needed that creates a magnetic field of magnitude $B$. You have a length of wire $L$ that can carry a maximum of current $I$ without its temperature rising too greatly. What is the radius of a solenoid that will use this length of wire and have a volume $V$?

**Solution:**

The volume of the solenoid will be the volume, $V$, of a cylinder of radius $R$ and length $S$:

$$V = \pi R^2 S$$

The magnetic field of the solenoid is given by

$$B = \mu_0 n I$$

The number of turns per length, $n$, will be the number of turns, $N$, divided by the length of the coil, $S$:

$$n = \frac{N}{S}$$

The number of turns on the coil will be the length of the wire divided by the circumference of the coil:

$$N = \frac{L}{2\pi R}$$

If we write the magnetic field in terms of $L$, $S$, and $R$, we have:

$$B = \mu_0 \frac{LI}{2\pi RS}$$

We can use this to write the relationship between the radius of the solenoid and the length of the solenoid:

$$S = \mu_0 \frac{LI}{2\pi BR}$$

Substituting this into the expression for the volume, we have

$$V = \pi R^2 \mu_0 \frac{LI}{2\pi BR} = \mu_0 \frac{LI}{2\pi B} R$$

If we solve this for the radius of the solenoid, we get

$$R = \frac{2\pi B}{LI\mu_0} V$$

Of course, this might not be a solenoid. Certainly if $V$ were large enough, the so-called solenoid would look more like a flat coil (or might not even have one complete loop). We would need to verify if the shape and turns per length satisfied the conditions of being a solenoid.

**Section 28-6 Biot-Savart Law**

The **Biot-Savart law** relates the distribution of currents in space to the magnetic field at any given point. A infinitesimal length element, $dl$, of current, $I$, contributes an infinitesimal component, $d\mathbf{B}$, to the magnetic field at a given point

$$d\mathbf{B} = \frac{\mu_0 I}{4\pi} \frac{dl \times \hat{\mathbf{r}}}{r^2}$$

To determine the magnetic field at a point in space, the contributions from each infinitesimal length of current must be added together:

$$\mathbf{B} = \int \frac{\mu_0}{4\pi} \frac{I\, d\mathbf{l} \times \hat{\mathbf{r}}}{r^2}$$

The magnitude of the magnetic field of a circular current loop with radius $R$ and current $I$ along the line through the center of the loop perpendicular to the plane of the loop is

$$B = \frac{\mu_0 IR^2}{2\left(R^2 + x^2\right)^{\frac{3}{2}}} = \frac{\mu_0 \mu}{2\pi\left(R^2 + x^2\right)^{\frac{3}{2}}}$$

where $\mu$ is the magnetic moment of the current loop and $x$ is the distance from the plane of the current loop to the point at which the magnetic field is evaluated. If the distance from the plane of the loop to the point at which the field is evaluated is much greater than the radius of the loop, $x \gg R$, then the magnitude of the magnetic field can be approximated by

$$B = \frac{\mu_0 \mu}{2\pi\, x^3}$$

### Section 28-7 Magnetic Materials–Ferromagnetism

Many atoms have magnetic dipole moments because of the motion of the electrons within the atoms and because the electrons themselves have magnetic moments. In most materials, these magnetic dipole moments lie in random directions and the net macroscopic field is negligible when the sum of the fields of all magnetic dipoles is found. However, in a certain class of materials called **ferromagnetic** materials, it is energetically favorable for the magnetic dipole moments of some of the electrons to align with each other. In this case the macroscopic magnetic field that results from the sum of the fields of the individual electron magnetic dipole moments can be relatively large. These are the materials that are attracted to magnets and some can make permanent magnets.

### Section 28-8 Electromagnets and Solenoids

A long cylindrical coil of wire of many loops of wire is called a **solenoid**:

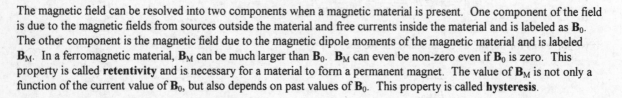

### Section 28-9 Magnetic Fields in Magnetic Materials; Hysteresis

The magnetic field can be resolved into two components when a magnetic material is present. One component of the field is due to the magnetic fields from sources outside the material and free currents inside the material and is labeled as $\mathbf{B}_0$. The other component is the magnetic field due to the magnetic dipole moments of the magnetic material and is labeled $\mathbf{B}_M$. In a ferromagnetic material, $\mathbf{B}_M$ can be much larger than $\mathbf{B}_0$. $\mathbf{B}_M$ can even be non-zero even if $\mathbf{B}_0$ is zero. This property is called **retentivity** and is necessary for a material to form a permanent magnet. The value of $\mathbf{B}_M$ is not only a function of the current value of $\mathbf{B}_0$, but also depends on past values of $\mathbf{B}_0$. This property is called **hysteresis**.

### Section 28-10 Paramagnetism and Diamagnetism

In materials that are non-ferromagnetic, the $\mathbf{B}_M$ component of the magnetic field is small compared to $\mathbf{B}_0$, except for a special class of materials called superconductors. The total magnetic field is the sum of the two component fields. For these types of materials, it is a good approximation to write the total field as a constant $\mu$ multiplied by $\mathbf{B}_0$:

$$\mathbf{B} = \mu\, \mathbf{B}_0 = K\mu_0\, \mathbf{B}_0$$

where $\mu$ is called the **magnetic permeability** of the material and $K_m$ is called the **relative magnetic permeability** of the material. This is also written in terms of the **magnetic susceptibility**, $\chi_m$, of the material. The magnetic susceptibility is defined as

$$\chi_m = K_m - 1$$

Paramagnetic substances have relative magnetic permeabilities greater than one and magnetic susceptibilities less than one. Diamagnetic materials have relative magnetic permeabilities less than one and magnetic susceptibilities greater than one.

In a paramagnetic material, the magnetic field within the material is slightly greater in magnitude than the applied magnetic field and in a diamagnetic field, the magnetic field is slightly less in magnitude than the applied magnetic field.

## Practice Quiz

1.  The magnetic force per length between two non-parallel wires separated by a distance $d$ where they are closest and out of parallel by an angle $\theta < 90°$ will be

    a)  equal to $\dfrac{\mu_0}{2\pi} \dfrac{I_1 I_2}{d}$ .

    b)  equal to $\dfrac{\mu_0}{2\pi} \dfrac{I_1 I_2}{d} \sin\theta$ .

    c)  equal to $\dfrac{\mu_0}{2\pi} \dfrac{I_1 I_2}{d} \cos\theta$ .

    d)  less than $\dfrac{\mu_0}{2\pi} \dfrac{I_1 I_2}{d} \cos\theta$ .

2.  You are located north of a wire carrying an upward vertical current. What direction is the magnetic field at your location due to the current?

    a)  north
    b)  south
    c)  east
    d)  west

3.  You walk around a vertical metal conduit carrying a magnetic compass. As you walk completely around the conduit in a clockwise direction, the compass always points in the direction you are moving. What can you conclude?

    a)  There is a permanent magnet inside the conduit.
    b)  The conduit has a net electrical charge at its surface.
    c)  The compass must be broken.
    d)  There is a current flowing through the conduit.

4.  To double the magnetic field inside a solenoid,

    a)  double the length of the solenoid.
    b)  double the diameter of the solenoid.
    c)  double the square of the radius of the solenoid.
    d)  double the current in the solenoid.

5.  The magnitude of the magnetic field a distance $d$ from a current-carrying wire is $B$. What is the magnitude of the magnetic field a distance $2d$ from the wire?

    a)  $2B$
    b)  $B/2$
    c)  $4B$
    d)  $B/4$

6.  A segment of wire length $L$ carries a current $I$ in the direction shown in the diagram. The point P is on the same line as the segment of wire and a distance $L$ from the wire. What is the magnitude of the magnetic field at the point P due to the current in the wire segment?

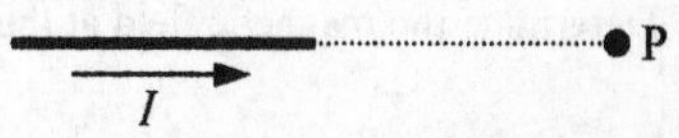

    a)  $\mu_0 I/L$
    b)  $\mu_0 I/2L$
    c)  $\mu_0 I/3L$
    d)  0

7.    A type of superconducting material always has zero magnetic field within it when it is in a superconducting state. What is the magnetic susceptibility of such a material?

   a) 0
   b) 1
   c) −1
   d) infinite

8.    If you look into the north pole end of a solenoid, which way does the current travel around the solenoid?

   a) clockwise
   b) counterclockwise
   c) east
   d) west

9.    A permanent magnet is brought near an unknown type of material.  No matter which pole of the permanent magnet is brought near the material, the material is repelled by the permanent magnet.  What type of material is this unknown material?

   a) diamagnetic
   b) paramagnetic
   c) ferromagnetic
   d) non-magnetic

10.    Which of the following properties of a material is necessary for a permanent magnet to be constructed from the material?

   a) diamagnetism
   b) paramagnetism
   c) ferromagnetism
   d) all of the above

11.    Determine the magnitude and direction of the magnetic field at a position 0.16 m East of a long vertical wire carrying an upward current 4.8 A.

12.    A square loop of wire with sides of length $L$ carries a current $I$ in the clockwise direction as shown in the diagram.  A long wire lies parallel to the left side of the square and is a distance $L$ away from the square.  The long wire also carries a current $I$ toward the top of the page as shown in the diagram.  What is the magnetic force on the loop?

13.    A solenoid is to be designed so that the magnetic field within the solenoid has a magnitude of 0.28 T.  The diameter of the coil needs to be 8.8 cm and the length needs to be 12 cm to accommodate the material that is to be placed within the solenoid.  How many turns of wire that can carry a maximum current of 3.0 A must be used?

14.    A long straight coaxial cable carries a current of 3.0 A in one direction on its inner conductor and a current of 5.0 A in the opposite direction on its outer conductor.  The inner conductor has a diameter of 0.50 mm and the outer conductor has a diameter of 6.0 mm.  Determine the magnitude of the magnetic field in the space between the two conductors and in the space outside of the outer conductor.

15.    Determine the magnetic field at the center of a square loop of wire that carries a current $I$.

## Problem Solutions

3.    The two currents in the same direction will be attracted with a force of
$$F = I_1(\mu_0 I_2/2\pi d)L = \mu_0 I_1 I_2 L/2\pi d$$
$$= (4\pi \times 10^{-7}\ \mathrm{T \cdot m/A})(35\ \mathrm{A})(35\ \mathrm{A})(45\ \mathrm{m})/2\pi(0.060\ \mathrm{m}) = 0.18\ \mathrm{N\ attraction.}$$

7.  We find the direction of the field for each wire from
    the tangent to the circle around the wire, as shown.
    For their magnitudes, we have

    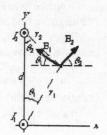

    $B_1$   $= (\mu_0/4\pi)2I/r_1$
        $= (10^{-7}\ T \cdot m/A)2(25\ A)/(0.120\ m) = 4.17 \times 10^{-5}\ T.$
    $B_2$   $= (\mu_0/4\pi)2I_B/r_2$
        $= (10^{-7}\ T \cdot m/A)2(25\ A)/(0.050\ m) = 1.00 \times 10^{-4}\ T.$

    We use the property of the triangle to find the angles shown:
    $r_2^2 = r_1^2 + d^2 - 2r_1d \cos \theta_1;$
    $(5.0\ cm)^2 = (12.0\ cm)^2 + (15.0\ cm)^2 - 2(12.0\ cm)(15.0\ cm) \cos \theta_1,$
    which gives $\cos \theta_1 = 0.956, \theta_1 = 17.1°;$
    $r_1^2 = r_2^2 + d^2 - 2r_2d \cos \theta_2;$
    $(12.0\ cm)^2 = (5.0\ cm)^2 + (15.0\ cm)^2 - 2(5.0\ cm)(15.0\ cm) \cos \theta_2,$
    which gives $\cos \theta_2 = 0.707, \theta_2 = 45.0°;$
    From the vector diagram, we have

    **B**  $= B_1(-\cos \theta_1\ \mathbf{i} + \sin \theta_1\ \mathbf{j}) + B_2 (\cos \theta_2\ \mathbf{i} + \sin \theta_2\ \mathbf{j})$
        $= (4.17 \times 10^{-5}\ T)(-\cos 17.1°\ \mathbf{i} + \sin 17.1°\ \mathbf{j}) + (1.00 \times 10^{-4}\ T)(\cos 45.0°\ \mathbf{i} + \sin 45.0°\ \mathbf{j})$
        $= (3.1 \times 10^{-5}\ T)\ \mathbf{i} + (8.3 \times 10^{-5}\ T)\ \mathbf{j}.$

    We find the direction from
    $\tan \alpha = B_y/B_x = (8.30 \times 10^{-5}\ T)/(3.09 \times 10^{-5}\ T) = 2.68, \alpha = 70.1°.$
    We find the magnitude from
    $B = B_x/\cos \alpha = (3.09 \times 10^{-5}\ T)/\cos 70.1° = 8.9 \times 10^{-5}\ T,\ 70°$ above horizontal.

11. (a) When the currents are in the same direction, the fields between
        the currents will be in opposite directions, so at the midpoint we have

        $B_a$  $= B_2 - B_1 = [(\mu_0/4\pi)2I_2/r] - [(\mu_0/4\pi)2I/r] = [(\mu_0/4\pi)2/r](I_2 - I)$
            $= (10^{-7}\ T \cdot m/A)2/(0.010\ m)(15\ A - I)$
            $= (2.0 \times 10^{-5}\ T/A)(15\ A - I)$ up, with the currents as shown.

    (b) When the currents are in opposite directions, the fields between
        the currents will be in the same direction, so at the midpoint we have

        $B_b$  $= B_2 + B_1 = [(\mu_0/4\pi)2I_2/r] + [(\mu_0/4\pi)2I/r] = [(\mu_0/4\pi)2/r](I_2 + I)$
            $= (10^{-7}\ T \cdot m/A)2/(0.010\ m)(15\ A + I)$
            $= (2.0 \times 10^{-5}\ T/A)(15\ A + I)$ down, with the currents as shown.

15. Because the currents are in the same direction, between the
    wires the fields will be in opposite directions.
    For the net field we have

    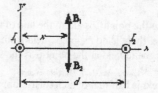

    **B**  $= \mathbf{B}_1 - \mathbf{B}_2 = [(\mu_0/4\pi)2I_1/x]\mathbf{j} - [(\mu_0/4\pi)2I_2/(d-x)]\mathbf{j}$
        $= (\mu_0/4\pi)2I\{[(d-x) - x]/x(d-x)\}\mathbf{j}$
        $= [(\mu_0/4\pi)2I(d-2x)/x(d-x)]\mathbf{j}.$

19. (a) The figure shows a view looking directly at the current.
        The sheet may be thought of as an infinite number of
        parallel wires. We choose a differential element of
        width $dx$ a distance $x$ from the center of the strip.
        This element has a current $dI = (I/L)\ dx$ which
        produces a magnetic field

        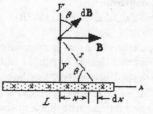

        $dB = (\mu_0/4\pi)2\ dI/r = (\mu_0I/2\pi L)\ dx/r$, in the direction shown.
        Because a differential element at $-x$ will produce a field
        of the same magnitude but below the horizontal, the
        symmetry means the resultant field will be parallel to the
        strip in the $x$-direction. We find the total field by adding
        (integrating) the $x$-components of the differential fields:

        $$B = \int \sin \theta\ dB = \int_{-L/2}^{L/2} \left(\frac{\mu_0 I}{2\pi L}\right)\left(\frac{y}{x^2 + y^2}\right)dx = \left(\frac{\mu_0 Iy}{2\pi L}\right)\left[\left(\frac{1}{y}\right)\tan^{-1}\left(\frac{x}{y}\right)\right]_{-L/2}^{L/2} = \left(\frac{\mu_0 I}{2\pi L}\right)\tan^{-1}\left(\frac{L}{2y}\right)$$

    (b) If $y \gg L$, or $L/2y \ll 1$, the angle is small and equal to the tangent, so we have
        $\tan^{-1}(L/2y) \approx L/2y$, and
        $B \approx (\mu_0I/\pi L))(L/2y) = (\mu_0/4\pi)(2I/y),$

which is the magnetic field produced by a long wire. This is what the strip will appear to be when we are far from the strip.

23. We use the results from Example 28–4.

   (*a*) The magnetic field at the surface of the wire is
   $$B_{surface} = \mu_0 I/2\pi R = (4\pi \times 10^{-7}\ \text{T} \cdot \text{m/A})(40\ \text{A})/2\pi(1.25 \times 10^{-3}\ \text{m}) = 6.4 \times 10^{-3}\ \text{T}.$$

   (*b*) Inside the wire we have
   $$B_{inside} = (\mu_0 I/2\pi)(r/R^2) = (4\pi \times 10^{-7}\ \text{T} \cdot \text{m/A})(40\ \text{A})(0.75 \times 10^{-3}\ \text{m})/2\pi(1.25 \times 10^{-3}\ \text{m})^2$$
   $$= 3.8 \times 10^{-3}\ \text{T}.$$

   (*c*) Outside the wire we have
   $$B_{outside} = \mu_0 I/2\pi r = (4\pi \times 10^{-7}\ \text{T} \cdot \text{m/A})(40\ \text{A})/2\pi(3.75 \times 10^{-3}\ \text{m}) = 2.1 \times 10^{-3}\ \text{T}.$$

27. The current densities in the wires are
   $$J_{inner} = I_0/\pi R_1^2 \quad \text{and} \quad J_{outer} = I_0/\pi(R_3^2 - R_2^2).$$
   Because of the cylindrical symmetry, we know that the magnetic fields will be circular. In each case we apply Ampere's law to a circular path of radius $r$.

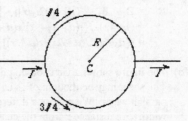

   (*a*) Inside the inner wire, $r < R_1$:
   $$\oint \mathbf{B} \cdot d\mathbf{s} = \mu_0 I_{enclosed};$$
   $$B2\pi r = \mu_0 J_{inner}\pi r^2 = \mu_0 I_0 \pi r^2/\pi R_1^2,$$
   which gives $B = (\mu_0 I_0/2\pi R_1^2)r$ circular CCW, $r < R_1$.

   (*b*) Between the wires, $R_1 < r < R_2$:
   $$\oint \mathbf{B} \cdot d\mathbf{s} = \mu_0 I_{enclosed};$$
   $B2\pi r = \mu_0 I_0$, which gives    $B = \mu_0 I_0/2\pi r$ circular CCW, $R_1 < r < R_2$.

   (*c*) Inside the outer wire, $R_2 < r < R_3$:
   $$\oint \mathbf{B} \cdot d\mathbf{s} = \mu_0 I_{enclosed};$$
   $$B2\pi r = \mu_0[I_0 - J_{outer}(\pi r^2 - \pi R_2^2)]$$
   $$= \mu_0[I_0 - I_0(\pi r^2 - \pi R_2^2)/\pi(R_3^2 - R_2^2)] = \mu_0 I_0[1 - (r^2 - R_2^2)/(R_3^2 - R_2^2)],$$
   which gives $B = (\mu_0 I_0/2\pi r)(R_3^2 - r^2)/(R_3^2 - R_2^2)$ circular CCW, $R_2 < r < R_3$.

   (*d*) Outside the outer wire, $R_3 < r$:
   $$\oint \mathbf{B} \cdot d\mathbf{s} = \mu_0 I_{enclosed};$$
   $B2\pi r = \mu_0(I_0 - I_0)$, which gives $B = 0$, $R_3 < r$.

31. Because the point C is along the line of the two straight segments of the wire, there is no magnetic field from these segments. The magnetic field at the point C is the sum of two fields:

   $\mathbf{B} = \mathbf{B}_{lower\ semicircle} + \mathbf{B}_{upper\ semicircle}.$

   Each field is half that of a circular loop, with the field of the lower semicircle out the page and that of the upper semicircle into the page, so we subtract the two magnitudes:
   $$B = \tfrac{1}{2}(\mu_0\tfrac{3}{4}I/2R) - \tfrac{1}{2}(\mu_0\tfrac{1}{4}I/2R) = \mu_0 I/8R \text{ out of the page.}$$

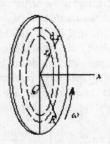

35. (*a*) We can treat the disk as an infinite number of rings. We choose a differential element of radius $r$ and thickness $dr$. The charge density on the disk is $\sigma = Q/\pi R^2$, so the current in the ring is
   $$dI = \sigma\, dA/T = (Q/\pi R^2)2\pi r\, dr/(2\pi/\omega) = Q\,\omega r\, dr/\pi R^2.$$
   We find the magnetic dipole moment by integration:
   $$\mu = \mathbf{i}\int_0^R \pi r^2\, dI = \frac{\pi Q\omega}{\pi R^2}\mathbf{i}\int_0^R r^3\, dr = \frac{Q\omega R^2}{4}\mathbf{i}$$

(b) The field from each of the rings is along the x-axis, so we integrate the magnitudes:

$$B = \int dB = \int_{r=0}^{r=R} \frac{\mu_0\, dI}{2} \frac{r^2}{\left(r^2 + x^2\right)^{3/2}}$$

$$= \int_0^R \frac{\mu_0 Q\omega}{2\pi R^2} \frac{r^3\, dr}{\left(r^2 + x^2\right)^{3/2}} = \frac{\mu_0 Q\omega}{2\pi R^2}\left[2\left(r^2+x^2\right)^{1/2} - \frac{r^2}{\left(r^2+x^2\right)^{1/2}}\right]_0^R$$

$$= \frac{\mu_0 Q\omega}{2\pi R^2}\left[2\left(R^2+x^2\right)^{1/2} - 2x - \frac{R^2}{\left(R^2+x^2\right)^{1/2}}\right] = \frac{\mu_0 Q\omega}{2\pi R^2}\left[\frac{R^2 - 2x^2 - 2x\left(R^2+x^2\right)^{1/2}}{\left(R^2+x^2\right)^{1/2}}\right]$$

Thus the magnetic field is

$$\mathbf{B} = \frac{\mu_0 Q\omega}{2\pi R^2}\left[\frac{R^2 - 2x^2 - 2x\left(R^2+x^2\right)^{1/2}}{\left(R^2+x^2\right)^{1/2}}\right]\mathbf{i}$$

(c) When $x \gg R$, we have

$$B = (\mu_0 Q\, \omega/2\pi R^2)x^2\{(R/x)^2 + 2 - 2[1 + (R/x)^2]^{1/2}\}/x[1 + (R/x)^2]^{1/2}$$
$$\approx (\mu_0 Q\, \omega/2\pi R^2)x\{2 + (R/x)^2 - 2[1 + (R/x)^2/2 - (R/x)^4/8]\}[1 - (R/x)^2/2]$$
$$\approx (\mu_0 Q\, \omega/2\pi R^2)x(R/x)^4/4 = \mu_0 Q\, \omega R^2/8\pi x^3 = \mu_0\mu/2\pi x^3.$$

This is Eq. 28–7b, so yes it does apply.

39. (a) The angle subtended by a side of the polygon at the center point P is
$\theta = 2\pi/n$.
The length of a side is
$L = 2R \sin(\theta/2) = 2R \sin(\pi/n)$.
The perpendicular distance from the center to the side is
$D = R \cos(\theta/2) = R \cos(\pi/n)$.
We use the result from Problem 36 to find the magnetic field of one side:
$B_1 = \mu_0 I_0 L/2\pi D(L^2 + 4D^2)^{1/2}$
$= \mu_0 I_0[2R \sin(\pi/n)]/2\pi[R \cos(\pi/n)][4R^2 \sin^2(\pi/n) + 4R^2 \cos^2(\pi/n)]^{1/2}$
$= (\mu_0 I_0/2\pi R) \tan(\pi/n)$.
Thus the field from the $n$ sides is
$B = nB_1 = (\mu_0 I_0/2\pi R)n \tan(\pi/n)$ into the page.

(b) If we let $n \to \infty$, the angle $\pi/n \to 0$, so $\tan(\pi/n) \to \pi/n$. Thus we have
$B \to (\mu_0 I_0/2\pi R)n(\pi/n) = \mu_0 I_0/2R$, which is the expression for a circular loop.

43. (a) If the iron bar is completely magnetized, the dipoles of all of the atoms are aligned. Thus the dipole moment will be
$\mu = N\mu_1 = (\rho V/M)N_A\mu_1$
$= (7.8 \times 10^3 \text{ kg/m}^3)(10^3 \text{ g/kg})(0.12 \text{ m})(1.2 \times 10^{-2} \text{ m})^2 \times$
$(6.02 \times 10^{23} \text{ atoms/mol})(1.8 \times 10^{-23} \text{ A} \cdot \text{m}^2)/(55.8 \text{ g/mol}) = 26 \text{ A} \cdot \text{m}^2$.

(b) Because the field is perpendicular to the dipole moment, the torque is maximal:
$\tau = \mu B = (26 \text{ A} \cdot \text{m}^2)(1.2 \text{ T}) = 31 \text{ m} \cdot \text{N}$.

47. Because the currents and the separations are the same, we find the force per unit length between any two wires from
$F/L = I_1(\mu_0 I_2/2\pi d) = \mu_0 I^2/2\pi d$
$= (4\pi \times 10^{-7} \text{ T} \cdot \text{m/A})(8.00 \text{ A})^2/2\pi(0.380 \text{ m})$
$= 3.37 \times 10^{-5} \text{ N/m}$.

The directions of the forces are shown on the diagram.

The symmetry of the force diagrams simplifies the vector addition, so we have

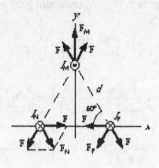

$$F_M/L = 2(F/L) \cos 30°$$
$$= 2(3.37 \times 10^{-5} \text{ N/m}) \cos 30° = 5.84 \times 10^{-5} \text{ N/m up.}$$
$$F_N/L = F/L$$
$$= 3.37 \times 10^{-5} \text{ N/m } 60° \text{ below the line toward P.}$$
$$F_P/L = F/L$$
$$= 3.37 \times 10^{-5} \text{ N/m } 60° \text{ below the line toward N.}$$

51. The sheet may be thought of as an infinite number of parallel wires. The figure shows a view looking directly at the current. If we consider a point above the sheet, the wire directly underneath produces a magnetic field parallel to the sheet. By considering a pair of wires symmetrically placed about the first one, we see that the net field will be parallel to the sheet. Below the sheet, the field will be in the opposite direction. We apply Ampere's law to the rectangular path shown in the diagram. For the sides perpendicular to the sheet, **B** is perpendicular to $d$s. For the sides parallel to the sheet, **B** is parallel to $d$s and constant in magnitude, because the upper and lower paths are equidistant from the sheet. We have

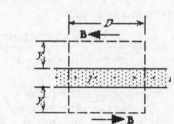

$$\oint B \cdot ds = \mu_0 I_{enclosed}$$

$$\int_{side \, s} \mathbf{B} \cdot d\mathbf{s} + \int_{length \, s} \mathbf{B} \cdot d\mathbf{s} = 0 + B \int_{length \, s} ds = B2D = \mu_0 jtD$$

This gives

$B = \mu_0 jt/2$ parallel to the sheet, perpendicular to the current (with opposite directions on the two sides).

55. We use the expression for the magnetic field on the axis of a circular loop:
$$B = (\mu_0 I/2)R^2/(R^2 + x^2)^{3/2} = (\mu_0 I/2)R_E^2/(R_E^2 + R_E^2)^{3/2} = \mu_0 I/4R_E\sqrt{2};$$
$$1 \times 10^{-4} \text{ T} = (4\pi \times 10^{-7} \text{ T} \cdot \text{m/A})I/4\sqrt{2}(6.4 \times 10^6 \text{ m}), \text{ which gives } I = 3 \times 10^9 \text{ A.}$$

59. Because the airplane is flying parallel to the wire, the circular magnetic field of the wire will be perpendicular to the velocity. Thus we have
$$qvB = ma, \text{ which gives}$$
$$a = qvB/m = qv[(\mu_0/4\pi)2I/r]/m$$
$$= (18.0 \times 10^{-3} \text{ C})(2.8 \text{ m/s})(10^{-7} \text{ T·m/A})2(30 \text{ A})/(0.086 \text{ m})(0.175 \text{ kg})(g/9.80 \text{ m/s}^2) = 2.1 \times 10^{-6} \, g.$$

63. At any instant the currents will be in opposite directions, with
$$I_{rms} = P/V = (40 \times 10^6 \text{ W})/(10 \times 10^3 \text{ V}) = 4.0 \times 10^3 \text{ A.}$$
The maximal current will be
$$I_{max} = I_{rms}\sqrt{2}.$$
We show the directions of each field on the diagram, with
$$\tan \theta = d/2H, \quad \text{and } \sin \theta = d/2r, \text{ where } r^2 = H^2 + \tfrac{1}{4}d^2.$$
Because the magnitudes of the two fields are the same, the net field will be down with a magnitude given by
$$B_{max} = 2B_1 \sin \theta = 2[(\mu_0/4\pi)2I_{max}/r](d/2r) = 2(\mu_0/4\pi)I_{max}d/r^2$$
$$= 2(10^{-7} \text{ T} \cdot \text{m/A})(4.0 \times 10^3 \text{ A}) \sqrt{2}(3 \text{ m})/[(30 \text{ m})^2 + \tfrac{1}{4}(3 \text{ m})^2]$$
$$= 4 \times 10^{-6} \text{ T.}$$
When we compare to the Earth's field, we get
$$B_{max}/B_E = (4 \times 10^{-6} \text{ T})/(5 \times 10^{-5} \text{ T}) \approx 0.1,$$
so it is about 10% of the Earth's field.

# Chapter 29: Electromagnetic Induction and Faraday's Law

## Chapter Overview and Objectives

This chapter introduces Faraday's law, Lenz's law, and magnetic flux. It discusses several of the implications and applications of Faraday's law.

After completing study of this chapter, you should:

- Know the definition of magnetic flux and be able to calculate magnetic flux for a given situation.
- Know Faraday's law and Lenz's law.
- Know how to calculate the emf induced in a moving conductor.
- Know what generators, dynamos, and alternators are.
- Know what a transformer is and how to relate primary and secondary voltages and currents.

## Summary of Equations

Magnetic flux for a uniform magnetic field:
$$\Phi_B = BA\cos\theta$$

General definition of magnetic flux:
$$\Phi_B = \int \mathbf{B} \cdot d\mathbf{A}$$

Faraday's law:
$$\varepsilon = -\frac{d\Phi_B}{dt}$$

Emf induced in a moving conductor:
$$\varepsilon = vLB$$

Emf induced in generator coil:
$$\varepsilon = NBA\omega\sin(\omega t + \phi)$$

Relationship between primary and secondary voltages in a transformer:

$$\frac{V_s}{V_p} = \frac{N_s}{N_p}$$

Relationship between primary and secondary currents in a transformer:

$$\frac{I_s}{I_p} = \frac{N_p}{N_s}$$

Faraday's law written in terms of electric field:
$$\oint E \cdot dl = -\frac{d\Phi_B}{dt}$$

## Chapter Summary

### Section 29-1 Induced EMF

A coil of wire will have an **induced emf** if the magnetic field within the coil is changing with time.

### Section 29-2 Faraday's Law of Induction; Lenz's Law

A more complete description of the cause of induced emf requires the notion of magnetic flux, $\Phi_B$. In a uniform magnetic field of magnitude $B$, the magnetic flux through an area $A$ of a plane with its normal at an angle $\theta$ to the direction of the magnetic field is

$$\Phi_B = BA\cos\theta$$

This can also be written as

$$\Phi_B = \mathbf{B} \cdot \mathbf{A}$$

If the magnetic field is not uniform, the magnetic flux must be determined from

$$\Phi_B = \int \mathbf{B} \cdot d\mathbf{A}$$

The dimensions of magnetic flux are dimensions of magnetic field × area. The SI unit of magnetic flux is the **weber**, abbreviated Wb. One Wb is equal to one T·m².

**Faraday's law of induction** states that the induced emf in a loop is equal to the negative of the time derivative of the flux through the loop:

$$\mathcal{E} = -\frac{d\Phi_B}{dt}$$

The minus sign means that the induced emf in the coil is in a direction that would drive a current through the coil in a direction that would create a magnetic field that creates a flux through the coil to oppose the change in flux that is taking place. This is called **Lenz's law**.

**Example 29-2-A**

A magnetic field is given by the expression

$$\mathbf{B}(x, y, z) = (3.4\,\text{T})\frac{(y+z)}{L}\mathbf{i} + (3.4\,\text{T})\frac{(x-z)}{L}\mathbf{j}$$

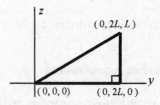

Determine the flux through the triangle shown in the *yz* plane.

**Solution:**

The magnetic field is not constant over the area for which the flux is being determined, so we must use the integral expression for the flux:

$$\Phi_B = \int \mathbf{B} \cdot d\mathbf{A}$$

We have the expression for **B**, but we need to determine the appropriate differential area elements, *d***A**. If we take as our differential area elements rectangles in the *yz* plane with sides *dy* and *dz*, we can cover the area of the triangle. A rectangle in the *yz* plane has its normal pointing either in the +**i** or −**i** direction. We will take the +**i** direction as the direction of our area elements, so *d***A** = *dy dz* **i**. We can write the integral for the flux as

$$\Phi_B = \int \mathbf{B} \cdot d\mathbf{A} = \int_{y=0}^{2L} \int_{z=0}^{y/2} \left[ (3.4\,\text{T})\frac{(y+z)}{L}\mathbf{i} + (3.4\,\text{T})\frac{(x-z)}{L}\mathbf{j} \right] \cdot dy\,dz\,\mathbf{i}$$

$$= \int_{y=0}^{2L} \int_{z=0}^{y/2} (3.4\,\text{T})\frac{(y+z)}{L}\,dy\,dz = \frac{(3.4\,\text{T})}{L} \int_{y=0}^{2L} \left[ \left( yz + \frac{z^2}{2} \right) \Bigg|_{z=0}^{y/2} \right] dy$$

$$= \frac{(3.4\,\text{T})}{L} \int_{y=0}^{2L} \frac{5}{8} y^2\,dy = \frac{5(3.4\,\text{T})}{8L} \frac{y^3}{3} \Bigg|_{y=0}^{2L} = \frac{5(3.4\,\text{T})L^2}{3} = (5.7\,\text{T})L^2$$

**Example 29-2-B**

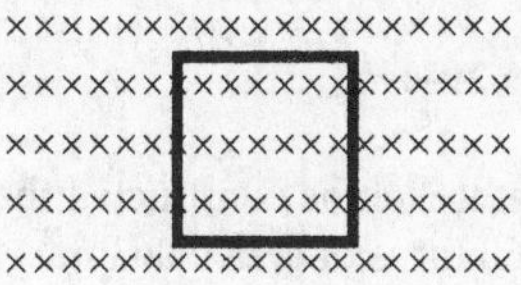

A square loop of wire with sides of length $L$ is in a uniform magnetic field pointing into the page as shown in the diagram. The magnitude of the magnetic field is $B$. At time $t = 0$, the square is stretched in such a way that it remains a rectangle, but two opposites sides grow at the rate $v$ and the other two sides shrink at the rate $v$ so that the perimeter of the rectangle remains constant. Determine the induced emf in the loop as a function of time.

**Solution:**

As the magnetic field remains constant in time and remains perpendicular to the loop, we recognize that any change in flux through the loop is due to the changing area of the loop. The long sides of the loop will have a length $L + vt$ and the short sides of the loop will have a length $L - vt$. The area of the loop will be

$$A = (L + vt)(L - vt) = L^2 - v^2 t^2$$

The flux through the loop is

$$\Phi_B = BA = B(L^2 - v^2 t^2)$$

The emf induced in the loop is

$$\mathcal{E} = -\frac{d\Phi_B}{dt} = -\frac{d}{dt}\left[B(L^2 - v^2 t^2)\right] = 2Bv^2 t$$

**Section 29-3 EMF Induced in a Moving Conductor**

If a conductor of length $L$ moves with a velocity $\mathbf{v}$ perpendicular to the length and perpendicular to a magnetic field $\mathbf{B}$, it will have an emf $\mathcal{E}$ induced along the length of the conductor such that

$$\mathcal{E} = vLB$$

where $v$ is the speed of the conductor and $B$ is the magnitude of the magnetic field.

**Example 29-3-A**

A metal ball of diameter 1.00 inches rolls from east to west in a region where the magnetic field points toward the north and has a magnitude of $1.3 \times 10^{-2}$ T. The emf induced across the diameter of the ball is 12.6 μV. How fast is the ball moving? Does the emf increase from top to bottom or bottom to top?

**Solution:**

The length of the ball perpendicular to the velocity and the magnetic field is the top-to-bottom diameter of the ball. We use the relationship between induced emf, $B$, $L$, and $v$ to solve for $v$:

$$\mathcal{E} = BLv \quad \Rightarrow \quad v = \frac{\mathcal{E}}{BL} = \frac{12.6 \times 10^{-6}\ \text{V}}{(1.3 \times 10^{-2}\ \text{T})(1.00\ \text{in} \times 0.0254\ \text{m/in})} = 3.8 \times 10^{-2}\ \text{m/s}$$

**Section 29-4 Electric Generators**

An application of Faraday's law is the conversion of mechanical energy into electrical energy. Generators, dynamos, and alternators are devices that utilize this principle. The most common form of these devices is a coil of wire that is rotated within a magnetic field with mechanical work. If a coil of wire consisting of $N$ loops of area $A$ is rotated in a magnetic field of magnitude $B$ at an angular speed $\omega$, an emf $\mathcal{E}$ will be induced in the coil with a time dependence

$$\mathcal{E} = NBA\omega \sin(\omega t + \phi)$$

where $\phi$ is a phase angle that depends on the orientation of the coil at time $t = 0$.

**Example 29-4-A**

A generator is to be constructed that has an output frequency $f = 60$ Hz and an output rms voltage of 120 V. The coil is to have 60 turns of area 0.063 m². What does the magnitude of the magnetic field inside the generator need to be to create the desired output voltage?

**Solution:**

The relationship

$$\mathcal{E} = NBA\omega \sin(\omega t + \phi)$$

gives the output emf of the generator as a function of time. The coefficient of the sine function is the amplitude of the output voltage from the generator. We know that the rms value of the voltage is the amplitude of voltage divided by the square root of two:

$$\mathcal{E}_{rms} = \frac{\mathcal{E}_p}{\sqrt{2}} = \frac{NBA\omega}{\sqrt{2}}$$

We are given the desired frequency of the voltage. The angular frequency is

$$\omega = 2\pi f = 2\pi \left(60 \text{ s}^{-1}\right) = 3.8 \times 10^2 \text{ s}^{-1}$$

We can calculate the required magnetic field:

$$B = \frac{\sqrt{2}\mathcal{E}_{rms}}{NA\omega} = \frac{\sqrt{2}(120 V)}{(60)(0.063 \text{ m}^2)(3.8 \times 10^2 \text{ s}^{-1})} = 0.12 \text{ T}$$

**Section 29-5 Counter EMF and Torque; Eddy Currents**

A motor turns a coil of wire within a magnetic field. This sounds just like the electric generators described above. In this case, the induced emf opposes the applied emf and reduces the current into the coil from the external emf source. This opposing emf is called a **back emf** or **counter emf**.

When a conductor moves through a magnetic field or is within a changing magnetic field, currents are produced within the conductor. These currents are called **eddy currents**.

**Section 29-6 Transformers and Transmission of Power**

Two coils that are located in close proximity and oriented in such a way that the magnetic field from currents in one coil cause a flux in the other coil form a functional unit called a **transformer**. One of the coils is called the **primary** coil with $N_p$ turns and the other is called the **secondary** coil with $N_s$ turns. In use, an AC voltage of amplitude $V_p$ is applied to the primary of the transformer and an AC voltage of amplitude $V_s$ appears at the secondary due to the induced emf in the secondary coil caused by the magnetic flux produced by the current in the primary coil. If the transformer is constructed in such a way that the magnetic flux through the primary and secondary coils is identical and the internal resistance of the coils is negligible, then

$$\frac{V_s}{V_p} = \frac{N_s}{N_p}$$

and the current in the primary coil $I_p$ and the current in the secondary coil $I_s$ are related by

$$\frac{I_s}{I_p} = \frac{N_p}{N_s}$$

Note that these relationships apply to both peak values and rms values as long as you are consistent within a given application of the relationships.

**Example 29-6-A**

A transformer is used to convert 120 V rms voltage to 5000 V rms voltage. The primary of the transformer has 180 turns of wire. The secondary circuit has a resistance of 180 kΩ. Determine the required number of turns in the secondary, the rms current in the secondary circuit and the primary circuit, and the power transformed by the transformer.

**Solution:**

First, we can determine the number of secondary turns required:

$$\frac{N_s}{N_p} = \frac{V_s}{V_p} \quad \Rightarrow \quad N_s = \frac{V_s}{V_p} N_p = \frac{5000\,\text{V}}{120\,\text{V}} 180 = 7500$$

To determine the secondary current, we use Ohm's law:

$$V_s = I_s R \quad \Rightarrow \quad I_s = \frac{V_s}{R} = \frac{5000\,V}{180 \times 10^3\,\Omega} = 2.7 \times 10^{-2}\,\text{A}$$

We can now find the primary current from the transformer relationship for currents:

$$\frac{I_s}{I_p} = \frac{N_p}{N_s} \quad \Rightarrow \quad I_p = \frac{N_s}{N_p} I_s = \frac{7500}{180} \left(2.7 \times 10^{-2}\,\text{A}\right) = 1.1\,\text{A}$$

The average power is the product of the rms voltage and the rms current in either the primary or the secondary circuit:

$$\overline{P} = V_s I_s = \left(5000\,\text{V}\right)\left(2.7 \times 10^{-2}\,\text{A}\right) = 135\,\text{W}$$

**Section 29-7 A Changing Magnetic Flux Produces an Electric Field**

Our discussion of Faraday's law to this point may make it seem as if a coil of wire is necessary to produce the emf due to the changing magnetic flux. The coil is not necessary, however. In general, if there is a changing magnetic flux through any closed path in space, there will be an emf induced around the path. If we write that the emf is the path integral of the electrical field, then we obtain a more general form of Faraday's law:

$$\oint E \cdot dl = -\frac{d\Phi_B}{dt}$$

**Example 29-7-A**

The electric field is integrated over a circular path of radius 16.2 cm. At each point along the path, the electric field points outward from the tangent to the circle (as shown in the diagram) at an angle of 30°. The magnitude of the electric field is 20 V/m at every point on the circle. What is the time rate of change of magnetic flux through the circle?

**Solution:**

We need to find the path integral

$$\oint \mathbf{E} \cdot d\mathbf{l}$$

What is $d\textbf{l}$ in this case? The length element $d\textbf{l}$ points in a direction tangent to the circular path. If we travel around the path in the counterclockwise direction, then $d\textbf{l}$ is always at an angle of $30°$ to the electric field. At every point along the path, the scalar product between $\textbf{E}$ and $d\textbf{l}$ will be

$$\textbf{E} \cdot d\textbf{l} = Edl \cos 30°$$

We integrate this expression around the closed circular path:

$$\oint \textbf{E} \cdot d\textbf{l} = E \cos 30° \oint dl = E \cos 30° \left(2\pi R\right)$$

where $R$ is the radius of the circle. Substituting the values of the magnitude of the electric field and radius of the circle,

$$\oint \textbf{E} \cdot d\textbf{l} = E \cos 30° \left(2\pi R\right) = 20 \, \text{V/m} \cos 30° \left(2\pi \, 0.162 \, \text{m}\right) = 18 \, \text{V}$$

Faraday's law states that this will also be minus the time rate of change of the flux through the circle:

$$\frac{d\Phi_B}{dt} = -\oint \textbf{E} \cdot d\textbf{l} = -18 \, \text{V} = -18 \frac{\text{T} \cdot \text{m}^2}{\text{s}}$$

What does the minus sign mean in this case? We followed the emf around in the counterclockwise direction. That is the direction that the induced electric field would push any induced current. Using the right hand rule, that current would produce a magnetic field that points out of the page through the loop as drawn in the diagram. According to Lenz's law, that field would oppose the change in flux taking place. The flux must be decreasing in the direction out of the page.

### Section 29-8 Applications of Induction: Sound Systems, Computer Memory, the Seismograph

There are many applications of magnetic induction, including microphones, magnetic recording, and seismographs.

## Practice Quiz

1.    A loop of area $A$ lies flat on the ground. A uniform magnetic field points horizontal and to the north with a magnitude of $B$. The magnetic field decreases uniformly to zero in $t$. What is the emf induced in the loop?

    a) $-BA/t$
    b) $+BA/t$
    c) $-Bat$
    d) zero

2.    A loop of area $A$ lies flat on the ground. A uniform magnetic field points vertically upward with a magnitude of $B$. The magnetic field decreases uniformly to zero in $t$. What is the magnitude of the emf induced in the loop?

    a) $BA/t$
    b) $B/At$
    c) $A/Bt$
    d) zero

3.    What is the direction of any induced current in the loop described in Question 2 above, as viewed looking down at the loop on the ground from above?

    a) clockwise
    b) counterclockwise
    c) depends on the resistance of the loop
    d) No emf is induced, so there is no direction.

4.    If the angular speed of a generator is doubled, what happens to the amplitude of the output voltage?

   a) doubles
   b) decreases by 50%
   c) amplitude doesn't change, only the frequency changes
   d) quadruples

5.    Considering emf, when is an electric motor operating most efficiently?  This means when is the motor putting the highest fraction of the input electrical energy into the mechanical work it is doing?

   a) when the motor is at rest
   b) when the motor is turning as slow as it goes without stalling
   c) when the motor is turning at full speed
   d) The motor is always converting energy to mechanical energy at the same efficiency.

6.    For a transformer in which the flux in the secondary coil is less than the flux through the primary coil for any reason, we can write the relationship between secondary and primary voltage as

$$\frac{V_s}{V_p} = \alpha \frac{N_s}{N_p}$$

where $\alpha$ is a number.  What must be true about the number $\alpha$?

   a) $\alpha$ is less than one.
   b) $\alpha$ is greater than one.
   c) $\alpha$ must be equal to one.
   d) $\alpha$ must be negative.

7.    An arrow is shot from a bow with a speed $v$ in a direction perpendicular to a magnetic field of magnitude $B$.  The arrow's shaft has a length $L$ and the diameter $D$.  What is the magnitude of the emf induced in the shaft of the arrow as it moves through the magnetic field?

   a) $BLv$
   b) $BDv$
   c) $Bv(L + D)/2$
   d) zero

8.    A DC voltage of 10.0 V is applied across the primary of a transformer that has 300 turns in its primary and 150 turns in its secondary.  What is the voltage across the secondary of the transformer?

   a) 20.0 V
   b) 5.0 V
   c) 1500 V
   d) zero

9.    What is the current in the primary circuit of the transformer in Question 8, if there are no voltage sources in the secondary circuit of the transformer, the resistances in the secondary circuit is 20 $\Omega$, and there is no resistance in the primary circuit?

   a) Infinite
   b) 0.50 A
   c) 0.25 A
   d) 0.75 A

10.    A coil of wire is in a magnetic field pointing perpendicular to and into the page as shown in the diagram.  The coil is being pulled from the magnetic field as shown by the arrow.  Which direction does induced current flow around the coil?

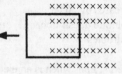

    a)  clockwise
    b)  counterclockwise
    c)  It depends on how fast the coil is pulled from the magnetic field.
    d)  No induced current flows because there is no induced emf.

11.    A circular loop of wire of diameter 15.6 cm rests in a uniform magnetic field of magnitude 1.19 T that is at an angle of 32° to the normal to the plane of the loop.  The resistance of the loop is 3.8 Ω.  The magnitude of the magnetic field goes to zero uniformly in 2.78 s.  What is the current induced in the loop while the magnetic field is changing?

12.    A magnetic field is given by

$$\mathbf{B}(x, y, z) = (1.0\,\text{T/m})\,x\,\mathbf{i} + (1.0\,\text{T/m})\,y\,\mathbf{j} - (2.0\,\text{T/m})(x + z)\mathbf{k}$$

Calculate the magnetic flux through the square that lies in the $xy$ plane with vertices at ($x = 0$, $y = 0$), ($x = 0$, $y = 1.0$ m), ($x = 1.0$ m, $y = 1.0$ m), and ($x = 1.0$ m, $y = 0$).

13.    A generator is made from a coil with 120 turns of wire that are rectangular and have a width of 12.8 cm and a length of 21.6 cm.  The coil turns in a uniform magnetic field of magnitude 1.93 T.  At what frequency must the coil turn to result in an output voltage with an amplitude of 10.0 V?

14.    An engineer wants to design a transformer that has a primary voltage of 120 V and a secondary voltage of 48 V.  If the transformer has 300 turns in the primary winding, how many turns are needed in the secondary winding?

15.    An engineer designs a transformer that requires a primary voltage of 120 V rms and the secondary circuit has a resistance of 10.0 Ω with a power requirement 360 W.  What ratio of secondary turns to primary turns must be used and what current must the primary and secondary windings be able to safely carry?

## Problem Solutions

3.    As the coil is pushed into the field, the magnetic flux increases into the page.  To oppose this increase, the flux produced by the induced current must be out of the page, so the induced current is counterclockwise.

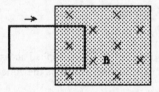

7.    (*a*)  The increasing current in the wire will cause an increasing field into the page through the loop.   To oppose this increase, the induced current in the loop will produce a flux out of the page, so the direction of the induced current will be counterclockwise.

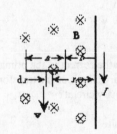

    (*b*)  The decreasing current in the wire will cause a decreasing field into the page through the loop.   To oppose this decrease, the induced current in the loop will produce a flux into the page, so the direction of the induced current will be clockwise.

    (*c*)  Because the current is constant, there will be no change in flux, so the induced current will be zero.

    (*d*)  The increasing current in the wire will cause an increasing field into the page through the loop.   To oppose this increase, the induced current in the loop will produce a flux out of the page, so the direction of the induced current will be counterclockwise.

11.    (*a*)  The magnetic flux through the loop is into the paper and decreasing, because the area is decreasing.  To oppose this decrease, the induced current in the loop will produce a flux

into the paper, so the direction of the induced current will be clockwise.

(b) We choose into the paper as the positive direction.  The average induced emf is

$$\varepsilon = -\Delta\Phi_B/\Delta t = -B\ \Delta A/\Delta t = -\pi B\ \Delta(r^2)/\Delta t$$
$$= -\pi(0.75\ \text{T})[(0.030\ \text{m})^2 - (0.100\ \text{m})^2]/(0.50\ \text{s}) = 4.3\times10^{-2}\ \text{V} = 43\ \text{mV}.$$

(c) We find the average induced current from

$$I = \varepsilon/R = (43\ \text{mV})/(2.5\ \Omega) = 17\ \text{mA}.$$

15.  For the resistance of the loop, we have

$$R = \rho L/A = \rho\pi D/\tfrac14\pi d^2 = 4\rho D/d^2.$$

The induced emf is

$$\varepsilon = -\Delta\Phi_B/\Delta t = -\tfrac14\pi D^2\ \Delta B/\Delta t;$$

so the induced current is

$$I = \varepsilon/R = -(\pi D d^2/16\rho)\ \Delta B/\Delta t.$$

In the time $\Delta t$ the amount of charge that will pass a point is

$$Q = I\ \Delta t$$
$$= -(\pi D d^2/16\rho)\ \Delta B = -[\pi(0.156\ \text{m})(2.05\times10^{-3}\ \text{m})^2/16(1.68\times10^{-8}\ \Omega\cdot\text{m})](0 - 0.550\ \text{T}) = 4.21\ \text{C}.$$

19.  The flux through the loop is

$$\Phi_B = BA.$$

The induced emf is

$$\varepsilon = -d\Phi_B/dt = -B\ dA/dt = -(0.48\ \text{T})(-3.50\times10^{-2}\ \text{m}^2/\text{s}) = 1.7\times10^{-2}\ \text{V}.$$

Because the area changes at a constant rate, this is the induced emf for both times.

23.  (a) Because the velocity is perpendicular to the magnetic field and the rod, we find the induced emf from

$$\varepsilon = Blv$$
$$= (0.35\ \text{T})(0.240\ \text{m})(1.8\ \text{m/s}) = 0.15\ \text{V}.$$

(b) We find the induced current from

$$I = \varepsilon/R = (0.151\ \text{V})/(2.2\ \Omega + 26.0\ \Omega) = 5.4\times10^{-3}\ \text{A}.$$

(c) The induced current in the rod will be down.  Because this current is in an outward magnetic field, there will be a magnetic force to the left.  To keep the rod moving, there must be an equal external force to the right, which we find from

$$F = IlB = (5.4\times10^{-3}\ \text{A})(0.240\ \text{m})(0.35\ \text{T}) = 4.5\times10^{-4}\ \text{N}.$$

27.  (a) At a distance $r$ from the long wire, the magnetic field is directed into the paper with magnitude

$$B = \mu_0 I/2\pi r.$$

Because the field is not constant over the short section, we find the induced emf by integration.  We choose a differential element $dr$ a distance $r$ from the long wire. The induced emf in this element is

$$d\varepsilon = Bv\ dr \text{ toward the long wire.}$$

We find the total emf by integrating:

$$\varepsilon = \int_b^{a+b} Bv\,dr = \frac{\mu_0 Iv}{2\pi}\int_b^{a+b}\frac{dr}{r} = \frac{\mu_0 Iv}{2\pi}\ln\!\left(\frac{a+b}{b}\right) \text{ toward long wire}$$

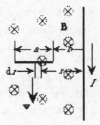

(b) If the current is in the opposite direction to $I$, the only change will be in the direction of the emf:

$$\varepsilon = \frac{\mu_0 Iv}{2\pi}\ln\!\left(\frac{a+b}{b}\right) \text{ away from long wire}$$

31. We find the peak emf from

$$\mathcal{E}_{peak} = NBA\omega = (350 \text{ turns})(0.45 \text{ T})\pi(0.050 \text{ m})^2(60 \text{ rev/s})(2\pi \text{ rad/rev}) = 466 \text{ V}.$$

The rms voltage output is

$$V_{rms} = \mathcal{E}_{peak}/\sqrt{2} = (466 \text{ V})/\sqrt{2} = 330 \text{ V} = 0.33 \text{ kV}.$$

If only the rotation frequency changes, to double the rms voltage output, we must double the rotation speed, so we have

$$f_2 = 2f_1 = 2(60 \text{ rev/s}) = 120 \text{ rev/s}.$$

35. Because the counter emf is proportional to the rotation speed, we find the new value from

$$\mathcal{E}_{back2}/\mathcal{E}_{back1} = \omega_2/\omega_1 ;$$

$$\mathcal{E}_{back2}/(108 \text{ V}) = (1/2), \text{ which gives } \mathcal{E}_{back2} = 54 \text{ V}.$$

We find the new current from

$$\mathcal{E} - \mathcal{E}_{back} = IR;$$
$$120 \text{ V} - 54 \text{ V} = I(5.0 \text{ }\Omega), \text{ which gives } I = 13 \text{ A}.$$

39. With 100% efficiency, the power on each side of the transformer is the same:

$$I_P V_P = I_S V_S , \text{ so we have}$$
$$I_S/I_P = V_P/V_S = (22 \text{ V})/(120) = 0.18.$$

43. We find the output voltage of this step-up transformer from

$$V_S/V_P = N_S/N_P ;$$
$$V_S/(120 \text{ V}) = (1510 \text{ turns})/(330 \text{ turns}), \text{ which gives } V_S = 549 \text{ V}.$$

We find the input current from

$$I_S/I_P = N_P/N_S ;$$
$$(15.0 \text{ A})/I_P = (330 \text{ turns})/(1510 \text{ turns}), \text{ which gives } I_P = 68.6 \text{ A}.$$

47. We find the electric field from

$$\mathcal{E}_{induced} = Blv = \int \mathbf{E} \cdot d\mathbf{l} = El, \text{ which gives}$$
$$E = Bv = (0.750 \text{ T})(0.250 \text{ m/s}) = 0.188 \text{ V/m}.$$

51. (*a*)  Because the current in the rod is constant, the potential difference across the rod must be constant: $V = IR$. The net electric field is the potential gradient along the rod:

$$E_{net} = V/l = IR/l \text{ (constant)}.$$

(*b*)  With a constant source emf $\mathcal{E}_0$, the potential difference across the rod will be

$$V = IR = \mathcal{E}_0 - Blv.$$

If we use the result for the velocity from Problem 26, we have

$$V = \mathcal{E}_0 - Blv = \mathcal{E}_0 - \mathcal{E}_0\left(1 - e^{-B^2l^2t/mR}\right) = \mathcal{E}_0 e^{-B^2l^2t/mR}$$

The net electric field is the potential gradient along the rod:

$$E_{net} = V/l = \frac{\mathcal{E}_0}{l}e^{-B^2l^2t/mR}$$

55. A side view of the rail and bar is shown in the figure. The component of the velocity of the bar that is perpendicular to the magnetic field is $v \cos \theta$, so the induced emf is

$$\mathcal{E} = Blv \cos \theta.$$

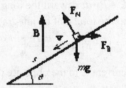

If we assume the resistance of the copper bar can be neglected, this produces a current in the bar

$$I = \mathcal{E}/R = (Blv \cos \theta)/R \text{ into the page.}$$

Because the current is perpendicular to the magnetic field, the force on the bar from the magnetic field will be horizontal, as shown, with magnitude

$F_B = IlB = (B^2l^2v\cos\theta)/R$.

For the bar to slide down at a steady speed, the net force must be zero. If we consider the components along the rail, we have

$F_B\cos\theta - mg\sin\theta = 0$,   or

$[(B^2l^2v\cos\theta)/R]\cos\theta = (B^2l^2v\cos^2\theta)/R = mg\sin\theta$;

$(0.55\text{ T})^2(0.30\text{ m})^2v(\cos^2 5.0°)/(0.60\ \Omega) = (0.040\text{ kg})(9.80\text{ m/s}^2)\sin 5.0°$, which gives $v = 0.76$ m/s.

59. The magnitude of the average induced emf is

$\mathcal{E} = \Delta\Phi_B/\Delta t$,

so the average current is

$I = \mathcal{E}/R = (1/R)\,\Delta\Phi_B/\Delta t$.

The charge moving past a fixed point is

$q = I\,\Delta t = \Delta\Phi_B/R = (B/R)\,\Delta A$.

The number of electrons is

$N = q/e = (B/eR)\,\Delta A$.

The maximum number of electrons occurs for the maximum change in area in a 90° rotation. The area perpendicular to the field varies sinusoidally. From the plot we see that the greatest change occurs when the coil moves from an angle of 45° with the field to an angle of 45° on the other side. Thus we have

$N_{max} = [(0.15\text{ T})\pi(0.030\text{ m})^2/(1.60\times 10^{-19}\text{ C})(0.025\ \Omega)][\cos 45° - (-\cos 45°)] = 1.5\times 10^{17}$.

63. (a) From the efficiency of the transformer, we have $P_S = 0.80P_P$.
    For the power input to the transformer, we have
    $P_P = I_PV_P$ ;
    $(75\text{ W})/0.80 = I_P(110\text{ V})$, which gives $I_P = 0.85$ A.
    (b) We find the secondary voltage from
    $P_S = V_S^2/R_S$ ;
    $75\text{ W} = V_S^2/(2.4\ \Omega)$, which gives $V_S = 13.4$ V.
    We find the ratio of the number of turns from
    $N_P/N_S = V_P/V_S = (110\text{ V})/(13.4\text{ V}) = 8.2$.

67. We can find the current in the transmission lines from the power transmitted to the user:
    $P_T = IV$,   or   $I = P_T/V$.
    The power loss in the lines is
    $P_L = I^2R_L = (P_T/V)^2R_L = (P_T)^2R_L/V^2$.

71. We choose a differential element $dr$ a distance $r$ from the center of rotation. The speed of this element is $v = \omega r$, so the radial differential induced emf is

$d\mathcal{E} = Bv\,dr = B\omega r\,dr$.

The electric field is the potential gradient:

$E = d\mathcal{E}/dr = B\omega r$, radially out from the axis.

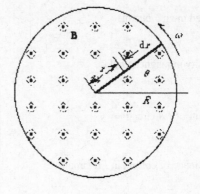

# Chapter 30: Inductance; and Electromagnetic Oscillations

## Chapter Overview and Objectives

This chapter introduces the concept of mutual and self-inductance. Faraday's law is rewritten in terms of inductance, a useful form for analysis of electronic circuits. The energy density of a magnetic field is discussed. Solutions to the time dependence of charge or current in $LR$, $LC$, and series $LRC$ circuits are developed.

After completing study of this chapter, you should:

- Know what mutual and self-inductance are.
- Know how mutual and self-inductance are related to Faraday's law.
- Know how to calculate induced emf's in electronic circuits.
- Know the energy density of a magnetic field.
- Know the characteristics of $LR$, $LC$, and $LRC$ circuits.
- Know how to calculate the time constant of an $LR$ circuit.
- Know how to calculate the natural frequencies of $LC$ and $LRC$ circuits.

## Summary of Equations

Definition of mutual inductance:
$$M_{21} = \frac{\Phi_{21} N_2}{I_1}$$

Emf induced in secondary coil:
$$\varepsilon_2 = -M_{21} \frac{dI_1}{dt}$$

Definition of self-inductance:
$$L = \frac{N\Phi_B}{I}$$

Emf induced by self-inductance:
$$\varepsilon = -L \frac{dI}{dt}$$

Power into an inductor:
$$P = I\varepsilon = LI \frac{dI}{dt}$$

Energy stored in an inductor:
$$U = \tfrac{1}{2} LI^2$$

Energy density in magnetic field:
$$u = \tfrac{1}{2} \frac{B^2}{\mu_0}$$

Time constant of $LR$ circuit:
$$\tau = \frac{L}{R}$$

Time dependence of current in $LR$ circuit:
$$I(t) = \frac{V_0}{R}\left(1 - e^{-Rt/L}\right) = \frac{V_0}{R}\left(1 - e^{-t/\tau}\right)$$

Natural frequency of oscillation of $LC$ circuit:
$$\omega = \sqrt{\frac{1}{LC}}$$

Time dependence of charge in $LC$ circuit:
$$Q(t) = Q_0 \cos(\omega t + \phi)$$

Natural frequency of oscillation in series $LRC$ circuit:
$$\omega = \sqrt{\frac{1}{LC} - \frac{R^2}{4L^2}}$$

Time dependence of charge in series *LRC* circuit:    $Q(t) = Q_0 e^{-\frac{R}{2L}t} \cos(\omega t + \phi)$

## Chapter Summary

### Section 30-1 Mutual Inductance

When two coils of wire are relatively close to each other, a changing current in one of the coils creates a changing magnetic field through the other coil, possibly creating a changing flux through the other coil and inducing an emf. The induced emf in the second coil, $\mathcal{E}_2$, is related to the rate of change of the flux through each loop of the second coil due to the magnetic field created by the first coil, $\Phi_{21}$, and the number of turns in the second coil, $N_2$, by Faraday's law:

$$\mathcal{E}_2 = -N_2 \frac{d\Phi_{21}}{dt}$$

For a given current in the first coil, the flux through the second coil depends on the geometry of the two coils and their position and orientation relative to each other. We do know that the flux will be proportional to the current in the first coil. We lump together all of the geometrical factors into a constant we call the mutual inductance of the coils and write the flux through each loop of the second coil as

$$\Phi_{21} = \frac{M_{21} I_1}{N_2}$$

The constant $M_{21}$ is called the **mutual inductance** of the two coils. We can write the induced emf in the second coil in terms o the mutual inductance of the two coils and the rate of change of current on the first coil:

$$\mathcal{E}_2 = -M_{21} \frac{dI_1}{dt}$$

A changing current in the second coil will also cause an induced emf in the first coil:

$$\mathcal{E}_1 = -M_{12} \frac{dI_2}{dt}$$

The mutual inductances $M_{12}$ and $M_{21}$ are the same, so usually the subscripts on $M$ are dropped. The dimensions of mutual inductance are voltage × time / current. The SI unit of inductance is the Henry, H. One Henry is one volt second per ampere ( 1 H = 1 V·s/A ).

### Example 30-1-A

Determine the mutual inductance of two small circular loops of wire of radius $r$ separated along a line joining the centers of the loops perpendicular to the plane of both loops. Assume the separation distance is much greater than the radii of the loops.

### Solution:

Return to Chapter 28 to find the magnetic field of a small circular loop far from the loop along a line that passes through the center of the loop. The magnetic field is given by

$$B = \frac{\mu_0}{2\pi} \frac{N_1 I_1 A_1}{x^3} \hat{n} = \frac{\mu_0}{2\pi} \frac{I_1 \pi r_1^2}{x^3} \hat{n}$$

where $I_1$ is the current in the first coil, $r_1$ is the radius of the first coil, $x$ is the distance between the centers of the coils, and $\hat{n}$ is the unit vector that points along the line joining the centers of the loops. Because the loop is small compared to the distance between the loops, we can approximate the magnetic field as being uniform over the area of the second loop. The flux through the second loop is then

$$\Phi_{21} = B \cdot A = \frac{\mu_0}{2\pi} \frac{I_1 \pi r_1^2 \pi\ r_2^2}{x^3} = \frac{\mu_0 \pi}{2x^3} r_1^2 r_2^2 I_1$$

where $r_2$ is the radius of the second coil. By inspection, we see that the mutual inductance is

$$M = \frac{\mu_0 \pi}{2x^3} r_1^2 r_2^2$$

## Section 30-2 Self-Inductance

A changing current through a coil causes a changing flux through that coil itself. The changing flux induces an emf in the coil by Faraday's law. In a similar manner to mutual inductance, we define the **self-inductance**, $L$, of a coil of wire:

$$L = \frac{N\Phi_B}{I}$$

where $N$ is the number of loops in the coil, $\Phi_B$ is the flux through the coil created by a current, $I$, in the coil. The induced emf in the coil can be written in terms of the self-inductance and the time rate of change of the current through the coil:

$$\varepsilon = -L\frac{dI}{dt}$$

The minus sign means that the induced emf is in a direction such that the induced emf would try to create a current that would oppose the change in current that is taking place. The dimensions of self-inductance are identical to those of mutual inductance; the SI unit Henry is also used for self-inductance.

An inductor is a coil of wire used in an electronic circuit for its self-inductance property. The symbol used for an inductor in an electronic circuit diagram is

—〰〰〰—

## Example 30-2-A

The current through an inductor with inductance $L$ as a function of time is $I_0 t^2$. What is the induced emf as a function of time?

**Solution:**

The induced emf is given by

$$\varepsilon = -L\frac{dI}{dt} = -L\frac{d}{dt}I_0 t^2 = -2LI_0 t$$

## Section 30-3 Energy Stored in a Magnetic Field

Power, $P$, is added to an inductor while its current is increasing in magnitude:

$$P = I\varepsilon = LI\frac{dI}{dt}$$

Note that the power into the inductor is positive when the magnitude of the current is increasing and negative when the magnitude of the current is decreasing. The power added to the inductor increases its potential energy, $U$:

$$U = \tfrac{1}{2}LI^2$$

If we replace the inductance and current by equivalent expressions involving the magnetic field, we find, equivalently, that

$$u = \tfrac{1}{2}\frac{B^2}{\mu_0}$$

where $u$ is the energy density or energy per volume associated with the magnetic field.

## Example 30-3-A

What is the total magnetic field energy within the cube shown in the diagram if the magnetic field is given by

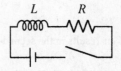

$$B(x, y, z) = B_0 \frac{x}{L}i - \tfrac{1}{2}B_0 \frac{y}{L}j - \tfrac{1}{2}B_0 \frac{z}{L}k$$

**Solution:**

The energy is given by

$$\int u^2 \, dV = \int \tfrac{1}{2}\frac{B^2}{\mu_0}dV = \frac{1}{2\mu_0}\int_0^L \int_0^L \int_0^L \mathbf{B} \cdot \mathbf{B} \, dxdydz$$

$$= \frac{1}{2\mu_0}\int_0^L \int_0^L \int_0^L \left(B_0^2 x^2 - \tfrac{1}{4}B_0^2 y^2 - \tfrac{1}{4}B_0^2 y^2\right)dxdydz$$

$$= \frac{B_0^2}{2\mu_0}\left[\left(\frac{x^3}{3}\right)\Big|_0^L + \tfrac{1}{4}\left(\frac{y^3}{3}\right)\Big|_0^L + \tfrac{1}{4}\left(\frac{z^3}{3}\right)\Big|_0^L\right]$$

$$= \frac{B_0^2 L^3}{4\mu_0}$$

## Example 30-3-B

A proposed energy storage mechanism is to store electrical energy in the magnetic field of a large inductor made of superconducting (R = 0) material. How long could you operate a 100 W light bulb with the energy stored in a 10 H inductor that carries a current of 40 A?

**Solution:**

The amount of energy stored in the inductor is

$$U = \tfrac{1}{2}LI^2 = \tfrac{1}{2}(10\,\text{H})(40\,\text{A})^2 = 1600\,\text{J}$$

The light bulb uses energy at the rate of 100 J/s, so this energy would last a time, $t$, given by

$$P = \frac{E}{t} \quad \Rightarrow \quad t = \frac{E}{P} = \frac{1600\,\text{J}}{100\,\text{W}} = 16\,\text{s}$$

Not very long!

## Section 30-4 *LR* Circuits

A series $RL$ circuit is shown in the diagram. When the switch is closed, applying Kirchoff's voltage rule to the circuit results in the equation:

$$V_0 - L\frac{dI}{dt} - IR = 0$$

where $I$ is a function of time. This equation has the general solution:

$$I(t) = \frac{V_0}{R} + I_1 e^{-Lt/R} = \frac{V_0}{R} + I_1 e^{-t/\tau}$$

where $I_1$ is a constant that depends on the current at time $t = 0$ and the **time constant**

$$\tau = \frac{L}{R}$$

For each increase in time of one time constant, the exponential function decreases by a factor $1/e$. At $t = 0$, the current is $V_0/R + I_1$, so we can write $I_1$ as

$$I_1 = I(0) - \frac{V_0}{R}$$

where $I(0)$ is the current at time $t = 0$. If the current is zero at time $t = 0$, then we can write the current in the circuit as

$$I(t) = \frac{V_0}{R}\left(1 - e^{-Rt/L}\right) = \frac{V_0}{R}\left(1 - e^{-t/\tau}\right)$$

**Example 30-4-A**

An $LR$ circuit has a potential difference source $V_0 = 5.0$ V, a resistance $R = 45\ \Omega$, an inductance $L = 34$ mH, and a switch in series. The switch is initially open. When the switch is closed, how long will it take the current in the circuit to reach 90 mA?

**Solution:**

We know the current in an $LR$ circuit has a time dependence given by

$$I = \frac{V_0}{R}\left(1 - e^{-Rt/L}\right)$$

To determine the time at which a particular current occurs, we solve this equation for the time

$$t = -\frac{L}{R}\ln\left(1 - \frac{IR}{V_0}\right) = -\frac{3.4 \times 10^{-2}\ \text{H}}{45\ \Omega}\ln\left(1 - \frac{\left(90 \times 10^{-3}\ \text{A}\right)\left(45\ \Omega\right)}{5.0\ \text{V}}\right) = 1.6 \times 10^{-4}\ \text{s}$$

**Section 30-5 *LC* Circuits and Electromagnetic Oscillations**

An $LC$ circuit has a capacitor and an inductor connected in series as shown in the diagram. Applying Kirchoff's voltage rule to this circuit, we get[1]

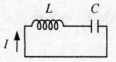

$$-L\frac{dI}{dt} + \frac{Q}{C} = 0$$

We can write the current $I$ in terms of the charge on the capacitor, $C$, as

$$I = -\frac{dQ}{dt}$$

This results in the second-order differential equation

---

[1] Note that the signs on these terms, as well as the sign on the relationship between $I$ and $dQ/dt$, depend on the direction of the current flow you define and which plate of the capacitor the charge $Q$ is on.

$$L\frac{d^2Q}{dt^2}+\frac{Q}{C}=0 \quad \Rightarrow \quad \frac{d^2Q}{dt^2}+\frac{Q}{LC}=0$$

This equation has the general solution for the charge on the capacitor as a function of time, $Q(t)$,

$$Q(t)=Q_0 \cos(\omega t+\phi)$$

where $Q_0$ and $\phi$ are constants that depend on the initial capacitor charge and the initial current. The angular frequency, $\omega$, is fixed by the circuit components and is given by

$$\omega=\sqrt{\frac{1}{LC}}$$

The current in the circuit is the negative of the time derivative of the charge on the capacitor:

$$I(t)=-\frac{dQ}{dt}=\frac{d}{dt}[Q_0\cos(\omega t+\phi)]=\omega Q_0\sin(\omega t+\phi)=I_0\sin(\omega t+\phi)$$

**Example 30-5-A**

Determine the charge on the capacitor as a function of time for an $LC$ circuit that has an inductor with inductance 365 mH and a capacitor with capacitance 534 nF. The charge on the capacitor is 0.365 μC and the current in the circuit is 2.45 mA and increasing at the time t = 1.76 μs.

**Solution:**

We know the general form of the solution of the charge as a function of time in an LC circuit:

$$Q(t)=Q_0\cos(\omega t+\phi)$$

and we know how to determine $\omega$ from the inductance and the capacitance:

$$\omega=\sqrt{\frac{1}{LC}}=\sqrt{\frac{1}{(365\times10^{-3}\text{ H})(534\times10^{-9}\text{ F})}}=2.27\times10^3\text{ s}^{-1}$$

The charge information at the given time tells us

$$Q(t)=Q_0\cos(\omega t+\phi) \quad \Rightarrow \quad 0.365\times10^{-6}\text{ C}=Q_0\cos(\omega t+\phi)$$

The information about the current at the given time tells us

$$I(t)=\omega Q_0\sin(\omega t+\phi) \quad \Rightarrow \quad 2.45\times10^{-3}\text{ A}=\omega Q_0\sin(\omega t+\phi)$$

If we divide the equation for the current condition by the equation for the charge condition, we get

$$\frac{2.45\times10^{-3}\text{ A}}{0.365\times10^{-6}\text{ C}}=\frac{\omega Q_0\sin(\omega t+\phi)}{Q_0\cos(\omega t+\phi)} \quad \Rightarrow$$

$$6.71\times10^3\text{ s}^{-1}=\omega\tan(\omega t+\phi) \quad \Rightarrow$$

$$\phi=\arctan\left(\frac{6.71\times10^3\text{ s}^{-1}}{2.27\times10^3\text{ s}^{-1}}\right)-(2.27\times10^3\text{ s}^{-1})(1.76\times10^{-3}\text{ s})=0.39 \text{ or } -2.75\,(\text{rad})$$

Your calculator will probably not give both results. Remember that the arctan function on the calculator only gives principle values between $-\pi/2$ and $\pi/2$. The second result is obtained by adding or subtracting $\pi$ from the result given by your calculator. Also, remember that angles different by any integer multiple of $2\pi$ are the same angle. Which result is the one we are looking for? We have a positive current in the problem, so we need to check which phase angle results in a positive current at the given time:

$$\text{For } \phi = 0.39 \quad \Rightarrow \quad \sin(\omega t + \phi) = \sin\left[(2.27 \times 10^3 \, \text{s}^{-1})(1.76 \times 10^{-3} \, \text{s}) + 0.39\right] = -0.95 < 0$$

$$\text{For } \phi = -2.75 \quad \Rightarrow \quad \sin(\omega t + \phi) = \sin\left[(2.27 \times 10^3 \, \text{s}^{-1})(1.76 \times 10^{-3} \, \text{s}) - 2.75\right] = +0.95 > 0$$

So the correct phase angle is $-2.75$ radians.

To find $Q_0$, we square the charge and current equations and add them after dividing the current equation by $\omega$ on both sides:

$$\left(0.365 \times 10^{-6} \, \text{C}\right)^2 + \left(\frac{2.45 \times 10^{-3} \, \text{A}}{\omega}\right)^2 = Q_0^2 \cos^2(\omega t + \phi) + Q_0^2 \sin^2(\omega t + \phi) = Q_0^2$$

Solving this for $Q_0$, we get

$$Q_0 = \sqrt{\left(0.365 \times 10^{-6} \, \text{C}\right)^2 + \left(\frac{2.45 \times 10^{-3} \, \text{C/s}}{2.27 \times 10^3 \, \text{s}^{-1}}\right)^2} = 1.14 \times 10^{-6} \, \text{C}$$

## Section 30-6 *LC* Oscillations with Resistance (*LRC* Circuit)

A series *LRC* circuit is shown in the circuit diagram. If we apply Kirchoff's voltage rule to this circuit, we get

$$-L\frac{dI}{dt} - IR + \frac{Q}{C} = 0$$

Again, we can use the relationship

$$I = -\frac{dQ}{dt}$$

to rewrite the equation as

$$L\frac{d^2Q}{dt^2} + \frac{dQ}{dt}R + \frac{Q}{C} = 0$$

This equation is of the same form as the damped harmonic oscillator equation of Chapter 14. Recall that the nature of the solution for the damped harmonic oscillator depended on the relative sizes of the damping constant, the mass, and the spring constant. In this case, the *LRC* circuit solutions are of the types shown below:

$$R^2 < \frac{4L}{C} \qquad\qquad \text{underdamped}$$

$$R^2 = \frac{4L}{C} \qquad\qquad \text{critically damped}$$

$$R^2 > \frac{4L}{C} \qquad\qquad \text{overdamped}$$

The solution for the underdamped case is

$$Q(t) = Q_0 e^{-\frac{R}{2L}t} \cos(\omega t + \phi)$$

where

$$\omega = \sqrt{\frac{1}{LC} - \frac{R^2}{4L^2}}$$

and $Q_0$ and $\phi$ are constants that depend on the initial charge on the capacitor and the initial current.

## Practice Quiz

1.    The magnetic field of a current loop is not zero anywhere. What can you do with a second current loop so that the mutual inductance between the two current loops is zero?

      a)  Place the second loop directly on top of the first current loop.
      b)  Form the second loop into a shape with as large an area as possible.
      c)  Orient the plane of the second loop so that it is parallel to the magnetic field of the first loop.
      d)  Apply an electric potential difference across the second loop.

2.    If you wanted to increase the time constant of an $LR$ circuit, you could

      a)  decrease the inductance.
      b)  increase the inductance.
      c)  increase the voltage.
      d)  either increase the inductance or increase the voltage.

3.    To help eliminate unwanted emf's induced in a circuit due to changing currents elsewhere, you should

      a)  make the area enclosed by the conductors of the circuit as large as possible.
      b)  make the area enclosed by the conductors of the circuit as small as possible.
      c)  place the circuit as close as possible to the other currents.
      d)  only use AC current in your circuit.

4.    The natural frequency of a damped harmonic oscillator is often approximately the same as the natural frequency of the undamped harmonic oscillator. Under what conditions is this a good approximation?

      a)  It is always a good approximation.
      b)  The smaller the resistance in the circuit, the better the approximation is.
      c)  The larger the resistance in the circuit, the better the approximation is.
      d)  It is a good approximation whenever you are solving a homework or test problem.

5.    An $LR$ circuit with no battery in the circuit has a current $I$ at time $t = 0$. At time $t = t_1$, the circuit has a current $I/2$. At what time does the circuit have a current $I/4$?

      a) 1.5 $t_1$
      b) 2.0 $t_1$
      c) 4.0 $t_1$
      d) The circuit will never have a current $I/4$.

6.    One way to tap a phone line without physical connection to the line is to place a coil of wire near the phone line. How should a coil be oriented to pick up the maximum emf from the signal on the phone line?

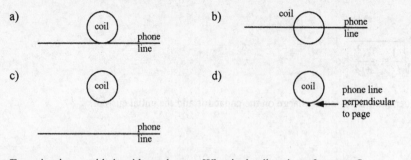

7.    Two circuits are side by side as shown. What is the direction of current flow induced in the right-hand circuit when the switch in the left-hand circuit is closed?

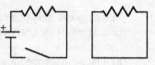

a)  clockwise
b)  counter-clockwise
c)  No current flow is induced in the coil on the right.
d)  It depends on how quickly the switch is closed.

8.    In the diagram for Question 7, if the switch has been closed for some time so that the current in the left-hand circuit is constant, what will be the direction of the induced current in the right-hand circuit?

a) clockwise
b) counter-clockwise
c) no current flow is induced in the coil on the right
d) it depends on how long the current has been constant

9.    Use dimensional analysis to determine the proportionality relationship between the inductance of a single circular loop of wire and its radius. The only dimensional parameters involved in determining the inductance are $\mu_0$, $I$, and $r$.

a) $L \propto r$
b) $L \propto r^2$
c) $L \propto 1/r$
d) $L \propto 1/r^2$

10.   Which of three elements in an *LRC* series circuit always has (or have) non-negative power put into it (or them) during the operation of the circuit?

a) inductor
b) resistor
c) capacitor
d) inductor and capacitor

11.   Two loops have a mutual inductance $M$. Loop #1 has a current as a function of time equal to $I_1(t) = I_0 \sin (2\pi ft)$. What is the induced emf in loop #2?

12.   A wire runs next to a rectangular loop of wire as shown in the diagram. The wire carries a current $I$. What is the average energy density of the magnetic field in the plane of the rectangular loop interior to the loop?

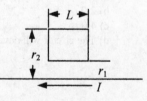

13.   In an *LR* circuit with a resistance of 123 $\Omega$, it takes the current 32.6 ms to reach 50.0% of its maximum value from zero. If the resistance is changed to 394 $\Omega$, how long would the current take to reach 98% of its maximum value starting from zero?

14.    You want to design an AM radio receiver to pick up a station operating on a frequency of 1040 kHz. You have an inductor that is a solenoid with a diameter of 5.78 cm, a length of 14.6 cm, and has 240 turns of wire. What value of capacitance do you need so that the natural frequency of an LC circuit made from these components is 1040 kHz?

15.    An $LRC$ series circuit has a frequency 340 kHz and loses 3.78% of its energy on each cycle. The resistor in the circuit has a resistance of 28.5 $\Omega$. What are the values of the inductance and the capacitance of the circuit?

## Problem Solutions

3.    The magnetic field inside the outer solenoid is
$$B = \mu_0 n_1 I_1.$$
The magnetic flux linked with the inner solenoid is
$$\Phi_{21} = \mu_0 n_1 I_1 A_2.$$
Thus the mutual inductance with the $N_2$ turns of the inner solenoid is
$$M_{21} = N_2 \Phi_{21}/I_1 = n_2 l \mu_0 n_1 A_2,$$ so the mutual inductance per unit length is
$$M/l = \mu_0 n_1 n_2 \pi r_2^2.$$

7.    We estimate the inductance by using the inductance of a solenoid:
$$L = \mu_0 N^2 A/l = (4\pi \times 10^{-7}\text{ T} \cdot \text{m/A})(20{,}000 \text{ turns})^2 \pi (1.85 \times 10^{-2}\text{ m})^2/(0.45\text{ m}) = 1.2\text{ H}.$$

11.    We use the result for the inductance from Ex. 30–4:
$$L/l = (\mu_0/2\pi)\ln(r_2/r_1);$$
$(2 \times 10^{-7}\text{ T} \cdot \text{m/A})\ln(3.0\text{ mm}/r_1) \le 40 \times 10^{-9}\text{ H/m}$, which gives $r_1 \ge 2.5$ mm.

15.    (a) For two inductors placed in series, the current through each inductor is the same. This current is also the current through the equivalent inductor, so the total emf is
$$\mathcal{E} = \mathcal{E}_1 + \mathcal{E}_2$$
$- L_{\text{series}}\, dI/dt = (- L_1\, dI/dt) + (- L_2\, dI/dt) = - (L_1 + L_2)\, dI/dt$, which gives $L_{\text{series}} = L_1 + L_2$.

(b) For two inductors placed in parallel, the potential difference across each inductor, which is the emf, is the same:
$$\mathcal{E} = \mathcal{E}_1 = \mathcal{E}_2 = - L_1\, dI_1/dt = - L_2\, dI_2/dt = - L_{\text{parallel}}\, dI/dt.$$
The total current through the equivalent inductor is
$I = I_1 + I_2$, so we have
$dI/dt = dI_1/dt + dI_2/dt$;
$- \mathcal{E}/L_{\text{parallel}} = - \mathcal{E}/L_1 - \mathcal{E}/L_2$, which gives $1/L_{\text{parallel}} = (1/L_1) + (1/L_2)$, or $L_{\text{parallel}} = L_1 L_2/(L_1 + L_2)$.

19.    (a) For the energy densities we have
$$u_E = \tfrac{1}{2}\,\varepsilon_0 E^2 = \tfrac{1}{2}(8.85 \times 10^{-12}\text{ F/m})(1.0 \times 10^4\text{ V/m})^2 = 4.4 \times 10^{-4}\text{ J/m}^3;$$
$$u_B = \tfrac{1}{2}B^2/\mu_0 = \tfrac{1}{2}(2.0\text{ T})^2/(4\pi \times 10^{-7}\text{ T} \cdot \text{m/A}) = 1.6 \times 10^6\text{ J/m}^3.$$
We see that $u_B \gg u_E$.

(b) We find the magnitude of the electric field from
$$u_E = \tfrac{1}{2}\,\varepsilon_0 E^2;$$
$1.6 \times 10^6\text{ J/m}^3 = \tfrac{1}{2}(8.85 \times 10^{-12}\text{ F/m})E^2$, which gives $E = 6.0 \times 10^8\text{ V/m}$.

23.    From Example 30–4, the magnetic field is
$$B = \mu_0 I/2\pi r.$$
The energy density of this field is
$$u = \tfrac{1}{2}B^2/\mu_0 = \mu_0 I^2/8\pi^2 r^2.$$
To find the total energy stored in the magnetic field, we integrate over the volume. For a differential element, we choose a ring at a radius $r$ with width $dr$ and length $l$:
$$\frac{U}{l} = \int \frac{u}{l}\,dV = \int_{r_1}^{r_2} \frac{\mu_0 I^2}{8\pi^2 r^2 l} 2\pi r l\, dr = \frac{\mu_0 I^2}{4\pi^2}\ln\!\left(\frac{r_2}{r_1}\right)$$
This is the same as found in Example 30–5.

27.    At $t = 0$ there is no voltage drop across the resistor, so we have
$$V_0 = L(dI/dt)_0,\text{ or }(dI/dt)_0 = V_0/L.$$
The maximum value of the current is reached after a long time, when there is no voltage across the inductor:

$V_0 = I_{max}R$, or $I_{max} = V_0/R$.

We find the time to reach maximum if the initial rate were maintained from

$V_0/R = (dI/dt)_0 t = (V_0/L)t$, which gives $t = L/R = \tau$.

31. (*a*)  The resonant frequency is given by

$f_0^2 = (1/2\pi)^2(1/LC)$.

When we form the ratio for the two stations, we get

$(f_{02}/f_{01})^2 = C_1/C_2$;

$(1600 \text{ kHz}/550 \text{ kHz})^2 = (1800 \text{ pF})/C_2$, which gives $C_2 = 213$ pF.

(*b*)  We find the inductance from the first frequency:

$f_{01} = (1/2\pi)(1/LC_1)^{1/2}$;

$550 \times 10^3 \text{ Hz} = (1/2\pi)[1/L(1800 \times 10^{-12} \text{ F})]^{1/2}$, which gives $L = 4.65 \times 10^{-5} \text{ H} = 46.5 \text{ μH}$.

35. (*a*)  The initial energy is the energy stored in the capacitor at $t = 0$:

$U_0 = \frac{1}{2}Q_0^2/C$.

When the capacitor has half the energy, we have

$U_C/U_0 = Q^2/Q_0^2 = \frac{1}{2}$, which gives $Q = Q_0/\sqrt{2}$.

(*b*)  The charge on the capacitor is

$Q = Q_0 \cos \omega t = Q_0/\sqrt{2}$.

This gives

$\omega t = \pi/4$;

$(2\pi/T)t = \pi/4$, which gives $t = T/8$.

39.  The charge is given by

$Q = Q_0 e^{-Rt/2L} \cos \omega' t$.

We find the current by differentiating:

$I = -dQ/dt = -Q_0(-R/2L)e^{-Rt/2L} \cos \omega' t + Q_0\omega' e^{-Rt/2L} \sin \omega' t$

$= +Q_0 e^{-Rt/2L}[\omega' \sin \omega' t + (R/2L) \cos \omega' t]$.

We form the triangle shown so we can express the terms in the bracket
as a combination of trig functions:

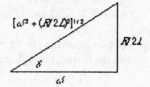

$I = Q_0 e^{-Rt/2L}[\omega'^2 + (R/2L)^2]^{1/2}(\{\omega'/[\omega'^2 + (R/2L)^2]^{1/2}\} \sin \omega' t + \{(R/2L)/[\omega'^2 + (R/2L)^2]^{1/2}\} \cos \omega' t)$

$= Q_0 e^{-Rt/2L}[\omega'^2 + (R/2L)^2]^{1/2}(\cos \delta \sin \omega' t + \sin \delta \cos \omega' t)$, where $\tan \delta = R/2L\omega'$.

If we use the expression for $\omega'$, we find

$[\omega'^2 + (R/2L)^2]^{1/2} = [(1/LC) - (R/2L)^2 + (R/2L)^2]^{1/2} = 1/(LC)^{1/2}$.

Thus we have

$I = [Q_0/(LC)^{1/2}] e^{-Rt/2L} \sin (\omega' t + \delta)$, with $\delta = \tan^{-1}(R/2L\omega')$.

43.  We use the inductance of a solenoid:

$L = \mu_0 AN^2/l = (4\pi \times 10^{-7} \text{ T} \cdot \text{m/A})\pi(1.25 \times 10^{-2} \text{ m})^2(3000 \text{ turns})^2/(0.282 \text{ m}) = 1.97 \times 10^{-2} \text{ H} = 20 \text{ mH}$.

If we form the ratio of inductances for the two conditions, we have

$L_2/L_1 = (\mu/\mu_0)(N_2/N_1)^2$;

$1 = (1000)[N_2/(3000 \text{ turns})]^2$, which gives $N_2 = 95$ turns.

47.  We use the inductance of a solenoid:

$L = \mu_0 AN^2/l$;

$25 \times 10^{-3} \text{ H} = (4\_ \times 10^{-7} \text{ T} \cdot \text{m/A})\_(1.1 \times 10^{-2} \text{ m})^2 N^2/(0.170 \text{ m})$, which gives $N = 3.0 \times 10^3$ turns.

If we form the ratio of inductances for the two conditions, we have

$L_2/L_1 = (\mu/\mu_0)(N_2/N_1)^2$;

$1 = (10^3)[N_2/(3.0 \times 10^3 \text{ turns})]^2$, which gives $N_2 = 95$ turns.

51. (*a*)  For two coils placed in series, the current through each coil is the same and is
also the current through the equivalent inductor.  The total induced emf is

$\varepsilon = \varepsilon_1 + \varepsilon_2 + \varepsilon_{21} + \varepsilon_{12}$

$-L_{series} \, dI/dt = (-L_1 \, dI/dt) + (-L_2 \, dI/dt) \pm [(-M \, dI/dt) + (-M \, dI/dt)]$

$= -(L_1 + L_2 \pm 2M) \, dI/dt$, which gives $L_{series} = L_1 + L_2 \pm 2M$.

The sign of the mutual inductance term is determined by the orientation of one winding with
respect to the other.  The upper sign is used when the emf from the mutual inductance is in the
same sense as the emf produced by the self-inductance.  The lower sign is used when the emf from
the mutual inductance is in the opposite sense as the emf produced by the self-inductance.

(b) To reduce the value of $M$ the flux linkage must be reduced. If the separation cannot be increased, the coils can be oriented so the magnetic field of one coil is parallel to the other coil. This can be achieved by positioning one coil perpendicular to the other.

(c) For two coils placed in parallel, the potential difference across each coil, which is the emf, is the same. With no mutual inductance we have

$\varepsilon = \varepsilon_1 = \varepsilon_2 = -L_1\, dI_1/dt = -L_2\, dI_2/dt = -L_{parallel}\, dI/dt$.

The total current through the equivalent inductor is

$I = I_1 + I_2$, so we have

$dI/dt = dI_1/dt + dI_2/dt$;

$-\varepsilon/L_{parallel} = -\varepsilon/L_1 - \varepsilon/L_2$, which gives $1/L_{parallel} = (1/L_1) + (1/L_2)$, or $L_{parallel} = L_1 L_2/(L_1 + L_2)$.

If we cannot ignore the mutual inductance, we have

$\varepsilon = \varepsilon_1 = -L_1\, dI_1/dt - M\, dI_2/dt = \varepsilon_2 = -L_2\, dI_2/dt - M\, dI_1/dt$.

When these equations are combined, we get

$dI_1/dt = -(L_2 - M)\,\varepsilon/(L_1 L_2 - M^2)$, and $dI_2/dt = -(L_1 - M)\,\varepsilon/(L_1 L_2 - M^2)$.

Because $I = I_1 + I_2$, we have

$dI/dt = dI_1/dt + dI_2/dt$;

$-\varepsilon/L_{parallel} = -(L_2 - M)\,\varepsilon/(L_1 L_2 - M^2) - (L_1 - M)\,\varepsilon/(L_1 L_2 - M^2)$,

which gives

$L_{parallel} = (L_1 L_2 - M^2)/(L_1 + L_2 - 2M)$.

The sign of the mutual inductance term in the denominator is determined by the orientation of one winding with respect to the other. The upper sign is used when the emf from the mutual inductance is in the same sense as the emf produced by the self-inductance. The lower sign is used when the emf from the mutual inductance is in the opposite sense as the emf produced by the self-inductance.

55. (a) For lightly damped motion,

$R \ll (4L/C)^{1/2}$, $\omega' \approx \omega = 1/(LC)^{1/2}$, $T = 2\pi/\omega' \approx 2\pi(LC)^{1/2}$, and $R/L\omega' \approx R(C/L)^{1/2} \ll 1$.

The charge on the capacitor is

$Q = Q_0 e^{-Rt/2L} \cos(\omega't + \phi)$.

We find the current by differentiating:

$I = -dQ/dt = -Q_0(-R/2L)e^{-Rt/2L}\cos(\omega't + \phi) + Q_0\omega' e^{-Rt/2L}\sin(\omega't + \phi)$

$= +Q_0 e^{-Rt/2L}[\omega'\sin(\omega't + \phi) + (R/2L)\cos(\omega't + \phi)]$.

Because the values for the sine and cosine functions range over $\pm 1$, and $\omega' \gg R/2L$, we neglect the second term to get

$I \approx Q_0 e^{-Rt/2L}\omega'\sin(\omega't + \phi)$.

The total energy is

$U = U_E + U_B = \frac{1}{2}(Q^2/C) + \frac{1}{2}LI^2$

$= \frac{1}{2}(Q_0{}^2/C)e^{-Rt/L}\cos^2(\omega't + \phi) + \frac{1}{2}LQ_0{}^2\omega'^2 e^{-Rt/L}\sin^2(\omega't + \phi)$.

When we use $\omega'^2 \approx 1/LC$, we get

$U = \frac{1}{2}(Q_0{}^2/C)e^{-Rt/L}[\cos^2(\omega't + \phi) + \sin^2(\omega't + \phi)] = \frac{1}{2}(Q_0{}^2/C)e^{-Rt/L}$.

(b) The rate at which the stored energy changes is

$dU/dt = \frac{1}{2}(Q_0{}^2/C)(-R/L)e^{-Rt/L} = -\frac{1}{2}(Q_0{}^2 R/LC)e^{-Rt/L}$.

The rate at which thermal energy is produced in the resistor is

$I^2R = Q_0{}^2\omega'^2 R\, e^{-Rt/L}\sin^2(\omega't + \phi) = (Q_0{}^2 R/LC)\,e^{-Rt/L}\sin^2(\omega't + \phi)$.

For lightly damped motion, the amplitude does not change significantly over a period, while the average value for $\sin^2(\omega't + \phi)$ is $\frac{1}{2}$. Thus we get

$I^2R = \frac{1}{2}(Q_0{}^2 R/LC)\,e^{-Rt/L} = -dU/dt$.

Thus the loss of stored energy becomes the thermal energy produced in the resistor.

# Chapter 31: AC Circuits

## Chapter Overview and Objectives

This chapter discusses the behavior of circuits with resistors, inductors, and capacitors when the circuits have ac voltage sources. It covers resonance of *LRC* series circuits impedance matching, and three-phase power.

After completing study of this chapter, you should:

- Know how to relate the ac current to the ac voltage in resistors, capacitors, and inductors.
- Know how to calculate reactance of inductors and capacitors.
- Know the current lags voltage in an inductor and current leads voltage in a capacitor.
- Know how to calculate the impedance of a series *LRC* circuit.
- Know how to determine the resonant frequency of a series *LRC* circuit.
- Know what impedance matching is.
- Know what three-phase power is.

## Summary of Equations

Definition of inductive reactance:  $X_L = \omega L$

Definition of capacitive reactance:  $X_C = \dfrac{1}{\omega C}$

Impedance of a series *LRC* circuit:  $Z = \sqrt{R^2 + (X_L - X_C)^2} = \sqrt{R^2 + \left(\omega L - \dfrac{1}{\omega C}\right)^2}$

Average power from source in series *LRC* circuit:  $\overline{P} = I_{rms}^2 Z \cos\phi = I_{rms} V_{rms} \cos\phi$

Resonant frequency of series *LRC* circuit:  $f_0 = \dfrac{\omega_0}{2\pi} = \dfrac{1}{2\pi\sqrt{LC}}$

Impedance matching condition:  $R_{load} = R_{source}$

## Chapter Summary

### Section 31-1 Introduction: AC Circuits

An ac circuit is an electrical circuit that has ac current or voltage sources in the circuit. To simplify our discussion of such circuits, we will assume that the current has a time dependence

$$I = I_0 \sin 2\pi\, ft = I_0 \sin \omega t$$

where $\omega = 2\pi f$.

### Section 31-2 AC Circuit Containing only Resistance *R*

Ohm's law applies at all instants of time in an ideal resistance. If an ac voltage source is connected to a resistance, then the current in the circuit is related to the voltage from the source as

$$V(t) = I(t)R$$

So, for our assumed ac current,

$$V(t) = I_0 R \sin \omega t = V_0 \sin \omega t$$

where $V_0 = I_0 R$ is the amplitude of the ac voltage. Because the phase angle of the voltage and the current is identical, we say the current and the voltage are in phase. The average electrical power input into the resistance is

$$\overline{P} = \tfrac{1}{2} I_0 V_0 = I_{rms}^2 R = V_{rms}^2 / R$$

## Section 31-3 AC Circuit Containing only Inductance $L$

Kirchhoff's voltage loop rule for a circuit consisting of an inductor connected to an ac voltage source is

$$V - L \frac{dI}{dt} = 0 \quad \Rightarrow \quad V = L \frac{dI}{dt}$$

With our assumed ac current,

$$V = L \frac{d}{dt} (I_0 \sin \omega t) = \omega I_0 L \cos \omega t = \omega I_0 L \sin(\omega t + 90°) = V_0 \sin(\omega t + 90°)$$

where

$$V_0 = \omega L I_0$$

is the amplitude of the voltage signal. We notice that the phase of the voltage is 90° greater than the current. We say **the current lags the voltage in an inductor by 90°**.

The previous equation looks similar to Ohm's law, with $\omega L$ taking the place of the resistance. It is not the same as Ohm's law, because it only relates the amplitudes of the current and voltage, not values at a particular time. It does not reflect the 90° phase shift between the current and the voltage. We call $\omega L$ the inductive reactance, $X_L$:

$$X_L = \omega L$$

Notice that the inductive reactance increases with increasing frequency.

## Example 31-3-A

An inductor is made from a solenoid with a length 10.0 cm and a diameter of 2.88 cm. When the solenoid is connected to a voltage source with an amplitude of 30.0 V and a frequency of 400 Hz, a current of amplitude 10.6 mA flows in the circuit. Determine the number of turns in the inductor.

## Solution:

We know that the inductance of the solenoid depends on the number of turns in the solenoid and that the relationship between the voltage and current amplitude in the circuit depends on the inductive reactance of the inductor. The inductance, $L$, of a solenoid was given in Chapter 30 of the text:

$$L = \frac{\mu_0 N^2 A}{l}$$

where $A$ is the cross-sectional area of the solenoid, $l$ is the length of the solenoid, and $N$ is the number of turns in the solenoid. The amplitude of the voltage source is related to the current amplitude in the circuit by

$$V_0 = I_0 \omega L$$

which implies

$$L = \frac{V_0}{\omega I_0}$$

Setting the two expressions for inductance equal to each other and solving for the number of turns in the solenoid, we have

$$N = \sqrt{\frac{V_0 l}{\mu_0 \omega I_0 A}} = \sqrt{\frac{(30.0 \text{ V})(0.0100 \text{ m})}{\left(4\pi \times 10^{-7} \text{ T·m/A}\right)\left(2\pi \, 400 \times 10^3 \text{ s}^{-1}\right)\left(106 \times 10^{-3} \text{ A}\right)\left(\frac{1}{4}\pi \left[0.0288 \text{ m}\right]^2\right)}} = 1376$$

### Section 31-4 AC Circuit Containing only Capacitance *C*

Kirchhoff's voltage loop rule for a circuit consisting of a capacitor connected to an ac voltage source is

$$V - \frac{Q}{C} = 0 \qquad V = \frac{Q}{C}$$

The current is the time derivative of the charge on the capacitor:

$$I = \frac{dQ}{dt} = \frac{d}{dt}CV = C\frac{d}{dt}V \quad \Rightarrow \quad dV = \frac{1}{C}I\,dt$$

Substituting our assumed ac current and integrating, we have

$$\int dV = \frac{1}{C}\int I_0 \sin \omega t \, dt \quad \Rightarrow \quad V = \frac{-I_0}{\omega C}\cos \omega t = V_0 \sin(\omega t - \pi/2)$$

where

$$V_0 = \frac{1}{\omega C}I_0$$

is the amplitude of the voltage signal. We notice that the phase of the voltage is 90° ($\pi/2$ radians) less than the current. We say **the current leads the voltage in a capacitor by 90°**.

The previous equation looks similar to Ohm's law, with $1/\omega C$ taking the place of the resistance. It is not the same as Ohm's law, because it only relates the amplitudes of the current and voltage, not values at a particular time. It does not reflect the 90° phase shift between the current and the voltage. We call $1/\omega C$ the **capacitive reactance**, $X_C$:

$$X_C = \frac{1}{\omega C}$$

Notice that the capacitive reactance decreases with increasing frequency.

### Example 31-4-A

The rms current in a capacitive circuit with a voltage source of rms voltage 12.0 V and frequency 1.20 kHz is 2.78 mA. What is the capcitance of the capcitor in the circuit?

### Solution:

We know the rms current and the rms voltage are related by

$$V_{rms} = I_{rms}X_C = \frac{I_{rms}}{\omega C} = \frac{I_{rms}}{2\pi \, fC}$$

If we solve for the capacitance, $C$, we get

$$C = \frac{I_{rms}}{2\pi \, f V_{rms}} = \frac{2.78 \times 10^{-3} \, A}{2\pi \left(1.20 \times 10^3 \, s^{-1}\right)\left(12.0 \, V\right)} = 3.07 \times 10^{-8} \, F = 30.7 \, nF$$

### Section 31-5 *LRC* Series AC Circuit

A series *LRC* circuit has an inductor, a resistor, and a capacitor in series as shown in the diagram. Again, we assume the current in the circuit has the form $I(t) = I_0 \sin \omega t$. By Kirchhoff's voltage rule, we know the sum of the voltage drops across each element will add up to the voltage of the source:

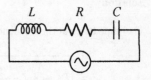

$$V = V_L + V_R + V_C$$

We know how to write the voltage across each element as a function of time when the current is given, so we can write:

$$V = \omega L I_0 \, \sin\left(\omega t + \pi/2\right) + R I_0 \, \sin\left(\omega t\right) + \frac{1}{\omega C}\sin\left(\omega t - \pi/2\right)$$

The sum of several sine functions that have the same frequency but different amplitudes and phases can always be written, using trigonometric identities, as a single sine function with the same frequency and some amplitude and phase. We would like to do that to the expression above. Using trigonometric identities or another method that makes use of writing trigonometric functions as the real or imaginary parts of exponential functions of complex numbers, called phasors, we find that

$$V = \sqrt{R^2 + \left(\omega L - \frac{1}{\omega C}\right)^2} \; I_0 \, \sin\left[\omega t + \phi\right]$$

where

$$\phi = \arctan\left(\frac{\omega L - \dfrac{1}{\omega C}}{R}\right)$$

The voltage can also be written as

$$V = \sqrt{R^2 + \left(X_L - X_C\right)^2} \; I_0 \, \sin\left[\omega t + \phi\right]$$

where

$$\phi = \arctan\left(\frac{X_L - X_C}{R}\right)$$

The coefficient of $I_0$ is called the impedance of the circuit and is denoted by $Z$:

$$Z = \sqrt{R^2 + \left(X_L - X_C\right)^2}$$

The average power in such a circuit is

$$\overline{P} = I_{rms}^2 Z \cos\phi = I_{rms}V_{rms}\cos\phi$$

The cos $\phi$ factor in the average power expression is called the **power factor**.

**Example 31-5-A**

A series *LRC* circuit has an inductor with an inductance of 34 mH, a resistor with a resistance of 2.8 kΩ, and a capacitor with a capacitance of 20 nF.  The voltage source has a voltage with an rms voltage of 4.8 V and a frequency of 36 kHz. What is the resonant frequency of the circuit?  What is the impedance of the circuit at the frequency of the voltage source?  What is the rms current in the circuit?  What is the power delivered to the circuit?

**Solution:**

The impedance at the frequency of the voltage source is

$$Z = \sqrt{R^2 + \left(\omega L - \frac{1}{\omega C}\right)^2}$$

$$= \sqrt{\left(2.8 \times 10^3 \ \Omega\right)^2 + \left(\frac{\left(36 \times 10^3 \ \text{s}^{-1}\right)\left(34 \times 10^3 \ \text{H}\right)}{2\pi} - \frac{2\pi}{\left(36 \times 10^3 \ \text{s}^{-1}\right)\left(20 \times 10^{-9} \ \text{F}\right)}\right)^2}$$

$$= 9.0 \times 10^3 \ \Omega$$

The rms current will be the rms voltage divided by the impedance:

$$I_{rms} = \frac{V_{rms}}{Z} = \frac{4.8 \ \text{V}}{9.0 \times 10^3 \ \Omega} = 5.3 \times 10^{-4} \ \text{A}$$

To calculate the power, we need the phase so that we can calculate the power factor:

$$\phi = \arctan\left(\frac{X_L - X_C}{R}\right)$$

$$= \arctan\left(\frac{\dfrac{\left(36 \times 10^3 \ \text{s}^{-1}\right)\left(34 \times 10^3 \ \text{H}\right)}{2\pi} - \dfrac{2\pi}{\left(36 \times 10^3 \ \text{s}^{-1}\right)\left(20 \times 10^{-9} \ \text{F}\right)}}{2.8 \times 10^3 \ \Omega}\right)$$

$$= -72°$$

The average power is equal to

$$\overline{P} = I_{rms} V_{rms} \cos\phi = \left(5.3 \times 10^{-4} \ \text{A}\right)\left(4.8 \ \text{V}\right)\cos\left(-72°\right) = 7.8 \times 10^{-4} \ \text{W}$$

**Section 31-6 Resonance in AC Circuits**

For a voltage source of fixed amplitude in an *LRC* circuit, we can look at how the amplitude of the current depends on frequency.  The current amplitude is

$$I_0 = \frac{V_0}{Z} = \frac{V_0}{\sqrt{R^2 + \left(\omega L - \dfrac{1}{\omega C}\right)^2}}$$

It is easy to see that this will reach a maximum value $V_0/R$ when the angular frequency is such that

$$\omega L - \frac{1}{\omega C} = 0$$

This occurs when

$$\omega = \frac{1}{\sqrt{LC}}$$

We give this angular frequency the symbol $\omega_0$. When the ac voltage source has this angular frequency, we call it resonance. The **resonant frequency** of the circuit is

$$f_0 = \frac{\omega_0}{2\pi} = \frac{1}{2\pi\sqrt{LC}}$$

## Example 31-6-A

Determine the resonant frequency of a series $LRC$ circuit that has an inductance of 66.8 mH, a resistance of 2.82 $\Omega$, and a capacitance of 92.6 nF. Determine the rms current in this series $LRC$ circuit when the source operates at the resonant frequency with an rms voltage of 12.0 V. Determine the rms voltage across the resistor, the inductor, and the capacitor with this source.

**Solution:**

The resonant frequency of the circuit is

$$f_0 = \frac{\omega_0}{2\pi} = \frac{1}{2\pi\sqrt{LC}} = \frac{1}{2\pi\sqrt{\left(66.8\times10^{-3}\text{ H}\right)\left(92.6\times10^{-9}\text{ F}\right)}} = 2.02\times10^3 \text{ Hz}$$

The impedance at the resonant frequency of the circuit is just the resistance of the circuit, 2.82 $\Omega$. The rms current and rms amplitude are related by

$$I_{rms} = \frac{V_{rms}}{Z} = \frac{12.0\text{ V}}{2.82\,\Omega} = 4.26 \text{ A}$$

The rms current is the same for each component of the circuit. We calculate the rms voltage across each circuit element using its impedance:

Resistor : $\quad V_{rms} = I_{rms}R = (4.26\text{ A})(2.82\,\Omega) = 12.0 \text{ V}$

Inductor : $\quad V_{rms} = I_{rms}X_L = I_{rms}\omega L = (4.26\text{ A})\left(2\pi\times2.02\times10^3\text{ s}^{-1}\right)\left(66.8\times10^{-3}\text{ H}\right) = 3.62\times10^3 \text{ V}$

Capacitor : $\quad V_{rms} = I_{rms}X_C = \dfrac{I_{rms}}{\omega C} = \dfrac{(4.26\text{ A})}{\left(2\pi\times2.02\times10^3\text{ s}^{-1}\right)\left(92.6\times10^{-9}\text{ F}\right)} = 3.62\times10^3 \text{ V}$

## Section 31-7 Impedance Matching

A source of electrical power can usually be modeled as an ideal voltage source in series with an impedance. The source is often connected to a load that has its own impedance as shown in the diagram. The total impedance is the sum of the source impedance and the load impedance. In some instances, we want to transfer the maximum possible power from the source. What should the load impedance be so that maximum power is transferred to the load? We can easily solve for the power delivered to the load in the case where the impedance is due only to resistance:

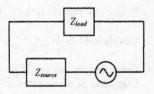

$$\overline{P} = I_{rms}^2 R_{load} = \left(\frac{V_{rms\,source}}{R_{total}}\right)^2 R_{load} = \left(\frac{V_{rms\,source}}{R_{source} + R_{load}}\right)^2 R_{load}$$

If we take the derivative of this expression with respect to $R_{load}$ and set it equal to zero, we can find the stationary point of this function:

$$\frac{d\overline{P}}{dR_{load}} = 2\frac{V_{rms\,source}^2}{(R_{source}+R_{load})^3}R_{load} + \left(\frac{V_{rms\,source}}{R_{source}+R_{load}}\right)^2 = 0$$

If we solve this for $R_{load}$, we get

$$R_{load} = R_{source}$$

Adjusting the load impedance to equal the source impedance so that maximum power is transferred is called **impedance matching**.

### Section 31-8 Three-Phase AC

A three-phase voltage source supplies three voltages of the same amplitude and frequency relative to a common potential (usually ground potential). The three voltages differ in phase by 120° ($2\pi/3$ radians) from each other:

$$V_1 = V_0 \sin \omega t$$
$$V_2 = V_0 \sin(\omega t + 2\pi/3)$$
$$V_3 = V_0 \sin(\omega t + 4\pi/3)$$

The biggest advantage to three-phase voltage sources over single-phase voltage sources is that the power is delivered at a constant rate (if the load is balanced). As an example, if the load on each phase is a resistor $R$ connected to the common potential, than the total power delivered as a function of time is

$$P = P_1 + P_2 + P_3 = \frac{V_1^2}{R} + \frac{V_2^2}{R} + \frac{V_3^2}{R}$$
$$= \frac{[V_0 \sin \omega t]^2}{R} + \frac{[V_0 \sin(\omega t + 2\pi/3)]^2}{R} + \frac{[V_0 \sin(\omega t + 4\pi/3)]^2}{R}$$

We can use the trigonometric identity $\sin a + b = \sin a \cos b + \sin b \cos a$ to rewrite the second two terms in the expression:

$$P = \frac{V_0^2}{R}\left[\sin^2 \omega t + (\sin \omega t \cos 2\pi/3 + \sin 2\pi/3 \cos \omega t)^2 + (\sin \omega t \cos 4\pi/3 + \sin 4\pi/3 \cos \omega t)^2\right]$$

$$= \frac{V_0^2}{R}\left[\sin^2 \omega t + \left(-\tfrac{1}{2}\sin \omega t + \tfrac{\sqrt{3}}{2}\cos \omega t\right)^2 + \left(-\tfrac{1}{2}\sin \omega t - \tfrac{\sqrt{3}}{2}\cos \omega t\right)^2\right]$$

$$= \frac{V_0^2}{R}\left[\sin^2 \omega t + \left(\tfrac{1}{4}\sin^2 \omega t - \tfrac{\sqrt{3}}{2}\sin \omega t \cos \omega t + \tfrac{3}{4}\cos \omega t\right) + \left(\tfrac{1}{4}\sin^2 \omega t + \tfrac{\sqrt{3}}{2}\sin \omega t \cos \omega t + \tfrac{3}{4}\cos \omega t\right)\right]$$

$$= \frac{V_0^2}{R}\left(\tfrac{3}{2}\sin^2 \omega t + \tfrac{3}{2}\cos^2 \omega t\right) = \frac{3V_0^2}{2R}$$

Notice that there is no time dependence to the expression for the power.

### Example 31-8-A

Determine the power delivered by a three-phase voltage source where the amplitude of each source is $V_0$ relative to a common potential and resistors of resistance $R$ connected between each pair of phases makes the load on the source as shown in the diagram.

**Solution:**

Although the resistances are identical, we label them differently to help keep track of which resistor is which. In this case, it should be easy to see that the voltage across each resistor is the difference between two of the phase voltages:

The voltage drop across $R_A$ is equal to $V_2 - V_1$:

$$V_{R_A} = V_0 \sin(\omega t + 2\pi/3) - V_0 \sin \omega t$$

The voltage drop across $R_B$ is equal to $V_3 - V_2$:

$$V_{R_B} = V_0 \sin(\omega t + 4\pi/3) - V_0 \sin(\omega t + 2\pi/3)$$

The voltage drop across $R_C$ is equal to $V_1 - V_3$:

$$V_{R_C} = V_0 \sin \omega t - V_0 \sin(\omega t + 4\pi/3)$$

The total power in the circuit is

$$P = \frac{V_{R_A}^2}{R_A} + \frac{V_{R_B}^2}{R_B} + \frac{V_{R_C}^2}{R_C}$$

$$= \frac{[V_0 \sin(\omega t + 2\pi/3) - V_0 \sin \omega t]^2}{R} + \frac{[V_0 \sin(\omega t + 4\pi/3) - V_0 \sin(\omega t + 2\pi/3)]^2}{R}$$

$$+ \frac{[V_0 \sin \omega t - V_0 \sin(\omega t + 4\pi/3)]^2}{R}$$

Using the trigonometric identities that were used previously and expanding the binomial squares, we can write this as

$$P = \frac{V_0^2}{R} \left\{ \left[ \frac{9}{4} \sin^2 \omega t - \frac{3\sqrt{3}}{2} \sin \omega t \cos \omega t + \frac{3}{4} \cos^2 \omega t \right] \right.$$

$$+ \left[ 3 \cos^2 \omega t \right] + \left. \left[ \frac{9}{4} \sin^2 \omega t + \frac{3\sqrt{3}}{2} \sin \omega t \cos \omega t + \frac{3}{4} \cos^2 \omega t \right] \right\}$$

$$= \frac{V_0^2}{R} \left\{ \frac{9}{2} \sin^2 \omega t + \frac{9}{2} \cos^2 \omega t \right\} = \frac{9V_0^2}{2R}$$

Again, we see that this circuit, called a delta connected three-phase circuit also has constant power.

## Practice Quiz

1.    An electric circuit component is of an unknown type of resistor, inductor, or capacitor. When connected to an ac source, the amplitude of the current increases with increasing frequency for a fixed voltage amplitude. What type of component is this?

a) resistor
b) inductor
c) capacitor
d) can't tell from the information given

2.    An electric circuit component is of an unknown type of resistor, inductor, or capacitor. When connected to an ac source of voltage with a time dependence $V(t) = V_0 \cos(\omega t + 3\pi/7)$, the current is $I(t) = I_0 \sin(\omega t + 3\pi/7)$. What type of component is this?

a) resistor
b) inductor
c) capacitor
d) can't tell from the information given

3.    At frequencies above the resonant frequency of a series *RLC* circuit

a)  The current leads the voltage source in phase.
b)  The current lags the voltage source in phase.
c)  The current is in phase with the voltage source.
d)  The current has no definite phase relationship to the voltage source.

4.    At the resonant frequency of a series *RLC* circuit, the amplitude of the voltage drop across the resistor is

a) equal to the amplitude of the source voltage.
b) greater than the amplitude of the source voltage.
c) less than the amplitude of the source voltage.
d) zero.

5.    At source frequencies below the resonant frequency of a series *RLC* circuit, the frequency of the voltage across the inductor is

a) less than the frequency of the voltage source.
b) greater than the frequency of the voltage source.
c) equal to the frequency of the voltage source.
d) zero

6.    An amplifier is designed to power some loudspeakers that have an impedance of 8 Ω.  Four of these speakers are connected in parallel to the amplifier.  What should be the output impedance of the amplifier so that maximum power is transferred to the loudspeakers?

a) 2 Ω
b) 8 Ω
c) 32 Ω
d) 64 Ω

7.    What is the average power into an inductor in an ac circuit with an rms current $I_{rms}$ and angular frequency $\omega$?

a) $I_{rms}^2 L$
b) $I_{rms}^2 \omega L$
c) $I_{rms}^2 / \omega L$
d) zero

8.    If you lower the load resistance below the voltage source resistance, the current from the voltage source increases, but according to our maximum power condition, the power to the load is smaller.  The voltage source is supplying greater power because voltage times current has increased.  Where is the additional power going?

a) The power disappears.
b) The power from the source actually decreases because the power factor drops.
c) The power goes into heating the wires that connect the source to the load.
d) The power goes into the source resistance.

9.    Consider Example 31-6-A in this study guide.  How can the amplitudes of the voltages across the capacitor and the inductor be over 3000 V when the voltage of the source is only 12.0 V and still satisfy Kirchhoff's voltage rule?

a) Kirchhoff's voltage rule does not apply to this circuit.
b) The calculations in Example 31-6-A are incorrect.
c) The circuit doesn't function as stated because of this problem.
d) The inductor voltage and the capacitor voltage are 180° out of phase and add to zero.

10.    What is one advantage of three-phase power?

       a) The power delivered to a balanced load is constant in time.
       b) The total voltage delivered to a balanced load is constant in time.
       c) The total current delivered to a balanced load is constant in time.
       d) If one phase of the supply fails, equipment will function properly on the two remaining phases.

11.    Determine the inductive reactance of an inductor with an inductance of 36.4 mH at a frequency of 320 Hz.  What is the amplitude of the current that flows in a circuit consisting of this inductor and a voltage source with an amplitude of 5.00 V and a frequency of 320 Hz?

12.    Determine the capacitance in a capacitive circuit in which the rms voltage of the ac source is 12.0 V and the frequency of the source is 1.80 kHz when there is a current amplitude of 78.2 mA in the circuit.

13.    Determine the resonant frequency of a series $LRC$ circuit that has an inductance of 10.8 mH, a resistance of 5.82 $\Omega$, and a capacitance of 45.6 nF.  Determine the rms current in this series $LRC$ circuit when the source operates at the resonant frequency with an rms voltage of 12.0 V.

14.    An LRC circuit has an inductance of 8.82 mH, a resistance of 1.88 $\Omega$, and a capacitance of 0.228 $\mu$F.  Determine the amplitude of the current in the circuit when it is connected to a voltage source with a frequency of 5.00 kHz and amplitude 12.0 V.  What are the amplitudes of the voltage across the inductor, the voltage across the resistor, and the voltage across the capacitor under these conditions?

15.    Determine the power delivered by a three phase power source in which each phase voltage has an amplitude of 280 V and the load consists of a 32.6 $\Omega$ resistance connected between each pair of voltages.

## Problem Solutions

3.   We find the frequency from
       $X_C = 1/2\pi f C$;
       $6.70 \times 10^3\ \Omega = 1/2\pi f(2.40 \times 10^{-6}\ \text{F})$, which gives $f = 9.90$ Hz.

7.   If there is no current in the secondary, there will be no induced emf from the mutual inductance.  We find the impedance from
       $Z = X_L = V_{rms}/I_{rms} = (110\ \text{V})/(2.2\ \text{A}) = 50\ \Omega$.
       We find the inductance from
       $X_L = 2\pi f L$;
       $50\ \Omega = 2\pi(60\ \text{Hz})L$, which gives $L = 0.13$ H.

11.  Because the capacitor and resistor are in parallel, their currents are
       $I_C = V/X_C$,   and   $I_R = V/R$.
       The total current is $I = I_C + I_R$ .
       (*a*)  The reactance of the capacitor is
              $X_{C1} = 1/2\pi f_1 C = 1/2\pi(60\ \text{Hz})(0.35 \times 10^{-6}\ \text{F}) = 7.58 \times 10^3\ \Omega = 7.58$ k$\Omega$.
              For the fraction of current that passes through $C$, we have
              fraction1    $= I_{C1}/(I_{C1} + I_R) = (1/X_{C1})/[(1/X_{C1}) + (1/R)] = R/(R + X_{C1})$
                           $= (0.400\ \text{k}\Omega)/(0.400\ \text{k}\Omega + 7.58\ \text{k}\Omega) = 0.050 = 5.0\%$.
       (*b*)  The reactance of the capacitor is
              $X_{C2} = 1/2\pi f_2 C = 1/2\pi(60,000\ \text{Hz})(0.35 \times 10^{-6}\ \text{F}) = 7.58\ \Omega$.
              For the fraction of current that passes through $C$, we have
              fraction2    $= I_{C2}/(I_{C2} + I_R) = (1/X_{C2})/[(1/X_{C2}) + (1/R)] = R/(R + X_{C2})$
                           $= (400\ \Omega)/(400\ \Omega + 7.58\ \Omega) = 0.98 = 98\%$.
              Thus most of the high-frequency current passes through the capacitor.

15.  (*a*)  The reactance of the capacitor is
              $X_C = 1/2\pi f C = 1/2\pi(60\ \text{Hz})(0.80 \times 10^{-6}\ \text{F}) = 3.32 \times 10^3\ \Omega = 3.32$ k$\Omega$.
              The impedance of the circuit is
              $Z = (R^2 + X_C^2)^{1/2} = [(6.0\ \text{k}\Omega)^2 + (3.32\ \text{k}\Omega)^2]^{1/2} = 6.86$ k$\Omega$.
              The rms current is
              $I_{rms} = V_{rms}/Z = (120\ \text{V})/(6.86\ \text{k}\Omega) = 18$ mA.

(b) We find the phase angle from
$$\cos \phi = R/Z = (6.0\text{ k}\Omega)/(6.86\text{ k}\Omega) = 0.875.$$
In an $RC$ circuit, the current leads the voltage, so $\phi = -29°$.

(c) The power dissipated is
$$P = I_{rms}^2 R = (17.5 \times 10^{-3}\text{ A})^2(6.0 \times 10^3\ \Omega) = 1.8\text{ W}.$$

(d) The rms readings across the elements are
$$V_R = I_{rms}R = (17.5\text{ mA})(6.0\text{ k}\Omega) = 105\text{ V};$$
$$V_C = I_{rms}X_C = (17.5\text{ mA})(3.32\text{ k}\Omega) = 58\text{ V}.$$
Note that, because the maximal voltages occur at different times, the two readings do not add to the applied voltage of 120 V.

19. We find the resistance from
$$Z = [R^2 + X_L^2]^{1/2};$$
$$335\ \Omega = [R^2 + (45.5\ \Omega)^2]^{1/2},\text{ which gives } R = 332\ \Omega.$$

23. The reactance of the capacitor is
$$X_C = 1/2\pi fC = 1/2\pi(10.0 \times 10^3\text{ Hz})(5000 \times 10^{-12}\text{ F}) = 3.18 \times 10^3\ \Omega = 3.18\text{ k}\Omega.$$
The reactance of the inductor is
$$X_L = 2\pi fL = 2\pi(10.0\text{ kHz})(0.0320\text{ H}) = 2.01\text{ k}\Omega.$$
The impedance of the circuit is
$$Z = [R^2 + (X_L - X_C)^2]^{1/2} = [(8.70\text{ k}\Omega)^2 + (2.01\text{ k}\Omega - 3.18\text{ k}\Omega)^2]^{1/2} = 8.78\text{ k}\Omega.$$
We find the phase angle from
$$\tan \phi = (X_L - X_C)/R = (2.01\text{ k}\Omega - 3.18\text{ k}\Omega)/(8.70\text{ k}\Omega) = -0.134,\text{ so } \phi = -7.66°.$$
The rms current is
$$I_{rms} = V_{rms}/Z = (800\text{ V})/(8.78\text{ k}\Omega) = 91.1\text{ mA}.$$

27. We find the capacitance from the new resonant frequency:
$$f_0 = (1/2\pi)(1/LC)^{1/2};$$
$$2(33.0 \times 10^3\text{ Hz}) = (1/2\pi)[1/(4.15 \times 10^{-3}\text{ H})C]^{1/2},\text{ which gives } C = 1.40 \times 10^{-9}\text{ F}.$$
At the applied frequency the reactances are
$$X_L = 2\pi fL = 2\pi(33.0 \times 10^3\text{ Hz})(4.15 \times 10^{-3}\text{ H}) = 860\ \Omega = 0.860\text{ k}\Omega.$$
$$X_C = 1/2\pi fC = 1/2\pi(33.0 \times 10^3\text{ Hz})(1.40 \times 10^{-9}\text{ F}) = 3.44 \times 10^3\ \Omega = 3.44\text{ k}\Omega.$$
The impedance is
$$Z = [R^2 + (X_L - X_C)^2]^{1/2} = [(0.220\text{ k}\Omega)^2 + (0.860\text{ k}\Omega - 3.44\text{ k}\Omega)^2]^{1/2} = 2.59\text{ k}\Omega.$$
The peak current is
$$I_0 = V_0/Z = (136\text{ V})/(2.59\text{ k}\Omega) = 52.5\text{ mA}.$$

31. (a) The average power dissipated is
$$\overline{P} = I_{rms}^2 Z \cos \phi = I_{rms}^2 R = \tfrac{1}{2}I_0^2 R.$$
If we use $I_0 = V_0/Z$, we get
$$\overline{P} = V_0^2 R/2Z^2 = V_0^2 R/2[R^2 + (\omega L - 1/\omega C)^2].$$

(b) We find the frequency to make $\overline{P}$ a maximum by setting the first derivative equal to zero:
$$\frac{d\overline{P}}{d\omega} = 0,\text{ or equivalently, } \frac{dZ^2}{d\omega} = 0;$$
$$\frac{dZ^2}{d\omega} = 2\left(\omega L - \frac{1}{\omega C}\right)\left(L + \frac{1}{\omega^2 C}\right) = 0$$
Because $(L + 1/\omega^2 C) > 0$, we have
$$\omega'L = 1/\omega'C,\text{ or } \omega'^2 = 1/LC.$$

(c) The peak for the average power, $P_0 = V_0^2 R/2Z_0^2$, occurs when $Z_0 = R$.
To get $\overline{P} = 406 V_0^2 R/2Z^2 = \tfrac{1}{2}P_0 = V_0^2 R/4Z_0^2$, we have
$$Z^2 = R^2 + (\omega L - 1/\omega C)^2 = 2Z_0^2 = 2R^2,\text{ or } R = \pm(\omega L - 1/\omega C),$$
which is a quadratic equation for $\omega$:
$$\pm L\omega^2 - R\omega - 1/C = 0,\text{ or } L\omega^2 - R\omega - 1/C = 0,\text{ with solutions } (\omega \text{ must be positive})$$
$$\omega = \pm R/2L + [(R^2/4L^2) + (1/LC)]^{1/2}.$$
Thus the width of the peak is
$$\Delta\omega = \omega_+ - \omega_- = R/L.$$

35. For the current and voltage to be in phase, the net reactance of the capacitor and inductor must be zero. Thus we have

$X_L = X_C$;

$2\pi f L = 1/2\pi f C$, or

$f = (1/2\pi)(1/LC)^{1/2}$;

3360 Hz $= (1/2\pi)[1/(0.230\text{ H})C]^{1/2}$, which gives $C = 9.76 \times 10^{-9}$ F $= 9.76$ nF.

39. At 60 Hz, the reactance of the inductor is

$X_{L1} = 2\pi f_1 L = 2\pi(60\text{ Hz})(0.620\text{ H}) = 234\ \Omega$.

The impedance of the circuit is

$Z_1 = (R^2 + X_{L1}^2)^{1/2} = [(3.5\text{ k}\Omega)^2 + (0.234\text{ k}\Omega)^2]^{1/2} = 3.51\text{ k}\Omega$.

Thus the impedance at the new frequency is

$Z_2 = 2Z_1 = 2(3.51\text{ k}\Omega) = 7.02\text{ k}\Omega$.

We find the new reactance from

$Z_2 = (R^2 + X_{L2}^2)^{1/2}$;

7.02 k$\Omega = [(3.5\text{ k}\Omega)^2 + X_{L2}^2]^{1/2}$, which gives $X_{L2} = 6.09\text{ k}\Omega$.

We find the new frequency from

$X_{L2} = 2\pi f_2 L$;

6.09 k$\Omega = 2\pi f_2(0.620\text{ H})$, which gives $f_2 = 1.6$ kHz.

43. We find the impedance from the power factor:

$\cos\phi = R/Z$;

$0.17 = (200\ \Omega)/Z$, which gives $Z = 1.18 \times 10^3\ \Omega$.

We get an expression for the reactances from

$Z^2 = R^2 + (X_L - X_C)^2$;

$(1.18 \times 10^3\ \Omega)^2 = (200\ \Omega)^2 + (X_L - X_C)^2$, which gives $X_L - X_C = \pm 1.16 \times 10^3\ \Omega$.

When we express this in terms of the inductance and capacitance, we get

$2\pi f L - (1/2\pi f C) = \pm 1.16 \times 10^3\ \Omega$;

$2\pi f(0.020\text{ H}) - [1/2\pi f(50 \times 10^{-9}\text{ F})] = \pm 1.16 \times 10^3\ \Omega$, which reduces to two quadratic equations:

$0.126 f^2 \pm (1.16 \times 10^3\text{ Hz})f - 3.183 \times 10^6\text{ Hz}^2 = 0$,

which have positive solutions of $f = 2.2 \times 10^3$ Hz, $1.1 \times 10^4$ Hz.

47. The impedance of the circuit to ac current is

$Z = [R^2 + X_C^2]^{1/2} = [R^2 + (1/\omega C)^2]^{1/2}$.

The ac current amplitude is

$I_0 = V_{20}/Z = V_{20}/[R^2 + (1/\omega C)^2]^{1/2}$.

We find the phase angle from

$\tan\phi = (X_L - X_C)/R = -X_C/R$.

If $R \gg X_C$, then $\tan\phi \approx 0$, and $\phi \approx 0$.

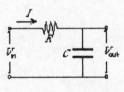

Thus the current is

$I = I_0 \sin\omega t$.

There will be no dc current through the capacitor, and thus no appreciable dc voltage drop across the resistor, so $V_{1\text{out}} = V_{1C} = V_1$, and $Q_1 = CV_1$.

At any instant the charge on the capacitor is

$Q = CV_{\text{out}} = CV_1 + CV_{2\text{out}}$,

and the current through the capacitor is

$I = dQ/dt = C\, dV_{2\text{out}}/dt = I_0 \sin\omega t$.

When we integrate this, we get

$$\int_0^{2\text{out}} dV'_{2\text{out}} = \int_0^t \frac{I_0}{C}\sin\omega t'\, dt'$$

$$V_{2\text{out}} = -\frac{I_0}{\omega C}\cos\omega t = \frac{X_C V_{20}}{[R^2 + X_c^2]^{1/2}}\sin(\omega t - 90^\circ)$$

If $R \gg X_C$, then

$Z = [R^2 + (1/\omega C)^2]^{1/2} \approx R$, and the current will be

$I_0 = V_{20}/Z = V_{20}/[R^2 + (1/\omega C)^2]^{1/2} \approx V_{20}/R$, which will be low.

If we form the ratio of output ac voltage to input ac voltage, we get

$V_{2\text{out}}/V_{20} = X_C/[R^2 + X_C^2]^{1/2} \approx X_C/R \ll 1$.

51. We can find the resistance from the power dissipated:

$P = I_{rms}^2 R$;

300 W = $(4.0 \text{ A})^2 R$, which gives $R = 19 \, \Omega$.

The impedance of the electromagnet is

$Z = V_{rms}/I_{rms} = (120 \text{ V})/(4.0 \text{ A}) = 30 \, \Omega$.

We can find the inductance from

$Z = (R^2 + X_L^2)^{1/2} = [R^2 + (2\pi f L)^2]^{1/2}$;

$\quad\quad$ 30 $\Omega$ = $\{(19 \, \Omega)^2 + [2\pi(60 \text{ Hz})L]^2\}^{1/2}$, which gives $L = 6.2 \times 10^{-2} \text{ H} = 62 \text{ mH}$.

# Chapter 32: Maxwell's Equations and Electromagnetic Waves

## Chapter Overview and Objectives

This chapter introduces changing electric fields as a source of magnetic fields and Gauss's law for magnetic fields. Electromagnetic waves are shown to be solutions of Maxwell's equations. Some properties of electromagnetic waves are discussed.

After completing study of this chapter, you should:

- Know that a changing electric flux produces a magnetic field.
- Know Gauss's law for magnetic fields.
- Know Maxwell's equations.
- Know that Maxwell's equations predict that electromagnetic waves exist.
- Know that the speed of light is predicted by Maxwell's equations.
- Know what the Poynting vector is, what it represents, and how to calculate it.
- Be able to calculate the intensity of electromagnetic waves given the amplitude or rms value of their electric or magnetic fields.

## Summary of Equations

Magnetic field due to a changing electric flux:
$$\oint \mathbf{B} \cdot d\mathbf{l} = \mu_0 \varepsilon_0 \frac{d\Phi_E}{dt}$$

Gauss's law for magnetic fields:
$$\oint \mathbf{B} \cdot d\mathbf{A} = 0$$

Maxwell's equations:
$$\oint \mathbf{E} \cdot d\mathbf{A} = \frac{Q}{\varepsilon_0}$$
$$\oint \mathbf{B} \cdot d\mathbf{A} = 0$$
$$\oint \mathbf{E} \cdot d\mathbf{l} = - \frac{d\Phi_B}{dt}$$
$$\oint \mathbf{B} \cdot d\mathbf{l} = \mu_0 I_{encl} + \mu_0 \varepsilon_0 \frac{d\Phi_E}{dt}$$

Electromagnetic wave equations:
$$\frac{\partial^2 E}{\partial t^2} = \frac{1}{\mu_0 \varepsilon_0} \frac{\partial^2 E}{\partial x^2}$$
$$\frac{\partial^2 B}{\partial t^2} = \frac{1}{\mu_0 \varepsilon_0} \frac{\partial^2 B}{\partial x^2}$$

Speed of electromagnetic waves:
$$v = \frac{1}{\sqrt{\mu_0 \varepsilon_0}} = 3.00 \times 10^8 \text{ m/s}$$

Speed of electromagnetic waves in a material:
$$v = \frac{1}{\sqrt{\mu \varepsilon}}$$

Definition of Poynting vector:
$$\mathbf{S} = \frac{1}{\mu_0} \mathbf{E} \times \mathbf{B}$$

Average magnitude of Poynting vector:
$$\overline{S} = \tfrac{1}{2} \varepsilon_0 c E_0^2 = \tfrac{1}{2} \frac{c}{\mu_0} B_0^2 = \frac{E_0 B_0}{2 \mu_0} = \frac{E_{rms} B_{rms}}{2 \mu_0}$$

Radiation pressure of electromagnetic waves traveling perpendicular to surface on perfect absorber:

$$P = \frac{\overline{S}}{c}$$

Radiation pressure of electromagnetic waves traveling perpendicular to surface on perfect reflector:

$$P = \frac{2\overline{S}}{c}$$

## Chapter Summary

### Section 32-1 Changing Electric Fields Produce Magnetic Fields; Ampère's Law and Displacement Current

Currents are not the only source of magnetic fields. In analogy to Faraday's law that an electric field is produced by a changing magnetic flux, induced magnetic fields are produced by changing **electric flux**. The mathematical relationship between the induced magnetic field and the changing electric flux has the same mathematical form as Faraday's law:

$$\oint \mathbf{B} \cdot dl = \mu_0 \varepsilon_0 \frac{d\Phi_E}{dt}$$

We can add this term to Ampere's law to write an equation for the total magnetic field

$$\oint \mathbf{B} \cdot dl = \mu_0 I_{encl} + \mu_0 \varepsilon_0 \frac{d\Phi_E}{dt}$$

### Example 32-1-A

A circular path in space has a magnetic field that points tangent to the circle at all points in a counterclockwise direction as seen in the diagram. The magnitude of the magnetic field is 3.98 μT at all points on the circle of radius 5.68 cm. A current of 450 mA flows out of the page through the circle. What is the time rate of change of electric flux through the circle? If the electric field is pointing out of the page through the circle, is the magnitude of the electric field increasing or decreasing with time?

### Solution:

First, we determine the path integral of the magnetic field. If we take our path of our integral to be counterclockwise around the circle, then *dl* and **B** are parallel everywhere and

$$\mathbf{B} \cdot dl = B\, dl$$

We can easily integrate this around the path:

$$\oint \mathbf{B} \cdot dl = \oint B\, dl = B 2\pi R = \left(3.98 \times 10^{-6}\ \text{T}\right) 2\pi (0.0568\ \text{m}) = 1.42 \times 10^{-6}\ \text{T} \cdot \text{m}$$

We relate this integral to the current through the loop and the rate of change of the electric flux through the loop:

$$\oint \mathbf{B} \cdot dl = \mu_0 I_{encl} + \mu_0 \varepsilon_0 \frac{d\Phi_E}{dt} \quad \Rightarrow \quad \frac{d\Phi_E}{dt} = \frac{1}{\mu_0 \varepsilon_0}\left(\oint \mathbf{B} \cdot dl - \mu_0 I_{encl}\right)$$

$$\frac{d\Phi_E}{dt} = \frac{\left(1.42 \times 10^{-6}\ \text{T} \cdot \text{m}\right)}{\left(4\pi \times 10^{-7}\ \text{T} \cdot \text{m/A}\right)\left(8.85 \times 10^{-12}\ \text{C}^2/\text{N} \cdot \text{m}^2\right)} - \frac{0.450\ \text{A}}{\left(8.85 \times 10^{-12}\ \text{C}^2/\text{N} \cdot \text{m}^2\right)}$$

$$= 7.68 \times 10^{10}\ \text{V} \cdot \text{m/s}$$

Note that we called the current a positive current. Why? By going around the circle counterclockwise, the right hand rule is consistent with the positive current direction defined as out-of-the-page through the loop. That implies that because our $d\Phi_E/dt$ is positive, the flux is increasing in the out of the page direction. That implies that the electric field is increasing in magnitude.

### Section 32-2 Gauss's Law for Magnetism

Because there is a symmetry between the electric field and the magnetic field in the way in which changing magnetic flux produces an electric field and changing electric flux produces a magnetic field, you might expect there to also be an analogy between Gauss's law for electric fields and a similar law for magnetic fields. There is, but because no magnetic charge has ever been observed, the term in the Gauss's law corresponding to the electric charge enclosed by the surface is identically zero (at least until a magnetic charge is observed). **Gauss's law for magnetism** is

$$\oint \mathbf{B} \cdot d\mathbf{A} = 0$$

### Section 32-3 Maxwell's Equations

The complete set of equations that relate the electric and magnetic fields to their sources is called **Maxwell's equations**. We restate them here:

$$\oint \mathbf{E} \cdot d\mathbf{A} = \frac{Q}{\varepsilon_0}$$

$$\oint \mathbf{B} \cdot d\mathbf{A} = 0$$

$$\oint \mathbf{E} \cdot dl = -\frac{d\Phi_B}{dt}$$

$$\oint \mathbf{B} \cdot dl = \mu_0 I_{encl} + \mu_0 \varepsilon_0 \frac{d\Phi_E}{dt}$$

### Section 32-4 Production of Electromagnetic Waves

Electric and magnetic fields change with time. The changes that take place do not propagate at infinite speed, so changes do not take place everywhere instantaneously. If we look at great distances from a source compared with the size of the spatial extent of the source, the solutions to the time-changing fields with time take a relatively simple approximate form. In this large distant limit, the changing electric and magnetic fields due to time dependent sources take the form of **radiation fields**. The general form of radiation fields are outward propagating waves of electric and magnetic field. The changing electric and magnetic fields are perpendicular to the direction of wave propagation. The magnitudes of the electric and magnetic fields decrease with distance as $1/r$, where $r$ is the distance from the source. The electric and magnetic fields have a time dependence of the same form as the source's time dependence. These waves are called **electromagnetic (EM) waves**. In general, electromagnetic waves are created by accelerating charges.

### Section 32-5 Electromagnetic Waves, and Their Speed, from Maxwell's Equations

Maxwell's equations are interdependent on each other. Changing magnetic field causes an electric field and changing electric field causes a magnetic field. If the equations are solved to separate out the interdependence, the results are wave equations for the electric and magnetic fields:

$$\frac{\partial^2 E}{\partial t^2} = \frac{1}{\mu_0 \varepsilon_0} \frac{\partial^2 E}{\partial x^2}$$

$$\frac{\partial^2 B}{\partial t^2} = \frac{1}{\mu_0 \varepsilon_0} \frac{\partial^2 B}{\partial x^2}$$

Both of these equations are one-dimensional wave equations, identical in form to the wave equation from Chapter 15 Section 5. From our study of this wave equation previously, we know the solutions to this equation are waves that propagate with a velocity, $v$:

$$v^2 = \frac{1}{\mu_0 \varepsilon_0} \quad \Rightarrow \quad v = \frac{1}{\sqrt{\mu_0 \varepsilon_0}}$$

The value of $v$ can be calculated to be

$$v = \frac{1}{\sqrt{\mu_0 \varepsilon_0}} = \frac{1}{\sqrt{\left(4\pi \times 10^{-7}\ \text{T} \cdot \text{m/A}\right)\left(8.85 \times 10^{-12}\ \text{C}^2/\text{N} \cdot \text{m}^2\right)}} = 3.00 \times 10^{-12}\ \text{m/s}$$

*3 × 10⁸ m/s*

## Section 32-6 Light as an Electromagnetic Wave and the Electromagnetic Spectrum

The waves predicted by Maxwell's equations have a predicted speed that corresponds to the measured speed of light. It is now known that light is the electromagnetic wave predicted by Maxwell's equations. There are several other phenomena that are the electromagnetic waves predicted by Maxwell's equations. Radio waves, microwaves, infrared light, visible light, ultraviolet light, X-rays, and gamma rays are all electromagnetic waves. The only difference between these different phenomena is the wavelength of the waves. They are classified differently because their different wavelengths cause them to interact with matter in different ways.

The waves predicted by Maxwell's equations can also propagate in materials. The wave equations are altered by the presence of charges and currents that are driven by the electric and magnetic fields of the electromagnetic waves, which in turn create additional electromagnetic waves. Maxwell's equations and the equations of motion of the charged particles must be solved consistently. In uniform, isotropic materials, the solution works out to be the same wave equation with the velocity in the material given by

*Velocity of EM wave thru a material*

$$v = \frac{1}{\sqrt{\mu \varepsilon}}$$

where $\mu$ is the magnetic permeability of the material and $\varepsilon$ is the electric permittivity of the material.

## Section 32-7 Energy in EM Waves; the Poynting Vector

The energy density, $u$, in space in the presence of either electric or magnetic fields has already been discussed in earlier chapters. In the case of electromagnetic waves, the magnitude of the electric and magnetic fields are related so we can write the energy density in terms of electric fields, magnetic fields, or both:

*energy density*

$$u = \varepsilon_0 E^2 = \frac{B^2}{\mu_0} = \sqrt{\frac{\varepsilon_0}{\mu_0}} EB$$

A vector quantity that has a magnitude equal to the energy per time per area that is carried by the electromagnetic wave, and a direction that is in the direction of the energy flow is the **Poynting vector, S**. The Poynting vector can be written in terms of the electric field and the magnetic field:

$$\mathbf{S} = \frac{1}{\mu_0} \mathbf{E} \times \mathbf{B}$$

The Poynting vector is time dependent at a given point in space because the electric and magnetic fields vary with time as the wave moves past the given point. A useful quantity for calculating the energy transfer by electromagnetic fields is the magnitude of the time-averaged Poynting vector. This is useful because we often look at the energy transferred over many periods of the electromagnetic wave. The magnitude of the time-averaged Poynting vector for sinusoidal electromagnetic waves is

$$\overline{S} = \tfrac{1}{2} \varepsilon_0 c E_0^2 = \tfrac{1}{2} \frac{c}{\mu_0} B_0^2 = \frac{E_0 B_0}{2\mu_0} = \frac{E_{rms} B_{rms}}{2\mu_0}$$

where $E_0$ and $B_0$ are the amplitudes of the electric and magnetic fields respectively and $E_{rms}$ and $B_{rms}$ are the rms values of the electric and magnetic fields respectively.  The final expression is valid regardless of whether the waves are sinusoidal or not.  The time-averaged Poynting vector is the **intensity** of the electromagnetic wave.

**Example 32-7-A**

An electromagnetic wave has a magnetic field that points to the East and an electric field that points to the North.  The magnitude of the magnetic field is 3.56 µT.  Determine the magnitude of the electric field of the electromagnetic wave.  Determine the Poynting vector of the electromagnetic wave when the electric and magnetic fields are at their maximum.  Determine the intensity of the wave.

**Solution:**

The electric field amplitude is related to the magnetic field amplitude by

$$B = \frac{E}{c} \quad \Rightarrow \quad E = Bc = \left(3.56 \times 10^{-6} \ T\right)\left(3.00 \times 10^8 \ m/s\right) = 1.1 \times 10^3 \ V/m$$

To determine the Poynting vector, set up the coordinate system shown.  We can then write

$$\mathbf{B} = \left(3.56 \times 10^{-6} \ T\right)\mathbf{i}$$

and

$$\mathbf{E} = \left(1.07 \times 10^3 \ V/m\right)\mathbf{j}$$

when the fields are at their maximum values.  We can then determine the Poynting vector, **S**,

$$\mathbf{S} = \frac{1}{\mu_0}\mathbf{E} \times \mathbf{B} = \frac{1}{\left(4\pi \times 10^{-7} \ T \cdot m/A\right)}\left(1.07 \times 10^3 \ V/m\right)\mathbf{j} \times \left(3.56 \times 10^{-6} \ T\right)\mathbf{i}$$
$$= -\left(3.03 \times 10^3 \ W/m^2\right)\mathbf{k}$$

The average intensity of the electromagnetic radiation is the magnitude of the average Poynting vector:

$$\bar{S} = \frac{1}{2}\frac{c}{\mu_0}B_0^2 = \frac{1}{2}\frac{\left(3 \times 10^8 \ m/s\right)}{\left(4\pi \times 10^{-7} \ T \cdot m/A\right)}\left(3.56 \times 10^{-6} \ T\right)^2 = 1.51 \times 10^3 \ W/m^2$$

For a check, this should be one-half the magnitude of the Poynting vector at maximum field, which it is.

**Section 32-8 Radiation Pressure**

Electromagnetic waves transfer momentum as well as energy.  **Radiation pressure** is the force per area on a surface due to the momentum transferred to the surface by the electromagnetic radiation incident upon it.  The radiation pressure, $P$, on a surface that completely absorbs the electromagnetic radiation incident upon it when the electromagnetic waves are traveling perpendicular to the surface is

$$P = \frac{\bar{S}}{c}$$

and the radiation pressure on a surface that completely reflects the electromagnetic radiation incident upon it when the electromagnetic waves are traveling perpendicular to the surface is

$$P = \frac{2\bar{S}}{c}$$

**Example 32-8-A**

A future design of space ships or airplanes could be that they emit electromagnetic radiation out of the ship to provide propulsion. What intensity of light would need to be emitted out the back of such a ship to provide a thrust of $1.00 \times 10^5$ N?

**Solution:**

Emission is the reverse of absorption, so that the appropriate expression to use for radiation pressure is

$$P = \frac{\overline{S}}{c}$$

The force is the pressure times the area, so

$$F = PA = \frac{\overline{S}A}{c} \quad \Rightarrow \quad \overline{S} = \frac{cF}{A} = \frac{\left(3.00 \times 10^8 \text{ m/s}\right)\left(1.00 \times 10^5 \text{ N}\right)}{\pi \left(1.00 \text{ m}\right)^2 / 4} = 3.8 \times 10^{13} \text{ W/m}^2$$

**Section 32-9 Radio and Television**

Electromagnetic wave propagation is used to carry information between two distant points. To allow multiple communications of information to take place simultaneously, users of electromagnetic waves for information transfer are assigned a **carrier frequency** and **bandwidth** to use. This means that all of the frequencies electromagnetic radiation sent out by an entity sending signals must fall between an upper and lower limit in frequency. The information is usually sent out by starting with a sinusoidal signal with the carrier frequency and then altering the signal slightly through a process called **modulation** to encode the information on the electromagnetic wave.

There are many methods of modulating the carrier to place the information on the signal. In amplitude modulation, the amplitude of the carrier wave is varied. In frequency modulation, the frequency of the wave is varied. In phase modulation, the phase of the wave is varied. Whether amplitude, frequency, or phase modulation is used, the modulation causes the electromagnetic wave to have a range of frequencies present in it. The modulating signal is limited so the frequencies in the electromagnetic wave to not spread beyond the bandwidth allowed.

A receiver of the electromagnetic waves must perform the process of **demodulation** to obtain the information from the modulated carrier.

## Practice Quiz

1.    There are no currents present, but there is a non-zero magnetic field. What else must be present?

    a) electric charge
    b) electric field
    c) changing electric field
    d) It is impossible to have a magnetic field without current.

2.    If the amplitude of the electric field in an electromagnetic wave is doubled, what happens to the intensity of the electromagnetic wave?

    a) doubles
    b) halves
    c) quadruples
    d) doesn't change

3.    Which travels faster in a vacuum, X-rays or radio waves?

a) X-rays
b) radio waves
c) It depends on whether the radio waves are AM or FM.
d) They travel at the same speed.

4.    An electromagnetic wave travels toward the north. At a given instant, the electric field in the electromagnetic wave is toward the east. What direction is the magnetic field in the electromagnetic wave at that instant?

a) west
b) south
c) up
d) down

5.    Transparent materials have a dielectric constant greater than one at the frequency of visible light. The magnetic susceptibility of these materials is much less than one. What can you say about the speed of light in these materials?

a) The speed of light in the material is the same as the speed of light in vacuum.
b) The speed of light in the material is greater than the speed of light in vacuum.
c) The speed of light in the material is less than the speed of light in vacuum.
d) The speed of light in the material might less than or greater than the speed of light in vacuum.

6.    If the amplitude of the electric field in an electromagnetic wave is doubled, what happens to the amplitude of the magnetic field in the electromagnetic wave?

a) doubles
b) halves
c) quadruples
d) doesn't change

7.    What direction is an electromagnetic wave traveling that has an electric field pointing east and a magnetic field pointing north?

a) upward
b) downward
c) northeast
d) southwest

8.    Two surfaces are of equal area and are illuminated by the same source of electromagnetic radiation. Surface A absorbs 30% of the electromagnetic energy incident on it and reflects the remainder and surface B absorbs 60% of the electromagnetic intensity incident on it and reflects the remainder. Which surface has a greater radiation pressure acting on it?

a) surface A
b) surface B
c) Both surfaces have the same radiation pressure acting on them.
d) There is insufficient information to answer the question.

9.    The magnetic field at all points on the circle shown in the diagram points tangent to the circle in the counterclockwise direction. No current flows through the circle, but there is an electric field pointing upward out of the page. Is the magnitude of the electric field increasing or decreasing?

a) increasing
b) decreasing
c) remaining constant
d) There is insufficient information to answer the question.

10.    The magnetic field at all points on the circle shown is equal to zero.  There is an electric field pointing out of the page and it is increasing in magnitude at a uniform rate.  What else must be true?

   a)  There is a constant current flowing into the page through the circle.
   b)  There is an increasing current flowing into the page through the circle.
   c)  There is a constant current flowing out of the page through the circle.
   d)  There is an increasing current flowing into the page through the circle.

11.    A circular area of space of radius 12 cm has a uniform electric field pointing through it perpendicular to the area with a magnitude of $4.0 \times 10^3$ V/m.  The electric field decreases uniformly in magnitude to $1.0 \times 10^3$ V/m in a time of 0.12 s.  What is the magnetic field along the perimeter of the circular area?

12.    Determine the radiation pressure on the Earth due to the Sun, assuming the radiation from the Sun is completely absorbed by the Earth.  The intensity of the light reaching the Earth from the Sun is 1.5 kW/m$^2$.

13.    Determine the rms electric field and rms magnetic field due to the electromagnetic radiation from a 100 W light bulb at a distance of 1.00 m from the bulb.  Assume all of the energy going into the bulb is radiated as electromagnetic radiation.

14.    Determine the wavelength of a radio station transmitting with a carrier frequency of 830 kHz in the AM radio band.

15.    Calculate the speed of an electromagnetic wave in a dielectric material with a dielectric constant 3.22.  Assume $\mu = \mu_0$.

## Problem Solutions

3.    The current in the wires must also be the displacement current in the capacitor.  We find the rate at which the electric field is changing from
$$I_D = \varepsilon_0 A \ (dE/dt);$$
$$1.8 \text{ A} = (8.85 \times 10^{-12} \text{ C}^2/\text{N} \cdot \text{m}^2)(0.0160 \text{ m})^2 \ dE/dt, \text{ which gives } dE/dt = 7.9 \times 10^{14} \text{ V/m} \cdot \text{s}.$$

7.    Gauss's law for electricity and Ampere's law will not change.
   From the analogy to Gauss's law for electric fields, where $Q$ is the source, $Q_m$ would be the source of the magnetic field, so we have
$$\oint \mathbf{B} \cdot d\mathbf{A} = \mu_0 Q_m.$$
   From the analogy to Ampere's law, we have an additional "current" contribution to Faraday's law.
$$\oint \mathbf{E} \cdot d\mathbf{l} = \mu_0 \ dQ_m/dt - d\Phi_B/dt.$$
   The $dQ_m/dt$ term corresponds to an electric field created by the "current" of magnetic monopoles.

11.    (a)  If we write the argument of the cosine function as $kz + \omega t = k(z + ct)$, we see that the wave is traveling in the $-z$-direction.
   Because **E** and **B** are perpendicular to each other and to the direction of propagation, and **E** is in the $x$-direction, **B** can have only a $y$-component, with magnitude $B_0 = E_0/c$.
   Because a rotation of **E** into **B** must give the direction of propagation, $-z$-direction, **B** must be in the $-y$-direction.
   (b)  The wave is traveling in the $-z$-direction.

15.    The wavelength of the wave is
$$\lambda = c/f = (3.00 \times 10^8 \text{ m/s})/(9.56 \times 10^{14} \text{ Hz}) = 3.14 \times 10^{-7} \text{ m} = 314 \text{ nm}.$$
   This wavelength is just outside the violet end of the visible region, so it is ultraviolet.

19.    The energy per unit area per unit time is
$$S = \tfrac{1}{2} c \varepsilon_0 E_0^2$$
$$= \tfrac{1}{2}(3.00 \times 10^8 \text{ m/s})(8.85 \times 10^{-12} \text{ C}^2/\text{N} \cdot \text{m}^2)(36.5 \times 10^{-3} \text{ V/m})^2 = 1.77 \times 10^{-6} \text{ W/m}^2.$$

23. For the energy density, we have
$u = \tfrac{1}{2}\varepsilon_0 E_0{}^2 = S/c = (1350 \text{ W/m}^2)/(3.00 \times 10^8 \text{ m/s}) = 4.50 \times 10^{-6} \text{ J/m}^3$.
The radiant energy is
$U = uV = (4.50 \times 10^{-6} \text{ J/m}^3)(1.00 \text{ m}^3) = 4.50 \times 10^{-6} \text{ J}$.

27. (a) We choose the axis of the circular plates as the $z$-axis.
We find the electric field from the charge density:
$\mathbf{E} = (\sigma/\varepsilon_0)\mathbf{k} = (Q/\pi R^2 \varepsilon_0)\mathbf{k}$.
From the cylindrical symmetry, we know that the magnetic
field will be circular, centered on the $z$-axis. We choose a
circular path with radius $r < R$ to apply Ampere's law:

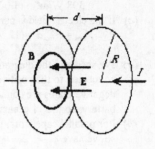

$\oint \mathbf{B} \cdot d\mathbf{s} = \mu_0 I_{enclosed} + \mu_0 \varepsilon_0 (d/dt) \oint \mathbf{E} \cdot d\mathbf{A}$;

$B2\pi r = 0 + \mu_0 \varepsilon_0 (d/dt)[(Q/\pi R^2 \varepsilon_0)\pi r^2]$, which gives
$B = (\mu_0 r/2\pi R^2) \, dQ/dt$ circular, for $r \leq R$.
The Poynting vector,
$\mathbf{S} = (1/\mu_0)\mathbf{E} \times \mathbf{B}$,
will be directed toward the axis of the plates:
$\mathbf{S} = -(1/\mu_0)(Q/\pi R^2 \varepsilon_0)(\mu_0 r/2\pi R^2)(dQ/dt) \, \hat{\mathbf{r}} = -(Qr/2\pi^2 R^4 \varepsilon_0)(dQ/dt) \, \hat{\mathbf{r}}$.
(b) The energy stored in the electric field is
$U = u_E \pi R^2 d = \tfrac{1}{2}\varepsilon_0 E^2 \pi R^2 d = \tfrac{1}{2}\varepsilon_0 (Q/\pi R^2 \varepsilon_0)^2 \pi R^2 d = \tfrac{1}{2}(d/\pi R^2 \varepsilon_0)Q^2$.
The rate at which this energy is being stored is
$dU/dt = (Qd/\varepsilon_0 \pi R^2) \, dQ/dt$.
To find the energy flow into the capacitor through the cylinder at $r = R$, we note that $\mathbf{S}$ is
perpendicular to the surface and has a constant magnitude. Thus
$P = \int \mathbf{S} \cdot d\mathbf{A} = S(\text{surface area}) = (Qr/2\pi^2 \varepsilon_0 R^4)(dQ/dt)(2\pi Rd) = (Qd/\varepsilon_0 \pi R^2) \, dQ/dt$,
which is the rate at which energy is being stored in the electric field.

31. The resonant frequency is given by
$f_0{}^2 = (1/2\pi)^2(1/LC)$.
When we form the ratio for the two stations, we get
$(f_{02}/f_{01})^2 = C_1/C_2$ ;
$(1550 \text{ kHz}/550 \text{ kHz})^2 = (2400 \text{ pF})/C_2$ , which gives $C_2 = 302 \text{ pF}$.

35. (a) The wavelength of the AM station is
$\lambda = c/f = (3.00 \times 10^8 \text{ m/s})/(680 \times 10^3 \text{ Hz}) = 441 \text{ m}$.
(b) The wavelength of the FM station is
$\lambda = c/f = (3.00 \times 10^8 \text{ m/s})/(100.7 \times 10^6 \text{ Hz}) = 2.979 \text{ m}$.

39. (a) The time for a signal to travel to the Moon is
$\Delta t = L/c = (3.84 \times 10^8 \text{ m})/(3.00 \times 10^8 \text{ m/s}) = 1.28 \text{ s}$.
(b) The time for a signal to travel to Mars at the closest approach is
$\Delta t = L/c = (78 \times 10^9 \text{ m})/(3.00 \times 10^8 \text{ m/s}) = 260 \text{ s} = 4.3 \text{ min}$.

43. The radiation from the Sun has the same intensity in all directions, so the rate at which it passes through a sphere
centered at the Sun is
$P = S4\pi R^2$.
The rate must be the same for the two spheres, one containing the Earth and one containing Mars. When we form
the ratio, we get
$P_{Mars}/P_{Earth} = (S_{Mars}/S_{Earth})(R_{Mars}/R_{Earth})^2$;
$1 = (S_{Mars}/1350 \text{ W/m}^2)(1.52)^2$, which gives $S_{Mars} = 584.3 \text{ W/m}^2$.
We find the rms value of the electric field from
$S_{Mars} = c\varepsilon_0 E_{rms}{}^2$;
$584.3 \text{ W/m}^2 = (3.00 \times 10^8 \text{ m/s})(8.85 \times 10^{-12} \text{ C}^2/\text{N} \cdot \text{m}^2)E_{rms}{}^2$, which gives $E_{rms} = 469 \text{ V/m}$.

47. (a) The radio waves have the same intensity in all directions, so the energy per unit area per unit
time over a sphere centered at the source with a radius of 100 m is
$S = P_0/A = P_0/4\pi r^2 = (50 \times 10^3 \text{ W})/4\pi(100 \text{ m})^2 = 0.398 \text{ W/m}^2$.
Thus the power through the area is
$P = SA = (0.398 \text{ W/m}^2)(1.0 \text{ m}^2) = 0.40 \text{ W}$.

(b) We find the rms value of the electric field from

$$S = c\varepsilon_0 E_{rms}^2;$$

$$0.398 \text{ W/m}^2 = (3.00 \times 10^8 \text{ m/s})(8.85 \times 10^{-12} \text{ C}^2/\text{N} \cdot \text{m}^2)E_{rms}^2, \text{ which gives } E_{rms} = 12 \text{ V/m}.$$

(c) If the electric field is parallel to the antenna, the voltage over the length of the antenna is

$$V_{rms} = E_{rms}d = (12 \text{ V/m})(1.0 \text{ m}) = 12 \text{ V}.$$

51. (a) We see from the diagram that all positive plates are connected to the positive side of the battery, and that all negative plates are connected to the negative side of the battery, so the 11 capacitors are connected in parallel.

(b) For parallel capacitors, the total capacitance is the sum, so we have

$$C_{min} = 11(\varepsilon_0 A_{min}/d)$$
$$= 11(8.85 \times 10^{-12} \text{ C}^2/\text{N} \cdot \text{m}^2)(1.0 \times 10^{-4} \text{ m}^2)/(1.1 \times 10^{-3} \text{ m}) = 8.85 \times 10^{-12} \text{ F} = 8.9 \text{ pF};$$

$$C_{max} = 11(\varepsilon_0 A_{max}/d)$$
$$= 11(8.85 \times 10^{-12} \text{ C}^2/\text{N} \cdot \text{m}^2)(9.0 \times 10^{-4} \text{ m}^2)/(1.1 \times 10^{-3} \text{ m}) = 79.7 \times 10^{-12} \text{ F} = 80 \text{ pF}.$$

Thus the range is $8.9 \text{ pF} \leq C \leq 80 \text{ pF}$.

(c) The lowest resonant frequency requires the maximum capacitance.
We find the inductance for the lowest frequency:

$$f_{01} = (1/2\pi)(1/L_1 C_{max})^{1/2};$$

$$550 \times 10^3 \text{ Hz} = (1/2\pi)[1/L_1(79.7 \times 10^{-12} \text{ F})]^{1/2}, \text{ which gives } L_1 = 1.05 \times 10^{-3} \text{ H} = 1.05 \text{ mH}.$$

We must check to make sure that the highest frequency can be reached.
We find the resonant frequency using this inductance and the minimum capacitance:

$$f_{0max} = (1/2\pi)(1/L_1 C_{min})^{1/2}$$
$$= (1/2\pi)[1/(1.05 \times 10^{-3} \text{ H})(8.85 \times 10^{-12} \text{ F})]^{1/2} = 1.65 \times 10^6 \text{ Hz} = 1650 \text{ kHz}.$$

Because this is greater than the highest frequency desired, the inductor will work.
We could also start with the highest frequency.
We find the inductance for the highest frequency:

$$f_{02} = (1/2\pi)(1/L_2 C_{min})^{1/2};$$

$$1600 \times 10^3 \text{ Hz} = (1/2\pi)[1/L_2(8.85 \times 10^{-12} \text{ F})]^{1/2}, \text{ which gives } L_2 = 1.12 \times 10^{-3} \text{ H} = 1.12 \text{ mH}.$$

We must check to make sure that the lowest frequency can be reached.
We find the resonant frequency using this inductance and the maximum capacitance:

$$f_{0min} = (1/2\pi)(1/L_2 C_{max})^{1/2}$$
$$= (1/2\pi)[1/(1.12 \times 10^{-3} \text{ H})(79.7 \times 10^{-12} \text{ F})]^{1/2} = 533 \times 10^5 \text{ Hz} = 533 \text{ kHz}.$$

Because this is less than the lowest frequency desired, this inductor will also work.
Thus the range of inductances is

$$1.05 \text{ mH} \leq L \leq 1.12 \text{ mH}.$$

# Chapter 33: Light: Reflection and Refraction

## Chapter Overview and Objectives

This chapter introduces the ray model of light. It discusses properties of the propagation of light as it interacts with boundaries between two materials. The laws of reflection and refraction are discussed, as is the phenomenon of total internal reflection. Image formation by reflecting and refracting surfaces is described.

After completing study of this chapter, you should:

- Know what the index of refraction of a material is.
- Know what the ray model of light is and when it applies.
- Know the law of reflection.
- Know the difference between specular and diffuse reflection.
- Know what an object and an image are.
- Know what a focal point is and what the corresponding focal length is.
- Know the image formation equation for a mirror.
- Know how to calculate the focal length of a spherical mirror.
- Know how to calculate the lateral magnification of an image.
- Know what real and virtual images are.
- Know what total internal reflection and critical angle are.
- Know what dispersion is.
- Know how to determine the image position formed by a refracting surface.

## Summary of Equations

Definition of index of refraction:
$$n = \frac{c}{v}$$

Law of reflection:
$$\theta_i = \theta_r$$

Relationship between focal length and radius of curvature of spherical mirror:

$$f = \frac{r}{2}$$

Mirror equation:
$$\frac{1}{f} = \frac{1}{d_o} + \frac{1}{d_i}$$

Lateral magnification of image:
$$m = \frac{h_i}{h_o} = -\frac{d_i}{d_o}$$

Snell's law or law of refraction:
$$n_1 \sin \theta_1 = n_2 \sin \theta_2$$

Critical angle for total internal reflection:
$$\theta_C = \arcsin \frac{n_2}{n_1}$$

Relationship between image distance, object distance, and radius of curvature of spherical refracting surface:

$$\frac{n_1}{d_o} + \frac{n_2}{d_i} = \frac{n_2 - n_1}{R}$$

# Chapter Summary

### Section 33-1 The Ray Model of Light

The **ray model** of light makes the assumption that light follows definite paths just as particles do. We call the paths **rays**. In uniform materials, these paths are straight lines. The ray model of light usually is applicable to understanding the behavior of light when all objects and apertures in the path of the light rays are large compared to the wavelength of the light. The field of **geometric optics** deals with problems concerning light in which the ray model is applicable.

### Section 33-2 The Speed of Light and Index of Refraction

The speed of light in a vacuum, usually denoted as $c$, is $2.99792458 \times 10^8$ m/s ($\sim 3.00 \times 10^8$ m/s). The speed of light in materials is less than the speed of light in vacuum. The ratio of the speed of light in vacuum, $c$, to the speed of light in a material, $v$, is called the **refractive index** of the material, $n$:

$$n = \frac{c}{v}$$

Because the speed of light in materials is less than the speed of light in vacuum, the refractive indices of materials are greater than one. The refractive index of a material depends on the wavelength of the light.

### Example 33-1-A

An optical fiber used for communication purposes carries a light signal a distance of 1.34 km in a time of 6.52 μs. What is the refractive index of the optical fiber?

**Solution:**

The speed of the light through the fiber is

$$v = \frac{d}{t} = \frac{1.34 \times 10^3 \text{ m}}{6.52 \times 10^{-6} \text{ s}} = 2.06 \times 10^8 \text{ m/s}$$

We can then calculate the refractive index:

$$n = \frac{c}{v} = \frac{3.00 \times 10^8 \text{ m/s}}{2.06 \times 10^8 \text{ m/s}} = 1.46$$

### Section 33-3 Reflection; Image Formation by a Plane Mirror

When light reaches a boundary between two materials, some of the energy of the light propagates back into the material from which it reached the boundary. This is called **reflection**.

To express the relationship between the direction the light travels toward the boundary and the direction the light travels after reflecting from the boundary, we need a convention for specifying the direction of travel of light rays relative to the boundary surface. The normal to the surface is the direction perpendicular to the surface. We define all of our directions of ray travel in our geometric optics laws relative to the normal of the boundary surface. This means light rays traveling perpendicular to a surface have angle zero and light rays traveling parallel to surface have angle 90°.

The light ray that travels toward the boundary surface is called the incident light ray. Its direction relative to the boundary surface normal is called the **angle of incidence** and is denoted $\theta_i$. The light ray that travels away from the surface after reflection from the surface is called the reflected light ray. Its direction relative to the boundary surface normal is called the **angle of reflection** and is denoted $\theta_r$. The **law of reflection** states **the angle of incidence is equal to the angle of reflection**:

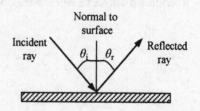

$$\theta_i = \theta_r$$

When a surface is rough, the normal to the surface changes direction by a large amount for a small displacement along the surface. Parallel light rays slightly displaced from each other are reflected into many different directions or scattered. This is called **diffuse reflection**. When the surface is very smooth, the normal to the surface points in almost the same direction for an extended area of the surface. Parallel light rays that are displaced slightly from each other are reflected into almost identical directions. This is called **specular reflection**. When you look at a surface from which the reflection is specular, you are able to see images of the sources of the light that is reflecting from the surface. This is why you see yourself when you look in a mirror. When you look at a surface from which the reflection is diffuse, you do not see images of the sources of the light that is reflecting from the surface. This is why you do not see yourself when you look at a piece of white paper even though the white paper reflects more light than a mirror does.

The image of an object in a flat mirror is the same distance behind the mirror, called the **image distance**, $d_i$ that the object is from the front of the mirror, called the **object distance**, $d_o$. The image is called a **virtual image** because the light rays do not travel from any point where the object is located, but follow a path after striking the mirror as if they had left the image behind the mirror. When a **real image** is formed, the light rays pass through the image point.

### Section 33-4 Formation of Images by Spherical Mirrors

A common shape for non-flat mirrors is a spherical shape. A mirror surface that is the inside surface of a sphere is called a **concave** mirror and a mirror surface that is the outside surface of a sphere is called a **convex** mirror.

A mirror can be characterized by its focal length. The **focal length,** $f$, is the distance from the center of the mirror to the point at which light rays that are parallel to the axis of the mirror converge to a point after reflecting from the mirror. This point is called the **focal point**, F.

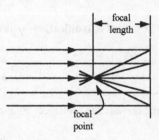

For a spherical mirror, the focal length, $f$, is one half the radius of curvature, $r$, of the mirror:

$$f = \frac{r}{2}$$

A spherical mirror will form an approximate point image of a point object or source of light. The reason the image is only approximately a point image is because of **spherical aberration**. We can locate the image formed by a mirror using a ray diagram. A ray diagram is a drawing of the mirror and object location used to determine the position of the image by drawing the paths of light rays. Unless we are interested in the spherical aberration, we only need draw two **principle rays** to determine the location of the image.

   1. Draw a line representing the optic axis of the mirror and a second perpendicular line at the position of the mirror to represent the mirror.

   2. Mark the position of the focal point on the optic axis a distance of one focal length away from the mirror.

   3. Mark the position of the object in the diagram.

   4. Draw a ray from the object to the mirror parallel to the optic axis. This ray leaves the mirror along a line that goes from the intersection of this ray with the mirror and passes through the focal point of the mirror.

   5. Draw a second ray from the object to the mirror passing through the focal point of the mirror. This ray then leaves the mirror from this point and continues parallel to the optic axis.

   6. The image is the point of common intersection of the reflected rays.

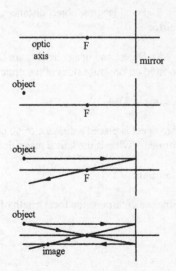

If the real light rays pass through this point, the image is called a **real image**. If the real light rays do not pass through their point of intersection, the image is a **virtual image**. Consider the ray diagram below, for a mirror with a negative focal length (a convex mirror). Note that the focal point F is on the back side of the mirror because of the negative focal length. Follow the instructions above to see how the ray diagram below was constructed for a negative focal length mirror.

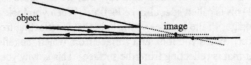

The distance of the object from the mirror parallel to the optic axis is called the **object distance**, $d_o$, and the distance of the image from the mirror parallel to the optic axis is called the **image distance**, $d_i$. The perpendicular distance of the object from the optic axis is called the **object height**, $h_o$, and the perpendicular distance of the image from the optic axis is called the **image height**, $h_i$. The mirror equation relates the object distance, the image distance, and the focal length of the mirror:

$$\frac{1}{f} = \frac{1}{d_o} + \frac{1}{d_i}$$

The **lateral magnification**, $m$, is defined as the ratio of the image height to the object height:

$$m = \frac{h_i}{h_o}$$

and is related to the object distance and the image distance by

$$m = \frac{h_i}{h_o} = -\frac{d_i}{d_o}$$

The image is called upright if the lateral magnification is positive and the image is called inverted if the lateral magnification is negative.

The mirror equation and lateral magnification relationship can be used for both concave and convex mirrors and for real and virtual images if the following sign conventions are observed:

   1. The focal length of concave mirrors is positive and the focal length of ~~concave~~ convex mirrors is negative

   2. Focal lengths, object distances, and image distances are positive in front of the mirror and negative behind the mirror.

   3. Object and image heights are the same sign if located on the same side of the optic axis and of opposite signs if located on opposite sides of the optic axis.

## Example 33-4-A

An object is placed a distance of 40 cm from a convex lens with a radius of curvature of 50 cm. Where is the image formed? What is the lateral magnification of the image? Is the image real or virtual? Is the image upright or inverted?

## Solution:

First we determine the focal length of the mirror:

$$f = \frac{r}{2} = \frac{50\,\text{cm}}{2} = 25\,\text{cm}$$

However, our sign convention tells us the focal length of a convex mirror is negative, so

$$f = -25 \text{ cm}$$

We then use the mirror equation to determine where the image is formed:

$$\frac{1}{f} = \frac{1}{d_o} + \frac{1}{d_i} \quad \Rightarrow \quad d_i = \frac{f d_o}{d_0 - f} = \frac{(-25 \text{ cm})(40 \text{ cm})}{(40 \text{ cm}) - (-25 \text{ cm})} = -15 \text{ cm}$$

We can determine the lateral magnification from the lateral magnification relationship to the object and image distances:

$$m = -\frac{d_i}{d_o} = -\frac{-15 \text{ cm}}{40 \text{ cm}} = 0.38$$

The image is upright because the magnification is positive and the image is virtual because the image distance is negative.

### Section 33-5 Refraction: Snell's Law

When light is transmitted across a boundary between two materials with different refractive indices, the light changes direction. This is called **refraction**. The direction of the refracted light ray, $\theta_2$, called the **angle of refraction**, is related to direction of the incident light ray, $\theta_1$, called the **angle of incidence**, and the indices of refraction of the materials on each side of the boundary. This relationship is called the **law of refraction** or **Snell's law**:

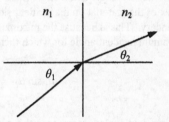

$$n_1 \sin \theta_1 = n_2 \sin \theta_2$$

### Example 33-5-A

A rectangular block of material has a refractive index 1.28. A light ray incident from air on the end of the block refracts through the end and reaches the side of the block, where it refracts again, passing back out of the block. Solve for the angle $\theta$ in the diagram for the ray exiting the block.

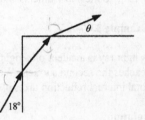

### Solution:

We use Snell's law to determine the direction of the light ray after passing through the first surface. We have to be careful about the incident angle. The incident angle is defined as measured from the normal of the surface to the direction of the light ray. That means the incident angle at the first surface is the complement of 18°, so the incident angle is 72° and Snell's law gives

$$n_1 \sin \theta_1 = n_2 \sin \theta_2 \quad \Rightarrow \quad \theta_2 = \arcsin\left(\frac{n_1}{n_2} \sin \theta_2\right) = \arcsin\left(\frac{1}{1.28} \sin 72°\right) = 48°$$

We use this information to determine the incident angle on the second surface. The perpendiculars to each surface and the light ray within the material form a right triangle. Because the two non-right angles in a right triangle are complementary, that means the incident angle at the second surface must be 90° − 48° = 42°. We can use this to solve for the refracted angle going out of the block

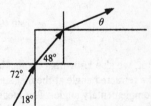

$$n_1 \sin \theta_1 = n_2 \sin \theta_2 \quad \Rightarrow \quad \theta_2 = \arcsin\left(\frac{n_1}{n_2} \sin \theta_2\right) = \arcsin\left(\frac{1.28}{1.00} \sin 42°\right) = 60°$$

The angle $\theta$ is the complement of the refracted angle, so

$$\theta = 90° - 60° = 30°$$

### Section 33-6 Visible Spectrum and Dispersion

Our eyes are sensitive to light with wavelengths from about 400 nm to 750 nm. Light that falls between these two wavelengths is part of the **visible spectrum**. Light with wavelengths greater than 750 nm is called **infrared light** and light with wavelengths less than 400 nm is called **ultraviolet light**.

Because the refractive index of a material depends on the wavelength of light, the direction of a refracted light ray also depends on the wavelength. This is what allows prisms to disperse light of different wavelengths in different directions. This process of separating different wavelengths into different directions is called **dispersion**.

### Section 33-7 Total Internal Reflection; Fiber Optics

If the incident angle is too large, there will not be a solution to Snell's law for the refracted angle when the refractive index of the material on the incident side of the boundary is larger than the refractive index on the refracted side of the boundary. This is because the maximum value of the sine function is one. We give the name **critical angle**, $\theta_C$, to the maximum incident angle for which there is a solution, when the sine of the refracted angle is one:

$$n_1 \sin\theta_C = n_2 \cdot 1 \quad \Rightarrow \quad \theta_C = \arcsin\frac{n_2}{n_1}$$

For incident angles greater than the critical angle, all of the light is reflected from the surface back into the material on the incident side of the boundary. This phenomenon is called **total internal reflection**. This behavior is taken advantage of in many optical instruments and in **fiber optics**.

### Example 33-7-A

A light ray is incident on a right-angle block in air as shown. It internally reflects when it reaches the second surface. What is the minimum refractive index of the block so that total internal reflection takes place at the second surface?

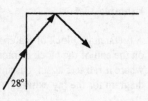

### Solution:

The direction of the ray after it is refracted at the first surface is given by Snell's law. The incident angle is the complement of the 28° angle given, $\theta_1 = 90° - 28° = 62°$. Using Snell's law, we get

$$n_1 \sin\theta_1 = n_2 \sin\theta_2 \quad \Rightarrow \quad \sin\theta_2 = \frac{1.00}{n_2}\sin 62°$$

We also know at the minimum refractive index, the angle will be the critical angle, so we then know

$$\sin\theta_C = \frac{1.00}{n_2}$$

But, we also know that the incident angle at the second surface, which must be the critical angle, $\theta_C$, is the complement of the refracted angle at the first surface or $\theta_C = 90° - \theta_2$. We also know that the cosine of an angle is the sine of the complementary angle. This means

$$\cos\theta_C = \sin\theta_2$$

If we use the trigonometric identity

$$\cos\theta = \sqrt{1 - \sin^2\theta}$$

we can write

$$\sqrt{1-\sin^2\theta_C} = \sin\theta_2 \quad \Rightarrow \quad \sqrt{1-\left(\frac{1.00}{n_2}\right)^2} = \frac{1.00}{n^2}\sin 62°$$

Solving this for $n_2$ we get

$$n_2 = \sqrt{1+\sin 62°} = 1.37$$

## Section 33-8 Refraction at a Spherical Surface

A spherical refractive surface can form an image. If the refractive index of the material where the point source of light is located is $n_1$ and the refractive index on the other side of the refracting surface of radius $R$ is $n_2$, then the object distance is related to the image distance by

$$\frac{n_1}{d_o} + \frac{n_2}{d_i} = \frac{n_2 - n_1}{R}$$

The sign convention for this image formation system is different than that for the mirror equation:

    1.  The surface is convex if it bulges toward the source and its radius of curvature is positive. The surface is concave if it bulges away from the source and its radius of curvature is negative.
    2.  The object distance is positive if located on the side of the surface that the light enters the surface and it is negative if it is on the side of the surface the light leaves the surface.
    3.  The image distance is positive if located on the side of the surface that the light leaves the surface and it is positive if it is on the side of the surface the light enters the surface.

### Example 33-8-A

Assume an eye has no lens and all of the refraction occurs at the curved surface of the cornea to form an image on the retina at the back of the eye. Water, which you can consider the material inside the eye to be, has a refractive index of 1.33. Assume an eye has a diameter of 3.2 cm. What must be the radius of curvature of the cornea to form an image on the retina if the object is located 50 cm from the eye?

### Solution:

We know the object distance is 50 cm and the image distance is 3.2 cm. The index of refraction on the source side of the surface is the index of refraction of air, $n_1 = 1$. The index of refraction on the image side of the surface is the index of refraction of water, $n_2 = 1.33$. We use the image formation equation above to determine the radius of curvature of the surface:

$$\frac{n_1}{d_o} + \frac{n_2}{d_i} = \frac{n_2 - n_1}{R} \quad \Rightarrow \quad R = \frac{(n_2-n_1)d_o d_i}{n_1 d_i - n_2 d_o} = \frac{(1.33-1.00)(50\,\text{cm})(3.2\,\text{cm})}{(1.00)(50\,\text{cm})-(1.33)(3.2\,\text{cm})} = 1.2\,\text{cm}$$

## Practice Quiz

1.    Under what conditions is the ray model of light applicable?

    a)  All dimensions in the system are small compared to the wavelength of light.
    b)  All dimensions in the system are large compared to the wavelength of light.
    c)  The light is of a single wavelength.
    d)  There are no reflecting surfaces in the system.

2.    Under what conditions does a concave mirror form a virtual image?

   a)  when the object is closer to the mirror than the radius of curvature of the mirror
   b)  when the object is farther from the mirror than the radius of curvature of the mirror
   c)  when the object is closer to the mirror than the focal length of the mirror
   d)  when the object is farther from the mirror than the focal length of the mirror

3.    Which of the following best describes the reason a prism sends different wavelengths of light in a different direction from incident white light?

   a)  total internal reflection
   b)  diffuse reflection
   c)  critical angle
   d)  dispersion

4.    Which of the following wavelengths of light is not visible to human eyes?

   a)  440 nm
   b)  20 µin
   c)  0.060 µm
   d)  $1.8 \times 10^{-6}$ ft

5.    In the diagram to the right, a light ray enters a material with refractive index $n_2$ from a material with a refractive index $n_1$.  Which statement is necessarily true about the refractive indices?

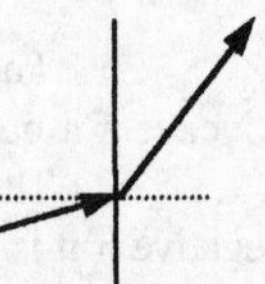

   a)  $n_2 > n_1$
   b)  $n_1 > n_2$
   c)  $n_2 < 1$ and $n_1 > 1$
   d)  $n_1 < 1$ and $n_2 > 1$

6.    Why can't you see your image in a reflection from a piece of paper?

   a)  The paper does not reflect enough light.
   b)  The reflection is a specular reflection.
   c)  The reflection is a diffuse reflection.
   d)  The dispersion that occurs separates all of the different colored images, making it impossible to see the image.

7.    When the sun is close to the horizon, what does refraction of light in the atmosphere do to the position of the sun? (The refractive index of the air is larger, the nearer to the surface of the earth you are.)

   a)  The position of the image of the Sun is closer to the horizon than the actual position of the Sun.
   b)  The position of the image of the Sun is farther from the horizon than the actual position of the Sun.
   c)  The position of the image of the Sun is the same as the actual position of the Sun.
   d)  The answer depends on whether its is morning or evening.

8.    In the ray diagram to the right, the vertical line represents a mirror.  The paths of two rays from an object are shown.  What type of mirror does the vertical line represent?

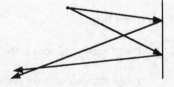

   a)  concave
   b)  convex
   c)  flat
   d)  broken

9.    What type of image is formed in Question 8 above?

   a) real and inverted
   b) virtual and inverted
   c) real and upright
   d) virtual and upright

10.   When you look at a pencil in a glass of water it appears bent.  Which phenomenon explains why the pencil appears bent?

   a) reflection
   b) refraction
   c) dispersion
   d) total internal reflection

11.   Determine the speed of light in a material that has a critical angle of 48° in air.

12.   A light ray is incident on a material's surface from air at an incident angle of 30°.  The direction of the light ray is 10° toward the normal to the surface after it passes through the surface.  What is the refractive index of the material?

13.   A convex mirror with a radius of curvature of 50.0 cm has an object located 25.0 cm in front of it.  Where is the image of this object formed?  What is the lateral magnification of the image?  Is the image upright or inverted?  Is the image real or virtual?

14.   A mirror is used to form an image of an object that is 30.0 cm from the mirror.  The image is formed 40.0 cm in front of the mirror and is inverted.  What is the radius of curvature of the mirror?  Is it a concave mirror or a convex mirror?

15.   A glass marble of diameter 0.75 in has a small seed of width 0.86 mm at its center.  The glass the marble is made of has a refractive index 1.55.  If you look at the seed inside the marble, where is its image formed?

## Problem Solutions

3.   The time for light to travel from the Sun to the Earth is
   $$\Delta t = L/c = (1.50 \times 10^{11} \text{ m})/(3.00 \times 10^8 \text{ m/s}) = 5.00 \times 10^2 \text{ s} = 8.33 \text{ min.}$$

7.   The eight-sided mirror would have to rotate 1/8 of a revolution for the succeeding mirror to be in position to reflect the light in the proper direction.  During this time the light must travel to the opposite mirror and back.  Thus the angular speed required is
   $$\omega = \Delta\theta/\Delta t = (2\pi \text{ rad}/8)/(2L/c) = (\pi \text{ rad})c/8L$$
   $$= (\pi \text{ rad})(3.00 \times 10^8 \text{ m/s})/8(35 \times 10^3 \text{ m}) = 3.4 \times 10^3 \text{ rad/s } ( = 3.2 \times 10^4 \text{ rev/min}).$$

11.   From the triangle formed by the mirrors and the first reflected ray, we have
   $$\theta + \alpha + \phi = 180°;$$
   $$40° + 135° + \phi = 180°, \text{ which gives } \phi = 5°.$$

15.   The rays from the Sun will be parallel, so the image will be at the focal point.  The radius is
   $$r = 2f = 2(18.2 \text{ cm}) = 36.4 \text{ cm.}$$

19.   We find the image distance from the magnification:
   $$m = h_i/h_o = -d_i/d_o ;$$
   $$+3 = -d_i/(1.5 \text{ m}), \text{ which gives } d_i = -4.5 \text{ m.}$$
   We find the focal length from
   $$(1/d_o) + (1/d_i) = 1/f,$$
   $$[1/(1.5 \text{ m})] + [1/(-4.5 \text{ m})] = 1/f, \text{ which gives } f = 2.25 \text{ m.}$$
   The radius of the concave mirror is
   $$r = 2f = 2(2.25 \text{ m}) = 4.5 \text{ m.}$$

23. (a) With $d_i = d_o$ , we locate the object from
$$(1/d_o) + (1/d_i) = 1/f,$$
$$(1/d_o) + (1/d_o) = 1/f,$$ which gives $d_o = 2f = r$.
The object should be placed at the center of curvature.
(b) Because the image is in front of the mirror, $d_i > 0$, it is real.
(c) The magnification is
$$m = -d_i/d_o = -d_o/d_o = -1.$$
Because the magnification is negative, the image is inverted.
(d) As found in part (c), $m = -1$.

27. From the ray diagram, we see that
$$\tan\theta = h_o/d_o = h_i/d_i \ ;$$
$$\tan\alpha = h_o/(d_o + r) = h_i/(r - d_i).$$
When we divide the two equations, we get
$$(d_o + r)/d_o = (r - d_i)/d_i;$$
$$1 + (r/d_o) = (r/d_i) - 1, \text{ or}$$
$$(r/d_o) - (r/d_i) = -2;$$
$$(1/d_o) - (1/d_i) = -2/r = -1/f, \text{ with } f = r/2.$$
From the ray diagram, we see that $d_i < 0$.
If we consider $f$ to be negative, we have
$$(1/d_o) + (1/d_i) = 1/f.$$

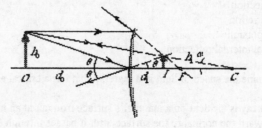

31. For a mirror we have
$$(1/d_o) + (1/d_i) = 1/f, \text{ or } d_i = d_o f/(d_o - f); \text{ and}$$
$$m = h_i/h_o = -d_i/d_o = -f/(d_o - f).$$
If we place the object so one end is at $d_{o1}$ and the
other is at $d_{o2} = d_{o1} - l$, the image distances to
the ends will be
$$d_{i1} = d_{o1}f/(d_{o1} - f) \text{ and } d_{i2} = d_{o2}f/(d_{o2} - f) = d_{i1} - l'.$$
Thus the length of the image is
$$l' = d_{i1} - d_{i2} = [d_{o1}f/(d_{o1} - f)] - [(d_{o1} - l)f/(d_{o1} - l - f)]$$
$$= [(d_{o1} - l - f)d_{o1}f - (d_{o1} - l)(d_{o1} - f)f]/(d_{o1} - f)(d_{o1} - l - f)$$
$$= -l f^2/(d_{o1} - f)(d_{o1} - l - f).$$
With $l \ll d_{o1}$, this becomes
$$l' \approx -l f^2/(d_{o1} - f)^2.$$
Thus the longitudinal magnification is
$$l'/l = -m^2.$$
The negative sign indicates that the image is reversed front to back, as shown in the diagram.

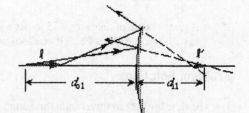

35. We find the incident angle in the air from
$$n_1 \sin\theta_1 = n_2 \sin\theta_2;$$
$$(1.00) \sin\theta_1 = (1.33) \sin 43.0°, \text{ which gives } \theta_1 = 65.1°.$$
Thus the angle above the horizon is
$$90.0° - \theta_1 = 90.0° - 65.1° = 24.9°.$$

39. Because all of the surfaces are parallel, the angle of refraction
from one surface is the angle of incidence at the next:
$$n_1 \sin\theta_1 = n_2 \sin\theta_2;$$
$$n_2 \sin\theta_2 = n_3 \sin\theta_3.$$
Thus we have
$$n_1 \sin\theta_1 = n_3 \sin\theta_3,$$
which is the same as refraction from the first medium directly
into the third medium.

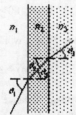

43. We find the speed of light from the index of refraction, $v = c/n$. For the change, we have
$$(v_{red} - v_{violet})/v_{violet} = [(c/n_{red}) - (c/n_{violet})]/(c/n_{violet})$$
$$= (n_{violet} - n_{red})/n_{red} = (1.662 - 1.613)/(1.613) = 0.030 = 3.0\%.$$

47. When the light in the material with a higher index is incident at the critical angle, the refracted angle is 90°:

$n_{\text{lucite}} \sin \theta_1 = n_{\text{water}} \sin \theta_2$ ;

$(1.51) \sin \theta_1 = (1.33) \sin 90°$, which gives $\theta_1 = 61.7°$.

Because lucite has the higher index, the light must start in lucite.

51. We find the angle of incidence from the distances:

$\tan \theta_1 = L/h = (7.0 \text{ cm})/(8.0 \text{ cm}) = 0.875$, so $\theta_1 = 41.2°$.

For the maximum incident angle for the refraction from liquid into air, we have

$n_{\text{liquid}} \sin \theta_1 = n_{\text{air}} \sin \theta_2$ ;

$n_{\text{liquid}} \sin \theta_{1\text{max}} = (1.00) \sin 90°$, which gives $\sin \theta_{1\text{max}} = 1/n_{\text{liquid}}$ .

Thus we have

$\sin \theta_1 \geq \sin \theta_{1\text{max}} = 1/n_{\text{liquid}}$ ;

$\sin 41.2° \geq 1/n_{\text{liquid}}$ , or $n_{\text{liquid}} \geq 1.5$.

55. We find the location of the image of a point on the bottom from the refraction from water to glass (with $R = \infty$):

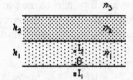

$(n_1/d_{o1}) + (n_2/d_{i1}) = (n_2 - n_1)/R = 0$;

$[(1.33)/(12.0 \text{ cm})] + [(1.50)/d_{i1}] = 0$, which gives $d_{i1} = -13.5$ cm.

Using this as the object for the refraction from glass to air, we find the location of the final image (with $R = \infty$):

$(n_2/d_{o2}) + (n_3/d_{i2}) = (n_3 - n_2)/R = 0$;

$[(1.50)/(25.5 \text{ cm})] + [(1.00)/d_{i2}] = 0$, which gives $d_{i2} = -17.0$ cm.

Thus the bottom appears to be 17.0 cm below the surface of the glass.

59. For a plane mirror each image is as far behind the mirror as the object is in front. Each reflection produces a front-to-back reversal. We show the three images and the two intermediate images that are not seen.

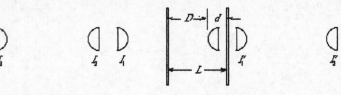

(a) The first image is from a single reflection, so it is

$d_1 = 2D = 2(1.5 \text{ m}) = 3.0$ m away.

The second image is from two reflections, so it is

$d_2 = L + d + D = 2.0 \text{ m} + 0.5 \text{ m} + 1.5 \text{ m} = 4.0$ m away.

The third image is from three reflections, so it is

$d_3 = 2L + D + D = 2(2.0 \text{ m}) + 1.5 \text{ m} + 1.5 \text{ m} = 7.0$ m away.

(b) We see from the diagram that

the first image is facing toward you;

the second image is facing away from you;

the third image is facing toward you.

63. The two students chose different signs for the magnification, i. e., one upright and one inverted.

The focal length of the concave mirror is $f = R/2 = (40 \text{ cm})/2 = 20$ cm.

We relate the object and image distances from the magnification:

$m = -d_i/d_o$ ;

$\pm 3 = -d_i/d_o$ , which gives $d_i = -3d_o$ .

When we use this in the mirror equation, we get

$(1/d_o) + (1/d_i) = 1/f$ ;

$(1/d_o) + [1/(-3d_o)] = 1/f$ , which gives $d_o = 2f/3, 4f/3 = 13.3$ cm, 26.7 cm.

The image distances are $= -40$ cm (virtual, upright), and $+ 80$ cm (real, inverted).

67. (a) The ray travels from point A, a distance $H_1$ from the surface, to point B, a distance $H_2$ from the surface, after reflecting from the surface. The component of the distance between A and B parallel to the surface is $L$. If $x$ is the component of the distance between A and the point P, where the ray meets the surface, parallel to

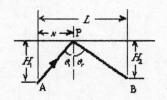

the surface, the time of travel is

$$t_{AB} = t_{AP} + t_{PB}$$
$$= [(x^2 + H_1^2)^{1/2}/c] + \{[(L - x)^2 + H_2^2]^{1/2}/c\}.$$

We find the value of $x$ for the minimum time from

$$dt_{AB}/dx = [(1/c)\tfrac{1}{2}(2x)/(x^2 + H_1^2)^{1/2}] + \{(1/c)\tfrac{1}{2}(-2)(L - x)/[(L - x)^2 + H_2^2]^{1/2}\} = 0,$$

which reduces to

$$x/(x^2 + H_1^2)^{1/2} = (L - x)/[(L - x)^2 + H_2^2]^{1/2}.$$

From the diagram, we see that

$$x/(x^2 + H_1^2)^{1/2} = \sin\theta_i \text{ and } (L - x)/[(L - x)^2 + H_2^2]^{1/2} = \sin\theta_r,$$

so we have $\sin\theta_i = \sin\theta_r$, or $\theta_i = \theta_r$.

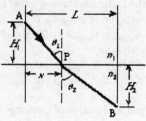

(b) The ray travels from point A in the top medium, a distance $H_1$ from the surface, to point B in the bottom medium, a distance $H_2$ from the surface. The component of the distance between A and B parallel to the surface is $L$. If $x$ is the component of the distance between A and the point P, where the ray meets the surface, parallel to the surface, the time of travel is

$$t_{AB} = t_{AP} + t_{PB}$$
$$= [(x^2 + H_1^2)^{1/2}/(c/n_1)] + \{[(L - x)^2 + H_2^2]^{1/2}/(c/n_2)\}.$$

We find the value of $x$ for the minimum time from

$$dt_{AB}/dx = [(n_1/c)\tfrac{1}{2}(2x)/(x^2 + H_1^2)^{1/2}] + \{(n_2/c)\tfrac{1}{2}(-2)(L - x)/[(L - x)^2 + H_2^2]^{1/2}\} = 0,$$

which reduces to

$$n_1 x/(x^2 + H_1^2)^{1/2} = n_2(L - x)/[(L - x)^2 + H_2^2]^{1/2}.$$

From the diagram, we see that

$$x/(x^2 + H_1^2)^{1/2} = \sin\theta_1 \quad \text{and} \quad (L - x)/[(L - x)^2 + H_2^2]^{1/2} = \sin\theta_2,$$

so we have $n_1 \sin\theta_1 = n_2 \sin\theta_2$.

# Chapter 34:Lenses and Optical Instruments

## Chapter Overview and Objectives

This chapter introduces the optics of thin lenses. It discusses how to locate the position of images formed by thin lenses and properties of the images. It discusses the various optical instruments, including the human eye.

After completing study of this chapter, you should:

- Know what a focal point of a lens is.
- Know what the focal length of a lens is defined to be.
- Know how to construct ray diagrams to locate images formed by lenses.
- Know how to use the lens equation to located images formed by lenses.
- Know how to determine the lateral magnification of an image.
- Know how to use the lensmaker's equation.
- Know how various optical instruments form images.
- Know that thin lenses are not perfect image-forming systems, but have aberrations and distortions.

## Summary of Equations

Definition of power of lens:

$$P = \frac{1}{f}$$

Relationship between image and object positions:

$$f = \frac{1}{d_o} + \frac{1}{d_i}$$

Definition of lateral magnification:

$$m = \frac{h_i}{h_o}$$

Relationship between lateral magnification and image and object positions:

$$m = \frac{h_i}{h_o} = -\frac{d_i}{d_o}$$

Lensmaker's equation:

$$\frac{1}{f} = (n-1)\left(\frac{1}{R_1} + \frac{1}{R_2}\right)$$

Definition of $f$-stop number:

$$f\text{-stop} = \frac{f}{D}$$

Definition of angular magnification:

$$M = \frac{\theta_i}{\theta_0}$$

Angular magnification of simple magnifier with image at an infinite distance:

$$M = \frac{N}{f}$$

Angular magnification of simple magnifier with image at near point distance:

$$M = \frac{N}{f} + 1$$

Approximate angular magnification of telescope:    $M \approx -\dfrac{f_o}{f_e}$

Approximate angular magnification of compound microscope:

$$M \approx \frac{N}{f_e}\frac{l-f_e}{d_o} \approx \frac{Nl}{f_e f_o}$$

## Chapter Summary

### Section 34-1 Thin Lenses; Ray Tracing

A lens has two surfaces that form images by refraction of light. It is considered a thin lens if the thickness of the lens is much smaller than any radius of curvature of the refracting surfaces. A lens can be characterized by its focal length. The **focal length,** $f$, is the distance from the center of the lens to the point at which parallel light rays that are parallel to the axis of the lens converge to a point after passing through the lens. This point is called the **focal point**, **F**. The reciprocal of the focal length is called the **power**, $P$, of the lens:

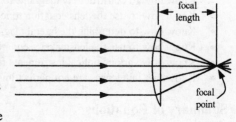

$$P = \frac{1}{f}$$

In this usage, the power of the lens is in diopters (D) if the focal length is in meters, 1 diopter = 1 m$^{-1}$.

If the lens has a positive focal length, it is called a **converging lens** because it causes the light rays to turn toward each other. If the lens has a negative focal length, it is called a **diverging lens** because it causes the light rays to turn away from each other.

A ray diagram can be used to determine the location of an image for a given object location. To draw a ray diagram we follow the following procedure:

    1. Draw a line representing the optic axis of the lens and second perpendicular line at the position of the lens to represent the lens.

    2. Mark the position of the two focal points on the optic axis, each a focal length away from the lens.

    3. Mark the position of the object in the diagram.

    4. Draw a ray from the object to the lens that is parallel to the optic axis. This ray exits the lens along a line that goes from the intersection of this ray with the lens and passes through the focal point of the lens.

    5. Draw a second ray that goes from the object through the intersection of the optic axis and the lens and continues straight through the lens.

    6. Draw a third ray from the object to the lens that passes through the focal point for rays from the opposite side of the lens to the lens. This ray then leaves the lens from this point and continues parallel to the optic axis.

    7. The image is the point of common intersection of the rays.

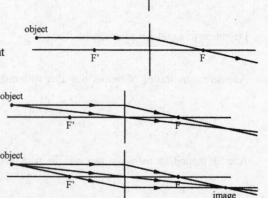

If the real light rays pass through this point, the image is called a **real image**. If the real light rays do not pass through their point of intersection, the image is a **virtual image**. Consider the ray diagram below, for a lens with a negative focal length. Note that the focal point F is on the same side of the lens as the object because of the negative focal length. Follow the instructions above to see how the ray diagram below was constructed for a negative focal length lens.

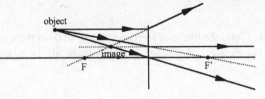

### Section 34-2 The Lens Equation

There is a simple relationship between the position of the object along the optic axis, $d_o$, the position of the image along the optic axis, $d_i$, and the focal length, $f$, of a thin lens. That relationship is

$$f = \frac{1}{d_o} + \frac{1}{d_i}$$

There is also a simple relationship between the position perpendicular to the optic axis for the object, $h_o$, the position perpendicular to the axis for the image, $h_i$, and the corresponding positions along the optic axis:

$$\frac{h_i}{h_o} = -\frac{d_i}{d_o}$$

The ratio of the positions perpendicular to the axis is called the lateral magnification, m:

$$m = \frac{h_i}{h_o} = -\frac{d_i}{d_o}$$

All of these equations apply to all situations involving lenses if the following sign conventions are used:

1.  The focal length of a converging lens is positive and the focal length of a diverging lens is negative.

2.  The object distance is positive if the object is on the side of the lens from which the light is approaching the lens.

3.  The image distance is positive if the image is on the side of the lens from which the light is leaving the lens.

4.  The image height is positive if the image is upright and the image height is negative if the image is inverted.

### Example 34-2-A

An object is placed a distance of 36.0 cm from a lens with a focal length of 20.0 cm. Where is the image formed? What is the lateral magnification of the image? Is the image upright or inverted? Is the image real or virtual?

**Solution:**

We use the lens equation to solve for the image distance:

$$\frac{1}{f} = \frac{1}{d_o} + \frac{1}{d_i} \qquad d_i = \frac{fd_o}{d_o - f} = \frac{(20.0\,\text{cm})(36.0\,\text{cm})}{36.0\,\text{cm} - 20.0\,\text{cm}} = 45.0\,\text{cm}$$

The lateral magnification of the image is

$$m = -\frac{d_i}{d_o} = -\frac{45.0\,\text{cm}}{36.0\,\text{cm}} = -1.25$$

The magnification is negative, so the image is inverted.

The image distance is positive, so the image is real.

### Example 34-2-B

A lens of focal length –12.5 cm is used to form the image of an object that is 40.0 cm from the lens. Where is the image located? What is the magnification of the image? Is the image upright or inverted? Is the image virtual or real?

**Solution:**

We use the lens equation to solve for the image distance:

$$\frac{1}{f} = \frac{1}{d_o} + \frac{1}{d_i} \qquad d_i = \frac{fd_o}{d_o - f} = \frac{(-12.5\,\text{cm})(40.0\,\text{cm})}{40.0\,\text{cm} - (-12.5\,\text{cm})} = -9.52\,\text{cm}$$

The lateral magnification of the image is

$$m = -\frac{d_i}{d_o} = -\frac{-9.52\,\text{cm}}{40.0\,\text{cm}} = +0.238$$

The magnification is positive, so the image is upright.

The image distance is negative, so the image is virtual.

### Section 34-3 Combinations of Lenses

If there is more than one lens in an optic system, the object for each succeeding lens is the image of the previous lens.

### Example 34-3-A

A lens with a focal length of +10.0 cm is placed 20.0 cm from a lens with a focal length of –25.0 cm. An object is placed 12.5 cm from the converging lens. Where is the final image of the two-lens system formed? What is the lateral magnification of the final image? Is the final image upright or inverted? Is the final image real or virtual?

**Solution:**

We use the lens equation to solve for the image distance of the first lens:

$$\frac{1}{f} = \frac{1}{d_o} + \frac{1}{d_i} \qquad d_i = \frac{fd_o}{d_o - f} = \frac{(10.0\,\text{cm})(12.5\,\text{cm})}{12.5\,\text{cm} - 10.0\,\text{cm}} = 50.0\,\text{cm}$$

We can also solve for the lateral magnification of the first lens:

$$m = -\frac{d_i}{d_o} = -\frac{50.0\,\text{cm}}{12.5\,\text{cm}} = -4.00$$

The image of the first lens is the object of the second lens. Because the lenses are 20.0 cm apart and the image of the first lens is 50.0 cm from the first lens, that means the image of the first lens is 30.0 cm past the second lens. The image is 30.0 cm from the second lens on the side the light leaves the lens. That means that the second object distance is –30.0 cm. We solve for the object distance:

$$\frac{1}{f'} = \frac{1}{d_o'} + \frac{1}{d_i'} \qquad d_i' = \frac{f'd_o'}{d_o' - f'} = \frac{(-25.0\,\text{cm})(-30.0\,\text{cm})}{-30.0\,\text{cm} - (-25.0\,\text{cm})} = -150\,\text{cm}$$

The magnification of the second image is

$$m' = -\frac{d_i'}{d_o'} = -\frac{-150\,\text{cm}}{-30.0\,\text{cm}} = -5\,\text{cm}$$

The overall magnification is the product of the two individual magnifications:

$$m_{total} = mm' = (-4.00)(-5.00) = 20.0$$

The overall magnification is positive, so the final image is upright.

The final image distance is negative, so the image is virtual.

### Section 34-4 Lensmaker's Equation

The focal length, $f$, of a thin lens in air or vacuum is given by the **lensmaker's equation**:

$$\frac{1}{f} = (n-1)\left(\frac{1}{R_1} + \frac{1}{R_2}\right)$$

where $n$ is the index of refraction of the lens material, $R_1$ is the radius of curvature of the first lens surface, $R_2$ is the radius of curvature of the second lens surface. A convex surface is defined to have a positive radius of curvature and a concave surface is defined to have a negative radius of curvature.

### Example 34-4-A

A lens with a focal length of +25.6 cm is needed. An existing lens has one surface that is convex with a radius of curvature of 40.8 cm and a second surface that is flat. Currently, the lens has a focal length of 80.0 cm. To what radius of curvature must the convex surface be ground to give the lens the necessary focal length? If instead of the convex surface being ground, the flat surface is ground, what radius of curvature should it be given to give the desired focal length?

### Solution:

First, we determine the refractive index of the material of the lens using the information about the existing lens. Using the lensmaker's equation, we solve for the refractive index:

$$n = \frac{1}{f}\left(\frac{1}{R_1} + \frac{1}{R_2}\right)^{-1} + 1 = \frac{1}{80.0\,\text{cm}}\left(\frac{1}{40.8\,\text{cm}} + 0\right)^{-1} + 1 = 1.51$$

Then we can solve for the radius of curvature needed for the desired focal length when one side of the lens is flat using the lensmaker's equation. We solve the lensmaker's equation for $R_1$:

$$R_1 = \left(\frac{1}{f(n-1)} - \frac{1}{R_2}\right)^{-1} = \left(\frac{1}{(25.6\,\text{cm})(1.51-1)} - 0\right)^{-1} = 13.1\,\text{cm}$$

We repeat this for the case when the flat side is ground and the other side has a 40.8 cm radius of curvature:

$$R_2 = \left(\frac{1}{f(n-1)} - \frac{1}{R_1}\right)^{-1} = \left(\frac{1}{(25.6\,\text{cm})(1.51-1)} - \frac{1}{40.8\,\text{cm}}\right)^{-1} = 19.2\,\text{cm}$$

## Section 34-5 Cameras

A camera is a device that has an optical system designed to form real images on either photographic film or an electronic detection system. There are three main adjustments on a better quality camera. These adjustments can be made automatically on better quality modern cameras.

The **shutter speed** of a camera is the length of time the film is exposed to light passing through the lenses of the camera to reach the film or electronic sensing device. In a camera that uses film, a mechanical shutter blocks the light path except for the specific length of time for which the shutter is open. In cameras that use electronic sensing devices to record the image, the shutter is electronic. The shutter speed is set to allow the correct amount of light energy to reach the film or electronic sensor or to limit the length of time light enters the lens to prevent moving subjects from forming blurred images.

The **aperture** of the lens is the effective area of the lens that is allowed to collect light. It is measured by a number called the *f*-**stop number**. The *f*-stop number is defined to be

$$f\text{-stop} = \frac{f}{D}$$

where *f* is the focal length of the lens and *D* is the diameter of the opening. The aperture is adjusted along with the shutter speed to control the amount of light reaching the film or electronic recording device. It is also adjusted to control the **depth of field** of the camera. The depth of field is the range of object distances that form a sharp image on the recording film or electronic sensing device. As the aperture becomes smaller to a point, the depth of field becomes greater. Once the aperture becomes too small, diffraction (see the chapters on the wave nature of light) decreases the sharpness of the image.

The **focusing** controls of a camera allow the user to ensure that the desired objects form a sharp image. The film or electronic sensing device must be at the image distance for the given object distance. Good quality cameras allow a means of adjusting the optical elements to place the image on the film or electronic sensing device.

## Section 34-6 The Human Eye; Corrective Lenses

The human eye forms a real image on the retina. The retina detects the light that falls on it. The eye is able to adjust the focal length of its image formation system so that objects located different distances from the eye can form a sharp image on the retina. This is called **accommodation**. The eye's accommodation ability is measured by determining the closest object for which the eye can form a sharp image on the retina, called the **near point**, and the farthest object for which the eye can form a sharp image on the retina, called the **far point**. A near point of 25 cm and a far point that is an infinite distance away are considered normal vision.

If the far point is not an infinite distance away, the eye cannot form a sharp image of distant objects on the retina. This condition is called **nearsightedness** or **myopia**. A negative focal length lens is needed to correct for nearsightedness.

If the far point is farther from the eye than 25 cm, the eye cannot form a sharp image of objects that are closer than the far point. This condition is called **farsightedness** or **hyperopia**. Everyday close-up activites such as reading normal print sizes are hindered by this condition, so it is desirable to correct vision with this condition. A positive focal length lens is required to correct this condition.

Astigmatism is a condition caused by a lens or cornea that is not spherical in shape, but has different radii of curvature in different planes passing perpendicularly through the surface. This implies that light rays lying in different planes have different focal lengths and, therefore, different image positions. A sharp image cannot be formed on the retina. A corrective lens for astigmatism is aspherical.

## Section 34-7 Magnifying Glass

A magnifying glass or simple magnifier is a single simple convex lens used to view objects so that an enlarged upright image of the object can be viewed by the eye. The important quantity that measures how enlarged an image looks to the eye is the angular magnification. The angular size, $\theta$, of an object or image to the eye will be

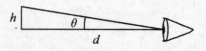

$$\theta = \arctan\frac{h}{d}$$

where $h$ is the actual height of the object of image and $d$ is the distance from the eye to the object or image. The angular magnification, $M$, is defined to be the ratio of angular sizes of the object without the instrument, $\theta_0$, and with the instrument, $\theta_i$:

$$M = \frac{\theta_i}{\theta_0}$$

The angular magnification of an object that is viewed with a simple magnifier is the angular size of the object viewed with the magnifier divided by the maximum angular size of the object viewed without the magnifier. The maximum angular size without a magnifier will be the angular size when the object is held at our near point distance, $N$. The angular magnification, $M$, of a simple lens when the image is formed an infinite distance from the eye is

$$M = \frac{N}{f}$$

where $f$ is the focal length of the magnifier. If the magnifier is used so that the image is formed at the near point of the eye, then the angular magnification is

$$M = \frac{N}{f} + 1$$

The magnification is greater when the image is formed at the near point, but the eye is not relaxed when viewing an image close to the near point.

**Example 34-7-A**

A simple magnifier is used with a power of +5.0 diopters. What is the angular magnification of this magnifier if it forms an image an infinite distance from the eye? What is the angular magnification if it forms an image at the near point of 25 cm?

**Solution:**

The angular magnification for the magnifier for an image formed an infinite distance from the eye is

$$M = \frac{N}{f} = (0.25\,\text{m})(5.0\,\text{m}^{-1}) = 1.25$$

The angular magnification for the magnifier for an image formed at the near point of the eye is

$$M = \frac{N}{f} + 1 = (0.25\,\text{m})(5.0\,\text{m}^{-1}) + 1 = 2.25$$

**Section 34-8 Telescopes**

A telescope is used to view objects that are far away. There are many different arrangements of optical elements used to make a telescope. Both lenses and mirrors are used to construct telescopes. To convey the basic principles of the design of a telescope, we consider the design of a **Keplerian** refracting telescope. This telescope contains two converging lenses, one called an **objective lens** and the other called an **eyepiece lens**. The objective lens forms a real image of the distant object and the eyepiece lens is used as a simple magnifier to increase the angular size of the real image formed by the objective lens. The approximate angular magnification of a telescope is

$$M \approx -\frac{f_o}{f_e}$$

where $f_o$ is the focal length of the objective lens and $f_e$ is the focal length of the eyepiece lens. The minus sign means that the image is inverted.

438    Giancoli, *Physics for Scientists & Engineers*: Study Guide

## Example 34-8-A

A Keplerian telescope has an angular magnification of –100. The distance between the objective lens and the eyepiece lens is 1.60 m. What are the focal lengths of the objective lens and the eyepiece lens?

**Solution:**

We know the relationship between the angular magnification and the focal lengths of the two lenses:

$$M \approx -\frac{f_o}{f_e}$$

We also know that the construction of the telescope requires the two lenses to be separated by a distance $L$ equal to the sum of the two focal lengths:

$$L = f_o + f_e$$

We solve for the focal length of the eyepiece lens using, from the first equation, $f_o = -Mf_e$:

$$L = -Mf_e + f_e \quad \Rightarrow \quad f_e = \frac{L}{1-M} = \frac{1.60\,m}{1-(-100)} = 1.58\times10^{-2}\,m$$

We can then solve for the focal length of the objective lens:

$$f_0 = -Mf_e = -(-100)\left(1.58\times10^{-2}\,m\right) = 1.58\,m$$

## Section 34-9 Compound Microscope

A microscope is used to form enlarged images of nearby objects. A compound microscope is made from two or more simple lenses. A two-lens compound microscope has an objective lens and an eyepiece lens like the telescope. Again the objective lens forms a real image of the object and the eyepiece lens acts as a simple magnifier of the real image. When the final image of a compound microscope is formed an infinite distance from the eye, so that the eye is relaxed when viewing the image, the angular magnification of the image is given approximately by

$$M \approx \frac{N}{f_e}\frac{l-f_e}{d_o}$$

where $N$ is the near point distance of the eye, $l$ is the distance between the objective lens and the eyepiece lens, $f_e$ is the focal length of the eyepiece lens, and $d_o$ is the distance from the objective lens to the object being viewed. In cases in which the objective lens focal length, $f_o$, is small compared to the separation of the lenses, this can be further approximated by

$$M \approx \frac{Nl}{f_e f_o}$$

## Example 34-9-A

For a compound microscope, determine the difference in the two approximate angular magnifications of the microscope. The microscope has an eyepiece lens with a focal length of 4.0 cm, an objective lens with a focal length of 4.8 mm, a separation between the two lenses of 18 cm, and the user has a near point distance of 25 cm. Assume the object is located a distance 4.9 mm from the objective.

**Solution:**

The first approximation gives

$$M \approx \frac{N}{f_e}\frac{l-f_e}{d_o} = \frac{25\,\text{cm}}{4.0\,\text{cm}}\frac{18\,\text{cm}-4.0\,\text{cm}}{0.49\,\text{cm}} = 179$$

The second approximation gives

$$M \approx \frac{Nl}{f_e f_o} = \frac{(25\,\text{cm})(18\,\text{cm})}{(4.0\,\text{cm})(0.48\,\text{cm})} = 234$$

### Section 34-10 Aberrations of Lenses and Mirrors

There are several different identifiable reasons why lenses do not form exact point images of point objects. These are called aberrations. Some of these aberrations arise when the object location is moved off of the optic axis. There are two common aberrations that occur for a point object, even if it is located on the optic axis. **Spherical aberration** is the result of using a spherically ground surface. Rays that are incident on the front surface of the lens for which the angle between the ray direction and the optic axis miss the image point of the rays near the axis by greater and greater amounts as the angle becomes larger and larger.

Another aberration that occurs is **chromatic aberration**. The refractive index of the lens material is, in general, dependent on the wavelength of the light. This causes images of different wavelengths to be in different positions and there is no single image point. This aberration is easily correctable by using two lenses such that one chromatic aberration is cancelled by the chromatic aberration of the other one. Such a pair of lenses is called an **achromatic doublet**.

Two of the off-axis aberrations are **coma** and **astigmatism**. Coma creates image shapes that are asymmetric shaped for small circular image shapes when the object is located off of the optic axis. Astigmatism cause rays in one plane to have a different image location than rays in a perpendicular plane.

**Distortions** are variations in the magnification of the image depending on the location of the object.

Much of the aberration and distortion can be removed by using compound lenses consisting of several simple lens elements. Spherical aberration can be removed and some other aberrations reduced by using **aspherical** lenses or mirrors.

## Practice Quiz

1.  A single-lens system produces a real image a distance of 30.0 cm from the lens. What does this imply about the lens?

    a) The lens has a positive focal length and a focal length less than or equal to 30.0 cm.
    b) The lens has a negative focal length and a focal length less than −30.0 cm.
    c) The lens has a positive focal length and a focal length greater than or equal to 30.0 cm.
    d) The lens has a negative focal length and a focal length greater than or equal to −30.0 cm.

2.  A positive focal length lens forms a real image. If the object is moved farther from the lens, the image

    a) moves farther from the lens and has greater magnification.
    b) moves farther from the lens and has less magnification.
    c) moves closer to the lens and has greater magnification.
    d) moves closer to the lens and has less magnification.

3.  Nearsighted people can use which type of lens to correct their vision?

    a) convex
    b) concave
    c) plano-convex
    d) all of the above

4.    What type of image is formed by a simple magnifier in its normal use?

    a)  inverted
    b)  real
    c)  virtual
    d)  No image is formed.

5.    For a negative focal length lens, if the object distance is positive, the image is

    a)  real and upright.
    b)  real and inverted.
    c)  virtual and upright.
    d)  virtual and inverted.

6.    If the object distance is positive, where must the object be located so that a converging lens forms a virtual image?

    a)  farther from the lens than two focal lengths
    b)  farther from the lens than one focal length
    c)  closer to the lens than one-half focal length
    d)  closer to the lens than one focal length

7.    A person is diagnosed with hyperopia.  What does that mean?

    a)  The person can't see far away objects sharply.
    b)  The person can't see nearby objects sharply.
    c)  The person can't sit still.
    d)  The person has bad-smelling feet.

8.    When you look through a thin lens at a distant object, you see an upright image that is smaller than its normal size. If you reverse the lens so that you are looking through it in the opposite direction, how will the image now appear?

    a)  inverted and larger than its normal size
    b)  inverted and smaller than its normal size
    c)  upright and larger than its normal size
    d)  upright and smaller than its normal size

9.    A lens is made from a material with a refractive index of 1.500 and has a focal length f.  If the lens were made identically, but from a material with a refractive index 2.000, what would its focal length be?

    a)  ⅓ f
    b)  ¾ f
    c)  2 f
    d)  ½ f

10.   Which aberration does a lens have that a mirror doesn't ?

    a)  spherical aberration
    b)  chromatic aberration
    c)  coma
    d)  astigmatism

11.   A +4.5 diopter lens forms a real image a distance of +56.4 cm from the lens.  Where is the object located?  What is the magnification of the image?

12.   A lens is needed with a focal length of –36 cm.  The lens blank has a refractive index of 1.59.  What should the radii of curvature of the lens surfaces be if the lens is ground so both surfaces have the same radius of curvature?

13.   A person is farsighted with a near point of 200 cm to be corrected so that the corrected near point is 25 cm.  What should the focal length of the lens be?

14.   A Keplerian telescope is to be built with an angular magnification of 80. The objective lens of the telescope has a focal length of 2.5 m. What should the focal length of the eyepiece lens be? How far apart do the eyepiece and objective lens need to be placed?

15.   What is the magnification of a compound microscope made with an objective lens with a focal length of +2.00 mm and an eyepiece lens with a focal length of +4.80 cm? The length of the microscope tube is 18.6 cm and the object is located so that the final image is 25 cm from the eye.

## Problem Solutions

3.   (a)   The power of the lens is
$$P = 1/f = 1/0.275 \text{ m} = 3.64 \text{ D, converging.}$$
(b)   We find the focal length of the lens from
$$P = 1/f,$$
$$-6.25 \text{ D} = 1/f, \text{ which gives } f = -0.160 \text{ m} = -16.0 \text{ cm, diverging.}$$

7.   (a)   We find the image distance from
$$(1/d_o) + (1/d_i) = 1/f,$$
$$(1/10.0 \times 10^3 \text{ mm}) + (1/d_i) = 1/80 \text{ mm, which gives } d_i = 81 \text{ mm.}$$
(b)   For an object distance of 3.0 m, we have
$$(1/3.0 \times 10^3 \text{ mm}) + (1/d_i) = 1/80 \text{ mm, which gives } d_i = 82 \text{ mm.}$$
(c)   For an object distance of 1.0 m, we have
$$(1/1.0 \times 10^3 \text{ mm}) + (1/d_i) = 1/80 \text{ mm, which gives } d_i = 87 \text{ mm.}$$
(d)   We find the smallest object distance from
$$(1/d_{omin}) + (1/120 \text{ mm}) = 1/80 \text{ mm, which gives } d_{omin} = 240 \text{ mm} = 24 \text{ cm.}$$

11.   (a)   We find the focal length of the lens from
$$(1/d_o) + (1/d_i) = 1/f,$$
$$(1/37.5 \text{ cm}) + [1/(-8.20 \text{ cm})] = 1/f, \text{ which gives } f = -10.5 \text{ cm (diverging).}$$
The image is in front of the lens, so it is virtual.
(b)   We find the focal length of the lens from
$$(1/d_o) + (1/d_i) = 1/f,$$
$$(1/37.5 \text{ cm}) + [1/(-46.0 \text{ cm})] = 1/f, \text{ which gives } f = +203 \text{ cm (converging).}$$
The image is in front of the lens, so it is virtual.

15.   We get an expression for the image distance from the lens equation:
$$(1/d_o) + (1/d_i) = 1/f,$$
$$1/d_i = (1/f) - (1/d_o), \quad \text{or} \quad d_i = fd_o/(d_o - f).$$
The magnification is
$$m = -d_i/d_o = -f/(d_o - f).$$
If the lens is converging, $f > 0$.
For a real object, $d_o > 0$.
When $d_o > f$, we have $(d_o - f) > 0$, so all factors in the expressions for $d_i$ and $m$ are positive; thus $d_i > 0$ (real), and $m < 0$ (inverted).
When $d_o < f$, we have $(d_o - f) < 0$, so the denominator in the expressions for $d_i$ and $m$ are negative; thus $d_i < 0$ (virtual), and $m > 0$ (upright).
For an object beyond the lens, $d_o < 0$.
When $-d_o > f$, we have $(d_o - f) < 0$, so both numerator and denominator in the expression for $d_i$ are negative; thus $d_i > 0$, so the image is real. The numerator in the expression for $m$ is negative; thus $m > 0$, so the image is upright.
When $0 < -d_o < f$, we have $(d_o - f) < 0$, so we get the same result: real and upright.

19.   (a)   The image from the first lens        (c)
is the same: $d_{i1} = +30.0$ cm.
This image is the object for the
second lens. Because it is
beyond the second lens, it has
a negative object distance:
$$d_{o2} = 20.0 \text{ cm} - 30.0 \text{ cm} = -10.0 \text{ cm.}$$

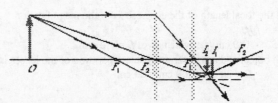

We find the image formed by the
refraction of the second lens:

$(1/d_{o2}) + (1/d_{i2}) = 1/f_2$ ;

$[1/(- 10.0 \text{ cm})] + (1/d_{i2}) = 1/25.0 \text{ cm}$, which gives $d_{i2} = + 7.14$ cm.

(b) The total magnification is the product of the magnifications for the two lenses:

$m = m_1 m_2 = (- d_{i1}/d_{o1})(- d_{i2}/d_{o2}) = d_{i1}d_{i2}/d_{o1}d_{o2}$

$= (+ 30.0 \text{ cm})(+ 7.14 \text{ cm})/(+ 60.0 \text{ cm})(- 10.0 \text{ cm}) = - 0.357$ (inverted).

23. We find the index from the lensmaker's equation:

$1/f = (n - 1)[(1/R_1) + (1/R_2)]$;

$1/28.9 \text{ cm} = (n - 1)[(1/31.0 \text{ cm}) + (1/31.0 \text{ cm})]$, which gives $n = 1.54$.

27. We find the radius from the lensmaker's equation:

$1/f = (n - 1)[(1/R_1) + (1/R_2)]$;

$+ 2.50 \text{ D} = (1.56 - 1)[(1/0.200 \text{ m}) + (1/R_2)]$, which gives $R_2 = - 1.87$ m (concave).

31. We find the f-number from

$f\text{-stop} = f/D = (14 \text{ cm})/(6.0 \text{ cm}) = f/2.3$.

35. The converging camera lens will form a real, inverted image. For the magnification, we have

$m = h_i/h_o = - d_i/d_o$ ;

$- (24 \times 10^{-3} \text{ m})/(32 \text{ m}) = - d_i/(55 \text{ m})$, or $d_i = 4.13 \times 10^{-2}$ m.

We find the focal length of the lens from

$(1/d_o) + (1/d_i) = 1/f$,

$(1/55 \text{ m}) + (1/4.13 \times 10^{-2} \text{ m}) = 1/f$, which gives $f = 4.1 \times 10^{-2}$ m = 41 mm.

39. With the contact lens, an object at infinity would have a virtual image at the far point of the eye.

We find the power of the lens from

$(1/d_o) + (1/d_i) = 1/f = P$, when distances are in m;

$(1/\infty) + (1/- 0.17 \text{ m}) = P = - 5.9$ D.

To find the new near point, we have

$(1/d_o) + (1/d_i) = 1/f = P$;

$(1/d_o) + (1/- 0.12 \text{ m}) = - 5.9$ D, which gives $d_o = 0.41$ m.

Glasses would be better, because they give a near point of 32 cm from the eye.

43. We find the far point of the eye by finding the image distance with the lens for an object at infinity:

$(1/d_{o1}) + (1/d_{i1}) = 1/f_1$ ;

$(1/\infty) + (1/d_{i1}) = 1/- 25.0 \text{ cm}$, which gives $d_{i1} = - 25.0$ cm from the lens, or 26.8 cm from the eye.

We find the focal length of the contact lens from

$(1/d_{o2}) + (1/d_{i2}) = 1/f_2$ ;

$(1/\infty) + (1/- 26.8 \text{ cm}) = 1/f_2$ , which gives $f_2 = - 26.8$ cm.

47. We find the focal length from

$M = N/f$;

$3.0 = (25 \text{ cm})/f$, which gives $f = 8.3$ cm.

51. (a) We find the image distance from

$(1/d_o) + (1/d_i) = 1/f$;

$(1/5.35 \text{ cm}) + (1/d_i) = 1/6.00 \text{ cm}$,

which gives $d_i = - 49.4$ cm.

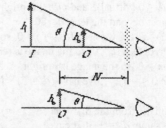

(b) From the diagram we see that the angular magnification is

$M = \theta'/\theta = (h_o/d_o)/(h_o/N) = N/d_o$

$= (25 \text{ cm})/(5.35 \text{ cm}) = 4.7\times$.

55. We find the focal length of the eyepiece from the magnification:

$M = - f_o/f_e$ ;

$- 25 = - (80 \text{ cm})/f_e$ , which gives $f_e = 3.2$ cm.

For both object and image far away, the separation of the lenses is

$L = f_o + f_e = 80 \text{ cm} + 3.2 \text{ cm} = 83$ cm.

59. For both object and image far away, we find the (negative) focal length of the eyepiece from the separation of the lenses:

$L = f_o + f_e$ ;

33.0 cm = 36.0 cm + $f_e$ , which gives $f_e$ = – 3.0 cm.

The magnification of the telescope is given by

$M = -f_o/f_e = -(36.0\text{ cm})/(-3.0\text{ cm}) = 12\times$.

63. We assume a prism binocular so the magnification is positive, but simplify the diagram by ignoring the prisms. We find the focal length of the eyepiece from the design magnification:

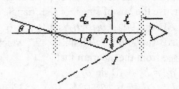

$M = f_o/f_e$ ;

7.0 = (26 cm)/$f_e$ , which gives $f_e$ = 3.71 cm.

We find the intermediate image formed by the objective:

$(1/d_{oo}) + (1/d_{oi}) = 1/f_o$ ;

(1/400 cm) + (1/$d_{oi}$) = 1/26 cm, which gives $d_{oi}$ = 27.8 cm.

With the final image at infinity (relaxed eye), the intermediate image will be at the focal point of the eyepiece lens. From the diagram the angle subtended by the object is

$\theta = h/d_{oi}$ ,

while the angle subtended by the image is

$\theta' = h/f_e$ .

Thus the angular magnification is

$M = \theta'/\theta = (h/f_e)/(h/d_{oi}) = d_{oi}/f_e = (27.8\text{ cm})/(3.71\text{ cm}) = 7.5\times$.

67. (a) Because the image from the objective is at the focal point of the eyepiece, the image distance for the objective is

$d_{io} = l - f_e = 16.0\text{ cm} - 1.8\text{ cm} = 14.2\text{ cm}$.

We find the object distance from the lens equation for the objective:

$(1/d_{oo}) + (1/d_{io}) = 1/f_o$ ;

(1/$d_{oo}$) + (1/14.2 cm) = 1/0.80 cm, which gives $d_{oo}$ = 0.85 cm.

(b) With the final image at infinity, the magnification of the eyepiece is

$M_e = N/f = (25.0\text{ cm})/(1.8\text{ cm}) = 13.9\times$.

The magnification of the objective is

$M_o = d_{io}/d_{oo} = (14.2\text{ cm})/(0.85\text{ cm}) = 16.7\times$.

The total magnification is

$M = M_oM_e = (16.7)(13.9) = 230\times$.

71. (a) We find the incident angle from

$\sin\theta_1 = h_1/R = (1.0\text{ cm})/(12.0\text{ cm}) = 0.0833$, so $\theta_1 = 4.78°$.

For the refraction at the curved surface, we have

$\sin\theta_1 = n\sin\theta_2$ ;

sin 4.78° = (1.50) sin $\theta_2$ , which gives sin $\theta_2$ = 0.0556,

so $\theta_2 = 3.18°$.

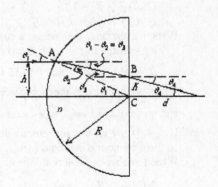

We see from the diagram that

$\theta_3 = \theta_1 - \theta_2 = 4.78° - 3.18° = 1.60°$.

For the refraction at the flat face, we have

$n\sin\theta_3 = \sin\theta_4$ ;

(1.50) sin 1.60° = sin $\theta_4$ , which gives sin $\theta_4$ = 0.0419,

so $\theta_4 = 2.40°$.

If we use the law of sines for the triangle ABC, we have

AB/sin (90° – $\theta_1$) = $h_1'$/sin $\theta_2$ ;

AB/sin 85.22° = $h_1'$/sin 3.18°, which gives AB = 17.96$h_1'$.

We see from the diagram that

$h_1' = h_1 - AB \sin\theta_3$ ;

$h_1'$ = 1.0 cm – 17.96$h_1'$ sin 1.60°, which gives $h_1'$ = 0.666 cm.

Thus the distance from the flat face to the point where the ray crosses the axis is

$d_1 = h_1'/\tan\theta_4 = (0.666\text{ cm})/\tan 2.40° = 15.9\text{ cm}$.

(b) We find the incident angle from

$\sin\theta_1 = h_2/R = (6.0\text{ cm})/(12.0\text{ cm}) = 0.500$, so $\theta_1 = 30.00°$.

For the refraction at the curved surface, we have

$\sin \theta_1 = n \sin \theta_2$ ;

$\sin 30.00° = (1.50) \sin \theta_2$ , which gives $\sin \theta_2 = 0.333$, so $\theta_2 = 19.47°$.

We see from the diagram that

$\theta_3 = \theta_1 - \theta_2 = 30.00° - 19.47° = 10.53°$.

For the refraction at the flat face, we have

$n \sin \theta_3 = \sin \theta_4$ ;

$(1.50) \sin 10.53° = \sin \theta_4$ , which gives $\sin \theta_4 = 0.274$, so $\theta_4 = 15.91°$.

If we use the law of sines for the triangle ABC, we have

$AB/\sin(90° - \theta_1) = h_2'/\sin \theta_2$ ;

$AB/\sin 60.00° = h_2'/\sin 19.47°$, which gives $AB = 2.598 h_2'$.

We see from the diagram that

$h_2' = h_2 - AB \sin \theta_3$ ;

$h_2' = 6.0 \text{ cm} - 2.598 h_2' \sin 10.53°$, which gives $h_2' = 4.07$ cm.

Thus the distance from the flat face to the point where the ray crosses the axis is

$d_2 = h_2'/\tan \theta_4 = (4.07 \text{ cm})/\tan 15.91° = 14.3$ cm.

(c) The separation of the "focal points" is

$\Delta d = d_1 - d_2 = 15.9 \text{ cm} - 14.3 \text{ cm} = 1.6$ cm.

(d) When $h_2 = 6.0$ cm, the rays focus closer to the lens, so they will form a circle at the "focal point" for $h_1 = 1.0$ cm.  We find the radius of this circle from similar triangles:

$h_2'/d_2 = r/(d_1 - d_2)$;

$(4.07 \text{ cm})/(14.3 \text{ cm}) = r/(1.6 \text{ cm})$, which gives $r = 0.46$ cm.

75. If we recognize that the image is inverted, for the magnification we have

$m = h_i/h_o = -d_i/d_o = -1$, so $d_i = d_o$ .

We find the object distance from

$(1/d_o) + (1/d_i) = 1/f$;

$(1/d_o) + (1/d_o) = 1/50$ mm, which gives $d_o = 100$ mm.

The distance between the object and the film is

$d = d_o + d_i = 100 \text{ mm} + 100 \text{ mm} = 200$ mm.

79. We find the object distance from

$(1/d_o) + (1/d_i) = 1/f$;

$(1/d_o) + (1/7.50 \times 10^3 \text{ mm}) = 1/100$ mm, which gives $d_o = 101 \text{ mm} = 0.101$ m.

We find the size of the image from

$m = h_i/h_o = -d_i/d_o$ ;

$h_i/(0.036 \text{ m}) = -(7.50 \text{ m})/(0.101 \text{ m})$, which gives $h_i = -2.7$ m.

83. (a) We use the lens equation with $d_o + d_i = d_T$ :

$(1/d_o) + (1/d_i) = 1/f$ ;

$(1/d_o) + [1/(d_T - d_o)] = 1/f$ .

When we rearrange this, we get a quadratic equation for $d_o$ :

$d_o^2 - d_T d_o + d_T f = 0$, which has the solution

$d_o = \frac{1}{2}[d_T \pm \sqrt{d_T^2 - 4 d_T f}]. [\text{dT} \pm]$

If $d_T > 4f$, we see that the term inside the square root is positive: $d_T^2 - 4 d_T f > 0$,

and $\sqrt{d_T^2 - 4 d_T f} < d_T$ , so we get two real, positive solutions for $d_o$ .

(b) If $d_T < 4f$, we see that the term inside the square root is negative: $d_T^2 - 4 d_T f < 0$,

so there are no real solutions for $d_o$ .

(c) When there are two solutions, the distance between them is

$\Delta d = d_{o1} - d_{o2} = \frac{1}{2}[d_T + \sqrt{d_T^2 - 4 d_T f}] - \frac{1}{2}[d_T - \sqrt{d_T^2 - 4 d_T f}] = \sqrt{d_T^2 - 4 d_T f}$ .

The image positions are given by

$d_i = d_T - d_o = \frac{1}{2}[d_T - \sqrt{d_T^2 - 4 d_T f}]$.

The ratio of image sizes is the ratio of magnifications:

$m = m_2/m_1 = (d_{i2}/d_{o2})/(d_{i1}/d_{o1}) = (d_{i2}/d_{o2})(d_{o1}/d_{i1})$

87. To find the new near point, we have

$(1/d_{o1}) + (1/d_{i1}) = 1/f_1 = P_1$ , when distances are in m;

$(1/0.35 \text{ m}) + (1/d_{i1}) = +2.5 \text{ D}$, which gives $d_{i1} = -2.8 \text{ m}$.

To give him a normal near point, we have

$(1/d_{o2}) + (1/d_{i2}) = 1/f_2 = P_2$ ;

$(1/0.25 \text{ m}) + (1/-2.8 \text{ m}) = P_2 = +3.6 \text{ D}$.

91. (a) The magnification of the telescope is given by

$M = -f_o/f_e = -P_e/P_o = -(5.0 \text{ D})/(2.0 \text{ D}) = -2.5\times$.

(b) To get a magnification greater than 1, for the eyepiece we use the lens with the smaller focal length, or greater power: 5.0 D.

# Chapter 35: The Wave Nature of Light; Interference

## Chapter Overview and Objectives

This chapter introduces the wave model of light and discusses several applications of the model that are not easily explained by Newton's particle model of light. Several cases of two-source interference are discussed, along with the concept of constructive and destructive interference.

After completing study of this chapter, you should:

- Know what Huygens' principle is.
- Know what constructive and destructive interference are.
- Know how to determine the position of the dark and bright fringes in a two-slit interference pattern.
- Know how to relate the intensity in a two-slit interference pattern to the position within the pattern.
- Know what thin film interference is and how to determine the conditions of constructive and destructive interference in a thin film
- Know what a Michelson interferometer is.

## Summary of Equations

Constructive interference condition in two-slit interference pattern:

$$d \sin \theta = m\lambda \qquad m \in \{0, 1, 2, ...\}$$

Destructive interference condition in two-slit interference pattern:

$$d \sin \theta = \left(m + \tfrac{1}{2}\right)\lambda \qquad m \in \{0, 1, 2, ...\}$$

Intensity in a two-slit interference pattern: $\qquad I = I_0 \cos^2\left(\dfrac{\pi d \sin \theta}{\lambda}\right) = I_0 \cos^2\left(\dfrac{\pi d}{\lambda L} y\right)$

Thin film interference conditions when $n_1 > n_2 > n_3$ or $n_1 < n_2 < n_3$

$$\text{Constructive:} \qquad 2t = m\lambda_{n_2} \qquad m \in \{0, 1, 2, ...\}$$

$$\text{Destructive:} \qquad 2t = \left(m + \tfrac{1}{2}\right)\lambda_{n_2} \qquad m \in \{0, 1, 2, ...\}$$

Thin film interference conditions when $n_1 > n_2$ and $n_3 > n_2$, or $n_2 > n_1$ and $n_2 > n_3$

$$\text{Constructive:} \qquad 2t = \left(m + \tfrac{1}{2}\right)\lambda_{n_2} \qquad m \in \{0, 1, 2, ...\}$$

$$\text{Destructive:} \qquad 2t = m\lambda_{n_2} \qquad m \in \{0, 1, 2, ...\}$$

## Chapter Summary

### Section 35-1 Huygens' Principle and Diffraction

Each point on the wave front of a wave is equivalent to a point source of spherical waves. The continuing wave is the sum of these outgoing spherical waves in the forward direction. This principle can be used to understand the phenomenon of **diffraction**. Diffraction is the property of waves to propagate some energy in the non-forward direction when they encounter a barrier that cuts off a wavefront.

### Section 35-2 Huygens' Principle and the Law of Refraction

Huygens' principle is consistent with the laws of reflection and refraction. In particular, the law of refraction is explained by the wave model of light consistently with the wave speed of light in the material.

### Section 35-3 Interference–Young's Double-Slit Experiment

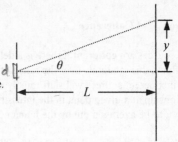

When two slits are illuminated by in-phase monochromatic coherent light of equal amplitude, a pattern of bright and dark fringes appears on a screen placed in front of the slits. The bright fringes occur where there is **constructive interference** and the dark fringes occur where there is **destructive interference**. The position of the bright fringes is given by

$$d \sin \theta = m\lambda \qquad m \in \{0, 1, 2, ...\}$$

and the position of the dark fringes is given by

$$d \sin \theta = \left(m + \tfrac{1}{2}\right)\lambda \qquad m \in \{0, 1, 2, ...\}$$

where $d$ is the distance between the slits, $\theta$ is the angular position of the fringe as shown in the diagram, $\lambda$ is the wavelength of the light, and $m$ is an integer. These expressions can also be written in terms of the position along the screen from the center of the pattern, $y$, and the distance from the slits to the screen, $L$, as

$$d\frac{y}{L} = m\lambda \qquad m \in \{0, 1, 2, ...\}$$

for constructive interference and by

$$d\frac{y}{L} = \left(m + \tfrac{1}{2}\right)\lambda \qquad m \in \{0, 1, 2, ...\}$$

All of these relationships are approximations. The approximations are good when the distance to the screen is much greater than the separation of the slits and when the angle $\theta$ is a small angle.

There is a general rule that applies to all two-source interference measurements, whether it is the two-slit interference described above or some other geometrical arrangement of two sources. Constructive interference occurs if the path length difference from the sources to the observation point is a multiple of the wavelength of the light. Destructive interference occurs if the path length difference from the sources to the observation point is an odd multiple of the one-half a wavelength of the light.

### Example 35-3-A

An interference pattern from a two-slit interference pattern is shown in the diagram. The slit separation used was 0.322 mm. The pattern is on a screen a distance 1.48 m from the slits. The distance shown on the diagram was measured on the screen. What is the wavelength of the light used to form the two-slit interference pattern?

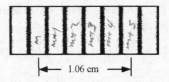

1.06 cm

### Solution:

This is a little bit tricky in counting, so we have to be careful. We don't know the order of any of the fringes. Let's say the fringe at the left edge of the measurement is order m, then the fringe at the right edge of the measurement is order m + 5 (or possibly m – 5, it doesn't make a difference). The error that is often made is that because the measurement spans six fringes, the last fringe is called the m + 6 order fringe. The distance measure is the difference between the positions of the $m^{\text{th}}$ fringe and the $m + 5^{\text{th}}$ fringe:

$$10.6 \, \text{cm} = \frac{(m+5)\lambda L}{d} - \frac{m\lambda L}{d} = \frac{5\lambda L}{d}$$

Solving this for the wavelength of the light, we get

$$\lambda = \frac{(10.6\,\text{cm})d}{5L} = \frac{\left(1.06\times10^{-2}\,\text{m}\right)\!\left(0.322\times10^{-3}\,\text{m}\right)}{5(1.48\,\text{m})} = 4.61\times10^{-7}\,\text{m}$$

## Section 35-4 Coherence

Two sources are **coherent** if there is a definite phase relationship between the two sources. If there is a phase relationship that fluctuates randomly over the range of $2\pi$ radians, the two sources are **incoherent**. A two-source interference pattern with incoherent sources will shift rapidly over the time scale that the phase fluctuations occur in because the relative phase relationship at a given point in the intensity pattern will fluctuate as well. This usually causes the interference intensity pattern to be averaged out by the human eye or a measuring instrument so that no interference pattern is distinguishable.

## Section 35-5 Intensity in the Double-Slit Interference Pattern

Rather then only looking for the positions in the intensity pattern of a two-slit interference pattern at which constructive and destructive interference occur, we can look at the intensity as a function of position. When the two slits are illuminated with equal intensity in phase-coherent light, the intensity as a function of position is given by

$$I = I_0 \cos^2\!\left(\frac{\pi d \sin\theta}{\lambda}\right) = I_0 \cos^2\!\left(\frac{\pi d}{\lambda L}y\right)$$

where $d$ is the separation of the two slits, $\lambda$ is the wavelength of the light, $\theta$ is the position in the pattern as shown in the diagram, and $I_0$ is the intensity of the light at the center of the intensity pattern ($\theta = 0$), $y$ is the distance from the center of the pattern to the position at which the intensity is being evaluated, and $L$ is the distance from the slits to the plane at which the intensity is being measured.

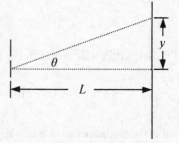

Again, these relationships are approximations. The approximations are good when the distance to the screen is much greater than the separation of the slits and when the angle $\theta$ is a small angle.

## Example 35-5-A

A two-slit interference pattern is formed using slits of separation 0.556 mm, light of wavelength 546 nm, and a screen that is 98.6 cm from the slits. The intensity at a position on the screen a distance of 1.00 cm from the center of the central bright fringe is 3.35 W/m². What is the intensity a distance of 1.25 cm from the central bright fringe?

## Solution:

This problem is solved with a direct application of the relationship between intensity and position in the interference pattern:

$$I_{1.00\,\text{cm}} = I_0 \cos^2\!\left(\frac{\pi d}{\lambda L}1.00\,\text{cm}\right)$$

$$I_{1.45\,\text{cm}} = I_0 \cos^2\!\left(\frac{\pi d}{\lambda L}1.45\,\text{cm}\right)$$

If we eliminate $I_0$ from these two equations and solve for the intensity at 1.25 cm from the central bright fringe, we get

$$I_{1.45\,\text{cm}} = \frac{\cos^2\!\left(\dfrac{\pi d}{\lambda L}1.45\,\text{cm}\right)}{\cos^2\!\left(\dfrac{\pi d}{\lambda L}1.00\,\text{cm}\right)}I_{1.00\,\text{cm}} = \frac{\cos^2\!\left(\dfrac{\pi\,0.556\times10^{-3}\,\text{m}}{\left(546\times10^{-9}\,\text{m}\right)\!\left(0.986\,\text{m}\right)}1.25\,\text{cm}\right)}{\cos^2\!\left(\dfrac{\pi\,0.556\times10^{-3}\,\text{m}}{\left(546\times10^{-9}\,\text{m}\right)\!\left(0.986\,\text{m}\right)}1.00\,\text{cm}\right)}3.35\,\text{W/m}^2 = 11.6\,\text{W/m}^2$$

**Section 35-6 Interference in Thin Films**

The two-source interference principle applies to thin films of materials that are transparent. A thin film of material will have a reflection from the front surface and from the back surface. These two reflections will usually be approximately equal in intensity and will interfere with each other. The interference condition depends on the thickness of the film, $t$, the wavelength of the light in the film, $\lambda_n$, and the relative size of the refractive indices of the material on each side of film and that of the film.

If the refractive indices of the materials are in increasing or decreasing order in going from incident material, $n_1$, to the film, $n_2$, to the material behind the film, $n_3$, (either $n_1 > n_2 > n_3$ or $n_1 < n_2 < n_3$), then constructive interference occurs for the condition

$$2t = m\lambda_{n_2} \qquad m \in \{0, 1, 2, ...\}$$

and destructive interference occurs when

$$2t = \left(m + \tfrac{1}{2}\right)\lambda_{n_2} \qquad m \in \{0, 1, 2, ...\}$$

where $\lambda_{n_2}$ is the wavelength of the light in the film and $m$ is a non-negative integer. The wavelength of the light in the film is given by

$$\lambda_{n_2} = \frac{\lambda}{n_2}$$

where $\lambda$ is the wavelength of the light in vacuum.

If the refractive indices of the materials are not in increasing or decreasing order in going from incident material, $n_1$, to the film, $n_2$, to the material behind the film, $n_3$, (either $n_1 > n_2$ and $n_3 > n_2$ or $n_2 > n_1$ and $n_2 > n_3$), then constructive interference occurs for the condition

$$2t = \left(m + \tfrac{1}{2}\right)\lambda_{n_2} \qquad m \in \{0, 1, 2, ...\}$$

and destructive interference occurs when

$$2t = m\lambda_{n_2} \qquad m \in \{0, 1, 2, ...\}$$

where $\lambda_{n_2}$ is the wavelength of the light in the film and $m$ is a non-negative integer.

**Example 35-6-A**

A thin film of oil of refractive index 1.28 floats on a pool of water with refractive index 1.33. Light of wavelength 623 nm in air is reflected brightly by the film. What are the three smallest possible non-zero thicknesses of the oil film?

**Solution:**

First, we determine what the orderings of the refractive indices are. The light is incident from air with refractive index $n_1 = 1.00$. The film has refractive index $n_2 = 1.28$. The water, the material behind the film, has a refractive index $n_3 = 1.33$. So, we know $n_1 < n_2 < n_3$. This means the condition for constructive interference is

$$2t = m\lambda_{n_2} \qquad m \in \{0, 1, 2, ...\}$$

The wavelength in the film is the wavelength of light in air divided by the refractive index of the film:

$$\lambda_{n_2} = \frac{\lambda}{n_2} = \frac{623\,\text{nm}}{1.28} = 486.7\,\text{nm}$$

We can then calculate the three non-zero film thicknesses

$$t = \frac{m\lambda_{n_2}}{2} \quad \Rightarrow$$

$$t_1 = \frac{(1)(486.7)}{2} = 243\ \text{nm}$$

$$t_2 = \frac{(2)(486.7)}{2} = 487\ \text{nm}$$

$$t_3 = \frac{(3)(486.7)}{2} = 730\ \text{nm}$$

### Section 35-7 Michelson Interferometer

A **Michelson interferometer** is an instrument that uses two-source interference and can be used to measure the wavelength of light, refractive indices of gases, or small changes in lengths. A **beam splitter** is used to create two beams of light that travel along two separate paths that return to the beam splitter where they interfere, as shown in the diagram.

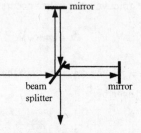

### Section 35-8 Luminous Intensity

The sensitive of our eye is not the same for all visible wavelengths of light. This means two sources of light of the same intensity, but different wavelengths, do not appear to be the same brightness. It is important for lighting engineers to have a measurement that reflects the effectiveness of a light source relative to the sensitivity of the human eye. **Luminous flux**, $F_l$, is the quantity defined to measure the brightness of a source and its unit is a **lumen (lm)**. An additional quantity that is useful is **luminous intensity**, which is the luminous flux per solid angle. The unit of luminous intensity is the **candela (cd)**. One candela is defined as one lumen per steradian. The **illuminance** of a surface is the luminous flux per unit area and is measured in units of lumens per square meter. Illuminance depends on the distance from the source.

## Practice Quiz

1.  As you are viewing a two-slit interference pattern, the wavelength of the source is changed. If the interference fringes move farther apart during this change, the wavelength has become

    a) greater.
    b) smaller.
    c) stayed the same.
    d) can't tell from this information.

2.  As the spacing of the slits in a two-slit interference experiment with a fixed wavelength become more closely spaced, the angle at which the first-order bright fringe occurs is

    a) greater.
    b) smaller.
    c) unchanged.
    d) can't tell from this information.

3.  In a two-slit interference experiment, the intensity of the light reaching the screen from either source individually would be $I_0$. What happens to the energy missing from the dark fringes?

    a) Energy is not conserved in wave optics.
    b) The energy is reaching the screen at the dark fringe; it just isn't seen as visible light.
    c) The intensity of the bright fringes is $4I_0$, so the missing energy is in the bright fringes.
    d) The energy never leaves the sources.

4.    A wedge-shaped film of air is formed between two glass slides.  What happens to the number of fringes seen if the index of refraction of the glass is increased?

a) The number of fringes increases.
b) The number of fringes decreases.
c) The number of fringes stays the same.
d) There is not enough information to determine the answer.

5.    A two-slit interference experiment gives an interference pattern shown in the top figure when the slits are illuminated with light of wavelength 420 nm. What is the wavelength of light illuminating the slits if the interference pattern is that shown in the bottom figure?

a) 280 nm
b) 630 nm
c) 840 nm
d) 1260 nm

6.    A monochromatic light source is illuminating one slit of a two-slit interference experiment and a second, independent, but identical, light source is illuminating the other slit.  The light passing through the slits does not create an interference pattern; why not?

a) The two sources are incoherent.
b) The two sources do not have a definite phase relationship.
c) The interference pattern is shifting too rapidly to see.
d) All of the above.

7.    An air gap between two glass slides gives a first-order bright reflection due to constructive interference at wavelength $\lambda_1$.  If the gap between the two plates is filled with a liquid that has a refractive index greater than the refractive index of the glass, what will the wavelength of the first-order bright reflection be?

a) $\lambda_1/2$
b) $2\lambda_1$
c) some wavelength less than $\lambda_{1/}$
d) some wavelength greater than $\lambda_1$

8.    What is the thinnest non-zero thickness film that will show constructive interference for visible light when there is a 180° phase change for reflections at both surfaces?

a) one half the wavelength of the most violet visible light
b) one half the wavelength of the most red visible light
c) one fourth the wavelength of the most violet visible light
d) one fourth the wavelength of the most red visible light

9.    The Michelson interferometer forms a two-source interference pattern.  The two sources that form this pattern are

a) a source being studied and a source within the human eye.
b) a single source that is split into two beams that travel along different paths.
c) a direct source and a second source that is reflected in a mirror.
d) a source and its image formed by a lens.

10.   If the distance from a light source is $r$ and the size of the source is small compared to $r$, how should the illuminance, $E_l$, depend on $r$?

a) $E_l \propto r$
b) $E_l \propto r^2$
c) $E_l \propto 1/r$
d) $E_l \propto 1/r^2$

11.  A two-slit interference pattern is formed on a screen a distance 1.56 m from two slits that are separated by a distance of 0.342 mm. The $5^{th}$ order bright fringe and the $12^{th}$ order bright fringe are separated by a distance of 1.87 cm. What is the wavelength of the light?

12.  A two-slit interference pattern is formed using a slit separation of 0.54 mm, a screen distance of 1.20 m, and a wavelength of 459 nm. How far from the center of the central bright fringe does the intensity first drop to one half the intensity at the center of the central bright spot?

13.  Two plates of glass illuminate by 560 nm wavelength light are in contact at one end and the other end is held apart by a very fine wire placed in between the plates as shown. If the number of bright fringes per length is 16 per cm and the glass plates are 2.54 cm long, what is the diameter of the wire?

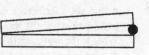

14.  A thin film of oil of refractive index 1.65 rests on a piece of plastic with refractive index 1.37. The thickness of the film is 1.18 µm. What are the three longest visible wavelengths of light in air that will have a bright reflection from this film due to constructive interference?

15.  What is the distance of separation between the $2^{nd}$ bright fringe away from the central bright fringe and the dark fringe which is the third dark fringe away from the central bright fringe in a two-slit interference pattern that has a slit separation of 0.78 mm, a wavelength of 433 nm, and a slit-to-screen distance of 1.44 m?

## Problem Solutions

3.  For constructive interference, the path difference is a multiple of the wavelength:
    $d \sin \theta = m\lambda$,   $m = 0, 1, 2, 3, \ldots$ .
    For the third order, we have
    $d \sin 18° = (3)(610 \times 10^{-9}$ m$)$, which gives $d = 5.9 \times 10^{-6}$ m $= 5.9$ µm.

7.  For constructive interference, the path difference is a multiple of the wavelength:
    $d \sin \theta = m\lambda$,   $m = 0, 1, 2, 3, \ldots$ .
    We find the location on the screen from
    $y = L \tan \theta$.
    For small angles, we have
    $\sin \theta \approx \tan \theta$, which gives $y = L(m\lambda/d) = mL\lambda/d$.
    For the second order of the two wavelengths, we have
    $\Delta y = mL \, \Delta\lambda/d = 2(1.0$ m$)[(720 - 660) \times 10^{-9}$ m$]/(0.58 \times 10^{-3}$ m$) = 2.07 \times 10^{-4}$ m $= 0.21$ mm.

11. To change the center point from constructive interference to destructive interference, the phase shift produced by the introduction of the plastic must be an odd multiple of half a wavelength, corresponding to the change in the number of wavelengths in the distance equal to the thickness of the plastic. The minimum thickness will be for a shift of a half wavelength:
    $N = (t/\lambda_{plastic}) - (t/\lambda) = (tn_{plastic}/\lambda) - (t/\lambda) = (t/\lambda)(n_{plastic} - 1) = \frac{1}{2}$;
    $[t/(640$ nm$)](1.60 - 1) = \frac{1}{2}$, which gives $t = 533$ nm.

15. (*a*)  If the sources have equal intensities, their electric fields will have the same magnitude:
    $E_{10} = E_{20} = E_{30} = E_0$.
    From the symmetry of the phasor diagram we see that $\phi = \delta$, where $\delta = 2\pi d \sin \theta/\lambda$ is the phase difference between adjacent slits. Thus the amplitude of the resultant field is
    $E_{\theta 0} = E_{10} \cos \delta + E_{20} + E_{30} \cos \delta = E_0(1 + 2 \cos \delta)$.
    At the center where $\theta = 0$, $\delta = 0$, we have
    $E_{00} = 3E_0$.
    The ratio of intensities is
    $I_\theta/I_0 = E_0^2(1 + 2 \cos \delta)^2/(3E_0)^2 = (1 + 4 \cos \delta + 4 \cos^2 \delta)/9$.

    (*b*)  The intensity will be maximal when $1 + 2 \cos \delta$ is maximal, which will be when $\cos \delta = 1$.
    This occurs when the three phasors are parallel. Thus we have
    $\delta = \ldots, - 2\pi, 0, 2\pi, 4\pi, \ldots = 2m\pi$, $m = 0, \pm1, \pm2, \ldots$ .

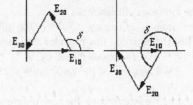

The angles for the maxima are given by

$\sin \theta_{max} = 2m\pi\lambda/2\pi d = m\lambda/d$, $m = 0, \pm1, \pm2, \dots$.

The intensity will be minimal when $1 + 2 \cos \delta$ is minimal (zero), which will be when $\cos \delta = -\frac{1}{2}$. This occurs when the three phasors form an equilateral triangle, as shown. Thus we have

$\delta = \dots, -2\pi/3, 2\pi/3, 4\pi/3, 8\pi/3, 10\pi/3, \dots$; or

$\delta = (m + \frac{1}{3}k)2\pi$; $k = 1, 2$; $m = 0, \pm1, \pm2, \dots$.

The angles for the minima are given by

$\sin \theta_{min} = (m + \frac{1}{3}k)\lambda/d$; $k = 1, 2$; $m = 0, \pm1, \pm2, \dots$.

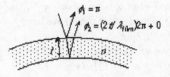

19. We equate a path difference of one wavelength with a phase difference of $2\pi$. With respect to the incident wave, the wave that reflects at the top surface from the higher index of the soap bubble has a phase change of $\phi_1 = \pi$.

With respect to the incident wave, the wave that reflects from the air at the bottom surface of the bubble has a phase change due to the additional path-length but no phase change on reflection: $\phi_2 = (2t/\lambda_{film})2\pi + 0$.

For destructive interference, the net phase change is

$\phi = (2t/\lambda_{film})2\pi - \pi = (m - \frac{1}{2})2\pi$, $m = 0, 1, 2, \dots$;   or   $t = \frac{1}{2}\lambda_{film}m = \frac{1}{2}(\lambda/n)m$, $m = 0, 1, 2, \dots$.

The minimum non-zero thickness is

$t_{min} = \frac{1}{2}[(480 \text{ nm})/(1.34)](1) = 179 \text{ nm}$.

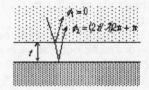

23. With respect to the incident wave, the wave that reflects from the air at the top surface of the air layer has a phase change of $\phi_1 = 0$.

With respect to the incident wave, the wave that reflects from the glass at the bottom surface of the air layer has a phase change due to the additional path-length and a change on reflection: $\phi_2 = (2t/\lambda)2\pi + \pi$.

For constructive interference, the net phase change is

$\phi = (2t/\lambda)2\pi + \pi - 0 = m2\pi$, $m = 1, 2, 3, \dots$;   or   $t = \frac{1}{2}\lambda(m - \frac{1}{2})$, $m = 1, 2, 3, \dots$.

The minimum thickness is

$t_{min} = \frac{1}{2}(480 \text{ nm})(1 - \frac{1}{2}) = 120 \text{ nm}$.

For destructive interference, the net phase change is

$\phi = (2t/\lambda)2\pi + \pi - 0 = (2m + 1)\pi$, $m = 0, 1, 2, \dots$;   or   $t = \frac{1}{2}m\lambda$, $m = 0, 1, 2, \dots$.

The minimum non-zero thickness is

$t_{min} = \frac{1}{2}(480 \text{ nm})(1) = 240 \text{ nm}$.

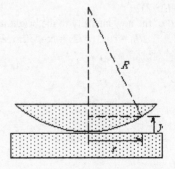

27. At a distance $r$ from the center of the lens, the thickness of the air space is $y$, and the phase difference for the reflected waves from the path-length difference and the reflection at the bottom surface is

$\phi = (2y/\lambda)2\pi + \pi$.

For the dark rings, we have

$\phi = (2y/\lambda)2\pi + \pi = (2m + 1)\pi$, $m = 0, 1, 2, \dots$;   or

$y = \frac{1}{2}m\lambda$, $m = 0, 1, 2, \dots$.

Because $m = 0$ corresponds to the dark center, $m$ represents the number of the ring. From the triangle in the diagram, we have

$r^2 + (R - y)^2 = R^2$,   or   $r^2 = 2yR - y^2 \approx 2yR$, when $y \ll R$,

which becomes

$r^2 = 2(\frac{1}{2}m\lambda)R = m\lambda R$, $m = 0, 1, 2, \dots$;   or   $r = (m\lambda R)^{1/2}$.

31. One fringe shift corresponds to a change in path length of $\lambda$. The number of fringe shifts produced by a mirror movement of $\Delta L$ is

$N = 2 \Delta L/\lambda$;

$750 = 2 \Delta L/(589 \times 10^{-9} \text{ m})$, which gives $\Delta L = 2.21 \times 10^{-4} \text{ m} = 0.221 \text{ mm}$.

454    Giancoli, *Physics for Scientists & Engineers*: Study Guide

35. (a) The luminous efficiency is

luminous efficiency = $F/P$ = (1700 lm)/(100 W) = 17 lm/W.

(b) If half the luminous flux provides the illuminance of the floor, we have

$E = \frac{1}{2}NP$(luminous efficiency)$/A$;

250 lm/m$^2$ = $\frac{1}{2}N$(40 W)(60 lm/W)/(25 m)(30 m), which gives $N = 156$.

39. For constructive interference, the path difference is a multiple of the wavelength:

$d \sin \theta = m\lambda$,   $m = 0, 1, 2, 3, \ldots$ .

We find the location on the screen from

$y = L \tan \theta$.

For small angles, we have

$\sin \theta \approx \tan \theta$, which gives   $y = L(m\lambda/d) = mL\lambda/d$.

For the second-order fringes we have

$y_1 = 2L\lambda_1/d$;   $y_2 = 2L\lambda_2/d$.

When we subtract the two equations, we get

$\Delta y = y_1 - y_2 = (2L/d)(\lambda_1 - \lambda_2)$;

(1.13 mm)(1 × 10$^{-3}$ m/mm) = [2(1.50 m)/(0.60 × 10$^{-3}$ m)](690 nm − $\lambda_2$), which gives $\lambda_2 = 464$ nm.

43. With respect to the incident wave, the wave that reflects
from the top surface of the coating has a phase change of

$\phi_1 = \pi$.

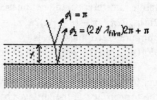

With respect to the incident wave, the wave that reflects
from the glass ($n \approx 1.5$) at the bottom surface of the coating
has a phase change due to the additional path-length and
a phase change of $\pi$ on reflection:

$\phi_2 = (2t/\lambda_{\text{film}})2\pi + \pi$.

For destructive interference, this phase difference
must be an odd multiple of $\pi$, so we have

$\phi = (2t/\lambda_{\text{film}})2\pi + \pi - \pi = (2m + 1)\pi$, $m = 0, 1, 2, \ldots$;   or   $t = \frac{1}{4}(2m + 1)\lambda_{\text{film}}$,  $m = 0, 1, 2, \ldots$ .

Thus the minimum thickness is

$t_{\text{min}} = \frac{1}{4}\lambda/n$.

(a) For the blue light we get

$t_{\text{min}} = \frac{1}{4}$ (450 nm)/(1.38) = 81.5 nm.

(b) For the red light we get

$t_{\text{min}} = \frac{1}{4}$ (700 nm)/(1.38) = 127 nm.

47. From the diagram in the text, we see that

$\sin \theta = v/v_p = c/nv_p$ = (3.00 × 10$^8$ m/s)/(1.52)(2.21 × 10$^8$ m/s) = 0.893, so $\theta = 63.3°$.

51. When the mirror is moved a distance $x$, the path length changes by $2x$. Thus the additional phase shift is

$\delta = (2x/\lambda)2\pi$.

The ratio of the new intensity to the bright maximum is

$I/I_0 = \cos^2(\delta/2) = \cos^2(2\pi x/\lambda)$.

# Chapter 36: Diffraction and Polarization

## Chapter Overview and Objectives

This chapter introduces diffraction and polarization. The diffraction pattern of a single slit is discussed in detail. The relationship between diffraction and the resolution of optical instruments is discussed. Diffraction gratings and their applications to spectrometers are introduced. Polarization is described and methods of polarizing light are discussed.

After completing study of this chapter, you should:

- Know how diffraction arises from the wave nature of light.
- Know how to determine where single-slit diffraction intensity minima are located.
- Know how to determine the intensity of light at a given position in a single-slit diffraction pattern.
- Know how the fact that the double-slit interference experiment results are altered by the fact that a double-slit experiment must use two finite-width single slits.
- Know where the first minimum intensity in the diffraction pattern of a circular aperture is located.
- Know what it means to resolve two point sources and Rayleigh's criteria.
- Know what a diffraction grating is.
- Know how to determine the resolution and resolving power of a diffraction grating.
- Know what polarization is.
- Know that a Polaroid transmits linearly polarized light.
- Know how to determine the intensity of light transmitted by a Polaroid.
- Know what Brewster's angle is and how to determine it from the refractive indices of the materials on either side of a reflecting surface.

## Summary of Equations

Position of intensity minima in single-slit diffraction pattern:

$$a \sin \theta = m\lambda \qquad m \in \{1, 2, 3, \ldots\}$$

Intensity as function of position in single-slit diffraction pattern:

$$I = I_0 \left[ \frac{\sin\left( \frac{\pi a \sin \theta}{\lambda} \right)}{\left( \frac{\pi a \sin \theta}{\lambda} \right)} \right]^2$$

Intensity as function of position in double-slit interference pattern including single-slit diffraction effects:

$$I_\theta = I_0 \left[ \frac{\sin\left( \frac{\pi a \sin \theta}{\lambda} \right)}{\left( \frac{\pi a \sin \theta}{\lambda} \right)} \right]^2 \cos^2\left( \frac{\pi d \sin \theta}{\lambda} \right)$$

Direction of first minimum in diffraction pattern of a circular aperture:

$$\theta = \frac{1.22\lambda}{D}$$

Resolving power of a microscope:

$$RP = \frac{1.22\lambda\, f}{D}$$

Angular width of diffraction peaks from a diffraction grating:

$$\Delta\theta_m = \frac{\lambda}{Nd\cos\theta_m}$$

Definition of resolving power:    $R = \dfrac{\lambda}{\Delta\lambda}$

Resolving power of a diffraction grating:    $R = Nm$

Bragg condition:    $m\lambda = 2d\sin\phi \qquad m \in \{0,1,2,3,\dots\}$

Transmitted intensity of Polaroid when incident light is unpolarized:

$$I = \tfrac{1}{2}I_0$$

Transmitted intensity of Polaroid when incident light is linearly polarized, Malus' law:

$$I = I_0 \cos^2\theta$$

Relationship between Brewster's angle and indices of refraction:

$$\tan\theta_p = \frac{n_2}{n_1}$$

## Chapter Summary

### Section 36-1 Diffraction by a Single Slit

Plane waves of light passing through a slit of width $a$ produce a diffraction pattern on a screen far from the slit such that there is a minimum (zero) intensity in the diffraction pattern whenever

$$a\sin\theta = m\lambda \qquad m \in \{1,2,3,\dots\}$$

where $\lambda$ is the wavelength of the light illuminating the slit and the angle $\theta$ is defined as shown in the diagram.

### Example 36-1-A

A single-slit diffraction pattern is formed on a screen a distance of 1.4 m from the slit. The third minimum from the center peak in the intensity pattern is a distance of 1.9 cm from the sixth minimum from the center peak in the intensity pattern. If the width of the slit is 0.16 mm, what is the wavelength of the light illuminating the slit?

**Solution:**

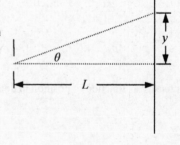

For small angles, we can approximate $\sin\theta \approx \theta$. The position of the minima in the diffraction pattern is given by:

$$a\sin\theta = m\lambda \qquad \Rightarrow \qquad y_m = \frac{m\lambda L}{a}$$

We know $y_6 - y_3 = 1.9$ cm. This implies

$$\frac{6\lambda L}{a} - \frac{3\lambda L}{a} = 1.9 \text{ cm} \quad \Rightarrow \quad \lambda = 1.9 \text{ cm} \frac{a}{3L} = \left(1.9 \times 10^{-2} \text{ m}\right) \frac{1.6 \times 10^{-4} \text{ m}}{3(1.4 \text{ m})} = 7.2 \times 10^{-7} \text{ m}$$

## Section 36-2 Intensity in Single-Slit Diffraction Pattern

The intensity of the diffraction pattern as a function of the direction $\theta$ shown in the diagram above is given by

$$I = I_0 \left[ \frac{\sin\left(\dfrac{\pi a \sin\theta}{\lambda}\right)}{\left(\dfrac{\pi a \sin\theta}{\lambda}\right)} \right]^2$$

where $I_0$ is the intensity in the $\theta = 0$ direction.

### Example 36-2-A

The intensity at the central peak of a single-slit diffraction pattern is 1.56 W/m$^2$, where the wavelength of the light is 640 nm and the width of the slit is 0.24 mm. What are the first three angles away from the central maximum at which the intensity is 0.012 W/m$^2$?

### Solution:

We start from the expression for intensity as a function of angle in the diffraction pattern:

$$I = I_0 \left[ \frac{\sin\left(\dfrac{\pi a \sin\theta}{\lambda}\right)}{\left(\dfrac{\pi a \sin\theta}{\lambda}\right)} \right]^2 \quad \Rightarrow \quad \frac{\sin x}{x} = \pm \sqrt{\frac{I}{I_0}} = \pm \sqrt{\frac{0.0120 \text{ W/m}^2}{1.56 \text{ W/m}^2}} = \pm 0.0877$$

where we have used $x$ to represent $\pi a \sin\theta/\lambda$. The equation above is transcendental, so we cannot solve it algebraically, but must use a numerical method. If we do so, we find that the three lowest solutions for $x$ are 2.89, 3.45, and 5.75. Using the values of $x$ to solve for $\theta$, we get

$$\theta = \arcsin \frac{\lambda x}{\pi a} = \arcsin \frac{\left(640 \times 10^{-9} \text{ m}\right)\left(2.89\right)}{\pi \left(0.24 \times 10^{-3} \text{ m}\right)} = 2.45 \times 10^{-3} \text{ rad}$$

$$= \arcsin \frac{\left(640 \times 10^{-9} \text{ m}\right)\left(3.45\right)}{\pi \left(0.24 \times 10^{-3} \text{ m}\right)} = 2.93 \times 10^{-3} \text{ rad}$$

$$= \arcsin \frac{\left(640 \times 10^{-9} \text{ m}\right)\left(5.75\right)}{\pi \left(0.24 \times 10^{-3} \text{ m}\right)} = 4.88 \times 10^{-3} \text{ rad}$$

## Section 36-3 Diffraction in the Double-Slit Experiment

The double-slit or two-slit interference pattern of Chapter 35 is the result of light passing through two closely spaced single slits. The discussion of Chapter 35 was incomplete because it ignored the single-slit diffraction of each of the two slits. The single slits cause lower light intensity in some directions relative to other directions. The two-slit intensity pattern is modified by the single-slit diffraction

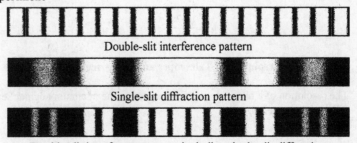

Double-slit interference pattern

Single-slit diffraction pattern

Double-slit interference pattern including single-slit diffraction pattern. The intensity patterns show the double-slit interference pattern at the top, the single-slit diffraction pattern in the middle, and the double-slit interference pattern when the single slit-diffraction is accounted for at the bottom. In this case, the center-to-center separation of the two slits is three times the slit width. The resulting two-slit interference pattern intensity is given by

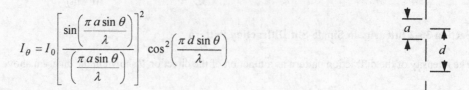

$$I_\theta = I_0 \left[ \frac{\sin\left(\dfrac{\pi a \sin\theta}{\lambda}\right)}{\left(\dfrac{\pi a \sin\theta}{\lambda}\right)} \right]^2 \cos^2\left(\frac{\pi d \sin\theta}{\lambda}\right)$$

where the symbols are as defined in the previous section and $d$ is the center-to-center spacing of the two slits as shown in the diagram. For the intensity patterns shown above, $d = 3a$.

### Example 36-3-A

A two-slit interference pattern is missing the $12^{th}$ order, $24^{th}$ order, and $36^{th}$ order constructive interference intensity peaks. What is the ratio of the separation of the slits to the width of each slit?

### Solution:

We know that the missing constructive interference peaks are the result of a single-slit diffraction minimum lying in the same direction as the two-slit interference intensity peak. Evidently, the first single-slit diffraction minimum is in the same position as the $12^{th}$ order two-slit interference minimum. Setting these conditions equal to each other, we get

$$\frac{1\lambda}{a} = \frac{12\lambda}{d} \quad \Rightarrow \quad \frac{d}{a} = 12$$

### Section 36-4 Limits of Resolution; Circular Apertures

A circular aperture produces a diffraction pattern, also. The diffraction pattern is circular and the location of the minima in the diffraction pattern are not uniformly spaced in distance as in the diffraction pattern of a slit. The first minimum in the diffraction is at an angle $\theta$ from the center of the central bright spot approximately given by

$$\theta = \frac{1.22\lambda}{D}$$

where $D$ is the diameter of the circular opening.

Most optical instruments have circular apertures. This means that the image formed by an optical instrument is not perfect, but can do no better than the limit given by the relationship above. The image of the point source cannot be smaller in angular width than the diffraction limit given above. This also limits how well the optical instrument can resolve two closely spaced point sources of light. A precise limit on the resolution of the instrument depends on how precisely the intensity of the light in the diffraction pattern is measured. Because many instruments are viewed with the eye, which may only see the intensity of the central bright spot of the diffraction pattern, an appropriate definition of resolution is the Rayleigh criterion. The Rayleigh criterion states that two sources can be resolved if the diffraction pattern central bright spot of one source falls on or outside of the first diffraction pattern minimum of the other light source. This implies that the angular separation of the sources of light must be

$$\theta = \frac{1.22\lambda}{D}$$

### Section 36-5 Resolution of Telescopes and Microscopes; the $\lambda$ Limit

The resolution of a telescope is usually stated as the minimum angular separation of the distant sources viewed by the telescope as given above.

The **resolving power**, **RP**, of a microscope is usually given as the minimum distance between two sources that are barely resolvable. If the objective lens is the smallest diameter aperture in the microscope, the resolving power of a microscope is given by

$$RP = \frac{1.22\lambda\, f}{D}$$

where $\lambda$ is the wavelength of the light, $f$ is the focal length of the objective lens, and $D$ is the diameter of the aperture formed by the objective lens.

The practical lower limit on the focal length of the objective lens is roughly the diameter of the lens divided by 2. This results in a practical rule of thumb regarding the limit of resolution of an optical instrument:

**The minimum distance between two sources of light that are resolvable by an optical instrument is about one wavelength of the light used.**

**Example 36-5-A**

Assume the opening of the pupil of the eye is 2.8 mm. What is the smallest distance between two light sources emitting light at a wavelength of 500 nm at a distance of 1.00 mile such that the eye can resolve the sources as two separate sources of light?

**Solution:**

We know that the angular separation between the two sources must be

$$\theta = \frac{1.22\lambda}{D}$$

where $\lambda$ is the wavelength of the light and $D$ is the diameter of the aperture of the eye. We know the wavelength is much smaller than the diameter of the pupil, so we can use the small angle approximation that $\theta$ is approximately the separation of the sources, $d$, divided by the distance to the sources, $L$. We then get

$$\frac{d}{L} = \frac{1.22\lambda}{D} \Rightarrow d = \frac{1.22\lambda L}{D} = \frac{1.22\left(500\times10^{-9}\,\text{m}\right)\left(1.00\,\text{mi}\times1.61\times10^{3}\,\text{m/mi}\right)}{2.8\times10^{-3}\,\text{m}} = 0.35\,\text{m}$$

**Section 36-6 Resolution and the Human Eye and Useful Magnification**

The limit of resolution of the human eye is about $5 \times 10^{-4}$ radians at best. This limits the greatest visible magnification of a microscope to about 500×.

**Section 36-7 Diffraction Grating**

An optical device that has many parallel, closely spaced transmitting or reflecting elements is called a **diffraction grating**. If the diffraction grating is made of transmitting elements, it is a **transmission grating** and if it is made of reflecting elements it is called a **reflection grating**. If the spacing between each element in the diffraction grating is $d$, the light that strikes the grating will interfere constructively, after it interacts with the grating, in directions given by

$$\sin\theta = \frac{m\lambda}{d} \qquad m \in \left\{0, 1, 2, 3, \dots\right\}$$

The intensity pattern is different from the intensity pattern of a double-slit interference pattern in that the width of the constructive interference peaks from the diffraction grating will be narrower depending on the number of elements illuminated by the incoming light.

**Section 36-8 The Spectrometer and Spectroscopy**

An instrument that is designed to determine the wavelength of light is called a spectrometer or spectroscope.  The basis of a spectrometer is an element that disperses the light in a direction that depends on the wavelength of the light. Spectrometers make use of either diffraction gratings or prisms to disperse the light.

**Section 36-9 Peak Widths and Resolving Power for a Diffraction Grating**

The angular width, $\Delta\theta_m$, of the constructive interference peaks for a diffraction grating is given by

$$\Delta\theta_m = \frac{\lambda}{Nd\cos\theta_m},$$

where $\theta_m$ is the angular position of the $m^{\text{th}}$ peak, $N$ is the number of diffraction grating elements illuminated by the light incident on the diffraction grating, and $d$ is the spacing of the diffraction grating elements.  The **resolving power**, $R$, of a diffraction grating is defined to be the ratio of the wavelength of the light, $\lambda$,  divided by the minimum difference in wavelengths, $\Delta\lambda$, that can be resolved by the diffraction grating:

$$R = \frac{\lambda}{\Delta\lambda}$$

It can be shown that the resolving power of a diffraction grating is given by

$$R = Nm,$$

where $m$ is the order of the constructive interference peak in the diffraction pattern.

**Example 36-9-A**

A transmission diffraction grating with a total width of 2.00 cm has slits with a separation of 180 μm.  Determine the angle in the diffraction pattern at which the first-order peak in the diffraction pattern for light of wavelength 532 nm occurs.  Determine the next largest wavelength that can be resolved from the 532 nm wavelength by this diffraction grating.

**Solution:**

We know the position of the diffraction grating peaks is given by

$$\sin\theta = \frac{m\lambda}{d} \quad\Rightarrow\quad \theta = \arcsin\frac{m\lambda}{d} = \arcsin\frac{1\left(532\times10^{-9}\text{ m}\right)}{180\times10^{-6}\text{ m}} = 5.16\times10^{-5}\text{ rad} = 0.00296° = 11''$$

The minimum difference in wavelengths that can be resolved with this grating is

$$\Delta\lambda = \frac{\lambda}{R} = \frac{\lambda}{Nm}$$

The number of slits that are illuminated by the light is equal to the width of the grating divided by the separation of the slits:

$$N = \frac{w}{d} = \frac{2.00\times10^{-2}\text{ m}}{180\times10^{-6}\text{ m}} = 111$$

Using this in the expression above for $\Delta\lambda$, we get

$$\Delta\lambda = \frac{\lambda}{Nm} = \frac{532\times10^{-9}\text{ m}}{(111)(1)} = 4.79\times10^{-9}\text{ m} = 4.79\text{ nm}$$

If this is the minimum difference in wavelengths that can be resolved, the next wavelength greater than 532 nm that can be resolved from 532 nm will be 532 nm + $\Delta\lambda$ = 532 nm + 5 nm = 537 nm.

## Section 36-10 X-Rays and X-Ray Diffraction

X-rays penetrate below the surface of materials and so diffraction of X-rays from crystal surfaces depends not only on the spacing of atoms along the surface that is illuminated by the X-rays, but also depends on the spacing of the atoms in depth from the surface. The directions of constructive interference from this three-dimensional array are given by

$$m\lambda = 2d \sin \phi \qquad m \in \{0, 1, 2, 3, \dots\}$$

where $d$ is the perpendicular distance between two planes of atoms in the array and $\phi$ is defined in the diagram to the right. This is called the **Bragg condition** for X-ray diffraction.

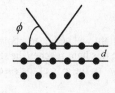

X-ray diffraction is used to determine the arrangement of atoms in sold materials.

## Section 36-11 Polarization

Because there are two independent directions that are perpendicular to the direction of propagation of a wave, transverse waves have two independent **polarizations**, or directions of the displacement of the wave. Electromagnetic waves are transverse waves and, therefore, have two polarizations. If the electric field of an electromagnetic wave has its electric field in a constant direction, it is said to be **plane-polarized** or **linearly polarized**. Most sources of light emit **unpolarized** light in which the electric field changes direction rapidly with time in a random fashion.

A **polarizer** can be used to create polarized light from unpolarized light. There are several different ways to construct a polarizer. A **Polaroid** is a sheet of semi-transparent material that transmits light of one linear polarization and absorbs light of the perpendicular polarization. The direction of the electric field of the linear polarized light that is transmitted through the Polaroid is called the **axis** of the Polaroid. The light transmitted by a Polaroid is linearly polarized in the direction of the axis of the Polaroid. The intensity of light transmitted by a polarizer depends on whether the light incident on the Polaroid is initially linearly polarized or not. If the light incident on the Polaroid is unpolarized the transmitted intensity, $I$, is related to the incident intensity, $I_0$, by

$$I = \tfrac{1}{2} I_0$$

If the light incident on the Polaroid is linearly polarized with a polarization direction at an angle $\theta$ to the axis of the polarizer, then the transmitted intensity $I$ is related to the incident intensity, $I_0$, by Malus' law:

$$I = I_0 \cos^2 \theta$$

A Polaroid can be used as an analyzer to determine whether light is linearly polarized or not and to determine the direction of the polarization if the light is linearly polarized.

Light is also polarized by reflection from surfaces if the incident angle is not zero. The polarization is not complete unless the angle of incidence is at a particular angle called Brewster's angle, $\theta_p$, where this angle is given by

$$\tan \theta_p = \frac{n_2}{n_1}$$

where $n_1$ is the refractive index of the material on the incident side of the reflecting surface and $n_2$ is the refractive index on the other side of the reflecting surface.

Some materials have a refractive index that depends on the direction of the polarization of light. This is called **birefringence**. This birefringence can be used to polarize light by refracting light in different directions depending on its polarization.

**Example 36-11-A**

Unpolarized light is incident on a Polaroid. The light that passes through the Polaroid is incident upon a second Polaroid with its axis tilted at an angle of 50° relative to the axis of the first Polaroid. What fraction of the initial intensity of the unpolarized light transmits through the second Polaroid?

**Solution:**

Let the intial intensity be $I_0$. Because this light is unpolarized, we know the intensity that passes through the first Polaroid, $I_1$, is given by

$$I_1 = \tfrac{1}{2} I_0$$

This light will be linearly polarized with the direction of polarization in the direction of the axis of the first polarizer. When this light reaches the second Polaroid, it is linearly polarized and its polarization direction will be at an angle of 50° relative to the axis of the second Polaroid. This implies that the second Polaroid will transmit an intensity $I_2$ given by

$$I_2 = I_1 \cos^2 50^0 = \tfrac{1}{2} I_0 \cos^2 50^0 = 0.207 \, I_0$$

The fraction of the initial intensity of the unpolarized light that transmits through the second Polaroid is 0.207.

**Example 36-11-B:**

Light reaching the surface of a piece of glass from air at an incident angle of 48° is refracted so the refracted angle is 33°. What is Brewster's angle for this surface?

**Solution:**

Brewster's angle is given by

$$\tan \theta_p = \frac{n_2}{n_1}$$

We know $n_1$, the refractive index of air, is 1.00. We can determine the refractive index of the material from the information given and Snell's law:

$$n_1 \sin \theta_1 = n_2 \sin \theta_2 \quad \Rightarrow \quad n_2 = n_1 \frac{\sin \theta_1}{\sin \theta_2} = 1.00 \frac{\sin 48^\circ}{\sin 33^\circ} = 1.36$$

Using this in the expression for Brewster's angle, we get

$$\tan \theta_p = \frac{n_2}{n_1} \quad \Rightarrow \quad \theta_p = \arctan \frac{n_2}{n_1} = \arctan \frac{1.36}{1.00} = 54^\circ$$

**Section 36-12 Scattering of Light by the Atmosphere**

Light is scattered by gas molecules because the light causes an acceleration of the electrons and accelerating electrons radiate electromagnetic waves. The polarization of the emitted waves is dependent on the direction of the electron's acceleration. No light is emitted in the direction that lies along the direction of the electron's acceleration. When light scatters off molecules in a gas, the light will be partially polarized. The polarization of the incoming light that lies along the line of sight to the molecule scattering the light to your eye will not emit light toward your eye and that polarization will be suppressed in the scattered light that you view.

## Practice Quiz

1.   Near the center of the single-slit diffraction pattern, the minima may appear to be equally spaced on one side of the central peak near the central peak.  What happens to the spacing between the minima on a flat screen as  the order number increases?

a)  The spacing remains the same.
b)  The spacing becomes smaller.
c)  The spacing becomes larger.
d)  The spacing can become larger or smaller depending on the wavelength of the light.

2.   The third order single-slit diffraction minimum for light of wavelength 540 nm falls on the fourth order minimum for light of a second unknown wavelength.  What is the unknown wavelength?

a)  720 nm
b)  405 nm
c)  640 nm
d)  360 nm

3.   For the double-slit interference pattern shown, what is the ratio of the center-to-center separation of the slits to the width of each slit?

a)  2
b)  3
c)  4
d)  8

4.   You are viewing objects with your eye.  Which wavelength should you use to be able to resolve the objects with the greatest amount of detail?

a)  200 nm
b)  450 nm
c)  600 nm
d)  750 nm

5.   Two Polaroids have their axes at right angles to one another so no light is transmitted by the second Polaroid. How can a third Polaroid be placed and oriented so that the maximum amount of light transmits through the three Polaroids?

a)  Place the third Polaroid after the second polarizer with its axis parallel to the first Polaroid's axis.
b)  Place the third Polaroid before the first Polaroid with its axis parallel to the second Polaroid's axis.
c)  Place the third Polaroid between the first two Polaroids with its axis at 45° to the other two axes.
d)  it is impossible to place the third Polaroid so that any light is transmitted through.

6.   A telescope with a primary mirror of diameter D is designed to be diffraction limited by a primary mirror.  A suspected binary star appears as one bright spot on photographs and CCD cameras used with the telescope.  In order to resolve the binary star as two stars, what improvement needs to be made?

a)  A better camera must be attached to the telescope.
b)  Film with better resolution must be used.
c)  A longer exposure time must be used.
d)  A telescope with a larger diameter primary mirror must used.

7.    A large concrete and steel building is directly between you and a radio tower.  The radio waves cannot pass through the building, but your radio receives the station just fine.  What is one possible reason your radio may still be receiving the station?

   a) The radio waves are diffracting around the building.
   b) The atoms in the building are spaced such that the Bragg scattering condition is satisfied for the wavelength of the radio station.
   c) The radio waves do not need to reach the radio receiver in order to receive the radio station.
   d) The radio waves are polarized parallel to the ground, but the building is standing vertically.

8.    What do the results of X-rays scattering off solid materials support as a model of solids?

   a) The atoms in all solids are randomly dispersed throughout the volume of the solid.
   b) Atoms in many materials are arranged in orderly arrays.
   c) There are no such thing as atoms.
   d) Solids are only solid on the surface, but are liquid inside.

9.    Whether the light that is incident on a Polaroid is polarized or unpolarized, what do you know about the light that is transmitted by the Polaroid?

   a) The transmitted light is lower in intensity than the incident light.
   b) The transmitted light is polarized in the direction of the axis of the polarizer.
   c) Both a) and b) are true.
   d) There is nothing you can conclude unless you know if the incident light was polarized or unpolarized.

10.   You measure the intensity of light passing through a Polaroid to be one-half the intensity of the light incident on the Polaroid.  What do you know?

   a) The light incident on the Polaroid was unpolarized.
   b) The light incident on the Polaroid was linearly polarized and its polarization direction was at an angle of 45° to the Polaroid's axis.
   c) Either a) or b) is true.
   d) The Polaroid is not functioning correctly.

11.   Two wavelengths, $\lambda_1$ and $\lambda_2$, illuminate a single slit.  The third minima in the diffraction pattern of $\lambda_1$ is located in the same position as the fourth minima of $\lambda_2$.  If $\lambda_1$ = 540 nm, what is $\lambda_2$?

12.   A single slit of width 0.180 mm is illuminated with light of wavelength 488 nm.  A diffraction pattern forms on a screen that is 1.25 m away.  How far away from the central peak of the single-slit diffraction pattern is the intensity first reduced to one-fourth the intensity at the center of the central peak?

13.   What is the minimum angular separation of two distant stars that can be resolved with a telescope with an effective aperture of 1.20 m when viewing the stars in light of wavelength 589 nm?

14.   Unpolarized light is incident on a Polaroid with its axis aligned in the vertical direction.  It then passes through a second Polaroid with its axis tipped clockwise in a direction 30° to the vertical.  It then passed though a third Polaroid with its axis tipped clockwise in a direction of 80° to vertical.  What fraction of the intensity of the initially unpolarized light is transmitted through the final Polaroid?

15.   Brewster's angle is determined to be 56° for the surface of a material when light is incident on the surface from air.  What is the refractive index of the material?

## Problem Solutions

3.  For constructive interference from the single slit, the path difference is
    $$a \sin \theta = (m + \tfrac{1}{2})\lambda, \quad m = 1, 2, 3, \dots .$$
    For the first fringe away from the central maximum, we have
    $$(3.50 \times 10^{-6}\ \text{m}) \sin \theta_1 = (\tfrac{3}{2})(550 \times 10^{-9}\ \text{m}), \text{ which gives } \theta_1 = 13.7°.$$
    We find the distance on the screen from
    $$y_1 = L \tan \theta_1 = (10.0\ \text{m}) \tan 13.7° = 2.4\ \text{m}.$$

7.  We find the angle to the first minimum from
    $$\sin \theta_{1\min} = m\lambda/a = (1)(400 \times 10^{-9}\ \text{m})/(0.0655 \times 10^{-3}\ \text{m}) = 6.11 \times 10^{-3}, \text{ so } \theta_{1\min} = 0.350°.$$
    We find the distance on the screen from
    $$y_1 = L \tan \theta_1 = (3.50\ \text{m}) \tan 0.350° = 2.14 \times 10^{-2}\ \text{m} = 2.14\ \text{cm}.$$
    Thus the width of the peak is
    $$\Delta y_1 = 2y_1 = 2(2.14\ \text{cm}) = 4.28\ \text{cm}.$$

11. To find the angular width at half-maximum, we find the phase
    at half-maximum:
    $$I = I_0[(\sin \tfrac{1}{2}\beta_h)/\tfrac{1}{2}\beta_h]^2 = \tfrac{1}{2}I_0, \quad \text{or} \quad \beta_h^{\ 2} = 8 \sin^2 \tfrac{1}{2}\beta_h.$$
    This equation can be solved graphically or numerically to get $\beta_h = 2.783$ rad.
    We find the corresponding angle from
    $$\beta_h = (2\pi a/\lambda) \sin \theta_h .$$
    Thus the angular width at half-maximum is
    $$\Delta \theta_h = 2\theta_h = 2 \sin^{-1}(\lambda\beta_h/2\pi a) = 2 \sin^{-1}(0.443\lambda/a).$$
    The angle will be small, so we have
    $$\Delta \theta_h \approx 2(0.443\lambda/a) = 0.886\lambda/a.$$
    For the given data we have
    $$\Delta \theta_h = 0.886(550 \times 10^{-9}\ \text{m})/(2.60 \times 10^{-6}\ \text{m}) = 0.187\ \text{rad} = 10.7°.$$

15. (a)  The maxima of the double slit are given by
    $$\sin \theta = m\lambda/d, \quad m = 0, \pm 1, \pm 2, \dots .$$
    The distance of a fringe on the screen from the center of the pattern is
    $$y = L \tan \theta.$$
    If the angles are small, we have
    $$y \approx L \sin \theta = L\lambda m/d.$$
    The separation of adjacent interference fringes will be
    $$\Delta y = (L\lambda\ \Delta m)/d = (1.0\ \text{m})(550 \times 10^{-9}\ \text{m})(1)/(0.030 \times 10^{-3}\ \text{m}) = 0.018\ \text{m} = 1.8\ \text{cm}.$$
    (b)  The minima of the single-slit pattern are given by
    $$\sin \theta = m_s\lambda/a, \quad m_s = \pm 1, \pm 2, \dots .$$
    The distance of the first minimum on the screen from the center of the pattern is
    $$y = L \tan \theta.$$
    If the angle is small, we have
    $$y \approx L \sin \theta = L\lambda/a = (1.0\ \text{m})(550 \times 10^{-9}\ \text{m})/0.010 \times 10^{-3}\ \text{m}) = 0.055\ \text{m} = 5.5\ \text{cm}.$$
    Thus the distance between the first minima on either side of the center is $2y = 11.0$ cm.

19. The minimum angular resolution is
    $$\theta = 1.22\lambda/D = (1.22)(500 \times 10^{-9}\ \text{m})/(100\ \text{in})(0.0254\ \text{m/in}) = 2.4 \times 10^{-7}\ \text{rad} = (1.4 \times 10^{-5})° = 0.050\text{"}.$$

23. We find the angle for the third order from
    $$d \sin \theta = m\lambda;$$
    $$(1.35 \times 10^{-5}\ \text{m}) \sin \theta = (3)(440 \times 10^{-9}\ \text{m}), \text{ which gives } \sin \theta = 9.78 \times 10^{-2}, \text{ so } \theta = 5.61°.$$

27. We find the wavelengths from
    $$d \sin \theta = m\lambda;$$
    $$[1/(10{,}000\ \text{lines/cm})](10^{-2}\ \text{m/cm}) \sin 29.8° = (1)\lambda_1, \text{ which gives } \lambda_1 = 4.97 \times 10^{-7}\ \text{m} = 497\ \text{nm};$$
    $$[1/(10{,}000\ \text{lines/cm})](10^{-2}\ \text{m/cm}) \sin 37.7° = (1)\lambda_2, \text{ which gives } \lambda_2 = 6.12 \times 10^{-7}\ \text{m} = 612\ \text{nm};$$
    $$[1/(10{,}000\ \text{lines/cm})](10^{-2}\ \text{m/cm}) \sin 39.6° = (1)\lambda_3, \text{ which gives } \lambda_3 = 6.37 \times 10^{-7}\ \text{m} = 637\ \text{nm};$$
    $$[1/(10{,}000\ \text{lines/cm})](10^{-2}\ \text{m/cm}) \sin 48.9° = (1)\lambda_4, \text{ which gives } \lambda_4 = 7.54 \times 10^{-7}\ \text{m} = 754\ \text{nm}.$$

31. We have the same average angles, but the path differences causing the interference must be measured in terms of the wavelengths in water. Thus the wavelengths calculated in Problem 30 are those in water. The wavelengths in air are

$\lambda_{1air} = \lambda_1 n_{water} = (467 \text{ nm})(1.33) = 621 \text{ nm};$

$\lambda_{2air} = \lambda_2 n_{water} = (684 \text{ nm})(1.33) = 909 \text{ nm}.$

Note that the second wavelength is not visible.

35. (*a*) The maximum angle is 90°, so we have

$d \sin \theta = m\lambda;$

$(1200 \text{ nm}) \sin 90° = m(580 \text{ nm})$, which gives $m = 2.07$.

Thus there are two orders on each side of the central maximum.

(*b*) The half width of an order is

$\Delta\theta_m = \lambda/Nd \cos \theta_m.$

Because $Nd = L$, the width of the grating, the full width is

$2 \Delta\theta_m = 2\lambda/L \cos \theta_m.$

For the principal maxima, we have

$m = 0$:

$\theta_0 = 0;$

$2 \Delta\theta_0 = 2(580 \times 10^{-9} \text{ m})/(1.80 \times 10^{-2} \text{ m}) \cos 0° = 6.44 \times 10^{-5} \text{ rad} = 13.3".$

$m = 1$:

$\sin \theta_1 = (1)(580 \text{ nm})/1200 \text{ nm}) = 0.483, \theta_1 = 28.9°;$

$2 \Delta\theta_1 = 2(580 \times 10^{-9} \text{ m})/(1.80 \times 10^{-2} \text{ m}) \cos 28.9° = 7.36 \times 10^{-5} \text{ rad} = 15.2".$

$m = 2$:

$\sin \theta_2 = (2)(580 \text{ nm})/1200 \text{ nm}) = 0.967, \theta_2 = 75.2°;$

$2 \Delta\theta_2 = 2(580 \times 10^{-9} \text{ m})/(1.80 \times 10^{-2} \text{ m}) \cos 75.2° = 2.52 \times 10^{-4} \text{ rad} = 52.0".$

39. The frequency is $f = c/\lambda$. If we approximate the changes as differentials, we have

$\Delta f = -(c/\lambda^2) \Delta\lambda.$

Disregarding the negative sign, we get

$\Delta f = (c/\lambda)(\Delta\lambda/\lambda) = fR = f/mN.$

43. If the initial intensity is $I_0$, through the two sheets we have

$I_1 = \tfrac{1}{2}I_0,$

$I_2 = I_1 \cos^2 \theta = \tfrac{1}{2}I_0 \cos^2 \theta$, which gives

$I_2/I_0 = \tfrac{1}{2}\cos^2 \theta = \tfrac{1}{2}\cos^2 75° = 0.033.$

47. If the original intensity is $I_0$, the first Polaroid sheet will reduce the intensity of the original beam to

$I_1 = \tfrac{1}{2}I_0.$

If the axis of the second Polaroid sheet is oriented at an angle $\theta$, the intensity is

$I_2 = I_1 \cos^2 \theta = \tfrac{1}{2}I_0 \cos^2 \theta.$

(*a*) $I_2 = \tfrac{1}{2}I_0 \cos^2 \theta = \tfrac{1}{3}I_0$, which gives $\theta = 35°$.

(*b*) $I_2 = \tfrac{1}{2}I_0 \cos^2 \theta = 0.10I_0$, which gives $\theta = 63°$.

51. Through the successive polarizers we have

$I_1 = \tfrac{1}{2}I_0;$

$I_2 = I_1 \cos^2 \theta_2 = \tfrac{1}{2}I_0 \cos^2 \theta_2;$

$I_3 = I_2 \cos^2 \theta_3 = \tfrac{1}{2}I_0 \cos^2 \theta_2 \cos^2 \theta_3;$

$I_4 = I_3 \cos^2 \theta_4 = \tfrac{1}{2}I_0 \cos^2 \theta_2 \cos^2 \theta_3 \cos^2 \theta_4;$

$I_5 = I_4 \cos^2 \theta_5 = \tfrac{1}{2}I_0 \cos^2 \theta_2 \cos^2 \theta_3 \cos^2 \theta_4 \cos^2 \theta_5 = \tfrac{1}{2}I_0 (\cos^2 45)^4 = I_0 /32.$

55. The wavelength of the sound is

$\lambda = v/f = (343 \text{ m/s})/(750 \text{ Hz}) = 0.457 \text{ m}.$

We find the angles of the minima from

$a \sin \theta = m\lambda, \quad m = 1, 2, 3, \dots;$

$(0.88 \text{ m}) \sin \theta_1 = (1)(0.457 \text{ m})$, which gives $\sin \theta_1 = 0.520$, so $\theta_1 = 31°;$

$(0.88 \text{ m}) \sin \theta_2 = (2)(0.457 \text{ m})$, which gives $\sin \theta_2 = 1.04$, so there is no $\theta_2$.

Thus the whistle would not be heard clearly at angles of 31° on either side of the normal.

59. The path difference between the top and bottom of the slit for the incident wave is

$a \sin \theta_i$ .

The path difference between the top and bottom of the slit for the diffracted wave is

$a \sin \theta$.

When $\theta = \theta_i$ , the net path difference is zero, and there will be constructive interference. There is a central maximum at $\theta = 20°$. When the net path difference is a multiple of a wavelength, there will be minima given by

$(a \sin \theta_i) - (a \sin \theta) = m\lambda, \quad m = \pm 1, \pm 2, \dots ,$   or

$\sin \theta = \sin 20° - (m\lambda/a)$, where $m = \pm 1, \pm 2, \dots .$

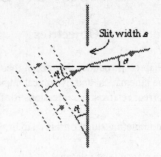

63. We find the angles for the first order from

$d \sin \theta = m\lambda = \lambda;$

$[1/(7600 \text{ lines/cm})](10^{-2} \text{ m/cm}) \sin \theta_\alpha = 656 \times 10^{-9}$ m, which gives $\sin \theta_\alpha = 0.499$, so $\theta_\alpha = 29.9°;$

$[1/(7600 \text{ lines/cm})](10^{-2} \text{ m/cm}) \sin \theta_\delta = 410 \times 10^{-9}$ m, which gives $\sin \theta_\delta = 0.312$, so $\theta_\delta = 18.2°.$

Thus the angular separation is

$\theta_\alpha - \theta_\delta = 29.9° - 18.2° = 11.7°.$

67. (a) If the initial intensity is $I_0$ , through the two sheets we have

$I_1 = \frac{1}{2}I_0 ;$

$I_2 = I_1 \cos^2 \theta = \frac{1}{2}I_0 \cos^2 90° = 0.$

(b) With the third polarizer inserted, we have

$I_1 = \frac{1}{2}I_0 ;$

$I_2 = I_1 \cos^2 \theta_1 = \frac{1}{2}I_0 \cos^2 60°;$

$I_3 = I_2 \cos^2 \theta_2 = \frac{1}{2}I_0 \cos^2 60° \cos^2 30° = 0.094I_0.$

(c) If the third polarizer is placed in front of the other two, we have the same situation as in (a), with $I_0$ being less. Thus no light gets transmitted.

71. We can write the electric field amplitude as

$\mathbf{E} = E_1\mathbf{i} + E_2\mathbf{j}$, so $E^2 = E_1^{\,2} + E_2^{\,2}$.

Because the intensity is proportional to $E^2$, we have

$I = I_1 + I_2,$

so the intensities add, with no interference.

# Chapter 37: Special Theory of Relativity

## Chapter Overview & Objectives

This chapter introduces the special theory of relativity. It discusses the principles and postulates of special relativity and the kinematical and dynamical implications of the postulates of special relativity. This chapter defines and discusses the correct relativistic expressions for familiar classical and dynamical quantities such as momentum and energy.

After completing study of this chapter, you should:

- Know the principle of relativity.
- Know the postulates of special relativity.
- Know how to calculate time dilation and length contraction.
- Know the Lorentz transformation equations.
- Know the Lorentz velocity transformation equations.
- Know the definition of relativistic momentum.
- Know the definition of relativistic energy.
- Know the definition of relativistic kinetic energy.
- Know what rest mass is.
- Know how to calculate the Doppler shift of light.

## Summary of Equations

Relativistic time dilation:

$$\Delta t = \frac{\Delta t_0}{\sqrt{1 - v^2/c^2}}$$

Relativistic length contraction:

$$L = L_0 \sqrt{1 - v^2/c^2}$$

Galilean transformation equations:

$$x' = x - vt$$
$$y' = y$$
$$z' = z$$
$$t' = t$$

Galilean velocity transformation equations:

$$u'_x = u_x - v$$
$$u'_y = u_y$$
$$u'_z = u_z$$

Lorentz transformation equations:

$$x' = \gamma(x - vt)$$
$$y' = y$$
$$z' = z$$
$$t' = \gamma\left(t - \frac{vx}{c}\right)$$

Lorentz velocity transformation equations:

$$u'_x = \frac{u_x - v}{1 - u_x v/c^2}$$
$$u'_y = \frac{u_y}{\gamma\left(1 - u_x v/c^2\right)}$$
$$u'_z = \frac{u_z}{\gamma\left(1 - u_x v/c^2\right)}$$

Definition of $\gamma$:

$$\gamma = \frac{1}{\sqrt{1 - v^2/c^2}}$$

Definition of (relativistic) momentum:

$$\mathbf{p} = \gamma\, m\mathbf{v} = \frac{m\mathbf{v}}{\sqrt{1 - v^2/c^2}}$$

Definition of relativistic mass:

$$m_{rel} = \gamma\, m = \frac{m}{\sqrt{1 - v^2/c^2}}$$

Definition of (relativistic) energy:

$$E = \gamma\, mc^2 = \frac{mc^2}{\sqrt{1 - v^2/c^2}}$$

Definition of (relativistic) kinetic energy:

$$K = E - mc^2 = \gamma\, mc^2 - mc^2 = (\gamma - 1)mc^2$$

Relationship between energy and momentum:

$$E^2 = p^2 c^2 + m^2 c^4$$

Relativistic Doppler shift:

$$f = f_0 \sqrt{\frac{c+v}{c-v}} \qquad \text{when source approaches observer at speed } v$$

$$f = f_0 \sqrt{\frac{c-v}{c+v}} \qquad \text{when source moves away from observer at speed } v$$

$$\lambda = \lambda_0 \sqrt{\frac{c-v}{c+v}} \qquad \text{when source approaches observer at speed } v$$

$$\lambda = \lambda_0 \sqrt{\frac{c+v}{c-v}} \qquad \text{when source moves away from observer at speed } v$$

## Chapter Summary

### Section 37-1 Galilean–Newtonian Relativity

The **relativity principle** is that the laws of physics are the same for all inertial observers. An inertial observer is an observer for which Newton's first law of motion holds true. In Galilean-Newtonian relativity, space and time are considered absolute. This means that the length of an object or the time between two events measured by one inertial observer will be the same as the length or time measured by any other inertial observer. In Galilean-Newtonian relativity, measurements of force, mass, and acceleration made by any two different inertial observers are also the same for both observers. This means that if Newton's second law applies to one inertial observer, it applies for any other inertial observer. A general statement of this principle is **all inertial reference frames are equivalent**.

The laws of electromagnetism predict the speed of light to be $c$, but with respect to what and as measured by which observer? We know that Galilean relativity implies that velocities measured by one inertial observer are different from those measured by a second inertial observer in motion relative to the first. That implies an observer in a special reference frame is the only one that measures the speed of light to be $c$ and other observers measure a different value. That is in violation of the relativity principle. The Michelson-Morley experiment attempted to measure the velocity of the Earth in its motion around the Sun relative to the special reference frame. They could not detect any difference in the speed of light in any direction.

### Section 37-2 The Michelson–Morley Experiment

The Michelson-Morley experiment used a Michelson interferometer to attempt to measure the difference in the speed of light in different directions. The Michelson interferometer creates interference between light beams that travel along paths in two perpendicular directions. The interference condition depends on the lengths of the two paths as well as the speed of light along the two paths. If the interference pattern is observed for a particular orientation of the paths, and then the orientation is changed, the interference condition should change if the speed of light along different paths is great enough. The Michelson-Morley experiment was easily sensitive enough to measure speed differences on the order of the speed of the Earth in its orbit about the Sun, but no changes in the interference condition were observed. The speed of light appears to be the same in all directions.

## Section 37-3 Postulates of the Special Theory of Relativity

Albert Einstein was able to reinstate the relativity principle by giving up the idea that space and time are absolutes. Instead of the absolute of space and time, he started with the principle that the speed of light measured by all inertial observers is the same and then derived the necessary properties of space and time from that principle. Einstein's **special theory of relativity** is based on two postulates:

*First postulate (the relativity principle)*: **The laws of physics have the same form in all inertial reference frames.**

*Second postulate (constancy of the speed of light)*: **Light propagates through empty space with a definite speed $c$ independent of the speed of the source or the observer.**

## Section 37-4 Simultaneity

The special theory of relativity has certain implications for the measurements of length and time. One of those implications is that simultaneity of events at different places is dependent on the observer. Two events that occur at different positions in space may be viewed as simultaneous by one inertial observer, but then the events will not necessarily be simultaneous to an inertial observer in motion relative to the first observer.

## Section 37-5 Time Dilation and the Twin Paradox

The time interval between two events measured by an observer in one inertial reference frame is, in general, different from the time interval between those same two events measured by an observer in a different inertial reference frame. There is a special inertial reference frame for measuring the time interval between two events (note that this special reference frame idea does not violate the principle of relativity because the events define the reference frame). The special inertial reference frame is the reference frame in which the two events occur at the same position. We call the time between the events in this special inertial reference frame the **proper time**, $\Delta t_0$. We might ask if it is always possible to find an inertial reference frame in which two events occur at the same place in space. The answer is it is not. In that case, it turns out that some inertial observers will not even agree to the order in which the events occur.

What does an inertial observer measure for the time between the two events if the inertial observer does not observe the two events at the same position? The observer measures a time $\Delta t$ between the two events that is related to the proper time, $\Delta t_0$, by

$$\Delta t = \frac{\Delta t_0}{\sqrt{1 - v^2/c^2}}$$

where $v$ is the speed of the observer's inertial reference frame relative to the inertial reference in which the two events appear at the same place. For $v$ less than $c$ in the above relationship, the denominator is real and less than one, so $\Delta t$ is greater than $\Delta t_0$. Because the time measured by the inertial observer is greater than the proper time, this is called **time dilation**.

Note that a clock always measures the proper time of its own inertial reference frame, because it is always located at the same place in its inertial reference frame. The time dilation effect then implies that if someone is watching a clock that is moving with respect to them, the clock appears to be running slower than it is supposed to be running. However, a person in the same inertial reference frame as the clock sees it running at its normal rate.

## Example 37-5-A

An approximate average human lifetime is 70 years. If an observer in a spaceship moving at a speed of $4.8 \times 10^7$ m/s relative to the Earth were watching the Earth, what would their measurement of an average human lifetime be?

**Solution:**

We use the time dilation relationship to solve for the time interval measured by the observers in the moving spaceship for the proper time of one average human lifetime:

$$\Delta t = \frac{\Delta t_0}{\sqrt{1-v^2/c^2}} = \frac{70\,\text{yr}}{\sqrt{1-\left(4.8\times10^7\,\text{m/s}\right)^2/\left(3.00\times10^8\,\text{m/s}\right)^2}} = 71\,\text{yr}$$

### Section 37-6  Length Contraction

The length of an object as measured by observers in one inertial reference frame is, in general, different from the length of the object measured by an observer in a different inertial reference frame. There is a special inertial reference frame for measuring the length of an object. The special inertial reference frame is the reference frame in which the object is at rest. We call the length of the object in this special inertial reference frame the **proper length**, $L_0$. What do inertial observers not at rest with respect to an object measure for the length of the object? They measure a length $L$ where $L$ is related to $L_0$ by

$$L = L_0\sqrt{1-v^2/c^2}$$

Again, for $v$ less than $c$, the square root is real and less than one. This implies that the length of an object measured by a person in motion relative to the object is shorter than the length measured by the observer at rest relative to the object. Because the length of the object measured by the observer in motion relative to the object is shorter, this is called **length contraction**.

### Example 37-6-A

You observe a spaceship moving past at a high speed and its length is 10.0 m. You know its length at rest is 45.0 m. At what speed is the spaceship passing by you?

**Solution:**

We use the length contraction relationship and solve for the speed $v$:

$$L = L_0\sqrt{1-v^2/c^2} \quad \Rightarrow \quad v = \sqrt{1-L^2/L_0^2}\;c = \sqrt{1-\left(10.0\,\text{m}\right)^2/\left(45.0\,\text{m}\right)^2}\;c = 0.975\,c$$

### Section 37-7 Four-Dimensional Space–Time

When an observer in one inertial reference frame measures positions and times and then relates those measurements of position and time to an observer in another inertial reference frame, the position measurements of the second observer are related to both the position and time measurements of the first observer. Also, the time measurements of the first observer are related to both the time and position measurements of the first observer. This means that space and time are linked together inseparably. Rather than treat space and time as separate quantities, it makes sense to treat them as different parts of the same mathematical structure. The mathematical space used to describe both space and time together is **four-dimensional space–time**. This structure has three spatial dimensions as does the familiar Galilean space, but an additional dimension for time.

### Section 37-8 Galilean and Lorentz Transformations

We can assign a position in space and a time to each event we observe. As we have learned, observers in different inertial reference frames will assign different positions and different times. There is a simple relationship called a transformation that relates the measurements of position and time of observers in one inertial reference frame to observers in another inertial reference frame. For the purposes of simplifying the relationship, we will choose the x-axis of both observers to be parallel to the direction of the relative velocity of the two observers. We will also place the coordinate systems so that at time equal to zero in each observer's reference frame, the origins of the coordinates are at the same place in space. Neither of these two conditions is necessary for choosing the two observers' coordinate systems, it just makes the form of the relationships between the two coordinate systems simpler.

472    Giancoli, *Physics for Scientists & Engineers*: Study Guide

**Galilean transformations** relate the observations of two inertial observers in a way that preserves Galilean relativity. The relationship of the measurements of position, $x, y, z$, and time $t$ of one observer are related to the measurements of position, $x', y', z'$, and time $t'$ of a second observer by

$$x' = x - vt$$
$$y' = y$$
$$z' = z$$
$$t' = t$$

where $v$ is the velocity of the second observer relative to the first. The Galilean transformation preserves time intervals and lengths of objects.

Galilean transformations can be used to determine how velocity measurements of two inertial observers are related to each other. The relationships are

$$u'_x = u_x - v$$
$$u'_y = u_y$$
$$u'_z = u_z$$

where $u_x, u_y$, and $u_z$ are the components of a velocity measured by one inertial observer and $u'_x, u'_y$, and $u'_z$ are the components measured by a second inertial observer in motion with respect to the first observer and $v$ is the velocity of the second observer with respect to the first. We have again chosen the direction of the relative velocity of the second observer with respect to the first along the common $x$-direction of the two observers. These relationships are called the Galilean velocity transformations. Notice that these tell us that the speed observed for light should be measured differently by different inertial observers.

If we start from Einstein's postulates of special relativity and try to find a linear transformation law that preserves the speed of light, we come up with a set of transformation equations called the **Lorentz transformation equations**. These transformation equations are

$$x' = \gamma(x - vt)$$
$$y' = y$$
$$z' = z$$
$$t' = \gamma\left(t - \frac{vx}{c}\right)$$

where

$$\gamma = \frac{1}{\sqrt{1 - v^2/c^2}},$$

the other symbols are the same as defined for the Galilean transformation, and $c$ is the speed of light.

The velocity transformation equations that result from the Lorentz transformation equations are

$$u'_x = \frac{u_x - v}{1 - u_x v/c^2}$$
$$u'_y = \frac{u_y}{\gamma\left(1 - u_x v/c^2\right)}$$
$$u'_z = \frac{u_z}{\gamma\left(1 - u_x v/c^2\right)}$$

where all of the symbols are the same as defined for the Galilean transformations.

**Example 37-8-A**

Our example will show that the order of events in time can be different for different observers when there is no inertial reference frame in which two events occur at the same place. Observer B moves relative to observer A with a velocity of $0.900c$ in the positive $x$ direction. Both events occur at $y = 0$ and $z = 0$ for both observers. Event number one occurs at a position $x = 200$ m at time $t_1 = 1.00$ μs according to observer A. Event number 2 occurs at a position x = –200 m at a time $t_2 = 2.00$ μs according to observer A. When and where do the two events occur for observer B?

**Solution:**

We will use unprimed symbols for observer A's measurements and primed symbols for observer B's measurements. First let's calculate $\gamma$, as we will use it several times:

$$\gamma = \frac{1}{\sqrt{1 - v^2/c^2}} = \frac{1}{\sqrt{1 - (0.900c)^2/c^2}} = 2.29$$

We determine observer B's position and time measurements from observers A's position and time measurements using the Lorentz transformation equations:

$$x_1' = \gamma(x_1 - vt_1) = 2.29[(200\,\text{m}) - (0.900\text{c})(1.00\,\mu\text{s})] = -160\,\text{m}$$

$$t_1' = \gamma\left(t_1 - \frac{vx_1}{c^2}\right) = 2.29\left[(1.00\,\mu\text{s}) - \frac{(0.900c)(200\,\text{m})}{c^2}\right] = 0.916\,\mu\text{s}$$

$$x_2' = \gamma(x_2 - vt_2) = 2.29[(600\,\text{m}) - (0.900\text{c})(2.00\,\mu\text{s})] = 137\,\text{m}$$

$$t_2' = \gamma\left(t_2 - \frac{vx_2}{c^2}\right) = 2.29\left[(2.00\,\mu\text{s}) - \frac{(0.900c)(600\,\text{m})}{c^2}\right] = 0.458\,\mu\text{s}$$

We see that the second of the two events according to observer A is the first of the two events according to observer B.

**Example 37-8-B**

You observe two spaceships passing by the Earth in the same direction. The first space ship passes the Earth with a speed of 0.800c. An observer on the first spaceship observers the second spaceship to be moving in the same direction the Earth is moving relative to the observer's spaceship with a speed of $0.700c$. What is the velocity of the second spaceship relative to the Earth?

**Solution:**

It is easy to confuse the $u$'s and the $v$'s in the Lorentz velocity transformation equations unless we are careful. Remember, we are always relating the velocity measurements of two observers in relative motion. In this case we are relating the two measurements of the velocity of the second spaceship made by the observer in the first spaceship, $u_x$, and the observer on the Earth, $u_x'$. It doesn't matter which one we call which, but we must be consistent with the relative velocity, $v$. The relative velocity $v$ is the velocity of the observer using the prime coordinates relative to the observer using the unprimed coordinates for the Lorentz transformation equation written in the form shown. If we choose the positive $x$ direction as the direction the observer on the first spaceship sees the Earth move, then $v = +0.800c$ and the second spaceship has a velocity relative to the first spaceship, $u_x = +0.700c$. We solve for the velocity of the second spaceship relative to the Earth using the Lorentz transformation equation:

$$u_x' = \frac{u_x - v}{1 - u_x v/c^2} = \frac{0.700\,c - 0.800\,c}{1 - (0.700c)(0.800c)/c^2} = -0.227c$$

## Section 37-9 Relativistic Momentum and Mass

Netwon's second law

$$F = \frac{d\mathbf{p}}{dt}$$

is valid for all inertial observers if we redefine **momentum** to be

$$\mathbf{p} = \gamma\, m\mathbf{v} = \frac{m\mathbf{v}}{\sqrt{1 - v^2/c^2}}$$

where $m$ is the mass of the particle as measured by an observer in the inertial frame of the mass and $\mathbf{v}$ is the velocity of the particle. In this case, although $\gamma$ has the same mathematical form as the other $\gamma$ used previously, it has a slightly different meaning because the $v$ is the velocity of the object, not the relative velocity of some other inertial reference frame.

Sometimes, the $\gamma$ is grouped together with the mass to define a quantity called **relativistic mass**:

$$m_{rel} = \gamma\, m = \frac{m}{\sqrt{1 - v^2/c^2}}$$

The quantity $m$ is called the **rest mass** of the object to make a clear distinction between the two defined masses.

## Section 37-10 The Ultimate Speed

No object can be accelerated from a speed below the speed of light to a speed greater than the speed of light. This makes the speed of light a universal speed limit.

## Section 37-11 Energy and Mass; $E = mc^2$

If we investigate the work-energy theorem for a system that obeys the Lorentz transformation equations, we find that the work-energy theorem is valid for all inertial observers if the quantity **energy** is defined to be

$$E = \gamma\, mc^2 = \frac{mc^2}{\sqrt{1 - v^2/c^2}}$$

For an object at rest, $\gamma = 1$, which implies

$$E = mc^2 = \frac{mc^2}{\sqrt{1 - v^2/c^2}}$$

for an object at rest. This amount of energy is called the **rest energy**. The object has energy even though it is at rest. This is consistent with the production of particles in high-energy physics experiments.

For a moving object, we call the excess energy above the rest energy the **relativistic kinetic energy**, $K$:

$$K = E - mc^2 = \gamma\, mc^2 - mc^2 = (\gamma - 1)mc^2$$

If we combine the definition of momentum and the definition of energy, we can derive a relationship between momentum and energy:

$$E^2 = p^2c^2 + m^2c^4$$

**Example 37-11-A**

A particle has a kinetic energy that is three times its rest energy. Determine the energy of the particle, the momentum of the particle, and the speed of the particle.

**Solution:**

We know that the kinetic energy is related to the rest mass by

$$K = (\gamma - 1)mc^2$$

From the problem statement, we then know that $\gamma - 1 = 3$ or $\gamma = 4$. Therefore, we know the energy $E$ is

$$E = \gamma mc^2 = 4mc^2$$

The momentum is related to the energy and the rest mass by

$$E^2 = p^2c^2 + m^2c^4 \quad \Rightarrow \quad p = \frac{1}{c}\sqrt{E^2 - m^2c^4} = \frac{1}{c}\sqrt{\gamma^2 m^2 c^4 - m^2 c^4}$$

$$p = \sqrt{(\gamma^2 - 1)}\, mc = \sqrt{(4^2 - 1)}\, mc = \sqrt{15}\, mc$$

We know $\gamma$, so from the definition of $\gamma$, we can solve for the speed, $v$:

$$\gamma = \frac{1}{\sqrt{1 - v^2/c^2}} \quad \Rightarrow \quad v = \sqrt{1 - \frac{1}{\gamma^2}}\, c = \sqrt{1 - \frac{1}{4^2}}\, c = \sqrt{\frac{15}{16}}\, c$$

**Section 37-12 Doppler Shift for Light**

The principle of the Doppler effect that was discussed in relationship to sound waves also applies to light waves. The mathematical relationship between the source frequency and the observed frequency is somewhat different than that previously derived because we must use the Lorentz transformations in relating the measurements of the source and observer because they are in different reference frames. The frequency of light measured by an observer, $f$, is related to the frequency of light emitted by the source, $f_0$, by

$$f = f_0\sqrt{\frac{c+v}{c-v}} \qquad \text{when source approaches observer at speed } v$$

$$f = f_0\sqrt{\frac{c-v}{c+v}} \qquad \text{when source moves away from observer at speed } v$$

and the wavelength of the light measured by the observer, $\lambda$, is related to the wavelength of light emitted by the source, $\lambda_0$, by

$$\lambda = \lambda_0\sqrt{\frac{c-v}{c+v}} \qquad \text{when source approaches observer at speed } v$$

$$\lambda = \lambda_0\sqrt{\frac{c+v}{c-v}} \qquad \text{when source moves away from observer at speed } v$$

**Example 37-12-A**

An FM radio station broadcasts with a carrier frequency of 101.5 MHz. What velocity relative to the transmitter would a receiver need to be moving so that this station would be received at a frequency of 101.7 MHz?

**Solution:**

We know that the receiver must be approaching the transmitter because the received frequency is higher than the transmitted frequency. We use the Doppler shift equation and solve for the speed of approach:

$$f = f_0 \sqrt{\frac{c+v}{c-v}} \quad \Rightarrow \quad c\frac{f^2 - f_0^2}{f^2 + f_0^2} = \left(3.00 \times 10^8 \text{ m/s}\right) \frac{(101.7 \text{ MHz})^2 - (101.5 \text{ MHz})^2}{(101.7 \text{ MHz})^2 + (101.5 \text{ MHz})^2}$$

$$f = 5.91 \times 10^5 \text{ m/s}$$

### Section 37-13 The Impact of Special Relativity

The results of the theory of special relativity are consistent with all experimental evidence gathered to test this theory. It is also consistent with Galilean relativity, which the universe seems to obey in our everyday life, at relative speeds small compared to the speed of light.

## Practice Quiz

1.  You are in a spaceship when a second spaceship passes by at a relative speed $v$. You observe a meter stick oriented along the direction of the relative velocity on the spaceship to be one-half meter in length as it passes by. What does an observer on the other spaceship observe for the length of a meter stick oriented along the direction of the relative velocity on your spaceship?

    a) one-half meter
    b) two meters
    c) one and a half meters
    d) need to know $v$ to answer that question

2.  Is the relativistic kinetic energy of a particle not at rest always less than, equal to, or greater than its Newtonian kinetic energy?

    a) less than
    b) equal to
    c) greater than
    d) sometime less than and sometimes greater than

3.  Two events for observer A occur simultaneously and at the same position in space. Observer B is in relative motion to observer A. What statement is true about what observer B sees for the location and time of the events observed by observer A?

    a) The events are not simultaneous for observer B.
    b) The events are not located in the same position for observer B.
    c) The events are simultaneous and located in the same position for observer B.
    d) More information is needed to answer this question.

4.  You are moving with a velocity **v** relative to another observer. What is the relativistically correct expression for the velocity of the other observer relative to you?

    a) $-\mathbf{v}$
    b) $-\sqrt{1 - v^2/c^2}\,\mathbf{v}$
    c) 0
    d) $-\gamma\mathbf{v}$

5.    You observe the lifetime of a particle at rest in the laboratory as $T$. If the particle is moving at high speed relative to the laboratory, what will be the observed lifetime of this particle?

a)  less than $T$
b)  greater than $T$
c)  equal to $T$
d)  The direction of the motion must be known to answer the question.

6.    Why are we still able to apply Newton's laws in their non-relativistic form and get results that agree with experiment even though we now know those relationships should be replaced by the relativistic expressions learned in this chapter?

a)  The relativistic expressions are very well approximated by the non-relativistic expressions for speeds much smaller than the speed of light.
b)  The relativistic expressions only apply when speeds are greater than $0.1c$.
c)  Special relativity applies to cases where light is involved and Galilean relativity applies to mechanics.
d)  You don't really think anyone really believes this science fiction stuff, do you?

7.    You are moving at a speed of $0.8c$ directly toward a star. What is the speed of the light from the star relative to you?

a)  $0.2c$
b)  $1.8c$
c)  $0.8c$
d)  $1.0c$

8.    When an object initially at rest is acted on by a constant force, what can you say about the Newtonian acceleration of the object as observed by an observer in an inertial reference frame?

a)  The acceleration remains constant.
b)  The acceleration increases with time.
c)  The acceleration decreases with time.
d)  The acceleration is zero.

9.    Is the relativistic kinetic energy less than, equal to, or greater than the square of the relativistic momentum divided by twice the rest mass of the object?

a)  less than
b)  equal to
c)  greater than
d)  sometime less than and sometimes greater than

10.    Two sources of light that emit identical frequencies are both moving relative to you at speed $v$. One is moving directly away from you and one is moving directly toward you. Which source will have its observed frequency at a larger difference from its source frequency?

a)  the source approaching you
b)  the source moving away from you
c)  Neither, the two differences will be the same magnitude.
d)  It depends on their relative distances to you.

11.    An Oxygen 15 nucleus decays at rest in an average time of 122 s. How fast would this nucleus need to be moving relative to us so that we observe the average time of this nucleus to decay to be 3.00 minutes?

12.    Two spaceships pass by the Earth in the same direction. To an observer on spaceship A, clocks on the Earth appear to be running at half the rate of his clock. To an observer on space ship B, clocks on the Earth appear to be running at one third the rate of her clock. What is the speed of each spaceship relative to the Earth and what is the relative velocity of spaceship B to spaceship A?

13.    A particle with rest mass $m$ has an energy of $6mc^2$. What are the kinetic energy and momentum of this particle?

14.    An observer moves with a velocity $0.782c$ in the positive $x$ direction relative to you. You see lightning strike the location $x = 232$ m, $y = 189$ m, $z = 0$ at a time $t = 3.56$ μs. At what location and time does the other observer see the lightening strike? Assume the observers are using coordinate systems with origins that coincided at time $t = 0$.

15.    The spectrum of a star that is moving along the line of sight to the star is observed. A hydrogen atom is known to emit light of a wavelength of 656 nm when it is stationary relative to the observer. The spectrum of the star shows the wavelength of this light to be 648 nm. What is the velocity of this star relative to the Earth?

## Problem Solutions

3.    We find the lifetime at rest from
$\Delta t = \Delta t_0/[1 - (v^2/c^2)]^{1/2}$;
$4.76 \times 10^{-6}$ s $= \Delta t_0/\{1 - [(2.70 \times 10^8$ m/s$)/(3.00 \times 10^8$ m/s$)]^2\}^{1/2}$, which gives $\Delta t_0 = 2.07 \times 10^{-6}$ s.

7.    For a 1.00 per cent change, the factor in the expressions for time dilation and length contraction must equal
$1 - 0.0100 = 0.9900$:
$[1 - (v/c)^2]^{1/2} = 0.9900$, which gives $v = 0.141c$.

11.    In the Earth frame, the average lifetime of the pion will be dilated:
$\Delta t = \Delta t_0/[1 - (v^2/c^2)]^{1/2}$.
The speed as a fraction of the speed of light is
$v/c = d/c\, \Delta t = d[1 - (v^2/c^2)]^{1/2}/c\, \Delta t_0$ ;
$v/c = (15$ m$)[1 - (v^2/c^2)]^{1/2}/(3.00 \times 10^8$ m/s$)(2.6 \times 10^{-8}$ s$)$,
which gives $v = 0.89c = 2.7 \times 10^8$ m/s.

15.    (*a*)  We choose the Earth for the S frame and spaceship 2 for the S′ frame, so $v = -0.50c$. The speed of spaceship 1 in the S frame is $0.50c$. We find the speed of spaceship 1 in S′ from the velocity transformation:
$u'$    $= (u - v)/[1 - (vu/c^2)]$
$= [0.50c - (-0.50c)]/[1 - (-0.50c)(0.50c)/c^2] = 0.80c.$
(*b*)  We could redefine our reference frames, but we know that the velocity of spaceship 2 relative to spaceship 1 must be $-0.80c$.

19.    We choose the Earth for the S frame and the spaceship for the S′ frame. The velocity components of the module in the S′ frame are $u_x' = 0$, $u_y' = 0.82c$. We find the velocity components of the module in the Earth frame from the velocity transformation:
$u_x = (u_x' + v)/[1 + (u_x'v/c^2)] = (0 + 0.66c)/(1 + 0) = 0.66c$;
$u_y = u_y'[1 - (v/c)^2]^{1/2}/[1 + (u_x'v/c^2)] = (0.82c)[1 - (0.66)^2]^{1/2}/(1 + 0) = 0.616c.$
The magnitude of the velocity is
$u = (u_x^2 + u_y^2)^{1/2} = [(0.66c)^2 + (0.616c)^2]^{1/2} = 0.90c = 2.7 \times 10^8$ m/s.
We find the angle the velocity makes with the $x$-axis from
$\tan \theta = u_y/u_x = (0.616c)/(0.66c) = 0.933$, so $\theta = 43°.$

23.    To an observer in the barn reference frame, if the boy runs fast enough, the measured contracted length of the pole will be less than 13.0 m, so the observer can say that the two ends of the pole were inside the barn simultaneously. We find the necessary speed for the contracted pole to fit inside the barn from
$L_{pole} = L_{0pole}[1 - (v/c)^2]^{1/2}$;
$10.0$ m $= (13.0$ m$)[1 - (v/c)^2]^{1/2}$, which gives $v = 0.64c.$
To the boy, the barn is moving and thus the length of the barn, as he would measure it, is less than the length of the pole:
$L_{barn} = L_{0barn}[1 - (v/c)^2]^{1/2} = (10.0$ m$)[1 - (0.64)^2]^{1/2} = 7.7$ m.
However, simultaneity is relative. Thus when the two ends are simultaneously inside the barn to the barn observer, those two events are not simultaneous to the boy. Thus he would claim that the observer in the barn determined that the ends of the pole were inside the barn at different times, which is also what the boy would say. It is not possible in the boy's frame to have both ends of the pole inside the barn simultaneously.

27. The two expressions for the momentum are
$p_{rel} = mv/[1 - (v^2/c^2)]^{1/2}$, and $p_c = mv$.
Thus the error is
$(p_{rel} - p_c)/p_{rel} = (\{mv/[1 - (v^2/c^2)]^{1/2}\} - mv)/\{mv/[1 - (v^2/c^2)]^{1/2}\} = 1 - [1 - (v^2/c^2)]^{1/2}$.
  (a) For the given speed we have
$(p_{rel} - p_c)/p_{rel} = 1 - [1 - (v^2/c^2)]^{1/2} = 1 - [1 - (0.10)^2]^{1/2} = 0.005 = 0.5\%$.
  (b) For the given speed we have
$(p_{rel} - p_c)/p_{rel} = 1 - [1 - (v^2/c^2)]^{1/2} = 1 - [1 - (0.50)^2]^{1/2} = 0.13 = 13\%$.

31. The rest energy of the electron is
$E = mc^2 = (9.109 \times 10^{-31} \text{ kg})(2.998 \times 10^8 \text{ m/s})^2 = 8.19 \times 10^{-14} \text{ J}$
$= (8.187 \times 10^{-14} \text{ J})/(1.602 \times 10^{-13} \text{ J/MeV}) = 0.511 \text{ MeV}$.

35. If the kinetic energy is equal to the rest energy, we have
$K = \{mc^2/[1 - (v^2/c^2)]^{1/2}\} - mc^2 = mc^2$, or
$1/[1 - (v^2/c^2)]^{1/2} = 2$, which gives $v = 0.866c$.

39. The total energy of the proton is
$E = K + mc^2 = 750 \text{ MeV} + 939 \text{ MeV} = 1689 \text{ MeV}$.
The relation between the momentum and energy is
$(pc)^2 = E^2 - (mc^2)^2$;
$p^2(3.00 \times 10^8 \text{ m/s})^2 = [(1689 \text{ MeV})^2 - (939 \text{ MeV})^2](1.60 \times 10^{-13} \text{ J/MeV})^2$,
which gives $p = 7.49 \times 10^{-19} \text{ kg} \cdot \text{m/s}$.

43. If $M$ is the mass of the new particle, for conservation of energy we have
$2(K + mc^2) = Mc^2$;
$2mc^2/[1 - (v^2/c^2)]^{1/2} = Mc^2$, which gives $M = 2m/[1 - (v^2/c^2)]^{1/2}$.
Because energy is conserved, there was no loss.
The final particle is at rest, so the kinetic energy loss is the initial kinetic energy of the two colliding particles:
$K_{loss} = 2K = (M - 2m)c^2 = 2mc^2(\{1/[1 - (v^2/c^2)]^{1/2}\} - 1)$.

47. The speed of the proton is
$v = (8.4 \times 10^7 \text{ m/s})/(3.00 \times 10^8 \text{ m/s}) = 0.280c$.
The kinetic energy is
$K = mc^2(\{1/[1 - (v/c)^2]^{1/2}\} - 1)$
$= (939 \text{ MeV})(\{1/[1 - (0.280)^2]^{1/2}\} - 1) = 39 \text{ MeV} (= 6.3 \times 10^{-12} \text{ J})$.
The momentum of the proton is
$p = mv/[1 - (v/c)^2]^{1/2}$
$= (1.67 \times 10^{-27} \text{ kg})(8.4 \times 10^7 \text{ m/s})\{1/[1 - (0.280)^2]^{1/2}\} = 1.46 \times 10^{-19} \text{ kg} \cdot \text{m/s} = 1.5 \times 10^{-19} \text{ kg} \cdot \text{m/s}$.
From the classical expressions, we get
$K_c = \frac{1}{2}mv^2 = \frac{1}{2}(1.67 \times 10^{-27} \text{ kg})(8.4 \times 10^7 \text{ m/s})^2 = 5.9 \times 10^{-12} \text{ J}$, with an error of
$(5.9 - 6.3)/(6.3) = -0.06 = -6\%$.
$p = mv = (1.67 \times 10^{-27} \text{ kg})(8.4 \times 10^7 \text{ m/s}) = 1.40 \times 10^{-19} \text{ kg} \cdot \text{m/s}$, with an error of
$(1.40 - 1.46)/(1.46) = -0.04 = -4\%$.

51. The total energy of the proton is
$E = m_{rel}c^2 = K + mc^2 = 900 \text{ GeV} + 0.938 \text{ GeV} = 901 \text{ GeV}$, so the relativistic mass is 901 GeV/$c^2$.
We find the speed from
$m_{rel} = m/[1 - (v^2/c^2)]^{1/2}$;
901 GeV/$c^2$ = (0.938 GeV/$c^2$)/$[1 - (v^2/c^2)]^{1/2}$, which gives $[1 - (v^2/c^2)]^{1/2} = 1.04 \times 10^{-3}$, so $v \approx 1.00c$.
The speed is constant so the relativistic mass is constant. The magnetic force provides the radial acceleration:
$qvB = m_{rel}v^2/r$, or
$B = m_{rel}v/qr = mv/qr[1 - (v^2/c^2)]^{1/2}$
$= (1.67 \times 10^{-27} \text{ kg})(3.00 \times 10^8 \text{ m/s})/(1.6 \times 10^{-19} \text{ C})(1.0 \times 10^3 \text{ m})(1.04 \times 10^{-3}) = 3.0 \text{ T}$.

55. (a) We let $m$ be the mass of the particle. Its velocity has an $x$-component only, $u_x$, so its momentum
components in frame S are
$p_x = mu_x/[1 - (u_x^2/c^2)]^{1/2}$, $p_y = 0$, $p_z = 0$.
The energy of the particle in frame S is
$E = K + mc^2 = mc^2/[1 - (u_x^2/c^2)]^{1/2}$.

The velocity in frame S' is

$u'_x = (u_x - v)/[1 - (u_x v/c^2)]$, $u'_y = 0$, $u'_z = 0$;

so the particle's momentum is

$p'_x = mu'_x/[1 - (u'^2_x/c^2)]^{1/2}$, $p'_y = 0 = p_y$, $p'_z = 0 = p_z$.

If we consider the denominator, we have

$$\sqrt{1 - \frac{u'^2_x}{c^2}} = \sqrt{1 - \frac{(u_x - v)^2}{\left[1 - u_x v/c^2\right]^2 c^2}} = \frac{1}{\left[1 - u_x v/c^2\right]} \sqrt{\left[1 - u_x v/c^2\right]^2 - \frac{\left(u_x^2 - 2u_x v + v^2\right)}{c^2}}$$

$$= \frac{1}{\left[1 - u_x v/c^2\right]} \sqrt{1 - 2u_x v/c^2 + \left(u_x v/c^2\right)^2 - u_x^2/c^2 + 2u_x v/c^2 - v^2/c^2}$$

$$= \frac{1}{\left[1 - u_x v/c^2\right]} \sqrt{\left[1 - u_x^2/c^2\right]\left[1 - v^2/c^2\right]}$$

When we use this and the velocity transformation in the expression for the momentum, we have

$$p'_x = \frac{mu'_x}{\sqrt{1 - \left(u'^2_x/c^2\right)}} = \frac{m(u_x - v)\left[1 - \left(u_x v/c^2\right)\right]}{\left[1 - \left(u_x v/c^2\right)\right]\sqrt{\left[1 - \left(u_x^2/c^2\right)\right]\left[1 - \left(v^2/c^2\right)\right]}} = \frac{1}{\sqrt{1 - \left(v^2/c^2\right)}} \frac{mu_x - mv}{\sqrt{1 - \left(u_x^2/c^2\right)}}$$

$$= \frac{p_x - \left(vE/c^2\right)}{\sqrt{1 - \left(v^2/c^2\right)}}$$

For the transformation of the energy we have

$$E' = \frac{mc^2}{\sqrt{1 - \left(u'^2_x/c^2\right)}} = \frac{mc^2\left[1 - \left(u_x v/c^2\right)\right]}{\sqrt{\left[1 - \left(u_x^2/c^2\right)\right]\left[1 - \left(v^2/c^2\right)\right]}} = \frac{1}{\sqrt{1 - \left(v^2/c^2\right)}} \frac{mc^2 - mu_x v}{\sqrt{1 - \left(u_x^2/c^2\right)}}$$

$$= \frac{E - p_x c}{\sqrt{1 - \left(v^2/c^2\right)}}$$

(b)  To simplify the expressions, we use $\gamma = 1/[1 - (v^2/c^2)]^{1/2}$. The transformations are

$x' = \gamma[x - (v/c)ct]$, $y' = y$, $z' = z$, $ct' = \gamma[ct - (vx/c)]$;

$p'_x = \gamma[p_x - (v/c)E/c]$, $p'_y = p_y$, $p'_z = p_z$, $E'/c = \gamma[(E/c) - (vp_x/c)]$.

Thus we see that $p_x$, $p_y$, $p_z$, $E/c$ transform in the same way as $x, y, z, ct$.

59.  For a source moving away from us the Doppler shift is

$\lambda = \lambda_0\{[1 + (v/c)]/[1 - (v/c)]\}^{1/2}$.

If $v \ll c$, we can use the approximation $1/(1 - x) \approx 1 + x$:

$\lambda \approx \lambda_0\{[1 + (v/c)]^2\}^{1/2} = \lambda_0[1 + (v/c)]$.

Thus the fractional change is

$\Delta\lambda/\lambda_0 = (\lambda - \lambda_0)/\lambda_0 = 1 + (v/c) - 1 = v/c$.

63.  The dependence of the relativistic mass on the speed is

$m_{rel} = m/[1 - (v^2/c^2)]^{1/2}$.

If we consider a box with sides $x_0$, $y_0$, and $z_0$, dimensions perpendicular to the motion, which we take to be the x-axis, do not change, but the length in the direction of motion will contract:

$x = x_0[1 - (v/c)^2]^{1/2}$.

Thus the density is

$\rho = m_{rel}/xyz = m/[1 - (v^2/c^2)]^{1/2} x_0[1 - (v^2/c^2)]^{1/2} y_0 z_0 = \rho_0/[1 - (v^2/c^2)]$.

67.  The minimum energy is required to produce the pair at rest:

$E_{min} = 2mc^2 = 2(0.511 \text{ MeV}) = 1.02 \text{ MeV} (1.64 \times 10^{-13} \text{ J})$.

71.  The kinetic energy comes from the decrease in mass:

$K = [m_n - (m_p + m_e + m_\nu)]c^2$

$= [1.008665 \text{ u} - (1.00728 \text{ u} + 0.000549 \text{ u} + 0)]c^2(931.5 \text{ MeV/u}c^2) = 0.78 \text{ MeV}$.

75. We convert the speed: $(110 \text{ km/h})/(3.6 \text{ ks/h}) = 30.6 \text{ m/s}$.
Because this is much smaller than $c$, the relativistic mass of the car is
$$m_{\text{rel}} = m/[1 - (v^2/c^2)]^{1/2} \approx m[1 + \tfrac{1}{2}(v/c)^2].$$
The fractional change in mass is
$$\begin{aligned}
(m_{\text{rel}} - m)/m &= [1 + \tfrac{1}{2}(v/c)^2] - 1 = \tfrac{1}{2}(v/c)^2 \\
&= \tfrac{1}{2}[(30.6 \text{ m/s})/(3.00 \times 10^8 \text{ m/s})]^2 = 5.19 \times 10^{-15} = 5.19 \times 10^{-13} \text{ \%}.
\end{aligned}$$

79. From the Lorentz transformation we have
$$x' = \gamma[x - (v/c)ct], \quad t' = \gamma[t - (vx/c^2)], \quad \text{or} \quad \Delta x' = \gamma[\Delta x - (v/c)c\,\Delta t], \quad c\Delta t' = \gamma[c\,\Delta t - (v\,\Delta x/c)].$$
Thus we have
$$\begin{aligned}
(c\,\Delta t')^2 - (\Delta x')^2 &= \gamma^2[(c\,\Delta t)^2 - 2v\,\Delta x\,\Delta t + v^2(\Delta x)^2/c^2] - \gamma^2[(\Delta x)^2 - 2v\,\Delta x\,\Delta t + v^2(c\,\Delta t)^2/c^2] \\
&= \gamma^2[(c\,\Delta t)^2(1 - v^2/c^2) - (\Delta x)^2(1 - v^2/c^2)] = \gamma^2(1 - v^2/c^2)[(c\,\Delta t)^2 - (\Delta x)^2]
\end{aligned}$$

# Chapter 38: Early Quantum Theory and Models of the Atom

## Chapter Overview and Objectives

This chapter describes the phenomena that could not be satisfactorily described by classical mechanics and how the introduction of what was then an arbitrary quantization condition gave theoretical results consistent with these phenomena. Some of these phenomena are blackbody radiation, the photoelectric effect, the Compton effect, and atomic spectra.

After completing study of this chapter, you should:

- Know what blackbody radiation is.
- Know Wien's law.
- Know what Planck's quantization condition is.
- Know what the photoelectric effect is and its properties.
- Know what Einstein's quantization condition on the electromagnetic field is.
- Know what Compton scattering is.
- Know what pair production is.
- Know what the de Broglie wavelength of a particle is.
- Know how to calculate the energy and wavelength of photons emitted from one-electron atoms.
- Know what the Bohr model of the atom is and what Bohr's quantization condition is.
- Know how to calculate orbit energies and radii for Bohr model orbits.

## Summary of Equations

Planck's quantum hypothesis:

$$E = nhf \qquad n \in \{0, 1, 2, 3, \ldots\}$$

Blackbody radiation intensity per unit wavelength:

$$I(\lambda, T) = \frac{2\pi hc^2 \lambda^{-5}}{e^{hc/\lambda kT} - 1}$$

Wien's law:

$$\lambda_P T = 2.90 \times 10^{-3} \text{ m} \cdot \text{K}$$

Photon energy:

$$E = hf$$

Maximum kinetic energy of photoelectron:

$$K_{\max} = hf - W_0$$

Compton shift:

$$\Delta\lambda = \lambda' - \lambda = \frac{h}{m_e c}(1 - \cos\theta)$$

Compton wavelength:

$$\lambda_C = \frac{h}{m_e c} = 2.43 \times 10^{-12} \text{ m}$$

de Broglie wavelength:

$$\lambda = \frac{h}{p} = \frac{h}{mv}$$

Wavelengths in the line spectrum of hydrogen:

$$\frac{1}{\lambda} = R\left(\frac{1}{n'^2} - \frac{1}{n^2}\right) \qquad n, n' \in \{1, 2, 3, \ldots\} \qquad n > n'$$

Bohr model quantization condition:

$$L = mv_n r_n = n\frac{h}{2\pi} \qquad n \in \{1, 2, 3, \ldots\}$$

Radii of Bohr model atomic orbits:
$$r_n = \frac{n^2 h^2 \varepsilon_0}{Z \pi m e^2} \approx \left(5.29 \times 10^{-11} \text{ m}\right) \frac{n^2}{Z}$$

Energy of Bohr model atomic orbits:
$$E_n = -\frac{m e^4}{8 \varepsilon_0^2 h^2} \frac{Z^2}{n^2} = \left(-13.6 \text{ eV}\right) \frac{Z^2}{n^2}$$

# Chapter Summary

## Section 38-1 Planck's Quantum Hypothesis

All bodies at non-zero absolute temperature radiate electromagnetic energy characteristic of the temperature. For an idealized object called a **blackbody**, the intensity per wavelength interval emitted by the body only depends on the absolute temperature of the body and the wavelength of the radiation. This radiation is called **blackbody radiation**. Classical electromagnetism and classical thermodynamics predict that the intensity of electromagnetic radiation should be more intense as the wavelength gets shorter for any blackbody at a non-zero temperature, with the intensity per unit wavelength going to infinity as the wavelength goes to zero. Experimental measurements show that the intensity per unit wavelength goes to zero as the wavelength goes to zero. To resolve this failure of electromagnetic and thermodynamics theory, Planck proposed that the energy of vibration, $E$, of atoms in the blackbody that create the blackbody radiation be restricted to vibrating with amplitudes that were multiples of a constant, $h$, times their frequency of vibration, $f$:

$$E = nhf \qquad n \in \{0, 1, 2, 3, \dots\}$$

The constant $h$ is now called Planck's constant and it has the value $6.626 \times 10^{-34}$ J·s.

The blackbody emission intensity per unit wavelength interval is given by the expression

$$I(\lambda, T) = \frac{2\pi h c^2 \lambda^{-5}}{e^{hc/\lambda kT} - 1}$$

where $\lambda$ is the wavelength of the light emitted, $c$ is the speed of light, $k$ is Boltzman's constant, and $T$ is the absolute temperature. This curve has a maximum at a wavelength $\lambda_p$, given by **Wien's law**:

$$\lambda_P T = 2.90 \times 10^{-3} \text{ m} \cdot \text{K}$$

where $T$ is the absolute temperature of the surface.

## Example 38-1-A

Metal workers often temper or anneal metals either by heating them to a given temperature and then either quenching them in a liquid coolant or allowing them to cool slowly. They often judge the temperature by looking at the color of light emitted by the hot metal. What is the peak wavelength emitted by a blackbody at a temperature of 900° C?

## Solution:

We use Wien's law. The absolute temperature of the surface is

$$T_K = T_C + 273.15 = 900 + 273 = 1.17 \times 10^3 \text{ K}$$

Using this temperature in Wien's law, we get

$$\lambda_P = \frac{2.90 \times 10^{-3} \text{ m} \cdot \text{K}}{1.17 \times 10^3 \text{ K}} = 2.48 \times 10^{-6} \text{ m}$$

Although the peak wavelength of emission is in the infrared part of the electromagnetic spectrum, the blackbody spectrum extends far enough into the visible that a reddish colored glow will be seen from an object at this temperature.

## Section 38-2 Photon Theory of Light and the Photoelectric Effect

Einstein proposed that not only are the energies of vibrations of the atoms in the walls of a blackbody radiator quantized, but so are the energies of the electromagnetic field. To conserve energy when a blackbody radiator radiates an energy $hf$ by changing from one energy level to the next lower energy level, the electromagnetic radiation that carries away the energy must carry energy $hf$ also. The electromagnetic field can only gain or lose energy in packets or quanta of energy, $E$, such that

$$E = hf$$

The electromagnetic field takes on a particle property, the property that particles come in countable units. This apparent electromagnetic particle is called a **photon**.

This quantization condition proposed by Einstein is consistent with the **photoelectric effect**. The photoelectric effect is the emission of electrons from a surface when the surface is illuminated with light. Classical eletromagnetism does predict that electrons in a material can have work done on them by the electromagnetic field to gain enough energy to escape from a material, but the details of the behavior of the photoelectric effect are inconsistent with classical electromagnetism. The observed properties of the photoelectric effect were that the light has a maximum wavelength, which will cause the electrons to be ejected by the material. This is inconsistent with classical electromagnetism, which predicts any wavelength should be able to eject the electrons if it has great enough intensity. A second property of the photoelectric effect is that the ejected electrons leave the material with a maximum kinetic energy for a given wavelength. This maximum is the same, regardless of the intensity of the light. In classical electromagnetic theory, the electron should be able to gain more kinetic energy from a more intense electric field. The maximum kinetic energy increases with decreasing wavelength of the light.

The quantization of the light into photons results in a behavior that is consistent with the observed results. If the wavelength of light is longer than the value for which $hf$ is less than the energy to remove the electron from the material, no electrons are emitted. If the intensity is increased, the number of photons per second that reach the material is larger, but a given electron still can only receive the same amount of energy, $hf$, so the maximum kinetic energy of the emitted electrons does not change. The maximum kinetic energy, $K_{max}$, is given by

$$K_{max} = hf - W_0$$

where $h$ is Planck's constant, $f$ is the frequency of the light illuminating the material, and $W_0$ is called the work function of the material. The work function of the material is the minimum required energy to remove an electron from the material.

## Example 38-2-A

Photons of wavelength 420 nm are incident on a material. The photoemitted electrons have a maximum kinetic energy of 2.2 eV. Determine the work function of the material.

### Solution:

The relationship between the maximum kinetic energy and the work function is

$$K_{max} = hf - W_0 = \frac{hc}{\lambda} - W_0$$

If we solve this for the work function, we get

$$W_0 = \frac{hc}{\lambda} - K_{max} = \frac{\left(6.63 \times 10^{-34}\ \text{J} \cdot \text{s}\right)\left(1\,\text{eV}/1.602 \times 10^{-19}\ \text{J}\right)\left(3.00 \times 10^8\ \text{m/s}\right)}{420 \times 10^{-9}\ \text{m}} - 2.2\,\text{eV} = 0.8\,\text{eV}$$

## Section 38-3 Photons and the Compton Effect

The **Compton effect** is the scattering of photons by a free or weakly bound electron. (By weakly bound, we mean the energy to free the electron from the material is very small compared to the energy of the photon.)  The scattered photon is found to have a different wavelength from the incident photon, and hence, a different frequency.  Classical physics is unable to describe a system in which the frequency of the outgoing electromagnetic field is different from the frequency of the incoming electromagnetic field.  However, the quantization of photons, together with conservation of energy and momentum, predicts the very shift in wavelength and frequency observed.  The predicted wavelength of a Compton scattered photon, $\lambda'$, is related to the incident wavelength, $\lambda$, by

$$\lambda' = \lambda + \frac{h}{m_e c}(1 - \cos\theta)$$

where $m_e$ is the rest mass of the electron, $c$ is the speed of light, and $\theta$ is the angle by which the photon is scattered.  The difference between the outgoing and incoming wavelengths, $\Delta\lambda$, is called the Compton shift:

$$\Delta\lambda = \lambda' - \lambda = \frac{h}{m_e c}(1 - \cos\theta)$$

The coefficient of the $1 - \cos\theta$ term is called the Compton wavelength, $\lambda_C$, of the electron:

$$\lambda_C = \frac{h}{m_e c} = 2.43 \times 10^{-12} \text{ m}$$

## Example 38-3-A

Compare the wavelength of a 788 keV photon Compton scattered from a free electron and the same photon Compton scattered from a free proton when they are scattered at an angle of 90°.

## Solution:

The Compton shift for the electron scattered photon is

$$\lambda' = \lambda + \frac{h}{m_e c}(1 - \cos\theta) = (1.57 \times 10^{-12} \text{ m}) + \frac{6.63 \times 10^{-34} \text{ J} \cdot \text{s}}{(9.11 \times 10^{-31} \text{ kg})(3.00 \times 10^8 \text{ m/s})}(1 - \cos 90°)$$

$$= 4.00 \times 10^{-12} \text{ m}$$

The Compton shift for the proton scattered photon is

$$\lambda' = \lambda + \frac{h}{m_p c}(1 - \cos\theta) = (1.57 \times 10^{-12} \text{ m}) + \frac{6.63 \times 10^{-34} \text{ J} \cdot \text{s}}{(1.67 \times 10^{-27} \text{ kg})(3.00 \times 10^8 \text{ m/s})}(1 - \cos 90°)$$

$$= 1.57 \times 10^{-12} \text{ m}$$

The Compton shift of the photon scattered from the proton is very small compared to the wavelength of the photon.

## Section 38-4 Photon Interactions; Pair Production

The energy of the electromagnetic field, a photon, can create a pair of particles in a process called **pair production**.  The easiest pair of particles to create are an electron and an anti-electron (also called a positron).  The particles must be produced in pairs to conserve electric charge.  To conserve energy, the minimum energy of the photon must be at least equal to the total energy of the rest mass of the particles created.  In this case, that would be $2m_e c^2 = 1.02$ MeV.  To also conserve momentum, there must be another particle present to take some of the momentum of the incoming photon.  The more massive this extra particle is, the less energy it takes to give it the required momentum.

**Example 38-4-A**

A photon passes near a nucleus of mass $4.48 \times 10^{-26}$ kg. It produces an electron-positron pair. If the electron and positron have negligible kinetic energy, what was the amount of energy given to the nucleus when the photon produced this pair?

**Solution:**

Because the electron and positron have negligible kinetic energy, they will also have negligible momentum after being produced. That implies that all the momentum of the photon must be in the nucleus. The nucleus will then have kinetic energy, which also must have come from the photon to conserve energy in the reaction. First, we set the momentum of the photon equal to the momentum of the nucleus after the pair production:

$$p_{photon} = p_{nucleus} \quad \Rightarrow \quad \frac{E_{photon}}{c} = \frac{1}{c}\sqrt{E_{nucleus}^2 - m_{nucleus}^2 c^4}$$

Next we write down the conservation of energy:

$$E_{photon} = 2m_e c^2 + K_{nucleus} = 2m_e c^2 + E_{nucleus} - m_{nucleus}c^2$$

Solving these two equations simultaneously for the energy of the photon, we get

$$E_{photon} = 2m_e c^2 \left( \frac{m_{nucleus} - m_e}{m_{nucleus} - 2m_e} \right)$$

and the kinetic energy of the nucleus is

$$K_{nucleus} = E_{photon} - 2m_e c^2 = 2m_e c^2 \left[ \left( \frac{m_{nucleus} - m_e}{m_{nucleus} - 2m_e} \right) - 1 \right] = 2m_e c^2 \left( \frac{m_e}{m_{nucleus} - 2m_e} \right)$$

Because of the relatively small mass of the electron compared to the nucleus, this will be a small energy compared to the rest mass energy of the electrons:

$$K_{nucleus} = 2\left(0.511 \times 10^6 \text{ eV}\right) \frac{\left(9.11 \times 10^{-31} \text{ kg}\right)}{\left(4.48 \times 10^{-26} \text{ kg}\right) - 2\left(9.11 \times 10^{-31} \text{ kg}\right)} = 20.8 \text{ eV}$$

This is very small compared to the rest mass energy of the two particles produced.

**Section 38-5 Wave-Particle Duality; The Principle of Complementarity**

The **wave-particle duality** of light is the property of light that sometimes it exhibits what is classical wave behavior and sometimes it exhibits what is classically a particle like behavior. The principle of complementarity states that based on the type of experiment that is performed, it can be understood either in terms of the wave nature of light or the particle nature of light, but both descriptions cannot be used simultaneously.

**Section 38-6 Wave Nature of Matter**

The de Broglie wavelength, $\lambda$, of a particle is

$$\lambda = \frac{h}{p} = \frac{h}{mv}$$

Giving a wave property to matter is analogous to giving a particle property of electromagnetic waves and extends the idea of wave-particle duality to what are classically considered particles.

**Example 38-6-A**

Determine the wavelength of a 100 kg person walking with a speed of 1.00 m/s. Determine the location of the first single-slit minimum away from the center peak of the diffraction pattern when a wave of this wavelength passes through a single-slit that has a width of 1.00 m (about the width of a doorway) after traveling 1.00 km.

**Solution:**

Using the de Broglie relationship, the wavelength of this person would be

$$\lambda = \frac{h}{mv} = \frac{6.63 \times 10^{-34} \text{ J} \cdot \text{s}}{(100 \text{ kg})(1.00 \text{ m/s})} = 6.63 \times 10^{-36} \text{ m}$$

The angular position of the first diffraction peak minimum is given by

$$\sin \theta = \frac{m\lambda}{a} = \frac{(1)(6.63 \times 10^{-36} \text{ m})}{1.00 \text{ m}} = 6.63 \times 10^{-36} \text{ rad}$$

This is a small angle, so we can approximate the tangent of the angle by the sine of the angle. The distance, $y$, that the first minimum is from the central peak divided by the distance, $L$, from the slits is the tangent of the angle. This implies

$$y = L \tan \theta \approx L \sin \theta = (1.00 \times 10^3 \text{ m})(6.63 \times 10^{-36}) = 6.63 \times 10^{-33} \text{ m}$$

This distance is immeasurably small and we see that it is impossible to use this as a test of the de Broglie wavelength relationship.

## Section 38-7 Electron Microscopes

A light microscope forms an image of light that has scattered from the object being viewed. There is no reason a microscope could not be constructed based on detecting electrons scattered from an object also. Because the wavelength of electrons can easily be made much smaller than the wavelength of light, a microscope based on electrons can have a much greater resolution than an optical microscope. Electron microscopes have been constructed and used that have better than atomic-sized resolution.

## Section 38-8 Early Models of the Atom

Although the concept of atoms of matter goes back to ancient Greece, the first model of the atom that is similar to the currently used model was put forward by Ernest Rutherford. Rutherford created his nuclear model of the atom to explain the results of scattering of alpha particles sent toward a thin gold foil.

## Section 38-9 Atomic Spectra: Key to the Structure of the Atom

The emission spectrum from hydrogen gas is a **line spectrum**. The light emitted does not have a continuum of wavelengths, but only a discrete set of wavelengths appears in the spectrum. Those wavelengths are given by the relationship

$$\frac{1}{\lambda} = R\left(\frac{1}{n'^2} - \frac{1}{n^2}\right) \qquad n, n' \in \{1, 2, 3, \dots\} \qquad n > n'$$

where $R$ is called the **Rydberg constant** and has a value of $1.097 \times 10^7 \text{ m}^{-1}$.

The lines can be classified in groups according to the value of $n'$. If $n' = 1$, then the group of emission lines is called the **Lyman series**. If $n' = 2$, then the group of emission lines is called the **Balmer series**. If $n' = 3$, then the group of emission lines is called the **Paschen series**.

## Example 38-9-A

Determine the five longest wavelengths in the series of emission lines from hydrogen that has $n' = 4$.

## Solution:

The longest wavelengths in a series correspond to the smallest values of $n$. The values of $n$ must be larger than the values of $n'$, so we want the wavelength for values of $n$ equal to 5, 6, 7, 8, and 9. Using the relationship above

$$\text{For } n = 5 \qquad \lambda = \left[ R\left( \frac{1}{n'^2} - \frac{1}{n^2} \right) \right]^{-1} = \left[ \left( 1.097 \times 10^7 \text{ m}^{-1} \right)\left( \frac{1}{4^2} - \frac{1}{5^2} \right) \right]^{-1} = 4.051 \times 10^{-6} \text{ m}$$

$$\text{For } n = 6 \qquad \lambda = \left[ R\left( \frac{1}{n'^2} - \frac{1}{n^2} \right) \right]^{-1} = \left[ \left( 1.097 \times 10^7 \text{ m}^{-1} \right)\left( \frac{1}{4^2} - \frac{1}{6^2} \right) \right]^{-1} = 2.625 \times 10^{-6} \text{ m}$$

$$\text{For } n = 7 \qquad \lambda = \left[ R\left( \frac{1}{n'^2} - \frac{1}{n^2} \right) \right]^{-1} = \left[ \left( 1.097 \times 10^7 \text{ m}^{-1} \right)\left( \frac{1}{4^2} - \frac{1}{7^2} \right) \right]^{-1} = 2.166 \times 10^{-6} \text{ m}$$

$$\text{For } n = 8 \qquad \lambda = \left[ R\left( \frac{1}{n'^2} - \frac{1}{n^2} \right) \right]^{-1} = \left[ \left( 1.097 \times 10^7 \text{ m}^{-1} \right)\left( \frac{1}{4^2} - \frac{1}{8^2} \right) \right]^{-1} = 1.945 \times 10^{-6} \text{ m}$$

$$\text{For } n = 9 \qquad \lambda = \left[ R\left( \frac{1}{n'^2} - \frac{1}{n^2} \right) \right]^{-1} = \left[ \left( 1.097 \times 10^7 \text{ m}^{-1} \right)\left( \frac{1}{4^2} - \frac{1}{9^2} \right) \right]^{-1} = 1.818 \times 10^{-6} \text{ m}$$

All of these wavelengths are in the infrared portion of the electromagnetic spectrum.

## Section 38-10 The Bohr Model

Bohr applied the quantization condition

$$L = mv_n r_n = n\frac{h}{2\pi} \qquad n \in \{1, 2, 3, \dots\}$$

to circular orbits of electrons in atoms. The value of $n$ is called the quantum number of the orbit. Imposing this condition on the orbits results in only certain allowed values of the radius of the orbits, $r_n$, corresponding allowed values of the speed of the electron in its orbit, $v_n$, and corresponding allowed values of the energy of the electron in its orbit, $E_n$. These values are given by

$$r_n = \frac{n^2 h^2 \varepsilon_0}{Z\pi m e^2} \approx \left( 5.29 \times 10^{-11} \text{ m} \right)\frac{n^2}{Z}$$

$$v_n = \frac{Ze^2}{2n\varepsilon_0 h} = \left( 2.19 \times 10^6 \text{ m/s} \right)\frac{Z}{n}$$

$$E_n = -\frac{me^4}{8\varepsilon_0^2 h^2}\frac{Z^2}{n^2} = \left( -13.6 \text{ eV} \right)\frac{Z^2}{n^2}$$

The difference in the energies of states with quantum numbers $n'$ and $n$ is

$$\Delta E = E_{n'} - E_n = \left( -13.6 \text{ eV} \right)\left( \frac{1}{n'^2} - \frac{1}{n^2} \right)Z^2$$

If we assume the energy lost by the electron in the atom making a transition from a higher quantum number state, $n$, to a lower quantum number state, $n'$, goes into creating a photon, the energy of the photon will be

$$E = (13.6\,\text{eV})\left(\frac{1}{n'^2} - \frac{1}{n^2}\right)Z^2$$

and the wavelength of the photon will be

$$\frac{1}{\lambda} = \frac{E}{hc} = \left(1.097 \times 10^7 \text{ m}^{-1}\right)\left(\frac{1}{n'^2} - \frac{1}{n^2}\right)Z^2$$

The quantity $1.097 \times 10^7$ m$^{-1}$ is called **Rydberg's constant** and the energy $-13.6$ eV is unit of energy called a **Rydberg**. The distance $5.29 \times 10^{-11}$ m is called the **Bohr radius**.

The spectrum of wavelengths that results from the relationship above is in excellent agreement with experimental measurements of the emission spectrum of hydrogen atoms.

**Example 38-10-A**

Determine the wavelength of light emitted for a transition between the $n = 2$ and $n = 1$ orbits of a single electron in orbit around an iron nucleus.

**Solution:**

Iron has an atomic number 26. The energy of the photon released in the given transition will be

$$E = (13.6\,\text{eV})\left(\frac{1}{n'^2} - \frac{1}{n^2}\right)Z^2 = (13.6\,\text{eV})\left(\frac{1}{1^2} - \frac{1}{2^2}\right)26^2 = 6.90 \times 10^3 \text{ eV}$$

We can calculate the wavelength of this photon:

$$\lambda = \frac{hc}{E} = \frac{1.240 \times 10^3 \text{ eV} \cdot \text{nm}}{6.90 \times 10^3 \text{ eV}} = 0.180\,\text{nm}$$

**Section 38-11 de Broglie's Hypothesis Applied to Atoms**

de Broglie's hypothesis that the wavelength, $\lambda$, of a particle is given by

$$\lambda = \frac{h}{p}$$

where $h$ is Planck's constant and $p$ is the momentum of the particle, can be applied to the electron in a one-electron atom, the results of the electron's allowed energies, orbital radii, and speeds is identical to Bohr's calculations if it is assumed a standing wave is created by fitting a whole number of wavelengths in the orbit's circumference. This reinforces the notion that some type of wave theory of matter may be the underlying physics of matter in atomic sized systems.

## Practice Quiz

1.    If the absolute temperature of a blackbody doubles, what happens to the wavelength of the peak of the intensity per unit wavelength spectrum of the blackbody radiation?

a) It doubles in wavelength.
b) It decreases to one-half the initial wavlength.
c) It stays the same.
d) It increases by a factor of $e^2$.

2.    A blackbody glows with an emission spectrum that appears red to the human visual system. The temperature is changed so that the glow now appears yellow to the human visual system. How has the temperature changed?

a) The temperature has increased.
b) The temperature has decreased.
c) The temperature has not changed.
d) You need to know the change in area of the surface to know how the temperature changed.

3.    A metal is being illuminated with light of a fixed wavelength and an intensity $I$. What change takes place in the electron emission if the intensity of the light is doubled to $2I$?

a) The maximum kinetic energy of the electrons doubles.
b) The sum of the maximum kinetic energy of the electrons and the work function of the material doubles.
c) The number of electrons emitted per second by the surface doubles.
d) No change occurs in the electron emission.

4.    Why can't two electrons and a positron be produced by a single photon?

a) Electric charge is a conserved quantity.
b) Energy is a conserved quantity.
c) Momentum is a conserved quantity.
d) There are only two polarizations of a photon, not three.

5.    When a photon Compton scatters off of a free electron, what happens to the energy lost by the photon?

a) The energy lost by the photon is gained by the electron.
b) The photon does not lose energy, it only changes wavelength.
c) An additional photon is emitted with an energy equal to the energy lost by the original photon.
d) Energy is not conserved during Compton scattering, so the energy is lost.

6.    If the de Broglie wavelength of a particle is given by $h/p$, how does the wavelength of a particle with speed $v$ calculated with the relativistic momentum expression compare to the wavelength calculated with the Newtonian momentum expression, $mv$?

a) The wavelength calculated with the relativistic expression for the momentum is always greater than the wavelength calculated with the Newtonian expression for the wavelength.
b) The wavelength calculated with the relativistic expression for the momentum is always shorter than the wavelength calculated with the Newtonian expression for the wavelength.
c) The two wavelengths are identical.
d) Sometimes one wavelength is longer, sometimes the other wavelength is longer.

7.    When an electron in a hydrogen atom goes from a lower $n$ value orbit to a higher $n$ value orbit, what happens to the electron's kinetic energy?

a) The kinetic energy increases.
b) The kinetic energy decreases.
c) The kinetic energy remains the same.
d) The kinetic energy might increase or decrease.

8.    What result of classical physics does the Bohr atom conflict with?

a) Angular momentum is a conserved quantity.
b) A centripetal force is necessary for an object to move on a circular path.
c) Newton's second law.
d) Accelerating charges radiate electromagnetic radiation.

9.    The reason an electron microscope can detect much finer detail than an optical microscope is

    a) Electrons are point masses, but light is a wave.
    b) The wavelength of electrons in the microscope is smaller than the wavelength of visible light.
    c) Electron lenses are better in quality than optical lenses.
    d) The electron detector has better resolution than the human eye.

10.   The $n = 1$ orbit of an electron in orbit around a hydrogen nucleus has a radius equal to one Bohr radius, $r_0$. What is the radius of the orbit of a single electron in the $n = 1$ orbit around a silicon nucleus?

    a) $(14)r_0$
    b) $(1/14)r_0$
    c) $(296)r_0$
    d) $(1/296)r_0$

11.   What is the Celsius temperature of a blackbody if its peak intensity per unit wavelength is at a photon energy of 2.5 eV?

12.   Determine the de Broglie wavelength of a 0.486 kg ball that is thrown with a speed of 40 mph.

13.   What is the scattering angle of a 662 keV photon that has an energy of 584 keV after being Compton scattered from a free electron?

14.   A material has a work function of 1.1 eV. What is the longest wavelength of light that will cause electrons to be emitted from the surface of this material?

15.   Determine the longest three wavelengths of light emitted in the Lyman series for a single electron in orbit around a lithium nucleus.

## Problem Solutions

3.    Because the energy is quantized, $E = nhf$, the difference in energy between adjacent levels is
$$\Delta E = hf = (6.63 \times 10^{-34} \text{ J} \cdot \text{s})(8.1 \times 10^{13} \text{ Hz}) = 5.4 \times 10^{-20} \text{ J} = 0.34 \text{ eV}.$$

7.    (a) To find the wavelength when $I(\lambda, T)$ is maximal at constant temperature, we set $dI/d\lambda = 0$:

$$\frac{dI}{d\lambda} = \frac{d}{d\lambda}\left(\frac{2\pi hc^2 \lambda^{-3}}{e^{hc/\lambda kT} - 1}\right)$$

$$= 2\pi hc^2 \left[\frac{-5\lambda^{-6}}{e^{hc/\lambda kT} - 1} - \frac{\lambda^{-5}\left(-hc/\lambda^2 kT\right)e^{hc/\lambda kT}}{\left(e^{hc/\lambda kT} - 1\right)^2}\right]$$

$$= \frac{2\pi hc^2}{\lambda^6}\left[\frac{\left(5 - hc/\lambda^2 kT\right)e^{hc/\lambda kT} - 5}{\left(e^{hc/\lambda kT} - 1\right)^2}\right] = 0 \quad \Rightarrow$$

$$\left[5 - \left(hc/\lambda_p kT\right)\right]e^{hc/\lambda_p kT} - 5 = 0 \quad \text{or} \quad 5 - \left(hc/\lambda_p kT\right) = 5e^{-hc/\lambda_p kT}$$

This equation will have a solution $\lambda_P T$ = constant, which is the Wien displacement law.

(b) To find the value of the constant, we let $x = hc/\lambda_P kT$,
so the transcendental equation is
    $5 - x = 5e^{-x}$.
One way to solve this equation is to plot each side against $x$.
We see from the plot that the solution is very close to $x = 5$.
If we let $\Delta = 5 - x$, we get
    $\Delta = 5e^{(\Delta - 5)} \approx 5e^{-5} = 0.034$, so $x = 4.966$.
Thus we have
    $\lambda_P T = hc/xk$ ;
    $2.90 \times 10^{-3} \text{ m} \cdot \text{K} = h(3.00 \times 10^8 \text{ m/s})/(4.966)(1.38 \times 10^{-23} \text{ J/K})$,
which gives $h = 6.63 \times 10^{-34} \text{ J} \cdot \text{s}$.

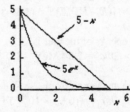

(c) For the rate at which energy is radiated per unit surface area for all wavelengths we have

$$\int_0^\infty I(\lambda, T)\,d\lambda = \int_0^\infty \frac{2\pi hc^2 \lambda^{-5}}{e^{hc/\lambda kT} - 1}\,d\lambda$$

If we change variable to $x = hc/\lambda kT$, then $d\lambda = -(hc/kTx^2)\,dx$, so we have

$$\int_0^\infty I(\lambda, T)\,d\lambda = \frac{2\pi}{h^3 c^2}(kT)^4 \int_0^\infty \frac{x^3}{e^x - 1}\,dx = \frac{2\pi k^4}{h^3 c^2}\left(\int_0^\infty \frac{x^3}{e^x - 1}\,dx\right)T^4 = \text{constant} \times T^4$$

11. The photon energy must be at least 0.1 eV. We find the minimum frequency from
   $E_{min} = hf_{min}$ ;
   $(0.1 \text{ eV})(1.60 \times 10^{-19} \text{ J/eV}) = (6.63 \times 10^{-34} \text{ J} \cdot \text{s})f_{min}$ , which gives $f_{min} = 2.4 \times 10^{13}$ Hz.
   The maximum wavelength is
   $\lambda_{max} = c/f_{min} = (3.00 \times 10^8 \text{ m/s})/(2.4 \times 10^{13} \text{ Hz}) = 1.2 \times 10^{-5}$ m.

15. (a) At the threshold wavelength, the kinetic energy of the photoelectrons is zero, so we have
    $K_{max} = hf - W_0 = 0$;
    $W_0 = hc/\lambda_{max} = (1.24 \times 10^3 \text{ eV} \cdot \text{nm})/(570 \text{ nm}) = 2.18$ eV.
    (b) The stopping voltage is the voltage that gives a potential energy change equal to the maximum
    kinetic energy:
    $K_{max} = eV_0 = hf - W_0$ ;
    $(1 \text{ e})V_0 = [(1.24 \times 10^3 \text{ eV} \cdot \text{nm})/(400 \text{ nm})] - 2.18 \text{ eV} = 3.10 \text{ eV} - 2.18 \text{ eV} = 0.92$ eV,
    so the stopping voltage is 0.92 V.

19. The energy required for the chemical reaction is provided by the photon:
    $E = hf = hc/\lambda = (1.24 \times 10^3 \text{ eV} \cdot \text{nm})/\lambda = (1.24 \times 10^3 \text{ eV} \cdot \text{nm})/(660 \text{ nm}) = 1.88$ eV.
    Each reaction takes place in a molecule, so we have
    $E$  = (1.88 eV/molecule)(6.02 × 10$^{23}$ molecules/mol)(1.60 × 10$^{-19}$ J/eV)/(4186 J/kcal)
        = 43.3 kcal/mol.

23. (a) The energy of a photon is
    $E = hf = hc/\lambda$.
    For the fractional loss, we have
    $(E - E')/E = [(1/\lambda) - (1/\lambda')]/(1/\lambda) = (\lambda' - \lambda)/\lambda'$.
    For 45° we get
    $(E - E'_a)/E = (\lambda'_a - \lambda)/\lambda'_a = (7.12 \times 10^{-13} \text{ m})/(0.120 \times 10^{-9} \text{ m} + 7.12 \times 10^{-13} \text{ m}) = 5.90 \times 10^{-3}$.
    For 90° we get
    $(E - E'_b)/E = (\lambda'_b - \lambda)/\lambda'_b = (2.43 \times 10^{-12} \text{ m})/(0.120 \times 10^{-9} \text{ m} + 2.43 \times 10^{-12} \text{ m}) = 1.98 \times 10^{-2}$.
    For 180° we get
    $(E - E'_c)/E = (\lambda'_c - \lambda)/\lambda'_c = (4.86 \times 10^{-12} \text{ m})/(0.120 \times 10^{-9} \text{ m} + 4.86 \times 10^{-12} \text{ m}) = 3.89 \times 10^{-2}$.
    (b) The energy of the incident photon is
    $E = hf = hc/\lambda = (1.24 \times 10^3 \text{ eV} \cdot \text{nm})/\lambda = (1.24 \times 10^3 \text{ eV} \cdot \text{nm})/(0.120 \text{ nm}) = 10.3 \times 10^3$ eV.
    From conservation of energy, the energy given to the scattered electron is the energy lost by the
    photon:
    $K = (E - E') = [(E - E')/E]E$.
    For 45° we get
    $K_a = [(E - E'_a)/E]E = (5.90 \times 10^{-3})(10.3 \times 10^3 \text{ eV}) = 60.8$ eV.
    For 90° we get
    $K_b = [(E - E'_b)/E]E = (1.98 \times 10^{-2})(10.3 \times 10^3 \text{ eV}) = 204$ eV.
    For 180° we get
    $K_c = [(E - E'_c)/E]E = (3.89 \times 10^{-2})(10.3 \times 10^3 \text{ eV}) = 401$ eV.

27. The kinetic energy of the pair is
    $K = hf - 2mc^2 = 2.84 \text{ MeV} - 2(0.511 \text{ MeV}) = 1.82$ MeV.

31. We find the wavelength from
    $\lambda = h/p = h/mv = (6.63 \times 10^{-34} \text{ J} \cdot \text{s})/(0.21 \text{ kg})(0.10 \text{ m/s}) = 3.2 \times 10^{-32}$ m.

35. Because all the energies are much less than $mc^2$, we can use $K = p^2/2m$, so

$\lambda = h/p = h/(2mK)^{1/2} = hc/(2mc^2K)^{1/2}$.

(a) $\lambda = hc/(2mc^2K)^{1/2} = (1.24 \times 10^3 \text{ eV} \cdot \text{nm})/[2(0.511 \times 10^6 \text{ eV})(10 \text{ eV})]^{1/2} = 0.39 \text{ nm}$.

(b) $\lambda = hc/(2mc^2K)^{1/2} = (1.24 \times 10^3 \text{ eV} \cdot \text{nm})/[2(0.511 \times 10^6 \text{ eV})(100 \text{ eV})]^{1/2} = 0.12 \text{ nm}$.

(c) $\lambda = hc/(2mc^2K)^{1/2} = (1.24 \times 10^3 \text{ eV} \cdot \text{nm})/[2(0.511 \times 10^6 \text{ eV})(1.0 \times 10^3 \text{ eV})]^{1/2} = 0.039 \text{ nm}$.

39. For diffraction, the wavelength must be of the order of the opening. We find the speed from

$\lambda = h/p = h/mv$;

$10 \text{ m} = (6.63 \times 10^{-34} \text{ J} \cdot \text{s})/(2000 \text{ kg})v$, which gives $v = 3.3 \times 10^{-38} \text{ m/s}$.

Not a good speed if you want to get somewhere.

At a speed of 30 m/s, $\lambda \ll 10$ m, so there will be no diffraction.

43. To ionize the atom means removing the electron, or raising it to zero energy:

$E_{\text{ion}} = 0 - E_n = (13.6 \text{ eV})/n^2 = (13.6 \text{ eV})/2^2 = 3.4 \text{ eV}$.

47. For the Rydberg constant we have

$R = e^4m/8\varepsilon_0^2h^3c$

$= (1.602177 \times 10^{-19} \text{ C})^4(9.109390 \times 10^{-31} \text{ kg})/$
$\qquad 8(8.854187 \times 10^{-12} \text{ C}^2/\text{N} \cdot \text{m}^2)^2(6.626076 \times 10^{-34} \text{ J} \cdot \text{s})^3(2.997925 \times 10^8 \text{ m/s})$

$= 1.0974 \times 10^7 \text{ m}^{-1}$.

51. Singly-ionized helium is like hydrogen, except that
there are two positive charges ($Z = 2$) in the nucleus.
The square of the product of the positive and negative
charges appears in the energy term for the energy levels.
We can use the results for hydrogen, if we replace $e^2$ by $Ze^2$:

$E_n = -Z^2(13.6 \text{ eV})/n^2 = -2^2(13.6 \text{ eV})/n^2 = -(54.4 \text{ eV})/n^2$.

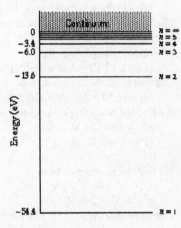

55. We find the velocity from the quantum condition:

$mvr_1 = nh/2\pi$;

$(9.11 \times 10^{-31} \text{ kg})v(0.529 \times 10^{-10} \text{ m}) = (1)(6.63 \times 10^{-34} \text{ J} \cdot \text{s})/2\pi$,

which gives $v = 2.18 \times 10^6 \text{ m/s} = 7.3 \times 10^{-3}c$.

The relativistic factor is

$[1 - (v/c)^2]^{1/2} \approx 1 - \frac{1}{2}(v/c)^2 = 1 - 2.7 \times 10^{-5}$.

Because this is essentially 1, the use of nonrelativistic formulas is justified.

59. (a) We find the frequency of the radiation for a jump from level $n$ to level $n - 1$ from

$hf = E_n - E_{n-1}$;

$f = [(Z^2e^4m/8\varepsilon_0^2h^2)/h]\{[1/(n-1)^2] - (1/n^2)\}$

$= (Z^2e^4m/8\varepsilon_0^2h^3)\{[n^2 - (n-1)^2]/n^2(n-1)^2\} \approx (Z^2e^4m/8\varepsilon_0^2h^3)(2n/n^4) = Z^2e^4m/4\varepsilon_0^2h^3n^3$.

From the quantum condition we have

$mvr_n = nh/2\pi$, or $v = nh/2\pi mr_n$.

Thus we get

$v/2\pi r_n = nh/4\pi^2 mr_n^2 = nh/4\pi^2m(n^2h^2\varepsilon_0/\pi mZe^2)^2 = Z^2e^4m/4\varepsilon_0^2h^3n^3$,

which is the same as the above frequency.

(b) From the classical theory for an electron revolving in a circular orbit, the time for one revolution is

$T = 2\pi r_n/v$,

so the frequency is

$f = 1/T = v/2\pi r_n$.

(c) Classically an accelerated charge radiates. For circular motion, the frequency of the radiation is
the orbital frequency. This agrees with the Bohr prediction for large values of $n$, consistent with
the correspondence principle.

63. For the energy of the photon, we have

$E = hf = hc/\square$

$= (6.63 \times 10^{-34} \text{ J} \cdot \text{s})(3.00 \times 10^8 \text{ m/s})/(1.60 \times 10^{-19} \text{ J/eV})\lambda = (1.24 \times 10^{-6} \text{ eV} \cdot \text{m})/\lambda$.

67. The energy of the photon is

$hf = (1.24 \times 10^3 \text{ eV} \cdot \text{nm})/\lambda = (1.24 \times 10^3 \text{ eV} \cdot \text{nm})/(550 \text{ nm}) = 2.25 \text{ eV}$.

Because the light radiates uniformly, the intensity at a distance $L$ is

$I = P/4\pi L^2$, so the rate at which energy enters the pupil is

$E/t = I\pi r^2 = Pr^2/4L^2$.

Thus the rate at which photons enter the pupil is

$n/t = (E/t)/hf = Pr^2/4L^2 hf$
$= (0.030)(100 \text{ W})(2.0 \times 10^{-3} \text{ m})^2/4(1.0 \times 10^3 \text{ m})^2(2.25 \text{ eV})(1.60 \times 10^{-19} \text{ J/eV})$
$= 8.3 \times 10^6 \text{ photons/s}$.

71. If we ignore the recoil motion of the gold nucleus, at the closest approach the kinetic energy of both particles is zero. The potential energy of the two charges must equal the initial kinetic energy of the $\alpha$ particle:

$K = Z_\alpha Z_{Au} e^2/4\pi\varepsilon_0 r_{min}$ ;

$(4.8 \text{ MeV})(1.60 \times 10^{-13} \text{ J/MeV}) = (2)(79)(1.60 \times 10^{-19} \text{ C})^2/4\pi(8.85 \times 10^{-12} \text{ C}^2/\text{N} \cdot \text{m}^2)r_{min}$ ,

which gives $r_{min} = 4.7 \times 10^{-14}$ m.

75. The ratio of the forces is

$F_g/F_e = (Gm_e m_p/r^2)/(e^2/4\pi\varepsilon_0 r^2) = 4\pi\varepsilon_0 Gm_e m_p/e^2$
$= 4\pi (8.85 \times 10^{-12} \text{ C}^2/\text{N} \cdot \text{m}^2)(6.67 \times 10^{-11} \text{ N} \cdot \text{m}^2/\text{kg}^2)(9.11 \times 10^{-31} \text{ kg})(1.67 \times 10^{-27} \text{ kg})/$
$(1.60 \times 10^{-19} \text{ C})^2 = 4.4 \times 10^{-40}$.

Yes, the gravitational force may be safely ignored.

79. The energy of the ultraviolet photon is

$hf_1 = (1.24 \times 10^3 \text{ eV} \cdot \text{nm})/\square = (1.24 \times 10^3 \text{ eV} \cdot \text{nm})/(290 \text{ nm}) = 4.28 \text{ eV}$.

The stopping potential is the potential difference that gives a potential energy change equal to the maximum kinetic energy:

$K_{max1} = eV_{01} = hf_1 - W_0$ ;

$(1 \text{ e})(2.10 \text{ V}) = 4.28 \text{ eV} - W_0$ , which gives $W_0 = 2.18 \text{ eV}$.

The energy of the blue photon is

$hf_1 = (1.24 \times 10^3 \text{ eV} \cdot \text{nm})/\lambda = (1.24 \times 10^3 \text{ eV} \cdot \text{nm})/(440 \text{ nm}) = 2.82 \text{ eV}$.

We find the new stopping potential from

$K_{max2} = eV_{02} = hf_2 - W_0$ ;

$(1 \text{ e})V_{02} = 2.82 \text{ eV} - 2.18 \text{ eV}$, which gives $V_{02} = 0.64 \text{ V}$.

83. We find the momentum from

$E^2 = (K + mc^2)^2 = p^2 c^2 + m^2 c^4$,  or
$p^2 c^2 = K^2 + 2mc^2 K$.

The wavelength is

$\lambda = h/p = hc/pc = hc/(K^2 + 2mc^2 K)^{1/2}$
$= (1.24 \times 10^3 \text{ eV} \cdot \text{nm})/[(60 \times 10^3 \text{ eV})^2 + 2(0.511 \times 10^6 \text{ eV})(60 \times 10^3 \text{ eV})]^{1/2} = 4.9 \times 10^{-3} \text{ nm}$.

The theoretical resolution limit is of the order of the wavelength, or $5 \times 10^{-12}$ m.

# Chapter 39: Quantum Mechanics

## Chapter Overview and Objectives

This chapter introduces the physics of quantum mechanics. It discusses the Heisenberg Uncertainty Principle and the Schrödinger wave equation. The application and solution of Schrödinger's equation to several situations is also discussed.

After completing study of this chapter, you should:

- Know Heisenberg's Uncertainty Principle in both the position-momentum and energy-time forms.
- Know the approximate value of Planck's constant.
- Know how the wave function is related to probability density.
- Know Schrödinger's time-dependent and time-independent equations in one dimension.
- Know what the normalization condition is.
- Know the form of the solutions to Schrödinger's equation for free particles.
- Know the solutions to Schrödeinger's equation for particles in an infinite square well.
- Know the continuity conditions on wave functions that are solutions to Schrödinger's equation.
- Know what quantum mechanical tunneling is and how to solve for the tunneling probability using the WKB approximation.

## Summary of Equations

Heisenberg Uncertainty Principle:
$$\Delta x \Delta p \geq \frac{h}{2\pi}$$

$$\Delta E \Delta t \geq \frac{h}{2\pi}$$

Schrödinger's time-dependent equation:
$$-\frac{\hbar^2}{2m}\frac{\partial^2 \Psi(x,t)}{dx^2} + U(x)\Psi(x,t) = i\hbar\frac{\partial \Psi(x,t)}{dt}$$

Schrödinger's time-independent equation:
$$-\frac{\hbar^2}{2m}\frac{\partial^2 \psi(x)}{dx^2} + U(x)\psi(x) = E\psi(x)$$

Normalization condition:
$$\int |\psi|^2 \, dV = 1$$

Relationship between energy and wave number for free particle:
$$k = \sqrt{\frac{2E}{m}}$$

Energy levels for particle in infinite square well:
$$E = \frac{h^2 n^2}{8mL^2}$$

Wave functions for particle in infinite square well:
$$\psi(x) = \sqrt{\frac{2}{L}} \sin\left(\frac{n\pi x}{L}\right)$$

Continuity conditions on wave function:
$$\psi_+ = \psi_-$$

$$\frac{d\psi_+}{dx} = \frac{d\psi_-}{dx} \quad \text{if} \quad -\infty < \Delta V < +\infty$$

WKB approximation to tunneling probability:
$$T \approx e^{-2GL} \quad \text{where} \quad G = \sqrt{\frac{2m(U_0 - E)}{\hbar^2}}$$

## Chapter Summary

### Section 39-1 Quantum Mechanics—A New Theory

Quantum mechanics is a self-consistent theory that accommodates the apparently contradictory dual wave-like and particle-like behaviors observed in nature.

### Section 39-2 The Wave Function and Its Interpretation; the Double-Slit Experiment

One approach to quantum mechanics is through a differential equation called a wave equation. The solution to the equation is a function of position and time we call a **wave function**, $\psi(x,t)$, but a wave of what? The wave function in quantum mechanics is a wave of **probability amplitude**. The interpretation of the wave is through its square magnitude, $|\psi|^2$. The probability that a measurement of the position of a particle results somewhere within a volume $dV$ around a position $x$ at time $t$ is equal to $|\psi(x, t)|^2 \, dV$. The function $|\psi(x, t)|^2$ is called the **probability density**.

### Section 39-3 The Heisenberg Uncertainty Principle

Simultaneous measurement of position and momentum of a particle are limited in resolution. The uncertainty in position, $\Delta x$, and the uncertainty in momentum, $\Delta p$, satisfy the inequality:

$$\Delta x \Delta p \geq \frac{h}{2\pi}$$

This inequality is called the **Heisenberg Uncertainty Principle.** The quantity $h$ is called Planck's constant and has the value

$$h = 6.626 \times 10^{-34} \text{ J} \cdot \text{s}$$

A second form of the uncertainty principle relates the uncertainty in the energy of a particle, $\Delta E$, to the length of time of observation of the particle, $\Delta t$:

$$\Delta E \Delta t \geq \frac{h}{2\pi}$$

Because the quantity $h/2\pi$ appears very often in quantum mechanics, it is given its own symbol, $\hbar$ :

$$\hbar \geq \frac{h}{2\pi} = 1.055 \times 10^{-34} \text{ J} \cdot \text{s}$$

### Example 39-3-A

A ball of mass 350 g is thrown straight through a doorway with a width 0.82 m. Determine the uncertainty in velocity of the ball parallel to the doorway after is passes through the doorway.

### Solution:

The doorway limits the uncertainty in the position in the direction parallel to the doorway to an uncertainty of 0.82 m. If we write the uncertainty in the momentum of the ball parallel to the doorway as

$$\Delta p = m \Delta v$$

then Heisenberg's Uncertainty Principle tells us

$$\Delta x (m \Delta v) \geq \frac{h}{2\pi} \quad \Rightarrow \quad \Delta v \geq \frac{h}{2\pi \, m \Delta x}$$

$$\Delta v \geq \frac{\left(6.63 \times 10^{-34} \text{ J} \cdot \text{s}\right)}{2\pi \left(0.350 \text{ kg}\right)\left(0.82 \text{ m}\right)} = 3.7 \times 10^{-34} \text{ m/s}$$

**Example 39-3-B**

The mass of an object is determined in a process that observes the mass for a time of 10 μs. What is the minimum uncertainty of the mass measurement?

**Solution:**

In this case, we need to make use of the special relativity relationship between mass and energy and the energy-time form of the Heisenberg Uncertainty Principle. The mass of the object is related to the energy by

$$m = \frac{E}{c^2}$$

which implies that

$$\Delta m = \frac{\Delta E}{c^2}$$

By the Heisenberg Uncertainty Principle, we know

$$\Delta E \Delta t \geq \frac{h}{2\pi} \quad \Rightarrow \quad \Delta E \geq \frac{h}{2\pi \, \Delta t}$$

Substituting this into the expression for $\Delta m$, we get

$$\Delta m \geq \frac{h}{2\pi \, c^2 \Delta t} = \frac{\left(6.63 \times 10^{-34} \text{ J} \cdot \text{s}\right)}{2\pi \left(3.00 \times 10^8 \text{ m/s}\right)^2 \left(10 \times 10^{-6} \text{ s}\right)} = 1.17 \times 10^{-46} \text{ kg}$$

This is a very small uncertainty in mass, but consider what happens if the lifetime of a particle becomes very small. The uncertainty in the mass could become on the same order of size as the mass of the particle.

**Section 39-4 Philosophic Implications; Probability versus Determinism**

In classical physics, if the complete physical conditions (positions and velocities of all particles) of a system are known at some time, then, in principle, the laws of classical physics can be used to determine the positions and velocities at all future times. Quantum mechanics is inherently different. Because of the fact that the wave function, the solution to quantum mechanical equations, is related to measurements as a probability amplitude, the implication is that quantum mechanics is not deterministic in the sense that which future events (values of measurements) happen are not determined by quantum mechanics.

**Section 39-5 The Schrödinger Equation in One Dimension—Time-Independent Form**

The wave function is a solution to Schrödinger's wave equation. In general, a wave equation has terms with derivatives of the wave function with respect to position and terms with derivatives of the wave function with respect to time. Some wave equations can be separated into two related equations, one with derivatives with respect to time and one with derivatives with respect to position. In general, these types of equations with derivatives with respect to one variable are easier to solve than equations with derivatives with respect to two different variables. We look first at such an equation, related to Schrödinger's wave equation, that only has derivatives with respect to position. This equation is called the **time-independent Schrödinger equation**:

$$-\frac{\hbar^2}{2m} \frac{\partial^2 \psi(x)}{dx^2} + U(x)\psi(x) = E\psi(x)$$

where $\psi(x)$ is a function related to the wave function $\Psi(x,t)$ and $E$ is a constant. It is consistent with classical mechanics for the constant $E$ to be interpreted as the total energy of the particle. The general form of Schrodinger's time-independent equation places some restrictions on the possible solutions $\psi(x)$. If the energy $E$ is to be finite, then $\psi(x)$ must be a continuous function everywhere and its derivative must be continuous except where there is an infinite discontinuity in $U(x)$.

Another restriction on the function $\psi(x)$ results from the interpretation of $|\psi|^2$ as the probability density. Because a measurement of the position of a particle must result in some location in space, then the integral of $|\psi|^2\, dV$ must be equal to one:

$$\int |\psi|^2\, dV = 1$$

This is called the **normalization condition.**

**Example 39-5-A**

A particle has a wave function given by

$$\psi(x) = Ae^{-b|x|}$$

Determine the constant $A$ in terms of $b$ so that the wave function is normalized.

**Solution:**

We need to find $A$ such that

$$\int |\psi|^2\, dx = 1 \quad \Rightarrow \quad \int_{-\infty}^{\infty} e^{-2b|x|} dx = 1$$

To deal with the absolute value sign in the integrand, we write the integrand in one form for $x < 0$ and another form for $x > 0$:

$$\int_{-\infty}^{\infty} A^2 e^{-2b|x|} dx = \int_{-\infty}^{0} A^2 e^{2bx} dx + \int_{0}^{\infty} A^2 e^{-2bx} dx = \frac{A^2}{2b} e^{2bx}\Big|_{-\infty}^{0} + \frac{A^2}{-2b} e^{-2bx}\Big|_{0}^{2b}$$

$$= \frac{A^2}{2b}(1-0) - \frac{A^2}{2b}(0-1) = \frac{A^2}{b}$$

Setting this equal to one, we easily solve for $A$:

$$A = \sqrt{b}$$

**Section 39-6 Time-Dependent Schrödinger Equation**

The complete Schrödinger equation is known as the time-dependent Schrödinger equation:

$$-\frac{\hbar^2}{2m} \frac{\partial^2 \Psi(x,t)}{dx^2} + U(x)\Psi(x,t) = i\hbar \frac{\partial \Psi(x,t)}{dt}$$

A comparison with the time-independent Schrödinger equation implies that solutions of the time-independent Schrödinger equation are solutions of the time-dependent equation when multiplied by a function of time:

$$\Psi(x,t) = \psi_E(x) e^{-i\frac{E}{\hbar}t}$$

where $\psi_E(x)$ is a solution to the time-independent Schrödinger equation with constant $E$.

**Section 39-7 Free Particles; Plane Waves and Wave Packets**

For free particles, the potential energy is zero everywhere.  Schrödinger's equation reduces to

$$-\frac{\hbar^2}{2m}\frac{\partial^2\psi(x)}{dx^2}+U(x)\psi(x)=E\psi(x)$$

Rearranging this, we have

$$\frac{d^2\psi(x)}{dx^2}=-\frac{2m}{\hbar^2}\psi(x)$$

The solutions to this equation have the form[1]:

$$\psi(x)=A\sin kx+B\cos kx$$

where the wave number $k$ is given by

$$k=\sqrt{\frac{2E}{m}}$$

This solution is not localized in space; the probability density is spread out over all space.  It has a definite momentum and so its uncertainty in position but must be infinite to satisfy the Heisenberg Uncertainty Principle.  In general, we are often interested in solutions to Schrödinger's equation that have some spatial localization.  We call these types of solutions **wave packets**.

**Section 39-8 Particle in an Infinitely Deep Square Well Potential (a Rigid Box)**

The infinitely deep square well potential energy function is

$$U(x)=0 \quad\quad \text{if} \quad\quad 0<x<L$$
$$U(x)=\infty \quad\quad \text{if} \quad\quad x\leq 0 \quad \text{or} \quad x\geq L$$

The solutions to Schrödinger's time-independent equation only exist for particular values of the energy, $E$.  These values are

$$E=\frac{h^2n^2}{8mL^2}$$

and the corresponding solutions to Schrödinger's time-independent equation are

$$\psi(x)=\sqrt{\frac{2}{L}}\sin\left(\frac{n\pi x}{L}\right)$$

**Section 39-9 Finite Potential Well**

The potential energy as a function of position for a finite potential well is given by

$$U(x)=0 \quad \text{if} \quad 0<x<L$$
$$U(x)=U_0 \quad \text{if} \quad x\leq 0 \quad \text{or} \quad x\geq L$$

In the regions where the potential energy is $U_0$, the solution to Schrödinger's equation can be written as

---

[1] The coefficients $A$ and $B$ cannot both be real numbers if this is a state with definite momentum; at least one of them must be a complex number.

$$\psi(x) = Ce^{Gx} + De^{-Gx}$$

where

$$G = \sqrt{\frac{2m(U_0 = E)}{\hbar^2}}$$

and there is a different value for $C$ and $D$ in each separate region. We also already know the expression for the solution in the region where the potential energy is equal to zero:

$$\psi(x) = A \sin kx + B \cos kx$$

where

$$k = \sqrt{\frac{2E}{m}}$$

The solutions to Schrödinger's equation must satisfy certain conditions. These conditions are not additional conditions, but are imposed by the form of Schrödinger's equation itself. The wave function must be continuous everywhere. We can write that as

$$\psi_+ = \psi_-$$

where we mean as we approach any given point from the positive side, the values of the wavefunction approach the same value that we get if we approach from the negative side and are equal to the value of the wave function at the given point. (Sorry, mathematicians.) The second condition on Schrödinger's equation is that the derivative of the wave function must be continuous everywhere except where there is an infinite discontinuity in the potential energy:

$$\frac{d\psi_+}{dx} = \frac{d\psi_-}{dx} \quad \text{where } U(x) \text{ does not make an infinite step}$$

If we apply these conditions to the solutions of Schrödinger's equation in the three separate regions in the finite square well problem, we have a set of equations that can be satisfied for only particular values of $E$ if $E < 0$ (or if $\psi = 0$ everywhere) and for all values of $E > 0$. Unfortunately, the equations are transcendental equations, equations that can not be solved by algebraic means, and we must use some approximate method of solving the equations.

### Section 39-10 Tunneling through a Barrier

The wave function solution of Schrödinger's equation will, in general, be non-zero in all regions where the potential energy is not infinite over some region of space. Of particular interest is the situation when a particle starts on one side of a potential energy barrier with less kinetic energy than the maximum potential energy of the barrier. In classical physics, the particle cannot cross the barrier. However, in quantum mechanics, this particle can appear on the opposite side of the barrier from which it started with a probability dependent on the amplitude of the wave function on the far side of the barrier. This process is called **quantum mechanical tunneling**. An approximation to the probability of a particle tunneling through the barrier is given by the WKB approximation as:

$$T \approx e^{-2GL}$$

where

$$G = \sqrt{\frac{2m(U_0 - E)}{\hbar^2}}$$

and $L$ is the thickness of the potential energy barrier, $U_0$ is the height of the potential energy barrier, $E$ is the energy of the particle, and $m$ is the mass of the particle.

## Example 39-10-A

An electron with an energy 32 eV tunnels through a potential energy barrier that is 1.6 nm thick with a probability of 0.312. What is the height of the potential energy barrier?

## Solution:

We know the probability for transmission, $T$, through the potential energy barrier is related to the thickness, $L$, of the barrier by

$$T \approx e^{-2GL}$$

Here, we can use the information given to calculate the value of $G$:

$$G = -\frac{\ln T}{2L} = -\frac{\ln(0.312)}{2(1.6 \times 10^{-9} \text{ m})} = 3.64 \times 10^9 \text{ m}^{-1}$$

We can now use the expression for $G$ to solve for $U_0$, the height of the potential barrier:

$$G = \sqrt{\frac{2m(U_0 - E)}{\hbar^2}} \quad \Rightarrow \quad U_0 = \frac{\hbar^2 G^2}{2m} + E$$

$$U_0 = \frac{\left(6.63 \times 10^{-34} \text{ J·s}\right)^2 \left(3.64 \times 10^9 \text{ m}^{-1}\right)^2}{2\left(9.11 \times 10^{-31} \text{ kg}\right)} + \left(32 \text{ eV}\right)\left(1.60 \times 10^{-19} \text{ J/eV}\right)$$

$$= 8.3 \times 10^{-18} \text{ J} = 52 \text{ eV}$$

# Practice Quiz

1.  The value of the wave function at point $x$ is equal to 2.00. What does this tell you?

    a)  The probabilty of finding the particle at point $x$ is 2.00.
    b)  The probabilty of finding the particle at point $x$ is 4.00.
    c)  The probability of finding the particle between $x$ and $x + dx$ is 4.00 $dx$.
    d)  The value of the wave function can't be bigger than 1.00 because of normalization.

2.  To be consistent with the Heisenberg Uncertainty Principle, if you improve an experiment to determine the position of a particle to a factor 10 smaller uncertainty, the smallest uncertainty in momentum you could possible measure simultaneously is

    a)  a factor 10 smaller, also.
    b)  a factor of 10 larger.
    c)  zero.
    d)  a factor 100 smaller.

3.  If you take a time $\Delta t$ to measure the energy of a particle, what is the minimum uncertainty in the energy measurement?

    a)  $1/\Delta t$
    b)  $1/\Delta t^2$
    c)  $h$
    d)  $\hbar/\Delta t$

4.     A particle in an infinite square well potential has a minimum energy $E$.  If the square well is doubled in length, what will the minimum energy be?

   a)  $2E$
   b)  $4E$
   c)  $E/2$
   d)  $E/4$

5.     From what assumption does the normalization principle come from?

   a)  The probability of a particle being somewhere in space must be equal to one.
   b)  The uncertainty in the momentum of a particle multiplied by the uncertainty in its position divided by $\hbar$ must be equal to 1.
   c)  The uncertainty in the momentum of a particle divided by the uncertainty in its position multiplied by $\hbar$ must be equal to 1.
   d)  There is no assumption, one is just a convenient number to work with.

6.     The time-independent Schrödinger equation has a wavefunction solution $\psi$ with an energy $E$.  What is the corresponding solution to the time-dependent Schrödinger equation?

   a)  $\psi$
   b)  $\psi t$
   c)  $\psi t^2$
   d)  $\psi e^{-iEt/\hbar}$

7.     Why don't we need to be concerned with the Heisenberg Uncertainty Principle in every day life when dealing with macroscopic objects?

   a)  The Heisenberg Uncertainty Principle only applies to microscopic objects.
   b)  The Heisenberg Uncertainty Principle only applies to electrons in atoms.
   c)  The uncertainties required by the uncertainty principle are much smaller than the uncertainities of position and momentum that we make on macroscopic objects.
   d)  The uncertainty principle only applies to measurements made by scattering light off of the object and we usually measure the position of macroscopic objects with a ruler.

8.     If a particle is placed in an infinite square well of length $L$ and a finite square well of length $L$, in which will it have the lower energy in the lowest energy state for that potential?

   a)  the infinite square well
   b)  the finite square well
   c)  Both energies will be identical.
   d)  need to know the depth of the finite square well to answer the question

9.     If the length of a barrier is doubled, by what factor must the difference between the energy and the barrier height change so that the tunneling probability remains the same?

   a)  4
   b)  2
   c)  ½
   d)  ¼

10.    What is the minimum height of a potential energy barrier for which a particle with energy $E$ will have zero probability of transmission through the barrier?

   a)  $E/2$
   b)  $E$
   c)  $2E$
   d)  $\infty$

11.    What is the minimum uncertainty of the momentum of an electron in a wire that is 5 μm long?

12.    You have measured the total energy of a system to within an uncertainty of $3 \times 10^{-5}$ eV. What is the minimum amount of time you had to make the measurement over to obtain this small of an uncertainty?

13.    The ratio of the energies of two consecutive energy levels of a particle in an infinite square well of length 0.645 μm is exactly 1.44. The lower energy of the two is $6.24 \times 10^{-32}$ J. What is the mass of the particle?

14.    Determine the probability that an electron with a kinetic energy of 1.5 eV is able to tunnel through a barrier of height 3.0 eV and a thickness of 0.12 μm.

15.    Determine the lowest four energy levels for an object of mass 3.8 g in a box 10 cm long. Treat the box as one-dimensional.

## Problem Solutions

3.    We find the uncertainty in the momentum:
$$\Delta p = m\Delta v = (1.67 \times 10^{-27} \text{ kg})(0.024 \times 10^5 \text{ m/s}) = 4.00 \times 10^{-24} \text{ kg} \cdot \text{m/s}.$$
We find the uncertainty in the proton's position from
$$\Delta x \geq \hbar / \Delta p = (1.055 \times 10^{-34} \text{ J} \cdot \text{s})/(4.00 \times 10^{-24} \text{ kg} \cdot \text{m/s}) = 2.6 \times 10^{-11} \text{ m}.$$
Thus the accuracy of the position is $\pm 1.3 \times 10^{-11}$ m.

7.    The uncertainty in the velocity is
$$\Delta v = (0.065/100)(75 \text{ m/s}) = 0.0488 \text{ m/s}.$$
For the electron, we have
$$\Delta x \geq \hbar / m\,\Delta v = (1.055 \times 10^{-34} \text{ J} \cdot \text{s})/(9.11 \times 10^{-31} \text{ kg})(0.0488 \text{ m/s}) = 2.4 \times 10^{-3} \text{ m}.$$
For the baseball, we have
$$\Delta x \geq \hbar / m\,\Delta v = (1.055 \times 10^{-34} \text{ J} \cdot \text{s})/(0.150 \text{ kg})(0.0488 \text{ m/s}) = 1.4 \times 10^{-32} \text{ m}.$$
The uncertainty for the electron is greater by a factor of $1.7 \times 10^{29}$.

11.    The electron has an initial momentum $p_x$ and a wavelength $\Box = \hbar / p_x$. For the maxima of the double-slit interference we have
$$d \sin \theta = n\lambda, \; n = 0, 1, 2, \ldots .$$
If the angles are small, the separation of maxima is
$$\Delta\theta_{\text{max}} = \lambda/d, \text{ so the angle between a maximum and a minimum is } \Delta\theta = \lambda/2d.$$
The separation on the screen will be
$$H = L\,\Delta\theta = \lambda L/2d.$$
The uncertainty in the $y$-position at the slits of $d/2$ produces an uncertainty in the $y$-momentum of
$$\Delta p_y \geq h/2\pi\Delta y = h/\pi d.$$
This produces an uncertainty of the $y$-position at the screen of
$$\Delta H = (\Delta p_y/p_x)L = (h/\pi d)L/(h/\lambda) = \lambda L/\pi d.$$
This is on the order of the separation of maxima and minima, so the pattern is destroyed.

15.    We form the wave packet from
$$\psi = \psi_1 + \psi_2 = A \sin k_1 x + A \sin k_2 x, \text{ where}$$
$$k_1 = 2\pi/\lambda_1, \text{ and } k_2 = 2\pi/\lambda_2.$$
If we use a trigonometric identity we get
$$\psi = A(\sin k_1 x + \sin k_2 x) = 2A \sin [\tfrac{1}{2}(k_1 + k_2)x] \cos [\tfrac{1}{2}(k_1 - k_2)x].$$
If the wavelengths are almost equal, we have
$$\Delta k = k_1 - k_2, \text{ and } k \approx \tfrac{1}{2}(k_1 + k_2); \text{ thus}$$
$$\psi = 2A \cos [\tfrac{1}{2}(\Delta k)x] \sin kx.$$
The width of the packet corresponds to the distance from one zero to the next zero of the cosine function:
$\tfrac{1}{2}(\Delta k)\Delta x = \pi$, which gives $\Delta x = 2\pi/\Delta k$.
The momentum is $p = \hbar k$, so $\Delta k = \Delta p/\hbar$. When we use this in the expression for the width, we have
$$\Delta x = 2\pi/(\Delta p/\hbar), \text{ or } \Delta x\, \Delta p = h.$$

19.    (a) The energy levels for the electron in an infinite square well are
$$E_n = n^2 h^2/8mL^2 = n^2 E_1.$$
The longest wavelength photon has the least energy, so the transition must be from

$n_2 = 2$ to $n_1 = 1$.  Thus we have

$hf = hc/\lambda = \Delta E = (n_2{}^2 - n_1{}^2)E_1$;

$(1.24 \times 10^3 \text{ eV} \cdot \text{nm})/\lambda = (2^2 - 1^2)(8.0 \text{ eV})$, which gives $\lambda = 52$ nm.

(b)  We find the width of the well from the ground state energy:

$E_1 = h^2/8mL^2$;

$(8.0 \text{ eV})(1.60 \times 10^{-19} \text{ J/eV}) = (6.63 \times 10^{-34} \text{ J} \cdot \text{s})^2/8(9.11 \times 10^{-31} \text{ kg})L^2$,

which gives $L = 2.2 \times 10^{-10}$ m $= 0.22$ nm.

23.  The energy levels for the electron in an infinite potential well are

$E_n = n^2h^2/8mL^2$

$= n^2(6.63 \times 10^{-34} \text{ J} \cdot \text{s})^2/8(9.11 \times 10^{-31} \text{ kg})(2.0 \times 10^{-9} \text{ m})^2(1.60 \times 10^{-19} \text{ J/eV}) = (9.42 \times 10^{-2} \text{ eV})n^2$.

The wave functions are

$\psi_n = (2/L)^{1/2} \sin(n\pi x/L) = (2/2.0 \text{ nm})^{1/2} \sin(n\pi x/2.0 \text{ nm}) = (1.00 \text{ nm}^{-1/2}) \sin(1.57 \text{ nm}^{-1} nx)$.

Thus we have

$E_1 = (9.42 \times 10^{-2} \text{ eV})(1)^2 = 0.094 \text{ eV}$;    $\psi_1 = (1.00 \text{ nm}^{-1/2}) \sin(1.57 \text{ nm}^{-1} x)$;

$E_2 = (9.42 \times 10^{-2} \text{ eV})(2)^2 = 0.38 \text{ eV}$;    $\psi_2 = (1.00 \text{ nm}^{-1/2}) \sin(3.14 \text{ nm}^{-1} x)$;

$E_3 = (9.42 \times 10^{-2} \text{ eV})(3)^2 = 0.85 \text{ eV}$;    $\psi_3 = (1.00 \text{ nm}^{-1/2}) \sin(4.71 \text{ nm}^{-1} x)$;

$E_4 = (9.42 \times 10^{-2} \text{ eV})(4)^2 = 1.51 \text{ eV}$;    $\psi_4 = (1.00 \text{ nm}^{-1/2}) \sin(6.28 \text{ nm}^{-1} x)$.

27.  (a)  Because the wavefunction is normalized, the probability is

$$P = \int_0^{L/4} |\psi_1|^2 \, dx + \int_{3L/4}^{L} |\psi_1|^2 \, dx = \frac{2}{L} \int_0^{L/4} \sin^2 \frac{\pi x}{L} dx + \frac{2}{L} \int_{3L/4}^{L} \sin^2 \frac{\pi x}{L} dx$$

If we change variable to $\theta = \pi x/L$, so $d\theta = (\pi/L) \, dx$, we have

$$P = \frac{2}{L} \frac{L}{\pi} \int_0^{\pi/4} \sin^2 \theta \, d\theta + \frac{2}{L} \frac{L}{\pi} \int_{3\pi/4}^{\pi} \sin^2 \theta \, d\theta$$

$$= \frac{2}{\pi} \left( \frac{1}{2}\theta - \frac{1}{4}\sin 2\theta \right)\Big|_0^{\pi/4} + \frac{2}{\pi} \left( \frac{1}{2}\theta - \frac{1}{4}\sin 2\theta \right)\Big|_{3\pi/4}^{\pi} = \frac{2}{\pi}\left( \frac{\pi}{4} - \frac{1}{2} \right)$$

$$= 0.18.$$

(b)  For the $n = 4$ state the probability is

$$P = \int_0^{L/4} |\psi_1|^2 \, dx + \int_{3L/4}^{L} |\psi_1|^2 \, dx = \frac{2}{L} \int_0^{L/4} \sin^2 \frac{4\pi x}{L} dx + \frac{2}{L} \int_{3L/4}^{L} \sin^2 \frac{4\pi x}{L} dx$$

If we change the variable to $\theta = 4\pi x/L$, so $d\theta = (4\pi/L) \, dx$, we have

$$P = \frac{2}{L} \frac{L}{4\pi} \int_0^{\pi} \sin^2 \theta \, d\theta + \frac{2}{L} \frac{L}{4\pi} \int_{3\pi}^{4\pi} \sin^2 \theta \, d\theta$$

$$= \frac{2}{4\pi} \left( \frac{1}{2}\theta - \frac{1}{4}\sin 2\theta \right)\Big|_0^{\pi} + \frac{2}{4\pi} \left( \frac{1}{2}\theta - \frac{1}{4}\sin 2\theta \right)\Big|_{3\pi}^{4\pi} = \frac{1}{2\pi}\left( \frac{\pi}{2} + \frac{\pi}{2} \right)$$

$$= 0.50.$$

(c)  Classically the electron has equal probability of being anywhere in the well.  Thus the classical prediction is

$[(\frac{1}{4}L - 0) + (L - \frac{3}{4}L)]/L = 0.50$.

We see that the probability approaches the classical value for large $n$.

31.  The wavefunction outside the well (negative $x$) is

$\psi_{\text{I}} = Ce^{Gx}$, where $G^2 = 2m(U_0 - E)/\hbar^2$.

We approximate the energy as that of the ground state of an infinite well:

$E \approx h^2/8mL^2 = (1)^2(6.63 \times 10^{-34} \text{ J} \cdot \text{s})^2/8(9.11 \times 10^{-31} \text{ kg})(0.10 \times 10^{-9} \text{ m})^2(1.60 \times 10^{-19} \text{ J/eV}) = 38 \text{ eV}$.

Because this is much less than $U_0$, this should be a good approximation.  We find $G$ from

$G^2 = 2(9.11 \times 10^{-31} \text{ kg})(1.00 \times 10^3 \text{ eV} - 38 \text{ eV})(1.60 \times 10^{-19} \text{ J/eV})/(1.055 \times 10^{-34} \text{ J} \cdot \text{s})^2$,

which gives $G = 1.59 \times 10^{11}$ m$^{-1}$.

Because of the continuity of the wavefunction, the value at the wall ($x = 0$) is $\psi_{\text{wall}} = C$.  Thus we have

$\psi_{\text{I}}/\psi_{\text{wall}} = e^{Gx}$;

$0.01 = e^{Gx}$,  or  $\ln(0.01) = (1.59 \times 10^{11} \text{ m}^{-1})x$, which gives $x = -2.9 \times 10^{-11}$ m, so $|x| = 0.03$ nm.

35.  (a)  We find the small change in $G$ from a small change in the barrier height $U_0$ :

$G^2 = 2m(U_0 - E)/\hbar^2$;

$2G \, dG = (2m/\hbar^2) \, dU_0$,  or  $dG = (mU_0/G\hbar^2) \, dU_0/U_0$ .

This produces a small change in the transmission coefficient:

$T = e^{-2GL}$;

$dT = -2L\,e^{-2GL}\,dG$, or $dT/T = -2L\,dG = -2(L^2 mU_0/GL^2)\,dU_0/U_0$ ;

$dT/T = -2[(0.10\times10^{-9}\text{ m})^2(9.11\times10^{-31}\text{ kg})(70\text{ eV})(1.60\times10^{-19}\text{ J/eV})/$
$(2.3)(1.055\times10^{-34}\text{ J}\cdot\text{s})^2](0.01) = -0.08.$

Thus $T$ decreases by 8%.

(b) We find the small change in $T$ from a small change in the barrier thickness $L$:

$T = e^{-2GL}$;

$dT = -2G\,e^{-2GL}\,dL$, or

$dT/T = -2G\,dL = -2GL\,(dL/L) = -2(2.3)(0.01) = -0.05.$

Thus $T$ decreases by 5%.

39. We use the radius as the uncertainty in position for the neutron. We find the uncertainty in the momentum from
$\Delta p \geq \hbar/\Delta x = (1.055\times10^{-34}\text{ J}\cdot\text{s})/(1.0\times10^{-15}\text{ m}) = 1.055\times10^{-19}\text{ kg}\cdot\text{m/s}.$
If we assume that the lowest value for the momentum is the least uncertainty, we estimate the lowest possible kinetic energy as
$E = (\Delta p)^2/2m = (1.055\times10^{-19}\text{ kg}\cdot\text{m/s})^2/2(1.67\times10^{-27}\text{ kg}) = 3.33\times10^{-12}\text{ J} = 21\text{ MeV}.$

43. We can relate the momentum to the radius of the orbit from the quantum condition:
$L = mvr = pr = n\hbar \geq \hbar$, so $p = n\hbar/r = \hbar/r_1$ for the ground state.
If we assume that this is the uncertainty of the momentum, the uncertainty of the position is
$\Delta x \geq \hbar/\Delta p = \hbar/(\hbar/r_1) = r_1$, which is the Bohr radius.

47. (a) When the pencil, assumed to be a uniform rod, makes an angle $\square$ with the vertical, the torque from its weight about the bottom point (assumed fixed) creates the angular acceleration:
$\tau = I\alpha$;
$Mg\tfrac{1}{2}l\sin\phi = \tfrac{1}{3}Ml^2\,d^2\square/dt^2.$
Classically, if $\phi = 0$, there is no torque and thus no rotation. From the uncertainty principle we have
$(\Delta L)(\Delta\phi) \geq \hbar$, or $\Delta\phi \geq \hbar/I\,\Delta\omega.$
Thus $\square$ will not always be zero. The resulting torque will cause the pencil to fall.
(b) We need the equation of motion to determine the time of fall. Most of the time will be taken while the angle is small. Thus while $\phi \ll 1$, we have
$d^2\phi/dt^2 = (3g/2l)\phi$, for which the solution is
$\phi = Ae^{\pm kt}.$
Because $\phi$ increases with time, we use
$\phi = Ae^{kt}$, where $k = (3g/2l)^{1/2} = [3(9.80\text{ m/s}^2)/2(0.20\text{ m})]^{1/2} = 8.6\text{ s}^{-1}.$
If we assume $\phi_0$ has the minimum value of $\Delta\phi$ at $t = 0$, then $\Delta\phi = A.$
From the uncertainty principle, the initial angular velocity will be
$\omega_0 = kA = \Delta\omega = \hbar/I\,\Delta\phi = \hbar/(\tfrac{1}{3}Ml^2)A$, which gives
$A^2 = 3\hbar/Ml^2k = 3(1.055\times10^{-34}\text{ J}\cdot\text{s})/(10\times10^{-3}\text{ kg})(0.20\text{ m})^2(8.6\text{ s}^{-1})$, which gives
$A = 3.0\times10^{-16}\text{ rad}.$
We find the time to fall through 0.1 rad from
$\phi = Ae^{kt}$, or $\ln(\phi/A) = kt$;
$\ln[(0.1\text{ rad})/(3.0\times10^{-16}\text{ rad})] = (8.6\text{ s}^{-1})t$, which gives $t = 4\text{ s}.$
Because most of the time will be taken while the angle is small, this should be within a factor of 2.

# Chapter 40: Quantum Mechanics of Atoms

## Chapter Overview and Objectives

This chapter gives the results of the application of quantum mechanics to one and many electron atoms. It gives some of the wave function solutions to Schrödinger's equation for the hydrogen atom and the energy and angular momentum of the electron for some of those solutions. This chapter also introduces some of the applied areas of atomic physics, such as lasers.

After completing study of this chapter, you should:

- Know the allowed energies and angular momentum values for an electron in a one-electron atom.
- Given a spherically symmetric hydrogen atom wave function, be able to calculate the probability of finding the electron in a given range of radii.
- Know what the Pauli exclusion principle is and how it applies to atoms.
- Be able to determine electron configurations for atoms.
- Know how to calculate the approximate energy and wavelength of an atom.
- Know how to calculate the magnetic dipole moment of an electron.
- Know how to calculate the possible total angular momentum quantum numbers of an electron of a given orbital angular momentum and spin angular momentum state.
- Know what fluorescence and phosphorescence are.

## Summary of Equations

Schrödinger equation for hydrogen atom electron:
$$-\frac{\hbar^2}{2m}\left(\frac{\partial^2\psi}{\partial x^2}+\frac{\partial^2\psi}{\partial y^2}+\frac{\partial^2\psi}{\partial z^2}\right)-\frac{1}{4\pi\varepsilon_0}\frac{e^2}{r}\psi=E\psi$$

Allowed energies for electron in hydrogen atom:
$$E_n=-\frac{13.6\,eV}{n^2}\qquad n\in\{1,2,3,...\}$$

Allowed magnitudes of orbital angular momentum for electron in hydrogen atom:
$$L=\hbar\sqrt{l(l+1)}\qquad l\in\{0,1,2,...,n-1\}$$

Allowed $z$ components of orbital angular momentum of electron in hydrogen atom:
$$L_z=m_l\hbar\qquad m_l\in\{-l,-l+1,...,l-1,l\}$$

Allowed $z$ components of spin angular momentum of an electron:
$$S_z=m_s\hbar\quad m_s\in\{+1/2,-1/2\}$$

Selection for single photon interaction:
$$\Delta l=\pm1$$

Ground state wave function solution to the hydrogen atom Schrödinger's equation:
$$\psi_{100}=\frac{1}{\sqrt{\pi r_0^3}}e^{-r/r_0}$$

Definition of Bohr radius:
$$r_0=\frac{h^2\varepsilon_0}{\pi me^2}=0.0529\,\text{nm}$$

Radial probability density:
$$P_r(r)=4\pi r^2|\psi(r)|^2$$

Magnetic moment of electron due to orbital motion:  $\boldsymbol{\mu} = -\dfrac{e}{2m}\mathbf{L}$

Definition of Bohr magneton:  $\mu_B = \dfrac{e\hbar}{2m} = 9.27 \times 10^{-27}\ \text{J/T}$

Mangetic moment of electron due to spin:  $\mu_z = -g\mu_B m_s$

Definition of total angular momentum of electron:  $\mathbf{J} = \mathbf{L} + \mathbf{S}$

## Chapter Summary

### Section 40-1 Quantum-Mechanical View of Atoms

When the time-independent Schrödinger's equation is solved for a one-electron atom, the result is a three-dimensional wave function centered on the nucleus of the atom. The electron does not follow a circular path as Bohr's theory of the atom assumes. The solution to Schrödinger's equation locates the electron only with a cloud of probability density around the nucleus. There is no definite position of the electron or trajectory that the electron follows.

### Section 40-2 Hydrogen Atom: Schrödinger Equation and Quantum Numbers

The potential energy function for the electron in the electrostatic field of a point-like positive charge +e is given by

$$U(r) = -\frac{1}{4\pi\varepsilon_0}\frac{e^2}{r}$$

where $r$ is the distance between the electron and the positive charge.

The time-independent Schrödinger equation for this system is

$$-\frac{\hbar^2}{2m}\left(\frac{\partial^2 \psi}{\partial x^2} + \frac{\partial^2 \psi}{\partial y^2} + \frac{\partial^2 \psi}{\partial z^2}\right) - \frac{1}{4\pi\varepsilon_0}\frac{e^2}{r}\psi = E\psi$$

The full solution of this equation lies beyond the scope of the textbook. We will discuss some of the properties of the solutions.

The solutions to the time-independent Schrödinger's equation for the one electron atom have the same energy levels as the solutions to Bohr's atom. The energy is determined by an integer $n$ associated with each solution. The value of $n$ is called the **principal quantum number** of the solution and can take on the value of any positive integer. The energy of the solution with principal quantum number $n$, $E_n$, is given by

$$E_n = -\frac{13.6\,eV}{n^2} \qquad n \in \{1,2,3,\ldots\}$$

A second quantum number, called the orbital quantum number, $l$, is related to the angular momentum of the electron. The magnitude of the angular momentum of the electron due to its translational motion, called the orbital angular momentum, is given by

$$L = \hbar\sqrt{l(l+1)} \qquad l \in \{0,1,2,\ldots,n-1\}$$

Another quantum number, the **magnetic quantum number**, $m_l$, is related to the component of the orbital angular momentum in a specified direction. The usual component specified is the component along the $z$ direction. If that is the case, then the $z$ component of the orbital angular momentum, $L_z$, is related to $m_l$ by

$$L_z = m_l\hbar \qquad m_l \in \{-l,-l+1,\ldots,l-1,l\}$$

The **spin quantum number**, $m_s$, is related to the component of the internal angular momentum of the electron, called the spin angular momentum, in a specified direction. Again the usual component specified is along the z direction. If that is the case, then the z component of the spin angular momentum, $S_z$, is related to $m_s$ by

$$S_z = m_s \hbar$$

The magnitude of the spin angular momentum for an electron is always $\hbar\sqrt{3}/2$.

Transitions between states that occur in interactions with the electromagnetic field in which a single photon is emitted or absorbed must satisfy a **selection rule**:

$$\Delta l = \pm 1$$

A transition that satisfies this rule is called an **allowed transition** and a rule that does not satisfy this condition is called a **forbidden transition**.[1]

**Example 40-2-A**

An electron in a hydrogen atom makes a transition from the $n = 3$, $l = 2$ state to the $n = 2$, $l = 3$. Is this an allowed or a forbidden transition? What is the change in energy of the electron making this transition?

**Solution:**

To determined whether this is an allowed or forbidden transition, we must determine whether $\Delta l = \pm 1$ or not:

$$\Delta l = l_{final} - l_{initial} = 3 - 2 = +1$$

So this is an allowed transition.

The change in energy of the electron is given by

$$\Delta E = E_{final} - E_{initial} = \left(-\frac{13.6\,\text{eV}}{n_{final}^2}\right) - \left(-\frac{13.6\,\text{eV}}{n_{initial}^2}\right)$$

$$= \left(-\frac{13.6\,\text{eV}}{2^2}\right) - \left(-\frac{13.6\,\text{eV}}{3^2}\right) = -1.89\,\text{eV}$$

**Section 40-3 Hydrogen Atom Wave Functions**

We label the different solutions to the time-independent Schrödinger's equation with subscripts $n$, $l$, and $m_l$ as

$$\psi_{nlm_l}$$

The solution with the lowest energy is

$$\psi_{100} = \frac{1}{\sqrt{\pi r_0^3}} e^{-r/r_0}$$

where $r_0$ is the Bohr radius:

$$r_0 = \frac{h^2 \varepsilon_0}{\pi m e^2} = 0.0529\,\text{nm}$$

---

[1] The "forbidden transitions" are not completely forbidden, but usually occur with a probability much lower (several orders of magnitude lower) than the allowed transitions.

The probability density is the absolute square of the wave function:

$$\left|\psi_{100}\right|^2 = \frac{1}{\pi r_0^3} e^{-2r/r_0}$$

Another function that is useful is the **radial probability distribution**, $P_r$. The radial probability function is the function such that $P_r(r)\,dr$ is the probability that a measurement of the location of the electron falls between a distance $r$ and $r + dr$ from the nucleus. For a wave function that only depends on distance from the nucleus and not on direction, the radial probability distribution function is given by

$$P_r(r) = 4\pi r^2 \left|\psi(r)\right|^2$$

and for the ground state wave function of the one-electron atom is given by

$$P_r(r) = 4\frac{r^2}{r_0^2} e^{-2r/r_0}$$

Some of the other hydrogen atom electron wave functions are

$$\psi_{200} = \frac{1}{\sqrt{32\pi r_0^3}}\left(2 - \frac{r}{r_0}\right)e^{-r/2r_0}$$

$$\psi_{210} = \frac{z}{\sqrt{32\pi r_0^3}}e^{-r/2r_0}$$

$$\psi_{211} = \frac{x+iy}{\sqrt{64\pi r_0^3}}e^{-r/2r_0}$$

$$\psi_{21-1} = \frac{x-iy}{\sqrt{64\pi r_0^3}}e^{-r/2r_0}$$

## Example 40-3-A

Determine the probability of finding an electron in the $\psi_{200}$ wave function closer to the nucleus then one Bohr radius. Determine the probability of finding an electron in the $\psi_{200}$ wave function farther from the nucleus than one Bohr radius, but closer to the nucleus than two Bohr radii.

## Solution:

In the first part, we need to evaluate

$$\int_{r=0}^{r_0} P_r(r)\,dr = 4\pi \int_{r=0}^{r_0} \left|\psi_{200}(r)\right|^2 r^2\,dr = 4\pi \int_{r=0}^{r_0}\left(\frac{1}{\sqrt{32\pi r_0^3}}\left(2-\frac{r}{r_0}\right)e^{-r/2r_0}\right)^2 r^2\,dr$$

$$= \frac{1}{8r_0^3}\int_{r=0}^{r_0}\left(4 - 4\frac{r}{r_0} + \frac{r^2}{r_0^2}\right)r^2 e^{-r/r_0}\,dr$$

$$= -\frac{1}{8r_0^3}\left[\left(\frac{r^4}{r_0} + 4r^2 r_0 + 8rr_0^2 + 8r_0^3\right)e^{-r/r_0}\right]_{r=0}^{r_0}$$

$$= 1 - \frac{21}{8}e^{-1} \approx 0.034$$

The second part is the same as the first part, but with the limits on the integral changed:

$$\int_{r=0}^{r_0} P_r(r)\, dr = -\frac{1}{8r_0^3}\left[\left(\frac{r^4}{r_0} + 4r^2 r_0 + 8rr_0^2 + 8r_0^3\right)e^{-r/r_0}\right]_{r=0}^{r_0}$$

$$= \frac{21}{8}e^{-1} - 7e^{-2} \approx 0.018$$

### Section 40-4 Complex Atoms; the Exclusion Principle

By complex atoms, we mean atoms or ions with more than one electron. The **atomic number**, $Z$, of an atom is the number of protons in the nucleus of the atom, which is the same as the number of electrons in the neutral atom. When there is more than one electron near a nucleus, the solution of Schrödinger's equation is much more difficult to determine because of the Coulomb potential energy term due to the interaction of any two electrons that must be included in the potential energy. Additional difficulty is created by the fact that all electrons are identical and we can't tell one from another. Despite these difficulties, we can still label electron wave functions with the same four quantum numbers, $n$, $l$, $m_l$, and $m_s$. The wave functions are different quantitatively, but this labeling is still useful.

The **Pauli exclusion principle** states that no two electrons can occupy the same quantum state. This means that one additional complete set of quantum numbers must be used for each additional electron in the atom. The structure of the **periodic table** can be understood on this basis.

### Section 40-5 The Periodic Table of Elements

The periodic table of the elements was originally arranged the elements by their atomic mass and by their similar chemical properties. Today, we know this arrangement also arranges them by their atomic number, $Z$. Most periodic tables of the elements give the average atomic mass of the atoms of that element found in nature. Some periodic tables of elements also give the electron configuration of the lowest energy state neutral atom. The periodic table of the elements on the inside back cover of the text includes this information by giving the configuration of the electrons in the outermost occupied **shell**. A shell is considered to be all of the wave functions for electrons that have the same principle quantum number $n$. The shells have been given names. The $n = 1$ shell is called the K shell, the $n = 2$ shell is called the L shell, and the $n = 3$ shell is called the M shell.

Each shell is divided into **subshells**. A given subshell consists of all the wave functions with the same principle quantum number and the same orbital quantum number, $l$. Rather than using the $l$ values to label the subshell, a corresponding letter is used. The letter s is used for $l = 0$, the letter p is used for $l = 1$, the letter d is used for $l = 2$, and the letter f is used for $l = 3$. There are $2l + 1$ independent spatial wave functions for any given value of $l$ and each spatial wavefunction can have both a spin up and a spin down electron making it possible for a subshell to have as many as $2(2l + 1)$ electrons in it. This means an s subshell can have two electrons, a p subshell can have six electrons, a d subshell can have ten electrons, and so on.

### Example 40-5-A

Determine the ground state electron configuration for a Germanium atom.

### Solution:

Germanium is atomic number 32. We need a ground state configuration with 32 electrons in it. The $n = 1$ shell has two electrons, the $n = 2$ shell has eight electrons (two $s$ electrons and six $p$ electrons), the 3s and 3p subshells adds another eight electrons, the 4s subshell adds two more, the 3d subshell adds ten more. This adds up to 30, leaving the final two electrons for the 4p subshell. So the complete electron configuration of the ground state Germanium atom is $1s^2 2s^2 2p^6 3s^2 3p^6 4s^2 3d^{10} 4p^2$.

### Section 40-6 X-Ray Spectra and Atomic Number

The emission and absorption spectra of the electron in the higher quantum number states in a many-electron atom are difficult to predict because the interaction energy between the electrons is on the same order of size as the transition energy and it is difficult to solve Schrödinger's equation with those interactions included. However, for electron transitions between an $n = 2$ state an $n = 1$ state, the effect of the electron interaction is still present, but is small compared to the size of the Coulomb interaction with the nucleus in large atomic number atoms. It also manifests itself in a relatively simple way. It effectively causes these transitions to have the same energy as for a one-electron atom with one less proton in the nucleus, $Z_{eff} = Z - 1$.

The emission lines from inner shell electron transitions have been given names corresponding to the final electron state with a subscript referring to the initial electron state. The $K_\alpha$ emission line is the emission line that results from the $n = 2$ shell to the $n = 1$ (K) shell. The $K_\beta$ emission line is the emission line that results from the $n = 3$ shell to the $n = 1$ (K) shell.

### Example 40-6-A

Determine the wavelength of the $K_\alpha$ emission from a gold atom.

### Solution:

The atomic number of gold is 79. We calculate the difference in energy of two states using the Bohr formula with $Z - 1$ replacing $Z$:

$$\Delta E = \left[ -\frac{(13.6 \, \text{eV})(Z-1)^2}{n_{final}^2} \right] - \left[ -\frac{(13.6 \, \text{eV})(Z-1)^2}{n_{initial}^2} \right]$$

$$= \left[ -\frac{(13.6 \, \text{eV})(79-1)^2}{1^2} \right] - \left[ -\frac{(13.6 \, \text{eV})(79-1)^2}{2^2} \right] = -62 \, \text{keV}$$

This means the emitted photon must have an energy of 62 keV and its wavelength must be

$$\lambda = \frac{hc}{E} = \frac{(6.63 \times 10^{-34} \, \text{J} \cdot \text{s})(3.00 \times 10^8 \, \text{m/s})}{(62 \times 10^3 \, \text{eV})(1.60 \times 10^{-19} \, \text{J/eV})} = 2.00 \times 10^{-11} \, \text{m}$$

### Section 40-7 Magnetic Dipole Moments; Total Angular Momentum

The magnetic dipole moment, $\mu$, of a classical electron moving in a circle at constant speed is related to its angular momentum, $\mathbf{L}$, by

$$\boldsymbol{\mu} = -\frac{e}{2m} \mathbf{L}$$

where $-e$ is the charge of the electron and $m$ is the mass of the electron. This same result applies to the spatial motion of the electron in quantum mechanics.

We already have learned about the potential energy, $U$, of a magnetic dipole in a magnetic field, $\mathbf{B}$:

$$U = -\boldsymbol{\mu} \cdot \mathbf{B}$$

If we take the direction of the magnetic field as the $z$ direction, then this expression reduces to

$$U = -\mu_z B = \frac{e}{2m} m_l \hbar B$$

where we have used $L_z = m_l \hbar$. If we define a **Bohr magneton**, $\mu_B$, to be

$$\mu_B = \frac{e\hbar}{2m} = 9.27 \times 10^{-27} \, \text{J/T} ,$$

so that the $z$ component of the magnetic dipole moment is given by

$$\mu_z = -\mu_B m_l$$

and the expression for the potential energy simplifies to

$$U = \mu_B m_l B .$$

There is also a magnetic dipole moment related to the spin quantum number, $m_s$, of the electron. The $z$ component of the spin magnetic dipole moment is given by

$$\mu_z = -g\mu_B m_s$$

where $g$ is called the gyromagnetic ratio of the electron and has an approximate value of 2.0023 for a free electron.

The total angular momentum, $\mathbf{J}$, is the sum of the orbital angular momentum, $\mathbf{L}$, and the spin angular momentum, $\mathbf{S}$:

$$\mathbf{J} = \mathbf{L} + \mathbf{S}$$

The magnitude of the total angular momentum is quantized and has a magnitude

$$J = \sqrt{j(j+1)}\,\hbar$$

where $j$ can be any integer between and including $|l - s|$ and $l + s$. The quantum number $j$ is called **the total angular momentum quantum number**. For the electron, $s = \frac{1}{2}$, so

$$j = l + \tfrac{1}{2} \quad \text{or} \quad j = l - \tfrac{1}{2} ,$$

unless $l = 0$, then $j = \frac{1}{2}$. In addition, the $z$ component of the total angular momentum, $J_z$, is given by

$$J_z = m_j \hbar \qquad m_j \in \left\{ -j, -j+1, ..., j-1, j \right\}$$

The electron state can be specified by spectroscopic notation. In this notation, the principal quantum number is given first, then the capitalized letter to specify the orbital quantum number, $l$, and the $s$ subscript equal to the total angular momentum quantum number, $j$.

The orbital motion of the electron creates a magnetic field in which the spin magnetic dipole moves. This adds an additional magnetic dipole interaction term that is proportional to the quantity $\mathbf{L}\cdot\mathbf{S}$ and is called the **spin-orbit interaction**.

### Section 40-8 Fluorescence and Phosphorescence

When light is absorbed by an atom, often it is re-emitted at very close to the same wavelength if the electron makes the same transition between the same two corresponding states during the absorption process and the emission process. It is also possible that the re-emission process could involve multiple photons being emitted because the electron makes transitions between states that are intermediate in energy between the initial and final states of the absorption process. The photons emitted in these multiple emissions will all have lower energy and longer wavelength than the photon initially absorbed by the atom. This type of process is called **fluorescence**.

Typical life times of atoms in excited states are on the order of $10^{-8}$ s. The emission of light by excited atoms takes place on that time scale. Certain states of atoms are metastable states. Atoms in these states can remain in these states for a much longer time on average, even many minutes or hours. When the atoms in these states do make a transition to a lower energy level and emit light in the process, it is called **phosphorescence**.

### Section 40-9 Lasers

A laser creates light through the process of stimulated emission. The word laser began as an acronym for light amplification by stimulated emission of radiation. Atoms in excited states are induced to radiate, making a transition to a lower energy state, by photons with the same energy as the emitted photon passing by the excited atoms. The photon emitted by stimulated emission is in phase and moves the same direction as the stimulating photons, thus amplifying the number of photons in the light traveling in the direction of the stimulating photons.

In a gas at thermal equilibrium, there are more atoms in the lower energy states than in the higher energy states. An atom in the lower energy state can absorb the same wavelength photons as the stimulated emission causes to be emitted from the higher energy state atoms. This process attenuates the light, rather than amplifying it. Because there are more atoms in the lower state than in the higher state in a gas in thermal equilibrium, more absorption than stimulated emission occurs as a light beam passes through a gas. The net result is overall absorption of the beam rather than amplification.

If, however, the gas can be taken out of equilibrium, so that there are more atoms in the higher energy states than in the lower energy states, the light can be amplified as it passes through the gas. The condition of more atoms in the higher energy state than in the lower energy state is called **population inversion**. The population inversion condition is obtained by different means in different types of lasers. If the energy input to cause the population inversion is continuous, the resulting laser is continuous laser. If the energy input to cause the population inversion is pulsed, the laser is called a **pulsed laser**.

### Section 40-10 Holography

A **hologram** is a recording of the interference pattern between a reference source of light and the light reflected from an object. Lasers make this possible because a definite phase relationship must exist between the reference light and the reflected light. The relatively long coherence length of lasers compared to ordinary sources of light makes this possible.

## Practice Quiz

1.  How many quantum numbers are necessary to completely specify an electron state in an atom?

    a) 1
    b) 2
    c) 3
    d) 4

2.  How many possible electron states are there in the $n = 4$ shell of the atom?

    a) 4
    b) 16
    c) 32
    d) 64

3.  Which of the following electron configurations is the correct ground state configuration for a fluorine atom?

    a) $1s^2 1p^6 2s^1$
    b) $1s^2 2s^2 2p^6$
    c) $1s^2 2s^2 3s^2 4s^2 5s^1$
    d) $1s^2 2s^2 2p^5$

4.  How many $m_l$ quantum numbers are the in the $l = 54$ subshell?

    a) 27
    b) 54
    c) 108
    d) 109

5.  Why is $Z - 1$ used in the Bohr formula in a many electron atom when calculating the energy or wavelength of a transition from an $n = 2$ state to an $n = 1$ state rather than $Z$?

    a) It is a relativistic correction because of the high kinetic energy of the 1s electron.
    b) The potential energy of the interaction with the other 1s electron cancels out the interaction of one proton.
    c) For larger atomic numbers, the atomic number is one greater than the number of protons in the nucleus.
    d) The proton charge decreases when it is in a large nucleus.

6.     The process when electromagnetic radiation is absorbed by an atom and then later re-emitted at longer wavelength is called

a) fluorescence.
b) spin-orbit interaction.
c) population inversion.
d) holography.

7.     The condition in a laser in which more atoms are in higher energy atomic states than in lower energy atomic states is called

a) fluorescence.
b) spin-orbit interaction.
c) population inversion.
d) holography.

8.     What is the approximate value of the gyromagnetic ratio of an electron?

a) $9.11 \times 10^{-31}$
b) $1.602 \times 10^{-19}$
c) 1
d) 2

9.     How many different total angular momentum quantum numbers are there for a particle with an orbital angular momentum quantum number $l = 2$ and a spin quantum number $s = 3/2$ ?

a) 2
b) 3
c) 4
d) 6

10.    Which is not a possible set of values of $j$ and $m_j$ resulting from an electron in an $l = 5$ state?

a) $j = 11/2$, $m_j = 11/2$
b) $j = 9/2$, $m_j = 11/2$
c) $j = 9/2$, $m_j = 9/2$
d) $j = 11/2$, $m_j = -1/2$

11.    Determine the probability that an electron in the $\psi_{100}$ wave function of the hydrogen atom is found farther from the nucleus than four Bohr radii.

12.    Determine the wavelength of the $K_\alpha$ X-ray emission from silicon ( $Z=14$ ).

13.    What element would have a $K_\alpha$ X-ray emission at a wavelength of approximately 0.12 nm?

14.    Determine the most probable distance from the nucleus at which you would find the electron in the $\psi_{200}$ wave function in a hydrogen atom.

15.    Determine the $j$ and $m_j$ quantum numbers all the possible states of total angular momentum resulting from an $l = 4$ state and an $s = 1$ state.

## Problem Solutions

3.     The value of $l$ can range from 0 to $n - 1$. Thus for $n = 4$, we have
$l = 0, 1, 2, 3$.
For each $l$ the value of $m_l$ can range from $-l$ to $+l$, or $2l + 1$ values. For each of these there are two values of $m_s$. Thus the total number for each $l$ is $2(2l + 1)$.
The number of states is
$N = 2(0 + 1) + 2(2 + 1) + 2(4 + 1) + 2(6 + 1) = 32$ states.

We start with $l = 0$, and list the quantum numbers in the order $(n, l, m_l, m_s)$;

$(4, 0, 0, -\frac{1}{2})$, $(4, 0, 0, +\frac{1}{2})$, $(4, 1, -1, -\frac{1}{2})$, $(4, 1, -1, +\frac{1}{2})$, $(4, 1, 0, -\frac{1}{2})$, $(4, 1, 0, +\frac{1}{2})$, $(4, 1, 1, -\frac{1}{2})$,
$(4, 1, 1, +\frac{1}{2})$, $(4, 2, -2, -\frac{1}{2})$, $(4, 2, -2, +\frac{1}{2})$, $(4, 2, -1, -\frac{1}{2})$, $(4, 2, -1, +\frac{1}{2})$, $(4, 2, 0, -\frac{1}{2})$, $(4, 2, 0, +\frac{1}{2})$,
$(4, 2, 1, -\frac{1}{2})$, $(4, 2, 1, +\frac{1}{2})$, $(4, 2, 2, -\frac{1}{2})$, $(4, 2, 2, +\frac{1}{2})$, $(4, 3, -3, -\frac{1}{2})$, $(4, 3, -3, +\frac{1}{2})$, $(4, 3, -2, -\frac{1}{2})$,
$(4, 3, -2, +\frac{1}{2})$, $(4, 3, -1, -\frac{1}{2})$, $(4, 3, -1, +\frac{1}{2})$, $(4, 3, 0, -\frac{1}{2})$, $(4, 3, 0, +\frac{1}{2})$, $(4, 3, 1, -\frac{1}{2})$, $(4, 3, 1, +\frac{1}{2})$,
$(4, 3, 2, -\frac{1}{2})$, $(4, 3, 2, +\frac{1}{2})$, $(4, 3, 3, -\frac{1}{2})$, $(4, 3, 3, +\frac{1}{2})$.

7. (a) The principal quantum number is $n = 6$.
   (b) The energy of the state is
   $E_6 = -(13.6 \text{ eV})/n^2 = -(13.6 \text{ eV})/6^2 = -0.378 \text{ eV}$.
   (c) From spdfgh, we see that the "g" subshell has $l = 4$. The magnitude of the angular momentum depends on $l$ only:
   $$L = \hbar[l(l+1)]^{1/2} = (1.055 \times 10^{-34} \text{ J} \cdot \text{s})[(4)(4+1)]^{1/2} = \hbar\sqrt{20} = 4.72 \times 10^{-34} \text{ kg} \cdot \text{m}^2/\text{s}.$$
   (d) For each $l$ the value of $m_l$ can range from $-l$ to $+l$: $m_l = -4, -3, -2, -1, 0, 1, 2, 3, 4$.

11. To see if the ground-state wave function is normalized, we integrate the radial probability density over all radii:
    $$\int_0^\infty 4\pi r^2 |\psi_{100}|^2 \, dr = \int_0^\infty 4\pi r^2 \frac{1}{\pi r_0^3} e^{-2r/r_0} \, dr$$
    We change variable to $x = 2r/r_0$ and use the result for the integration in Ex. 40–4:
    $$\frac{1}{2}\int_0^\infty x^2 e^{-x} dx = \left(-\frac{1}{2}x^2 - x - 1\right)e^{-x}\Big|_0^\infty = 0 - \left(-1e^0\right) = 1$$

15. We form the ratio of the radial probability densities:
    $$\frac{P_r(r_0)}{P_r(2r_0)} = \frac{\dfrac{4\pi r_0^2}{\pi r_0^3} 2^{-2r_0/r_0}}{\dfrac{4\pi (2r_0)^2}{\pi r_0^3} 2^{-4r_0/r_0}} = \frac{e^{-2}}{4e^{-4}} = 1.85$$

19. To find the probability for the electron to be within a sphere of radius $r$, we integrate the radial probability density:
    $$P = \int_0^r \frac{4\pi r'^2}{\pi r_0^3} 2^{-2r'/r_0} \, dr'.$$
    We change variable to $x' = 2r'/r_0$ and use the result for the integration in Ex. 40–4:
    $$P = \frac{1}{2}\int_0^x x'^2 e^{-x'} dx' = \left(-\frac{1}{2}x'^2 - x' - 1\right)e^{-x'}\Big|_0^x = 1 - \left(-\frac{1}{2}x^2 - x - 1\right)e^{-x}$$
    (a) For the probability to be 50% we have
    $(-\frac{1}{2}x^2 - x - 1)e^{-x} = 0.50$.
    A numerical calculation gives $x = 2.68$, so $r = 1.34r_0$.
    (b) For the probability to be 90% we have
    $(-\frac{1}{2}x^2 - x - 1)e^{-x} = 0.90$.
    A numerical calculation gives $x = 5.32$, so $r = 2.7r_0$.
    (c) For the probability to be 99% we have
    $(-\frac{1}{2}x^2 - x - 1)e^{-x} = 0.99$.
    A numerical calculation gives $x = 8.40$, so $r = 4.2r_0$.

23. We find the mean value of $r$ from
    $$\bar{r} = \int_0^\infty r|\psi_{100}|^2 4\pi r^2 \, dr = \int_0^\infty 4\pi r^3 \frac{1}{\pi r_0^3} e^{-2r/r_0} \, dr.$$
    We change variable to $x = 2r/r_0$:
    $$\bar{r} = \frac{r_0}{4}\int_0^\infty x^3 e^{-x} dx = \frac{r_0}{4} 3! = \frac{3}{2} r_0$$

27. (a) The $n = 3$, $l = 0$ wave function is

    $$\psi_{300} = \frac{1}{\sqrt{27\pi r_0^3}}\left(1 - \frac{2r}{3r_0} + \frac{2r^2}{27r_0^2}\right)e^{-r/3r_0}$$

    (b)

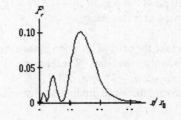

The radial probability distribution is

$$P_r(r) = 4\pi r^2 |\psi_{300}|^2 = \frac{4r^2}{27r_0^3}\left(1 - \frac{2r}{3r_0} + \frac{2r^2}{27r_0^2}\right)^2 e^{-2r/3r_0}$$

(c)  To find the most probable distance from the nucleus, we find the maxima by setting the first derivative equal to zero.  If we change variable to $x = r/r_0$, we have

$$P_r(x) = \frac{4}{27r_0}\left(1 - \frac{2x}{3} + \frac{2x^2}{27}\right)^2 x^2 e^{-2x/3}$$

If we suppress the constant in front, the derivative is

$$\frac{dP_r}{dx} = \left(1 - \frac{2x}{3} + \frac{2x^2}{27}\right)^2\left(2x - \frac{2x^2}{3}\right)e^{-2x/3} + 2x^2 e^{-2x/3}\left(1 - \frac{2x}{3} + \frac{2x^2}{27}\right)\left(-\frac{2}{3} + \frac{4x}{27}\right)$$

$$= 2xe^{-2x/3}\left(1 - \frac{2x}{3} + \frac{2x^2}{27}\right)\left(1 - \frac{5x}{3} + \frac{12x^2}{27} - \frac{2x^3}{81}\right) = 0$$

Thus we have

$x = 0$, which is the minimum at $r = 0$;

$$\left(1 - \frac{2x}{3} + \frac{2x^2}{27}\right) = 0$$

Because this is a factor in $\psi$, its two solutions are the other minima at $x = 1.90, 7.10$.

$$\left(1 - \frac{5x}{3} + \frac{12x^2}{27} - \frac{2x^3}{81}\right) = 0$$

which can be solved numerically for the three solutions,
which are $x = 0.74, 4.19, 13.07$.  These correspond to the three peaks in the distribution.
The highest peak is the most probable distance: $r = 13r_0$.

31.  (a)  Selenium has $Z = 34$:
     $1s^2 2s^2 2p^6 3s^2 3p^6 3d^{10} 4s^2 4p^4$.
     (b)  Gold has $Z = 79$:
     $1s^2 2s^2 2p^6 3s^2 3p^6 3d^{10} 4s^2 4p^6 4d^{10} 4f^{14} 5s^2 5p^6 5d^{10} 6s^1$.
     (c)  Uranium has $Z = 92$:
     $1s^2 2s^2 2p^6 3s^2 3p^6 3d^{10} 4s^2 4p^6 4d^{10} 4f^{14} 5s^2 5p^6 5d^{10} 6s^2 6p^6 5f^3 6d^1 7s^2$.

35.  In a filled subshell, we have $2(2l + 1)$ electrons.  All of the $m_l$ values
     $-l, -l + 1, \dots, 0, \dots, l - 1, l$
     are filled, so their sum is zero.  For each $m_l$ value, both values of $m_s$ are filled, so their sum is also zero.  Thus the total angular momentum is zero.

39.  The energy of the photon with the shortest wavelength must equal the maximum kinetic energy of an electron:
     $hf_0 = hc/\lambda_0 = eV$,  or
     $\lambda_0 = hc/eV = (6.63 \times 10^{-34}\text{ J} \cdot \text{s})(3 \times 10^8\text{ m/s})(10^9\text{ nm/m})/(1.60 \times 10^{-19}\text{ J/eV})(1\text{ e})V$
     $= (1.24 \times 10^3\text{ V} \cdot \text{nm})/V$.

43.  The $K_\alpha$ line is from the $n = 2$ to $n = 1$ transition.  We use the energies of the hydrogen atom with $Z$ replaced by $Z - 1$.  Thus we have
     $hf = \Delta E \propto (Z - 1)^2$,  so  $\lambda \propto 1/(Z - 1)^2$.
     When we form the ratio for the two materials, we get
     $\lambda_X/\lambda_{iron} = (Z_{iron} - 1)^2/(Z_X - 1)^2$;
     $(229\text{ pm})/(194\text{ pm}) = (26 - 1)^2/(Z_X - 1)^2$, which gives $Z_X = 24$,
     so the material is chromium.

47.  (a)  We take the original direction as the $x$-axis and the direction of the force from the magnetic field gradient as the $z$-axis.  The force is
     $F_z = \mu_z\, dB/dz = -g\mu_B m_s\, dB/dz$.
     This constant force will produce a constant acceleration.  The time to traverse the field is

$t = x/v_0 = (4.0 \times 10^{-2} \text{ m})/(700 \text{ m/s}) = 5.71 \times 10^{-5} \text{ s}.$

so the deflection of one of the beams is

$\begin{aligned} z &= \tfrac{1}{2}at^2 = \tfrac{1}{2}(F_z/m)t^2 = \tfrac{1}{2}[(g\mu_B m_s \, dB/dz)/m]t^2 \\ &= \tfrac{1}{2}[(2.0023)(9.27 \times 10^{-24} \text{ J/T})(\tfrac{1}{2})(1.5 \times 10^3 \text{ T/m})/(107.9)(1.66 \times 10^{-27} \text{ kg})](5.71 \times 10^{-5} \text{ s})^2 \\ &= 1.26 \times 10^{-4} \text{ m}. \end{aligned}$

Thus the separation is

$2z = 2.52 \times 10^{-4} \text{ m} = 0.25 \text{ mm}.$

(b) Because the separation is proportional to $g$, we have

$2z' = (1/2.0023)(2.52 \times 10^{-4} \text{ m}) = 1.26 \times 10^{-4} \text{ m} = 0.13 \text{ mm}.$

51. (a) The additional energy term for the spin-orbit interaction is

$U = -\mu_s B_n = -g\mu_B m_s B_n \, .$

Thus the separation of energy levels is

$\Delta U = \Delta m_s \, g\mu_B B_n \, ;$

$(5 \times 10^{-5} \text{ eV})(1.60 \times 10^{-19} \text{ J/eV}) = [\tfrac{1}{2} - (-\tfrac{1}{2})](2.0023)(9.27 \times 10^{-24} \text{ J/T})B_n,$ which gives $B_n = 0.4 \text{ T}.$

(b) If we consider the nucleus to be a charge $e$ revolving in a circle of radius $r$, the effective current is

$I = e/(2\pi r/v) = ev/2\pi r = m_e evr/2\pi m_e r^2 = eL/2\pi m_e r^2,$

where $L$ is the orbital angular momentum of the electron, because the nucleus has the same $v$ and $r$. The electron is at the center of this circular current, so the magnetic field is

$B = \mu_0 I/2r = \mu_0 eL/4\pi m_e r^3.$

If we use the Bohr quantization, $L = n\hbar$, and $r = n^2 r_0$, we have

$\begin{aligned} B &= \mu_0 e n \hbar/4\pi m_e n^6 r_0^3 = (\mu_0/4\pi)e\hbar m_e n^5 r_0^3 \\ &= (10^{-7} \text{ T} \cdot \text{m/A})(1.60 \times 10^{-19} \text{ C})(1.055 \times 10^{-34} \text{ J} \cdot \text{s})/(9.11 \times 10^{-31} \text{ kg})(2)^5(0.529 \times 10^{-10} \text{ m})^3 \\ &= 0.4 \text{ T}. \end{aligned}$

This is consistent with the result from part (a).

55. We find the temperature from

$\dfrac{N_2}{N_0} = e^{-\Delta E/kT} \, ;$

$\dfrac{1}{2} = e^{-(2.2 \text{ eV})(1.60 \times 10^{-19} \text{ J/eV})/(1.38 \times 10^{-23} \text{ J/K})(300 \text{ K})},$ which gives $T = 3.7 \times 10^4 \text{ K}$

59. (a) For $Z = 27$ we start with hydrogen and fill the levels as indicated in the periodic table:

$1s^2 2s^2 2p^6 3s^2 3p^6 3d^7 4s^2.$

Note that the $4s^2$ level is filled before the 3d level is completed.

(b) For $Z = 36$ we have

$1s^2 2s^2 2p^6 3s^2 3p^6 3d^{10} 4s^2 4p^6.$

(c) For $Z = 38$ we have

$1s^2 2s^2 2p^6 3s^2 3p^6 3d^{10} 4s^2 4p^6 5s^2.$

Note that the $5s^2$ level is filled before the 4d level is started.

63. The $n = 2$, $l = 0$ wave function is

$$\psi_{200} = \frac{1}{\sqrt{32\pi r_0^3}}\left(2 - \frac{r}{r_0}\right)e^{-r/2r_0}$$

The radial probability density is

$$P_r(r) = 4\pi r^2 |\psi_{200}|^2 = \frac{r^2}{8r_0^3}\left(2 - \frac{r}{r_0}\right)^2 e^{-r/r_0}$$

To find the most probable distance from the nucleus, we find the maxima by setting the first derivative equal to zero. If we change variable to $x = r/r_0$, we have

$$P_r(r) = \frac{1}{8r_0}x^2(2 - x)^2 e^{-x}$$

If we suppress the constant in front, the derivative is

$$\frac{dP_r}{dx} = 2x(2-x)^2 e^{-x} - 2x^2(2-x)e^{-x} - x^2(2-x)^2 e^{-x}$$

$$= x(2-x)(x^2 - 6x + 4)e^{-x} = 0$$

Thus we have

$x = 0$, which is the minimum at $r = 0$;

$2 - x = 0$.

Because this is a factor in $\psi$, this is the other minimum at $x = 2$.

$x^2 - 6x + 4 = 0$, with solutions $x = 0.76, 5.24$. These correspond to the two peaks in the distribution.

The highest peak is the most probable distance: $r = 5.24r_0$.

67. (*a*) The additional energy term in a magnetic field is

$U = -\mu_s B_n = -g\mu_B m_s B_n$ .

Thus the separation of energy levels is

$\Delta U = \Delta m_s\, g\mu_B B_n$

$= [\tfrac{1}{2} - (-\tfrac{1}{2})](2.0023)(9.27 \times 10^{-24}\ \text{J/T})(1.0\ \text{T}) = 1.9 \times 10^{-23}\ \text{J} = 1.2 \times 10^{-4}\ \text{eV}.$

(*b*) We find the wavelength from

$\lambda = (1.24 \times 10^3\ \text{eV} \cdot \text{nm})/\Delta U = (1.24 \times 10^3\ \text{eV} \cdot \text{nm})/(1.2 \times 10^{-4}\ \text{eV}) = 1.07 \times 10^7\ \text{nm} = 1.1\ \text{cm}.$

(*c*) From the periodic table we see that the participating electron is in the $5s^1$ state. Thus the splitting for both atoms is of a single *s*-state electron, so there will be no difference.

71. Each shell with quantum number $n$ can contain $2n^2$ electrons. Thus the maximum number of electrons in the shells from $n = 1$ to $n = 6$ is

$$N = \sum_{n=1}^{6} 2n^2 = 2\left(\tfrac{1}{6}\right)n(n+1)(2n+1)\Big|_{n=6} = 182$$

Because each electron corresponds to one proton in an atom, there would be a maximum of 182 elements.

# Chapter 41: Molecules and Solids

## Chapter Overview and Objectives

This chapter covers topics related to the interaction between atoms. Atoms can bond to each other to form molecules and solids. This chapter discusses the mechanisms responsible for bonding and some of the properties expected of atoms bound together in molecules and solids.

After completing study of this chapter, you should:

- Know what bonding is and the different types of bonds between atoms.
- Know the general form of the potential energy function for the interaction of two molecules.
- Know what van der Waals bonding is.
- Know what contributes to the vibrational-rotational spectra of a diatomic molecule.
- Know what the free electron theory of metals is.
- Know what band theory of solids is.
- Know what conditions determine whether a material will be a conductor, an insulator, or a semiconductor
- Know how semiconductor materials can be used in useful semiconductor electronic components

## Summary of Equations

Simple model of interatomic potential energy:
$$U = U_{repulse} + U_{attract} = -\frac{A}{r^m} + \frac{B}{r^n}$$

Van der Waals interaction potential:
$$U_{van\,der\,Waals} = -\frac{A}{r^6}$$

Rotational energy levels:
$$E_{rot} = \frac{\hbar^2 l(l+1)}{2I} \qquad l \in \{0, 1, 2, ...\}$$

Definition of reduced mass:
$$\mu = \frac{m_1 m_2}{m_1 + m_2}$$

Vibrational energy levels:
$$E_{vib} = \left(v + \tfrac{1}{2}\right)hf \qquad v \in \{0, 1, 2, ...\}$$

Free electron density of states:
$$g(E) = \frac{8\sqrt{2}\pi\, m^{3/2}}{h^3} E^{1/2}$$

Fermi-Dirac distribution function:
$$f(E) = \frac{1}{e^{(E-E_F)/kT} + 1}$$

## Chapter Summary

### Section 41-1 Bonding in Molecules

When two one-electron atoms are near each other, the potential energy function for either electron is altered by the presence of the other nucleus and electron. Because of the altered potential energy function, the solutions to Schrödinger's equation also change. In general, the sum of the energies for the lowest energy states for the two electrons when two nuclei are present is lower in energy than the sum of the two energies for two isolated atoms, unless the nuclei approach too close together. Because the energy is lower when the two nuclei are near each other, it requires the addition of energy to separate the nuclei. The energy that must be added to separate the atoms when they are in their minimum energy configuration is called the **binding energy**.

Sometimes the type of bond that is formed between the atoms is classified by how the new wave function of the electrons is distributed around the two nuclei. If the redistribution of the probabilities of finding the electrons is distributed equally around the two nuclei, the bond is called a **covalent bond**. If the new wave function transfers most of the probability of finding the electrons from one atom to the other atom, the bond is called an **ionic bond**. Bonds between dissimilar atoms are, in general, ionic to some degree. One of the physically important effects of the probability redistribution is the electric dipole moment of the molecule that results. A molecule that has an electric dipole moment is called a **polar** molecule.

### Section 41-2 Potential-Energy Diagrams for Molecules

The Schrödinger equation for two or more atoms in close proximity is very difficult to solve. However, we can make some simple qualitative statements about the interactions between atoms that should apply in all cases. We look at the case of two atoms interacting, but the ideas can be extended to apply to any number of atoms interacting.

For relatively large distances between two atoms, there is always an attractive force (see the next section). A simple model potential energy function of an attractive force that increases in magnitude as the distance decreases is

$$U_{attract} = -\frac{A}{r^m}$$

where $A$ and $m$ are positive numbers and r is the distance between the two atoms. For example, for the attractive Coulomb potential between two charges of opposite sign, $m = 1$.

As atoms become closer together, it is easy to see that electrons must remain between the two nuclei or the repulsive Coulomb force of the two nuclei will push the atoms apart. However, as the distance between the two nuclei decreases, the electrons between the two nuclei have a small volume of space to move in and still remain between the two nuclei. The uncertainty principle forces the uncertainty in momentum, and thus, the kinetic energy of the electrons to increase. This raises the total energy of the system. The conclusion is that the effective potential energy of the molecule begins to increase as the atoms approach close enough. A simple model potential-energy function for a repulsive force that increases in magnitude as the distance decreases is given by

$$U_{repulse} = \frac{B}{r^n}$$

where $B$ and $n$ are positive numbers and $r$ is the distance between the two atoms. Again, a particular potential that has this form is the repulsive Coulomb potential energy of two charges with the same sign of charge. In that case, $m = 1$.

The total model effective potential-energy function of two atoms as a function of the separation of the two atoms is given by

$$U = U_{repulse} + U_{attract} = -\frac{A}{r^m} + \frac{B}{r^n}$$

To ensure that this function is repulsive for small $r$ and attractive for large $r$, the value of $n$ must be larger than the value of $m$. A function of this form, with the given condition on $n$ and $m$, has the general behavior shown in the graph. The potential energy goes to zero for infinite distances. The potential energy reaches a minimum as the distance decreases and then the potential energy increases as the distance decreases further.

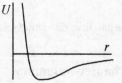

Remember that this is an *effective* potential-energy function. It includes the kinetic energy of the electrons and the Coulomb potential energies. We have also ignored the minimum momentum of the nuclei required by the uncertainty principle at this point. Also, remember that we have based the model on the behavior at large distances and small distances. At intermediate distances, the behavior could be more complicated.

**Example 41-2-A**

For a potential energy function given by

$$U(r) = -\frac{A}{r^m} + \frac{B}{r^n}$$

determine the equilibrium position in such a potential and determine the potential energy at the equilibrium separation.

**Solution:**

The equilibrium position is located where the derivative of the potential with respect to position is zero:

$$\left.\frac{dU}{dr}\right|_{r_0} = 0$$

where $r_0$ is the equilibrium position. We take the derivative of the potential energy function and determine where it is zero:

$$\frac{dU}{dr} = m\frac{A}{r^{m+1}} - n\frac{B}{r^{n+1}} = 0 \quad \Rightarrow \quad r_0 = \left(\frac{nB}{mA}\right)^{\frac{1}{(n-m)}}$$

If we substitute this distance into the expression for the potential energy, we get

$$U(r_0) = -\frac{A}{\left(\frac{nB}{mA}\right)^{\frac{m}{n-m}}} + \frac{B}{\left(\frac{nB}{mA}\right)^{\frac{n}{n-m}}}$$

## Section 41-3 Weak (van der Waals) Bonds

All atoms are attracted to each other by an induced dipole interaction called the **van der Waals force**. Fluctuations in the electron density of an atom create an electric dipole moment. These fluctuating dipole moments create induced dipole moments in nearby atoms caused by the electric field of the first atom. The force of one electric dipole on the other is attractive. We can write a potential energy function for this attractive force and it has a $1/r^6$ distance dependence[1]:

$$U_{van\,der\,Waals} = -\frac{A}{r^6}$$

The value of $A$ depends on the polarizability of the atoms. This is the greatest contribution to the attractive force of inert gas molecules, even at close distances.

## Section 41-4 Molecular Spectra

The electrons in molecules can make electronic transitions, just as the electrons in atoms do. In addition, molecules can have rotational and vibrations energy levels. The molecules can make transitions between these levels, also. If the transition is from a higher energy state to a lower energy state, electromagnetic energy can be emitted during the transition. If the transition is from a lower energy to state to a higher energy state, it may be the result of the absorbtion of electromagnetic radiation.

---

[1] If the separation becomes large enough, a relativistic effect, called the Casimir effect, changes the dependence to $1/r^7$. The Casimir effect, in simple terms, can be seen as the result of the finite speed of light. The electric field of one fluctuating dipole takes so much time to get to the other atom that it polarizes the other atom after the first atom has already changed its own polarization.

The rotational energy levels of molecules are given by

$$E_{rot} = \frac{L^2}{2I}$$

where $L$ is the magnitude of the angular momentum of the molecule and $I$ is the moment of inertia of the molecule about the rotation axis.  The angular momentum is quantized, so the energy levels are quantized.  If we use the angular momentum quantization condition:

$$L = \hbar\sqrt{l(l+1)}\,\hbar \qquad l \in \{0, 1, 2, ...\}$$

Writing the the energy levels in terms of the angular momentum quantum number $l$, we get

$$E_{rot} = \frac{\hbar^2 l(l+1)}{2I}$$

Transitions between rotational energy levels that involve the emission or absorption of a single photon must satisfy the selection rule:

$$\Delta l = \pm 1$$

If we calculate the transition energies allowed by this selection rule, we get the difference between two energy levels with a difference of one in their angular momentum quantum numbers.  If we call the greater angular momentum quantum number $l$, then the other angular momentum quantum number will be $l - 1$ and we get

$$\Delta E_{rot} = \frac{\hbar^2 l(l+1)}{2I} - \frac{\hbar^2 (l-1)(l-1+1)}{2I} = \frac{\hbar^2}{I} l$$

The moment of inertia of a diatomic molecule can be approximated by two point masses separated by the equilibrium distance between the atoms in the molecule.  The moment of inertia about an axis through the center of mass perpendicular to the line joining the two masses is given by

$$I = \frac{m_1 m_2}{m_1 + m_2} r^2$$

where $r$ is the equilibrium distance between the two nuclei, and $m_1$ and $m_2$ are the masses of the two nuclei.  We define the reduced mass, $\mu$, of the system by

$$\mu = \frac{m_1 m_2}{m_1 + m_2}$$

and write the moment of inertia as

$$I = \mu r^2$$

The vibrational energy levels can be approximated by the energy levels of a simple harmonic oscillator.  These energy levels are given by

$$E_{vib} = \left(v + \tfrac{1}{2}\right)hf \qquad v \in \{0, 1, 2, ...\}$$

where $v$ is called the **vibrational quantum number**.  For transitions involving the absorption or emission of a single photon, there is a selection rule:

$$\Delta v = \pm 1$$

When a molecule changes its vibrational or rotational energies level because of absorption or emission of a single photon, both the rotational and vibrational selection rules must be satisfied. This implies that the changes in energy of the molecule for the absorption of a photon are

$$\Delta E = \Delta E_{vib} + \Delta E_{rot}$$

$$= hf + (l+1)\frac{\hbar^2}{I} \quad \text{or} \quad hf - l\frac{\hbar^2}{I}$$

where $l$ is the angular momentum quantum number of the initial state of the molecule. The changes in energy of the molecule for the emission of a photon are

$$\Delta E = \Delta E_{vib} + \Delta E_{rot}$$

$$= -hf + (l+1)\frac{\hbar^2}{I} \quad \text{or} \quad -hf - l\frac{\hbar^2}{I}$$

where $l$ is the angular momentum quantum number of the initial state of the molecule.

### Example 41-4-A

A diatomic molecule is in the $l = 2$ angular momentum quantum state. It absorbs photons of wavelength 2.4921 μm and wavelength 2.4936 μm. What are the vibrational frequency and the reduced mass of the diatomic molecule?

**Solution:**

The two absorption wavelengths correspond to transition energies

$$\Delta E_1 = \frac{hc}{\lambda_1} = \frac{1.240\,\text{eV} \cdot \mu\text{m}}{2.4921\,\mu\text{m}} = 0.49757\,\text{eV} = 7.9711 \times 10^{-20}\,\text{J}$$

and

$$\Delta E_2 = \frac{hc}{\lambda_2} = \frac{1.240\,\text{eV} \cdot \mu\text{m}}{2.4936\,\mu\text{m}} = 0.49727\,\text{eV} = 7.9663 \times 10^{-20}\,\text{J}$$

We use the two solutions for the absorbed photon energies to determine the reduced mass of the molecule:

$$\Delta E_1 = hf + (l+1)\frac{\hbar^2}{I} \quad \text{and} \quad \Delta E_2 = hf - l\frac{\hbar^2}{I}$$

We solve these two equations for the moment of inertia of the molecule and the vibrational frequency of the molecule:

$$I = \frac{(2l+1)\ \hbar^2}{\Delta E_1 - \Delta E_2} = \frac{(2 \cdot 2 + 1)(6.63 \times 10^{-34}\,\text{J} \cdot \text{s})^2}{7.9711 \times 10^{-20}\,\text{J} - 7.9663 \times 10^{-20}\,\text{J}} = 4.6 \times 10^{-44}\,\text{kg} \cdot \text{m}^2$$

$$f = \frac{l\Delta E_1 + (l+1)\Delta E_2}{(2l+1)h} = \frac{2(7.9711 \times 10^{-20}\,\text{J}) + (2+1)(7.9663 \times 10^{-20}\,\text{J})}{(2 \cdot 2 + 1)(6.63 \times 10^{-34}\,\text{J} \cdot \text{s})} = 1.20 \times 10^{14}\,\text{Hz}$$

### Section 41-5 Bonding in Solids

A collection of positive and negative ions can bond together into a solid. The attractive potential energy can be modeled as the Coulomb potential energy. To find the total Coulomb potential energy of a collection of ions we need to find the sum

$$U = \sum \frac{kQ_i Q_j}{r_{ij}}$$

where $r_{ij}$ is the distance between ions with charges $Q_i$ and $Q_j$ respectively. The sum is over every pair of the ions in the collection. If the ions are in orderly array, called a crystal, we can rewrite the sum in terms of a characteristic length of the array, $r$. In that case, we can write the potential energy function as a function of $r$:

$$U = -\frac{\alpha\,ke^2}{r}$$

The constant $\alpha$ is called the **Madelung constant** and its value depends on the way in which the atoms are arranged in the solid.

Atoms in metals are bonded by **metallic bonds**. Metallic bonds can be considered as an extreme case of covalent bonds. The electrons are essentially shared by all of the atoms in the solid.

### Section 41-6 Free-Electron Theory of Metals

As a simple approximation, a metal can be treated as a relatively deep square potential well in which the least tightly bound electrons move. Within the well, the potential energy is zero everywhere. The electrons are free to move about inside the well. Because the number of these electrons is very large in any macroscopic sized piece of material and the electron states are very closely spaced in energy, the energy of the states can be treated as a continuous variable, even though they are still discrete states. The **density of states**, $g(E)$, is the number of states per unit energy level. The number of states, $dN$, between energy $E$ and energy $E + dE$ is related to the density of states by

$$dN = g(E)dE$$

It is relatively easy to show that the density of states for the model of free electrons in a square well is given by

$$g(E) = \frac{8\sqrt{2}\pi\,m^{3/2}}{h^3}E^{1/2}$$

where $m$ is the electron mass.

At zero temperature, the electrons in a metal would be in the lowest energy state. Because of the Pauli exclusion principle, the electrons fill up states to some finite energy, called the **Fermi level**, above the lowest energy state in the metal. The energy of this highest filled state at zero temperature is called the **Fermi energy**, $E_F$. At temperatures above absolute zero, the probability that an electron state is occupied is given by the **Fermi-Dirac distribution function**

$$f(E) = \frac{1}{e^{(E-E_F)/kT} + 1}$$

This function replaces the Boltzman distribution function that we used in the kinetic theory of gases. The reason the probability distribution function is different is because the Pauli exclusion principle applies to electrons. The **density of occupied states**, $n_o$, is the number of electrons in states in an energy interval between $E$ and $E + dE$:

$$n_o(E) = g(E)f(E) = \frac{8\sqrt{2}\pi\,m^{3/2}}{h^3}\frac{E^{1/2}}{e^{(E-E_F)/kT} + 1}$$

### Example 41-6-A

Determine the probability that states at energies $E_F - 10kT$, $E_F - kT$, $E_F - 0.1kT$, $E_F$, $E_F + 0.1kT$, $E_F + kT$, and $E_F + 2kT$ are occupied.

### Solution:

The probability that a given state is occupied is given by the Fermi-Dirac distribution function:

$$f(E_F - 10kT) = \frac{1}{e^{(E_F - 2kT - E_F)/kT} + 1} = \frac{1}{e^{-10} + 1} = 0.99995$$

$$f(E_F - kT) = \frac{1}{e^{(E_F - kT - E_F)/kT} + 1} = \frac{1}{e^{-1} + 1} = 0.731$$

$$f(E_F - 0.1kT) = \frac{1}{e^{(E_F - 0.1kT - E_F)/kT} + 1} = \frac{1}{e^{-0.1} + 1} = 0.525$$

$$f(E_F - 10kT) = \frac{1}{e^{(E_F - E_F)/kT} + 1} = \frac{1}{e^0 + 1} = 0.500$$

$$f(E_F + 0.1kT) = \frac{1}{e^{(E_F + 0.1kT - E_F)/kT} + 1} = \frac{1}{e^{0.1} + 1} = 0.475$$

$$f(E_F + 1kT) = \frac{1}{e^{(E_F + kT - E_F)/kT} + 1} = \frac{1}{e^1 + 1} = 0.269$$

$$f(E_F + 10kT) = \frac{1}{e^{(E_F + 10kT - E_F)/kT} + 1} = \frac{1}{e^{10} + 1} = 4.5 \times 10^{-5}$$

## Section 41-7 Band Theory of Solids

The discrete electronic energy levels of isolated atoms are spread into many states that have energies that are very close in energy to each other when the atoms are assembled into a solid. The energies are spread into this range of closely spaced energies because of the interaction of the electrons with the other atoms in the solid. These states are now called a band. The conductivity of a material can be understood in terms of the band theory of solids. A material that has a partially filled band in its highest occupied band is a **conductor**. The electrons in a partially filled band have plenty of empty states that the electrons can be excited into so that an electric field applied to the material can change the momentum of the electrons.

An insulator has a filled highest occupied band. To give some momentum to an electron in an insulator, the applied electric field would need to transfer an energy equal to the energy from the top of the filled band, called the **valence band**, up to the bottom of the next higher empty band, called the **conduction band**. This is more energy than the electric field can transfer to the electron, so no electrons can move. The difference in energy between the highest energy of the valence band and the lowest energy of the conduction band is called the **band gap** of the material.

If the band gap is not too large, the finite temperature of a semiconductor will give a reasonable probability of electrons having an energy that will place them in the conduction band. Such a material is called a **semiconductor**. The electrons that are thermally excited into the conduction band can gain momentum from an applied electric field because there are plenty of empty states available in the conduction band at nearby energies. Likewise, the empty states left behind in the valence allow the other electrons still in the valence to gain momentum from the electric field. The behavior of the motion of the charges remaining in the valence band can be understood by treating the empty states as positively charged particles called **holes**.

## Section 41-8 Semiconductors and Doping

The Fermi level and the conductivity of a semiconductor can be manipulated by the addition of impurities to the semiconductor. The addition of impurities is called doping. Atoms with five valence electrons, called **donor impurities**, added to the semiconductor have a high probability of their fifth electron being excited to the conduction band of the semiconductor, raising its conductivity in proportion to the number of impurity atoms added, as long as the impurity concentration remains small. It also raises the Fermi energy of the semiconductor toward the conduction band. When a semiconductor is doped in this way, it is called an **n-type semiconductor**. The $n$ is because the charge carriers are negative electrons.

Atoms with three valence electrons, called **acceptor impurities**, added to the semiconductor have a high probability of an electron from the valence band of the semiconductor to be thermally excited into their valence shell. This leaves an empty state or hole in the valence band of the semiconductor that is relatively free to move, again raising the conductivity of the semiconductor in proportion to the number of impurity atoms added, as long as the impurity concentration remains small. It also lowers the Fermi energy of the semiconductor toward the valence band. When a semiconductor is doped in this way, it is called a **p-type semiconductor**. The $p$ is because the charge carriers are effectively the holes in the valence band and their dynamics can be understood as if they were positively charged particles.

## Section 41-9 Semiconductor Diodes

A semiconductor that consists of both *p* and *n* type materials in contact forms what is called a **pn junction diode**. The boundary between the two types of materials is called a **pn junction**. Such a structure conducts when an applied potential difference across the diode is in a direction to cause conventional current to flow in the direction from p-type to the n-type material, but the structure does not conduct when the potential difference is reversed.

## Section 41-10 Transistors and Integrated Circuits

Transistors are semiconductor devices that have the ability to control either a large voltage or current with a small voltage or current. The function they are commonly used for is amplifying signals. The signal to be amplified is the small voltage or current that controls the output signal that is the large voltage or current.

Transistors can be divided into two general categories, junction transistors and field effect transistors. These two types of transistors are constructed differently, function on a different basis, and have different electrical characteristics. A bipolar junction transistor (**BJT**) is made from a semiconductor that has been doped so that an n-type semiconductor layer of material is between two p-type semiconductor layers.[2] This is called a pnp transistor. Alternatively, the bipolar junction transistor can be made so that a p-type semiconductor layer lies between two n-type semiconductor layers. This is called an npn transistor. A terminal is connected to each of the three regions of the semiconductor. The terminals are named the emitter, base, and collector terminals. The base is always the middle layer of the structure. In a circuit, these transistors have a relatively large current that flows from the emitter terminal to the collector terminal. This current passes through the base region of the transistor. A small percentage of the carriers from the emitter recombine with the opposite type of carriers in the base as the carriers travel from the emitter to the collector. This results in a relatively small base current. If the base current is controlled, it will control the larger emitter to collector current as the percentage of the emitter to collector current that ends up as base current remains relatively constant.

Field effect transistors could also be called voltage-controlled resistors. The structure of a field effect transistor can be modeled as a slab of semiconductor of one type with a terminal connected to it at each end. These terminals are called the source and the drain. Along the length of this slab of material lies a region of the opposite type of semiconductor called the gate. The pn junction that is formed, effectively reduces the cross-sectional area of the source-to-drain slab of material and increases it resistance. When the pn junction is reverse biased, the pn junction expands, further reducing the conductive cross-sectional area of the source-drain slab. This increases the resistance of the source-drain slab. Very little current flows through the reverse biased pn junction, so the gate current is much smaller than the base current of bipolar junction transistor. In another type of field effect transistor, there is a thin insulating layer placed between the gate terminal and the pn junction, which reduces the gate current even further. The field effect transistor without the insulating layer on the gate connection is called a junction field effect transistor (**JFET**) and the field effect transistor with the insulating layer on the gate is called an insulated gate field effect transistor (**IGFET**). A common construction method for making IGFETs is to form the insulating layer by oxidizing the semiconductor. In this case the field effect transistor is called a metal oxide semiconductor field effect transistor (**MOSFET**).

**Integrated circuits** contain many transistors and other circuit components on a single piece of semiconductor.

## Practice Quiz

1.    Consider a diatomic molecule made from two atoms from the same row but two different columns from the periodic table of elements. From the combinations listed, which would be the most ionic bond (i.e. most polar molecule)?

      a) I and VII
      b) II and VI
      c) III and V
      d) IV and IV

---

[2] Note: Don't think of this structure as separate pieces of material that have been sandwiched together. Although, it might be possible to physically put three separate slabs of material together and have them function as a transistor, it would be very difficult to do so and the electrical characteristics of such a transistor would probably be poor. Transistors are usually made from a single slab of material in which the impurities have been introduced in a manner to create the desired structure.

2.    The reduced mass of a two-body system with different masses is always

      a) greater than both masses.
      b) less than both masses.
      c) greater than the smaller mass and less than the larger mass.
      d) less than the smaller mass and greater than the larger mass.

3.    At absolute zero temperature, what is the probability of a state with an energy greater than the Fermi energy being occupied?

      a) 1
      b) ½
      c) 0
      d) It depends on how much greater than the Fermi energy the energy of the state is.

4.    The highest occupied energy band of a material has $2.98 \times 10^{24}$ states for electrons. The material has $1.48 \times 10^{24}$ electrons in this band. The energy gap to the next higher band is 5.8 eV. At room temperature, this material is a

      a) conductor.
      b) insulator.
      c) semiconductor.
      d) either an insulator or a semiconductor.

5.    At any temperature, the probability of an electron occupying a state with an energy located near the Fermi energy is

      a) 1
      b) ½
      c) 0
      d) ¼

6.    The selection laws that the change in the angular momentum quantum number and the vibrational quantum number of a molecule change by ±1 apply to

      a) all transitions of the molecule involving rotation or vibration.
      b) only collisions of molecules with other molecules.
      c) only transitions involving the emission or absorption of a single photon.
      d) all transitions involving only rotational transitions.

7.    A semiconductor that has been doped with an acceptor impurity is a

      a) n-type semiconductor.
      b) p-type semiconductor.
      c) pn junction diode.
      d) intrinsic semiconductor.

8.    The force between any two neutral atoms at distances greater than many atomic diameters is always

      a) repulsive.
      b) attractive.
      c) exactly zero.
      d) larger than when the atoms are closer to each other.

9.    When will a material that is a semiconductor at room temperature become an insulator?

      a) when the temperature becomes low enough
      b) when the temperature becomes high enough
      c) never
      d) when it has a few impurity atoms added to it

10.    What is the largest force that holds a solid mass of ions together?

      a) gravitational force
      b) van der Waals force
      c) electrostatic force
      d) magnetic force

11.    Determine the electrostatic potential energy of a pair of ions, one with charge $+e$ and one with charge $-e$, that are distance of 0.12 nm apart.

12.    A metal has a Fermi energy of 3.2 eV. Determine the probability that a state with an energy of 3.4 eV is occupied at a temperature of 20° C. Determine the speed of electrons that have the Fermi energy.

13.    Determine the absorption wavelengths of a diatomic molecule that has a moment of inertia of $4.6 \times 10^{-46}$ kg·m$^2$, a vibrational frequency of $1.817 \times 10^{14}$ Hz, and initially has an angular momentum quantum number $l = 3$.

14.    Determine the Fermi energy of a free electron model material with an electron density of $1.2 \times 10^{24}$ electrons per cubic centimeter.

15.    Determine the equilibrium separation of two ions, one of charge $+e$ and one of charge $-e$, when they have a repulsive potential energy function given by

$$U_{repulsive} = \frac{1.88 \times 10^{-58} \, J \cdot m^4}{r^4}$$

## Problem Solutions

3.    With the repulsion of the electron clouds, the binding energy is
      Binding energy $= -U - U_{clouds}$ ;
      4.43 eV $= 5.1$ eV $- U_{clouds}$ , which gives $U_{clouds} = 0.7$ eV.

7.    The reduced mass of the molecule is
      $\mu = m_1 m_2/(m_1 + m_2)$.
      (a) Using data from the periodic table, for NaCl we have
         $\mu = m_{Na} m_{Cl}/(m_{Na} + m_{Cl}) = (22.9898 \text{ u})(35.4527 \text{ u})/(22.9898 \text{ u} + 35.4527 \text{ u}) = 13.941$ u.
      (b) For $N_2$ we have
         $\mu = m_N m_N/(m_N + m_N) = (14.0067 \text{ u})(14.0067 \text{ u})/(14.0067 \text{ u} + 14.0067 \text{ u}) = 7.0034$ u.
      (c) For HCl we have
         $\mu = m_H m_{Cl}/(m_H + m_{Cl}) = (1.00794 \text{ u})(35.4527 \text{ u})/(1.00794 \text{ u} + 35.4527 \text{ u}) = 0.9801$ u.

11.    (a) The moment of inertia of $O_2$ about its CM is
      $I = 2m_O(r/2)^2 = m_O r^2/2$.
    We find the characteristic rotational energy from

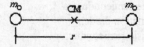

      $\hbar^2/2I = \hbar^2/m_O r^2$
             $= (1.055 \times 10^{-34} \, J \cdot s)^2/(16.0 \text{ u})(1.66 \times 10^{-27} \text{ kg/u})(0.121 \times 10^{-9} \text{ m})^2$
             $= 2.86 \times 10^{-23}$ J $= 1.79 \times 10^{-4}$ eV.
      (b) The rotational energy is
      $E_{rot} = L(L + 1)(\hbar^2/2I)$.
    Thus the energy of the emitted photon from the $L = 2$ to $L = 1$ transition is
      $hf = \Delta E_{rot} = [(2)(2 + 1) - (1)(1 + 1)](\hbar^2/2I) = 4(\hbar^2/2I) = 4(1.79 \times 10^{-4} \text{ eV}) = 7.16 \times 10^{-4}$ eV.
    The wavelength is
      $\lambda = c/f = hc/hf = (1.24 \times 10^3 \text{ eV} \cdot \text{nm})/(7.16 \times 10^{-4} \text{ eV}) = 1.73 \times 10^6$ nm $= 1.73$ mm.

15.    The ionic cohesive energy is
      $U_0 = -(\alpha e^2/4\pi\varepsilon_0 r)[1 - (1/m)]$
         $= -[(1.75)(2.30 \times 10^{-28} \, J \cdot m)/(0.28 \times 10^{-9} \text{ m})(1.60 \times 10^{-19} \text{ J/eV})][1 - (1/8)] = -7.9$ eV.

19. (a)  The potential energy is

$$U = -\alpha \frac{e^2}{4\pi\varepsilon_0 r} + \frac{\beta}{r^m}.$$

The separation at equilibrium occurs at the minimum in the energy.  We can relate this distance $r_0$ to the value of $\beta$ by setting $dU/dr = 0$:

$$\frac{dU}{dr} = \alpha \frac{e^2}{4\pi\varepsilon_0 r^2} - \frac{m\beta}{r^{m+1}} = 0 \quad \text{or} \quad \alpha \frac{e^2}{4\pi\varepsilon_0 r_0^2} = \frac{m\beta}{r_0^{m+1}} \quad \Rightarrow \quad \beta = \frac{\alpha e^2 r_0^{m-1}}{4\pi\varepsilon_0 m}.$$

The ionic cohesive energy is the value of $U$ at the equilibrium distance:

$$U = -\alpha \frac{e^2}{4\pi\varepsilon_0 r} + \frac{\alpha e^2 r_0^{m-1}}{4\pi\varepsilon_0 m r_0^m} = \frac{\alpha e^2}{4\pi\varepsilon_0 r_0}\left(1 - \frac{1}{m}\right).$$

(b)  For NaI we have

$$U_0 = -\frac{(1.75)(2.30\times10^{-28}\text{ J}\cdot\text{m})}{(1.60\times10^{-19}\text{ J/eV})(0.33\times10^{-9}\text{ m})}\left(1 - \frac{1}{10}\right) = -6.9\text{ eV}$$

(c)  If we assume the same value for the Madelung constant, for MgO we have

$$U_0 = -\frac{(1.75)(2.30\times10^{-28}\text{ J}\cdot\text{m})}{(1.60\times10^{-19}\text{ J/eV})(0.21\times10^{-9}\text{ m})}\left(1 - \frac{1}{10}\right) = -10.8\text{ eV}$$

(d)  If we use a new value for $m$, the fractional change is

$$\frac{\Delta U_0}{U_0} = -\frac{\left(1 - \frac{1}{10}\right) - \left(1 - \frac{1}{10}\right)}{\left(1 - \frac{1}{10}\right)} = 0.028 = 3\%$$

23. (a)  The value of $kT$ is
$kT = (1.38 \times 10^{-23}\text{ J/K})(300\text{ K})/(1.60 \times 10^{-19}\text{ J/eV}) = 0.0259\text{ eV}.$
We find the energy from
$$f = 1/\left(e^{(E-E_F)/kT} + 1\right);$$
$0.90 = 1/\left[e^{(E-7.0\,eV)/(0.0259eV)} + 1\right],$ which gives $E = 6.9$ eV.

(b)  The value of $kT$ is
$kT = (1.38 \times 10^{-23}\text{ J/K})(1200\text{ K})/(1.60 \times 10^{-19}\text{ J/eV}) = 0.1035\text{ eV}.$
We find the energy from
$$f = 1/\left(e^{(E-E_F)/kT} + 1\right);$$
$0.90 = 1/\left[e^{(E-7.0\,eV)/(0.1035eV)} + 1\right],$ which gives $E = 6.8$ eV.

27.  Because each sodium atom contributes one conduction electron, the density of conduction electrons is
$n = \rho N_A/M = (0.97 \times 10^3\text{ kg/m}^3)(6.02 \times 10^{23}/\text{mol})/(23.0 \times 10^{-3}\text{ kg/mol}) = 2.539 \times 10^{28}\text{ m}^{-3}.$
We find the Fermi energy from

$$E_F = \frac{h^2}{8m}\left(\frac{3n}{\pi}\right)^{2/3}$$

$$= \frac{(6.63\times10^{-34}\text{ J}\cdot\text{s})^2}{8(9.11\times10^{-31}\text{ kg})}\left[\frac{3(2.539\times10^{28}\text{ m}^{-3})}{\pi}\right]^{2/3} = 5.05\times10^{-19}\text{ J} = 3.2\text{ eV}$$

We find the Fermi speed as the speed which gives a kinetic energy equal to the Fermi energy:
$E_F = \tfrac{1}{2} m v_F^2;$
$5.05 \times 10^{-18}\text{ J} = \tfrac{1}{2}(9.11 \times 10^{-31}\text{ kg})v_F^2,$ which gives $v_F = 1.05 \times 10^6$ m/s.

31.  The maximum energy of an electron at $T = 0$ K is $E_F$.  All states above $E_F$ are unoccupied, while all states with $E \le E_F$ are occupied.  We find the average energy of an electron from

$$\bar{E} = \frac{E_{tot}}{N} = \frac{\int_0^{E_F} E n_0(E)\,dE}{\int_0^{E_F} n_0(E)\,dE} = \frac{\left(V 8\pi\sqrt{2}\,m^{3/2}/h^3\right)\int_0^{E_F} E^{3/2}\,dE}{\left(V 8\pi\sqrt{2}\,m^{3/2}/h^3\right)\int_0^{E_F} E^{1/2}\,dE} = \frac{(2/5)E_F^{5/2}}{(2/3)E_F^{3/2}} = \tfrac{3}{5}E_F$$

35. If we consider the cube to be a three-dimensional infinite well, we can apply the boundary conditions separately to each dimension. Each dimension gives a quantum number, which we label $n_1$, $n_2$, $n_3$.

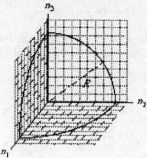

Thus there is a contribution to the energy from each dimension equal to the energy from the one-dimensional well:

$E_1 = (h^2/8mL^2)n_1^2$; $E_2 = (h^2/8mL^2)n_2^2$; $E_3 = (h^2/8mL^2)n_3^2$, $n_1$, $n_2$, $n_3 = 1, 2, 3, \ldots$.

The energy of a state specified by the three quantum numbers is

$E = (h^2/8mL^2)(n_1^2 + n_2^2 + n_3^2)$.

If we create a three-dimensional space with the axes labeled by $n_1$, $n_2$, $n_3$, each state corresponds to a point in the lattice. When we construct a sphere of radius $R$, where $R^2 = n_1^2 + n_2^2 + n_3^2$, each point within the octant corresponding to positive values for $n_1$, $n_2$, $n_3$ represents a state with energy between 0 and $E = h^2R^2/8mL^2$. Because the density of points is one and there can be two electrons in each state, the number of electrons with energy between 0 and $E$ is

$N = 2(1/8)\frac{4}{3}\pi R^3 = \pi R^3/3 = (\pi/3)(8mL^2E/h^2)^{3/2}$.

We find the density of states with energy $E$ from

$g(E) = dn/dE = (1/V)\, dN/dE = \frac{3}{2}(1/L^3)(\pi/3)(8mL^2E/h^2)^{3/2}E^{1/2} = (8\pi\sqrt{2}\; m^{3/2}/h^3)E^{1/2}$.

39. The photon with the longest wavelength or minimum frequency for conduction must have an energy equal to the energy gap:

$\lambda = c/f = hc/hf = hc/E_g = (1.24 \times 10^3 \text{ eV} \cdot \text{nm})/(1.1 \text{ eV}) = 1.1 \times 10^3 \text{ nm} = 1.1\ \mu\text{m}$.

43. The photon will have an energy equal to the energy gap:

$E_g = hf = hc/\lambda = (1.24 \times 10^3 \text{ eV} \cdot \text{nm})/(650 \text{ nm}) = 1.91 \text{ eV}$.

47. (a) For a half-wave rectifier without a capacitor, the current is zero for half the time. Thus the average current is

$I_{av} = \frac{1}{2} V_{rms}/R = \frac{1}{2}(120 \text{ V})/(28 \text{ k}\Omega) = 2.1 \text{ mA}$.

(b) For a full-wave rectifier without a capacitor, the current is positive all the time. Thus the average current is

$I_{av} = V_{rms}/R = (120 \text{ V})/(28 \text{ k}\Omega) = 4.3 \text{ mA}$.

51. The output voltage is the voltage across the resistor:

$V = i_C R_C = \beta i_B R_C$;

$0.40 \text{ V} = (100)(1.0 \times 10^{-6} \text{ A})R_C$, which gives $R_C = 4.0 \times 10^3\ \Omega = 4.0 \text{ k}\Omega$.

55. (a) We find the temperature from

$K = \frac{3}{2} kT$;

$(4.5 \text{ eV})(1.60 \times 10^{-19} \text{ J/eV}) = \frac{3}{2}(1.38 \times 10^{-23} \text{ J/K})T$, which gives $T = 3.5 \times 10^4 \text{ K}$.

(b) We find the temperature from

$K = \frac{3}{2} kT$;

$(0.15 \text{ eV})(1.60 \times 10^{-19} \text{ J/eV}) = \frac{3}{2}(1.38 \times 10^{-23} \text{ J/K})T$, which gives $T = 1.2 \times 10^3 \text{ K}$.

59. From the Boltzmann factor the population of a state with energy $E$ is $e^{-E/kT}$. The rotational energy of a state is

$E = (\hbar^2/2I)L(L + 1)$.

The selection rule requires $\Delta L = \pm 1$.

States with higher values of $L$ are less likely to be occupied and thus less likely to absorb a photon.

For example, there is a greater probability for absorption from $L = 1$ to $L = 2$ than from $L = 2$ to $L = 3$. The molecule is not rigid and thus $I$ will depend on $L$, which will affect the spacing.

63. The photon with the longest wavelength has the minimum energy, so the energy gap must be

$E_g = hc/\lambda = (1.24 \times 10^3 \text{ eV} \cdot \text{nm})/(1000 \text{ nm}) = 1.24 \text{ eV}$.

67. In a dielectric, Coulomb's law becomes

$F = e^2/4\pi K\varepsilon_0 r^2$.

Thus where $\varepsilon_0$ appears in an equation, we insert $K$. If the "extra" electron is outside the arsenic ion, the effective $Z$ will be 1, and we can use the hydrogen results.

(a) The energy of the electron is

$E = -Z^2 e^4 m/8K^2\varepsilon_0^2 h^2 n^2 = -(13.6 \text{ eV})Z^2/K^2 n^2 = -(13.6 \text{ eV})(1)^2/(12)^2(1)^2 = -0.094 \text{ eV}$.

Thus the binding energy is 0.094 eV.

(b) The radius of the electron orbit is

$r = K\varepsilon_0 h^2 n^2/\pi Z e^2 m = Kn^2 r_0/Z = Kn^2(0.0529 \text{ nm})/Z = (12)(1)^2(0.0529 \text{ nm})/(1)^2 = 0.63 \text{ nm}$.

Note that this result justifies the assumption that the electron is outside the arsenic ion.

# Chapter 42: Nuclear Physics and Radioactivity

## Chapter Overview and Objectives

This chapter describes the structure of the atomic nucleus and radioactivity. It discusses alpha, beta, and gamma decay of nuclei. It derives the time dependence of the number of nuclei in a given sample.

After completing study of this chapter, you should:

- Know that nuclei of atoms are made up of protons and neutrons.
- Know the definitions of atomic number, neutron number, and atomic mass number and the relationship between these quantities.
- Know the nuclear radius approximation.
- Know what a Bohr magneton is.
- Know how to calculate the binding energy of a nucleus.
- Know what alpha, beta, and gamma decay are.
- Know the definitions of decay constant, mean lifetime, and half-life.
- Know how to calculate the number of nuclei and the decay rate of a sample of radioactive material given the decay constant, mean lifetime, or half-life of the nucleus.
- Know the law of conservation of nucleon number.
- Know what radioactive dating is.
- Know how radiation is detected.

## Summary of Equations

Relationship between atomic number, neutron number, and atomic mass number:

$$A = Z + N$$

Approximate radius of nucleus:

$$r \approx \left(1.2 \times 10^{-13}\,\text{m}\right)A^{1/3}$$

Definition of a nuclear magneton:

$$\mu_N = \frac{e\hbar}{2m_p}$$

Binding energy of nucleus:

$$\text{Binding energy} = \left[Zm_p + (A - Z)m_n - m_{nucleus}\right]c^2$$

Alpha decay:

$$^A_Z N \rightarrow {}^{A-4}_{Z-2} N' + {}^4_2 \alpha$$

Beta decay:

$$^A_Z N \rightarrow {}^{A}_{Z+1} N' + {}^{0}_{-1} e + \bar{\nu} \quad \text{or} \quad {}^A_Z N \rightarrow {}^{A}_{Z-1} N' + {}^{0}_{+1} e + \nu$$

Electron capture:

$$^A_Z N + {}^{0}_{-1} e \rightarrow {}^{A}_{Z-1} N' + \nu$$

Gamma decay:

$$^A_Z N^* \rightarrow {}^A_Z N + \gamma$$

Radioactive decay time dependence:

$$N = N_0 e^{-\lambda t}$$

$$\frac{dN}{dt} = \left(\frac{dN}{dt}\right)_0 e^{-\lambda t} = -\lambda N_0 e^{-\lambda t}$$

Relationship between half-life, mean lifetime, and decay constant:

$$T_{1/2} = \frac{\ln 2}{\lambda} = \tau \ln 2$$

# Chapter Summary

### Section 42-1 Structure and Properties of the Nucleus

The nucleus of an atom consists of **protons** and **neutrons**. The proton has an electric positive charge of the same magnitude as the electron charge and it has a mass $1.67262 \times 10^{-27}$ kg. The neutron has zero electric charge and has a mass $1.67493 \times 10^{-27}$ kg. Both protons and neutrons are called **nucleons**.

The number of protons in a nucleus is called the **atomic number** of the nucleus and is designated by the symbol $Z$. The number of neutrons in a nucleus is called the **neutron number** of the nucleus and is designated by $N$. The total number of nucleons is called the **atomic mass number** of the nucleus and is designated by $A$. Be careful to distinguish between the similar phrases "atomic number" and "atomic mass number". The relationship between these three quantities is given by

$$A = Z + N$$

This information is commonly expressed along with the chemical symbol of the element that the nucleus belongs to in the form

$$^{A}_{Z}X$$

where $X$ is the chemical symbol of the element, $Z$ is its atomic number, and $A$ is its atomic mass number. The $Z$ and the $X$ are redundant because the atomic mass number determines which element the atom containing the nucleus belongs to, so the $Z$ is often omitted. The nuclei with different values of atomic mass number but the same atomic number are called **isotopes**. Most elements exist in nature in several different isotopes.

The shape of nuclei is approximately spherical and their approximate radius, $r$, depends on their atomic mass number, $A$, and is given by

$$r \approx \left(1.2 \times 10^{-13}\,\text{m}\right)A^{1/3}$$

Nuclear masses are often given in unified **atomic mass units (u)** or units of $\text{MeV}/c^2$. The unified atomic mass unit is defined in terms of the mass of a $^{12}_{6}\text{C}$ atom. The mass of the $^{12}_{6}\text{C}$ is defined to be 12.000000 u. The relationship between u, kg, and $\text{MeV}/c^2$ is given by

$$1\,\text{u} = 1.6605 \times 10^{-27}\,\text{kg} = 931.5\,\text{MeV}/c^2$$

The proton mass is

$$m_p = 1.67262 \times 10^{-27}\,\text{kg} = 1.007276\,\text{u} = 938.27\,\text{MeV}/c^2$$

and the neutron mass is

$$m_n = 1.67493 \times 10^{-27}\,\text{kg} = 1.007825\,\text{u} = 939.57\,\text{MeV}/c^2$$

The nucleons are, like the electron, spin ½ particles. The spin of the nucleons, along with the translational angular momentum of the nucleons within the nucleus, contributes to the **nuclear angular moment** of the nucleus. The nucleons have magnetic moments and so does the nucleus. A convenient unit for magnetic moments of nucleons and nuclei is the **nuclear magneton**, $\mu_N$:

$$\mu_N = \frac{e\hbar}{2m_p}$$

The magnetic moment of the proton is

$$\mu_p = 2.7928\,\mu_N$$

The magnetic moment of the neutron is

$$\mu_n = -1.9135\,\mu_N$$

The minus sign in this expression means that the magnetic moment is in the opposite direction to the spin in the neutron.

### Section 42-2 Binding Energy and Nuclear Forces

The mass of a stable nucleus is always less than the mass of its protons and neutrons if they were isolated from each other. The total binding energy of a nucleus is the difference between the mass of the protons and neutrons that make up the nucleus and the mass of the nucleus all multiplied by the square of the speed of light:

$$\text{Binding energy} = \left[Zm_p + (A-Z)m_n - m_{nucleus}\right]c^2$$

This is the energy required to separate the nucleus into its component protons and neutrons.

The **average binding energy per nucleon** is the binding energy divided by the number of nucleons in the nucleus. It is greatest for the $^{56}_{26}\text{Fe}$ nucleus. This means that the $^{56}_{26}\text{Fe}$ is the most stable nucleus.

The force that is responsible for holding nuclei together is the strong nuclear force. The strong nuclear force is an attractive force between nucleons. The strong nuclear force is the same for neutrons and protons. It is a **short range** force. At distances greater than about $10^{-15}$ m, its magnitude drops off very quickly.

There is a second type of nuclear force called the **weak nuclear force**. The weak nuclear force is involved in the decay of neutrons.

### Section 42-3 Radioactivity

The decay of nuclei along with the emission of a particle or electromagnetic radiation from the nucleus is called radioactivity. There are three different mechanisms of nuclear decay, which will be discussed in the following sections.

### Section 42-4 Alpha Decay

Some nuclei decay by emitting a $^4_2\text{He}$ nucleus. The $^4_2\text{He}$ nucleus emitted in the decay process is called an **alpha particle** and decay by this mechanism is called **alpha decay**. The decaying nucleus, called the **parent nucleus**, has its atomic number reduced by two and its atomic mass number reduced by four. The resulting nucleus is called the **daughter nucleus**. The general alpha decay can be written as

$$^A_Z N \rightarrow\, ^{A-4}_{Z-2} N' + ^4_2\text{He} \qquad \text{or} \qquad ^A_Z N \rightarrow\, ^{A-4}_{Z-2} N' + ^4_2\alpha$$

where $N$ is the parent nucleus element and $N'$ is the daughter nucleus element. The mass of the parent nucleus will always be greater than the mass of the daughter nucleus added to the mass of the alpha particle. This means that there will be energy released, also. The energy released is called the **disintegration energy** or **Q-value** of the decay:

$$Q = M_P c^2 - (M_D + M_\alpha)c^2$$

where $M_P$ is the mass of the parent nucleus, $M_D$ is the mass of the daughter nucleus, and $M_\alpha$ is the mass of the alpha particle.

### Example 42-4-A

A Californium 251 nucleus alpha decays. What is the daughter nucleus? What is the disintegration energy?

**Solution:**

The alpha decay is

$$^{251}_{98}Cf \rightarrow ^{247}_{96}N' + ^{4}_{2}\alpha$$

where $N'$ is the chemical symbol of the daughter nucleus. The chemical name of an atom with atomic number 96 is Curium (Cm). To find the disintegration energy, we determine the difference in the rest mass of the parent and the daughter and alpha particle and multiply by $c^2$:

$$Q = (M_P - M_D - m_a)c^2 = (251.079580\,u - 247.070346\,u - 4.002603\,u)\left(\frac{931.5\,MeV/c^2}{u}\right)c^2$$

$$= 6.176\,MeV$$

The masses given in the table of masses of isotopes are the masses for the neutral atoms, including the orbital electrons. Have we included the mass of the electrons correctly above? There were 98 electrons in the parent Californium atom. For the products of the reaction, we had a Curium atom, which would have 96 electrons when neutral, and an alpha particle. The alpha particle is usually ejected with no electrons bound to it, but we used the mass of a neutral helium atom in calculating the disintegration energy. Did we include the mass of two extra electrons in the product masses? No, because there were 98 electrons around the Californium nucleus, and there are 98 electrons after the decay. Even though we used the mass of a neutral helium atom when the alpha particle had no electrons, the electrons were present after the decay. The electrons were just not bound to the alpha particle.

### Section 42-5 Beta Decay

A **beta decay** is the emission of an electron and a **neutrino** from a nucleus. The emitted electron is called a **beta particle**, symbolized by $\beta^-$. The neutrino is a particle that is difficult to detect, has either no mass or a mass that is very small compared to the electron mass, is electrically neutral, and has an intrinsic spin of one-half. The neutrino is symbolized by $\nu$. The general beta decay can be written as

$$^{A}_{Z}N \rightarrow ^{A}_{Z+1}N' + ^{0}_{-1}e + \overline{\nu}$$

Notice the atomic mass number of the nucleus does not change; the same number of nucleons are present in the daughter nucleus as in the parent nucleus. Also notice that there is a bar placed over the neutrino because the neutrino that is emitted in a beta decay is called an **anti-neutrino**. Nuclei that beta decay have too many neutrons in the nucleus to be stable.

Neutrons that have too few neutrons in the nucleus to be stable decay by a related decay process, called $\beta^+$ **decay**. In $\beta^+$ decay, an **anti-electron**, also called a **positron**, is emitted, instead of an electron:

$$^{A}_{Z}N \rightarrow ^{A}_{Z-1}N' + ^{0}_{+1}e + \nu$$

Note that a neutrino, rather than an anti-neutrino, is emitted in a $\beta^+$ decay.

A nucleus that has too few neutrons to be stable can capture an inner shell electron in the form of inverse beta decay called **electron capture**. In electron capture a neutrino is emitted. The general electron capture reaction can be written as

$$^{A}_{Z}N + ^{0}_{-1}e \rightarrow ^{A}_{Z-1}N' + \nu$$

### Section 42-6 Gamma Decay

A nucleus can be in an excited state just as atoms can be in an excited state. Usually, a nucleus is left in an excited state after it undergoes an alpha or beta decay. Just as excited atoms often lose their excess energy by emitting electromagnetic radiation, nuclei can also lose their excess energy by emitting electromagnetic radiation. The electromagnetic radiation emitted by excited nuclei is roughly in the range of 10 keV to 10 MeV. The process of a nucleus losing its excess energy by emitting electromagnetic radiation is called **gamma decay** and the electromagnetic radiation emitted in the process is called **gamma radiation**.  The gamma decay process can be written as

$$_Z^A N^* \rightarrow {_Z^A} N + \gamma$$

where $N^*$ means an excited state of the nucleus $N$.

## Section 42-7 Conservation of Nucleon Number and Other Conservation Laws

All nuclear decays conserve energy, linear momentum, angular momentum, electric charge, and nucleon number (atomic mass number).  The conservation of nucleon number is called **the law of conservation of nucleon number**.

## Section 42-8 Half-Life and Rate of Decay

The time at which a given nucleus will decay is unpredictable.  The best that can be done is to state the probability that a given nucleus will decay in a unit time interval.  It appears to be true that the probability of decay of a nucleus per unit time interval is constant, regardless of the amount of time the nucleus has already existed.  Contrast this with the process of life.  A human that has existed for 90 years has a much higher probability of expiring within the next 10 year period than a human that has only existed for 10 years.  A nucleus that has existed for 9 million years has the same probability of decaying in the next million years as it had for decaying during its first million years.  The probability that a given nucleus will decay in a unit time interval is called the **decay constant** of the nucleus and is symbolized by $\lambda$.

It also appears to be true that in a collection of radioactive nuclei, the time of decay of each nucleus is independent of the decay time of the other nuclei.  If a collection of nuclei is large enough in number, these properties of radioactive decay lead to a rule about the change in number of parent nuclei, $\Delta N$, that will occur in a time interval $\Delta t$:

$$\Delta N = -\lambda N \Delta t$$

or in the limit where $\Delta t$ goes to zero:

$$dN = -\lambda N dt$$

$\Delta N$ and $dN$ are negative because the number of parent nuclei decreases in time as the nuclei decay.

If we solve this equation for $N$, we get

$$N = N_0 e^{-\lambda t}$$

and if we solve for the decay rate, $dN/dt$, we get

$$\frac{dN}{dt} = \left(\frac{dN}{dt}\right)_0 e^{-\lambda t} = -\lambda N_0 e^{-\lambda t}$$

Sometimes the exponential dependencies are written as

$$e^{-\lambda t} = e^{-t/\tau}$$

where $\tau$ is called the **mean lifetime** of the nucleus and is equal to one over the decay constant:

$$\tau = \frac{1}{\lambda}$$

or sometimes the exponential dependence is written in terms of the **half-life** of the nucleus, $T_{1/2}$ :

$$e^{-\lambda t} = \left(\frac{1}{2}\right)^{-t/T_{1/2}}$$

The half-life is the time for one-half of the nuclei present at the start of the time interval to decay. The relationship between the half-life and the decay constant and between the half-life and the mean lifetime is given by

$$T_{1/2} = \frac{\ln 2}{\lambda} = \tau \ln 2$$

## Example 42-8-A

The decay rate of a sample of a particular nucleus is measured to be $2.87 \times 10^4$ decays per second. 3 hours and 27 minutes later, the decay rate of the sample is measured to be 980 decays per second. What are the half-life and decay constant of the nucleus?

### Solution:

We know the dependence of the activity with time is given by

$$\frac{dN}{dt} = \left(\frac{dN}{dt}\right)_0 \left(\tfrac{1}{2}\right)^{t/T_{1/2}}$$

If we solve this for $T_{1/2}$ we get

$$T_{1/2} = \frac{-t \ln 2}{\ln\left[(dN/dt)/(dN/dt)_0\right]} = \frac{-(207\,\text{min})\ln 2}{\ln\left(980\,s^{-1}/2.87 \times 10^4\,s^{-1}\right)} = 42.5\,\text{min}$$

## Section 42-9 Decay Series

When a radioactive nucleus decays, the daughter nucleus is not necessarily stable and may itself undergo radioactive decay. This may also be true of the daughter of this nucleus, and so on. The list of isotopes that the decays successively go through until a stable isotope is reached is called a **decay series**.

## Section 42-10 Radioactive Dating

Ratios of the concentrations of different isotopes that are related by radioactive decay that are contained in a sample of material can be used to determine the age of the sample. Various isotopes can be used to carry out this **radioactive dating** determination of age.

Carbon 14 dating is used to date the remains of living creatures. During the lifetime of living creatures, a fraction, $1.3 \times 10^{-12}$, of their carbon nuclei is the $^{14}_{6}C$ isotope of carbon. This fraction is the same fraction as the carbon nuclei in the atmosphere. The reason the fraction of the nuclei in the living creature is the same as in the atmosphere is because the living creature is constantly exchanging its carbon nuclei with the environment. When a creature expires, it stops exchanging $^{14}_{6}C$ nuclei with the environment and the fraction of $^{14}_{6}C$ nuclei in the creature's remains begins to decrease because of the 5730-year half-life of these nuclei. The remains of a living creature can be dated by finding the fraction of $^{14}_{6}C$ nuclei in the remains and determining how many half-lives have passed since the creature expired.

## Example 42-10-A

Determine the age of a sample that has a fraction of $1.65 \times 10^{-14}$ of its carbon nuclei that are $^{14}_{6}C$ nuclei.

**Solution:**

We use the time dependence of radioactive nuclei in a sample:

$$N = N_0 \left(\tfrac{1}{2}\right)^{t/T_{1/2}}$$

and solve for $t$:

$$t = \frac{T_{1/2}}{-\ln 2} \ln \frac{N}{N_0} = \frac{5730 \text{ yr}}{-\ln 2} \ln \frac{1.65 \times 10^{-14}}{1.3 \times 10^{-12}} = 3.6 \times 10^4 \text{ yr}$$

### Section 42-11 Detection of Radiation

There are several types of detectors used for the detection of the products of radioactive decay. All of these devices function by utilizing the energy of the decay products to excite electrons in materials.

A **Geiger-Mueller tube** contains a gas and two electrodes held at different electric potentials. Radiation entering the tube has a finite probability of interacting with the gas atoms and causing them to ionize by losing an electron. Once a gas atom is ionized, it is accelerated by the electric field due to the potential difference between the two electrodes. Also, the electron separated from the atom is accelerated in the opposite direction. Either the current from the directly ionized atoms can be detected or, if the electric potential difference between the electrodes is great enough, the ions can gain enough kinetic energy from the electric field that collisions with other atoms cause the other atoms to ionize as well.

A **scintillator** is an optically transparent solid or liquid material that gives off visible light when a radioactive decay product interacts with the material. The number of visible photons is approximately proportional to the energy of the detected radioactive decay product. This light can be detected by a photomultiplier tube to produce an electrical signal that is approximately proportional in amplitude to the energy of the original radioactive decay product. By measuring the signal amplitude, knowledge of the energy of the detected radioactive decay product is gained.

A **semiconductor detector** utilizes a reverse biased pn junction to detect radioactive decay products. A reversed biased pn junction has only a small leakage current passing through it because the junction has no free charge carriers within it. When a radioactive decay product enters the junction and its energy is absorbed by electrons so that free charge carriers are created within the pn junction, these charge carriers move and create a current that can be detected. Like in the scintillator, the amplitude of the signal is approximately proportional to the energy of the detected radioactive decay product, so a semiconductor detector can be used to measure the energy of the detected radioactive decay product also.

## Practice Quiz

Necessary information about half-lives and nuclear atomic masses can be found in Appendix D of the text.

1.     Which of the following is a conserved quantity in nuclear decay processes?

   a) number of protons
   b) number of neutrons
   c) number of neutrinos
   d) number of nucleons

2.     Which of the following is not a conserved quantity in nuclear decay processes?

   a) angular momentum
   b) energy
   c) linear momentum
   d) number of electrons

3.   The decay rate of a sample of radioactive material has an initial activity of 1000 decays per second.  Five minutes later, it has an activity of 500 decays per second.  What will its activity be at the end of another five minutes of time?

a)  zero
b)  125 decays per second
c)  250 decays per second
d)  500 decays per second

4.   A $_{88}^{226}$Ra  nucleus alpha decays.  What is the daughter nucleus?

a)  $_{89}^{226}$Ac

b)  $_{87}^{224}$Fr

c)  $_{86}^{222}$Rn

d)  $_{84}^{222}$Po

5.   If you start with a sample of $N$ nuclei of a given isotope, there are 20% remaining after 10 minutes pass by.  What fraction of the nuclei are remaining after 10 minutes if you start with a sample of size $2N$?

a)  10%
b)  20%
c)  40%
d)  80%

6.   A nucleus with atomic number $Z$ and atomic mass number $A$ has a binding energy $BE$.  What is the binding energy per nucleon of this nucleus?

a)  $BE$
b)  $BE/Z$
c)  $BE/A$
d)  $BE/(A + Z)$

7.   Almost all stable nuclei with atomic number greater than 50 have

a)  no binding energy.
b)  more protons than neutrons.
c)  more neutrons than protons.
d)  more neutrons than nucleons.

8.   What is one assumption made when carbon 14 dating is used to determine when a living organism expired?

a)  The ratio of carbon 12 to carbon 14 in the environment at the time the creature lived was the same as it is today.
b)  The half-life of carbon 14 is decreasing at a uniform rate.
c)  The exchange of the carbon with the environment in the remains of the creature has continued since the time it expired.
d)  Both a) and c) above are assumed to be true.

9.   If nucleus A has a probability P of decaying in time t and nucleus B has a probability 2P of decaying in time $t$, which statement is true about the half-lives of nucleus A and nucleus B?

a)  The half-life of nucleus A is twice the half-life of nucleus B.
b)  The half-life of nucleus B is twice the half-life of nucleus A.
c)  The half-life of nucleus A is the same as the half-life of nucleus B.
d)  The half-life of nucleus A is $e^2$ times greater than the half-life of nucleus B.

10.    A particular nucleus has a half-life of 10 minutes. The decay rate of a given sample is 100 decays per second. What was the decay rate of the sample 20 minutes ago?

   a)  25 decays per second
   b)  200 decays per second
   c)  300 decays per second
   d)  400 decays per second

11.    A sample of $^{60}_{27}$Co contains $2.78 \times 10^{16}$ nuclei. How many nuclei remain after 20.0 years?

12.    A promethium 145 nucleus decays to a praseodymium 141 nucleus by alpha decay:

$$^{145}_{61}\text{Pm} \rightarrow \ ^{141}_{59}\text{Pr} \ + ^{4}_{2}\alpha$$

   What is the $Q$-value of this reaction?

13.    Determine the binding energy per nucleon for $^{28}_{14}$Si and for $^{31}_{14}$Si. Which of these nuclei would you expect to be more stable? Does your answer agree with the actual stability?

14.    What is the decay rate of a sample of 100.0 g of carbon from the remains from a creature that expired 16,000 years ago?

15.    A strontium 90 nucleus, $^{191}_{76}$Os, decays by beta decay. What is the daughter nucleus of this decay? How much energy is released in this decay?

## Problem Solutions

3.    The $\alpha$ particle is a helium nucleus:
      $r = (1.2 \times 10^{-15} \text{ m})A^{1/3} = (1.2 \times 10^{-15} \text{ m})(4)^{1/3} = 1.9 \times 10^{-15} \text{ m} = 1.9$ fm.

7.    We find the radii of the two nuclei from
      $r = r_0 A^{1/3}$;
      $r_\alpha = (1.2 \text{ fm})(4)^{1/3} = 1.9$ fm;
      $r_U = (1.2 \text{ fm})(238)^{1/3} = 7.4$ fm.
      If the two nuclei are just touching, the Coulomb potential energy must be the initial kinetic energy of the $\alpha$ particle:
      $K = U = Z_\alpha Z_U e^2 / 4\pi\varepsilon_0(r_\alpha + r_U)$
            $= (2)(92)(1.44 \text{ MeV} \cdot \text{fm})/(1.9 \text{ fm} + 7.4 \text{ fm}) = 28$ MeV.

11.  (a)  From Figure 42–1, we see that the average binding energy per nucleon at $A = 238$ is 7.5 MeV.
          Thus the total binding energy for $^{238}$U is
             $(238)(7.5 \text{ MeV}) = 1.8 \times 10^3$ MeV.
     (b)  From Figure 42–1, we see that the average binding energy per nucleon at $A = 84$ is 8.7 MeV.
          Thus the total binding energy for $^{84}$Kr is
             $(84)(8.7 \text{ MeV}) = 7.3 \times 10^2$ MeV.

15.  (a)  $^6$Li consists of three protons and three neutrons. We find the binding energy from the masses:
          Binding energy $= [3M(^1\text{H}) + 3m(^1\text{n}) – M(^6\text{Li})]c^2$
               $= [3(1.007825 \text{ u}) + 3(1.008665 \text{ u}) – (6.015122 \text{ u})]c^2(931.5 \text{ MeV}/uc^2) = 32.0$ MeV.
          Thus the binding energy per nucleon is
             $(32.0 \text{ MeV})/6 = 5.33$ MeV.
     (b)  $^{208}$Pb consists of 82 protons and 126 neutrons. We find the binding energy from the masses:
          Binding energy $= [82M(^1\text{H}) + 126m(^1\text{n}) – M(^{208}\text{Pb})]c^2$
               $= [82(1.007825 \text{ u}) + 126(1.008665 \text{ u}) – (207.976635 \text{ u})]c^2(931.5 \text{ MeV}/uc^2)$
               $= 1636$ MeV.
          Thus the binding energy per nucleon is
             $(1636 \text{ MeV})/208 = 7.87$ MeV.

19. (a)  We find the binding energy from the masses:
$$\text{Binding energy} = [2M(^4\text{He}) - M(^8\text{Be})]c^2$$
$$= [2(4.002603 \text{ u}) - (8.005305 \text{ u})]c^2(931.5 \text{ MeV/u}c^2) = -0.092 \text{ MeV}.$$
Because the binding energy is negative, the nucleus is unstable.
(b)  We find the binding energy from the masses:
$$\text{Binding energy} = [3M(^4\text{He}) - M(^{12}\text{C})]c^2$$
$$= [3(4.002603 \text{ u}) - (12.000000 \text{ u})]c^2(931.5 \text{ MeV/u}c^2) = +7.3 \text{ MeV}.$$
Because the binding energy is positive, the nucleus is stable.

23.  If $^{22}_{11}\text{Na}$ were a $\beta^-$ emitter, the resulting nucleus would be $^{22}_{12}\text{Mg}$, which has too few neutrons relative to the number of protons to be stable.  Thus we have a $\beta^+$ emitter.
For the reaction $^{22}_{11}\text{Na} \to ^{22}_{10}\text{Ne} + \beta^+ + \nu$, if we add 11 electrons to both sides in order to use atomic masses, we see that we have two extra electron masses on the right.  The kinetic energy of the $\beta^+$ will be maximum if no neutrino is emitted.  If we ignore the recoil of the neon, the maximum kinetic energy is
$$K = [M(^{22}\text{Na}) - M(^{22}\text{Ne}) - 2m(\text{e})]c^2$$
$$= [(21.994437 \text{ u}) - (21.991386 \text{ u}) - 2(0.00054858 \text{ u})]c^2(931.5 \text{ MeV/u}c^2) = 1.82 \text{ MeV}.$$

27. (a)  We find the final nucleus by balancing the mass and charge numbers:
$$Z(X) = Z(P) - Z(e) = 15 - (-1) = 16;$$
$$A(X) = A(P) - A(e) = 32 - 0 = 32, \text{ so the final nucleus is } ^{32}_{16}\text{S}.$$
(b)  If we ignore the recoil of the sulfur, the maximum kinetic energy of the electron is
$$K = [M(^{32}\text{P}) - M(^{32}\text{S})]c^2;$$
$$1.71 \text{ MeV} = [(31.973907 \text{ u}) - M(^{32}\text{S})]c^2(931.5 \text{ MeV/u}c^2),$$
which gives $M(^{32}\text{S}) = 31.97207 \text{ u}$.

31.  The total kinetic energy of the daughter and the $\alpha$ particle is
$$K_\alpha + K_{\text{Pb}} = Q = [M(^{210}\text{Po}) - M(^{206}\text{Pb}) - M(^4\text{He})]c^2$$
$$= [(209.982857 \text{ u}) - (205.974449 \text{ u}) - (4.002603 \text{ u})]c^2(931.5 \text{ MeV/u}c^2) = 5.41 \text{ MeV}.$$
If the polonium nucleus is at rest when it decays, for momentum conservation we have
$$p_\alpha = p_{\text{Pb}}.$$
The kinetic energy of the lead nucleus is
$$K_{\text{Pb}} = p_{\text{Pb}}^2/2m_{\text{Pb}} = p_\alpha^2/2m_{\text{Pb}} = (m_\alpha/m_{\text{Pb}})K_\alpha.$$
Thus we have
$$K_\alpha + (m_\alpha/m_{\text{Pb}})K_\alpha = [1 + (4 \text{ u})/(206 \text{ u})]K_\alpha = 5.41 \text{ MeV}, \text{ which gives } K_\alpha = 5.31 \text{ MeV}.$$

35.  We find the decay constant from
$$\lambda N = \lambda N_0 e^{-\lambda t};$$
$$320 \text{ decays/min} = (1280 \text{ decays/min}) e^{-\lambda(6 \text{ h})}, \text{ which gives } \lambda = 0.231/\text{h}.$$
Thus the half-life is
$$T_{1/2} = 0.693/\lambda = 0.693/(0.231/\text{h}) = 3.0 \text{ h}.$$
Note that in 6.0 h the decay rate was reduced to $\frac{1}{4}$ the original rate.  This means the elapsed time was 2 half-lives.

39.  We find the number of nuclei from the activity of the sample:
$$\Delta N/\Delta t = \lambda N;$$
$$875 \text{ decays/s} = [(0.693)/(4.468 \times 10^9 \text{ yr})(3.16 \times 10^7 \text{ s/yr})]N, \text{ which gives } N = 1.78 \times 10^{20} \text{ nuclei}.$$

43.  The decay constant is
$$\lambda = 0.693/T_{1/2} = 0.693/(30.8 \text{ s}) = 0.0225 \text{ s}^{-1}.$$
(a)  The initial number of nuclei is
$$N_0 = (9.8 \times 10^{-6} \text{ g})/(124 \text{ g/mol})(6.02 \times 10^{23} \text{ atoms/mol}) = 4.8 \times 10^{16} \text{ nuclei}.$$
(b)  When $t = 2.0$ min, the exponent is
$$\lambda t = (0.0225 \text{ s}^{-1})(2.0 \text{ min})(60 \text{ s/min}) = 2.7,$$
so we get
$$N = N_0 e^{-\lambda t} = (4.8 \times 10^{16}) e^{-2.7} = 3.2 \times 10^{15} \text{ nuclei}.$$
(c)  The activity is
$$\lambda N = (0.0225 \text{ s}^{-1})(2.5 \times 10^{15}) = 7.2 \times 10^{13} \text{ decays/s}.$$

(*d*)  We find the time from

$$\lambda N = \lambda N_0 e^{-\lambda t};$$

1 decay/s $= (0.0225 \text{ s}^{-1})(4.8 \times 10^{16}) \, e^{-(0.0225 \text{ /s}) t}$, which gives $t = 1.54 \times 10^3 \text{ s} = 26 \text{ min}$.

47.  We find the number of half-lives from

$$(\Delta N/\Delta t)/(\Delta N/\Delta t)_0 = (\tfrac{1}{2})^n;$$

$1/10 = (\tfrac{1}{2})^n$, or $n \log 2 = \log 10$, which gives $n = 3.32$.

Thus the half-life is

$$T_{1/2} = t/n = (8.6 \text{ min})/3.32 = 2.6 \text{ min}.$$

51.  The decay rate is

$$\Delta N/\Delta t = \lambda N.$$

If we assume equal numbers of nuclei decaying by $\alpha$ emission, we have

$$(\Delta N/\Delta t)_{218}/(\Delta N/\Delta t)_{214} = \lambda_{218}/\lambda_{214} = T_{1/2,214}/T_{1/2,218}$$
$$= (1.6 \times 10^{-4} \text{ s})/(3.1 \text{ min})(60 \text{ s/min}) = 8.6 \times 10^{-7}.$$

55.  The number of radioactive nuclei decreases exponentially:

$$N = N_0 e^{-\lambda t}.$$

Every radioactive nucleus that decays becomes a stable daughter nucleus, so we have

$$N_D = N_0 - N = N_0(1 - e^{-\lambda t}).$$

59.  If the $^{40}$K nucleus in the excited state is at rest, the gamma ray and the nucleus must have equal and opposite momenta:

$$p_K = p_\gamma = E_\gamma/c, \quad \text{or} \quad p_K c = E_\gamma = 1.46 \text{ MeV}.$$

The kinetic energy of the nucleus is

$$K = p_K^2/2m = (p_K c)^2/2mc^2 = (1.46 \text{ MeV})^2/2(40 \text{ u})(931.5 \text{ MeV/u}c^2)c^2 = 2.86 \times 10^{-5} \text{ MeV} = 28.6 \text{ eV}.$$

63.  If we assume a body has 70 kg of water, the number of water molecules is

$$N_{\text{water}} = [(70 \times 10^3 \text{ g})/(18 \text{ g/mol})](6.02 \times 10^{23} \text{ atoms/mol}) = 2.34 \times 10^{27} \text{ molecules}.$$

The number of protons in a water molecule ($H_2O$) is $2 + 8 = 10$, so the number of protons is

$$N_0 = 2.34 \times 10^{28} \text{ protons}.$$

If we assume that the time is much less than the half-life, the rate of decay is constant, so we have

$$\Delta N/\Delta t = \lambda N = (0.693/T_{1/2})N;$$

$(1 \text{ proton})/\Delta t = [(0.693)/(10^{32} \text{ yr})](2.34 \times 10^{28} \text{ protons})$, which gives $\Delta t = 6 \times 10^3 \text{ yr}$.

67.  If we use $^{14}_{6}\text{C} \rightarrow {}^{14}_{7}\text{N} + e^- + \bar{\nu}$ as the decay, we see that $^{14}$C and $^{14}$N each have an even number of nucleons with spin ½, so their total spin must be an integer.  Because $e^-$ has spin ½, $\bar{\nu}$ must have spin ½ to conserve angular momentum.

71.  We use an average nuclear radius of $5 \times 10^{-15}$ m.

We use the radius as the uncertainty in position for the electron to find the uncertainty in the momentum from

$$\Delta p \geq \hbar/\Delta x = (1.055 \times 10^{-34} \text{ J} \cdot \text{s})/2(5 \times 10^{-15} \text{ m}) = 2.1 \times 10^{-20} \text{ kg} \cdot \text{m/s}.$$

If we assume that the lowest value for the momentum is the least uncertainty, we estimate the minimum possible energy as

$$\begin{aligned} E &= K + mc^2 = (p^2 c^2 + m^2 c^4)^{1/2} = [(\Delta p)^2 c^2 + m^2 c^4]^{1/2} \\ &= [(2.1 \times 10^{-20} \text{ kg} \cdot \text{m/s})^2(3.00 \times 10^8 \text{ m/s})^2 + (9.11 \times 10^{-31} \text{ kg})^2(3.00 \times 10^8 \text{ m/s})^4]^{1/2} \\ &= 6.3 \times 10^{-12} \text{ J} = 40 \text{ MeV}. \end{aligned}$$

Because this is $\gg mc^2$, the electron is unlikely to be found in the nucleus.

75.  Because there are so many low-energy electrons available, the reaction $e^- + p \rightarrow n + \nu$ would turn most of the protons into neutrons, which would eliminate chemistry, and thus life.

The $Q$-value of the reaction is

$$Q = [M(^1\text{H}) - m(^1\text{n})]c^2 = [(1.007825 \text{ u}) - (1.008665 \text{ u})]c^2(931.5 \text{ MeV/u}c^2) = -0.782 \text{ MeV}.$$

The percentage increase in the proton's mass to make the $Q$-value $= 0$ is

$$(\Delta m/m)(100) = [(0.782 \text{ MeV}/c^2)/(938.3 \text{ MeV}/c^2)](100) = 0.083\%.$$

79. (a) If the speed of the $\alpha$ particle within the nucleus is $v_{in}$, the time to traverse the nucleus is $2R_0/v_{in}$.

The frequency of a collision with the Coulomb barrier is $v_{in}/2R_0$. From Ch. 39 the probability of passing through the barrier each time is $T = e^{-2GL}$, where $L$ is the thickness of a square barrier, and

$G = [2m(U_0 - E)/\hbar^2]^{1/2}$. The decay constant is the probability of the $\alpha$ particle escaping from the nucleus:

$$\lambda = (v_{in}/2R_0)T = (v_{in}/2R_0)e^{-2GL}.$$

The Coulomb barrier decreases as $1/r$. For a square barrier that approximates this we will use the maximum height found in Problem 34 for the height of the barrier. We find the value of $G$ from

$G^2 = 2m(U_0 - E)/\hbar^2$

$= 2(4)(1.67 \times 10^{-27} \text{ kg})(28 \text{ MeV} - 4.5 \text{ MeV})(1.60 \times 10^{-13} \text{ J/MeV})/(1.055 \times 10^{-34} \text{ J} \cdot \text{s})^2$,

which gives $G = 2.1 \times 10^{15} \text{ m}^{-1} = 2.1 \text{ fm}^{-1}$.

If we choose the potential inside the well to be zero, the kinetic energy of the $\alpha$ particle inside is the same as the kinetic energy of the $\alpha$ particle after the decay. We find the speed inside the well from

$K = \frac{1}{2}mv^2$;

$(4.49 \text{ MeV})(1.60 \times 10^{-13} \text{ J/MeV}) = \frac{1}{2}(4)(1.67 \times 10^{-27} \text{ kg})v_{in}^2$, which gives $v_{in} = 1.47 \times 10^7$ m/s.

The frequency of striking the barrier is

$v_{in}/2R_0 = (1.47 \times 10^7 \text{ m/s})/2(7.4 \times 10^{-15} \text{ m}) = 9.9 \times 10^{20} \text{ s}^{-1}$.

To account for the decrease with $r$ of the Coulomb barrier, we use a square barrier with a thickness less than the width found in Problem 34.

(b) If we use a thickness of $\frac{1}{2}$ the width of 50 fm, we have

$$\lambda = \left(\frac{v_{in}}{2R_0}\right)e^{-2GL} = \left(9.9 \times 10^{20} \text{ s}^{-1}\right)e^{-2\left(2.1\text{fm}^{-1}\right)(25\text{ fm})} = 2.5 \times 10^{-25} \text{ s}^{-1}$$

Thus the half-life is

$T_{1/2} = (\ln 2)/\lambda = 0.693/(2.5 \times 10^{-25} \text{ s}^{-1}) = 2.8 \times 10^{24} \text{ s} \approx 10^{17}$ yr.

(c) If we use a thickness of $\frac{1}{3}$ the width of 50 fm, we have

$$\lambda = \left(\frac{v_{in}}{2R_0}\right)e^{-2GL} = \left(9.9 \times 10^{20} \text{ s}^{-1}\right)e^{-2\left(2.1\text{ fm}^{-1}\right)(16.7\text{ fm})} = 3.4 \times 10^{-25} \text{ s}^{-1}$$

Thus the half-life is

$T_{1/2} = (\ln 2)/\lambda = 0.693/(3.4 \times 10^{-10} \text{ s}^{-1}) = 2.0 \times 10^9 \text{ s} \approx 60$ yr.

Note that the result is very sensitive to the value used for the barrier thickness.

(d) To find the width that gives us the known half-life, we have

$$T_{1/2} = (\ln 2)/\lambda = \frac{(\ln 2)e^{2GL}}{(v_{in}/2R_0)}, \text{ or } e^{2GL} = [(v_{in}/2R_0)/(\ln 2)]T_{1/2}$$

$e^{2GL} = [(9.9 \times 10^{20} \text{ s}^{-1})/(\ln 2)](2 \times 10^7 \text{ yr})(3.16 \times 10^7 \text{ s/yr})$, or $2GL = 83$;

$L = (83)/2(2.1 \text{ fm}^{-1}) = 20$ fm, which is 0.4 of the width found in Problem 34.

# Chapter 43: Nuclear Energy; Effects and Uses of Radiation

## Chapter Overview and Objectives

This chapter covers the physics of sources of nuclear energy and uses of radiation. It covers nuclear fission and nuclear fusion, the two types of nuclear reactions that could be used in large scale nuclear power sources. This chapter also covers the measurements used in determining the amount of damage received from exposure to ionizing radiation.

After completing study of this chapter, you should:

- Know what nuclear fusion and nuclear fission are.
- Know what a cross section is.
- Know how to calculate the $Q$ value of a nuclear reaction.
- Know the common and SI units for measurements of activity, absorbed dose, and effective dose.
- Know what tracers, tomography, and nuclear magnetic resonance are.

## Summary of Equations

$Q$-value of nuclear reaction:
$$Q = \sum M_R c^2 - \sum M_P c^2$$

Total cross section:
$$\sigma_T = \sigma_{el} + \sigma_{inel} + \sigma_R$$

Relationship between decay rate and decay constant:
$$\frac{dN}{dt} = -\lambda N$$

## Chapter Summary

### Section 43-1 Nuclear Reactions and the Transmutation of Elements

A nuclear reaction occurs when a nucleus interacts with another entity such as another nucleus, a particle, or a photon so that the entities that exist after the interaction are different from those that existed before the interaction. An example of such a reaction is

$$^{1}_{0}n + ^{131}_{53}I \rightarrow ^{132}_{53}I$$

These reactions must conserve proton number and nucleon number. This means that the sums of the lower numbers (proton number) must be the same on each side of the equation and the sums of the upper numbers (nucleon number) must be the same on both sides of the equation.

In addition to nuclear reactions conserving proton number and nucleon number, the laws of conservation of energy and conservation of momentum must both be satisfied. The **reaction energy** or **$Q$-value** of a nuclear reaction is defined to be

$$Q = \sum M_R c^2 - \sum M_P c^2$$

where the $M_R$ are the masses of the reacting particles and nuclei and the $M_P$ are the masses of the product particles and nuclei. This is just the difference in rest mass energy of the reactants and the products.

The $Q$-value will be the difference in kinetic energies of the products and the reactants

$$Q = \sum K_P - \sum K_R$$

**Example 43-1-A**

Determine the $Q$-value for the following reaction:

$$^{39}_{19}K + ^{1}_{1}p \rightarrow ^{40}_{20}Ca$$

**Solution:**

The atomic masses and the mass of the proton are:

$$^{39}_{19}K: \quad 38.962591\,u$$

$$^{40}_{20}Ca: \quad 39.963707\,u$$

$$^{1}_{1}H: \quad 1.007825\,u$$

Note that we have found the mass of a hydrogen atom, rather than the mass of a bare proton. The reason we have done that is the masses in the table of isotopes in the textbook are atomic masses. That means they include the mass of the electrons needed to make the atom a neutral object. The neutral calcium atom has one more electron than the neutral potassium atom. We need to add an electron mass to the reacting particles to account for the extra electron and a convenient way to do this is to use the mass of the hydrogen atom rather than just a proton mass.

$$Q = \sum M_R c^2 - \sum M_P c^2 = (38.962591\,u + 1.007825\,u - 39.963707\,u)(931.5\,MeV/c^2)c^2 = 6.24\,MeV$$

**Section 43-2 Cross Section**

If projectile particles are sent at a target nucleus, some of the projectiles might not encounter the nucleus at all and others may be elastically scattered, inelastically scattered, or absorbed. The **total cross section**, $\sigma_T$, for scattering from a nucleus is the area a hard-edged disk would have that would scatter the projectiles at the same rate that the nucleus does. This area depends on the type of incoming projectile and the energy of the incoming projectile.

The total cross section can be written as a sum of cross sections for each of the three processes mentioned above. The **elastic scattering cross section**, $\sigma_{el}$, is the cross section for scattering processes in which the sum of the initial kinetic energies of the projectile and the target are the same before and after the scattering process occurs. The **inelastic scattering cross section**, $\sigma_{inel}$, is the cross section for scattering processes in which the sum of the initial kinetic energies of the projectile and the target are not the same before and after the scattering process. Some of the initial kinetic energy goes into raising the projectile or the target to an excited state in an inelastic collision. A nuclear reaction may occur during a collision so that the objects that exist after the collision are not the same set of objects that existed before the collision. This results in a **reaction cross section**. There may be several possible nuclear reactions that occur in the collision and the sum of the cross sections for each of the reactions is called the **total reaction cross section**, $\sigma_R$. The total cross section is related to all of the cross sections for different types of processes by

$$\sigma_T = \sigma_{el} + \sigma_{inel} + \sigma_R$$

A unit of a barn is used to measure nuclear cross sections. One barn is $10^{-28}\,m^2$.

**Example 43-2-A**

A detector detects particles directly from a source at a rate of $2.68 \times 10^4$ per second. An aluminum block of thickness 2.00 cm is placed between the source and the detector. The rate of detection of the particles drops to $2.32 \times 10^4$ per second. What is the total cross section of a single aluminum atom for scattering the particles in the incident beam.

**Solution:**

The density of aluminum is $2.70 \times 10^3\,kg/m^3$ and its average atomic mass is 26.9815 u. We use this information to calculate the number density, $n$, of nuclei in the target:

$$n = \frac{\text{mass density}}{\text{atomic mass}} = \frac{2.70 \times 10^3 \text{ kg/m}^3}{(26.9815u)(1.6605 \times 10^{-27} \text{ kg/u})} = 6.03 \times 10^{28} \text{ m}^{-3}$$

The rate at which either scattering or absorption occurs, $R$, will be the difference between the rate of particle detection with and without the scattering material in position:

$$R = R_0 - R_{\text{with scattering}} = 2.68 \times 10^4 \text{ s}^{-1} - 2.32 \times 10^4 \text{ s}^{-1} = 5.6 \times 10^3 \text{ s}^{-1}$$

We can now calculate the total reaction cross section:

$$R = R_0 nt\sigma \quad \Rightarrow \quad \sigma = \frac{R}{R_0 nt} = \frac{5.6 \times 10^3 \text{ s}^{-1}}{(2.68 \times 10^4 \text{ s}^{-1})(6.03 \times 10^{28})(2.00 \times 10^{-2} \text{ m})} = 1.7 \times 10^{-28} \text{ m}^2$$

### Section 43-3 Nuclear Fission; Nuclear Reactors

A few large nuclei will break apart into two approximately equal size nuclei when the nuclei are bombarded by neutrons. This process is called **nuclear fission** and the product nuclei are called **fission fragments**. One or more neutrons may also be released in the fragmentation process. The binding energy per nucleon for fissionable nuclei is smaller than the binding energy per nuclei in the fission fragments and energy is released during the fission process. The fission of a $^{235}_{92}$U nucleus releases about 200 MeV of energy.

Neutrons released in the fission process can collide with other fissionable nuclei and cause further fissions to occur. These secondary fissions release neutrons that can cause further reactions to occur, and so on in a **chain reaction**. If at least one neutron on average from each fission that occurs goes on to cause another fission, the reaction is a **self-sustaining chain reaction**. A **nuclear reactor** is a device used to contain and control a self-sustaining chain reaction.

Constructing a nuclear reactor is not as simple as just throwing some fissionable material together in a container. There are some other factors that must be taken care of to ensure the reaction is self-sustaining. One of these factors is that the neutrons must have a relatively small kinetic energy to cause a nucleus to fission. Neutrons with relatively large kinetic energies are not absorbed by nuclei as easily as neutrons with relatively small kinetic energies. The neutrons released in a fission have relatively large kinetic energies, so they must be slowed down to increase the probability that they will cause another fission. A material called a moderator is included inside the reactor to slow down the neutrons without absorbing them. We know from the study of elastic collisions that more kinetic energy is lost in a collision with a similar mass object than either smaller or larger masses. Moderator materials should contain nuclei with masses as similar as possible to the mass of the neutron, but not have a large cross section for absorbing the neutron. Obviously, hydrogen nuclei with only a proton meet the similar mass requirement better than any other nuclei. However, a proton has a high probability for an inelastic collision in which the neutron binds to the proton to form a deuterium nucleus. Deuterium is not as efficient at removing kinetic energy from the neutrons, but it does not absorb as many neutrons either and so is a better nucleus for a moderator. **Heavy water** is water in which the hydrogen atoms are deuterium.

Another difficulty is created by the fact that not all isotopes are fissionable and the non-fissionable nuclei may absorb neutrons. The material may need to have the abundance of the fissionable isotope increased above the naturally occurring abundance of that isotope so that there are more fissionable targets for the neutrons to collide with before the neutrons are absorbed by the non-fissionable nuclei. This is called **enrichment**.

Neutrons can also escape from the surface of the fissionable material before colliding with a fissionable nucleus. This effect is reduced by making the volume-to-surface area ratio of the fissionable material larger. For a given geometrical shape of the fissionable material, this is often stated in terms of the minimum mass of material needed for sustainable fission. This mass is called the **critical mass**.

The average number of secondary fissions caused by the neutrons released in a fission is called the **multiplication factor**, $f$. If $f$ is greater than one, the rate of fissions increases exponentially. If $f$ is equal to one, then fissions occur at a constant rate. If $f$ is less than one, then the rate of fissions decreases with time. A fission power reactor that is to produce a constant output power should have $f$ equal to one. Control rods that absorb neutrons are used in nuclear reactor to adjust the rate of fissions to produce the desired output energy when $f$ is equal to one. An important aspect of design of nuclear power plants is **stability**. Stability means if the rate of fissions increases slightly for some reason, the change in the multiplication factor caused by this increase fission rate would be negative. This would tend to decrease the fission rate back to the desired level. If the reactor is stable, it decreases the chances of a nuclear reactor producing too much energy and damaging itself because of the internal temperature rising too high.

**Example 43-3-A**

Determine the $Q$-value of the fission of a $^{239}_{94}$Pu nucleus into a $^{90}_{38}$Sr nucleus, $^{138}_{56}$Ba nucleus, and 11 neutrons.

**Solution:**

The atomic masses and the mass of the neutrons are:

$$^{239}_{94}\text{Pu}: \quad 239.052157\,\text{u}$$
$$^{90}_{38}\text{Sr}: \quad 89.907737\,\text{u}$$
$$^{138}_{56}\text{Ba}: \quad 137.905241\,\text{u}$$
$$^{1}_{0}\text{n} \quad \quad 1.008665\,\text{u}$$

In this case, we do not need to add any electron masses because the total number of electrons in the neutral plutonium is the same as in the neutral product atoms.

$$Q = \sum M_R c^2 - \sum M_P c^2 = (239.052157\,\text{u} - 89.907737\,\text{u} - 39.963707\,\text{u} - 11 \times 1.008665\,\text{u})(931.5\,\text{MeV}/c^2)c^2$$
$$= 134\,\text{MeV}$$

**Section 43-4 Fusion**

The process of building nuclei of larger atomic mass from those of smaller atomic mass and individual protons and neutrons is called **nuclear fusion**. The mass of a nucleus is less than the mass of its constituent protons and neutrons, so energy is usually released in the nuclear fusion processes. However, nuclei with atomic mass numbers greater than about 60 have less binding energy per nucleon than does iron 56. If these heavier nuclei are made from a fusion process involving lighter nuclei, the reaction usually requires some additional energy besides the rest energy of the fusing nuclei.

**Example 43-4-A**

Calculate the $Q$ value of the following fusion reaction:

$$^{32}_{15}\text{P} + ^{32}_{15}\text{P} \rightarrow ^{64}_{30}\text{Zn}$$

**Solution:**

The atomic masses are:

$$^{32}_{15}\text{P}: \quad 31.973907\,\text{u}$$
$$^{64}_{30}\text{Zn}: \quad 63.929147\,\text{u}$$

In this case, we do not need to add any electron masses because the total number of electrons in the neutral phosphorus atoms is the same as in the neutral zinc atom.

$$Q = \sum M_R c^2 - \sum M_P c^2 = (2 \times 31.973907\,\text{u} - 63.929147\,\text{u})(931.5\,\text{MeV}/c^2)c^2 = 17.4\,\text{MeV}$$

**Section 43-5 Passage of Radiation Through Matter; Radiation Damage**

Radiation passing through matter causes damage by the energy it can give to atoms within the matter. Radiation that has enough energy is called ionizing radiation because it can ionize atoms that it interacts with. Ionization of an atom can cause chemical reactions to occur which are potentially disruptive to biological processes. Structural materials can be damaged by the disruption of their atomic structure because of collisions with high energy particles.

## Section 43-6 Measurement of Radiation—Dosimetry

One measurement of a radioactive source is its activity. Activity is defined as the number of radioactive decays that occur per unit time. A commonly used unit of activity is the Curie (Ci) and is defined as

$$1\,Ci = 3.70 \times 10^{10}\ \text{disintegration / s}$$

The SI unit for activity is the becquerel (Bq) and is defined as

$$1\,Bq = 1\,\text{disintegration} / s$$

A fixed collection of radioactive nuclei has a time-dependent activity. The activity of a source is $-dN/dt$ where $N$ is the number of radioactive nuclei in the collection. From Chapter 42, we know that $dN/dt$ is proportional to the number of nuclei in the sample:

$$\frac{dN}{dt} = -\lambda\,N$$

where $\lambda$ is the **decay constant** of the particular nuclei. As nuclei decay, the total number of nuclei decreases and so does the total decay rate.

The damage done by radiation depends on the amount of energy absorbed from the radiation. There are various measurements of the energy absorbed, called the **absorbed dose**, and the effectiveness of the absorbed energy in doing damage. The **roentgen** (R) is the amount of radiation that deposits 8.78 mJ of energy per kilogram of air. Another measurement that is in more common use today is the rad. One **rad** is the amount of radiation that deposits 10 mJ of energy into any absorbing material. The SI unit of absorbed dose is the gray (Gy). One **gray** is the amount of radiation that deposits one joule of energy per kilogram of absorbing material. One gray is equal to one hundred rad.

For a given amount of kinetic energy, more massive particles travel more slowly. This results in this energy being deposited in a smaller volume of the absorbing material. The damage in this smaller volume is greater than if the same amount of energy were to be deposited spread out over a larger volume. A measurement of the effectiveness in doing damage is useful, so the effective dose is defined. Each type of radiation is given a **relative biological effectiveness** (RBE) or **quality factor** (QF). The **effective dose** is defined as the quality factor multiplied by the absorbed dose. The **rad equivalent man** (rem) is the absorbed dose in rad multiplied by quality factor. The SI unit for effective dose is the **sievert** (Sv). One sievert is the absorbed dose in gray multiplied by the quality factor.

## Example 43-6-A

What is the effective dose received by an individual exposed to a gamma ray source that emits successive gamma rays with photon energies of 1.12 MeV and 0.465 MeV for 6.8 hours at a distance of 3.6 m? The activity of the source is 15 mCi. Assume 2.8% of the 1.12 MeV gamma rays are absorbed by the body and 4.3% of the 0.465 MeV gamma rays are absorbed by the body. Assume the effective area of the body to absorb the radiation is 0.85 m$^2$.

### Solution:

First we calculate the exposure rate of the individual at the distance the individual is from the source. The source activity is 15 mCi, which is

$$15\,mCi \times \frac{3.7 \times 10^{10}\ \text{decay/s}}{1\,Ci} = 5.55 \times 10^{8}\ \text{decays/s}$$

The rate of energy released by the source will be the activity multiplied by the energy released per decay. This decay releases two photons, one of energy 1.12 MeV and one of energy 0.465 MeV. The rate of energy released of each photon will be

$$\frac{E_{1.12\,\text{MeV}}}{t} = \left(1.12\,\text{MeV}\right)\!\left(5.55 \times 10^{8}\ s^{-1}\right) = 6.22 \times 10^{8}\ \text{MeV/s}$$

$$\frac{E_{0.465\,MeV}}{t} = (0.465\,MeV)(5.55\times10^8\ s^{-1}) = 2.58\times10^7\ MeV/s$$

The radiation that reaches the body will be the fraction of the area of the body divided by the area of a sphere that has a radius equal to the distance the body is from the source:

$$\text{fraction of energy reaching body} = \frac{\text{Area of body}}{4\pi r^2} = \frac{0.85\,m^2}{4\pi\ (3.6\,m^2)} = 5.22\times10$$

The energy absorbed by the body of each photon type will be the rate the energy is released by the source multiplied by the fraction of energy reaching the body multiplied by the fraction of the radiation absorbed by the body multiplied by the time of exposure:

$$E_{absorbed\ 1.12\,MeV} = (6.22\times10^8\ MeV/s)(5.22\times10^{-3})(0.028)\left(6.8\,hr\times\frac{3600\,s}{1\,hr}\right) = 2.22\times10^8\ MeV$$

$$E_{absorbed\ 0.465\,MeV} = (2.58\times10^7\ MeV/s)(5.22\times10^{-3})(0.043)\left(6.8\,hr\times\frac{3600\,s}{1\,hr}\right) = 1.42\times10^8\ MeV$$

The total absorbed energy is the sum of these two:

$$E_{absorbed} = (2.22\times10^8\ MeV + 1.42\times10^8\ MeV)\left(\frac{1.60\times10^{-13}\ J}{1\,MeV}\right)\left(\frac{1\,rad}{1.00\times10^{-2}\ J}\right) = 5.82\times10^{-3}\ rad$$

The effective dose is the absorbed dose multiplied by the relative biological effectiveness (rbe) of the radiation. Both gamma rays, the rbe is approximately 1, so the effective dose is $5.82\times10^{-3}$ rem or 5.82 mrem.

**Section 43-7 Radiation Therapy**

The ionization capability of radiation can be used to destroy cancerous cells within the human body.

**Section 43-8 Tracers**

Radioactive isotopes can be used in making chemical compounds that are biologically active. These compounds are introduced into biological systems and the radioactive nature of the isotopes can be used to determine what happens to the compound within the biological system. The isotope is called a **tracer** in this usage.

**Section 43-9 Imaging by Tomography: CAT Scans and Emission Tomography**

The familiar X-ray taken at a doctor's office is called a shadowgraph. It simply records, by varying density of the silver in the film, the amount of X-ray energy reaching the film that passes through the object. The density of silver grains in the film is a function of the total absorption of X-ray energy along the path through the object. There is no information in the film recording about how much energy was absorbed at a particular point along the path.

By recording shadowgraphs through a given cross section of an object from many different directions, there will be enough information to reconstruct the absorption density as a function of position within the cross section. To facilitate this, the shadowgraph is recorded electronically rather than on film. A computer is used to calculate the adsorption density as a function of position within the cross section. This is called **computed axial tomography (CAT)**.

Rather than sending X-rays through the body, as in a CAT scan, another imaging technique makes use of radioactive tracers absorbed within the body. If tracers absorbed within the body emit radiation that can be detected outside the body, information about the location of the tracer can be determined in a manner related to the techniques used in a CAT scan. Two methods that make use of this technique are **single photon emission tomography (SPET)** and **positron emission tomography (PET)**.

### Section 43-10 Nuclear Magnetic Resonance (NMR) and Magnetic Resonance Imaging (MRI)

We know that the potential energy of a magnetic dipole with dipole moment $\mu$ in a magnetic field **B** is given by

$$U = -\mu \cdot \mathbf{B}$$

Nuclei have magnetic dipole moments, so there is an energy that depends on the orientation of the dipole moment in a magnetic field. From the chapter on the quantum mechanics of atoms, we learned that the orientations are not continuous, but are quantized. That means that there will be discrete energy levels associated with the orientation of the nuclear magnetic dipole within a magnetic field. Transitions between energy levels can occur by absorption or emission of electromagnetic radiation.

In the simplest case, a hydrogen atom has a single proton in its nucleus. The magnetic dipole moment of the proton has a magnitude $\mu_p$. Like the electron, the proton is a spin ½ particle. Its angular momentum and, hence, its magnetic dipole moment can only be in one of two directions relative to the direction of the magnetic field; spin up or spin down (where up or down mean parallel to or opposite to the direction of the magnetic field). The difference in energy between the two orientations of the magnetic dipole moment is

$$\Delta E = U_{down} - U_{up} = \left(\mu_p B\right) - \left(-\mu_p B\right) = 2\mu_p B$$

The magnetic dipole can interact with an electromagnetic field, so photons with the energy difference between the two states can be absorbed by nuclei in the lowest energy state and emitted by nuclei in the higher energy state. This is in the radio frequency band of the electromagnetic spectrum.

There are several ways to utilize the interaction between the nuclear dipole moment and an electromagnetic field to make useful measurements. A simple measurement of the energy absorbed from an applied electromagnetic field at the resonant frequency will be proportional to the density of the nuclei with that resonant frequency. There is a slightly different magnetic field at the nucleus of an atom depending on to what other atoms it is chemically bound. By making a high resolution measurement of the relative amount of emission at different frequencies from nuclei excited to their higher energy states, the relative concentrations of the different chemical bonds can be determined in a sample. If the sample is placed in a spatially varying magnetic field, then nuclear concentration as a function of position can be determined. This is called **magnetic resonance imaging (MRI)**.

## Practice Quiz

1.     Which of the following reactions is impossible?

   a) $^{132}_{54}\text{Xe} + ^{1}_{1}\text{p} \rightarrow ^{133}_{55}\text{Cs}$

   b) $^{59}_{27}\text{Co} + ^{1}_{0}\text{n} \rightarrow ^{60}_{27}\text{Co}$

   c) $^{40}_{19}\text{K} + e^{-} \rightarrow ^{40}_{20}\text{Ca}$

   d) $^{31}_{14}\text{Si} + ^{32}_{15}\text{P} \rightarrow ^{63}_{29}\text{Cu}$

2.     One of the reactions below has an error in it. Which one?

   a) $^{35}_{17}\text{Cl} + ^{4}_{2}\text{He} \rightarrow ^{39}_{19}\text{K}$

   b) $^{56}_{26}\text{Fe} + ^{1}_{1}\text{p} \rightarrow ^{57}_{27}\text{Co}$

   c) $^{235}_{92}\text{U} \rightarrow ^{60}_{27}\text{Co} + ^{159}_{65}\text{Tb} + 6\,^{1}_{0}\text{n}$

   d) $^{235}_{92}\text{U} \rightarrow ^{102}_{42}\text{Mo} + ^{124}_{50}\text{Sn} + 9\,^{1}_{0}\text{n}$

3.     If radiation of energy 1.64 MeV is incident on the human body, which form of radiation will cause the most damage to internal organs?

   a) gamma radiation
   b) beta radiation
   c) protons
   d) alpha radiation

4.      Which of the following statements is true about nuclear fusion?

a) The fusion of any two nuclei is always accompanied by the release of energy.
b) The fusion of any two stable nuclei results in a stable nucleus.
c) The fusion of any two nuclei results in a nucleus with a greater binding energy per nucleon than the two fusing nuclei.
d) Any fusion process of stable lighter nuclei to form the $^{56}_{26}$Fe nucleus is accompanied by the release of energy.

5.      Which step is the most effective at reducing the exposure you receive from radiation?

a) Double your distance from the source of radiation.
b) Decrease your exposure time to one-half the time.
c) Turn sideways to the source so less area of your body receives radiation.
d) Close your eyes.

6.      Which is not a possible pair of fission products of a plutonium nucleus?

a) technetium, antimony
b) xenon, zirconium
c) cesium, yttrium
d) promethium, bromine

7.      Which pair of nuclei could fuse to form an $^{40}_{18}$Ar nucleus?

a) $^{22}_{10}$Na and $^{18}_{8}$O
b) $^{14}_{6}$C and $^{26}_{14}$Si
c) $^{26}_{12}$Mg and $^{14}_{8}$O
d) $^{39}_{19}$K and $^{1}_{1}$H

8.      For a given amount of energy deposited in the body, which of the following causes the most damage?

a) X-rays
b) beta particles
c) protons
d) alpha particles

9.      Nucleus A has a magnetic moment twice that of nucleus B. The wavelength of radio waves that are absorbed by nucleus A to change the direction of its magnetic moment are

a) twice the wavelength necessary to change the direction of nucleus B's magnetic moment
b) half the wavelength necessary to change the direction of nucleus B's magnetic moment
c) the same as the wavelength necessary to change the direction of nucleus B's magnetic moment
d) any wavelength.

10.     Estimate the total cross section for a baseball scattering off of a baseball bat.

a) 10 bn
b) $10^{-26}$ bn
c) $10^{26}$ bn
d) 0

11.     Calculate the $Q$-value of the following nuclear reaction:

$$^{51}_{23}V + ^{1}_{1}p \rightarrow ^{52}_{24}Cr$$

12.    Determine the effective dose an 84 kg person receives if their body absorbs 26 mJ of energy from beta particles.

13.    How much energy is released in the following fusion reaction?

$$^{10}_{5}B + ^{6}_{3}Li \rightarrow ^{16}_{8}O$$

14.    A source that undergoes beta decay has an activity of 3.24 mCi. The beta particle leaves the source with an energy of 0.385 MeV. The source is a distance of 1.4 m from an individual of mass 60 kg that has an effective absorption area of 0.92 $m^2$. What is the effective dose received by this individual in 4.0 hours? Assume the person absorbs all of the energy in the beta particles that reach the person's body.

15.    A hypothetical nucleus has spin ½ and a magnetic dipole moment of magnitude 3.421 $\mu_N$. Determine the resonant frequency of this nucleus in a magnetic field of magnitude 1.684 T.

## Problem Solutions

3.    For the reaction $^{238}_{92}U(n, \gamma)^{239}_{92}U$ with slow neutrons, whose kinetic energy is negligible, we find the difference of the initial and the final masses:
$$\Delta M = M(^{238}U) + m(n) - M(^{239}U)$$
$$= (238.050782 \text{ u}) + (1.008665 \text{ u}) - (239.054287 \text{ u}) = +0.005160 \text{ u}.$$
Thus no threshold energy is required, so the reaction is possible.

7.    (a)  For the reaction $^{7}_{3}Li(p, \alpha)^{4}_{2}He$, we determine the $Q$-value:
$$Q = [M(^{7}Li) + M(^{1}H) - M(^{4}He) - M(^{4}He)]c^2$$
$$= [(7.016004 \text{ u}) + (1.007825 \text{ u}) - 2(4.002603 \text{ u})]c^2(931.5 \text{ MeV}/uc^2) = +17.35 \text{ MeV}.$$
Because $Q > 0$, the reaction can occur.
       (b)  The kinetic energy of the products is
$$K = K_i + Q = 2.500 \text{ MeV} + 17.35 \text{ MeV} = 19.85 \text{ MeV}.$$

11.   (a)  We find the product nucleus by balancing the mass and charge numbers:
$$Z(X) = Z(^{6}Li) + Z(^{2}H) - Z(^{1}H) = 3 + 1 - 1 = 3;$$
$$A(X) = A(^{6}Li) + A(^{2}H) - A(^{1}H) = 6 + 2 - 1 = 7, \text{ so the product nucleus is } ^{7}_{3}Li.$$
       (b)  It is a "stripping" reaction because a neutron is stripped from the deuteron.
       (c)  For the reaction $^{6}_{3}Li(d, p)^{7}_{3}Li$, we determine the $Q$-value:
$$Q = [M(^{6}Li) + M(^{2}H) - M(^{1}H) - M(^{7}Li)]c^2$$
$$= [(6.015122 \text{ u}) + (2.014102 \text{ u}) - (1.007825 \text{ u}) - (7.016004 \text{ u})]c^2(931.5 \text{ MeV}/uc^2) = +5.025 \text{ MeV}.$$
Because $Q > 0$, the reaction is exothermic.

15.   From the figure we see that a collision will occur if
$$d \leq (R_1 + R_2).$$
Thus the area of the effective circle presented
by $R_2$ to the center of $R_1$ is
$$\sigma = \pi(R_1 + R_2)^2.$$

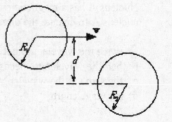

19.   We assume a 1% reaction rate allows us to treat the target as thin. The density of Cd atoms is
$$n = [(8650 \text{ kg/m}^3)(10^3 \text{ g/kg})/(113.9 \text{ g/mol})](6.02 \times 10^{23} \text{ atoms/mol}) = 4.57 \times 10^{28} \text{ m}^{-3}.$$
       (a)  From Fig. 43–3 we find that the cross section for 0.1-eV neutrons is 3000 bn. Thus we have
$$R/R_0 = n\sigma x;$$
$$0.01 = (4.57 \times 10^{28} \text{ m}^{-3})(3000 \times 10^{-28} \text{ m}^2)x, \text{ which gives } x = 7 \times 10^{-7} \text{ m} = 0.7 \text{ } \mu\text{m}.$$
       (b)  From Fig. 43–3 we find that the cross section for 10-eV neutrons is 2 bn. Thus we have
$$R/R_0 = n\sigma t;$$
$$0.01 = (4.57 \times 10^{28} \text{ m}^{-3})(2 \times 10^{-28} \text{ m}^2)x, \text{ which gives } x = 1 \times 10^{-3} \text{ m} = 1 \text{ mm}.$$

23. With an efficiency of 100% we find the number of fissions from
$P = E/t$;
300 W = (200 MeV)(1.60 × $10^{-13}$ J/MeV)$n$/(3.16 × $10^7$ s), which gives $n$ = 2.96 × $10^{20}$ fissions.
Each fission uses one uranium atom, so the required mass is
$m$ = [(2.96 × $10^{20}$ atoms)/(6.02 × $10^{23}$ atoms/mol)](235 g/mol) = 0.116 g.

27. When $^{236}$U decays by $\alpha$ emission, the resulting nucleus is $^{232}$Th. Thus the radii of the two particles are
$r_\alpha$ = (1.2 fm)(4)$^{1/3}$ = 1.90 fm;
$r_{Th}$ = (1.2 fm)(232)$^{1/3}$ = 7.37 fm.
At the instant of separation, the two particles are in contact, so the Coulomb energy is
$U_\alpha = Z_\alpha Z_{Th} e^2 / 4\pi\varepsilon_0 (r_\alpha + r_{Th})$.
If we assume fission into equal parts, the resulting nucleus has $A$ = 118. Thus each radius is
$r_f$ = (1.2 fm)(118)$^{1/3}$ = 5.89 fm.
At the instant of separation, the two particles are in contact, so the Coulomb energy is
$U_f = Z_f^2 e^2 / 4\pi\varepsilon_0 (2r_f)$.
The ratio is
$U_\alpha/U_f = Z_\alpha Z_{Th}(2r_f)/Z_f^2(r_\alpha + r_{Th})$ = (2)(90)(2)(5.89 fm)/(46)$^2$(1.90 fm + 7.37 fm) = 0.11.

31. For the reaction $^2_1$H + $^2_1$H → $^3_2$He + n , we determine the $Q$-value:
$Q$ = [$M(^2$H) + $M(^2$H) – $m$(n) – $M(^3$He)]$c^2$
= [2(2.014102 u) – (1.008665 u) – (3.016029 u)]$c^2$(931.5 MeV/u$c^2$) = + 3.27 MeV.
Thus 3.27 MeV is released.

35. Because the kinetic energies « $mc^2$, we can use a non-relativistic treatment: $K = mv^2/2 = p^2/2m$.
We assume the kinetic energy of the deuterium and the tritium can be neglected, so for momentum conservation we have
$p_{He} = p_n$.
The kinetic energy of the particles is
$Q = K_{He} + K_n = (p_{He}^2/2m_{He}) + (p_n^2/2m_n) = p_{He}^2(m_n + m_{He})/2m_{He}m_n$ ;
17.59 MeV = $p_{He}^2$(1 u + 4 u)/2(4 u)(1 u), which gives $p_{He}^2$ = 28.1 MeV · u.
The kinetic energy of $^4$He is
$K_{He} = p_{He}^2/2m_{He}$ = (28.1 MeV · u)/2(4 u) = 3.5 MeV.
The kinetic energy of n is
$K_n = p_n^2/2m_n$ = (28.1 MeV · u)/2(1 u) = 14 MeV.
This result is not independent of the plasma temperature, which is a measure of the initial kinetic energies.

39. (a) To initiate the reaction, the kinetic energy of the initial nuclei must be sufficient to overcome the
Coulomb barrier. The radii of the two initial nuclei for the first reaction of the carbon cycle are
$r_C$ = (1.2 fm)(12)$^{1/3}$ = 2.75 fm;
$r_p$ = (1.2 fm)(1)$^{1/3}$ = 1.2 fm.
We assume the total kinetic energy equals the Coulomb energy when the nuclei are in contact:
$K_{C-p} = U_{C-p} = Z_C Z_p e^2 / 4\pi\varepsilon_0 (r_C + r_p)$.
The radii of the two initial nuclei for the deuteron-tritium reaction are
$r_d$ = (1.2 fm)(2)$^{1/3}$ = 1.51 fm;
$r_t$ = (1.2 fm)(3)$^{1/3}$ = 1.73 fm.
We assume the total kinetic energy equals the Coulomb energy when the nuclei are in contact:
$K_{d-t} = U_{d-t} = Z_d Z_t e^2 / 4\pi\varepsilon_0 (r_d + r_t)$.
Thus the ratio is
$K_{C-p}/K_{d-t} = Z_C Z_p(r_d + r_t)/Z_d Z_t(r_C + r_p)$ = (6)(1)(1.51 fm + 1.73 fm)/(1)(1)(2.75 fm + 1.2 fm) = 4.9×.
(b) Because the kinetic energy is proportional to the temperature, we have
$K_{C-p}/K_{d-t} = T_{C-p}/T_{d-t}$ ;
4.9 = $T_{C-p}$/(3 × $10^8$ K), which gives $T_{C-p}$ = 1.5 × $10^9$ K.

43. Because the quality factor for slow neutrons is ≈ 3 and for fast neutrons is ≈ 10, we have
effective dose (rem) = effective dose (rad) × QF;
rad$_{slow}$ × 3 = rad$_{fast}$ × 10;
rad$_{slow}$ = (50 rad)(10)/(3) = 167 rad.

47. If we start with the current definition of the roentgen, we get

$$1 \text{ R} = (0.878 \times 10^{-2} \text{ J/kg})/(1.60 \times 10^{-19} \text{ J/eV})(1000 \text{ g/kg})(35 \text{ eV/pair})$$
$$= 1.57 \times 10^{12} \text{ pairs/g} \approx 1.6 \times 10^{12} \text{ pairs/g}.$$

51. Because each decay gives one gamma ray, the rate at which energy is emitted is

$$P = (2000 \times 10^{-12} \text{ Ci/L})(3.7 \times 10^{10} \text{ decays/s} \cdot \text{Ci})(1.5 \text{ MeV/decay})(1.60 \times 10^{-13} \text{ J/MeV})$$
$$= 1.78 \times 10^{-11} \text{ J/s} \cdot \text{L}.$$

If 10% of the energy is absorbed by the body for half a year (12 h/day), the total absorbed energy rate is

$$\text{rate} = (0.10)(1.78 \times 10^{-11} \text{ J/s} \cdot \text{L})(0.5 \text{ L})\tfrac{1}{2}(3.16 \times 10^{7} \text{ s/yr}) = 1.4 \times 10^{-5} \text{ J/yr}.$$

For beta particles and gamma rays, QF = 1.

(a) For an adult, the total dose in one year is

$$\text{dose} = (1.4 \times 10^{-5} \text{ J/yr})(1)/(1.00 \text{ J/kg} \cdot \text{Gy})(50 \text{ kg})$$
$$= 2.8 \times 10^{-7} \text{ Sv} = 2.8 \times 10^{-5} \text{ rem} = 0.03 \text{ mrem} \approx 0.006\% \text{ of allowed dose.}$$

(b) For a baby, the total dose in one year is

$$\text{dose} = (1.4 \times 10^{-5} \text{ J/yr})(1)/(1.00 \text{ J/kg} \cdot \text{Gy})(5 \text{ kg})$$
$$= 2.8 \times 10^{-6} \text{ Sv} = 2.8 \times 10^{-4} \text{ rem} = 0.3 \text{ mrem} \approx 0.06\% \text{ of allowed dose.}$$

55. (a) We find the other nucleus by balancing the mass and charge numbers:

$$Z(\text{X}) = Z(^{9}\text{Be}) + Z(^{4}\text{He}) - Z(\text{n}) = 4 + 2 - 0 = 6;$$
$$A(\text{X}) = A(^{9}\text{Be}) + A(^{4}\text{He}) - A(\text{n}) = 9 + 4 - 1 = 12, \text{ so the other nucleus is } ^{12}_{6}\text{C}.$$

(b) For the reaction $^{9}_{4}\text{Be} + ^{4}_{2}\text{He} \rightarrow ^{12}_{6}\text{C} + ^{1}_{0}\text{n}$, we determine the $Q$-value:

$$Q = [M(^{9}\text{Be}) + M(^{4}\text{He}) - M(^{12}\text{C}) - m(\text{n})]c^2$$
$$= [(9.012182 \text{ u}) + (4.002603 \text{ u}) - (12.000000 \text{ u}) - (1.008665 \text{ u})]c^2(931.5 \text{ MeV/u}c^2) = +5.70 \text{ MeV}.$$

59. The effective dose in rem = effective dose (rad) × QF. For the two radiations, we get

$$\text{dose (rem)} = \Sigma \text{ dose (rad)} \times \text{QF} = (21 \text{ mrad/yr})(1) + (3.0 \text{ mrad/yr})(10) = 51 \text{ mrem/yr}.$$

63. Because the reaction $^{18}_{8}\text{O}(\text{p}, \text{n})^{18}_{9}\text{F}$ requires an input of energy, the $Q$-value is negative:

$$Q = [M(^{18}\text{O}) + M(^{1}\text{H}) - m(\text{n}) - M(^{18}\text{F})]c^2;$$
$$-2.453 \text{ MeV} = [(17.999160 \text{ u}) + (1.007825 \text{ u}) - (1.008665 \text{ u}) - M(^{18}\text{F})]c^2(931.5 \text{ MeV/u}c^2),$$

which gives $M(^{18}\text{F}) = 18.000953 \text{ u}$.

67. In the net proton-proton cycle, four protons produce two neutrinos. If we use the result from Problem 66 for the rate at which protons are consumed, for the rate at which neutrinos are produced we have

$$n_\nu = (3.8 \times 10^{38} \text{ protons/s})(2 \text{ neutrinos/4 protons})(3.16 \times 10^{7} \text{ s/yr}) = 6.0 \times 10^{45} \text{ neutrinos/yr}.$$

The neutrinos are spread uniformly over a sphere centered at the Sun, so the number that would pass through an area of 100 m$^2$ at the Earth that is always perpendicular to the neutrino flux is

$$N_{\nu 0} = [(6.0 \times 10^{45} \text{ neutrinos/yr})/4\pi \ (1.50 \times 10^{11} \text{ m})^2](100 \text{ m}^2) = 2.1 \times 10^{24} \text{ neutrinos/yr}.$$

At a latitude of 40° the ceiling is not always perpendicular to the neutrino flux, so the number is reduced by a factor of cos 40°, which we assume as an average value for the annual variation in the elevation of the Sun. During the daily rotation of the Earth, the neutrino flux will vary sinusoidally from zero to the maximum. (We take both directions through the ceiling as positive.) To average this variation, we use a factor of cos 45°, so we have

$$N_\nu = N_{\nu 0} \cos 40° \cos 45° = (2.1 \times 10^{24} \text{ neutrinos/yr}) \cos 40° \cos 45° \approx 1 \times 10^{24} \text{ neutrinos/yr}.$$

71. (a) The rate of decay is

$$\text{Activity} = (0.10 \times 10^{-6} \text{ Ci})(3.7 \times 10^{10} \text{ decays/s} \cdot \text{Ci}) = 3.7 \times 10^{3} \text{ decays/s}.$$

(b) Because the QF for beta particles is 1, the dose in Sv is the dose in Gy:

$$\text{dose rate} = (3.7 \times 10^{3} \text{ decays/s})(1.4 \text{ MeV/decay})(3.16 \times 10^{7} \text{ s/yr})(1.6 \times 10^{-13} \text{ J/MeV})/$$
$$(50 \text{ kg})(1.00 \text{ J/kg} \cdot \text{Gy}) = 5.2 \times 10^{-4} \text{ Sv/yr} \approx 0.15 \text{ background}.$$

# Chapter 44: Elementary Particles

## Chapter Overview and Objectives

This chapter introduces the physics of elementary particles. It briefly discusses the experimental tools of high energy particle physics and introduces the current models of elementary particles.

After completing study of this chapter, you should:

- Know what particle accelerators are and why they are necessary to probe small length scales.
- Know how forces can be modeled by the exchange of particles.
- Know the quantum numbers that are conserved in elementary particle interactions.
- Know that all elementary particles can be classified as leptons, hadrons, or gauge bosons.
- Know what the quark model of hadrons is.
- Know what quantum chromodynamics and electroweak theory are.

## Summary of Equations

De Broglie wavelength:
$$\lambda = \frac{h}{p}$$

## Chapter Summary

### Section 44-1 High-Energy Particles

To investigate structure on a given size scale, probes must have a wavelength of the same size. Using the de Broglie wavelength relationship,

$$\lambda = \frac{h}{p},$$

we know that we must use energy of greater momentum, and, therefore, greater kinetic energy to probe details at smaller size scales. Particle accelerators are devices that give particles large kinetic energies so that they can be used to probe small size scales.

### Example 44-1-A

Compare the de Broglie wavelength of electrons with 2.0 GeV of kinetic energy with the de Broglie wavelength of protons with 2.0 GeV of kinetic energy.

### Solution:

We know that the momentum of a particle is related to its kinetic energy by

$$K = \sqrt{m^2c^4 + p^2c^2} - mc^2 \quad \Rightarrow \quad p = \frac{1}{c}\sqrt{\left(K + mc^2\right)^2 - m^2c^4}$$

We can solve for the de Broglie wavelength of the electron:

$$\lambda = \frac{h}{p} = \frac{hc}{\sqrt{\left(K + mc^2\right)^2 - m^2c^4}} = \frac{\left(6.63 \times 10^{-34}\,\text{J}\cdot\text{s}\right)\left(3.00 \times 10^8\,\text{m/s}\right)}{\sqrt{\left(2000\,\text{Mev} + 0.511\,\text{MeV}\right)^2 - \left(0.511\,\text{MeV}\right)^2}\left(1.602 \times 10^{-13}\,\text{J/MeV}\right)}$$

$$= 6.21 \times 10^{-16}\,\text{m}$$

We can solve for the de Broglie wavelength of the proton:

$$\lambda = \frac{h}{p} = \frac{hc}{\sqrt{(K+mc^2)^2 - m^2c^4}} = \frac{(6.63\times10^{-34}\,\text{J}\cdot\text{s})(3.00\times10^8\,\text{m/s})}{\sqrt{(2000\,\text{Mev}+938.3\,\text{MeV})^2 - (938.3\,\text{MeV})^2}\,(1.602\times10^{-13}\,\text{J/MeV})}$$

$$= 4.461\times10^{-16}\,\text{m}$$

Even though the particles have very different masses, their momenta are not very different because they are both relativistic. The very relativistic limit of the momentum becomes $p = E/c$ regardless of the mass of the particle.

### Section 44-2 Particle Accelerators and Detectors

There are several basic types of particle accelerators that can be used to accelerate particles to very large kinetic energies. Van de Graaff generators, cyclotrons, synchrotrons, and linear accelerators are all types of particle accelerators.

Particle detectors are used to determine the nature of particles that are the result of particle interactions that take place. The detector is used to determine the trajectories of the products of the interactions that occur. Information such as the charge-to-mass ratios of the products can be determined from the trajectories of the particles. Some of the types of detectors that are used or have been used are photographic emulsions, cloud chambers, bubble chambers, and wire drift chambers.

### Section 44-3 Beginnings of Elementary Particle Physics—Particle Exchange

An alternate to the representation of forces between particles being caused by fields, such as the electric field, is to treat force as the result of the exchange of particles. The result of an interaction between two objects is the transfer of momentum and energy between the two objects. Particle transfer from one object to another also results in the transfer of momentum and energy from one object to another. The transferred particle carries the momentum and energy from one object to the other. For example, rather than understanding electric forces by considering an electric field existing around charged particles, it is possible to treat the electric force as being the result of the exchange of photons between two charged particles.

Each fundamental force has a corresponding force carrier or mediator. The **photon** is the mediator of the electromagnetic force. The **pi meson** or **pion**, as it is more commonly called, is the mediator of the strong nuclear force. The $\mathbf{W^+}$, $\mathbf{W^-}$, and the **Z bosons** are the mediators of the weak nuclear force. The **graviton** is the mediator of the gravitational force. All of the force mediators are whole integer spin particles. They can carry intrinsic or spin angular momentum as well as momentum and energy.

### Section 44-4 Particles and Antiparticles

Every particle has a corresponding antiparticle. The antiparticle has the opposite electric charge as the particle, but is identical in mass and intrinsic spin. Some uncharged particles are thier own antiparticles, such as the photon, while other uncharged particles have a distinct antiparticle, such as the neutron.

### Section 44-5 Particle Interactions and Conservation Laws

We already know some of the conservation laws of nature. We know about conservation of charge, conservation of energy, conservation of translational momentum and conservation of angular momentum. There are some additional conservation laws that only become evident when investigating interactions between fundamental particles. Each particle has a **baryon number**, $B$. For example, the nucleons have a baryon number $B = +1$ and the antiparticles of the nucleons have a baryon number $B = -1$. Electrons and neutrinos have baryon number $B = 0$. Baryon number appears to be conserved during particle interactions.

Three other conserved quantities are the three lepton numbers, $L_e, L_\mu, L_\tau$. The electron lepton number, $L_e$, is one for the electron and neutrino and zero for all other particles. The electron lepton number, $L_e$, is one for the electron, $e$, and (electron) neutrino, $\nu_e$, negative one for their antiparticles, and zero for all other particles. The muon lepton number, $L_\mu$, is one for the muon, $\mu^-$, and muon neutrino, $\nu_\mu$, negative one for their antiparticles, and zero for all other particles. The tao lepton number, $L_\tau$, is one for the tao lepton, $\tau^-$, and tao neutrino, $\nu_\tau$, negative one for their antiparticles, and zero for all other particles. The muon and tao lepton are particles that are identical to the electron, but more massive.

**Example 44-5-A**

Which of the following forbidden particle reactions are not allowed because they violate conservation of baryon number?

1. $\tau^- \to \pi^- + \bar{\nu}_\tau$    3. $n \to p + \pi^-$
2. $\tau^- \to e^- + \bar{\nu}_\tau$    4. $\Lambda^0 \to \bar{p} + p + \pi^0$

**Solution:**

1. The baryon number of the $\tau^-$ is zero, the baryon number of the $\pi^-$ is zero, and the baryon number of the tau anti-neutrino are all zero, so baryon number is conserved.

2. The baryon number of the $\tau^-$ is zero, the baryon number of the $e^-$ is zero, and the baryon number of the tau anti-neutrino are all zero, so baryon number is conserved.

3. The baryon number of the neutron is one, the baryon number of the proton is one, and the baryon number of the $\pi^-$ is zero, so baryon number is conserved.

4. The baryon number of the $\Lambda^0$ is $+1$, the baryon number of the antiproton is $-1$, the baryon number of the proton is $+1$, and the baryon number of the $\pi^0$ is zero, so baryon number is not conserved.

**Example 44-5-B**

Which of the forbidden particle reactions in Example 44-5-A are not allowed because they violate conservation of a lepton number? Determine which lepton number is not conserved.

**Solution:**

1. The tau lepton number of the $\tau^-$ is $+1$, the tau lepton number of the $\pi^-$ is zero, and the tau lepton number of the tau neutrino is $-1$, so tau lepton number is not conserved.

2. The tau lepton number of the $\tau^-$ is $+1$, the tau lepton number of the $e^-$ is 0, and the tau lepton number of the tau anti-neutrino is $+1$, so tau lepton number is conserved. The electron lepton number of the $\tau^-$ is 0, the electron lepton number of the $e^-$ is $+1$, and the electron lepton number of the tau anti-neutrino is 0, so electron lepton number is not conserved.

3. All of these particles have zero lepton numbers so all of the lepton numbers are conserved.

4. All of these particles have zero lepton numbers so all of the lepton numbers are conserved.

**Section 44-6 Particle Classification**

The particles can be divided into categories. First of all, particles can be divided into two groups according to whether they are integer spin particles or odd half-integer spin particles. The particles with integer spins are called **bosons** and the particles with odd half-integer spins are called **fermions**. The bosons obey **Bose-Einstein statistics** and the fermions obey **Fermi-Dirac statistics**. The **gauge bosons** are particles which are the force mediators. The **leptons**, which are fermions, are the electron, muon, and tao leptons and their corresponding neutrinos and their antiparticles. Leptons do not interact through the strong nuclear force. **Hadrons** are the particles that interact through the strong nuclear force. The hadrons can be subdivided into **mesons** and **baryons**. Mesons have baryon number $B = 0$ and baryons have a non-zero baryon number.

**Section 44-7 Particle Stability and Resonances**

Most particles are unstable. They decay into two or more lighter mass particles. The lifetime of a particle is characteristic of the fundamental force which is responsible for the decay. Decays due to weak force interactions have typical lifetimes greater than $10^{-13}$ s. Decays due to strong force interactions are typically in the range of $10^{-19}$ to $10^{-16}$ s. Some particles have such a short lifetime they are called **resonances** rather than particles.

### Section 44-8 Strange Particles

The conservation of baryon number and the conservation of the lepton numbers allow us to see why some otherwise possible particle interactions do not occur in nature. However, the set of conservation laws we have looked at so far is incomplete. Some interactions that are allowed by the conservation laws studied to this point do not occur in nature. Introducing a new property called **strangeness**, along with a **conservation of strangeness law**, appears to be consistent with the observed behavior. Particles that have non-zero strangeness are called **strange particles**. Strangeness is conserved during strong interactions, but is not conserved during weak interactions.

### Section 44-9 Quarks

The leptons show no internal structure down to the smallest size scales probed so far. However, it has long been apparent that hadrons have some internal structure and are not elementary particles, but could be built from some "simpler" particles. In addition, the large number of different baryons appears to make nature unusually complex if each of the hadrons is an elementary particle. The **quark** model of hadrons constructs all the hadrons from six different quarks and their antiparticles. The six types or **flavors** of quarks are **up** (u), **down** (d), **strange** (s), **charmed** (c), **bottom** (b), and **top** (t). All of the quarks are spin ½ particles. Three new quantum numbers (charm, bottomness, and topness) are needed to completely specify a quark. The quantum numbers of the quarks are

| Quark | Spin | Charge | Baryon Number | Strangeness | Charm | Bottomness | Topness |
|-------|------|--------|---------------|-------------|-------|------------|---------|
| u | ½ | $+\frac{2}{3}e$ | ⅓ | 0 | 0 | 0 | 0 |
| d | ½ | $-\frac{1}{3}e$ | ⅓ | 0 | 0 | 0 | 0 |
| s | ½ | $-\frac{1}{3}e$ | ⅓ | −1 | 0 | 0 | 0 |
| c | ½ | $+\frac{2}{3}e$ | ⅓ | 0 | +1 | 0 | 0 |
| b | ½ | $-\frac{1}{3}e$ | ⅓ | 0 | 0 | −1 | 0 |
| t | ½ | $+\frac{2}{3}e$ | ⅓ | 0 | 0 | 0 | +1 |

Each quark has a corresponding anti-quark. The anti-quark has the negative value of each quantum number of the corresponding quark.

In the quark model, mesons are constructed from a quark-anti-quark pair and baryons are constructed from three quarks or anti-quarks. Arbitrary pairs or triplets are not allowed. The quantum numbers of the pairs or triplets are the sums of the quantum numbers of the quarks. The charge of the combination of quarks must be an integer multiple of an electron charge. The baryon number of the combination of quarks must be an integer.

### Section 44-10 The "Standard Model": Quantum Chromodynamics (QCD) and the Electroweak Theory

The quark model of matter discussed in Section 9 is a model of the structure of baryons, but a theory of the interaction between the quarks is also needed to complete the theory. The model of the quark interaction is somewhat more complex than the electromagnetic force. The standard model of quark interactions is called **quantum chromodynamics**. In this model, each quark type comes in one of three different possible **color charges**. The gauge boson that mediates the **color force** between two quarks is called a **gluon**.

**Electroweak theory** unifies the electromagnetic force and the weak force into two different manifestations of the same fundamental interaction.

Quantum chromodynamics together with electroweak theory are called the **standard model** of particle structure and interaction.

### Section 44-11 Grand Unified Theories

Theories that attempt to treat the electromagnetic force, the weak force, and the strong force as different manifestations of the same fundamental interaction are called **grand unified theories**.

## Practice Quiz

1.  Which of the following is not a conserved quantity?

    a) baryon number
    b) lepton number
    c) hadron number
    d) energy

2.  Which of the conservation laws does the reaction $p + n \rightarrow p + p + p + \pi^- + \pi^-$ violate?

    a) conservation of baryon number
    b) conservation of hadron number
    c) conservation of a lepton number
    d) conservation of charge

3.  Which of the following is a meson with charge +e, strangeness +1, bottomness 0, and topness 0?

    a) $ud\bar{s}$
    b) $u\bar{s}$
    c) $\bar{s}c$
    d) $us\bar{u}$

4.  Which particle is a force carrier or mediator of the weak force?

    a) photon
    b) hadron
    c) $W^+$ boson
    d) neutrino

5.  Which of the following statements is true?

    a) All particles that have integer spins are gauge bosons.
    b) All gauge particles have integer spin.
    c) All particles have integer spin.
    d) All gauge particles are massless.

6.  How can the quark model allow the uncharged neutron to have a magnetic moment?

    a) The quark model cannot account for the non-zero magnetic moment of the neutron.
    b) The intrinsic magnetic moments of the quarks that a neutron is composed of do not add up to zero.
    c) The orbital motion of the charged quarks within the nucleus result in a magnetic moment of the nucleus.
    d) Both reasons in b) and c) above can contribute to the magnetic moment of the neutron.

7.  Which of the following particles is not a hadron?

    a) muon
    b) pion
    c) neutron
    d) proton

8.  Which type of force does not affect a $\mu$ particle?

    a) electromagnetic force
    b) strong force
    c) weak force
    d) gravitational force

9.    Which quantity is conserved during strong interactions, but not in weak interactions?

a) baryon number
b) charge
c) lepton number
d) strangeness

10.   What property must a particle have to be its own anti-particle?

a) no spin
b) no mass
c) no charge
d) no magnetic moment

11.   What is the difference between the speed of light and the speed of a 2.0 GeV electron?

12.   A beam of electrons collides with a beam of protons with kinetic energy 12 GeV traveling in the opposite direction as the electrons. If the total momentum of the electrons and protons is zero, what is the kinetic energy of the electrons?

13.   A cyclotron is constructed with a radius of 5.82 cm. The magnetic field is 2.2 T. What frequency must the accelerating voltage must be so that the cyclotron can accelerate protons? What is the kinetic energy of the protons when they reach the outside radius of the cyclotron?

14.   If two protons collide with equal kinetic energies traveling in opposite directions, what is the minimum kinetic energy of each proton necessary for the reaction

$$p + p \rightarrow p + p + \pi^0 + \pi^0$$

to occur?

15.   A $\Sigma^-$ is at rest in the laboratory and decays into a neutron and a $\pi^-$. What are the momentum and energy of the two decay products?

## Problem Solutions

3.    We find the magnetic field from the cyclotron frequency:
$f = qB/2\pi m$;
$2.8 \times 10^7$ Hz $= (1.60 \times 10^{-19}$ C$)B/2\pi(1.67 \times 10^{-27}$ kg), which gives $B = 1.8$ T.

7.    The size of a nucleon is
$d \approx 2(1.2 \times 10^{-15}$ m$) = 2.4 \times 10^{-15}$ m.
Because 30 MeV $\ll mc^2$, we find the momentum from
$K = p^2/2m$, so the wavelength is
$\lambda = h/p = h/(2mK)^{1/2}$.
For the $\alpha$ particle we have
$\lambda_\alpha = (6.63 \times 10^{-34}$ J $\cdot$ s$)/[2(4)(1.67 \times 10^{-27}$ kg$)(30$ MeV$)(1.60 \times 10^{-13}$ J/MeV$)]^{1/2}$
$= 2.6 \times 10^{-15}$ m $\approx$ size of nucleon.
For the proton we have
$\lambda_p = (6.63 \times 10^{-34}$ J $\cdot$ s$)/[2(1.67 \times 10^{-27}$ kg$)(30$ MeV$)(1.60 \times 10^{-13}$ J/MeV$)]^{1/2}$
$= 5.2 \times 10^{-15}$ m $\approx$ 2(size of nucleon).
Thus the $\alpha$ particle is better.

11.   The number of revolutions is
$n = \Delta E/2eV = (900$ GeV $- 8.0$ GeV$)(10^3$ MeV/GeV$)/(1$ e$)(2.5$ MV$) = 3.57 \times 10^5$ rev.
The total distance traveled is
$d = n2\pi R = (3.57 \times 10^5)2\pi(1.0$ km$) = 2.2 \times 10^6$ km.
Very-high-energy protons will have a speed $v \approx c$. Thus the time is
$t = d/v = (2.2 \times 10^9$ m$)/(3.00 \times 10^8$ m/s$) = 7.5$ s.

15. For the reaction $\pi^+ \to \mu^+ + \nu_\mu$, we determine the $Q$-value:

$$Q = [m(\pi^+) - m(\mu^+)]c^2$$
$$= [(139.6 \text{ MeV}/c^2) - (105.7 \text{ MeV}/c^2)]c^2 = 33.9 \text{ MeV}.$$

Thus 33.9 MeV is released.

19. Because two protons are present before and after the process, and the total momentum is zero, the minimum kinetic energy will produce all three particles at rest. Thus the total initial kinetic energy must provide the rest energy of the $\pi^0$ meson:

$$2K = m_{\pi 0}c^2 = 135 \text{ MeV, which gives } K = 67.5 \text{ MeV}.$$

23. (a) For the reaction $e^- \to e^- + \gamma$, the isolated electron is at rest. For the photon, we have $E_\gamma = p_\gamma c$.
    For energy conservation we have
    $$m_e c^2 = [(p_e c)^2 + (m_e c^2)^2]^{1/2} + E_\gamma.$$
    For momentum conservation we have
    $$0 = p_e - p_\gamma.$$
    When we eliminate $p_e$ in the energy equation and rearrange, we have
    $$m_e c^2 - E_\gamma = [E_\gamma^2 + (m_e c^2)^2]^{1/2}.$$
    When we square both sides, we have
    $$(m_e c^2)^2 + E_\gamma^2 - 2E_\gamma m_e c^2 = E_\gamma^2 + (m_e c^2)^2, \text{ which gives } E_\gamma = 0.$$
    Thus no photon is emitted.

    (b) For the photon exchange in Fig. 44–10, the photon exists for such a short time that the uncertainty principle allows energy to not be conserved during the exchange.

27. The total kinetic energy after the decay of the stationary $\Xi^-$ is the $Q$-value:

$$K_\Lambda + K_\pi = Q = [m(\Xi^-) - m(\Lambda^0) - m(\pi^-)]c^2$$
$$= [(1321.3 \text{ MeV}/c^2) - (1115.7 \text{ MeV}/c^2) - (139.6 \text{ MeV}/c^2)]c^2 = 66.0 \text{ MeV}.$$

For energy conservation we have
$$m_\Xi c^2 = E_\Lambda + E_\pi, \quad \text{or} \quad E_\pi = m_\Xi c^2 - E_\Lambda.$$

For momentum conservation we have
$$0 = p_\pi - p_\Lambda, \quad \text{or} \quad (p_\Lambda c)^2 = (p_\pi c)^2 = E_\Lambda^2 - (m_\Lambda c^2)^2 = E_\pi^2 - (m_\pi c^2)^2.$$

When we combine this with the result from energy conservation, we get
$$E_\Lambda = [(m_\Xi c^2)^2 + (m_\Lambda c^2)^2 - (m_\pi c^2)^2]/2m_\Xi c^2$$
$$= [(1321.3 \text{ MeV})^2 + (1115.7 \text{ MeV})^2 - (139.6 \text{ MeV})^2]/2(1321.3 \text{ MeV}) = 1124.3 \text{ MeV}.$$

For the kinetic energies we have
$$K_\Lambda = E_\Lambda - m_\Lambda c^2 = 1124.3 \text{ MeV} - 1115.7 \text{ MeV} = 8.6 \text{ MeV};$$
$$K_\pi = Q - K_\Lambda = 66.0 \text{ MeV} - 8.6 \text{ MeV} = 57.4 \text{ MeV}.$$

31. We estimate the lifetime from
$$\Delta t = h/2\pi\Delta E = hc/2\pi c \, \Delta E = (1.24 \times 10^{-12} \text{ MeV} \cdot \text{m})/2\pi(3.00 \times 10^8 \text{ m/s})(0.088 \text{ MeV}) = 7.5 \times 10^{-21} \text{ s}.$$

35. (a) For the neutron we must have charge, strangeness, charm, bottomness, and topness = 0.
    For the baryon number to be 1, we need three quarks: n = d d u.

    (b) For the antineutron we have $\bar{n} = \bar{d}\bar{d}\bar{u}$.

    (c) For the $\Lambda^0$ we must have charge, charm, bottomness, and topness = 0. To get strangeness = –1, we need an s quark. To get a baryon number of 1, we need three quarks. To get a charge = 0, we have
    $$\Lambda^0 = \text{u d s}.$$

    (d) For the $\Sigma^0$, we must have charge, charm, bottomness, and topness = 0. To get strangeness = +1, we need an $\bar{s}$ quark. To get a baryon number of –1, we need three antiquarks. To get a charge = 0, we have $\Sigma^0 = \bar{u}\bar{d}\bar{s}$.

39. (a)

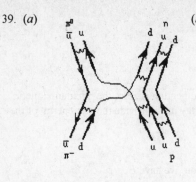

(b) From Table 44–3 we see that the $\pi^0$ could be either $u\bar{u}$ or $d\bar{d}$. Actually it is a combination of the two, but we will simplify the diagram by assuming that one $\pi^0$ is $u\bar{u}$ and the other is $d\bar{d}$.

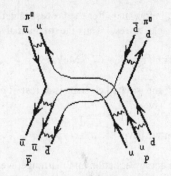

We see that the intermediate particle is $\pi^+$.

We see that the intermediate particle is p. Note that high-energy $p\bar{p}$ collisions usually produce many particles.

43. Because 7.0 TeV $\gg m_pc^2$, we have
$$K = E = pc.$$
We use the result from Problem 13:
$$E(eV) = Brc;$$
$$(7.0 \text{ TeV})(1.60 \times 10^{12} \text{ eV/TeV}) = B(4.25 \times 10^3 \text{ m})(3.00 \times 10^8 \text{ m/s}), \text{ which gives } B = 5.5 \text{ T}.$$

47. We estimate the energy from
$$\Delta x \geq h/2\pi\Delta p = hc/2\pi\Delta pc = hc/2\pi\Delta E;$$
$$10^{-18} \text{ m} = (1.24 \times 10^{-12} \text{ MeV} \cdot \text{m})/2\pi\Delta E, \text{ which gives } \Delta E = 2 \times 10^5 \text{ MeV} = 2 \times 10^2 \text{ GeV}.$$
This is on the order of the 80 GeV rest energy of the W particles.

51. We consider the quarks and leptons as the fundamental fermions. A water molecule consists of two hydrogen atoms and one oxygen atom, or $2 + 8 = 10$ protons, $2 + 8 = 10$ electrons, and 8 neutrons. Each of the protons and neutrons is made up of three quarks, so the number of fundamental fermions is
$$N = (10 + 8)(3) + 10 = 64.$$

55. We find the speed of the tau-lepton from its kinetic energy:
$$K = \{[1/(1 - v^2/c^2)^{1/2}] - 1\}mc^2;$$
$$450 \text{ MeV} = \{[1/(1 - v^2/c^2)^{1/2}] - 1\}(1777 \text{ MeV}), \text{ which gives } 1/(1 - v^2/c^2)^{1/2} = 1.25, \text{ and } v = 0.603c.$$
In the lab the lifetime of the tau-lepton will be dilated:
$$\Delta t = \Delta t_0/(1 - v^2/c^2)^{1/2}.$$
Thus the length of the track will be
$$\Delta x = v\,\Delta t = (0.603)(3.00 \times 10^8 \text{ m/s})(2.91 \times 10^{-13} \text{ s})(1.25) = 6.58 \times 10^{-5} \text{ m}.$$

# Chapter 45: Astrophysics and Cosmology

## Chapter Overview and Objectives

This chapter introduces the applications of the laws of physics to the universe on scales of size beyond our solar system. It provides an introduction to the ideas behind the theory of general relativity. It discusses the formation and evolution of stars. It also introduces the standard model of the creation and evolution of the universe.

After completing study of this chapter, you should:

- Be familiar with the units of measurement of distance used for astronomical size scales.
- Know that stars go through different evolutionary processes that depend on the masses of the stars.
- Know what the principle of equivalence is.
- Know that the general theory of relativity explains gravitational forces using curved space-time.
- Know that the universe is expanding and how Hubble's law describes that expansion.
- Know what the cosmic microwave background radiation is.
- Know what the big bang model of the universe is.
- Know what the critical density of the universe is and how it affects the distant future of the universe.

## Summary of Equations

Relationship between light-year and meter:
$$1\,\text{ly} = \left(2.998 \times 10^8 \text{ m/s}\right)\left(3.156 \times 10^7 \text{ s}\right) = 9.46 \times 10^{15} \text{ m}$$

Relationship between parsec, light-year, and meter:  $1\,\text{pc} = 3.26\,\text{ly} = 3.08 \times 10^{16} \text{ m}$

Relationship between apparent brightness and absolute luminosity:

$$l = \frac{L}{4\pi d^2}$$

Schwarschild radius:
$$R = \frac{2GM}{c^2}$$

Hubble's law:
$$v = Hd$$

Critical density of the universe:
$$\rho_C \approx 10^{-26} \text{ kg/m}^3$$

## Chapter Summary

### Section 45-1 Stars and Galaxies

Astronomical distances are so great, we define a more convenient unit to measure distances called the light-year. One **light-year (ly)** is the distance light travels in one year:

$$1\,\text{ly} = \left(2.998 \times 10^8 \text{ m/s}\right)\left(3.156 \times 10^7 \text{ s}\right) = 9.46 \times 10^{15} \text{ m}$$

Visible stars appear to be organized in large groups called **galaxies**. Galaxies appear to be organized in larger groups called **galactic clusters**. Galactic clusters appear to be organized in larger groups called **superclusters**.

The distance to an object can be determined from difference in viewing direction from two spatially separated viewing positions. The **parallax** angle is defined in the diagram to the right. It is one-half the angular difference in viewing directions of the object. For small angles, the distance to the object is inversely proportional to the parallax angle and the distance between the two direction measurements. We define a unit of distance called a parsec, which is the distance

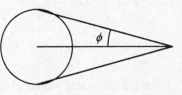

to an object that has a parallax angle of one second of arc when the two viewing locations are separated by the mean diameter of the Earth's orbit about the Sun. One parsec is equal to 3.26 light-years or $3.08 \times 10^{16}$ m:

$$1\,\text{pc} = 3.26\,\text{ly} = 3.08 \times 10^{16}\,\text{m}$$

### Example 45-1-A

The human visual system uses parallax to help determine the distance to an object by the difference in direction of the light source to the two eyes. We learned in Chapter 36 that the angular resolution of the human eye was about $5 \times 10^{-4}$ rad. Use this to determine the maximum possible distance of objects for which the visual system could use parallax to assist in determining distance. The separation of the two eyes is about 8 cm.

### Solution:

Our eyes cannot determine angular differences smaller than the angular resolution of the eye, so parallax angles smaller than the angular resolution of the eye are undetectable. We want to know the distance of an object such that the parallax angle is equal to the angular resolution of the eye. The angular resolution of the eye is a small angle, so we can use the small angle approximation:

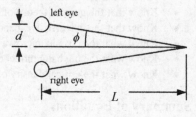

$$L \approx \frac{d}{\phi} = \frac{4 \times 10^{-2}\,\text{m}}{5 \times 10^{-4}\,\text{rad}} = 80\,\text{m}$$

The actual limiting distance for useful parallax is probably somewhat less than this, but you can easily verify that parallax is detectable to a distance of several tens of meters. How does the human visual system judge distances greater than that for which eye-to-eye parallax is useful?

### Section 45-2 Stellar Evolution: The Birth and Death of Stars

The brightness of an observed source of light depends on its radiated power output and its distance from the observer. The **absolute luminosity**, $L$, of a radiating source is defined to be the total power radiated by the source. The apparent brightness, $l$, is defined as the intensity of the radiated energy at the position of the observer. Assuming the energy is radiated isotropically by the source and none of the radiation is absorbed by intervening material, the apparent brightness is related to the absolute luminosity by

$$l = \frac{L}{4\pi d^2}$$

The luminosity of a star is related to its mass, with more massive stars having greater luminosity. There is a strong correlation between the luminosity of a star and its surface temperature. A Hertzprung-Russell diagram is a plot of stars' luminosity versus their surface temperature that shows this correlation. Most stars are located near a curve called the **main sequence** on this diagram.

### Example 45-2-A

An object has an apparent brightness of $1.75 \times 10^{-7}$ W/m$^2$ and a parallax angle of 38 seconds of arc. What is the absolute luminosity of the object, assuming it radiates isotropically and there is no scattering or absorbing material between the object and its observer?

### Solution:

The distance to the object is

$$d = 38'' \frac{3.08 \times 10^{16}\,\text{m}}{\text{arc sec}} = 1.17 \times 10^{18}\,\text{m}$$

We use the relationship between absolute luminosity and apparent brightness to solve for the absolute luminosity:

$$l = \frac{L}{4\pi d^2} \quad \Rightarrow \quad L = 4\pi d^2 I = 4\pi \left(1.17 \times 10^{18} \text{ m}\right)^2 \left(1.75 \times 10^{-7} \text{ W/m}^2\right) = 3.01 \times 10^{30} \text{ W}$$

A current model of star creation is that stars condense out of a cloud of hydrogen gas. A higher than average density region of hydrogen gas gravitationally attracts nearby hydrogen, increasing the denisty further. As the density increases, the temperature of the gas rises. If enough gas collects, the temperature rises far enough to begin fusion of hydrogen. The energy generated by hydrogen fusion increases the temperature and pressure in the core of the star until it balances out the pressure due to the weight of the hydrogen above it.

Most stars continue in this relatively stable dynamic equilibrium for a length of time that depends on the mass of the star. Eventually, most of the hydrogen in the core is fused into helium and there is not enough hydrogen available for enough further fusion to continue to keep the temperature and pressure high enough to prevent gravitational collapse. The star begins to collapse again. The collapse continues until the temperature rises enough to begin the fusion of helium nuclei. The temperature must rise before the helium fusion begins because the repulsive force of two helium nuclei is greater than the repulsive force of two hydrogen nuclei because their charge is twice as great.

If the mass of the star is great enough, this cycling of fusion of nuclei until exhaustion of the nuclei and then collapse continues until the product of the fusion is iron-56 nuclei. Further fusions of the nuclei are endothermic and the star begins its final gravitational collapse. The result of the final gravitational collapse depends on the mass of the star. Low mass stars end up as **white dwarfs**, radiating from the temperature rise caused by gravitational collapse. Stars greater than about 1.4 solar masses collapse into **neutron stars**. The gravitational interaction is strong enough to hold the matter into one large nucleus. During the collapse, a tremendous amount of gravitational potential energy is liberated in an explosion called a **supernova**. The neutron stars usually have a relatively rapid rate of rotation when formed because the moment of inertia of the star is greatly decreased during the collapse. The electromagnetic radiation of the neutron star combined with its rapid rate of rotation results in a **pulsar**. If the mass of the neuron star is great enough, the gravitational interaction will cause the star to continue to collapse endlessly. A **black hole** is formed from which not even light can escape.

**Section 45-3 General Relativity: Gravity and the Curvature of Space**

The **principle of equivalence** states that it is impossible to determine from local measurements whether an observer is accelerating or moving at constant velocity in a uniform gravitational field. Based on this principle, Albert Einstein developed the **theory of general relativity**. The theory of general relativity explains gravitational interactions by assuming that the presence of mass changes the geometry of space-time. The familiar geometry that is studied in high school is Euclidean geometry. The rules of Euclidean geometry describe what is called flat space. In the general theory of relativity, the presence of mass causes space-time to be curved. One of the results of the curvature of space-time is that the shortest distance between two points is not necessarily a Euclidean straight line, but can be a curve. The general name for the shortest distance path between two points is **geodesic**. In the ray model of light, light follows geodesics. If the curvature of space-time is great enough, light may be unable to escape from a given region of space. We can think of this as happening when the escape velocity of a given mass becomes greater than the speed of light. Using Newton's law of gravitation, this would happen at the surface of a mass when the radius of the mass is given by

$$R = \frac{2GM}{c^2}$$

This radius is called the **Schwarschild radius** of the mass $M$. The theory of general relativity predicts that if a mass collapses to within its Schwarschild radius, then it collapses to a single point called a **singularity**.

**Section 45-4 The Expanding Universe**

If spectra of light from distant galaxies are observed and compared to the known emission spectra of atoms, one finds the spectral lines shifted toward longer wavelengths. This **redshift**, as it is called, is caused by the Doppler effect. The velocity of the galaxy relative to the Earth can be calculated from the redshift. The velocity of galaxies relative to the Earth, in the direction away from the Earth, $v$, is found to be proportional to the distance, $d$, of the galaxy from the Earth. This is known as Hubble's law:

$$v = Hd$$

where the constant $H$ is known as the Hubble parameter and its value is approximately 20 km/s/$10^6$ ly.

If all galaxies are moving away from one another in the manner described by Hubble's law, we can infer that there may have been a time at which they were all in the same location. If we assume that the expansion rate has remained constant, we can solve for the time at which all of the galaxy positions were the same. This time works out to be $15 \times 10^9$ yr. We might take this as an estimate of the age of the universe.

### Section 45-5 The Big Bang and the Cosmic Microwave Background

No matter which direction in the sky one observes, there is a black-body spectrum of radiation reaching the Earth. The temperature of a black body with the observed radiation spectrum is about 3 K. The peak of this radiation is in the microwave portion of the electromagnetic spectrum, so the radiation is called the **cosmic microwave background radiation**. The cosmic microwave background radiation is consistent with an expanding universe and the big bang model of the universe.

### Section 45-6 The Standard Cosmological Model: the Early History of the Universe

In the big bang model of the universe, the universe expands outward from either a zero or very small radius. As the universe expands outward, the temperature drops with time. The types of matter and interactions that dominated the universe changed as the universe expanded and cooled. The early universe can be divided into different eras, based on the type of matter and interactions that dominated the universe at those times and temperatures.

### Section 45-7 The Future of the Universe?

Depending on the average density of the universe, the universe may go one expanding forever or it may reach a maximum size and then start contracting. If the density is greater than the critical density, $\rho_C$, then the universe will stop expanding at some point in the future and start contracting. If the density of the universe is less than the critical density, then the universe will go on expanding forever. If the critical density is equal to the critical velocity, the universe will go on expanding forever, but the rate of expansion will go to zero as the time increases to infinity. The critical density estimated using the current observed rate of expansion of the universe is

$$\rho_C \approx 10^{-26} \, \text{kg/m}^3$$

The amount of directly detectable matter in the universe appears to have a density significantly lower than the critical density. There is an unknown amount of matter that does not emit enough radiation to be detectable on the Earth. This is often called **dark matter**. There is some evidence that there is a significant amount of dark matter because the rotation rate of galaxies appears greater than the rate predicted if the visible mass of the galaxies is the entire mass of the galaxies.

## Practice Quiz

1.      A star has a mass of 1.5 solar masses. Which object will the star eventually end up as?

      a) white dwarf
      b) neutron star
      c) black hole
      d) green giant

2.      There are stars that can be observed that are close enough to determine their distance using parallax angle. If we calculate their absolute luminosity from their apparent brightness and their distance, their luminosity falls far below the main sequence for their surface temperature. What could cause the absolute luminosity of the star to be so low?

      a) There may be dust between the star and the Earth that absorbs or scatters much of the light from the star.
      b) The star may be moving toward the Earth at a relatively high speed.
      c) The star might be a red giant.
      d) The star's gravity might be so great that is is preventing the radiation from escaping efficiently.

3.    What should happen to the temperature of the cosmic background radiation as the universe continues expand?

a) The temperature of the cosmic background radiation should drop as the universe expands.
b) The temperature of the cosmic background radiation should rise as the universe expands.
c) The temperature of the cosmic background radiation should remain the same as the universe expands.
d) The change in temperature as the universe expands depends on whether the universe has greater or less than the critical density.

4.    If the universe is closed and begins to contract into the big crunch, which era would happen before the other?

a) The radiation era would occur before the lepton era.
b) The hadron era would occur before the radiation era.
c) The lepton era would occur before the matter-dominated era.
d) The hadron era would occur before the lepton era.

5.    What is the approximate age of the universe based on Hubble's law?

a) 1.5 million years
b) 15 million years
c) 1.5 billion years
d) 15 billion years

6.    Which of the following is a result of the universe having greater than the critical density?

a) the big crunch
b) a supernova
c) nuclear fission of hydrogen nuclei
d) stellar ignition

7.    Light cannot escape from a

a) white dwarf.
b) black dwarf.
c) red giant.
d) black hole.

8.    The reason nothing can be determined about the universe before a time of about $10^{-43}$ seconds after the big bang is

a) No one was alive at that time.
b) The universe did not exist prior to that time.
c) A theory of quantum gravity is needed to understand processes that occurred before that time.
d) No particles have a lifetime that short.

9.    In the standard cosmological model of the universe, the reason the hadron era preceded the lepton era is

a) Hadrons decay in a shorter time than leptons.
b) Hadrons were created before leptons.
c) All leptons were created from the decay of hadrons.
d) Hadron masses are greater than the electron mass.

10.    Which of the following is evidence that there is a non-negligible amount of dark matter?

a) The luminosity of distance galaxies is lower than expected because dark matter absorbs much of the energy before it reaches the Earth.
b) The rotation rate of galaxies is greater than expected for the amount of visible matter observed in the galaxies.
c) The sky is darker than expected without the presence of dark matter.
d) Many galaxies have been observed that do not emit any electromagnetic radiation.

11.    Determine the distance to an object that has an absolute luminosity of $2.4 \times 10^{20}$ W and an apparent brightness of $1.32 \times 10^{-5}$ W/m$^2$.

12.    Calculate the Schwarzschild radius of the Moon.

13.    What is the approximate redshift of the 122 nm emission line from hydrogen from a galaxy that is $1.8 \times 10^9$ ly away from the Earth?

14.    The product $\rho_c c^2$ is an energy density. It is the critical energy density. How many photons per cubic meter with the peak wavelength of the cosmic background radiation would be necessary to give the critical energy density? What would the aproximate intensity of this electromagnetic radiation be?

15.    Determine the parallax angle of an object located a distance of 34.6 light-years from the Earth.

## Problem Solutions

3.    We find the distance from
$$D = 1/\phi = [1/(0.00019°)(3600"/°)](3.26 \text{ ly/pc}) = 4.8 \text{ ly}.$$

7.    The star farther away will subtend a smaller angle, so the parallax angle will be less.
From $D = 1/\phi$, we see that
$$D_1/D_2 = 2 = \phi_2/\phi_1, \text{ or } \phi_1/\phi_2 = \tfrac{1}{2}.$$

11.    If we assume negligible mass change, as a red giant, the density of the Sun will be
$$\rho = M/V = M/\tfrac{4}{3}\pi r^3;$$
$$= (2 \times 10^{30} \text{ kg})/\tfrac{4}{3}\pi (1.5 \times 10^{11} \text{ m})^3 = 1.4 \times 10^{-4} \text{ kg/m}^3.$$

15.    For the reaction $^4_2\text{He} + {}^4_2\text{He} \rightarrow {}^8_4\text{Be} + \gamma$, we determine the $Q$-value:
$$Q = [M(^4\text{He}) + M(^4\text{He}) - M(^8\text{Be})]c^2$$
$$= [2(4.002603 \text{ u}) - (8.005305 \text{ u})]c^2(931.5 \text{ MeV}/uc^2) = -0.0922 \text{ MeV} = -92.2 \text{ keV}.$$
For the reaction $^4_2\text{He} + {}^8_4\text{Be} \rightarrow {}^{12}_6\text{C} + \gamma$, we determine the $Q$-value:
$$Q = [M(^4\text{He}) + M(^8\text{Be}) - M(^{12}\text{C})]c^2$$
$$= [(4.002603 \text{ u}) + (8.005305 \text{ u}) - (12.000000 \text{ u})]c^2(931.5 \text{ MeV}/uc^2) = 7.366 \text{ MeV}.$$
Note that the total $Q$-value is 7.27 MeV.

19.    From Wien's displacement law we can compare the temperatures of the two stars:
$$\lambda_1 T_1 = \lambda_2 T_2, \text{ or}$$
$$T_2/T_1 = \lambda_1/\lambda_2 = (500 \text{ nm})/(700 \text{ nm}) = 5/7.$$
If $r$ is the radius of a star, the radiating area is $A = 4\pi r^2$. From the Stefan-Boltzmann law we have
$$L = l4\pi D^2 \propto AT^4 = 4\pi r^2 T^4.$$
If we form the ratio for the two stars, we get
$$l_2/l_1 = (r_2/r_1)^2(T_2/T_1)^4;$$
$$1/0.091 = (r_2/r_1)^2(5/7)^4, \text{ which gives } r_2/r_1 = 6.5, \text{ so } d_2/d_1 = 6.5.$$

23.    If we consider a triangle with its three vertices on a great circle, such as one through the North pole as shown in the diagram, we see that the sum of the angles is 540°.

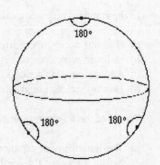

27.    We estimate the speed of the galaxy from Hubble's law:
$$v = Hd = [(70 \times 10^3 \text{ m/s/Mpc})/(3.26 \text{ ly/pc})](12 \times 10^3 \text{ Mly}) = 2.6 \times 10^8 \text{ m/s}$$
$$= 0.86c.$$

31.    We find the peak wavelength from
$$\lambda T = 2.90 \times 10^{-3} \text{ m} \cdot \text{K};$$
$$\lambda(2.7 \text{ K}) = 2.90 \times 10^{-3} \text{ m} \cdot \text{K}, \text{ which gives } \lambda = 1.1 \times 10^{-3} \text{ m} = 1.1 \text{ mm}.$$

35. The absolute luminosity depends on the radius of the star and the temperature:
$$L \propto AT^4 = 4\pi r^2 T^4.$$
   (a) For case A we have
   > temperature increases, luminosity is constant, and size decreases.
   (b) For case B we have
   > temperature is constant, luminosity decreases, and size decreases.
   (c) For case C we have
   > temperature decreases, luminosity increases, and size increases.

39. The kinetic energy of the neutron star is
$$K = \tfrac{1}{2} I\omega^2 = \tfrac{1}{2}\left(\tfrac{2}{5}Mr^2\right)\omega^2$$
$$= \tfrac{1}{2}\left(\tfrac{2}{5}\right)(1.5)(2.0 \times 10^{30}\text{ kg})(5.0 \times 10^3\text{ m})^2[(1.0\text{ rev/s})(2\pi\text{ rad/rev})]^2 = 5.92 \times 10^{38}\text{ J}.$$

   The power output is
$$P = E/t = (10^{-9}\text{ day}^{-1})(5.92 \times 10^{38}\text{ J})/(86{,}400\text{ s/day}) = 7 \times 10^{24}\text{ W}.$$

43. We find half the subtended angle from
$$\phi = 1/30\text{ pc} = (1/30)''/(3600''/^\circ) \approx (1 \times 10^{-5})^\circ.$$
   Thus the minimum subtended angle is $\approx (2 \times 10^{-5})^\circ$.

47. Because Venus has a more negative value, Venus is brighter.
   We write the logarithmic scale as
$$m = k \log l.$$
   We find the value of $k$ from
$$m_2 - m_1 = k(\log l_2 - \log l_1) = k \log(l_2/l_1);$$
$$+ 5 = k \log(1/100), \text{ which gives } k = -2.5.$$
   For Venus and Sirius, we have
$$m_V - m_S = k \log(l_V/l_S);$$
$$- 4.4 - (-1.4) = -2.5 \log(l_V/l_S), \text{ which gives } l_V/l_S = 16.$$

51. If we have only $N = M/m_n$ neutrons, the Fermi energy is
$$E_n = \left(\tfrac{3}{5}E_F\right)N = \tfrac{3}{5}N\frac{h}{8m_n}\left(\frac{3N}{\pi V}\right)^{2/3} = \frac{3Nh^2}{160m_nR^2}\left(\frac{18N}{\pi^2}\right)^{2/3}$$

   The gravitational energy will be the same, so the total energy is
$$E = E_n + E_g = \frac{3Nh^2}{160m_nR^2}\left(\frac{18N}{\pi^2}\right)^{2/3} - \frac{3GM^2}{5R}$$

   We find the equilibrium radius from $dE/dR = 0$:
$$\frac{dE}{dR} = -\frac{3Nh^2}{80m_nR^3}\left(\frac{18N}{\pi^2}\right)^{2/3} + \frac{3GM^2}{5R^2} = 0, \text{ which gives } R = \frac{Nh^2}{16m_nGM^2}\left(\frac{9N}{\pi^2}\right)^{2/3}$$

   In terms of the mass, this is
$$R = \frac{h^2}{16m_n^{8/3}GM^{1/3}}\left(\frac{18}{\pi^2}\right)^{2/3}$$

   For a mass of 1.5 solar masses, we get
$$R = \frac{\left(6.63 \times 10^{-34}\text{ J}\cdot\text{s}\right)^2}{16\left(1.67 \times 10^{-27}\text{ kg}\right)^{8/3}\left(6.67 \times 10^{-11}\text{ N}\cdot\text{m}^2/\text{kg}^2\right)\left[(1.5)\left(2.0 \times 10^{30}\text{ kg}\right)\right]^{1/3}}\left(\frac{18}{\pi^2}\right)^{2/3}$$
$$= 1.1 \times 10^4\text{ m} = 11\text{ km}.$$

# Quiz Answers

**Chapter 1**

1. c
2. a
3. a
4. c
5. c
6. b
7. b
8. b
9. b
10. a
11. 27.25
12. 0.2%
13. 5.7%
14. 100,000
15. $B^2CD$

**Chapter 2**

1. c
2. a
3. d
4. a
5. c
6. b
7. b
8. b
9. c
10. b
11. 32.1 m
12. 5.03 s
13. First car wins
14. 40 mph
15. 3.0 s

**Chapter 3**

1. d
2. d
3. b
4. c
5. d
6. b
7. d
8. b
9. a
10. a
11. 8.9, 4.6° N of E
12. 9.48, 88.7° S of W
13. 1.56 s, 1.57 m
14. $5.92 \times 10^{-3}$ m/s
15. 61.3 km/hr
    63.8° W of N

**Chapter 4**

1. c
2. b
3. d
4. c
5. a
6. b
7. b
8. d
9. b
10. d
11. 98 m/s$^2$
    17° W of S
12. 560 m
13. 98.8 N

14. 149 N
15. $2.88 \times 10^3$ N

**Chapter 5**

1. a
2. b
3. d
4. c
5. d
6. c
7. d
8. a
9. a
10. a
11. $1.41 \times 10^3$ N
    down the bank
12. 0.991 m/s$^2$
13. 13.3 kg
14. 19.2 s
15. 5.37 m/s$^2$ down

**Chapter 6**

1. c
2. d
3. b
4. a
5. b
6. c
7. d
8. d
9. a
10. a

11. $1.00 \times 10^4$ s
12. $r/(1 + 1/\sqrt{3})$
13. $d(M/m)^{1/3}$
14. $6.70 \times 10^7$ m
15. $1.20 \times 10^5$ N

**Chapter 7**

1. c
2. d
3. c
4. b
5. d
6. c
7. c
8. d
9. b
10. c
11. 7.0 N
12. 5.93 m/s
13. 246 J, 55°
14. 134 N
15. $c[(\sqrt{35}) - 1]/6$

## Chapter 8

1. We know there are nonconservative forces acting on the automobile because the automobile slows down if the engine does not continue to provide energy to the motion of the car. Nonconservative forces that act on the car are the frictional force of the air on the car, the force of friction between the tires and the road.

2. The change in gravitational potential energy only depends on the initial and final height. It does not depend on the path taken.

3. This is not enough information to determine the displacement of the end of the spring. There are two different displacements with the same potential energy, one with the spring compressed and one with the spring stretched.

4. For a constant force, the potential energy function increases linearly with position in the direction directly opposite to the direction of the force.

5. Energy conservation, in this context, means to conserve the energy in a useful form. We will learn more about what that means in Chapter 20.

6. No, power is proportional to speed, but the energy used is also proportional to the time the power is in use. The travel time of the second car is one-half the travel time of the first car, so the total energy used will be the same for each car.

7. Aristotle's law of motion implies that the kinetic energy of an object is always eventually reduced to zero. This is consistent with a statement that all objects have nonconservative forces acting on them.

8. The conditions for equilibrium are that the vector sum of the forces and the vector sum of the torques are equal to zero. The stability conditions are that a small displacement in any direction or rotation about any axis results in an increase in the potential energy of the system.

9. The more energy required to displace or rotate a system to a position in which it is no longer stable, the greater the stability of the system.

10. If there is greater output power than input power, the process does not conserve energy. All real machines have some nonconservative forces that are responsible for negative work done on the system. This means that for a real machine, the output power will always be less than the input power.

11. 8.85 mm
12. $U(x) = \frac{1}{2}kx^2 + \frac{1}{4}qx^4$
13. 5.48 cm above initial position
14. 11.0 N
15. Equilibrium positions at $x = 1 + \sqrt{(5/3)}$ and $x = 1 - \sqrt{(5/3)}$. The first equilibrium is unstable and the second equilibrium is stable.

# Chapter 9

1. The change in the momentum of the earth is equal in magnitude and opposite in direction to the change in momentum of the car.

2. The piece with less mass will fly out at the highest speed.

3. The two pieces must fly out in exactly opposite directions.

4. If the impulse is applied in the shortest time possible, the train will travel with the shortest time between stops, but the acceleration will be the most uncomfortable to the passengers. If the impulse is applied over a long time, the train will take longer to travel between stations, but the passengers will be the most comfortable for the passengers.

5. In two and three dimensions, there are more than two unknown components of velocity after the collision. More information must be given to reduce the number of unknowns to the number of relationships between those unknowns. For example, in two dimensions, there are four unknown velocity components after the collision. There are only three relationships between the initial and final velocities: the conservation of momentum in the two perpendicular directions and the conservation of kinetic energy. One additional piece of information must be known to solve the problem.

6. Yes. A horseshoe and a donut are examples of such objects.

7. Many collisions occur in a short time interval with the collision force much larger in magnitude than the external forces that the effect of the external forces during the collision can be ignored.

8. The padding on the surface extends the time that the collision occurs. For a given change in momentum, this implies a reduced force during the collision.

9. The collapse of the car implies that the driver moves for a greater amount of time before stopping than if the car remained intact and rigid. The same impulse is given to the driver regardless of what happens to the car, but the destructible car allows the impulse to occur over a greater time, reducing the force.

10. A bullet that bounces off of an object has a greater momentum change than a bullet that sticks in the object. This implies that the momentum change of the object struck by the bullet also has a greater momentum change when the bullet bounces off the object.

11. 170 N·s

12.    $5.65 \times 10^{-23}$ m/s
13.    15.0 m/s
14.    0.209 m from end with the sphere attached
15.    11.8 m/s 29.7° N of W

# Chapter 10

1.    $x = R \cos \omega t$ and $y = R \sin \omega t$

2.    No, unless $R = 0$.

3.    Yes.

4.    The angular velocity is directly into the face of the clock.

5.    The velocity of the top of the wheel relative to the ground is twice the velocity of the center of the wheel relative to the ground.

6.    The moment of inertia is undefined until the axis of rotation is defined.

7.

8.    The argument ignores the fact that the slowing of the rotation is caused by an external torque. If there is an external torque, the angular momentum is not conserved.

9.    For a given direction of rotation axis, the moment of inertia is minimum when the rotation axis passes through the center of mass of the object.

10.    If the knob is in the center of the door, the moment arm is shorter than if the knob is at the edge of the door opposite the hinge. This reduces the torque applied to the door for a given magnitude of force.

11.    $23.0$ rad/s$^2$
12.    $15.1$ rad/s$^2$
13.    $r_{cm} = 0.50$ m $\mathbf{i}$ + $0.33$ m $\mathbf{j}$, $I_{cm} = 5.20$ kg·m$^2$
14.    57.5 rev/min
15.    329 rad/s

# Chapter 11

1. d
2. b
3. a
4. d
5. a
6. d
7. c
8. d
9. c
10. b
11. $5\mathbf{i} + \mathbf{j} + 4\mathbf{k}$
12. $7\mathbf{k}$ N·m
13. $5.5 \times 10^{-14}$ rad
14. $2.20 \times 10^{-2}$ kg·m$^2$/s
15. 0.057 rad/s

# Chapter 12

1. d
2. d
3. a
4. d
5. c
6. b
7. b
8. c
9. c
10. d
11. 45 kg
12. 67 N
13. 12 kg
14. 0.42
15. $9.9 \times 10^{-4}$ m

# Chapter 13

1. b
2. c

3. d
4. b
5. c
6. d
7. d
8. c
9. d
10. d
11. 680 kg/m$^3$
12. $3.89 \times 10^3$ s
13. $1.56 \times 10^5$ N/m$^2$
14. 47.6 N
15. 13.0 m

# Chapter 14

1. d
2. d
3. c
4. c

5. d
6. b
7. d
8. b
9. b
10. d
11. $1.74 \times 10^4$ N/m
    939 kg
12. $5.13 \times 10^{-2}$ s
13. 161 s
14. $\pi/2$, 0, $-\pi/2$
15. 2.38 cm

# Chapter 15

1. b
2. c
3. c
4. c
5. d

6. a
7. a
8. b
9. d
10. b
11. $9.4 \times 10^{10}$ N/m$^2$
12. $2.38 \times 10^5$ J
13. 3.2 cm, 5.2 cm
    1.1 Hz, 0.92 s
    5.7 cm/s
14. 109 Hz
15. 9.56 m/s

## Chapter 16

1. b
2. b
3. d
4. a
5. b
6. c
7. d
8. c
9. d
10. d
11. 82 Hz, 164 Hz
    246 Hz
12. 90 dB
13. 2.2 m/s
14. 584 m/s
15. 2.1%

## Chapter 17

1. a
2. d
3. b
4. c
5. c
6. c
7. c
8. a
9. c
10. b
11. 0.051 in
12. -40° C
13. $8.67 \times 10^6$ N/m$^2$
14. 173 moles
15. $3.2 \times 10^6$

## Chapter 18

1. c
2. a
3. c
4. d
5. d
6. b

7. d
8. d
9. a
10. d
11. $4.03 \times 10^{-9}$ K
12. $\dfrac{2N}{\sqrt{\pi}} \dfrac{\sqrt{E}}{(kT)^{3/2}} e^{-E/kT} \, dE$
13. 20 g
14. $6.25 \times 10^{18}$ m$^{-3}$
15. $a = 50.6$ N·m$^4$
    $b = 1.00 \times 10^{-2}$ m$^3$

## Chapter 19

1. b
2. d
3. c
4. d
5. b
6. d
7. c
8. c
9. d
10. a
11. 55.2°
12. 135 s
13. $-3.85 \times 10^5$ J
14. $5.88 \times 10^{24}$ J
15. $1.26 \times 10^5$ N/m$^2$
    378 K

## Chapter 20

1. b
2. b
3. a
4. d
5. b
6. c
7. b
8. c
9. b
10. d
11. 50.5%
12. $5.49 \times 10^4$ J/s
13. $1.9 \times 10^{-2}$ kcal/K
14. $1.2 \times 10^{-3}$ kcal/K
15. 

| macro | micro |
|---|---|
| 2 | 1-1 |
| 3 | 1-2 |
|  | 2-1 |
| 4 | 1-3 |
|  | 2-2 |
|  | 3-1 |
| 5 | 1-4 |
|  | 2-3 |
|  | 3-2 |
|  | 4-1 |

| | |
|---|---|
| 6 | 1-5 |
|  | 2-4 |
|  | 3-3 |
|  | 4-2 |
|  | 5-1 |
| 7 | 1-6 |
|  | 2-5 |
|  | 3-4 |
|  | 4-3 |
|  | 5-2 |
|  | 6-1 |
| 8 | 2-6 |
|  | 3-5 |
|  | 4-4 |
|  | 5-3 |
|  | 6-2 |
|  | 7-1 |
| 9 | 3-6 |
|  | 4-5 |
|  | 5-4 |
|  | 6-3 |
| 10 | 4-6 |
|  | 5-5 |
|  | 6-4 |
| 11 | 5-6 |
|  | 6-5 |
| 12 | 6-6 |

## Chapter 21

1. a
2. a
3. d
4. a
5. d
6. d
7. a
8. c
9. d
10. d
11. 0.21 N west
12. 0.26 N
    133° CCW from +$x$-axis
13. $(-1.11\mathbf{i} + 0.792\mathbf{j} + 5.14\mathbf{k}) \times 10^4$ N·m
14. $1.29 \times 10^3$ m
15. $(-0.31\mathbf{i} + 0.37\mathbf{j} + 0.46\mathbf{k})$N·m

## Chapter 22

1. d
2. b
3. b
4. b
5. a
6. c
7. a

8. $1.99 \times 10^{-7}$ N·m$^2$/C
9. $-4.17 \times 10^{-15}$ C
10. $E_{r<R} =$
    $\dfrac{\rho_0}{\varepsilon_0}\left( \dfrac{r}{3} - \dfrac{r^2}{2R} + \dfrac{r^3}{5R^2} \right)$
    $E_{r<R} = \dfrac{\rho_0 R^3}{30\varepsilon_0 r^2}$
11. $2\varepsilon_0 E_0 x$
12. $4E_0 L^2/\pi^2$

## Chapter 23

1. d
2. c
3. a
4. b
5. d
6. b
7. c
8. b
9. c
10. b
11. 0.775 J
12. 0.768 J
13. 480 eV
    $1.30 \times 10^7$ m/s
14. At $l$: -25%
    At 10$l$: -0.25%
15. $E_x = -V_0\left( \dfrac{y}{z^2} - \dfrac{y}{x^2} \right)$
    $E_y = -V_0\left( \dfrac{x}{z^2} - \dfrac{1}{x} \right)$
    $E_z = V_0\left( \dfrac{2xy}{z^3} \right)$

## Chapter 24

1. c
2. b
3. a
4. b
5. a
6. c
7. b
8. a
9. d
10. a
11. $3.8 \times 10^{-9}$ F
12. 22.8 μF
13. $Q_{12} = 144$ μC
    $E_{12} = 864$ μJ
    $Q_{27} = 130$ μC
    $E_{27} = 313$ μJ
    $Q_{18} = 130$ μC
    $E_{18} = 469$ μJ
14. $\frac{1}{2}(K_1 + K_2)C$

15. $\left[\dfrac{K_1 + K_2}{2K_1K_2} - 1\right] \frac{1}{2}CV^2$

13. $(E_0/B_0)\mathbf{i}$
14. $5.48 \times 10^{-6}$ m
15. $1.32 \times 10^{23}$ m$^{-3}$

13. 57.4 s
14. 18.0 pF
15. $1.01 \times 10^{-10}$ F
    $2.18 \times 10^{-3}$ H

12. 2.88
13. 12.5 cm behind
    mirror, 0.50,
    upright, virtual
14. 34.3 cm concave
15. 0.18 in inside

## Chapter 25

1. d
2. b
3. c
4. c
5. c
6. c
7. b
8. b
9. c
10. b
11. 190 mA, 2.28 W
12. ≈ \$12
13. $V_0/\sqrt{3}$
14. ≈ 3 Ω
15. $5.2 \times 10^{-5}$ m/s

## Chapter 26

1. b
2. c
3. c
4. b
5. b
6. a
7. b
8. b
9. b
10. a
11. 8.9 Ω
12. 65.1 Ω
13. $I_{R1} = 0.19$ A
    $I_{R2} = 0.12$ A
    $I_{R3} = 0.07$ A
14. 71 ms
15. $2.50 \times 10^{-2}$ Ω

## Chapter 27

1. b
2. d
3. d
4. d
5. c
6. d
7. d
8. c
9. c
10. b
11. 7.87 N/m $\mathbf{k}$
12. $(-5.95\mathbf{i} + 4.67\mathbf{j}$
    $-3.58\mathbf{k}) \times 10^{13}$ m/s$^2$

## Chapter 28

1. c
2. d
3. d
4. d
5. b
6. d
7. c
8. b
9. a
10. c
11. $6.0 \times 10^{-6}$ T
12. $\mu_0 I^2/4\pi$
    toward wire
13. $8.9 \times 10^3$
14. $(6.0 \times 10^{-7}$ T·m)/$r$
    $(4.0 \times 10^{-7}$ T·m)/$r$
15. $4\mu_0 I/\pi a\sqrt{2}$

## Chapter 29

1. d
2. a
3. b
4. a
5. c
6. a
7. b
8. d
9. d
10. a
11. 1.83 mA
12. $-1.0$ T·m$^2$
13. 29.7 Hz
14. 120
15. ½, $I_{\text{primary}} = 3.0$ A
    $I_{\text{secondary}} = 6.0$ A

## Chapter 30

1. c
2. a
3. b
4. b
5. b
6. a
7. a
8. c
9. a
10. b
11. $2\pi f M I_0 \cos 2\pi ft$
12. $\mu_0 I^2/4\pi r_1 r_2$

## Chapter 31

1. c
2. b
3. b
4. a
5. c
6. a
7. d
8. d
9. d
10. a
11. 73 Ω, 68.3 mA
12. $4.07 \times 10^{-7}$ F
13. $7.17 \times 10^3$ Hz
    2.06 A
14. $V_L = 24.2$ V
    $V_R = 0.164$ V
    $V_C = 12.2$ V
15. $1.08 \times 10^4$ W

## Chapter 32

1. c
2. c
3. d
4. d
5. c
6. a
7. a
8. a
9. a
10. a
11. $5.6 \times 10^{-15}$ T
12. $5.0 \times 10^{-6}$ N/m$^2$
13. 77.4 V/m
    $2.58 \times 10^{-7}$ T
14. 361 m
15. $1.67 \times 10^8$ m/s

## Chapter 33

1. c
2. c
3. d
4. c
5. b
6. c
7. b
8. a
9. a
10. b
11. $2.23 \times 10^8$ m/s

## Chapter 34

1. a
2. d
3. b
4. c
5. c
6. b
7. b
8. d
9. d
10. b
11. 36.7 cm, $-1.54$
12. $-42.5$ cm
13. $+28.6$ cm
14. 3.1 cm, 2.53 m
15. ≈ 480

## Chapter 35

1. a
2. a
3. c
4. c
5. b
6. d
7. d
8. a
9. b
10. d
11. 410 nm
12. 0.255 mm
13. 11.4 μm
14. 514 nm, 1.54 μm
    2.57 μm
15. 0.40 mm

## Chapter 36

1. c
2. b
3. c
4. b
5. c
6. d
7. a
8. b
9. b
10. c
11. 405 nm

12. 1.96 mm
13. $2.17 \times 10^{-4}$ rad
14. 0.155
15. 1.48

## Chapter 37

1. a
2. c
3. c
4. a
5. b
6. a
7. d
8. c
9. c
10. a
11. $0.735c$
12. $v_A = 0.866c$
 $v_B = 0.943c$
 $v_{AB} = 0.420c$
13. $K = 5mc^2$
 $p = 5.91mc$
14. $x' = $ -967 m
 $y' = 189$ m
 $z' = 0$
 $t' = 4.74$ μs
15. $0.0122c$
 toward Earth

## Chapter 38

1. b
2. a
3. c
4. a
5. a
6. b
7. b
8. d
9. b
10. b
11. $5.8 \times 10^3$ K
12. $7.6 \times 10^{-35}$ m
13. 26.2°
14. 1.1 μm
15. 13.5 nm, 11.4 nm
 1.08 nm

## Chapter 39

1. c
2. b
3. d
4. d
5. a
6. d
7. c
8. b

9. d
10. d
11. $2.1 \times 10^{-27}$ kg·m/s
12. $2.18 \times 10^{-13}$ s
13. $5.31 \times 10^{-24}$ kg
14. 0.22
15. $1.45 \times 10^{-64}$ J
 $5.80 \times 10^{-64}$ J
 $1.31 \times 10^{-63}$ J
 $2.32 \times 10^{-63}$ J

## Chapter 40

1. d
2. c
3. d
4. d
5. b
6. a
7. c
8. d
9. c
10. b
11. $41e^{-8} \approx 0.014$
12. 0.54 nm
13. $Z = 27$, Cobalt
14. $(3 + \sqrt{5})r_0$
15. $j = 3$,
 $m_j = \{-3, -2, \ldots, 3\}$;
 $j = 4$,
 $m_j = \{-4, -3, \ldots, 4\}$;
 $j = 5$,
 $m_j = \{-5, -4, \ldots, 5\}$;

## Chapter 41

1. a
2. b
3. c
4. a
5. b
6. c
7. b
8. b
9. a
10. c
11. $1.9 \times 10^{-18}$ J
12. $3.6 \times 10^{-4}$
 $1.06 \times 10^6$ m/s
13. 1.652 μm
 1.650 μm
14. $6.60 \times 10^{-18}$ J
15. $1.48 \times 10^{-10}$ m

## Chapter 42

1. d
2. d
3. c
4. c
5. b
6. c
7. c
8. a
9. a
10. d
11. $2.0 \times 10^{15}$
12. 2.32 MeV
13. $^{28}_{14}$Si : 8.45 MeV
 $^{31}_{14}$Si : 8.46 MeV
 Expect $^{31}_{14}$Si to be more
 stable, but it isn't.
14. 29.9 decays/s
15. 0.313 MeV

## Chapter 43

1. c
2. c
3. a
4. d
5. a
6. d
7. a
8. d
9. b
10. c
11. 10.5 MeV
12. $3.1 \times 10^{-2}$ rem
13. 30.9 MeV
14. $6.6 \times 10^{-3}$ rem
15. 87.83 MHz

## Chapter 44

1. c
2. a
3. b
4. c
5. b
6. d
7. a
8. b
9. d
10. c
11. 9.8 m/s
12. 12.9 GeV
13. $3.37 \times 10^7$ Hz
 $1.26 \times 10^{-13}$ J
14. 270 MeV
15. $K_\pi = 98.6$ MeV
 $p_\pi = 193$ MeV/c

$K_n = 19.6$ MeV
$P_n = 193$ MeV/c

## Chapter 45

1. b
2. a
3. a
4. a
5. d
6. a
7. d
8. c
9. d
10. b
11. $1.20 \times 10^{12}$ m
12. $1.08 \times 10^{-4}$ m
13. $\Delta\lambda = 15.6$ nm
14. $4.81 \times 10^{12}$
 $10^{-18}$ W/m²
15. $4.58 \times 10^{-7}$ rad